STUDENT'S SOLUTIONS MANUAL

JEFFERY A. COLE

Anoka-Ramsey Community College

INTERMEDIATE ALGEBRA

NINTH EDITION

Margaret L. Lial

American River College

John Hornsby

University of New Orleans

Terry McGinnis

PEARSON

Addison Wesley

Boston San Francisco New York
London Toronto Sydney Tokyo Singapore Madrid
Mexico City Munich Paris Cape Town Hong Kong Montreal

Reproduced by Pearson Addison-Wesley from electronic files supplied by the author.

Publishing as Pearson Addison-Wesley, 75 Arlington Street, Boston, MA 02116

ISBN 0-321-15911-X

1 2 3 4 5 6 CRS 06 05 04 03

Table of Contents

Preface

This *Student's Solutions Manual* contains solutions to selected exercises in the text *Intermediate Algebra, Ninth Edition* by Margaret L. Lial, John Hornsby, and Terry McGinnis. It contains solutions to the odd-numbered exercises in each section, as well as solutions to all the exercises in the review sections, the chapter tests, and the cumulative review sections.

This manual is a text supplement and should be read along *with* the text. You should read all exercise solutions in this manual because many concept explanations are given and then used in subsequent solutions. All concepts necessary to solve a particular problem are not reviewed for every exercise. If you are having difficulty with a previously covered concept, refer back to the section where it was covered for more complete help.

A significant number of today's students are involved in various outside activities, and find it difficult, if not impossible, to attend all class sessions; this manual should help meet the needs of these students. In addition, it is my hope that this manual's solutions will enhance the understanding of all readers of the material and provide insights to solving other exercises.

I appreciate feedback concerning errors, solution correctness or style, and manual style. Any comments may be sent directly to me at the address below, at jeff.cole@anokaramsey.edu, or in care of the publisher, Pearson Addison-Wesley.

I would like to thank Ken Grace, of Anoka-Ramsey Community College, for typesetting the manuscript and providing invaluable help with many features of the manual; Marv Riedesel and Mary Johnson, of Inver Hills Community College, for their careful accuracy checking; Jolene Lehr and Suzanne Alley, of Pearson Addison-Wesley, for facilitating the production process; and the authors and Maureen O'Connor, of Pearson Addison-Wesley, for entrusting me with this project.

Jeffery A. Cole
Anoka-Ramsey Community College
11200 Mississippi Blvd. NW
Coon Rapids, MN 55433

CHAPTER 1 REVIEW OF THE REAL NUMBER SYSTEM

Section 1.1

1. $\{x \mid x$ is a natural number less than $6\}$
The set of natural numbers is $\{1, 2, 3, \dots\}$, so the set of natural numbers less than 6 is $\{1, 2, 3, 4, 5\}$.

3. $\{z \mid z$ is an integer greater than $4\}$
The set of integers is $\{\dots, -3, -2, -1, 0, 1, 2, 3, \dots\}$, so the set of integers greater than 4 is $\{5, 6, 7, 8, \dots\}$.

5. $\{z \mid z$ is an integer less than or equal to $4\}$
The set of integers is $\{\dots, -3, -2, -1, 0, 1, 2, 3, \dots\}$, so the set of integers less than or equal to 4 is $\{\dots, -1, 0, 1, 2, 3, 4\}$.

7. $\{a \mid a$ is an even integer greater than $8\}$
The set of even integers is $\{\dots, -2, 0, 2, 4, \dots\}$, so the set of even integers greater than 8 is $\{10, 12, 14, 16, \dots\}$.

9. $\{x \mid x$ is an irrational number that is also rational$\}$
Irrational numbers cannot also be rational numbers. The set of irrational numbers that are also rational is $\emptyset$.

11. $\{p \mid p$ is a number whose absolute value is $4\}$
This is the set of numbers that lie a distance of 4 units from 0 on the number line. Thus, the set of numbers whose absolute value is 4 is $\{-4, 4\}$.

In Exercises 13–15, we give one possible answer.

13. $\{2, 4, 6, 8\}$ can be described by $\{x \mid x$ is an even natural number less than or equal to $8\}$.

15. $\{4, 8, 12, 16, \dots\}$ can be described by $\{x \mid x$ is a multiple of 4 greater than $0\}$.

17. Yes, the choice of variables is not important. The sets have the same description so they represent the same set.

19. Graph $\{-3, -1, 0, 4, 6\}$.
Place dots for $-3, -1, 0, 4$, and 6 on a number line.

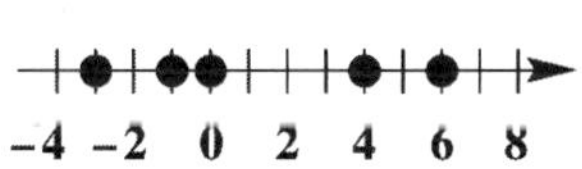

21. Graph $\left\{-\frac{2}{3}, 0, \frac{4}{5}, \frac{12}{5}, \frac{9}{2}, 4.8\right\}$.
Place dots to indicate the points
$-\frac{2}{3} = -.\overline{6}\,, 0, \frac{4}{5} = .8, \frac{12}{5} = 2.4, \frac{9}{2} = 4.5,$ and 4.8
on a number line.

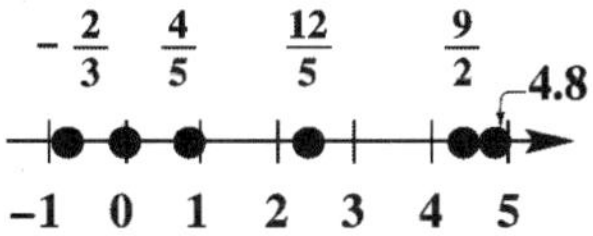

23. The graph of a number is a point on the number line. The coordinate of a point on the number line is the number that corresponds to the point.

25. $\left\{-8, -\sqrt{5}, -.6, 0, \frac{3}{4}, \sqrt{3}, \pi, 5, \frac{13}{2}, 17, \frac{40}{2}\right\}$

(a) The elements $5, 17,$ and $\frac{40}{2}$ (or 20) are natural numbers.

(b) The elements $0, 5, 17,$ and $\frac{40}{2}$ are whole numbers.

(c) The elements $-8, 0, 5, 17,$ and $\frac{40}{2}$ are integers.

(d) The elements $-8, -.6, 0, \frac{3}{4}, 5, \frac{13}{2}, 17,$ and $\frac{40}{2}$ are rational numbers.

(e) The elements $-\sqrt{5}\,, \sqrt{3}\,,$ and π are irrational numbers.

(f) All the elements are real numbers.

27. The statement "Every integer is a whole number" is *false*. Some integers are whole numbers, but the negative integers are not whole numbers.

29. The statement "Every irrational number is an integer" is *false*. Irrational numbers have decimal representations that neither terminate nor repeat, so no irrational numbers are integers.

31. The statement "Every natural number is a whole number" is *true*. The whole numbers consist of the natural numbers and zero.

33. The statement "Some rational numbers are whole numbers is *true*. Every whole number is rational.

35. The statement "The absolute value of any number is the same as the absolute value of its additive inverse" is *true*. The distance on a number line from 0 to a number is the same as the distance from 0 to its additive inverse.

37. **(a)** The additive inverse of 6 is -6.

(b) $6 > 0$, so $|6| = 6$.

39. **(a)** The additive inverse of -12 is $-(-12) = 12$.

(b) $-12 < 0$, so $|-12| = -(-12) = 12$.

41. **(a)** The additive inverse of $\frac{6}{5}$ is $-\frac{6}{5}$.

(b) $\frac{6}{5} > 0$, so $\left|\frac{6}{5}\right| = \frac{6}{5}$.

43. $|-8|$
Use the definition of absolute value.

$$-8 < 0, \text{ so } |-8| = -(-8) = 8$$

45. $\frac{3}{2} > 0$, so $\left|\frac{3}{2}\right| = \frac{3}{2}$.

47. $-|5| = -(5) = -5$

49. $-|-2| = -[-(-2)] = -(2) = -2$

51. $-|4.5| = -(4.5) = -4.5$

53. $|-2| + |3| = 2 + 3 = 5$

55. $|-9| - |-3| = 9 - 3 = 6$

57. $$\begin{aligned} |-1| + |-2| - |-3| &= 1 + 2 - 3 \\ &= 3 - 3 \\ &= 0 \end{aligned}$$

59. **(a)** The greatest absolute value is $|-10.6| = 10.6$. Therefore, Philadelphia had the greatest change in population. The population declined 10.6%.

(b) The smallest absolute value is $|.6| = .6$. Therefore, Chicago had the smallest change in population. The population increased .6%.

61. Compare the depths of the bodies of water. The deepest, that is, the body of water whose depth has the greatest absolute value, is the Pacific Ocean ($|-12,925| = 12,925$) followed by the Indian Ocean, the Caribbean Sea, the South China Sea, and the Gulf of California.

63. True; the absolute value of the depth of the Pacific Ocean is

$$|-12,925| = 12,925$$

which is greater than the absolute value of the depth of the Indian Ocean,

$$|-12,598| = 12,598.$$

65. True; since -6 is to the left of -2 on a number line, -6 is less than -2.

67. False; since -4 is to the left of -3 on a number line, -4 is *less* than -3, not greater.

69. True; since 3 is to the right of -2 on a number line, 3 is greater than -2.

71. True; since $-3 = -3$, -3 is greater than *or equal to* -3.

73. The inequality $6 > 2$ can also be written $2 < 6$. In each case, the inequality symbol points toward the smaller number so both inequalities are true.

75. $-9 < 4$ is equivalent to $4 > -9$.

77. $-5 > -10$ is equivalent to $-10 < -5$.

79. $0 < x$ is equivalent to $x > 0$.

81. "7 is greater than y" can be written as $7 > y$.

83. "5 is greater than or equal to 5" can be written as $5 \geq 5$.

85. "$3t - 4$ is less than or equal to 10" can be written as $3t - 4 \leq 10$.

87. "$5x + 3$ is not equal to 0" can be written as $5x + 3 \neq 0$.

89. "t is between -3 and 5" can be written as $-3 < t < 5$.

91. "$3x$ is between -3 and 4, including -3 and excluding 4" can be written as $-3 \leq 3x < 4$.

93. $-6 < 7 + 3$?
$-6 < 10$ *True*

The last statement is true since -6 is to the left of 10 on a number line.

95. $2 \cdot 5 \geq 4 + 6$?
$10 \geq 10$ *True*

97. $-|-3| \geq -3$?
$-3 \geq -3$ *True*

99. $-8 > -|-6|$?
$-8 > -6$ *False*

101. $\{x \mid x > -1\}$ includes all numbers greater than -1, written $(-1, \infty)$. Place a parenthesis at -1 since -1 is not an element of the set. The graph extends from -1 to the right.

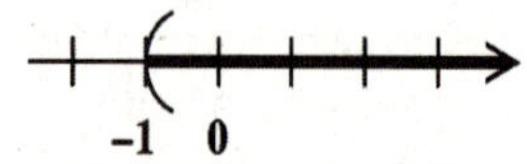

103. $\{x \mid x \leq 6\}$ includes all numbers less than or equal to 6, written $(-\infty, 6]$. Place a bracket at 6 since 6 is an element of the set. The graph extends from 6 to the left.

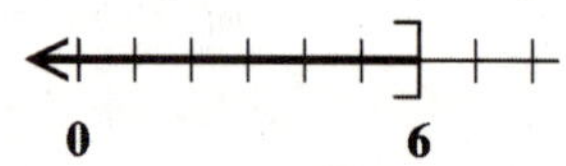

105. $\{x \mid 0 < x < 3.5\}$ includes all numbers between 0 and 3.5, written $(0, 3.5)$. Place parentheses at 0 and 3.5 since these numbers are not elements of the set. The graph goes from 0 to 3.5.

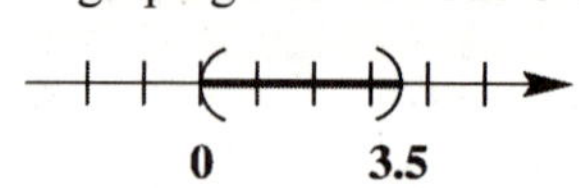

107. $\{x \mid 2 \leq x \leq 7\}$ includes all numbers from 2 to 7, written $[2, 7]$. Place brackets at 2 and 7 since these numbers are elements of the set. The graph goes from 2 to 7.

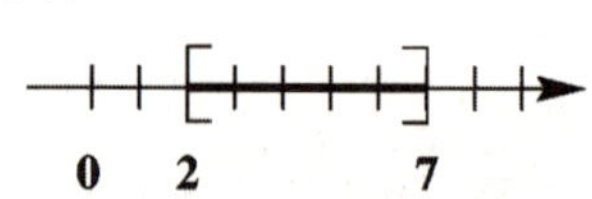

109. $\{x \mid -4 < x \leq 3\}$ includes all numbers between -4 and 3, excluding -4, but including 3, written $(-4, 3]$. Place a parenthesis at -4 and a bracket at 3 to show that -4 is not included, but 3 is. The graph goes from -4 to 3.

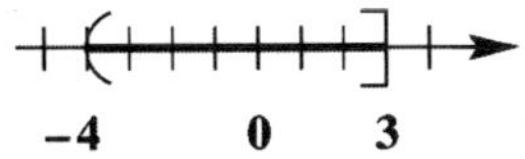

111. $\{x \mid 0 < x \leq 3\}$ includes all numbers between 0 and 3, excluding 0, but including 3, written $(0, 3]$. Place a parenthesis at 0 and a bracket at 3 to show that 0 is not included, but 3 is. The graph goes from 0 to 3.

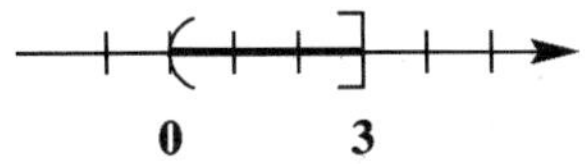

113. 1999: OH $= 681$, IA $= 548$, CA $= 532$
In 1999, Ohio (OH), Iowa (IA), and California (CA) had production greater than 500 million eggs.

115. 1999: TX $= x = 365$, OH $= y = 681$
Since $365 < 681$, $x < y$ is true.

117. The *natural numbers* are the numbers with which we count. Some examples are 1, 2, and 3. The *whole numbers* are formed by including 0 with the natural numbers. Three examples are 0, 10, and 100. The *integers* are formed by including the negatives of the natural numbers with the whole numbers, such as -1, -2, and -3. The *rational numbers*, such as $\frac{1}{2}$, .75, and $-\frac{3}{4}$, are formed by quotients of integers (nonzero denominator). The *irrational numbers* include positive or negative numbers that are not rational, such as π, $\sqrt{2}$, and $-\sqrt{5}$. The *real numbers* include all positive numbers, negative numbers, and 0, such as $-\pi$, 0, and $\sqrt{2}$.

Section 1.2

1. The sum of a positive number and a negative number is 0 if the numbers are additive inverses. For example, $4 + (-4) = 0$.

3. The sum of two negative numbers is a negative number. For example, $-7 + (-21) = -28$.

5. The sum of a positive number and a negative number is positive if the positive number has a larger absolute value. For example, $15 + (-2) = 13$.

7. The difference between two negative numbers is negative if the number with smaller absolute value is subtracted from the one with larger absolute value. For example, $-15 - (-3) = -12$.

9. The product of two numbers with unlike signs is negative. For example, $-5(15) = -75$.

11. $-6 + (-13) = -(6 + 13) = -19$

13. $13 + (-4) = 13 - 4 = 9$

15. $-\frac{7}{3} + \frac{3}{4} = -\frac{28}{12} + \frac{9}{12} = -\frac{19}{12}$

17. The difference between 2.3 and .45 is 1.85. The number with the larger absolute value, -2.3, is negative, so the answer is negative. Thus, $-2.3 + .45 = -1.85$.

19. $-6 - 5 = -6 + (-5) = -(6 + 5) = -11$

21. $8 - (-13) = 8 + 13 = 21$

23. $-16 - (-3) = -16 + 3 = -13$

25.
$$\begin{aligned} -12.31 - (-2.13) &= -12.31 + 2.13 \\ &= -(12.31 - 2.13) \\ &= -10.18 \end{aligned}$$

27. $\frac{9}{10} - \left(-\frac{4}{3}\right) = \frac{9}{10} + \frac{4}{3} = \frac{27}{30} + \frac{40}{30} = \frac{67}{30}$

29. $|-8 - 6| = |-14| = -(-14) = 14$

31. $-|-4 + 9| = -|5| = -5$

33. $-2 - |-4| = -2 - 4 = -2 + (-4) = -6$

35. $-7 + 5 - 9 = (-7 + 5) - 9 = -2 - 9 = -11$

37. $6 - (-2) + 8 = 6 + 2 + 8 = 8 + 8 = 16$

39.
$$\begin{aligned} -9 - 4 - (-3) + 6 &= (-9 - 4) + 3 + 6 \\ &= -13 + 3 + 6 \\ &= -10 + 6 \\ &= -4 \end{aligned}$$

41.
$$\begin{aligned} -8 - (-12) - (2 - 6) &= -8 + 12 - (-4) \\ &= -8 + 12 + 4 \\ &= 4 + 4 \\ &= 8 \end{aligned}$$

43.
$$\begin{aligned} -.382 + 4 - .6 &= 3.618 - .6 \\ &= 3.018 \end{aligned}$$

45.
$$\begin{aligned} \left(-\frac{5}{4} - \frac{2}{3}\right) + \frac{1}{6} &= \left(-\frac{15}{12} - \frac{8}{12}\right) + \left(\frac{2}{12}\right) \\ &= -\frac{23}{12} + \frac{2}{12} \\ &= -\frac{21}{12} \text{ or } -\frac{7}{4} \end{aligned}$$

47.
$$\begin{aligned} -\frac{3}{4} - \left(\frac{1}{2} - \frac{3}{8}\right) &= -\frac{6}{8} - \left(\frac{4}{8} - \frac{3}{8}\right) \\ &= -\frac{6}{8} - \frac{1}{8} \\ &= -\frac{7}{8} \end{aligned}$$

49.
$$\begin{aligned} &|-11| - |-5| - |7| + |-2| \\ &= 11 - 5 - 7 + 2 \\ &= 6 - 7 + 2 \\ &= -1 + 2 = 1 \end{aligned}$$

51. $A = -4$ and $B = 2$. The distance between A and B is the absolute value of their difference.

$$|-4 - 2| = |-6| = 6$$

53. $D = -6$ and $F = \frac{1}{2}$. The distance between D and F is the absolute value of their difference.

$$\begin{aligned}\left|-6 - \frac{1}{2}\right| &= \left|-\frac{12}{2} - \frac{1}{2}\right| \\ &= \left|-\frac{13}{2}\right| \\ &= \frac{13}{2} \text{ or } 6\frac{1}{2}\end{aligned}$$

55. Answers will vary. One example is

$$-4 - (-9) = -4 + 9 = 5.$$

The sign is determined by choosing the sign of the number with the larger absolute value after subtraction has been changed to addition. In this example, 9 has a larger absolute value than -4, so the sign of the answer is the same as the sign of 9, which is positive.

57. It is true for multiplication (and division). It is false for addition and for subtraction when the number to be subtracted has the smaller absolute value. A more precise statement is, "The product or quotient of two negative numbers is positive."

59. The product of two numbers with *different* signs is *negative*, so

$$5(-7) = -35.$$

61. The product of two numbers with the *same* sign is *positive*, so

$$-8(-5) = 40.$$

63. The product of two numbers with the *same* sign is *positive*, so

$$-10\left(-\frac{1}{5}\right) = 2.$$

65. The product of two numbers with *different* signs is *negative*, so

$$\frac{3}{4}(-16) = -\frac{3}{4} \cdot 4 \cdot 4 = -12.$$

67. The product of two numbers with the *same* sign is *positive*, so

$$-\frac{5}{2}\left(-\frac{12}{25}\right) = \frac{5 \cdot 2 \cdot 6}{2 \cdot 5 \cdot 5} = \frac{6}{5}.$$

69. The product of two numbers with the *same* sign is *positive*, so

$$-\frac{3}{8}\left(-\frac{24}{9}\right) = \frac{3 \cdot 3 \cdot 8}{8 \cdot 9} = 1.$$

71. $-2.4(-2.45) = 5.88$

73. $3.4(-3.14) = -10.676$

75. The quotient of two nonzero real numbers with *different* signs is *negative*, so

$$\frac{-14}{2} = -14 \cdot \frac{1}{2} = -2 \cdot 7 \cdot \frac{1}{2} = -7.$$

77. The quotient of two nonzero real numbers with the *same* sign is *positive*, so

$$\frac{-24}{-4} = 24 \cdot \frac{1}{4} = 4 \cdot 6 \cdot \frac{1}{4} = 6.$$

79. The quotient of two nonzero real numbers with *different* signs is *negative*, so

$$\frac{100}{-25} = -100 \cdot \frac{1}{25} = -4 \cdot 25 \cdot \frac{1}{25} = -4.$$

81. $\frac{0}{-8} = 0 \cdot \left(-\frac{1}{8}\right) = 0$

83. Division by 0 is undefined, so $\frac{5}{0}$ is undefined.

85. The quotient of two nonzero real numbers with the *same* sign is *positive*, so

$$-\frac{10}{17} \div \left(-\frac{12}{5}\right) = \frac{10}{17} \cdot \frac{5}{12} = \frac{2 \cdot 5 \cdot 5}{17 \cdot 2 \cdot 6} = \frac{25}{102}.$$

87. $$\frac{\frac{12}{13}}{-\frac{4}{3}} = \frac{12}{13} \div \left(-\frac{4}{3}\right) = \frac{12}{13}\left(-\frac{3}{4}\right)$$

$$= -\frac{3 \cdot 4 \cdot 3}{13 \cdot 4} = -\frac{9}{13}$$

89. $-\frac{27.72}{13.2} = -2.1$

91. $$\frac{-100}{-.01} = \frac{100}{\frac{1}{100}} = 100 \div \frac{1}{100} = 100 \cdot 100$$

$$= 10{,}000$$

93. To find the difference between these two temperatures, subtract the lowest temperature from the highest temperature.

$$\begin{aligned}90° - (-22°) &= 90° + 22° \\ &= 112°\end{aligned}$$

The difference is 112° F.

95. $37.58 - (-9.10) = 37.58 + 9.10 = 46.68$
The difference is 46.68%.

97.

Year	Difference (in billions)
2000	\$538 – \$409 = \$129
2010	\$916 – \$710 = \$206
2020	\$1479 – \$1405 = \$74
2030	\$2041 – \$2542 = –\$501

99. The difference between the January and February changes is

$$-6,439 - 5,039$$
$$= -(6,439 + 5,039) = -11,478.$$

101. $-5588 - 1377 \approx -7000$
The answer is **C**.

103. Choose values for a and b such that $a < b$, for example, $a = 1$ and $b = 2$. Then the difference $a - b$ is

$$1 - 2 = 1 + (-2) = -1.$$

104. The answer in Exercise 103, -1, is less than 0.

105. This time let $a = -2$ and $b = -1$. Then the difference $a - b$ is

$$-2 - (-1) = -2 + 1 = -1.$$

106. Again, the answer, -1, is less than 0. Based on these exercises, if $a < b$, then $a - b \underline{\ <\ } 0$.

Section 1.3

1. $-4^6 = (-4)^6$ is *false*.
$-4^6 = -(4 \cdot 4 \cdot 4 \cdot 4 \cdot 4 \cdot 4) = -4096$
whereas
$(-4)^6 = (-4)(-4)(-4)(-4)(-4)(-4)$
$= 4096.$

3. The statement "$\sqrt{16}$ is a positive number" is *true*. The symbol $\sqrt{\ }$ always gives a positive square root.

5. The statement "$(-2)^7$ is a negative number" is *true*. $(-2)^7$ gives an odd number of negative factors, so the product is negative.

7. The statement "The product of 8 positive factors and 8 negative factors is positive" is *true*. The product of an even number of negative factors is positive.

9. The statement "In the exponential -3^5, -3 is the base" is *false*. The base is 3, not -3. If the problem were written $(-3)^5$, then -3 would be the base.

11. **(a)** $8^2 = 64$

(b) $-8^2 = -(8 \cdot 8) = -64$

(c) $(-8)^2 = (-8)(-8) = 64$

(d) $-(-8)^2 = -(64) = -64$

13. $10 \cdot 10 \cdot 10 \cdot 10 = 10^4$

15. $\frac{3}{4} \cdot \frac{3}{4} \cdot \frac{3}{4} \cdot \frac{3}{4} \cdot \frac{3}{4} = \left(\frac{3}{4}\right)^5$

17. $(-9)(-9)(-9) = (-9)^3$

19. $z \cdot z \cdot z \cdot z \cdot z \cdot z \cdot z = z^7$

21. $4^2 = 4 \cdot 4 = 16$

23. $.28^3 = (.28)(.28)(.28) = .021952$

25. $\left(\frac{1}{5}\right)^3 = \frac{1}{5} \cdot \frac{1}{5} \cdot \frac{1}{5} = \frac{1}{125}$

27. $\left(\frac{4}{5}\right)^4 = \left(\frac{4}{5}\right)\left(\frac{4}{5}\right)\left(\frac{4}{5}\right)\left(\frac{4}{5}\right)$
$= \frac{4 \cdot 4 \cdot 4 \cdot 4}{5 \cdot 5 \cdot 5 \cdot 5} = \frac{256}{625}$

29. $(-5)^3 = (-5)(-5)(-5) = -125$

31. $(-2)^8 = (-2)(-2)(-2)(-2)(-2)(-2)(-2)(-2)$
$= 256$

33. $-3^6 = -(3 \cdot 3 \cdot 3 \cdot 3 \cdot 3 \cdot 3) = -729$

35. $-8^4 = -(8 \cdot 8 \cdot 8 \cdot 8) = -4096$

37. $\sqrt{81} = 9$ since 9 is positive and $9^2 = 81$.

39. $\sqrt{169} = 13$ since 13 is positive and $13^2 = 169$.

41. $-\sqrt{400} = -\left(\sqrt{400}\right) = -(20) = -20$

43. $\sqrt{\frac{100}{121}} = \frac{10}{11}$ since $\frac{10}{11}$ is positive and $\left(\frac{10}{11}\right)^2 = \frac{100}{121}$.

45. $-\sqrt{.49} = -\sqrt{(.7)^2} = -(.7) = -.7$

47. There is no real number whose square is negative, so $\sqrt{-36}$ is not real.

49. **(a)** $\sqrt{144} = 12$; choice **B**

(b) $\sqrt{-144}$ is not a real number; choice **C**

(c) $-\sqrt{144} = -12$; choice **A**

51. If a is a positive number, then $-a$ is a negative number. Therefore, $\sqrt{-a}$ is not a real number, so $-\sqrt{-a}$ *is not a real number.*

53. $12 + 3 \cdot 4 = 12 + 12$ *Multiply.*
$= 24$ *Add.*

55. $6 \cdot 3 - 12 \div 4 = 18 - 12 \div 4$ *Multiply.*
$= 18 - 3$ *Divide.*
$= 15$ *Subtract.*

57. $10 + 30 \div 2 \cdot 3 = 10 + 15 \cdot 3$ *Divide.*
$= 10 + 45$ *Multiply.*
$= 55$ *Add.*

59. $-3(5)^2 - (-2)(-8)$
$= -3(25) - (-2)(-8)$ *Evaluate power.*
$= -75 - 16$ *Multiply.*
$= -91$ *Subtract.*

61. $5 - 7 \cdot 3 - (-2)^3$

$= 5 - 7 \cdot 3 - (-8)$ *Evaluate power.*

$= 5 - 21 + 8$ *Multiply and subtract.*

$= -16 + 8$ *Subtract.*

$= -8$ *Add.*

63. $-7\left(\sqrt{36}\right) - (-2)(-3)$

$= -7(6) - (-2)(-3)$ *Evaluate root.*

$= -42 - 6$ *Multiply.*

$= -48$ *Subtract.*

65. $6|4 - 5| - 24 \div 3$

$= 6|-1| - 24 \div 3$ *Simplify within absolute value bars.*

$= 6(1) - 24 \div 3$ *Take absolute value.*

$= 6 - 8$ *Multiply and divide.*

$= -2$ *Subtract.*

67. $|-6 - 5|(-8) + 3^2$

$= |-11|(-8) + 3^2$ *Simplify within absolute value bars.*

$= 11(-8) + 3^2$ *Take absolute value.*

$= 11(-8) + 9$ *Simplify power.*

$= -88 + 9$ *Multiply.*

$= -79$ *Add.*

69. $6 + \frac{2}{3}(-9) - \frac{5}{8} \cdot 16 = 6 + (-6) - 10$ *Multiply.*

$= 0 - 10$ *Add.*

$= -10$ *Subtract.*

71. $-14\left(-\frac{2}{7}\right) \div (2 \cdot 6 - 10)$

$= 4 \div (12 - 10)$ *Multiply.*

$= 4 \div 2$ *Work inside parentheses.*

$= 2$ *Divide.*

73. $\dfrac{\left(-5 + \sqrt{4}\right)(-2^2)}{-5 - 1}$

$= \dfrac{(-5 + 2)(-4)}{-6}$ *Evaluate root and power; subtract.*

$= \dfrac{(-3)(-4)}{-6}$ *Work inside parentheses.*

$= \dfrac{12}{-6}$ *Multiply.*

$= -2$ *Divide.*

75. $\dfrac{2(-5) + (-3)(-2)}{-8 + 3^2 - 1}$

$= \dfrac{-10 + 6}{-8 + 9 - 1}$ *Evaluate power; multiply.*

$= \dfrac{-4}{1 - 1}$ *Add.*

$= \dfrac{-4}{0}$ *Subtract.*

Since division by 0 is undefined, the given expression is *undefined.*

77. $\dfrac{5 - 3\left(\frac{-5 - 9}{-7}\right) - 6}{-9 - 11 + 3 \cdot 7}$

$= \dfrac{5 - 3\left(\frac{-14}{-7}\right) - 6}{-9 - 11 + 21}$ *Work in numerator and denominator separately.*

$= \dfrac{5 - 3(2) - 6}{-20 + 21}$

$= \dfrac{5 - 6 - 6}{1}$

$= \dfrac{-7}{1}$

$= -7$ *Write in lowest terms.*

In Exercises 79–85, $a = -3$, $b = 64$, and $c = 6$.

79. $3a + \sqrt{b} = 3(-3) + \sqrt{64}$

$= 3(-3) + 8$

$= -9 + 8$

$= -1$

81. $\sqrt{b} + c - a = \sqrt{64} + 6 - (-3)$

$= 8 + 6 + 3$

$= 14 + 3 = 17$

83. $4a^3 + 2c = 4(-3)^3 + 2(6)$

$= 4(-27) + 12$

$= -108 + 12 = -96$

85. $\dfrac{2c + a^3}{4b + 6a} = \dfrac{2(6) + (-3)^3}{4(64) + 6(-3)}$

$= \dfrac{12 + (-27)}{256 + (-18)}$

$= \dfrac{-15}{238} = -\dfrac{15}{238}$

87. $(v \times .5485 - 4850) \div 1000 \times 31.44$
$= (100{,}000 \times .5485 - 4850) \div 1000 \times 31.44$
$= (54{,}850 - 4850) \div 1000 \times 31.44$
$= 50{,}000 \div 1000 \times 31.44$
$= 50 \times 31.44$
$= 1572$

The owner would pay \$1572 in property taxes.

89. $(v \times .5485 - 4850) \div 1000 \times 31.44$
$= (200{,}000 \times .5485 - 4850) \div 1000 \times 31.44$
$= (109{,}700 - 4850) \div 1000 \times 31.44$
$= 104{,}850 \div 1000 \times 31.44$
$= 104.85 \times 31.44$
$= 3296.484 \approx 3296$

The owner would pay \$3296 in property taxes.

91. number of oz $\times$ % alcohol $\times .075 \div$ body weight in lb $-$ hr of drinking $\times .015$

$= 36 \times 4.0 \times .075 \div 135 - 3 \times .015$
$= 144 \times .075 \div 135 - 3 \times .015$
$= 10.8 \div 135 - 3 \times .015$
$= .08 - 3 \times .015$
$= .08 - .045$
$= .035$

93. $2.2023x - 4356.6$

(a) $2.2023(1980) - 4356.6 \approx 4.0\%$

(b) $2.2023(1990) - 4356.6 \approx 26.0\%$

(c) $2.2023(2000) - 4356.6 = 48\%$

(d) The percent increased dramatically (more than ten times) from 1980 to 2000.

Section 1.4

1. The identity element for addition is 0 since, for any real number a, $a + 0 = 0 + a = a$.
Choice **B** is correct.

3. The additive inverse of a is $-a$ since, for any real number a, $a + (-a) = 0$ and $-a + a = 0$.
Choice **A** is correct.

5. The multiplication property of 0 states that the product of 0 and any real number is 0.

7. The associative property is used to change the grouping of three terms or factors.

9. When simplifying an expression, only like terms can be combined.

11. Using the distributive property,

$$2(m + p) = 2m + 2p.$$

13. Using the distributive property,

$$\begin{aligned} -12(x - y) &= -12[x + (-y)] \\ &= -12(x) + (-12)(-y) \\ &= -12x + 12y. \end{aligned}$$

15. Using the second form of the distributive property,

$$\begin{aligned} 5k + 3k &= (5 + 3)k \\ &= 8k. \end{aligned}$$

17. $$\begin{aligned} 7r - 9r &= 7r + (-9r) \\ &= [7 + (-9)]r \\ &= -2r \end{aligned}$$

19. $-8z + 4w$
Since there is no common variable factor here, we cannot use the distributive property to simplify the expression.

21. Using the identity property, then the distributive property,

$$\begin{aligned} a + 7a &= 1a + 7a \\ &= (1 + 7)a \\ &= 8a. \end{aligned}$$

23. $$\begin{aligned} -(2d - f) &= -1(2d - f) \\ &= -1(2d) + (-1)(-f) \\ &= -2d + f \end{aligned}$$

25. $-12y + 4y + 3 + 2y$
$= -12y + 4y + 2y + 3$
$= (-12 + 4 + 2)y + 3$
$= -6y + 3$

27. $-6p + 5 - 4p + 6 + 11p$
$= -6p - 4p + 11p + 5 + 6$
$= (-6 - 4 + 11)p + 11$
$= 1p + 11$ or $p + 11$

29. $3(k + 2) - 5k + 6 + 3$
$= 3k + 6 - 5k + 6 + 3$
$= 3k - 5k + 6 + 6 + 3$
$= (3 - 5)k + 6 + 6 + 3$
$= -2k + 15$

31. $-2(m + 1) - (m - 4)$
$= -2m - 2 - m + 4$
$= -2m - m - 2 + 4$
$= (-2 - 1)m + 2$
$= -3m + 2$

33. $.25(8 + 4p) - .5(6 + 2p)$
$= .25(8) + .25(4p) + (-.5)(6) + (-.5)(2p)$
$= 2 + p - 3 - p$
$= p - p + 2 - 3$
$= (1 - 1)p + 2 - 3$
$= 0p - 1$
$= -1$

35. $-(2p+5)+3(2p+4)-2p$
$= -2p-5+6p+12-2p$
$= (-2+6-2)p+(-5+12)$
$= 2p+7$

37. $2+3(2z-5)-3(4z+6)-8$
$= 2+6z-15-12z-18-8$
$= 6z-12z+2-15-26$
$= (6-12)z-13-26$
$= -6z-39$

39. $5x+8x=(5+8)x=13x$ *Distributive property*

41. $5(9r)=(5\cdot 9)r=45r$ *Associative property*

43. $5x+9y=9y+5x$ *Commutative property*

45. $1\cdot 7=7$ *Identity property*

47. $-\frac{1}{4}ty+\frac{1}{4}ty=0$ *Inverse property*
A number plus its opposite equals 0.

49. $8(-4+x)=8(-4)+8x$ *Distributive property*
$= -32+8x$

51. $0(.875x+9y-88z)=0$ *Multiplication property of 0*
Zero times any quantity equals 0.

53. Answers will vary. Commutative: one example is washing your face and brushing your teeth. The activities can be carried out in either order.
Not commutative: one example is putting on your socks and putting on your shoes.

55. $96\cdot 19+4\cdot 19=(96+4)19$
$= (100)19$
$= 1900$

57. $58\cdot\frac{3}{2}-8\cdot\frac{3}{2}=(58-8)\frac{3}{2}$
$= (50)\frac{3}{2}=\frac{50}{1}\cdot\frac{3}{2}$
$= \frac{150}{2}=75$

59. $4.31(69)+4.31(31)=(69+31)4.31$
$= (100)4.31$
$= 431$

61. The terms have been grouped using the associative property of addition.

62. The terms have been regrouped using the associative property of addition.

63. The order of the terms inside the parentheses has been changed using the commutative property of addition.

64. The terms have been regrouped using the associative property of addition.

65. The common factor, x, has been distributed over the expression in parentheses using the distributive property.

66. The numbers in parentheses have been added to simplify the expression.

67. Answers will vary but should include the following properties: distributive, inverse (2), identity (2), associative (2), commutative (2), and multiplying by 0. See the summary in the Quick Review for Section 1.4 on page 43 of the text.

69. Is $a+(b\cdot c)=(a+b)(a+c)$?
No. One example is

$$7+(5\cdot 3)=(7+5)(7+3),$$

which is false since

$$7+(5\cdot 3)=7+15=22$$

and

$$(7+5)(7+3)=12(10)=120.$$

Chapter 1 Review Exercises

1. $\left\{-4,-1,2,\frac{9}{4},4\right\}$

Place dots for $-4,-1,2,\frac{9}{4}=2.25$, and 4 on a number line.

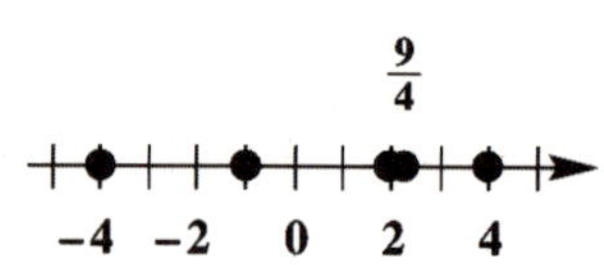

2. $\left\{-5,-\frac{11}{4},-.5,0,3,\frac{13}{3}\right\}$

Place dots for $-5,-\frac{11}{4}=-2.75,-.5,0,3$, and $\frac{13}{3}=4.\overline{3}$ on a number line.

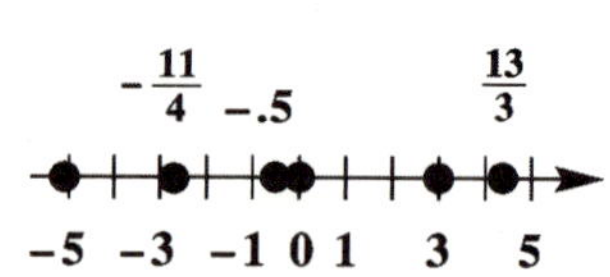

3. $|-16|=-(-16)=16$

4. $-|-4|=-[-(-4)]=-(4)=-4$

5. $|-8|-|-3|=-(-8)-[-(-3)]$
$= 8-[3]$
$= 5$

In Exercises 6–9,

$$S=\left\{-9,-\frac{4}{3},-\sqrt{4},-.25,0,.\overline{35},\frac{5}{3},\sqrt{7},\sqrt{-9},\frac{12}{3}\right\}.$$

6. The elements 0 and $\frac{12}{3}$ (or 4) are whole numbers.

7. The elements -9, $-\sqrt{4}$ (or -2), 0, and $\frac{12}{3}$ (or 4) are integers.

8. The elements $-9, -\frac{4}{3}, -\sqrt{4}$ (or -2), $-.25, 0, .\overline{35}, \frac{5}{3}$, and $\frac{12}{3}$ (or 4) are rational numbers. (Remember that terminating and repeating decimals are rational numbers.)

9. All the elements in the set are real numbers except $\sqrt{-9}$.

10. $\{x \mid x$ is a natural number between 3 and 9$\}$
The natural numbers between 3 and 9 are $4, 5, 6, 7,$ and 8. Therefore, the set is $\{4, 5, 6, 7, 8\}$.

11. $\{y \mid y$ is a whole number less than 4$\}$
The whole numbers less than 4 are $0, 1, 2,$ and 3. Therefore, the set is $\{0, 1, 2, 3\}$.

12. $4 \cdot 2 \leq |12 - 4|$
$8 \leq |8|$
$8 \leq 8$ *True*

13. $2 + |-2| > 4$
$2 + 2 > 4$
$4 > 4$ *False*

14. $4(3 + 7) > -|40|$
$4(10) > -40$
$40 > -40$ *True*

15. The longest bar represents the greatest change. Hyundai had the greatest change of 50%.

16. The shortest bar represents the smallest change. General Motors had the smallest change of -5%.

17. Honda: $|11| = 11$

Toyota: $|-11| = 11$

These are equal, not greater than. The answer is *false*.

18. Hyundai $>$ 4(Honda) ?
$50 > 4(11)$?
$50 > 44$ *True*

19. $\{x \mid x < -5\}$
In interval notation, $x < -5$ is written as $(-\infty, -5)$.

The parenthesis at -5 indicates that -5 is not included. The graph extends from -5 to the left.

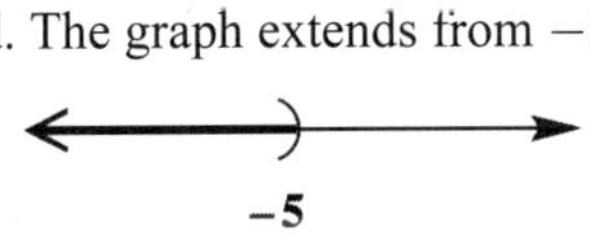

20. $\{x \mid -2 < x \leq 3\}$
In interval notation, $-2 < x \leq 3$ is written as $(-2, 3]$. The parenthesis indicates that -2 is not included, while the bracket indicates that 3 is included. The graph goes from -2 to 3.

21. $$-\frac{5}{8} - \left(-\frac{7}{3}\right) = -\frac{5}{8} + \frac{7}{3} = -\frac{15}{24} + \frac{56}{24} = \frac{41}{24}$$

22. $$-\frac{4}{5} - \left(-\frac{3}{10}\right) = -\frac{4}{5} + \frac{3}{10} = -\frac{8}{10} + \frac{3}{10} = -\frac{5}{10} = -\frac{1}{2}$$

23. $-5 + (-11) + 20 - 7$
$= -16 + 20 - 7$
$= 4 - 7$
$= -3$

24. $-9.42 + 1.83 - 7.6 - 1.9$
$= -7.59 - 7.6 - 1.9$
$= -15.19 - 1.9$
$= -17.09$

25. $-15 + (-13) + (-11)$
$= -28 + (-11)$
$= -39$

26. $-1 - 3 - (-10) + (-7)$
$= -4 + 10 + (-7)$
$= 6 + (-7)$
$= -1$

27. $$\frac{3}{4} - \left(\frac{1}{2} - \frac{9}{10}\right) = \frac{3}{4} - \left(\frac{5}{10} - \frac{9}{10}\right) = \frac{3}{4} - \left(-\frac{4}{10}\right) = \frac{3}{4} + \frac{4}{10} = \frac{15}{20} + \frac{8}{20} = \frac{23}{20}$$

28. $-|-12| - |-9| + (-4) - |10|$
$= -(12) - 9 + (-4) - 10$
$= -21 + (-4) - 10$
$= -25 - 10$
$= -35$

29. $11{,}049 - (-282) = 11{,}049 + 282$
$= 11{,}331$

The difference is 11,331 feet.

30. $2(-5)(-3)(-3) = (-10)(-3)(-3)$
$= (30)(-3)$
$= -90$

31. $-\frac{3}{7}\left(-\frac{14}{9}\right) = \frac{3}{7} \cdot \frac{2 \cdot 7}{3 \cdot 3} = \frac{2}{3}$

32. $\frac{75}{-5} = 15 \cdot 5\left(\frac{1}{-5}\right) = -15$

33. $\frac{-2.3754}{-.74} = 3.21$

34. $\frac{5}{7-7} = \frac{5}{0}$, which is undefined.

$\frac{7-7}{5} = \frac{0}{5} = 0$

35. $10^4 = 10 \cdot 10 \cdot 10 \cdot 10 = 10{,}000$

36. $\left(\frac{3}{7}\right)^3 = \frac{3}{7} \cdot \frac{3}{7} \cdot \frac{3}{7} = \frac{27}{343}$

37. $(-5)^3 = (-5)(-5)(-5) = -125$

38. $-5^3 = -(5 \cdot 5 \cdot 5) = -125$

39. $\sqrt{400} = 20$, because 20 is positive and $20^2 = 400$.

40. $\sqrt{\frac{64}{121}} = \frac{8}{11}$ since $\frac{8}{11}$ is positive and $\left(\frac{8}{11}\right)^2 = \frac{64}{121}$.

41. $-\sqrt{.81} = -.9$ since $-.9^2 = -.81$.

42. $\sqrt{-64}$ is not a real number.

43. $-14\left(\frac{3}{7}\right) + 6 \div 3 = -2(3) + (6 \div 3)$

$= -6 + 2 = -4$

44. $-\frac{2}{3}[5(-2) + 8 - 4^3]$

$= -\frac{2}{3}[5(-2) + 8 - 64]$ *Evaluate the power.*

$= -\frac{2}{3}[-10 + 8 - 64]$

$= -\frac{2}{3}[-66] = \frac{2}{3}(22 \cdot 3) = 44$

45. $\frac{-5(3^2) + 9\left(\sqrt{4}\right) - 5}{6 - 5(-2)}$

$= \frac{-5(9) + 9(2) - 5}{6 + 10}$

$= \frac{-45 + 18 - 5}{16}$

$= \frac{-32}{16} = -2$

In Exercises 46–48, let $k = -4$, $m = 2$, and $n = 16$.

46. $4k - 7m = 4(-4) - 7(2)$

$= -16 - 14$

$= -30$

47. $-3\sqrt{n} + m + 5k$

$= -3\left(\sqrt{16}\right) + 2 + 5(-4)$

$= -3(4) + 2 - 20$

$= -12 - 18$

$= -30$

48. $\frac{4m^3 - 3n}{7k^2 - 10} = \frac{4(2)^3 - 3(16)}{7(-4)^2 - 10}$

$= \frac{4(8) - 3(16)}{7(16) - 10}$

$= \frac{32 - 48}{112 - 10}$

$= \frac{-16}{102} = -\frac{8}{51}$

49. **(a)** 6 ft 3 in. $= (6 \cdot 12 + 3)$ in. $= 75$ in.

$704 \times$ (weight in pounds) $\div$ (height in inches)^2

$= 704 \times 195 \div 75^2$

$= 704 \times 195 \div 5625$

$= 137{,}280 \div 5625$

$\approx 24.41 \approx 24$

Derek Jeter's BMI is 24.

(b) Answers will vary.

50. $2q + 19q$

$= (2 + 19)q$ *Distributive property*

$= 21q$

51. $13z - 17z$

$= (13 - 17)z$ *Distributive property*

$= -4z$

52. $-m + 6m$

$= -1m + 6m$ *Identity property*

$= (-1 + 6)m$ *Distributive property*

$= 5m$

53. $5p - p$

$= 5p + (-1)p$ *Identity property*

$= [5 + (-1)]p$ *Distributive property*

$= 4p$

54. $-2(k + 3)$

$= -2(k) + (-2)(3)$ *Distributive property*

$= -2k - 6$

55. $6(r+3)$

$= 6(r) + 6(3)$ *Distributive property*

$= 6r + 18$

56. $9(2m+3n)$

$= 9(2m) + 9(3n)$ *Distributive property*

$= 18m + 27n$

57. $-(-p+6q)-(2p-3q)$

$= -1(-p+6q)+(-1)(2p-3q)$

$= -1(-p)+(-1)(6q)+(-1)(2p)+(-1)(-3q)$

$= p-6q-2p+3q$

$= p-2p-6q+3q$

$= -p-3q$

58. $-3y+6-5+4y$

$= -3y+4y+6-5$

$= y+1$

59. $2a+3-a-1-a-2$

$= 2a-a-a+3-1-2$

$= 0$

60. $-3(4m-2)+2(3m-1)-4(3m+1)$

$= -12m+6+6m-2-12m-4$

$= -12m+6m-12m+6-2-4$

$= -18m$

61. $2x+3x=(2+3)x=5x$ *Distributive property*

62. $-4\cdot 1=-4$ *Identity property*

63. $2(4x)=(2\cdot 4)x=8x$ *Associative property*

64. $-3+13=13+(-3)=10$ *Commutative property*

65. $-3+3=0$ *Inverse property*

66. $5(x+z)=5x+5z$ *Distributive property*

67. $0+7=7$ *Identity property*

68. $8\cdot\frac{1}{8}=1$ *Inverse property*

69. 1996

$|19.07-21.39|=|-2.32|=2.32$

The company experienced a loss of \$2.32 million (in the red).

70. 1997

$|67.41-90.29|=|-22.88|=22.88$

The company experienced a loss of \$22.88 million (in the red).

71. 1998

$|203.20-177.61|=|25.59|=25.59$

The company made a profit of \$25.59 million (in the black).

72. $\left(-\frac{4}{5}\right)^4=\left(-\frac{4}{5}\right)\left(-\frac{4}{5}\right)\left(-\frac{4}{5}\right)\left(-\frac{4}{5}\right)=\frac{256}{625}$

73. $-\frac{5}{8}(-40)=-\frac{5}{8}\cdot\frac{-40}{1}$

$= 5\cdot 5=25$

74. $-25\left(-\frac{4}{5}\right)+3^3-32\div\sqrt{4}$

$= -25\left(-\frac{4}{5}\right)+27-32\div 2$

$= 20+27-16$

$= 31$

75. $-8+|-14|+|-3|=-8+14+3$

$= 9$

76. $\dfrac{6\cdot\sqrt{4}-3\cdot\sqrt{16}}{-2\cdot 5+7(-3)-10}$

$= \dfrac{6\cdot 2-3\cdot 4}{-2\cdot 5+7(-3)-10}$

$= \dfrac{12-12}{-10-21-10}$

$= \dfrac{0}{-41}=0$

77. $-\sqrt{25}=-(5)=-5$

78. $-\frac{10}{21}\div-\frac{5}{14}=-\frac{10}{21}\cdot-\frac{14}{5}$

$= \frac{2\cdot 5}{3\cdot 7}\cdot\frac{2\cdot 7}{5}$

$= \frac{2\cdot 2}{3}=\frac{4}{3}$

79. $.8-4.9-3.2+1.14$

$= -4.1-3.2+1.14$

$= -7.3+1.14$

$= -6.16$

80. $-3^2=-(3\cdot 3)=-9$

81. $\frac{-38}{-19}=38\left(\frac{1}{19}\right)=2\cdot 19\left(\frac{1}{19}\right)=2$

82. $-2(k-1)+3k-k$

$= -2k+2+3k-k$

$= -2k+3k-k+2$

$= (-2+3-1)k+2$

$= 0k+2=2$

83. Since there is no real number whose square is -100, $-\sqrt{-100}$ is not a real number.

84. $-(3k - 4h)$
$= -1(3k - 4h)$ *Identity property*
$= -1(3k) + (-1)(-4h)$
$= -3k + 4h$

85. $-4.6(2.48) = -11.408$

86. $-\frac{2}{3}(-15) + (2^4 - 8 \div 4)$
$= 10 + [16 - (8 \div 4)]$
$= 10 + (16 - 2)$
$= 10 + 14 = 24$

87. $-2x + 5 - 4x - 1$
$= -2x - 4x + 5 - 1$
$= -6x + 4$

88. $-\frac{2}{3} - \left(\frac{1}{6} - \frac{5}{9}\right) = -\frac{2}{3} - \left(\frac{3}{18} - \frac{10}{18}\right)$
$= -\frac{2}{3} - \left(-\frac{7}{18}\right)$
$= -\frac{2}{3} + \frac{7}{18}$
$= -\frac{12}{18} + \frac{7}{18} = -\frac{5}{18}$

89. $-m(3k^2 + 5m) = -2[3(-4)^2 + 5(2)]$
Let k= –4, m=2
$= -2[3(16) + 5(2)]$
$= -2[48 + 10]$
$= -2[58]$
$= -116$

90. In order to evaluate $(3 + 2)^2$, you should work within the parentheses first.

Chapter 1 Test

1. $\left\{-3, .75, \frac{5}{3}, 5, 6.3\right\}$

Place dots at $-3, .75, \frac{5}{3} = 1.\overline{6}, 5,$ and 6.3.

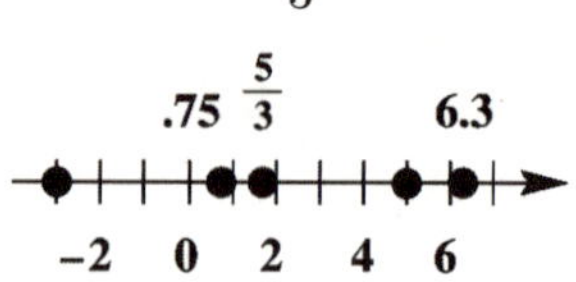

In Exercises 2–5,

$$A = \left\{-\sqrt{6}, -1, -.5, 0, 3, \sqrt{25}, 7.5, \frac{24}{2}, \sqrt{-4}\right\}.$$

2. The elements $0, 3, \sqrt{25}$ (or 5), and $\frac{24}{2}$ (or 12) are whole numbers.

3. The elements $-1, 0, 3, \sqrt{25}$ (or 5), and $\frac{24}{2}$ (or 12) are integers.

4. The elements $-1, -.5, 0, 3, \sqrt{25}$ (or 5), 7.5, and $\frac{24}{2}$ (or 12) are rational numbers.

5. All the elements in the set are real numbers except $\sqrt{-4}$.

6. $\{x \mid x < -3\}$
In interval notation, $x < -3$ is written as $(-\infty, -3)$. The parenthesis at -3 indicates that -3 is not included.. The graph extends from -3 to the left.

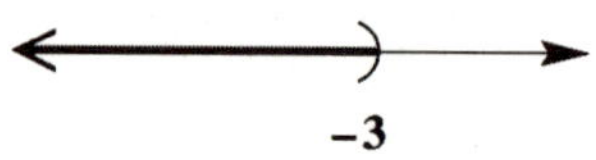

7. $\{y \mid -4 < y \le 2\}$
In interval notation, $-4 < y \le 2$ is written as $(-4, 2]$. The parenthesis indicates that -4 is not included, while the bracket indicates that 2 is included. The graph goes from -4 to 2.

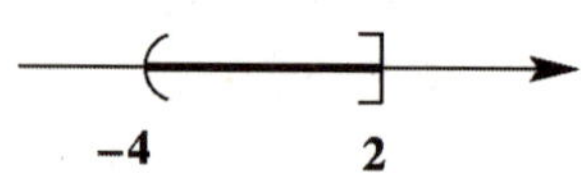

8. $-6 + 14 + (-11) - (-3)$
$= 8 + (-11) + 3$
$= -3 + 3 = 0$

9. $10 - 4 \cdot 3 + 6(-4)$
$= 10 - 12 + (-24)$
$= -2 + (-24) = -26$

10. $7 - 4^2 + 2(6) + (-4)^2$
$= 7 - 16 + 12 + 16$
$= 19$

11. $\frac{10 - 24 + (-6)}{\sqrt{16}(-5)}$
$= \frac{-14 + (-6)}{4(-5)}$
$= \frac{-20}{-20} = 1$

12. $\frac{-2[3 - (-1 - 2) + 2]}{\sqrt{9}(-3) - (-2)}$
$= \frac{-2[3 - (-3) + 2]}{3(-3) - (-2)}$
$= \frac{-2[8]}{-9 - (-2)}$
$= \frac{-16}{-7} = \frac{16}{7}$

13. $\dfrac{8\cdot 4-3^2\cdot 5-2(-1)}{-3\cdot 2^3+1}$

$$=\frac{8\cdot 4-9\cdot 5-2(-1)}{-3\cdot 8+1}$$
$$=\frac{32-45+2}{-24+1}$$
$$=\frac{-11}{-23}=\frac{11}{23}$$

14. $17{,}400-(-32{,}995)=17{,}400+32{,}995$
$$=50{,}395$$

The difference between the height of Mt. Foraker and the depth of the Philippine Trench is 50,395 feet.

15. $14{,}110-(-23{,}376)=14{,}110+23{,}376$
$$=37{,}486$$

The difference between the height of Pikes Peak and the depth of the Java Trench is 37,486 feet.

16. $-24{,}721-(-23{,}376)=-24{,}721+23{,}376$
$$=-1345$$

The Cayman Trench is 1345 feet deeper than the Java Trench.

17. $\sqrt{196}=14$, because 14 is positive and $14^2=196$.

18. $-\sqrt{225}=-\left(\sqrt{225}\right)=-(15)=-15$

19. Since there is no real number whose square is -16, $\sqrt{-16}$ is not a real number.

20. **(a)** If a is positive, then $\sqrt{a}$ will represent a positive number.

(b) If a is negative, then $\sqrt{a}$ will not represent a real number.

(c) If a is 0, then $\sqrt{a}$ will be 0.

21. $\dfrac{8k+2m^2}{r-2}=\dfrac{8(-3)+2(-3)^2}{25-2}$

Let k= –3, m= –3, r=25

$$=\frac{8(-3)+2(9)}{23}$$
$$=\frac{-24+18}{23}$$
$$=\frac{-6}{23}\text{ or }-\frac{6}{23}$$

22. $-3(2k-4)+4(3k-5)-2+4k$

$$=-3(2k)+(-3)(-4)+4(3k)$$
$$+4(-5)-2+4k$$
$$=-6k+12+12k-20-2+4k$$
$$=-6k+12k+4k+12-20-2$$
$$=10k-10$$

23. When simplifying

$$(3r+8)-(-4r+6),$$

the subtraction sign in front of $(-4r+6)$ changes the sign of the terms $-4r$ and 6.

$$(3r+8)-(-4r+6)$$
$$=3r+8-(-4r)-6$$
$$=3r+8+4r-6$$
$$=3r+4r+8-6$$
$$=7r+2$$

24. $6+(-6)=0$

The answer is **B**, *Inverse property.*
The sum of 6 and its inverse, -6, equals zero.

25. $-2+(3+6)=(-2+3)+6$

The answer is **D**, *Associative property.*
The order of the terms is the same, but the grouping has changed.

26. $5x+15x=(5+15)x$

The answer is **A**, *Distributive property.*
This is the second form of the distributive property.

27. $13\cdot 0=0$

The answer is **F**, *Multiplication property of 0.*
Multiplication by 0 always equal 0.

28. $-9+0=-9$

The answer is **C**, *Identity property.*
The addition of 0 to any number does not change the number.

29. $4\cdot 1=4$

The answer is **C**, *Identity property.*
Multiplication of any number by 1 does not change the number.

30. $(a+b)+c=(b+a)+c$

The answer is **E**, *Commutative property.*
The order of the terms a and b is reversed.

CHAPTER 2 LINEAR EQUATIONS, INEQUALITIES, AND APPLICATIONS

Section 2.1

1. **A.** $3x + x - 1 = 0$ can be written as $4x = 1$, so it is linear.

   C. $6x + 2 = 9$ is in linear form.

3. $3(x+4) = 5x$ *Original equation*
   $3(6+4) = 5 \cdot 6$? *Let x=6.*
   $3(10) = 30$? *Add.*
   $30 = 30$ *True*
   Since a true statement is obtained, 6 is a solution.

5. Suppose your last name is Lincoln. Then $x = 7$ and both sides are evaluated as -48. The equation is an identity, so any number is a solution.

7. **(a)** $3x = 6$ is an *equation* because it contains an equals sign.

   (b) $3x + 6$ is an *expression* because it does not contain an equals sign.

   (c) $5x + 6(x-3) = 12x + 6$ is an *equation* because it contains an equals sign.

   (d) $5x + 6(x-3) - (12x+6)$ is an *expression* because it does not contain an equals sign.

In the following exercises, we do not show the checks of the solutions. To be sure that your solutions are correct, check them by substituting into the original equations.

9. $7x + 8 = 1$
   $7x + 8 - 8 = 1 - 8$ *Subtract 8.*
   $7x = -7$
   $\frac{7x}{7} = \frac{-7}{7}$ *Divide by 7.*
   $x = -1$
   Solution set: $\{-1\}$

11. $7x - 5x + 15 = x + 8$
    $2x + 15 = x + 8$ *Combine terms.*
    $2x = x - 7$ *Subtract 15.*
    $x = -7$ *Subtract x.*
    Solution set: $\{-7\}$

13. $12w + 15w - 9 + 5 = -3w + 5 - 9$
    $27w - 4 = -3w - 4$ *Combine terms.*
    $30w - 4 = -4$ *Add 3w.*
    $30w = 0$ *Add 4.*
    $w = 0$ *Divide by 30.*
    Solution set: $\{0\}$

15. $2(x+3) = -4(x+1)$
    $2x + 6 = -4x - 4$ *Remove parentheses.*
    $6x + 6 = -4$ *Add 4x.*
    $6x = -10$ *Subtract 6.*
    $x = \frac{-10}{6} = -\frac{5}{3}$ *Divide by 6.*
    Solution set: $\left\{-\frac{5}{3}\right\}$

17. $3(2w+1) - 2(w-2) = 5$
    $6w + 3 - 2w + 4 = 5$ *Remove parentheses.*
    $4w + 7 = 5$ *Combine terms.*
    $4w = -2$ *Subtract 7.*
    $w = \frac{-2}{4} = -\frac{1}{2}$ *Divide by 4.*
    Solution set: $\left\{-\frac{1}{2}\right\}$

19. $2x + 3(x-4) = 2(x-3)$
    $2x + 3x - 12 = 2x - 6$
    $5x - 12 = 2x - 6$
    $3x = 6$
    $x = \frac{6}{3} = 2$
    Solution set: $\{2\}$

21. $6p - 4(3-2p) = 5(p-4) - 10$
    $6p - 12 + 8p = 5p - 20 - 10$
    $14p - 12 = 5p - 30$
    $9p = -18$
    $p = -2$
    Solution set: $\{-2\}$

23. $-[2z - (5z+2)] = 2 + (2z+7)$
    $-[2z - 5z - 2] = 2 + 2z + 7$
    $-2z + 5z + 2 = 2 + 2z + 7$
    $3z + 2 = 2z + 9$
    $z = 7$
    Solution set: $\{7\}$

25. $-3m + 6 - 5(m-1) = 4m - (2m-4) - 9m + 5$
    $-3m + 6 - 5m + 5 = 4m - 2m + 4 - 9m + 5$
    $-8m + 11 = -7m + 9$
    $-m = -2$
    $m = 2$
    Solution set: $\{2\}$

27.
$$\begin{aligned} -[3x-(2x+5)] &= -4-[3(2x-4)-3x] \\ -[3x-2x-5] &= -4-[6x-12-3x] \\ -[x-5] &= -4-[3x-12] \\ -x+5 &= -4-3x+12 \\ -x+5 &= -3x+8 \\ 2x &= 3 \\ x &= \frac{3}{2} \end{aligned}$$

Solution set: $\left\{\frac{3}{2}\right\}$

29.
$$\begin{aligned} -(9-3a)-(4+2a)-3 &= -(2-5a)+(-a)+1 \\ -9+3a-4-2a-3 &= -2+5a-a+1 \\ a-16 &= 4a-1 \\ -15 &= 3a \\ -5 &= a \end{aligned}$$

Solution set: $\{-5\}$

31. The denominators of the fractions are 3, 4, and 1. The least common denominator is $(3)(4)(1)=12$, since it is the smallest number into which each denominator can divide without a remainder.

33. Yes, the coefficients will be larger, but you will get the correct solution. As long as you multiply both sides of the equation by the *same* nonzero number, the resulting equation is equivalent and the solution does not change.

35.
$$\begin{aligned} -\frac{5}{9}k &= 2 \\ -5k &= 18 && \textit{Multiply by 9.} \\ k &= \frac{18}{-5} = -\frac{18}{5} && \textit{Divide by } -5. \end{aligned}$$

Solution set: $\left\{-\frac{18}{5}\right\}$

37.
$$\begin{aligned} \frac{6}{5}x &= -1 \\ 6x &= -5 && \textit{Multiply by 5.} \\ x &= \frac{-5}{6} = -\frac{5}{6} && \textit{Divide by 6.} \end{aligned}$$

Solution set: $\left\{-\frac{5}{6}\right\}$

39.
$$\frac{m}{2}+\frac{m}{3}=5$$

Multiply by the LCD, 6.

$$\begin{aligned} 6\left(\frac{m}{2}+\frac{m}{3}\right) &= 6(5) \\ 6\left(\frac{m}{2}\right)+6\left(\frac{m}{3}\right) &= 30 && \textit{Distributive property} \\ 3m+2m &= 30 \\ 5m &= 30 && \textit{Add.} \\ m &= 6 && \textit{Divide by 5.} \end{aligned}$$

Solution set: $\{6\}$

41.
$$\frac{x-10}{5}+\frac{2}{5}=-\frac{x}{3}$$

Multiply both sides by the LCD, 15.

$$\begin{aligned} 15\left(\frac{x-10}{5}+\frac{2}{5}\right) &= 15\left(-\frac{x}{3}\right) \\ 3(x-10)+3(2) &= -5x \\ 3x-30+6 &= -5x \\ 8x &= 24 \\ x &= \frac{24}{8} = 3 \end{aligned}$$

Solution set: $\{3\}$

43.
$$\frac{4t+1}{3}=\frac{t+5}{6}+\frac{t-3}{6}$$

Multiply both sides by the LCD, 6.

$$\begin{aligned} 6\left(\frac{4t+1}{3}\right) &= 6\left(\frac{t+5}{6}+\frac{t-3}{6}\right) \\ 2(4t+1) &= (t+5)+(t-3) \\ 8t+2 &= 2t+2 \\ 6t &= 0 \\ t &= 0 \end{aligned}$$

Solution set: $\{0\}$

45. $.05x+.12(x+5000)=940$

Multiply both sides by 100.

$$\begin{aligned} 5x+12(x+5000) &= 100(940) \\ 5x+12x+60,000 &= 94,000 \\ 17x &= 34,000 \\ x &= 2000 \end{aligned}$$

Solution set: $\{2000\}$

47. $.02(50)+.08r=.04(50+r)$

Multiply both sides by 100.

$$\begin{aligned} 2(50)+8r &= 4(50+r) \\ 100+8r &= 200+4r \\ 4r &= 100 \\ r &= 25 \end{aligned}$$

Solution set: $\{25\}$

49. $.05x+.10(200-x)=.45x$

Multiply both sides by 100.

$$\begin{aligned} 5x+10(200-x) &= 45x \\ 5x+2000-10x &= 45x \\ 2000-5x &= 45x \\ 2000 &= 50x \\ 40 &= x \end{aligned}$$

Solution set: $\{40\}$

51. By dividing the equation $8x=7x$ by x, he possibly divided the equation by 0. He should have subtracted $7x$ from both sides of the equation to get $x=0$. The solution set is $\{0\}$.

53. $$\begin{aligned} -2p+5p-9 &= 3(p-4)-5 \\ 3p-9 &= 3p-12-5 \\ 3p-9 &= 3p-17 \\ -9 &= -17 \quad \textit{False} \end{aligned}$$
The equation is a *contradiction*.
Solution set: $\emptyset$

55. $$\begin{aligned} 6x+2(x-2) &= 9x+4 \\ 6x+2x-4 &= 9x+4 \\ 8x-4 &= 9x+4 \\ -8 &= x \end{aligned}$$
This is a *conditional* equation.
Solution set: $\{-8\}$

57. $$\begin{aligned} -11m+4(m-3)+6m &= 4m-12 \\ -11m+4m-12+6m &= 4m-12 \\ -m-12 &= 4m-12 \\ 0 &= 5m \\ 0 &= m \end{aligned}$$
This is a *conditional* equation.
Solution set: $\{0\}$

59. $$\begin{aligned} 7[2-(3+4r)]-2r &= -9+2(1-15r) \\ 7[2-3-4r]-2r &= -9+2-30r \\ 7[-1-4r]-2r &= -7-30r \\ -7-28r-2r &= -7-30r \\ -7-30r &= -7-30r \end{aligned}$$
The equation is an *identity*.
Solution set: {all real numbers}

61. $5x=10$ and $\dfrac{5x}{x+2}=\dfrac{10}{x+2}$
Solve the first equation.
$$\begin{aligned} 5x &= 10 \\ x &= 2 \end{aligned}$$
Solution set: $\{2\}$
Solve the second equation.
$$\frac{5x}{x+2}=\frac{10}{x+2}$$
Since the denominators are the same, the numerators must be equal.
$$\begin{aligned} 5x &= 10 \\ x &= 2 \end{aligned}$$
Solution set: $\{2\}$
Since these two equations have the same solution set, they *are equivalent* equations.

63. $x=-3$ and $\dfrac{x}{x+3}=\dfrac{-3}{x+3}$
The solution set of the first equation is $\{-3\}$. Since -3 would cause the denominators in the second equation to become zero, it cannot be used as a solution. Therefore, the equations *are not equivalent*.

65. $k=4$ and $k^2=16$
The solution set of the first equation is $\{4\}$. Since $(-4)^2=16$ and $4^2=16$, the solution set of the second equation is $\{-4,4\}$.
Since these two equations do not have the same solution set, they *are not equivalent* equations.

Section 2.2

1. **(a)** $$\begin{aligned} x &= \frac{5x+8}{3} \\ 3x &= 3\left(\frac{5x+8}{3}\right) \\ 3x &= 5x+8 \end{aligned}$$
(b) $$\begin{aligned} t &= \frac{bt+k}{c}\ (c\neq 0) \\ ct &= c\left(\frac{bt+k}{c}\right) \\ ct &= bt+k \end{aligned}$$

2. **(a)** $$\begin{aligned} 3x-5x &= 5x+8-5x \\ 3x-5x &= 8 \end{aligned}$$
(b) $$\begin{aligned} ct-bt &= bt+k-bt \\ ct-bt &= k \end{aligned}$$

3. Use the distributive property in each case.
(a) $-2x=8$ **(b)** $(c-b)t=k$

4. **(a)** $$\begin{aligned} \frac{-2x}{-2} &= \frac{8}{-2} \\ x &= -4 \end{aligned}$$
(b) $$\begin{aligned} \frac{(c-b)t}{c-b} &= \frac{k}{c-b} \\ t &= \frac{k}{c-b} \end{aligned}$$

5. The restriction $b\neq c$ must be applied. If $b=c$, the denominator becomes 0 and division by 0 is undefined.

6. To solve an equation for a particular variable, such as solving the second equation for t, go through the same steps as you would in solving for x in the first equation. Treat all other variables as constants.

7. Solve $I=prt$ for r.
$$\begin{aligned} I &= prt \\ \frac{I}{pt} &= \frac{prt}{pt} \\ \frac{I}{pt} &= r \quad \text{or} \quad r=\frac{I}{pt} \end{aligned}$$

9. Solve $P=2L+2W$ for L.
$$\begin{aligned} P &= 2L+2W \\ P-2W &= 2L \\ \frac{P-2W}{2} &= \frac{2L}{2} \\ \frac{P-2W}{2} &= L \quad \text{or} \quad L=\frac{P}{2}-W \end{aligned}$$

11. Solve for $V = LWH$ for W.

$$V = LWH$$
$$\frac{V}{LH} = \frac{LWH}{LH}$$
$$\frac{V}{LH} = W$$

13. Solve $C = 2\pi r$ for r.

$$C = 2\pi r$$
$$\frac{C}{2\pi} = \frac{2\pi r}{2\pi} \quad \textit{Divide by } 2\pi.$$
$$\frac{C}{2\pi} = r$$

15. Solve $A = \frac{1}{2}(B+b)h$ for B.

$$2A = (B+b)h \quad \text{or} \quad 2A = Bh + bh$$
$$\frac{2A}{h} = B + b \qquad 2A - bh = Bh$$
$$\frac{2A}{h} - b = B \quad \text{or} \quad \frac{2A - bh}{h} = B$$

17. Solve $F = \frac{9}{5}C + 32$ for C.

$$F - 32 = \frac{9}{5}C$$
$$\frac{5}{9}(F - 32) = \frac{5}{9}\left(\frac{9}{5}C\right)$$
$$\frac{5}{9}(F - 32) = C$$

19. $A = \frac{1}{2}bh$

$$2A = 2\left(\frac{1}{2}bh\right) \quad \textit{Multiply by 2.}$$
$$2A = bh$$
$$\frac{2A}{b} = \frac{bh}{b} \quad \textit{Divide by b.}$$
$$\frac{2A}{b} = h$$
$$\frac{2A}{b} = \frac{2}{1} \cdot \frac{A}{b}$$
$$= 2\left(\frac{A}{b}\right) \quad \textit{This choice is } \textbf{A}.$$
$$= 2A\left(\frac{1}{b}\right) \quad \textit{This is choice } \textbf{B}.$$

To get choice **C**, multiply $\frac{2A}{b}$ by $1 = \frac{\frac{1}{2}}{\frac{1}{2}}$.

$$\frac{\frac{1}{2}}{\frac{1}{2}} \cdot \frac{2A}{b} = \frac{A}{\frac{1}{2}b}$$

Choice **D**, $h = \frac{\frac{1}{2}A}{b}$, can be multiplied by $\frac{2}{2}$ on the right side to get $h = \frac{A}{2b}$, so it is *not* equivalent to $h = \frac{2A}{b}$. Therefore, the correct answer is **D**.

21. Solve $2k + ar = r - 3y$ for r.
Get the "r-terms" on one side and the other terms on the other side.

$$ar - r = -2k - 3y$$
$$(a-1)r = -2k - 3y \quad \textit{Distributive property}$$
$$r = \frac{-2k - 3y}{a - 1} \quad \textit{Divide by a–1.}$$

The answer can also be written as

$$r = \frac{2k + 3y}{1 - a},$$

which would occur if you took the r-terms to the right side in your first step.

23. Solve $w = \frac{3y - x}{y}$ for y.

$$wy = 3y - x \quad \textit{Multiply by y.}$$
$$x = 3y - wy \quad \textit{Get y-terms on one side.}$$
$$x = (3 - w)y \quad \textit{Distributive property}$$
$$\frac{x}{3 - w} = y \quad \textit{Divide by 3–w.}$$

Equivalently, we have

$$y = \frac{-x}{w - 3}.$$

25. Solve $d = rt$ for t.

$$t = \frac{d}{r}$$

To find t, substitute $d = 600$ and $r = 136.424$.

$$t = \frac{600}{136.424} \approx 4.398$$

His time was about 4.398 hours.

27. Solve $d = rt$ for r.

$$r = \frac{d}{t}$$
$$r = \frac{520}{10} = 52 \quad \textit{Let d=520, t=10.}$$

Her rate was 52 mph.

29. Use the formula $F = \frac{9}{5}C + 32$.

$$\begin{aligned} F &= \frac{9}{5}(40) + 32 \quad \textit{Let } C = 40. \\ &= 72 + 32 \\ &= 104 \end{aligned}$$

The corresponding temperature is 104°F.

31. Solve $P = 4s$ for s.

$$s = \frac{P}{4}$$

To find s, substitute 920 for P.

$$s = \frac{920}{4} = 230$$

The length of each side is 230 m.

33. Use the formula $C = 2\pi r$.

$$\begin{aligned} 480\pi &= 2\pi r \quad \textit{Let } C = 480\pi. \\ \frac{480\pi}{2\pi} &= r \quad \textit{Divide by } 2\pi. \end{aligned}$$

So the radius of the circle is 240 inches and the diameter is twice that length, that is, 480 inches.

35. Solve $V = LWH$ for L.

$$L = \frac{V}{WH}$$

To find L, substitute $V = 128$, $W = 4$, and $H = 4$.

$$L = \frac{128}{(4)(4)} = 8$$

The stack is 8 ft long.

37. The mixture is 36 oz and that part which is alcohol is 9 oz. Thus, the percent of alcohol is

$$\frac{9}{36} = \frac{1}{4} = \frac{25}{100} = 25\%.$$

The percent of water is

$$100\% - 25\% = 75\%.$$

39. Find what percent \$6300 is of \$210,000.

$$\frac{6300}{210,000} = .03 = 3\%$$

The agent received a 3% rate of commission.

In Exercises 41–43, use the rule of 78.

$$u = f \cdot \frac{k(k+1)}{n(n+1)}$$

41. Substitute 700 for f, 4 for k, and 36 for n.

$$\begin{aligned} u &= 700 \cdot \frac{4(4+1)}{36(36+1)} \\ &= 700 \cdot \frac{4(5)}{36(37)} \approx 10.51 \end{aligned}$$

The unearned interest is \$10.51.

43. Substitute 380.50 for f, 8 for k, and 24 for n.

$$\begin{aligned} u &= (380.50) \cdot \frac{8(8+1)}{24(24+1)} \\ &= (380.50) \cdot \frac{8(9)}{24(25)} \approx 45.66 \end{aligned}$$

The unearned interest is \$45.66.

45. **(a)** Chicago:

$$\text{Pct.} = \frac{W}{W+L} = \frac{87}{87+58} = \frac{87}{145} = .600$$

(b) Cleveland:

$$\text{Pct.} = \frac{W}{W+L} = \frac{77}{77+65} = \frac{77}{142} \approx .542$$

(c) Detroit:

$$\text{Pct.} = \frac{W}{W+L} = \frac{71}{71+74} = \frac{71}{145} \approx .490$$

47. In 1982, 3% of the television audience watched cable.

$$\begin{aligned} &3\% \text{ of } 50,000 \\ &= .03(50,000) \\ &= 1,500 \end{aligned}$$

49. In 1999, 41% of the television audience watched cable.

$$\begin{aligned} &41\% \text{ of } 35,000 \\ &= .41(35,000) \\ &= 14,350 \end{aligned}$$

51. $.33(160{,}140) = 52{,}846.20$

To the nearest dollar, \$52,846 will be spent to provide housing.

Section 2.3

1. **(a)** 12 more than a number $\underline{x + 12}$

(b) 12 is more than a number $\underline{12 > x}$

3. **(a)** 4 smaller than a number $\underline{x - 4}$

(b) 4 is smaller than a number $\underline{4 < x}$

5. 20% can be written as $.20 = .2 = \frac{20}{100} = \frac{2}{10} = \frac{1}{5}$, so "20% of a number" can be written as $.20x$, $.2x$, or $\frac{x}{5}$. We see that "20% of a number" cannot be written as $20x$, choice **D**.

7. Twice a number, decreased by 13 $\underline{2x - 13}$

9. 12 increased by three times a number $\underline{12 + 3x}$

11. The product of 8 and 12 less than a number $\underline{8(x - 12)}$

13. The quotient of three times a number and 7 $\underline{\frac{3x}{7}}$

15. The sentence "the sum of a number and 6 is −31" can be translated as

$$\begin{aligned} x + 6 &= -31. \\ x &= -37 \quad \textit{Subtract 6.} \end{aligned}$$

The number is −37.

17. The sentence "if the product of a number and −4 is subtracted from the number, the result is 9 more than the number" can be translated as

$$\begin{aligned} x - (-4x) &= x + 9. \\ x + 4x &= x + 9 \\ 4x &= 9 \\ x &= \tfrac{9}{4} \end{aligned}$$

The number is $\frac{9}{4}$.

19. The sentence "when $\frac{2}{3}$ of a number is subtracted from 12, the result is 10" can be translated as

$$\begin{aligned} 12 - \frac{2}{3}x &= 10. \\ 36 - 2x &= 30 \quad \textit{Multiply by 3.} \\ -2x &= -6 \quad \textit{Subtract 36.} \\ x &= 3 \quad \textit{Divide by } -2. \end{aligned}$$

The number is 3.

21. $5(x + 3) - 8(2x - 6)$ is an *expression* because there is no equals sign.

23. $5(x + 3) - 8(2x - 6) = 12$ has an equals sign, so this represents an *equation.*

25. $\frac{r}{2} - \frac{r+9}{6} - 8$ is an *expression* because there is no equals sign.

27. *Step 1* Read the problem.
Step 2 Assign a variable.
Step 3 Write an equation.
Step 4 Solve the equation.
Step 5 State the answer.
Step 6 Check the answer.

29. *Step 2*
Let W = the width of the base. Then $2W - 65$ is the length of the base.

Step 3
The perimeter of the base is 860 feet. Using $P = 2L + 2W$ gives us

$$2(2W - 65) + 2W = 860.$$

Step 4

$$\begin{aligned} 4W - 130 + 2W &= 860 \\ 6W - 130 &= 860 \\ 6W &= 990 \\ W &= \frac{990}{6} = 165 \end{aligned}$$

Step 5
The width of the base is 165 feet and the length of the base is $2(165) - 65 = 265$ feet.

Step 6
$2L + 2W = 2(265) + 2(165) = 530 + 330 = 860$, which is the perimeter of the base.

31. *Step 2*
Let x = the length of the middle side. Then the shortest side is $x - 75$ and the longest side is $x + 375$.

Step 3
The perimeter of the Bermuda Triangle is 3075 miles. Using $P = a + b + c$ gives us

$$x + (x - 75) + (x + 375) = 3075.$$

Step 4

$$\begin{aligned} 3x + 300 &= 3075 \\ 3x &= 2775 \quad \textit{Subtract 300.} \\ x &= 925 \quad \textit{Divide by 3.} \end{aligned}$$

Step 5
The length of the middle side is 925 miles. The length of the shortest side is $x - 75 = 925 - 75 = 850$ miles. The length of the longest side is $x + 375 = 925 + 375 = 1300$ miles.

Step 6
The answer checks since $925 + 850 + 1300 = 3075$ miles, which is the correct perimeter.

33. *Step 2*
Let x = the amount of revenue for Wal-Mart. Then $x - 28$ is the amount of revenue for Exxon Mobil (in millions).

Step 3
The total revenue was \$412 million, so

$$x + (x - 28) = 412.$$

Step 4

$$\begin{aligned} 2x - 28 &= 412 \\ 2x &= 440 \quad \textit{Add 28.} \\ x &= 220 \quad \textit{Divide by 2.} \end{aligned}$$

Step 5
The amount of revenue for Wal-Mart was \$220 million. The amount of revenue for Exxon Mobil was $x - 28 = 220 - 28 = \$192$ million.

Step 6
The answer checks since $220 + 192 = \$412$ million, which is the correct total revenue.

35. *Step 2*
Let $x =$ votes for Dole. Then
$x + 220 =$ votes for Clinton.

Step 3
There were 538 total electoral votes, so

$$x + (x + 220) = 538.$$

Step 4

$$\begin{aligned} 2x + 220 &= 538 \\ 2x &= 318 \\ x &= 159 \end{aligned}$$

Step 5
Dole received 159 votes, so Clinton received $159 + 220 = 379$ votes.

Step 6
379 is 220 more than 159 and the total is $159 + 379 = 538$.

37. Let $x =$ the percent increase.

$$\begin{aligned} x &= \frac{\text{amount of increase}}{\text{base score}} \\ &= \frac{21.0 - 20.6}{20.6} = \frac{.4}{20.6} \approx .019 = 1.9\% \end{aligned}$$

The percent increase was approximately 1.9%.

39. Let $x =$ the approximate cost in 2000.
Since x is 85% more than the 1990 cost,

$$\begin{aligned} x &= 2035 + .85(2035) \\ &= 2035 + 1729.75 = 3764.75. \end{aligned}$$

To the nearest dollar, the cost was \$3765.

41. Let $x =$ the amount of the receipts excluding tax. Since the sales tax is 9% of x, the total amount is

$$\begin{aligned} x + .09x &= 2725 \\ 1x + .09x &= 2725 \\ 1.09x &= 2725 \\ x &= \frac{2725}{1.09} = 2500 \end{aligned}$$

Thus, the tax was $.09(2500) = \$225$.

43. Let $x =$ the amount invested at 3%. Then
$12,000 - x =$ the amount invested at 4%.
Complete the table. Use $I = prt$ with $t = 1$.

Principal	Rate (as a decimal)	Interest
x	.03	$.03x$
$12,000 - x$	.04	$.04(12,000 - x)$
12,000	← Total →	440

The last column gives the equation.

$$\begin{array}{ccccc} \text{Interest at 3\%} & + & \text{interest at 4\%} & = & \text{total interest.} \\ .03x & + & .04(12,000 - x) & = & 440 \end{array}$$

$$\begin{aligned} 3x + 4(12,000 - x) &= 44,000 \quad \textit{Multiply by 100.} \\ 3x + 48,000 - 4x &= 44,000 \\ -x &= -4000 \\ x &= 4000 \end{aligned}$$

He should invest \$4000 at 3% and
$12,000 - 4000 = \$8000$ at 4%.

Check: \$4000 @ 3% = \$120 and
\$8000 @ 4% = \$320; \$120 + \$320 = \$440.

45. Let $x =$ the amount invested at 4.5%. Then
$2x - 1000 =$ the amount invested at 3%.
Use $I = prt$ with $t = 1$. Make a table.

Principal	Rate (as a decimal)	Interest
x	.045	$.045x$
$2x - 1000$	.03	$.03(2x - 1000)$
	Total →	1020

The last column gives the equation.

$$\begin{array}{ccccc} \text{Interest at 4.5\%} & + & \text{interest at 3\%} & = & \text{total interest.} \\ .045x & + & .03(2x - 1000) & = & 1020 \end{array}$$

$$\begin{aligned} 45x + 30(2x - 1000) &= 1{,}020{,}000 \quad \textit{Multiply by 1000.} \\ 45x + 60x - 30,000 &= 1{,}020{,}000 \\ 105x &= 1{,}050{,}000 \\ x &= \frac{1{,}050{,}000}{105} = 10{,}000 \end{aligned}$$

He invested \$10,000 at 4.5% and
$2x - 1000 = 2(10,000) - 1000 = \$19,000$ at 3%.

Check: \$10,000 @ 4.5% = \$450 and
\$19,000 @ 3% = \$570; \$450 + \$570 = \$1020.

47. Let $x =$ the amount of additional money to be invested at 2%.
Use $I = prt$ with $t = 1$. Make a table.
Use the fact that the total return on the two investments is 3%.

Principal	Rate (as a decimal)	Interest
29,000	.05	.05(29,000)
x	.02	$.02x$
$29,000 + x$	.03	$.03(29,000 + x)$

The last column gives the equation.

$$\begin{array}{ccccc} \text{Interest at 5\%} & + & \text{interest at 2\%} & = & \text{interest at 3\%}. \\ .05(29,000) & + & .02x & = & .03(29,000 + x) \end{array}$$

continued

$$\begin{aligned} 5(29{,}000) + 2x &= 3(29{,}000 + x) && \textit{Multiply by 100.} \\ 145{,}000 + 2x &= 87{,}000 + 3x \\ 58{,}000 &= x \end{aligned}$$

They should invest \$58,000 at 2%.

Check: \$29,000 @ 5% = \$1450 and \$58,000 @ 2% = \$1160; \$1450 + \$1160 = \$2610, which is the same as (\$29,000 + \$58,000) @ 3%.

49. Let $x =$ the number of liters of 10% acid solution needed. Make a table.

Liters of Solution	Percent (as a decimal)	Liters of Pure Acid
10	.04	$.04(10) = .4$
x	.10	$.10x$
$x + 10$	.06	$.06(x + 10)$

Acid in 4% + acid in 10% = acid in 6%.

$$.4 + .10x = .06(x + 10)$$

$$\begin{aligned} .4 + .10x &= .06x + .6 && \textit{Distributive property} \\ .04x &= .2 && \textit{Subtract .06x and .4.} \\ x &= 5 && \textit{Divide by .04.} \end{aligned}$$

Five liters of the 10% solution are needed.

Check: 4% of 10 is .4 and 10% of 5 is .5; $.4 + .5 = .9$, which is the same as 6% of $(10 + 5)$.

51. Let $x =$ the number of liters of the 20% alcohol solution. Make a table.

Liters of Solution	Percent (as a decimal)	Liters of Pure Alcohol
12	.12	$.12(12) = 1.44$
x	.20	$.20x$
$x + 12$	.14	$.14(x + 12)$

Alcohol in 12% + alcohol in 20% = alcohol in 14%.

$$1.44 + .20x = .14(x + 12)$$

$$\begin{aligned} 144 + 20x &= 14(x + 12) && \textit{Multiply by 100.} \\ 144 + 20x &= 14x + 168 && \textit{Distributive property} \\ 6x &= 24 && \textit{Subtract 14x and 144.} \\ x &= 4 && \textit{Divide by 6.} \end{aligned}$$

4L of 20% alcohol solution are needed.

Check: 12% of 12 is 1.44 and 20% of 4 is .8; $1.44 + .8 = 2.24$, which is the same as 14% of $(12 + 4)$.

53. Let $x =$ the amount of pure dye used (pure dye is 100% dye). Make a table.

Gallons of Solution	Percent (as a decimal)	Gallons of Pure Dye
x	1	$1x = x$
4	.25	$.25(4) = 1$
$x + 4$	.40	$.40(x + 4)$

Write the equation from the last column in the table.

$$\begin{aligned} x + 1 &= .4(x + 4) \\ x + 1 &= .4x + 1.6 && \textit{Distributive property} \\ .6x &= .6 && \textit{Subtract .4x and 1.} \\ x &= 1 && \textit{Divide by .6.} \end{aligned}$$

One gallon of pure (100%) dye is needed.

Check: 100% of 1 is 1 and 25% of 4 is 1; $1 + 1 = 2$, which is the same as 40% of $(1 + 4)$.

55. Let $x =$ the amount of \$6 per lb nuts. Make a table.

Pounds of nuts	Cost per lb	Total Cost
50	\$2	$2(50) = 100$
x	\$6	$6x$
$x + 50$	\$5	$5(x + 50)$

The total value of the \$2 per lb nuts and the \$6 per lb nuts must equal the value of the \$5 per lb nuts.

$$\begin{aligned} 100 + 6x &= 5(x + 50) \\ 100 + 6x &= 5x + 250 \\ x &= 150 \end{aligned}$$

He should use 150 lb of \$6 nuts.

Check: 50 pounds of the \$2 per lb nuts are worth \$100 and 150 pounds of the \$6 per lb nuts are worth \$900; \$100 + \$900 = \$1000, which is the same as $(50 + 150)$ pounds worth \$5 per lb.

57. We cannot expect the final mixture to be worth more than each of the ingredients. Answers will vary.

59. **(a)** Let $x =$ the amount invested at 5%.
$800 - x =$ the amount invested at 10%.

(b) Let $y =$ the amount of 5% acid used.
$800 - y =$ the amount of 10% acid used.

60. Organize the information in a table.

(a)

Principal	Percent (as a decimal)	Interest
x	.05	$.05x$
$800 - x$	.10	$.10(800 - x)$
800	.0875	$.0875(800)$

The amount of interest earned at 5% and 10% is found in the last column of the table, $.05x$ and $.10(800 - x)$.

(b)

Liters of Solution	Percent (as a decimal)	Liters of Pure Acid
y	.05	$.05y$
$800 - y$	.10	$.10(800 - y)$
800	.0875	$.0875(800)$

The amount of pure acid in the 5% and 10% mixtures is found in the last column of the table, $.05y$ and $.10(800-y)$.

61. Refer to the tables for Exercise 60. In each case, the last column gives the equation.

(a) $.05x+.10(800-x)=.0875(800)$

(b) $.05y+.10(800-y)=.0875(800)$

62. In both cases, multiply by 10,000 to clear the decimals.

(a)
$$\begin{aligned} .05x+.10(800-x)&=.0875(800)\\ 500x+1000(800-x)&=875(800)\\ 500x+800,000-1000x&=700,000\\ -500x&=-100,000\\ x&=200\end{aligned}$$

Jack invested \$200 at 5% and $800-x=800-200=\$600$ at 10%.

(b)
$$\begin{aligned} .05y+.10(800-y)&=.0875(800)\\ 500y+1000(800-y)&=875(800)\\ 500y+800,000-1000y&=700,000\\ -500y&=-100,000\\ y&=200\end{aligned}$$

Jill used 200 L of 5% acid solution and $800-y=800-200=600$ L of 10% acid solution.

63. The processes used to solve Problems A and B were virtually the same. Aside from the variables chosen, the problem information was organized in similar tables and the equations solved were the same.

Section 2.4

1. The total amount is

$$\begin{aligned} 38(.05)+26(.10)&=1.90+2.60\\ &=\$4.50.\end{aligned}$$

3. Use $d=rt$, or $r=\frac{d}{t}$. Substitute 300 for d and 5 for t.

$$r=\frac{300}{5}=60$$

His rate was 60 mph.

5. The problem asks for the distance Jeff travels to the workplace, so we must multiply the rate, 10 mph, by the time, $\frac{3}{4}$ hr, to get the distance, 7.5 mi.

7. No, the answers must be whole numbers because they represent the number of coins.

9. Let x = the number of pennies. Then x is also the number of dimes, and $44-2x$ is the number of quarters.

Number of Coins	Denomination	Value
x	.01	$.01x$
x	.10	$.10x$
$44-2x$	.25	$.25(44-2x)$
44	← Total →	4.37

The sum of the values must equal the total value.

$$\begin{aligned} .01x+.10x+.25(44-2x)&=4.37\\ x+10x+25(44-2x)&=437\\ &\textit{Multiply by 100.}\\ x+10x+1100-50x&=437\\ -39x+1100&=437\\ -39x&=-663\\ x&=17\end{aligned}$$

There are 17 pennies, 17 dimes, and $44-2(17)=10$ quarters.

Check: The number of coins is $17+17+10=44$ and the value of the coins is $\$.01(17)+\$.10(17)+\$.25(10)=\$4.37$, as required.

11. Let x = the number of quarters. Then $47-x$ is the number of half-dollars.

Number of Coins	Denomination	Value
x	.25	$.25x$
$47-x$	.50	$.50(47-x)$
47	← Total →	17.00

The sum of the values must equal the total value.

$$\begin{aligned} .25x+.50(47-x)&=17.00\\ 25x+50(47-x)&=1700\\ &\textit{Multiply by 100.}\\ 25x+2350-50x&=1700\\ -25x&=-650\\ x&=26\end{aligned}$$

She has 26 quarters and $47-26=21$ half-dollars.

Check: The value of the coins is $\$.25(26)+\$.50(21)=\$17.00$, as required.

13. Let x = the number of \$10 coins. Then $41-x$ is the number of \$20 coins.

Number of Coins	Denomination	Value
x	10	$10x$
$41-x$	20	$20(41-x)$
41	← Total →	540

The sum of the values must equal the total value.

continued

$$\begin{aligned} 10x + 20(41 - x) &= 540 \\ 10x + 820 - 20x &= 540 \\ -10x &= -280 \\ x &= 28 \end{aligned}$$

He has 28 \$10 coins and $41 - 28 = 13$ \$20 coins.

Check: The value of the coins is $\$10(28) + \$20(13) = \$540$, as required.

15. Let $x =$ the number of floor tickets. Then $550 - x =$ the number of balcony tickets.

Cost of Ticket	Number Sold	Amount Collected
\$40	x	$40x$
\$28	$550 - x$	$28(550 - x)$
Total	550	\$20, 800

$$\begin{aligned} 40x + 28(550 - x) &= 20,800 \\ 40x + 15,400 - 28x &= 20,800 \\ 12x &= 5,400 \\ x &= 450 \end{aligned}$$

450 floor tickets and $550 - 450 = 100$ balcony tickets were sold.

Check: The amount collected was $\$40(450) + \$28(100) = \$20{,}800$, as required.

17. $d = rt$, so

$$r = \frac{d}{t} = \frac{100}{12.65} \approx 7.91$$

Her rate was about 7.91 m/sec.

19. $d = rt$, so

$$r = \frac{d}{t} = \frac{400}{47.50} \approx 8.42$$

His rate was about 8.42 m/sec.

21. Let $t =$ the time until they are 110 mi apart. Use the formula $d = rt$. Complete the table.

	Rate	Time	Distance
First Steamer	22	t	$22t$
Second Steamer	22	t	$22t$
			110

The total distance traveled is the sum of the distances traveled by each steamer, since they are traveling in opposite directions. This total is 110 mi.

$$\begin{aligned} 22t + 22t &= 110 \\ 44t &= 110 \\ t &= \frac{110}{44} = \frac{5}{2} \text{ or } 2\frac{1}{2} \end{aligned}$$

It will take them $2\frac{1}{2}$ hr.

Check: Each steamer traveled $22(2.5) = 55$ miles for a total of $2(55) = 110$ miles, as required.

23. Let $t =$ Mulder's time. Then $t - \frac{1}{2} =$ Scully's time.

	Rate	Time	Distance
Mulder	65	t	$65t$
Scully	68	$t - \frac{1}{2}$	$68\left(t - \frac{1}{2}\right)$

The distances are equal.

$$\begin{aligned} 65t &= 68\left(t - \frac{1}{2}\right) \\ 65t &= 68t - 34 \\ -3t &= -34 \\ t &= \frac{34}{3} \text{ or } 11\frac{1}{3} \end{aligned}$$

Mulder's time will be $11\frac{1}{3}$ hr. Since he left at 8:30 A.M., $11\frac{1}{3}$ hr or 11 hr 20 min later is 7:50 P.M.

Check: Mulder's distance was $65\left(\frac{34}{3}\right) = 736\frac{2}{3}$ miles. Scully's distance was $68\left(\frac{34}{3} - \frac{1}{2}\right) = 68\left(\frac{65}{6}\right) = 736\frac{2}{3}$, as required.

25. Let $x =$ Latrella's speed when she rides. Then $x - 10 =$ her speed walking.

	Rate	Time	Distance
Riding	x	$\frac{15}{60} = \frac{1}{4}$ hr	$\frac{1}{4}x$
Walking	$x - 10$	$\frac{45}{60} = \frac{3}{4}$ hr	$\frac{3}{4}(x - 10)$

The distances are equal.

$$\begin{aligned} \frac{1}{4}x &= \frac{3}{4}(x - 10) \\ x &= 3(x - 10) \quad \textit{Multiply by 4.} \\ x &= 3x - 30 \\ -2x &= -30 \\ x &= 15 \end{aligned}$$

Her speed when she rides is 15 mph.
Check: The distance she travels to school when riding is $\frac{1}{4}(15) = 3.75$ miles. The distance when walking is $\frac{3}{4}(15 - 10) = 3.75$ miles, as required.

27. Let $x =$ Anne's time. Then $x + \frac{1}{2} =$ Johnny's time.

	Rate	Time	Distance
Anne	60	x	$60x$
Johnny	50	$x+\frac{1}{2}$	$50\left(x+\frac{1}{2}\right)$

The total distance is 80.

$$\begin{aligned} 60x+50\left(x+\frac{1}{2}\right)&=80\\ 60x+50x+25&=80\\ 110x&=55\\ x&=\frac{55}{110}=\frac{1}{2}\end{aligned}$$

They will meet $\frac{1}{2}$ hr after Anne leaves.

Check: Anne travels $60\left(\frac{1}{2}\right)=30$ miles. Johnny travels $50\left(\frac{1}{2}+\frac{1}{2}\right)=50$ miles. The sum of the distances is 80 miles, as required.

29. The sum of the measures of the three angles of a triangle is 180°.

$$\begin{aligned}(x-30)+(2x-120)+\left(\frac{1}{2}x+15\right)&=180\\ \frac{7}{2}x-135&=180\\ 7x-270&=360\\ &\textit{Multiply by 2.}\\ 7x&=630\\ x&=90\end{aligned}$$

With $x=90$, the three angle measures become

$$\begin{aligned}&(90-30)^\circ=60^\circ,\\ &[2(90)-120]^\circ=60^\circ,\\ \text{and}\quad&\left[\frac{1}{2}(90)+15\right]^\circ=60^\circ.\end{aligned}$$

31. The sum of the measures of the three angles of a triangle is 180°.

$$\begin{aligned}(3x+7)+(9x-4)+(4x+1)&=180\\ 16x+4&=180\\ 16x&=176\\ x&=11\end{aligned}$$

With $x=11$, the three angle measures become

$$\begin{aligned}&(3\cdot 11+7)^\circ=40^\circ,\\ &(9\cdot 11-4)^\circ=95^\circ,\\ \text{and}\quad&(4\cdot 11+1)^\circ=45^\circ.\end{aligned}$$

33. The sum of the measures of the angles of a triangle is 180°.

$$\begin{aligned}x+2x+60&=180\\ 3x+60&=180\\ 3x&=120\\ x&=40\end{aligned}$$

The measures of the unknown angles are 40° and $2x=80°$.

34. Supplementary angles sum to 180°, so $180°-60°=120°$. The measure of the unknown angle is 120°.

35. The sum of the measures of the unknown angles in Exercise 33 is $40°+80°=120°$. This is equal to the measure of the angle in Exercise 34.

36. The sum of the measures of angles ① and ② is equal to the measure of angle ③ .

37. Vertical angles have equal measure.

$$\begin{aligned}8x+2&=7x+17\\ x&=15\end{aligned}$$

$$8\cdot 15+2=122\quad\text{and}\quad 7\cdot 15+17=122.$$

The angles are both 122°.

39. The sum of the two angles is 90°.

$$\begin{aligned}(5x-1)+2x&=90\\ 7x-1&=90\\ 7x&=91\\ x&=13\end{aligned}$$

The angles are $[5(13)-1]^\circ=64^\circ$ and $[2(13)]^\circ=26^\circ$.

41. Let $x=$ the first consecutive integer. Then $x+1$ will be the second consecutive integer, and $x+2$ will be the third consecutive integer.

The sum of the first and twice the second is 17 more than twice the third.

$$\begin{aligned}x+2(x+1)&=2(x+2)+17\\ x+2x+2&=2x+4+17\\ 3x+2&=2x+21\\ x&=19\end{aligned}$$

Since $x=19$, $x+1=20$, and $x+2=21$.
The three consecutive integers are 19, 20, and 21.

43. Let $x=$ the current age. Then $x+1$ will be the age next year.
The sum of these ages will be 103 years.

$$\begin{aligned}x+(x+1)&=103\\ 2x+1&=103\\ 2x&=102\\ x&=51\end{aligned}$$

If my current age is 51, in 10 years I will be

$51+10=61$ years old.

Summary Exercises on Solving Applied Problems

1. Let $x =$ the width of the rectangle. Then $x + 3$ is the length of the rectangle.

If the length were decreased by 2 inches and the width were increased by 1 inch, the perimeter would be 24 inches. Use the formula $P = 2L + 2W$, and substitute 24 for P, $(x + 3) - 2$ or $x + 1$ for L, and $x + 1$ for W.

$$\begin{aligned} P &= 2L + 2W \\ 24 &= 2(x + 1) + 2(x + 1) \\ 24 &= 2x + 2 + 2x + 2 \\ 24 &= 4x + 4 \\ 20 &= 4x \\ 5 &= x \end{aligned}$$

The width of the rectangle is 5 inches, and the length is $5 + 3 = 8$ inches.

3. Let $x =$ the regular price of the videodisc player. The sale price after a discount of 40% (or .40) was \$255, so an equation is

$$\begin{aligned} x - .40x &= 255. \\ .60x &= 255 \\ x &= 425 \end{aligned}$$

The regular price of the videodisc player was \$425.

5. Let $x =$ the amount invested at 4%. Then $2x$ is the amount invested at 5%. Use $I = prt$ with $t = 1$ yr. Make a table.

Principal	Rate (as a decimal)	Interest
x	.04	$.04x$
$2x$	.05	$.05(2x) = .10x$
	Total →	112

The last column gives the equation.

$$\begin{array}{ccccc} \text{Interest at 4\%} & + & \text{interest at 5\%} & = & \text{total interest.} \\ .04x & + & .10x & = & 112 \end{array}$$

$$\begin{aligned} 4x + 10x &= 11{,}200 \quad \textit{Multiply by 100.} \\ 14x &= 11{,}200 \\ x &= 800 \end{aligned}$$

\$800 is invested at 4% and 2(\$800) = \$1600 at 5%.

Check: \$800 @ 4% = \$32 and \$1600 @ 5% = \$80; \$32 + \$80 = \$112

7. Let $x =$ the number of votes for Willkie. Then $x + 367 =$ the number of votes for Roosevelt. Together they received 531 votes.

$$\begin{aligned} x + (x + 367) &= 531 \\ 2x + 367 &= 531 \\ 2x &= 164 \\ x &= 82 \end{aligned}$$

Willkie received 82 electoral votes, and Roosevelt received $82 + 367 = 449$ electoral votes.

Check: 449 is 367 more than 82 and $82 + 449 = 531$.

9. Let $t =$ the time it will take until John and Pat meet. Use $d = rt$ and make a table.

	Rate	Time	Distance
John	60	t	$60t$
Pat	28	t	$28t$

The total distance is 440 miles.

$$\begin{aligned} 60t + 28t &= 440 \\ 88t &= 440 \\ t &= 5 \end{aligned}$$

It will take 5 hours for John and Pat to meet.

Check: John traveled $60(5) = 300$ miles and Pat traveled $28(5) = 140$ miles; $300 + 140 = 440$, as required.

11. Let $x =$ the number of liters of the 5% drug solution.

Liters of Solution	Percent (as a decimal)	Liters of Pure Drug
20	.10	$20(.10) = 2$
x	.05	$.05x$
$20 + x$	.08	$.08(20 + x)$

$$\begin{array}{ccccc} \text{Drug in 10\%} & + & \text{drug in 5\%} & = & \text{drug in 8\%.} \\ 2 & + & .05x & = & .08(20 + x) \end{array}$$

$$\begin{aligned} 200 + 5x &= 8(20 + x) \quad \textit{Multiply by 100.} \\ 200 + 5x &= 160 + 8x \\ 40 &= 3x \\ x &= \frac{40}{3} \text{ or } 13\frac{1}{3} \end{aligned}$$

The pharmacist should add $13\frac{1}{3}$ L.

Check: 10% of 20 is 2 and 5% of $\frac{40}{3}$ is $\frac{2}{3}$; $2 + \frac{2}{3} = \frac{8}{3}$, which is the same as 8% of $\left(20 + \frac{40}{3}\right)$.

13. Let $x =$ the number of \$5 bills. Then $126 - x$ is the number of \$10 bills.

Number of Bills	Denomination	Value
x	5	$5x$
$126 - x$	10	$10(126 - x)$
126	← Total →	840

The sum of the values must equal the total value.

$$\begin{aligned} 5x + 10(126 - x) &= 840 \\ 5x + 1260 - 10x &= 840 \\ -5x &= -420 \\ x &= 84 \end{aligned}$$

There are 84 \$5 bills and $126 - 84 = 42$ \$10 bills.

Check: The value of the bills is $\$5(84) + \$10(42) = \$840$, as required.

15. The sum of the measures of the three angles of a triangle is 180°.

$$\begin{aligned} x + (6x - 50) + (x - 10) &= 180 \\ 8x - 60 &= 180 \\ 8x &= 240 \\ x &= 30 \end{aligned}$$

With $x = 30$, the three angle measures become

$$(6 \cdot 30 - 50)^\circ = 130^\circ,$$
$$(30 - 10)^\circ = 20^\circ, \text{ and } 30^\circ.$$

17. Let $x =$ the smallest integer. Then $x + 1$ is the middle integer and $x + 2$ is the largest integer.

"The sum of the smallest and largest of three consecutive integers is 32 more than the middle integer" translates to

$$\begin{aligned} x + (x + 2) &= 32 + (x + 1). \\ 2x + 2 &= x + 33 \\ x &= 31 \end{aligned}$$

The three consecutive integers are 31, 32, and 33.

Check: The sum of the smallest and largest integers is $31 + 33 = 64$, which is the same as 32 more than the middle integer.

19. Let $x =$ the length of the shortest side. Then $2x$ is the length of the middle side and $3x - 2$ is the length of the longest side.

The perimeter is 34 inches. Using $P = a + b + c$ gives us

$$\begin{aligned} x + 2x + (3x - 2) &= 34. \\ 6x - 2 &= 34 \\ 6x &= 36 \\ x &= 6 \end{aligned}$$

The lengths of the three sides are 6 inches, $2(6) = 12$ inches, and $3(6) - 2 = 16$ inches.

Check: The sum of the lengths of the three sides is $6 + 12 + 16 = 34$ inches, as required.

Section 2.5

1. $x \le 3$ can be written in interval notation as $(-\infty, 3]$. Choice **D**

3. $x < 3$ represents all numbers less than 3, which is graphed in choice **B**.

5. $-3 \le x \le 3$ can be written in interval notation as $[-3, 3]$. Choice **F**

7. If an endpoint is an element of the solution set, then this is shown on the graph using a bracket. If an endpoint is *not* an element of the solution set, then this is shown on the graph using a parenthesis.

9. $x + 4 \ge 20$

$x \ge 16$ *Subtract 4.*

Solution set: $[16, \infty)$

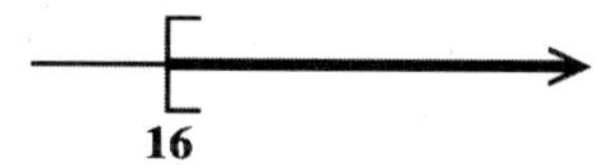

11. $3k - 1 > 20$

$3k > 21$ *Add 1.*

$k > 7$ *Divide by 3.*

Solution set: $(7, \infty)$

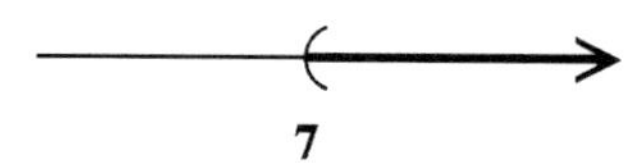

13. $4x < 16$

$x < 4$ *Divide by 4.*

Solution set: $(-\infty, 4)$

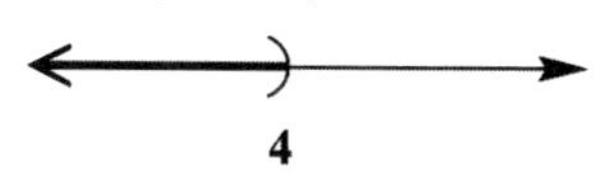

15. $$-\frac{3}{4}r \ge 30$$

Multiply both sides by $-\frac{4}{3}$, and reverse the inequality symbol.

$$-\frac{4}{3}\left(-\frac{3}{4}r\right) \le -\frac{4}{3}(30)$$
$$r \le -40$$

Solution set: $(-\infty, -40]$

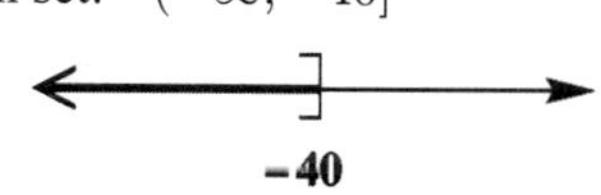

17. $$-1.3m \ge -5.2$$

Divide both sides by -1.3, and reverse the inequality symbol.

$$\frac{-1.3m}{-1.3} \le \frac{-5.2}{-1.3}$$
$$m \le 4$$

Solution set: $(-\infty, 4]$

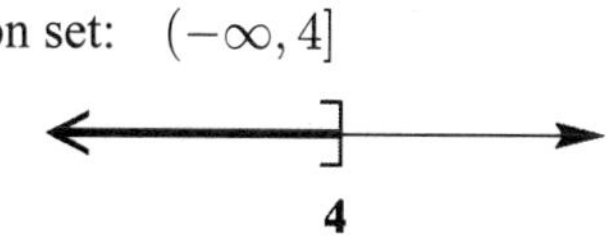

19. $$\frac{2k-5}{-4} > 5$$

Multiply both sides by -4, and reverse the inequality symbol.

$$-4\left(\frac{2k-5}{-4}\right) < -4(5)$$
$$2k - 5 < -20$$
$$2k < -15 \quad \textit{Add 5.}$$
$$k < -\frac{15}{2} \quad \textit{Divide by 2.}$$

Solution set: $\left(-\infty, -\frac{15}{2}\right)$

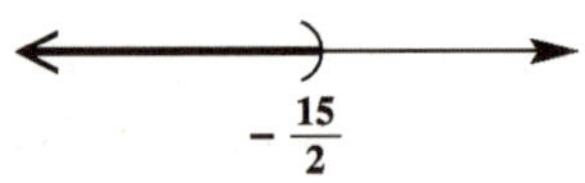

21. $6x - 4 \geq -2x$

$$8x - 4 \geq 0 \quad \textit{Add 2x.}$$
$$8x \geq 4 \quad \textit{Add 4.}$$
$$x \geq \frac{4}{8} = \frac{1}{2} \quad \textit{Divide by 8.}$$

Solution set: $\left[\frac{1}{2}, \infty\right)$

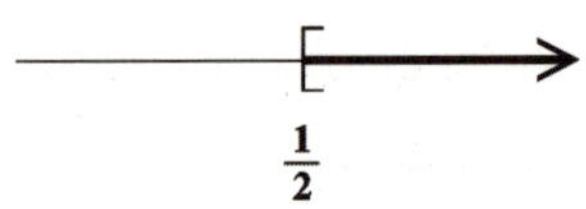

23. $-(4 + r) + 2 - 3r < -14$

$$-4 - r + 2 - 3r < -14 \quad \textit{Distributive property}$$
$$-4r - 2 < -14 \quad \textit{Combine terms.}$$
$$-4r < -12 \quad \textit{Add 2.}$$

Divide by -4, and reverse the inequality symbol.

$$r > 3$$

Solution set: $(3, \infty)$

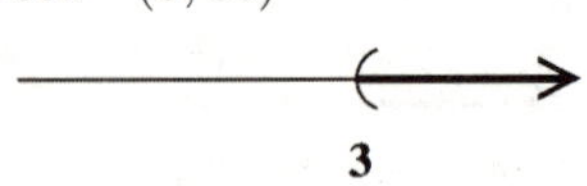

25. $-3(z - 6) > 2z - 2$

$$-3z + 18 > 2z - 2 \quad \textit{Distributive property}$$
$$-5z > -20 \quad \textit{Subtract 2z; subtract 18.}$$

Divide by -5, and reverse the inequality symbol.

$$z < 4$$

Solution set: $(-\infty, 4)$

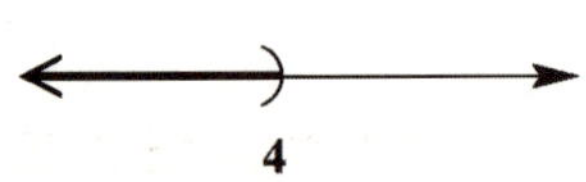

27. $\frac{2}{3}(3k - 1) \geq \frac{3}{2}(2k - 3)$

Multiply both sides by 6 to clear the fractions.

$$6 \cdot \frac{2}{3}(3k-1) \geq 6 \cdot \frac{3}{2}(2k-3)$$
$$4(3k - 1) \geq 9(2k - 3)$$
$$12k - 4 \geq 18k - 27 \quad \textit{Distributive property}$$
$$-6k \geq -23 \quad \textit{Subtract 18k; add 4.}$$

Divide by -6, and reverse the inequality symbol.

$$k \leq \frac{23}{6}$$

Solution set: $\left(-\infty, \frac{23}{6}\right]$

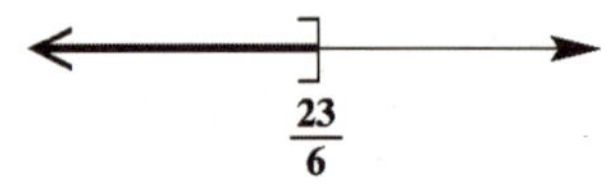

29. $-\frac{1}{4}(p + 6) + \frac{3}{2}(2p - 5) < 10$

Multiply each term by 4 to clear the fractions.

$$-1(p + 6) + 6(2p - 5) < 40$$
$$-p - 6 + 12p - 30 < 40$$
$$11p - 36 < 40$$
$$11p < 76$$
$$p < \frac{76}{11}$$

Solution set: $\left(-\infty, \frac{76}{11}\right)$

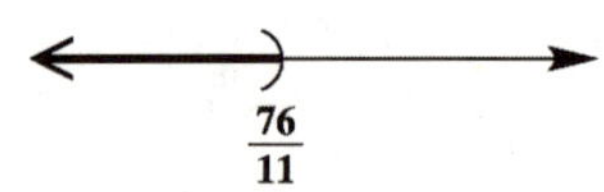

31. $3(2x - 4) - 4x < 2x + 3$

$$6x - 12 - 4x < 2x + 3$$
$$2x - 12 < 2x + 3$$
$$-12 < 3 \quad \textit{True}$$

The statement is true for all values of x. Therefore, the original inequality is true for any real number.

Solution set: $(-\infty, \infty)$

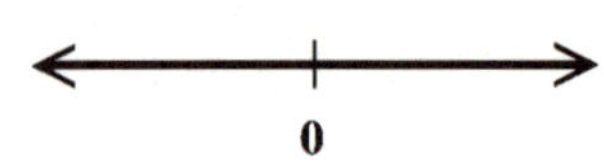

33. $8\left(\frac{1}{2}x + 3\right) < 8\left(\frac{1}{2}x - 1\right)$

$$4x + 24 < 4x - 8$$
$$24 < -8 \quad \textit{False}$$

This is a false statement, so the inequality is a contradiction.

Solution set: $\emptyset$

35. $5(x + 3) - 2(x - 4) = 2(x + 7)$

$$5x + 15 - 2x + 8 = 2x + 14$$
$$3x + 23 = 2x + 14$$
$$x = -9$$

Solution set: $\{-9\}$

The graph is the point -9 on a number line.

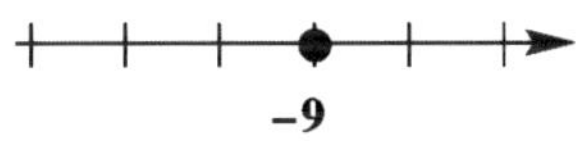

36. $5(x+3) - 2(x-4) > 2(x+7)$

$$5x + 15 - 2x + 8 > 2x + 14$$
$$3x + 23 > 2x + 14$$
$$x > -9$$

Solution set: $(-9, \infty)$

The graph extends from -9 to the right on a number line; -9 is not included in the graph.

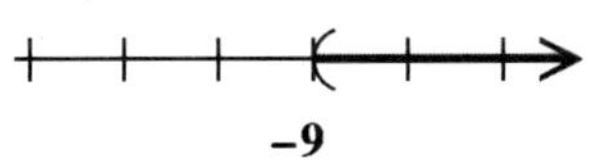

37. $5(x+3) - 2(x-4) < 2(x+7)$

$$5x + 15 - 2x + 8 < 2x + 14$$
$$3x + 23 < 2x + 14$$
$$x < -9$$

Solution set: $(-\infty, -9)$

The graph extends from -9 to the left on a number line; -9 is not included in the graph.

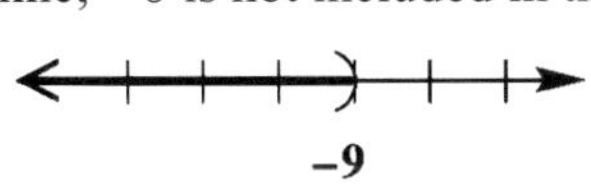

38. If we graph all the solution sets from Exercises 35–37; that is, $\{-9\}$, $(-9, \infty)$, and $(-\infty, -9)$, on the same number line, we will have graphed the set of all real numbers.

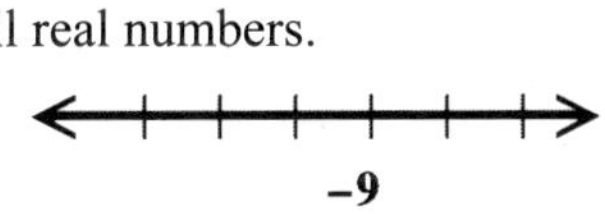

39. The solution set of the given equation is the point -3 on a number line. The solution set of the first inequality extends from -3 to the right (toward ∞) on the same number line. Based on Exercises 35–37, the solution set of the second inequality should then extend from -3 to the left (toward $-\infty$) on the number line. Complete the statement with $\underline{(-\infty, -3)}$.

41. $-4 < x - 5 < 6$

Add 5 to each part of the inequality to isolate the variable x.

$$-4 + 5 < x - 5 + 5 < 6 + 5$$
$$1 < x < 11$$

Solution set: $(1, 11)$

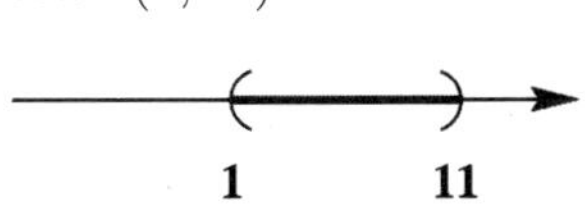

43. $-9 \le k + 5 \le 15$

Subtract 5 from each part.

$$-9 - 5 \le k + 5 - 5 \le 15 - 5$$
$$-14 \le k \le 10$$

Solution set: $[-14, 10]$

45. $-6 \le 2z + 4 \le 16$

$-10 \le 2z \le 12$ *Subtract 4.*

$-5 \le z \le 6$ *Divide by 2.*

Solution set: $[-5, 6]$

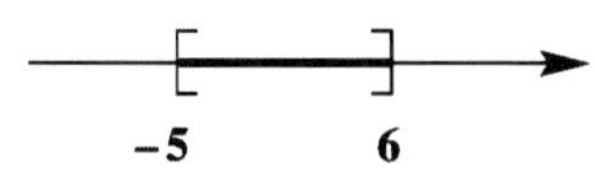

47. $-19 \le 3x - 5 \le 1$

$-14 \le 3x \le 6$ *Add 5.*

$-\frac{14}{3} \le x \le 2$ *Divide by 3.*

Solution set: $\left[-\frac{14}{3}, 2\right]$

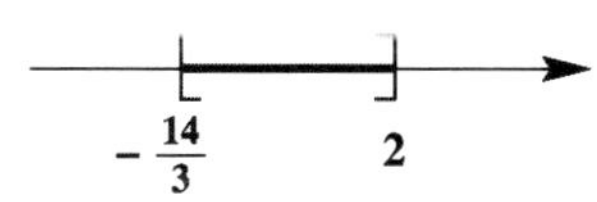

49. $-1 \le \frac{2x-5}{6} \le 5$

$-6 \le 2x - 5 \le 30$ *Multiply by 6.*

$-1 \le 2x \le 35$ *Add 5.*

$-\frac{1}{2} \le x \le \frac{35}{2}$ *Divide by 2.*

Solution set: $\left[-\frac{1}{2}, \frac{35}{2}\right]$

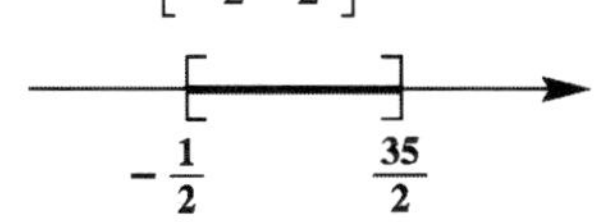

51. $4 \le 5 - 9x < 8$

$-1 \le -9x < 3$ *Subtract 5.*

Divide each part by -9; reverse the inequalities.

$$\frac{1}{9} \ge x > -\frac{1}{3}$$

This inequality may be written

$$-\frac{1}{3} < x \le \frac{1}{9}.$$

Solution set: $\left(-\frac{1}{3}, \frac{1}{9}\right]$

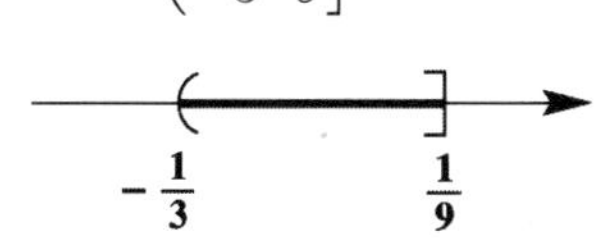

53. Six times a number is between -12 and 12.

$$\begin{aligned} -12 &< 6x < 12 \\ -2 &< x < 2 \quad \textit{Divide by 6.} \end{aligned}$$

This is the set of all numbers between -2 and 2, that is, $(-2, 2)$.

55. When 1 is added to twice a number, the result is greater than or equal to 7.

$$\begin{aligned} 2x + 1 &\geq 7 \\ 2x &\geq 6 \quad \textit{Subtract 1.} \\ x &\geq 3 \quad \textit{Divide by 2.} \end{aligned}$$

This is the set of all numbers greater than or equal to 3, that is, $[3, \infty)$.

57. One third of a number is added to 6, giving a result of at least 3.

$$\begin{aligned} 6 + \frac{1}{3}x &\geq 3 \\ \frac{1}{3}x &\geq -3 \quad \textit{Subtract 6.} \\ x &\geq -9 \quad \textit{Multiply by 3.} \end{aligned}$$

This is the set of all numbers greater than or equal to -9, that is, $[-9, \infty)$.

59. Draw a horizontal line at the 90°F mark. It intersects the *upper* boundary of the forecasted highs in two places. The temperature is expected to be at least 90°F from about 2:30 P.M. to 6:00 P.M.

61. Locate the dot at 5:39 P.M. and draw a vertical line through it. The line will intersect the lower and upper boundaries of the forecasted highs at about 84°F and 91°F, so the range of predicted temperatures is 84°F–91°F.

63. Let $x =$ her score on the third test.
Her average must be at least 84 (≥ 84). To find the average of three numbers, add them and divide by 3.

$$\begin{aligned} \frac{90 + 82 + x}{3} &\geq 84 \\ \frac{172 + x}{3} &\geq 84 \quad \textit{Add.} \\ 172 + x &\geq 252 \quad \textit{Multiply by 3.} \\ x &\geq 80 \quad \textit{Subtract 172.} \end{aligned}$$

She must score at least 80 on her third test.

65. Let $x =$ the number of miles driven.
The cost of renting from Avis is \$35 plus the mileage cost of $.14x$, while the cost of renting from Enterprise is \$34 plus the mileage cost of $.16x$.

$$\begin{aligned} \text{Cost from Enterprise} &> \text{Cost from Avis} \\ 34 + .16x &> 35 + .14x \\ 3400 + 16x &> 3500 + 14x \quad \textit{Multiply by 100.} \\ 2x &> 100 \\ x &> 50 \end{aligned}$$

After 50 mi, the price to rent from Enterprise exceeds the price to rent from Avis.

67. Cost $C = 20x + 100$; Revenue $R = 24x$
The business will show a profit only when $R > C$. Substitute the given expressions for R and C.

$$\begin{aligned} R &> C \\ 24x &> 20x + 100 \\ 4x &> 100 \\ x &> 25 \end{aligned}$$

The company will show a profit upon selling 26 tapes.

69. $4 < x < 1$ is an improper statement since $4 \not< 1$. There is no such number x.

71. $$\text{BMI} = \frac{704 \times (\text{weight in pounds})}{(\text{height in inches})^2}$$

(a) Let the height equal 72.

$$\begin{aligned} 19 \leq &\text{ BMI } \leq 25 \\ 19 \leq &\frac{704w}{72^2} \leq 25 \\ 19(72^2) \leq &\ 704w \leq 25(72^2) \\ \frac{19(72^2)}{704} \leq &\ w \leq \frac{25(72^2)}{704} \\ (\approx 139.91) \leq &\ w \leq (\approx 184.09) \end{aligned}$$

According to the BMI formula, the healthy weight range (rounded to the nearest pound) for a person that is 72 inches tall is 140 to 184 pounds.

(b) Answers will vary.

Section 2.6

1. This statement is *true*. The solution set of $x + 1 = 5$ is $\{4\}$. The solution set of $x + 1 > 5$ is $(4, \infty)$. The solution set of $x + 1 < 5$ is $(-\infty, 4)$. Taken together we have the set of real numbers. (See Section 2.5, Exercises 35–39, for a discussion of this concept.)

3. This statement is *false*. The union is $(-\infty, 8) \cup (8, \infty)$. The only real number that is *not* in the union is 8.

5. This statement is *false* since 0 is a rational number but not an irrational number. The sets of rational numbers and irrational numbers have no common elements so their intersection is $\emptyset$.

In Exercises 7–13, let

$A = \{1, 2, 3, 4, 5, 6\}$, $B = \{1, 3, 5\}$, $C = \{1, 6\}$, and $D = \{4\}$.

7. The intersection of sets B and A contains only those elements in both sets B and A.

$$B \cap A = \{1, 3, 5\} \text{ or set } B$$

9. The intersection of sets A and D is the set of all elements in both set A and D. Therefore,

$$A \cap D = \{4\} \text{ or set } D.$$

11. The intersection of set B and the set of no elements (empty set), $B \cap \emptyset$, is the set of no elements or $\emptyset$.

13. The union of sets A and B is the set of all elements that are in either set A or set B or both sets A and B. Since all numbers in set B are also in set A, the set $A \cup B$ will be the same as set A.

$$A \cup B = \{1, 2, 3, 4, 5, 6\} \text{ or set } A$$

15. Answers will vary. One example is: The intersection of two streets is the region common to *both* streets.

17. The first graph represents the set $(-\infty, 2)$. The second graph represents the set $(-3, \infty)$. The intersection includes the elements common to both sets, that is, $(-3, 2)$.

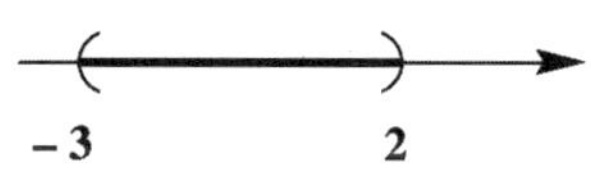

19. The first graph represents the set $(-\infty, 5]$. The second graph represents the set $(-\infty, 2]$. The intersection includes the elements common to both sets, that is, $(-\infty, 2]$.

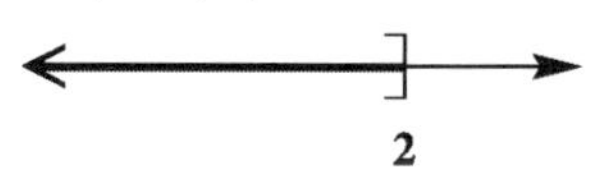

21. $x < 2$ and $x > -3$

The graph of the solution set will be all numbers that are both less than 2 and greater than -3. The solution set is $(-3, 2)$.

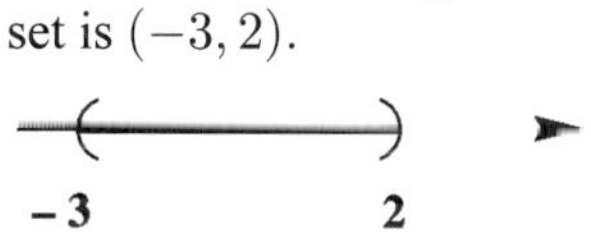

23. $x \leq 2$ and $x \leq 5$

The graph of the solution set will be all numbers that are both less than or equal to 2 and less than or equal to 5. The overlap is the numbers less than or equal to 2. The solution set is $(-\infty, 2]$.

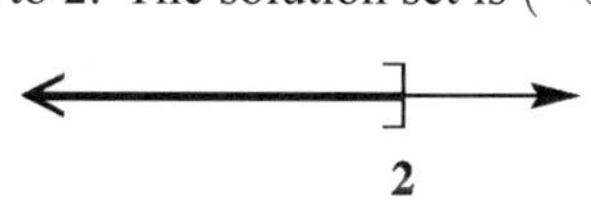

25. $x \leq 3$ and $x \geq 6$

The graph of the solution set will be all numbers that are both less than or equal to 3 and greater than or equal to 6. There are no such numbers. The solution set is $\emptyset$.

27. $x - 3 \leq 6$ and $x + 2 \geq 7$
$x \leq 9$ and $x \geq 5$
The graph of the solution set is all numbers that are both less than or equal to 9 and greater than or equal to 5. This is the intersection. The elements common to both sets are the numbers between 5 and 9, including the endpoints. The solution set is $[5, 9]$.

29. $-3x > 3$ and $x + 3 > 0$
$x < -1$ and $x > -3$
The graph of the solution set is all numbers that are both less than -1 and greater than -3. This is the intersection. The elements common to both sets are the numbers between -3 and -1, not including the endpoints. The solution set is $(-3, -1)$.

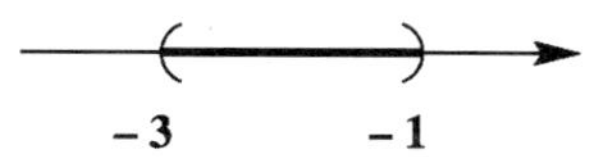

31. $3x - 4 \leq 8$ and $-4x + 1 \geq -15$
$3x \leq 12$ and $-4x \geq -16$
$x \leq 4$ and $x \leq 4$
Since both inequalities are identical, the graph of the solution set is the same as the graph of one of the inequalities. The solution set is $(-\infty, 4]$.

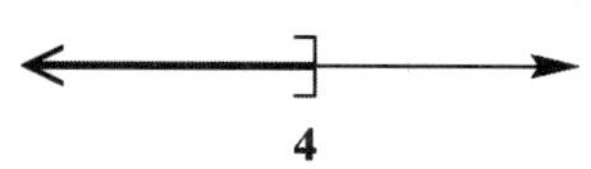

33. The first graph represents the set $(-\infty, 2]$. The second graph represents the set $[4, \infty)$. The union includes all elements in either set, or in both, that is, $(-\infty, 2] \cup [4, \infty)$.

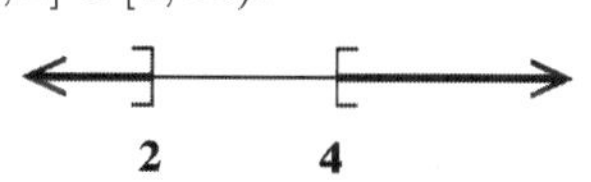

35. The first graph represents the set $[1, \infty)$. The second graph represents the set $(-\infty, 8]$. The union includes all elements in either set, or in both, that is, $(-\infty, \infty)$.

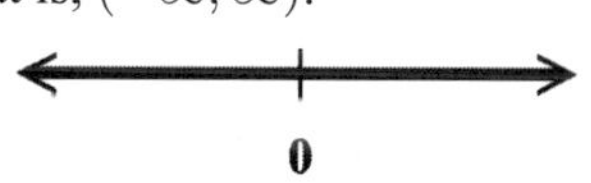

37. $x \leq 1$ or $x \leq 8$

The graph of the solution set will be all numbers that are either less than or equal to 1 or less than or equal to 8. The solution set is $(-\infty, 8]$.

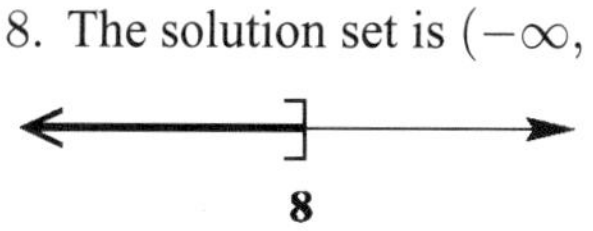

39. $x \geq -2$ or $x \geq 5$

The graph of the solution set will be all numbers that are either greater than or equal to -2 or greater than or equal to 5.
The solution set is $[-2, \infty)$.

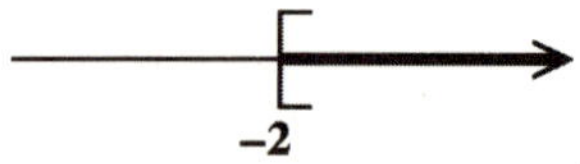

41. $x \geq -2$ or $x \leq 4$

The graph of the solution set will be all numbers that are either greater than or equal to -2 or less than or equal to 4. This is the set of all real numbers. The solution set is $(-\infty, \infty)$.

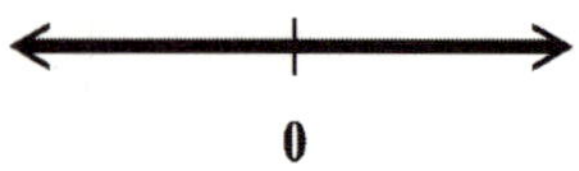

43. $x + 2 > 7$ or $1 - x > 6$
$-x > 5$
$x > 5$ or $x < -5$
The graph of the solution set is all numbers either greater than 5 or less than -5. This is the union. The solution set is $(-\infty, -5) \cup (5, \infty)$.

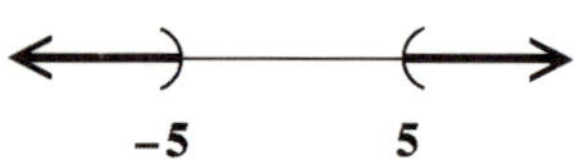

45. $x + 1 > 3$ or $-4x + 1 > 5$
$-4x > 4$
$x > 2$ or $x < -1$
The graph of the solution set is all numbers either less than -1 or greater than 2. This is the union. The solution set is $(-\infty, -1) \cup (2, \infty)$.

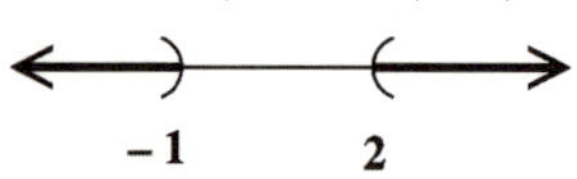

47. $(-\infty, -1] \cap [-4, \infty)$
The intersection is the set of numbers less than or equal to -1 and greater than or equal to -4. The numbers common to both original sets are between, and including, -4 and -1. The simplest interval form is $[-4, -1]$.

49. $(-\infty, -6] \cap [-9, \infty)$
The intersection is the set of numbers less than or equal to -6 and greater than or equal to -9. The numbers common to both original sets are between, and including, -9 and -6. The simplest interval form is $[-9, -6]$.

51. $(-\infty, 3) \cup (-\infty, -2)$
The union is the set of numbers that are either less than 3 or less than -2, or both. This is all numbers less than 3. The simplest interval form is $(-\infty, 3)$.

53. $[3, 6] \cup (4, 9)$
The union is the set of numbers between, and including, 3 and 6, or between, but not including, 4 and 9. This is the set of numbers greater than or equal to 3 and less than 9. The simplest interval form is $[3, 9)$.

55. $x < -1$ and $x > -5$
The word "and" means to take the intersection of both sets. $x < -1$ and $x > -5$ is true only when

$$-5 < x < -1.$$

The graph of the solution set is all numbers greater than -5 *and* less than -1. This is all numbers between -5 and -1, not including -5 or -1. The solution set is $(-5, -1)$.

57. $x < 4$ or $x < -2$
The word "or" means to take the union of both sets. The graph of the solution set is all numbers that are either less than 4 *or* less than -2, or both. This is all numbers less than 4. The solution set is $(-\infty, 4)$.

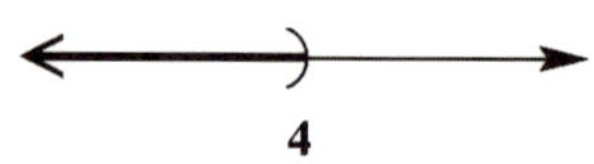

59. $-3x \leq -6$ or $-3x \geq 0$
$x \geq 2$ or $x \leq 0$
The word "or" means to take the union of both sets. The graph of the solution set is all numbers that are either greater than or equal to 2 *or* less than or equal to 0. The solution set is $(-\infty, 0] \cup [2, \infty)$.

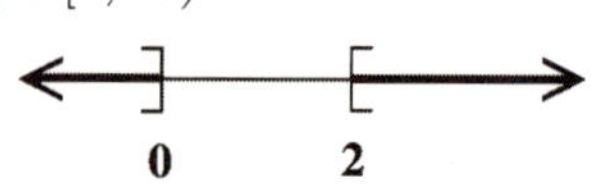

61. $x + 1 \geq 5$ and $x - 2 \leq 10$
$x \geq 4$ and $x \leq 12$
The word "and" means to take the intersection of both sets. The graph of the solution set is all numbers that are both greater than or equal to 4 *and* less than or equal to 12. This is all numbers between, and including, 4 and 12. The solution set is $[4, 12]$.

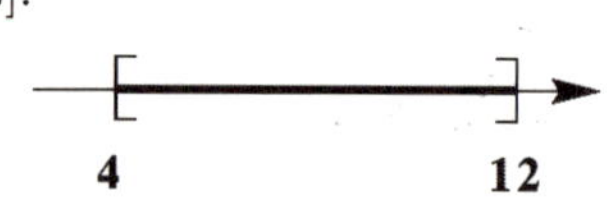

63. Find "the yard can be fenced *and* the yard can be sodded."

A yard that can be fenced has $P \leq 150$. Maria and Joe qualify.

A yard that can be sodded has $A \leq 1400$. Again, Maria and Joe qualify.

Find the intersection. Maria's and Joe's yards are common to both sets, so Maria and Joe can have their yards both fenced and sodded.

64. Find "the yard can be fenced *and* the yard cannot be sodded."

A yard that can be fenced has $P \leq 150$. Maria and Joe qualify.

A yard that cannot be sodded has $A > 1400$. Luigi and Than qualify.

Find the intersection. There are no yards common to both sets, so none of them qualify.

65. Find "the yard cannot be fenced *and* the yard can be sodded."

A yard that cannot be fenced has $P > 150$. Luigi and Than qualify.

A yard that can be sodded has $A \leq 1400$. Maria and Joe qualify.

Find the intersection. There are no yards common to both sets, so none of the qualify.

66. Find "the yard cannot be fenced *and* the yard cannot be sodded."

A yard that cannot be fenced has $P > 150$. Luigi and Than qualify.

A yard that cannot be sodded has $A > 1400$. Again, Luigi and Than qualify.

Find the intersection. Luigi's and Than's yards are common to both sets, so Luigi and Than qualify.

67. Find "the yard can be fenced *or* the yard can be sodded." From Exercise 63, Maria's and Joe's yards qualify for both conditions, so the union is Maria and Joe.

68. Find "the yard cannot be fenced *or* the yard can be sodded." From Exercise 65, Luigi's and Than's yards cannot be fenced, and Maria's and Joe's yards can be sodded. The union includes all of them.

69. Since the number of players with 40 or more home runs never exceeds 45, there are no years in which the number of players with 30–39 home runs exceeds 20 *and* the number of players with 40 or more home runs exceeds 45.

Section 2.7

1. $|x| = 5$ has two solutions, $x = 5$ or $x = -5$. The graph is Choice **E**.

$|x| < 5$ is written $-5 < x < 5$. Notice that -5 and 5 are not included. The graph is Choice **C**, which uses parentheses.

$|x| > 5$ is written $x < -5$ or $x > 5$. The graph is Choice **D**, which uses parentheses.

$|x| \leq 5$ is written $-5 \leq x \leq 5$. This time -5 and 5 are included. The graph is Choice **B**, which uses brackets.

$|x| \geq 5$ is written $x \leq -5$ or $x \geq 5$. The graph is Choice **A**, which uses brackets.

3. When solving an absolute value equation or inequality of the form

$$|ax+b| = k, \quad |ax+b| < k, \quad \text{or} \quad |ax+b| > k,$$

where k is a positive number, use *or* for the equality statement and the $>$ statement. Use *and* for the $<$ statement.

5. $|x| = 12$

$x = 12$ or $x = -12$

Solution set: $\{-12, 12\}$

7. $|4x| = 20$

$4x = 20$ or $4x = -20$

$x = 5$ or $x = -5$

Solution set: $\{-5, 5\}$

9. $|y-3| = 9$

$y - 3 = 9$ or $y - 3 = -9$

$y = 12$ or $y = -6$

Solution set: $\{-6, 12\}$

11. $|2x-1| = 11$

$2x - 1 = 11$ or $2x - 1 = -11$

$2x = 12$ $\quad$ $2x = -10$

$x = 6$ or $x = -5$

Solution set: $\{-5, 6\}$

13. $|4r-5| = 17$

$4r - 5 = 17$ or $4r - 5 = -17$

$4r = 22$ $\quad$ $4r = -12$

$r = \frac{22}{4} = \frac{11}{2}$ or $r = -3$

Solution set: $\left\{-3, \frac{11}{2}\right\}$

15. $|2y+5| = 14$

$2y + 5 = 14$ or $2y + 5 = -14$

$2y = 9$ $\quad$ $2y = -19$

$y = \frac{9}{2}$ or $y = -\frac{19}{2}$

Solution set: $\left\{-\frac{19}{2}, \frac{9}{2}\right\}$

17. $\left|\frac{1}{2}x + 3\right| = 2$

$\frac{1}{2}x + 3 = 2$ or $\frac{1}{2}x + 3 = -2$

$\frac{1}{2}x = -1$ $\quad$ $\frac{1}{2}x = -5$

$x = -2$ or $x = -10$

Solution set: $\{-10, -2\}$

19. $\left|1+\frac{3}{4}k\right|=7$

$1+\frac{3}{4}k=7$ or $1+\frac{3}{4}k=-7$

Multiply all sides by 4.

$4+3k=28$ or $4+3k=-28$

$3k=24$ $\quad 3k=-32$

$k=8$ or $k=\frac{-32}{3}$

Solution set: $\left\{-\frac{32}{3},8\right\}$

21. $|x|>3$

$x>3$ or $x<-3$

Solution set: $(-\infty,-3)\cup(3,\infty)$

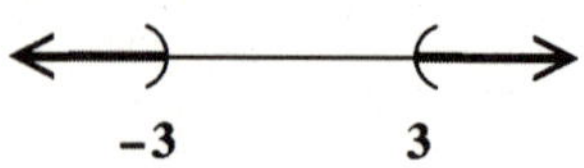

23. $|k|\geq 4$

$k\geq 4$ or $k\leq -4$

Solution set: $(-\infty,-4]\cup[4,\infty)$

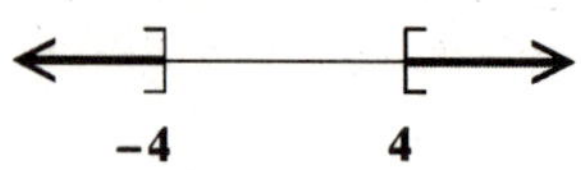

25. $|r+5|\geq 20$

$r+5\leq -20$ or $r+5\geq 20$

$r\leq -25$ or $r\geq 15$

Solution set: $(-\infty,-25]\cup[15,\infty)$

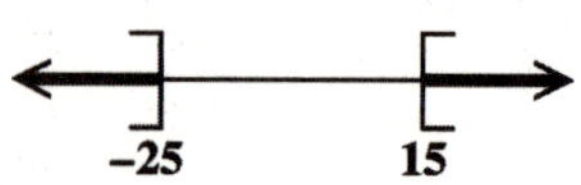

27. $|t+2|>10$

$t+2>10$ or $t+2<-10$

$t>8$ or $t<-12$

Solution set: $(-\infty,-12)\cup(8,\infty)$

29. $|3-x|>5$

$3-x>5$ or $3-x<-5$

$-x>2$ or $-x<-8$

Multiply by -1,
and reverse the inequality symbols.

$x<-2$ or $x>8$

Solution set: $(-\infty,-2)\cup(8,\infty)$

31. $|-5x+3|\geq 12$

$-5x+3\geq 12$ or $-5x+3\leq -12$

$-5x\geq 9$ $\quad -5x\leq -15$

$x\leq -\frac{9}{5}$ or $x\geq 3$

Solution set: $\left(-\infty,-\frac{9}{5}\right]\cup[3,\infty)$

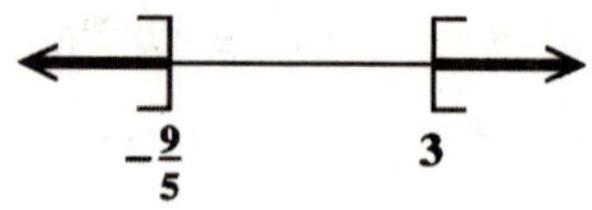

33. **(a)** $|2x+1|<9$

The graph of the solution set will be all numbers between -5 and 4, since the absolute value is less than 9.

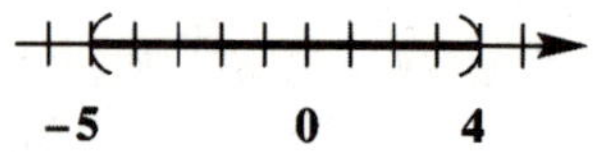

(b) $|2x+1|>9$

The graph of the solution set will be all numbers less than -5 or greater than 4, since the absolute value is greater than 9.

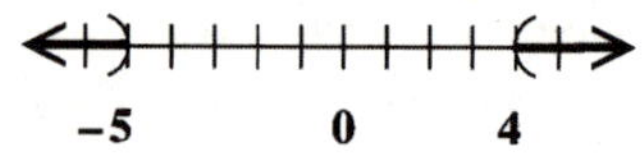

35. $|x|\leq 3$

$-3\leq x\leq 3$

Solution set: $[-3,3]$

37. $|k|<4$

$-4<k<4$

Solution set: $(-4,4)$

39. $|r+5|\leq 20$

$-20\leq r+5\leq 20$

$-25\leq r\leq 15$ *Subtract 5.*

Solution set: $[-25,15]$

41. $|t+2|\leq 10$

$-10\leq t+2\leq 10$

$-12\leq t\leq 8$

Solution set: $[-12,8]$

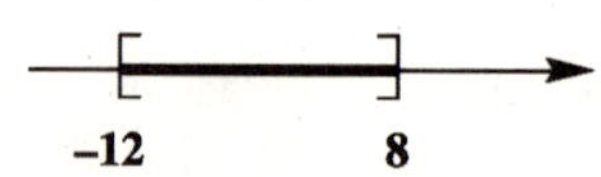

43. $|3-x| \le 5$
$-5 \le 3-x \le 5$
$-8 \le -x \le 2$
Multiply by -1, and reverse the inequality symbols.
$8 \ge x \ge -2$ or $-2 \le x \le 8$
Solution set: $[-2, 8]$

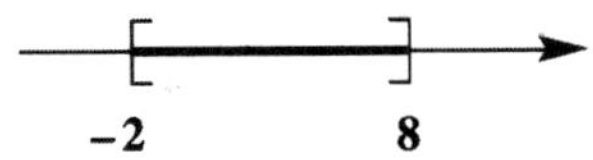

45. $|-5x+3| \le 12$

$-12 \le -5x+3 \le 12$
$-15 \le \quad -5x \quad \le 9$

Divide by -5, and reverse the inequality symbols.

$3 \ge x \ge -\frac{9}{5}$ or $-\frac{9}{5} \le x \le 3$

Solution set: $\left[-\frac{9}{5}, 3\right]$

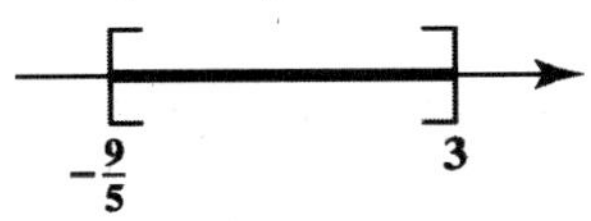

47. $|-4+k| > 9$
$-4+k > 9$ or $-4+k < -9$
$k > 13$ or $k < -5$

Solution set: $(-\infty, -5) \cup (13, \infty)$

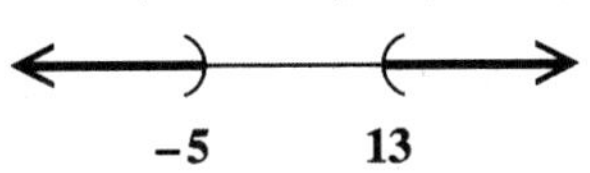

49. $|r+5| > 20$
$r+5 > 20$ or $r+5 < -20$
$r > 15$ or $r < -25$

Solution set: $(-\infty, -25) \cup (15, \infty)$

51. $|7+2z| = 5$
$7+2z = 5$ or $7+2z = -5$
$2z = -2$ $\qquad 2z = -12$
$z = -1$ or $z = -6$

Solution set: $\{-6, -1\}$

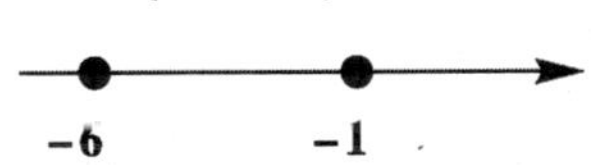

53. $|3r-1| \le 11$
$-11 \le 3r-1 \le 11$
$-10 \le 3r \le 12$
$-\frac{10}{3} \le r \le 4$

Solution set: $\left[-\frac{10}{3}, 4\right]$

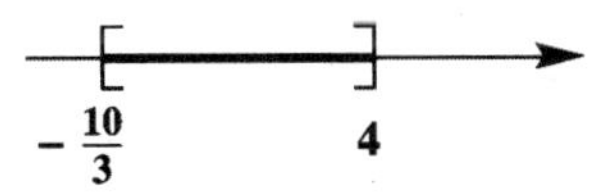

55. $|-6x-6| \le 1$
$-1 \le -6x-6 \le 1$
$5 \le -6x \le 7$
Divide by -6, and reverse the inequality symbols.
$-\frac{5}{6} \ge x \ge -\frac{7}{6}$ or $-\frac{7}{6} \le x \le -\frac{5}{6}$

Solution set: $\left[-\frac{7}{6}, -\frac{5}{6}\right]$

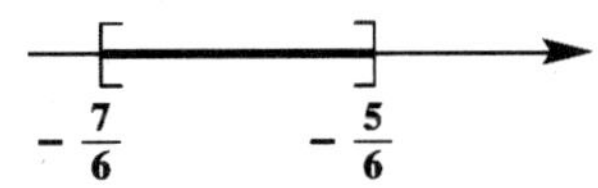

57. $|2x-1| \ge 7$

$2x-1 \ge 7$ or $2x-1 \le -7$
$2x \ge 8$ or $2x \le -6$
$x \ge 4$ or $x \le -3$

Solution set: $(-\infty, -3] \cup [4, \infty)$

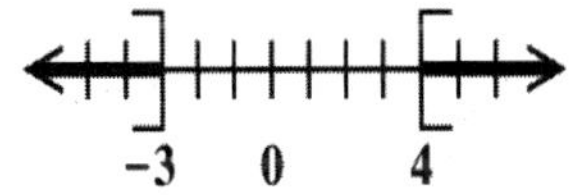

59. $|x|-1 = 4$
$|x| = 5$

$x = 5$ or $x = -5$

Solution set: $\{-5, 5\}$

61. $|x+2| = 3$

$x+2 = 3$ or $x+2 = -3$
$x = 1$ or $x = -5$

Solution set: $\{-5, 1\}$

63. $|x+4|+1 = 2$
$|x+4| = 1$
$x+4 = 1$ or $x+4 = -1$
$x = -3$ or $x = -5$

Solution set: $\{-5, -3\}$

65. $|2x+1|+3 > 8$
$|2x+1| > 5$
$2x+1 > 5$ or $2x+1 < -5$
$2x > 4$ $\qquad 2x < -6$
$x > 2$ or $x < -3$

Solution set: $(-\infty, -3) \cup (2, \infty)$

67. $|x+5|-6 \le -1$
$|x+5| \le 5$
$-5 \le x+5 \le 5$
$-10 \le x \le 0$

Solution set: $[-10, 0]$

69. $|2-x|>3$

$$\begin{aligned} 2-x&>3 &\text{or}\quad 2-x&<-3\\ -x&>1 & -x&<-5\\ x&<-1 &\text{or}\quad x&>5 \end{aligned}$$

Solution set: $(-\infty,-1)\cup(5,\infty)$

71. $|3x+1|=|2x+4|$

$$\begin{aligned} 3x+1&=2x+4 &\text{or}\quad 3x+1&=-(2x+4)\\ && 3x+1&=-2x-4\\ && 5x&=-5\\ x&=3 &\text{or}\quad x&=-1 \end{aligned}$$

Solution set: $\{-1,3\}$

73. $\left|m-\frac{1}{2}\right|=\left|\frac{1}{2}m-2\right|$

$$m-\frac{1}{2}=\frac{1}{2}m-2 \text{ or } m-\frac{1}{2}=-\left(\frac{1}{2}m-2\right)$$

Multiply by 2. $\qquad m-\frac{1}{2}=-\frac{1}{2}m+2$

$$\begin{aligned} 2m-1&=m-4 & 2m-1&=-m+4\\ && 3m&=5\\ m&=-3 &\text{or}\quad m&=\frac{5}{3} \end{aligned}$$

Solution set: $\left\{-3,\frac{5}{3}\right\}$

75. $|6x|=|9x+1|$

$$\begin{aligned} 6x&=9x+1 &\text{or}\quad 6x&=-(9x+1)\\ -3x&=1 & 6x&=-9x-1\\ && 15x&=-1\\ x&=-\frac{1}{3} &\text{or}\quad x&=-\frac{1}{15} \end{aligned}$$

Solution set: $\left\{-\frac{1}{3},-\frac{1}{15}\right\}$

77. $|2p-6|=|2p+11|$

$$\begin{aligned} 2p-6&=2p+11 &\text{or}\quad 2p-6&=-(2p+11)\\ -6&=11\quad \textit{False} & 2p-6&=-2p-11\\ && 4p&=-5\\ &\textit{No solution} &\text{or}\quad p&=-\frac{5}{4} \end{aligned}$$

Solution set: $\left\{-\frac{5}{4}\right\}$

79. $|12t-3|=-8$

Since the absolute value of an expression can never be negative, there are no solutions for this equation.

Solution set: $\emptyset$

81. $|4x+1|=0$

The expression $4x+1$ will equal 0 *only* for the solution of the equation

$$\begin{aligned} 4x+1&=0.\\ 4x&=-1\\ x&=\frac{-1}{4}\text{ or }-\frac{1}{4} \end{aligned}$$

Solution set: $\left\{-\frac{1}{4}\right\}$

83. $|2q-1|=-6$

Since the absolute value of an expression can never be negative, there are no solutions for this equation.

Solution set: $\emptyset$

85. $|x+5|>-9$

Since the absolute value of an expression is always nonnegative (positive or zero), the inequality is true for any real number x.

Solution set: $(-\infty,\infty)$

87. $|7x+3|\le 0$

The absolute value of an expression is always nonnegative (positive or zero), so this inequality is true only when

$$\begin{aligned} 7x+3&=0\\ 7x&=-3\\ x&=-\frac{3}{7}. \end{aligned}$$

Solution set: $\left\{-\frac{3}{7}\right\}$

89. $|5x-2|=0$

The expression $5x-2$ will equal 0 *only* for the solution of the equation

$$\begin{aligned} 5x-2&=0.\\ 5x&=2\\ x&=\frac{2}{5} \end{aligned}$$

Solution set: $\left\{\frac{2}{5}\right\}$

91. $|10z+7|+3<1$

$$|10z+7|<-2$$

There is no number whose absolute value is less than -2, so this inequality has no solution.

Solution set: $\emptyset$

93. Let x represent the calcium intake for a specific female. For x to be within 100 mg of 1000 mg, we must have

$$|x-1000|\le 100.$$

$$\begin{aligned} -100\le\ &x-1000\le 100\\ 900\le\ &\quad x\quad\ \le 1100 \end{aligned}$$

95. Add the given heights with a calculator to get 4729. There are 10 numbers, so divide the sum by 10.

$$\frac{4729}{10} = 472.9$$

The average height is 472.9 ft.

96. $|x - k| < 50$
Substitute 472.9 for k and solve the inequality.

$$|x - 472.9| < 50$$

$$-50 < x - 472.9 < 50$$
$$422.9 < x < 522.9$$

The buildings with heights between 422.9 ft and 522.9 ft are the 1201 Walnut, Fidelity Bank and Trust Building, City Hall, Kansas City Power and Light, and the Hyatt Regency.

97. $|x - k| < 75$
Substitute 472.9 for k and solve the inequality.

$$|x - 472.9| < 75$$

$$-75 < x - 472.9 < 75$$
$$397.9 < x < 547.9$$

The buildings with heights between 397.9 ft and 547.9 ft are City Center Square, Commerce Tower, Federal Office Building, 1201 Walnut, Fidelity Bank and Trust Building, City Hall, Kansas City Power and Light, and Hyatt Regency.

98. **(a)** This would be the opposite of the inequality in Exercise 97, that is,

$$|x - 472.9| \geq 75.$$

(b) $|x - 472.9| \geq 75$

$x - 472.9 \geq 75$ or $x - 472.9 \leq -75$
$x \geq 547.9$ or $x \leq 397.9$

(c) The buildings that are not within 75 ft of the average have height less than or equal to 397.9 or greater than or equal to 547.9. They are AT&T Town Pavilion and One Kansas City Place.

(d) The answer makes sense because it includes all the buildings *not* listed earlier which had heights within 75 ft of the average.

Summary Exercises on Solving Linear and Absolute Value Equations and Inequalities

1. $4z + 1 = 49$
$4z = 48$
$z = 12$
Solution set: $\{12\}$

3. $6q - 9 = 12 + 3q$
$3q = 21$
$q = 7$
Solution set: $\{7\}$

5. $|a + 3| = -4$
Since the absolute value of an expression is always nonnegative, there is no number that makes this statement true. Therefore, the solution set is $\emptyset$.

7. $8r + 2 \geq 5r$
$3r \geq -2$
$r \geq -\frac{2}{3}$
Solution set: $\left[-\frac{2}{3}, \infty\right)$

9. $2q - 1 = -7$
$2q = -6$
$q = -3$
Solution set: $\{-3\}$

11. $6z - 5 \leq 3z + 10$
$3z \leq 15$
$z \leq 5$
Solution set: $(-\infty, 5]$

13. $9x - 3(x + 1) = 8x - 7$
$9x - 3x - 3 = 8x - 7$
$6x - 3 = 8x - 7$
$4 = 2x$
$2 = x$
Solution set: $\{2\}$

15. $9x - 5 \geq 9x + 3$
$-5 \geq 3$ *False*
This is a false statement, so the inequality is a contradiction.
Solution set: $\emptyset$

17. $|q| < 5.5$
$-5.5 < q < 5.5$
Solution set: $(-5.5, 5.5)$

19. $\frac{2}{3}x + 8 = \frac{1}{4}x$
$8x + 96 = 3x$ *Multiply by 12.*
$5x = -96$
$x = -\frac{96}{5}$
Solution set: $\left\{-\frac{96}{5}\right\}$

21. $\frac{1}{4}p < -6$
$4\left(\frac{1}{4}p\right) < 4(-6)$
$p < -24$
Solution set: $(-\infty, -24)$

23. $\frac{3}{5}q - \frac{1}{10} = 2$

$6q - 1 = 20$ *Multiply by 10.*

$6q = 21$

$q = \frac{21}{6} = \frac{7}{2}$

Solution set: $\left\{\frac{7}{2}\right\}$

25. $r + 9 + 7r = 4(3 + 2r) - 3$

$8r + 9 = 12 + 8r - 3$

$8r + 9 = 8r + 9$

$0 = 0$ *True*

The last statement is true for any real number r.
Solution set: $(-\infty, \infty)$

27. $|2p - 3| > 11$

$2p - 3 > 11$ or $2p - 3 < -11$

$2p > 14$ $\qquad 2p < -8$

$p > 7$ or $p < -4$

Solution set: $(-\infty, -4) \cup (7, \infty)$

29. $|5a + 1| \le 0$

The expression $|5a + 1|$ is never less than 0 since an absolute value expression must be nonnegative. However, $|5a + 1| = 0$ if

$5a + 1 = 0$

$5a = -1$

$a = \frac{-1}{5} = -\frac{1}{5}$

Solution set: $\left\{-\frac{1}{5}\right\}$

31. $-2 \le 3x - 1 \le 8$

$-1 \le 3x \le 9$

$-\frac{1}{3} \le x \le 3$

Solution set: $\left[-\frac{1}{3}, 3\right]$

33. $|7z - 1| = |5z + 3|$

$7z - 1 = 5z + 3$ or $7z - 1 = -(5z + 3)$

$2z = 4$ $\qquad 7z - 1 = -5z - 3$

$12z = -2$

$z = 2$ or $z = \frac{-2}{12} = -\frac{1}{6}$

Solution set: $\left\{-\frac{1}{6}, 2\right\}$

35. $|1 - 3x| \ge 4$

$1 - 3x \ge 4$ or $1 - 3x \le -4$

$-3x \ge 3$ $\qquad -3x \le -5$

$x \le -1$ or $x \ge \frac{5}{3}$

Solution set: $(-\infty, -1] \cup \left[\frac{5}{3}, \infty\right)$

37. $-(m + 4) + 2 = 3m + 8$

$-m - 4 + 2 = 3m + 8$

$-m - 2 = 3m + 8$

$-10 = 4m$

$m = \frac{-10}{4} = -\frac{5}{2}$

Solution set: $\left\{-\frac{5}{2}\right\}$

39. $-6 \le \frac{3}{2} - x \le 6$

$-12 \le 3 - 2x \le 12$

$-15 \le -2x \le 9$

$\frac{15}{2} \ge x \ge -\frac{9}{2}$ or $-\frac{9}{2} \le x \le \frac{15}{2}$

Solution set: $\left[-\frac{9}{2}, \frac{15}{2}\right]$

41. $|x - 1| \ge -6$

The absolute value of an expression is always nonnegative, so the inequality is true for any real number x.
Solution set: $(-\infty, \infty)$

43. $8q - (1 - q) = 3(1 + 3q) - 4$

$8q - 1 + q = 3 + 9q - 4$

$9q - 1 = 9q - 1$ *True*

This is an identity.
Solution set: $(-\infty, \infty)$

45. $|r - 5| = |r + 9|$

$r - 5 = r + 9$ or $r - 5 = -(r + 9)$

$-5 = 9$ *False* $\qquad r - 5 = -r - 9$

$2r = -4$

No solution or $r = -2$

Solution set: $\{-2\}$

47. $2x + 1 > 5$ or $3x + 4 < 1$

$2x > 4$ $\qquad 3x < -3$

$x > 2$ or $x < -1$

Solution set: $(-\infty, -1) \cup (2, \infty)$

Chapter 2 Review Exercises

1. $-(8 + 3z) + 5 = 2z + 6$

$-8 - 3z + 5 = 2z + 6$

$-3z - 3 = 2z + 6$

$-5z = 9$

$z = -\frac{9}{5}$

Solution set: $\left\{-\frac{9}{5}\right\}$

2. $-\frac{3}{4}x = -12$

$-3x = -48$ *Multiply by 4.*

$x = 16$

Solution set: $\{16\}$

3. $\frac{2q+1}{3} - \frac{q-1}{4} = 0$

$4(2q+1) - 3(q-1) = 0$ *Multiply by 12.*

$8q + 4 - 3q + 3 = 0$

$5q + 7 = 0$

$5q = -7$

$q = -\frac{7}{5}$

Solution set: $\left\{-\frac{7}{5}\right\}$

4. $5(2x-3) = 6(x-1) + 4x$

$10x - 15 = 6x - 6 + 4x$

$10x - 15 = 10x - 6$

$-15 = -6$ *False*

This is a false statement, so the equation is a contradiction.

Solution set: $\emptyset$

5. $7r - 3(2r-5) + 5 + 3r = 4r + 20$

$7r - 6r + 15 + 5 + 3r = 4r + 20$

$4r + 20 = 4r + 20$

$20 = 20$ *True*

This equation is an *identity*.

Solution set: $(-\infty, \infty)$

6. $8p - 4p - (p-7) + 9p + 6 = 12p - 7$

$8p - 4p - p + 7 + 9p + 6 = 12p - 7$

$12p + 13 = 12p - 7$

$13 = -7$ *False*

This equation is a *contradiction*.

Solution set: $\emptyset$

7. $-2r + 6(r-1) + 3r - (4-r) = -(r+5) - 5$

$-2r + 6r - 6 + 3r - 4 + r = -r - 5 - 5$

$8r - 10 = -r - 10$

$9r = 0$

$r = 0$

This equation is a *conditional* equation.

Solution set: $\{0\}$

8. Solve $V = LWH$ for H.

$$\frac{V}{LW} = \frac{LWH}{LW}$$

$$\frac{V}{LW} = H \quad \text{or} \quad H = \frac{V}{LW}$$

9. Solve $A = \frac{1}{2}(B+b)h$ for b.

$2A = (B+b)h$ or $2A = Bh + bh$

$\frac{2A}{h} = B + b$ $\qquad 2A - Bh = bh$

$\frac{2A}{h} - B = b$ or $\frac{2A - Bh}{h} = b$

10. Solve $M = -\frac{1}{4}(x+3y)$ for x.

$4M = -1(x+3y)$ *Multiply by 4.*

$4M = -x - 3y$

$4M + 3y = -x$

$x = -4M - 3y$

11. Solve $P = \frac{3}{4}x - 12$ for x.

$P + 12 = \frac{3}{4}x$ *Add 12.*

$\frac{4}{3}(P+12) = x$ *Multiply by 4/3.*

$x = \frac{4}{3}(P+12)$ or $x = \frac{4}{3}P + 16$

12. Solve $-2x + 5 = 7$.

Begin by subtracting 5 from each side. Then divide each side by -2.

13. Use the formula $V = LWH$ and substitute 180 for V, 6 for L, and 5 for W.

$180 = 6(5)H$

$180 = 30H$

$6 = H$

The height is 6 feet.

14. Divide the amount of decrease by the original amount.

$$\frac{58{,}443 - 48{,}269}{58{,}443} = \frac{10{,}174}{58{,}443} \approx .174$$

The percent decrease is about 17.4%.

15. Use the formula $I = prt$, and solve for r.

$$\frac{I}{pt} = \frac{prt}{pt}$$

$$\frac{I}{pt} = r$$

Substitute 30,000 for p, 7800 for I, and 4 for t.

$$r = \frac{7800}{30,000(4)} = \frac{7800}{120,000} = .065$$

The rate is 6.5%.

16. Use the formula $C = \frac{5}{9}(F-32)$ and substitute 77 for F.

$C = \frac{5}{9}(77 - 32)$

$= \frac{5}{9}(45) = 25$

The Celsius temperature is 25°.

17. The number of people in the Hispanic category is

$.117(148{,}847{,}000) = 17{,}415{,}099,$

or approximately 17,415,000.

18. The number of people in the "Asian and other" category is

$$.049(148{,}847{,}000) = 7{,}293{,}503,$$

or approximately 7,294,000.

19. "One-third of a number, subtracted from 9" is written

$$9 - \tfrac{1}{3}x.$$

20. "The product of 4 and a number, divided by 9 more than the number" is written

$$\frac{4x}{x+9}.$$

21. Let $x =$ the width of the rectangle.
Then $2x - 3 =$ the length of the rectangle.

Use the formula $P = 2L + 2W$ with $P = 42$.

$$\begin{aligned} 42 &= 2(2x-3) + 2x \\ 42 &= 4x - 6 + 2x \\ 48 &= 6x \\ 8 &= x \end{aligned}$$

The width is 8 meters and the length is $2(8) - 3 = 13$ meters.

22. Let $x =$ the length of each equal side. Then $2x - 15 =$ the length of the third side.

Use the formula $P = a + b + c$ with $P = 53$.

$$\begin{aligned} 53 &= x + x + (2x - 15) \\ 53 &= 4x - 15 \\ 68 &= 4x \\ 17 &= x \end{aligned}$$

The lengths of the three sides are 17 inches, 17 inches, and $2(17) - 15 = 19$ inches.

23. Let $x =$ the number of kilograms of peanut clusters. Then $3x$ is the number of kilograms of chocolate creams.
The clerk has a total of 48 kg.

$$\begin{aligned} x + 3x &= 48 \\ 4x &= 48 \\ x &= 12 \end{aligned}$$

The clerk has 12 kilograms of peanut clusters.

24. Let $x =$ the number of liters of the 20% solution. Make a table.

Liters of Solution	Percent (as a Decimal)	Liters of Pure Chemical
x	.20	$.20x$
15	.50	$.50(15) = 7.5$
$x + 15$	.30	$.30(x + 15)$

The last column gives the equation.

$$\begin{aligned} .20x + 7.5 &= .30(x + 15) \\ .20x + 7.5 &= .30x + 4.5 \\ 3 &= .10x \\ 30 &= x \end{aligned}$$

30 L of the 20% solution should be used.

25.

Liters of Solution	Percent (as a Decimal)	Liters of Pure Acid
30	.40	$.40(30) = 12$
x	.00	$.00(x) = 0$
$30 + x$	.30	$.30(30 + x)$

The last column gives the equation.

$$\begin{aligned} 12 + 0 &= .30(30 + x) \\ 12 &= 9 + .3x \\ 3 &= .3x \\ 10 &= x \end{aligned}$$

10 L of water should be added.

26. Let $x =$ the amount invested at 6%. Then $x - 4000 =$ the amount invested at 4%.

Principal	Rate (as a Decimal)	Interest
x	.06	$.06x$
$x - 4000$	.04	$.04(x - 4000)$
	Total →	\$840

The last column gives the equation.

$$\begin{aligned} .06x + .04(x - 4000) &= 840 \\ 6x + 4(x - 4000) &= 84{,}000 \quad \textit{Multiply by 100.} \\ 6x + 4x - 16{,}000 &= 84{,}000 \\ 10x &= 100{,}000 \\ x &= 10{,}000 \end{aligned}$$

Anna should invest \$10,000 at 6% and \$10,000 − \$4000 = \$6000 at 4%.

27. Let $x =$ the number of quarters. Then $2x - 1$ is the number of dimes.

Number of coins	Denomination	Value
x	.25	$.25x$
$2x - 1$	.10	$.10(2x - 1)$
	Total →	3.50

The sum of the values equals the total value.

$$.25x + .10(2x - 1) = 3.50$$

Multiply by 100.

$$\begin{aligned} 25x + 10(2x - 1) &= 350 \\ 25x + 20x - 10 &= 350 \\ 45x &= 360 \\ x &= 8 \end{aligned}$$

There are 8 quarters and $2(8) - 1 = 15$ dimes.

Check: $8(.25) + 15(.10) = 3.50$

28. Let $x =$ the number of nickels.
Then $19 - x$ is the number of dimes.

Number of Coins	Denomination	Value
x	.05	$.05x$
$19 - x$	.10	$.10(19 - x)$
	Total →	1.55

The sum of the values equals the total value.

$$.05x + .10(19 - x) = 1.55 \quad \textit{Multiply by 100.}$$
$$5x + 10(19 - x) = 155$$
$$5x + 190 - 10x = 155$$
$$-5x = -35$$
$$x = 7$$

He had 7 nickels and $19 - 7 = 12$ dimes.

Check: $7(.05) + 12(.10) = 1.55$

29. Use the formula $d = rt$ or $r = \frac{d}{t}$. Here, d is about 400 mi and t is about 8 hr. Since $\frac{400}{8} = 50$, the best estimate is choice **A**.

30. Use the formula $d = rt$.

(a) Here, $r = 53$ mph and $t = 10$ hr.

$$d = 53(10) = 530$$

The distance is 530 miles.

(b) Here, $r = 164$ mph and $t = 2$ hr.

$$d = 164(2) = 328$$

The distance is 328 miles.

31. Let $x =$ the time it takes for the trains to be 297 mi apart.

Use the formula $d = rt$.

	Rate	Time	Distance
Passenger Train	60	x	$60x$
Freight Train	75	x	$75x$
			297

The total distance traveled is the sum of the distances traveled by each train.

$$60x + 75x = 297$$
$$135x = 297$$
$$x = 2.2$$

It will take the trains 2.2 hours before they are 297 miles apart.

32. Let $x =$ the speed of the faster car and $x - 15 =$ the speed of the slower car.
Make a table.

	Rate	Time	Distance
Faster Car	x	2	$2x$
Slower Car	$x - 15$	2	$2(x - 15)$
			230

The total distance traveled is the sum of the distances traveled by each car.

$$2x + 2(x - 15) = 230$$
$$2x + 2x - 30 = 230$$
$$4x = 260$$
$$x = 65$$

The faster car travels at $65 \frac{\text{km}}{\text{hr}}$, while the slower car travels at $65 - 15 = 50 \frac{\text{km}}{\text{hr}}$.

Check: $2(65) + 2(50) = 230$

33. Let $x =$ amount of time spent averaging 45 miles per hour. Then $4 - x =$ amount of time at 50 mph.

	Rate	Time	Distance
First Part	45	x	$45x$
Second Part	50	$4 - x$	$50(4 - x)$
Total			195

From the last column:

$$45x + 50(4 - x) = 195$$
$$45x + 200 - 50x = 195$$
$$-5x = -5$$
$$x = 1$$

The automobile averaged 45 mph for 1 hour.

Check: 45 mph for 1 hour $= 45$ miles and 50 mph for 3 hours $= 150$ miles; $45 + 150 = 195$.

34. Let $x =$ the average speed for the first hour. Then $x - 7 =$ the average speed for the second hour. Using $d = rt$, the distance traveled for the first hour is $x(1)$, for the second hour is $(x - 7)(1)$, and for the whole trip, 85.

$$x + (x - 7) = 85$$
$$2x - 7 = 85$$
$$2x = 92$$
$$x = 46$$

The average speed for the first hour was 46 mph.

Check: 46 mph for 1 hour $= 46$ miles and $46 - 7 = 39$ mph for 1 hour $= 39$ miles; $46 + 39 = 85$.

35. The sum of the angles in a triangle is 180°.

$$\begin{aligned}(3x+7)+(4x+1)+(9x-4)&=180\\16x+4&=180\\16x&=176\\x&=11\end{aligned}$$

The first angle is $3(11)+7=40°$.
The second angle is $4(11)+1=45°$.
The third angle is $9(11)-4=95°$.

36. The marked angles are supplements which have a sum of 180°.

$$\begin{aligned}(15x+15)+(3x+3)&=180\\18x+18&=180\\18x&=162\\x&=9\end{aligned}$$

The angle measures are
$15(9)+15=150°$ and $3(9)+3=30°$.

37. $-\frac{2}{3}k<6$

$-2k<18$ *Multiply by 3.*

Divide by -2; reverse the inequality symbol.

$k>-9$

Solution set: $(-9,\infty)$

38. $-5x-4\geq 11$

$-5x\geq 15$

Divide by -5; reverse the inequality symbol.

$x\leq -3$

Solution set: $(-\infty,-3]$

39. $\frac{6a+3}{-4}<-3$

Multiply by -4; reverse the inequality symbol.

$6a+3>12$

$6a>9$

$a>\frac{9}{6}=\frac{3}{2}$

Solution set: $\left(\frac{3}{2},\infty\right)$

40. $5-(6-4k)\geq 2k-7$

$$\begin{aligned}5-6+4k&\geq 2k-7\\4k-1&\geq 2k-7\\2k&\geq -6\\k&\geq -3\end{aligned}$$

Solution set: $[-3,\infty)$

41. $8\leq 3z-1<14$

$9\leq 3z<15$

$3\leq z<5$

Solution set: $[3,5)$

42. $\frac{5}{3}(m-2)+\frac{2}{5}(m+1)>1$

$25(m-2)+6(m+1)>15$

Multiply by 15.

$$\begin{aligned}25m-50+6m+6&>15\\31m-44&>15\\31m&>59\\m&>\frac{59}{31}\end{aligned}$$

Solution set: $\left(\frac{59}{31},\infty\right)$

43. Let $x=$ the length of the rectangle. The width of the rectangle is 22 and the perimeter can be no greater than 120.

$$\begin{aligned}P&\leq 120\\2L+2W&\leq 120\\2(x)+2(22)&\leq 120\\2x+44&\leq 120\\2x&\leq 76\\x&\leq 38\end{aligned}$$

The length must be 38 meters or less.

44. Let $x=$ the number of tickets that can be purchased. The total cost of the tickets is $10.59x$. Including the \$50 discount and staying within the available \$1000, we have

$$\begin{aligned}10.59x-50&\leq 1000.\\10.59&\leq 1050\\x&\lesssim 99.15\end{aligned}$$

The group can purchase 99 tickets or less.

45. Let $x=$ the student's score on the fifth test. The average of the five test scores must be at least 70. The inequality is

$$\begin{aligned}\frac{75+79+64+71+x}{5}&\geq 70.\\75+79+64+71+x&\geq 350\\289+x&\geq 350\\x&\geq 61\end{aligned}$$

The student will pass algebra if any score greater than or equal to 61% on the fifth test is achieved.

46. The result, $-8<-13$, is a false statement. There are no real numbers that make this inequality true. The solution set is $\emptyset$.

For Exercises 47–50, let $A=\{\text{a, b, c, d}\}$, $B=\{\text{a, c, e, f}\}$, and $C=\{\text{a, e, f, g}\}$.

47. $A\cap B=\{\text{a, b, c, d}\}\cap\{\text{a, c, e, f}\}$
$=\{\text{a, c}\}$

48. $A\cap C=\{\text{a, b, c, d}\}\cap\{\text{a, e, f, g}\}$
$=\{\text{a}\}$

49. $B\cup C=\{\text{a, c, e, f}\}\cup\{\text{a, e, f, g}\}$
$=\{\text{a, c, e, f, g}\}$

50. $A\cup C=\{\text{a, b, c, d}\}\cup\{\text{a, e, f, g}\}$
$=\{\text{a, b, c, d, e, f, g}\}$

51. $x > 6$ and $x < 9$

The graph of the solution set will be all numbers which are both greater than 6 and less than 9. The overlap is the numbers between 6 and 9, not including the endpoints.

Solution set: $(6, 9)$

52. $x + 4 > 12$ and $x - 2 < 12$
$x > 8$ and $x < 14$

The graph of the solution set will be all numbers between 8 and 14, not including the endpoints.

Solution set: $(8, 14)$

53. $x > 5$ or $x \leq -3$

The graph of the solution set will be all numbers that are either greater than 5 or less than or equal to -3.

Solution set: $(-\infty, -3] \cup (5, \infty)$

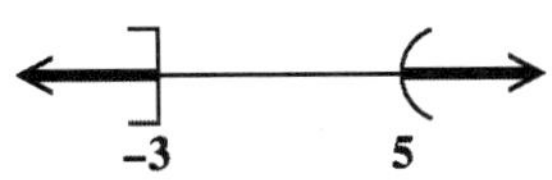

54. $x \geq -2$ or $x < 2$

The graph of the solution set will be all numbers that are either greater than or equal to -2 or less than 2. All real numbers satisfy these criteria.

Solution set: $(-\infty, \infty)$

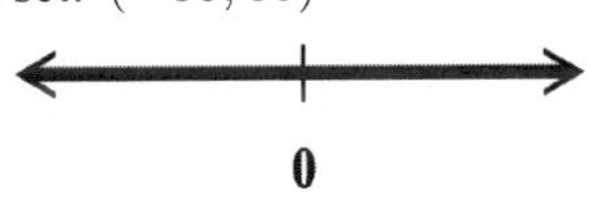

55. $x - 4 > 6$ and $x + 3 \leq 10$
$x > 10$ and $x \leq 7$

The graph of the solution set will be all numbers that are both greater than 10 and less than or equal to 7. There are no real numbers satisfying these criteria.

Solution set: $\emptyset$

56. $-5x + 1 \geq 11$ or $3x + 5 \geq 26$
$-5x \geq 10$ $\quad$ $3x \geq 21$
$x \leq -2$ or $x \geq 7$

The graph of the solution set will be all numbers that are either less than or equal to -2 or greater than or equal to 7.

Solution set: $(-\infty, -2] \cup [7, \infty)$

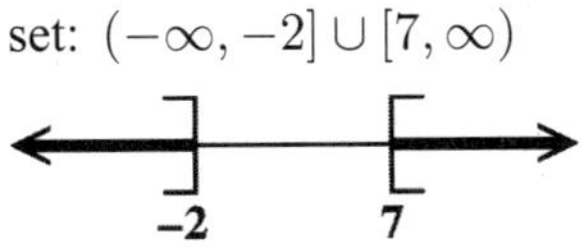

57. $(-3, \infty) \cap (-\infty, 4)$
$(-3, \infty)$ includes all real numbers greater than -3.
$(-\infty, 4)$ includes all real numbers less than 4. Find the intersection. The numbers common to both sets are greater than -3 and less than 4.

$$-3 < x < 4$$

Solution set: $(-3, 4)$

58. $(-\infty, 6) \cap (-\infty, 2)$
$(-\infty, 6)$ includes all real numbers less than 6.
$(-\infty, 2)$ includes all real numbers less than 2.
Find the intersection. The numbers common to both sets are less than 2.

Solution set: $(-\infty, 2)$

59. $(4, \infty) \cup (9, \infty)$
$(4, \infty)$ includes all real numbers greater than 4.
$(9, \infty)$ includes all real numbers greater than 9.
Find the union. The numbers in the first set, the second set, or in both sets are all the real numbers that are greater than 4.
Solution set: $(4, \infty)$

60. $(1, 2) \cup (1, \infty)$
$(1, 2)$ includes the real numbers between 1 and 2, not including 1 and 2.
$(1, \infty)$ includes all real numbers greater than 1.
Find the union. The numbers in the first set, the second set, or in both sets are all real numbers greater than 1.

Solution set: $(1, \infty)$

61. $|x| = 7$
$x = 7$ or $x = -7$
Solution set: $\{-7, 7\}$

62. $|x + 2| = 9$

$x + 2 = 9$ or $x + 2 = -9$
$x = 7$ or $x = -11$

Solution set: $\{-11, 7\}$

63. $|3k - 7| = 8$
$3k - 7 = 8$ or $3k - 7 = -8$
$3k = 15$ $\quad$ $3k = -1$
$k = 5$ or $k = -\frac{1}{3}$

Solution set: $\left\{-\frac{1}{3}, 5\right\}$

64. $|z - 4| = -12$
Since the absolute value of an expression can never be negative, there are no solutions for this equation.
Solution set: $\emptyset$

65. $|2k-7|+4=11$

$$|2k-7|=7$$

$$2k-7=7 \quad \text{or} \quad 2k-7=-7$$

$$2k=14 \qquad\qquad 2k=0$$

$$k=7 \quad \text{or} \quad k=0$$

Solution set: $\{0,7\}$

66. $|4a+2|-7=-3$

$$|4a+2|=4$$

$$4a+2=4 \quad \text{or} \quad 4a+2=-4$$

$$4a=2 \qquad\qquad 4a=-6$$

$$a=\frac{2}{4} \qquad\qquad a=-\frac{6}{4}$$

$$a=\frac{1}{2} \quad \text{or} \quad a=-\frac{3}{2}$$

Solution set: $\left\{-\frac{3}{2},\frac{1}{2}\right\}$

67. $|3p+1|=|p+2|$

$$3p+1=p+2 \quad \text{or} \quad 3p+1=-(p+2)$$

$$2p=1 \qquad\qquad 3p+1=-p-2$$

$$4p=-3$$

$$p=\frac{1}{2} \quad \text{or} \quad p=-\frac{3}{4}$$

Solution set: $\left\{-\frac{3}{4},\frac{1}{2}\right\}$

68. $|2m-1|=|2m+3|$

$$2m-1=2m+3 \quad \text{or} \quad 2m-1=-(2m+3)$$

$$-1=3 \quad \textit{False} \qquad 2m-1=-2m-3$$

$$4m=-2$$

$$\textit{No solution} \quad \text{or} \quad m=-\frac{2}{4}=-\frac{1}{2}$$

Solution set: $\left\{-\frac{1}{2}\right\}$

69. $|p|<14$

$$-14<p<14$$

Solution set: $(-14,14)$

70. $|-t+6|\le 7$

$$-7\le -t+6\le 7$$

$$-13\le -t\le 1$$

Multiply by -1; reverse the inequality symbols.

$$13\ge t\ge -1 \quad \text{or} \quad -1\le t\le 13$$

Solution set: $[-1,13]$

71. $|2p+5|\le 1$

$$-1\le 2p+5\le 1$$

$$-6\le 2p\le -4$$

$$-3\le p\le -2$$

Solution set: $[-3,-2]$

72. $|x+1|\ge -3$

Since the absolute value of an expression is always nonnegative (positive or zero), the inequality is *true* for any real number x.

Solution set: $(-\infty,\infty)$

73. $5-(6-4k)>2k-5$

$$5-6+4k>2k-5$$

$$-1+4k>2k-5$$

$$2k>-4$$

$$k>-2$$

Solution set: $(-2,\infty)$

74. Solve $ak+bt=6t-sk$ for k.

Get the "k-terms" on one side.

$$ak+sk=6t-bt$$

$$(a+s)k=6t-bt$$

$$k=\frac{6t-bt}{a+s}$$

If we take the k-terms to the right side in the first step, we get

$$k=\frac{bt-6t}{-a-s}.$$

75. $x<3$ and $x\ge -2$

The real numbers that are common to both sets are the numbers greater than or equal to -2 and less than 3.

$$-2\le x<3$$

Solution set: $[-2,3)$

76. $\dfrac{4x+2}{4}+\dfrac{3x-1}{8}=\dfrac{x+6}{16}$

Clear fractions by multiplying by the LCD, 16.

$$4(4x+2)+2(3x-1)=x+6$$

$$16x+8+6x-2=x+6$$

$$22x+6=x+6$$

$$21x=0$$

$$x=0$$

Solution set: $\{0\}$

77. $|3k+6|\ge 0$

The absolute value of an expression is always nonnegative, so the inequality is true for any real number k.

Solution set: $(-\infty,\infty)$

78. $-5r\ge -10$

$$r\le 2 \qquad \textit{Divide by } -5\textit{; reverse inequality.}$$

Solution set: $(-\infty,2]$

79. Use the formula $V = LWH$, and solve for H.

$$\frac{V}{LW} = \frac{LWH}{LW}$$

$$\frac{V}{LW} = H \quad \text{or} \quad H = \frac{V}{LW}$$

Substitute 1.5 for W, 5 for L, and 75 for V.

$$H = \frac{75}{5(1.5)} = \frac{75}{7.5} = 10$$

The height of the box is 10 ft.

80. Let $x =$ the smallest consecutive integer. Then $x + 1 =$ the middle consecutive integer and $x + 2 =$ the largest consecutive integer.
The sum of the smallest and largest integers is 47 more than the middle integer, so an equation is

$$x + (x + 2) = 47 + (x + 1).$$
$$2x + 2 = 48 + x$$
$$x = 46$$

Then $x + 1 = 47$, and $x + 2 = 48$.
The integers are 46, 47, and 48.

81. $|3x + 2| + 4 = 9$

$|3x + 2| = 5$

$3x + 2 = 5$ or $3x + 2 = -5$

$3x = 3$ $\quad$ $3x = -7$

$x = 1$ or $x = -\frac{7}{3}$

Solution set: $\left\{-\frac{7}{3}, 1\right\}$

82. $.05x + .03(1200 - x) = 42$

$5x + 3(1200 - x) = 4200$ *Multiply by 100.*

$5x + 3600 - 3x = 4200$

$2x = 600$

$x = 300$

Solution set: $\{300\}$

83. $|m + 3| \le 13$

$-13 \le m + 3 \le 13$

$-16 \le m \le 10$

Solution set: $[-16, 10]$

84. $\frac{3}{4}(a - 2) - \frac{1}{3}(5 - 2a) < -2$

$9(a - 2) - 4(5 - 2a) < -24$

Multiply by 12.

$9a - 18 - 20 + 8a < -24$

$17a - 38 < -24$

$17a < 14$

$a < \frac{14}{17}$

Solution set: $\left(-\infty, \frac{14}{17}\right)$

85. $-4 < 3 - 2k < 9$

$-7 < -2k < 6$

Divide by -2; reverse the inequality symbols.

$\frac{7}{2} > k > -3$ or $-3 < k < \frac{7}{2}$

Solution set: $\left(-3, \frac{7}{2}\right)$

86. $-.3x + 2.1(x - 4) \le -6.6$

$-3x + 21(x - 4) \le -66$

Multiply by 10.

$-3x + 21x - 84 \le -66$

$18x - 84 \le -66$

$18x \le 18$

$x \le 1$

Solution set: $(-\infty, 1]$

87. Let $x =$ the angle. Then $90 - x$ is its complement and $180 - x$ is its supplement.
The complement of an angle measures 10° less than one-fifth of its supplement.

$90 - x = \frac{1}{5}(180 - x) - 10$

$450 - 5x = 180 - x - 50$ *Multiply by 5.*

$450 - 5x = 130 - x$

$320 = 4x$

$80 = x$

The measure of the angle is 80°.

88. Use the rule of 78.

$$u = f \cdot \frac{k(k + 1)}{n(n + 1)}$$

Let $f = 450$, $k = 5$, and $n = 24$.

$$u = 450 \cdot \frac{5(6)}{24(25)} = 22.50$$

The unearned interest is \$22.50.

89. Let $x =$ the employee's earnings during the fifth month. The average of the five months must be at least \$1000.

$$\frac{900 + 1200 + 1040 + 760 + x}{5} \ge 1000$$
$$900 + 1200 + 1040 + 760 + x \ge 5000$$
$$3900 + x \ge 5000$$
$$x \ge 1100$$

Any amount greater than or equal to \$1100 will qualify the employee for the pension plan.

90. $|5r - 1| > 14$

$5r - 1 > 14$ or $5r - 1 < -14$

$5r > 15$ $\quad$ $5r < -13$

$r > 3$ or $r < -\frac{13}{5}$

Solution set: $\left(-\infty, -\frac{13}{5}\right) \cup (3, \infty)$

91. $x \geq -2$ or $x < 4$
The solution set includes all numbers either greater than or equal to -2 or all numbers less than 4. This is the union and is the set of all real numbers. The solution set is $(-\infty, \infty)$.

92. Let $x =$ the number of liters of the 20% solution. Then $x + 10$ is the number of liters of the resulting 40% solution.

Liters of Solution	Percent (as a decimal)	Liters of Mixture
x	.20	$.20x$
10	.50	$.50(10) = 5$
$x + 10$	.40	$.40(x + 10)$

From the last column:

$$\begin{aligned} .20x + 5 &= .40(x + 10) \\ .20x + 5 &= .40x + 4 \\ 1 &= .20x \\ 5 &= x \end{aligned}$$

5 L of the 20% solution should be used.

93. $|m - 1| = |2m + 3|$

$$\begin{aligned} m - 1 = 2m + 3 \quad &\text{or} \quad m - 1 = -(2m + 3) \\ & \qquad\quad m - 1 = -2m - 3 \\ & \qquad\qquad 3m = -2 \\ -4 = m \quad &\text{or} \qquad\quad m = -\frac{2}{3} \end{aligned}$$

Solution set: $\left\{-4, -\frac{2}{3}\right\}$

94. $\frac{3x}{5} - \frac{x}{2} = 3$

$6x - 5x = 30$ *Multiply by 10.*

$x = 30$

Solution set: $\{30\}$

95. $|m + 3| \leq 1$

$-1 \leq m + 3 \leq 1$

$-4 \leq m \leq -2$

Solution set: $[-4, -2]$

96. $|3k - 7| = 4$

$$\begin{aligned} 3k - 7 = 4 \quad &\text{or} \quad 3k - 7 = -4 \\ 3k = 11 \quad & \qquad\quad 3k = 3 \\ k = \frac{11}{3} \quad &\text{or} \qquad\quad k = 1 \end{aligned}$$

Solution set: $\left\{1, \frac{11}{3}\right\}$

97. Let $x =$ the number of electoral votes for George W. Bush. Then $x - 5$ is the number of votes for Al Gore. There were a total of 537 votes.

$$\begin{aligned} x + (x - 5) &= 537 \\ 2x - 5 &= 537 \\ 2x &= 542 \\ x &= 271 \end{aligned}$$

Bush received 271 votes and Gore received $271 - 5 = 266$ votes.

98. $5(2x - 7) = 2(5x + 3)$

$10x - 35 = 10x + 6$

$-35 = 6$ *False*

This equation is a *contradiction.*

Solution set: $\emptyset$

99. $x > 6$ and $x < 8$
The graph of the solution set is all numbers both greater than 6 *and* less than 8. This is the intersection. The elements common to both sets are the numbers between 6 and 8, not including the endpoints. The solution set is $(6, 8)$.

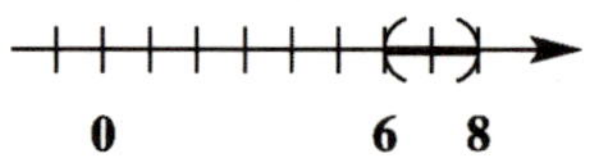

100.

$$\begin{aligned} -5x + 1 \geq 11 \quad &\text{or} \quad 3x + 5 \geq 26 \\ -5x \geq 10 \quad & \qquad\quad 3x \geq 21 \\ x \leq -2 \quad &\text{or} \qquad\quad x \geq 7 \end{aligned}$$

The graph of the solution set is all numbers either less than or equal to -2 *or* greater than or equal to 7. This is the union. The solution set is $(-\infty, -2] \cup [7, \infty)$.

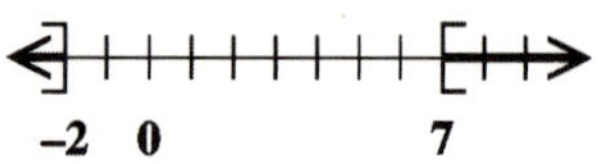

101. **(a)** There are *no* occupations with median earnings for men less than \$900 and for women greater than \$500.

(b) *All* of the listed occupations have median earnings for men greater than \$600 or for women less than \$400. They are: Managerial/Professional, Technical/Sales/Administrative Support, Service, Operators/Fabricators/Laborers.

Chapter 2 Test

1. $3(2x - 2) - 4(x + 6) = 3x + 8 + x$

$$\begin{aligned} 6x - 6 - 4x - 24 &= 4x + 8 \\ 2x - 30 &= 4x + 8 \\ -2x &= 38 \\ x &= -19 \end{aligned}$$

Solution set: $\{-19\}$

2. $.08x + .06(x + 9) = 1.24$

$8x + 6(x + 9) = 124$

Multiply by 100.

$8x + 6x + 54 = 124$

$$14x + 54 = 124$$
$$14x = 70$$
$$x = 5$$

Solution set: $\{5\}$

3. $$\frac{x+6}{10} + \frac{x-4}{15} = \frac{x+2}{6}$$
$$3(x+6) + 2(x-4) = 5(x+2)$$
Multiply by 30.
$$3x + 18 + 2x - 8 = 5x + 10$$
$$5x + 10 = 5x + 10 \quad \textit{True}$$
This is an *identity*.

Solution set: $(-\infty, \infty)$

4. $$3x - (2 - x) + 4x + 2 = 8x + 3$$
$$3x - 2 + x + 4x + 2 = 8x + 3$$
$$8x = 8x + 3$$
$$0 = 3 \quad \textit{False}$$
The false statement indicates that the equation is a *contradiction*.

Solution set: $\emptyset$

5. Solve $-16t^2 + vt - S = 0$ for v.
$$vt = S + 16t^2$$
$$v = \frac{S + 16t^2}{t}$$

6. Solve $ar + 2 = 3r - 6t$ for r.
Get the "r-terms" on one side.
$$ar - 3r = -2 - 6t$$
$$r(a-3) = -2 - 6t$$
$$r = \frac{-2-6t}{a-3}$$
If we take the r-terms to the right side in the first step, we get
$$r = \frac{2+6t}{3-a}.$$

7. Solve $d = rt$ for t and substitute 500 for d and 148.295 for r.
$$t = \frac{d}{r} = \frac{500}{148.295} \approx 3.372$$
Gordon's time was about 3.372 hr.

8. Use $I = Prt$ and substitute \$2281.25 for I, \$36,500 for P, and 1 for t.
$$2281.25 = 36{,}500r(1)$$
$$r = \frac{2281.25}{36{,}500} = .0625$$
The rate of interest is 6.25%.

9. $$\frac{27{,}952}{38{,}159} \approx .733$$
73.3% were classified as post offices.

10. Let $x =$ the amount invested at 3%. Then $28{,}000 - x =$ the amount invested at 5%.

Principal	Rate (as a decimal)	Interest
x	.03	$.03x$
$28,000 - x$	.05	$.05(28,000 - x)$
\$28,000	← Total →	\$1240

From the last column:
$$.03x + .05(28{,}000 - x) = 1240$$
$$3x + 5(28{,}000 - x) = 124{,}000$$
Multiply by 100.
$$3x + 140{,}000 - 5x = 124{,}000$$
$$-2x = -16{,}000$$
$$x = 8000$$
He invested \$8000 at 3% and \$28,000 − \$8000 = \$20,000 at 5%.

11. Let $x =$ the speed of the slower car. Then $x + 15 =$ the speed of the faster car.

Use the formula $d = rt$.

	Rate	Time	Distance
Slower Car	x	6	$6x$
Faster Car	$x + 15$	6	$6(x+15)$
			630

The total distance traveled is the sum of the distances traveled by each car.
$$6x + 6(x+15) = 630$$
$$6x + 6x + 90 = 630$$
$$12x = 540$$
$$x = 45$$
The slower car traveled at 45 mph, while the faster car traveled at $45 + 15 = 60$ mph.

12. The sum of the three angle measures is 180°.
$$(2x + 20) + x + x = 180$$
$$4x + 20 = 180$$
$$4x = 160$$
$$x = 40$$
The three angle measures are 40°, 40°, and $(2 \cdot 40 + 20)° = 100°$.

13. $$4 - 6(x+3) \le -2 - 3(x+6) + 3x$$
$$4 - 6x - 18 \le -2 - 3x - 18 + 3x$$
$$-6x - 14 \le -20$$
$$-6x \le -6$$
Divide by −6, and reverse the inequality symbol.
$$x \ge 1$$
Solution set: $[1, \infty)$

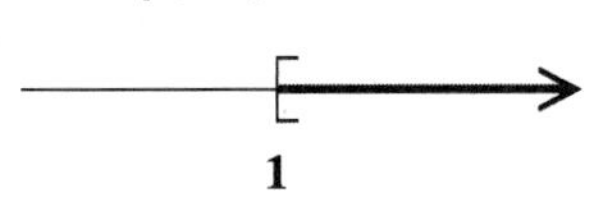

14. $-\frac{4}{7}x > -16$

$-4x > -112$ *Multiply by 7.*

Divide by -4, and reverse the inequality symbol.

$x < 28$

Solution set: $(-\infty, 28)$

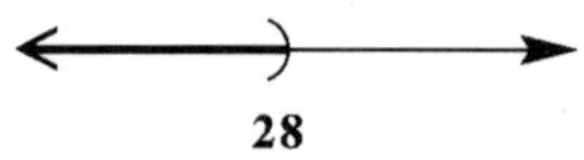

15. $-6 \le \frac{4}{3}x - 2 \le 2$

$-18 \le 4x - 6 \le 6$ *Multiply by 3.*

$-12 \le 4x \le 12$ *Add 6.*

$-3 \le x \le 3$ *Divide by 4.*

Solution set: $[-3, 3]$

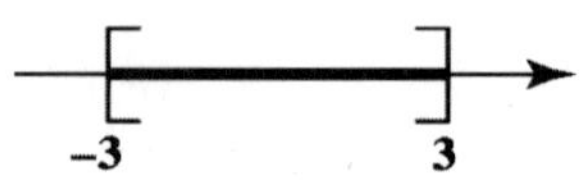

16. Let $x =$ the score on the fourth test.

$$\frac{83+76+79+x}{4} \ge 80$$
$$\frac{238+x}{4} \ge 80$$
$$238 + x \ge 320$$
$$x \ge 82$$

The minimum score must be 82 to guarantee a B.

17. $C = 50x + 5000$; $R = 60x$

$$R \ge C$$
$$60x \ge 50x + 5000$$
$$10x \ge 5000$$
$$x \ge 500$$

The solution set is $[500, \infty)$.

18. **(a)** $A \cap B = \{1,2,5,7\} \cap \{1,5,9,12\}$
$= \{1,5\}$

(b) $A \cup B = \{1,2,5,7\} \cup \{1,5,9,12\}$
$= \{1,2,5,7,9,12\}$

19. **(a)** $3k \ge 6$ and $k - 4 < 5$

$k \ge 2$ and $k < 9$

The solution set is all numbers both greater than or equal to 2 *and* less than 9. This is the intersection. The numbers common to both sets are between 2 and 9, including 2 but not 9. The solution set is $[2, 9)$.

(b) $-4x \le -24$ or $4x - 2 < 10$

$4x < 12$

$x \ge 6$ or $x < 3$

The solution set is all numbers less than 3 or greater than or equal to 6. This is the union. The solution set is $(-\infty, 3) \cup [6, \infty)$.

20. $|4x - 3| = 7$

$4x - 3 = 7$ or $4x - 3 = -7$

$4x = 10$ $\quad 4x = -4$

$x = \frac{10}{4} = \frac{5}{2}$ or $x = -1$

Solution set: $\left\{-1, \frac{5}{2}\right\}$

21. $|5 - 6x| > 12$

$5 - 6x > 12$ or $5 - 6x < -12$

$-6x > 7$ $\quad -6x < -17$

$x < -\frac{7}{6}$ or $x > \frac{17}{6}$

Solution set: $\left(-\infty, -\frac{7}{6}\right) \cup \left(\frac{17}{6}, \infty\right)$

22. $|7 - x| \le -1$

Since the absolute value of an expression is always nonnegative (positive or zero), the inequality is *false* for any real number x.

Solution set: $\emptyset$

23. $|3 - 5x| = |2x + 8|$

$3 - 5x = 2x + 8$ or $3 - 5x = -(2x + 8)$

$-7x = 5$ $\quad 3 - 5x = -2x - 8$

$-3x = -11$

$x = -\frac{5}{7}$ or $x = \frac{11}{3}$

Solution set: $\left\{-\frac{5}{7}, \frac{11}{3}\right\}$

24. $|-3x + 4| - 4 < -1$

$|-3x + 4| < 3$

$-3 < -3x + 4 < 3$

$-7 < -3x < -1$

$\frac{7}{3} > x > \frac{1}{3}$ or $\frac{1}{3} < x < \frac{7}{3}$

Solution set: $\left(\frac{1}{3}, \frac{7}{3}\right)$

25. **(a)** $|5x + 3| < k$

If $k < 0$, then $|5x + 3|$ would be less than a negative number. Since the absolute value of an expression is always nonnegative (positive or zero), the solution set is $\emptyset$.

(b) $|5x + 3| > k$

If $k < 0$, then $|5x + 3|$ would be greater than a negative number. Since the absolute value of an expression is always nonnegative (positive or zero), the solution set is the set of all real numbers, $(-\infty, \infty)$.

(c) $|5x + 3| = k$
If $k < 0$, then $|5x + 3|$ would be equal to a negative number. Since the absolute value of an expression is always nonnegative (positive or zero), the solution set is $\emptyset$.

Cumulative Review Exercises Chapters 1–2

Exercises 1–6 refer to set A.

Let $A = \left\{-8, -\frac{2}{3}, -\sqrt{6}, 0, \frac{4}{5}, 9, \sqrt{36}\right\}$.

Simplify $\sqrt{36} = 6$.

1. The elements 9 and 6 are natural numbers.

2. The elements 0, 9, and 6 are whole numbers.

3. The elements -8, 0, 9, and 6 are integers.

4. The elements $-8, -\frac{2}{3}, 0, \frac{4}{5}, 9,$ and 6 are rational numbers.

5. The element $-\sqrt{6}$ is an irrational number.

6. All the elements in set A are real numbers.

7. $$-\frac{4}{3} - \left(-\frac{2}{7}\right) = -\frac{4}{3} + \frac{2}{7} = -\frac{28}{21} + \frac{6}{21} = -\frac{22}{21}$$

8. $$|-4| - |2| + |-6| = 4 - 2 + 6 = 2 + 6 = 8$$

9. $(-2)^4 + (-2)^3 = 16 + (-8) = 8$

10. $\sqrt{25} - 5(-1)^0 = 5 - 5(1) = 5 - 5 = 0$

11. $(-3)^5 = (-3)(-3)(-3)(-3)(-3) = -243$

12. $\left(\frac{6}{7}\right)^3 = \frac{6}{7} \cdot \frac{6}{7} \cdot \frac{6}{7} = \frac{216}{343}$

13. $\left(-\frac{2}{3}\right)^3 = \left(-\frac{2}{3}\right)\left(-\frac{2}{3}\right)\left(-\frac{2}{3}\right) = -\frac{8}{27}$

14. $-4^6 = -(4 \cdot 4 \cdot 4 \cdot 4 \cdot 4 \cdot 4) = -4096$

15. $-\sqrt{36} = -(6) = -6$
$\sqrt{-36}$ is not a real number.

16. $\frac{4-4}{4+4} = \frac{0}{8} = 0$
$\frac{4+4}{4-4} = \frac{8}{0}$, which is undefined.

For Exercises 17–19, let $a = 2$, $b = -3$, and $c = 4$.

17. $$-3a + 2b - c = -3(2) + 2(-3) - 4 = -6 - 6 - 4 = -16$$

18. $$-8(a^2 + b^3) = -8[2^2 + (-3)^3] = -8[4 + (-27)] = -8(-23) = 184$$

19. $$\frac{3a^3 - b}{4 + 3c} = \frac{3(2)^3 - (-3)}{4 + 3(4)} = \frac{3(8) - (-3)}{4 + 3(4)} = \frac{24 + 3}{4 + 12} = \frac{27}{16}$$

20. $$-7r + 5 - 13r + 12 = -7r - 13r + 5 + 12 = (-7 - 13)r + (5 + 12) = -20r + 17$$

21. $$-(3k + 8) - 2(4k - 7) + 3(8k + 12) = -3k - 8 - 8k + 14 + 24k + 36 = -3k - 8k + 24k - 8 + 14 + 36 = 13k + 42$$

22. $(a + b) + 4 = 4 + (a + b)$
The order of the terms $(a + b)$ and 4 have been reversed. This is an illustration of the commutative property.

23. $4x + 12x = (4 + 12)x$
The common variable, x, has been removed from each term. This is an illustration of the distributive property.

24. $$\begin{aligned} -4x + 7(2x + 3) &= 7x + 36 \\ -4x + 14x + 21 &= 7x + 36 \\ 10x + 21 &= 7x + 36 \\ 3x &= 15 \\ x &= 5 \end{aligned}$$
Solution set: $\{5\}$

25. $$\begin{aligned} -\frac{3}{5}x + \frac{2}{3}x &= 2 \\ 3(-3x) + 5(2x) &= 15(2) \quad \textit{Multiply by 15.} \\ -9x + 10x &= 30 \\ x &= 30 \end{aligned}$$
Solution set: $\{30\}$

26. $$\begin{aligned} .06x + .03(100 + x) &= 4.35 \\ 6x + 3(100 + x) &= 435 \quad \textit{Multiply by 100.} \\ 6x + 300 + 3x &= 435 \\ 9x &= 135 \\ x &= 15 \end{aligned}$$
Solution set: $\{15\}$

27. Solve $P = a + b + c$ for b.

$$P - a - c = b \quad \text{or} \quad b = P - a - c$$

28.
$$\begin{aligned} 4(2x-6)+3(x-2) &= 11x+1 \\ 8x-24+3x-6 &= 11x+1 \\ 11x-30 &= 11x+1 \\ -30 &= 1 \quad \textit{False} \end{aligned}$$

Solution set: $\emptyset$

29.
$$\frac{2}{3}x+\frac{5}{8}x=\frac{31}{24}x$$
Multiply by the LCD, 24.
$$\begin{aligned} 8(2x)+3(5x) &= 31x \\ 16x+15x &= 31x \\ 31x &= 31x \quad \textit{True} \end{aligned}$$

Solution set: {all real numbers}

30.
$$\begin{aligned} 3-2(x+7) &\le -x+3 \\ 3-2x-14 &\le -x+3 \\ -2x-11 &\le -x+3 \\ -x &\le 14 \end{aligned}$$
Multiply by -1, and reverse the inequality symbol.
$$x \ge -14$$
Solution set: $[-14, \infty)$

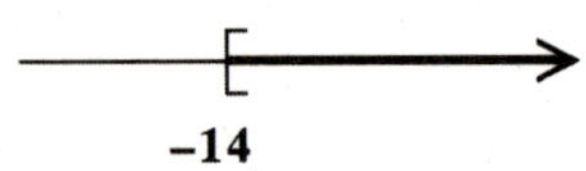

31.
$$\begin{aligned} -4 < 5-3x &\le 0 \\ -9 < -3x &\le -5 \end{aligned}$$
Divide by -3, and reverse the inequality symbol.
$$3 > x \ge \frac{5}{3} \quad \text{or} \quad \frac{5}{3} \le x < 3$$
Solution set: $\left[\frac{5}{3}, 3\right)$

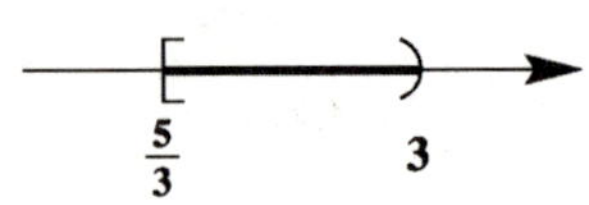

32.
$$\begin{aligned} 2x+1 > 5 \quad &\text{or} \quad 2-x > 2 \\ 2x > 4 \quad & \quad\quad -x > 0 \\ x > 2 \quad &\text{or} \quad x < 0 \end{aligned}$$
Solution set: $(-\infty, 0) \cup (2, \infty)$

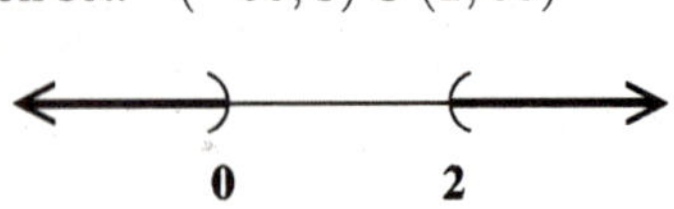

33.
$$|-7k+3| \ge 4$$
$$\begin{aligned} -7k+3 \ge 4 \quad &\text{or} \quad -7k+3 \le -4 \\ -7k \ge 1 \quad & \quad\quad -7k \le -7 \\ k \le -\frac{1}{7} \quad &\text{or} \quad k \ge 1 \end{aligned}$$
Solution set: $\left(-\infty, -\frac{1}{7}\right] \cup [1, \infty)$

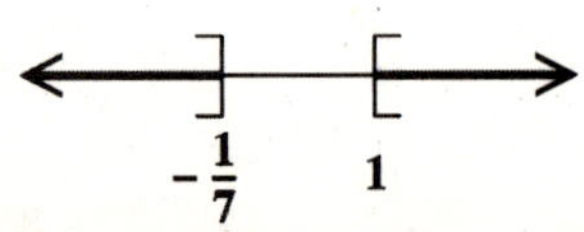

34. Let $x =$ the amount invested at 7% and at 10%. The total amount invested is $2x$, and the total interest is
$$.1(2x)-150 = .2x-150.$$
$$\begin{array}{ccccc} \text{Interest at 7\%} & + & \text{interest at 10\%} & = & \text{total interest.} \\ .07x & + & .10x & = & .2x-150 \\ & & .17x & = & .20x-150 \\ & & 150 & = & .03x \\ & & 5000 & = & x \end{array}$$
He invested \$5000 at each rate.

35. Let $x =$ the number of grams of food C. Then $2x$ is the number of grams of food A.
There are 5 grams of food B and the total is at most 24 grams.
$$\begin{aligned} x+2x+5 &\le 24 \\ 3x &\le 19 \\ x &\le \frac{19}{3} \text{ or } 6\frac{1}{3} \end{aligned}$$
He may use at most $6\frac{1}{3}$ grams of food C.

36. Let $x =$ the grade the student must make on the third test.
To find the average of the three tests, add them and divide by 3. This average must be at least 80.
$$\begin{aligned} \frac{88+78+x}{3} &\ge 80 \\ \frac{166+x}{3} &\ge 80 \\ 166+x &\ge 240 \\ x &\ge 74 \end{aligned}$$
She must score at least 74 on her third test.

37. Let $x =$ the amount of pure alcohol that should be added.

Liters of Solution	Percent (as a decimal)	Liters of Pure Alcohol
x	1.00	$1.00x$
7	.10	$.10(7) = .7$
$x+7$	.30	$.30(x+7)$

From the last column:
$$\begin{aligned} 1.00x + .7 &= .30(x+7) \\ 10x+7 &= 3(x+7) \quad \textit{Multiply by 10.} \\ 10x+7 &= 3x+21 \\ 7x &= 14 \\ x &= 2 \end{aligned}$$
2 L of pure alcohol should be added to the solution.

38. Let $x =$ the number of nickels. Then $x - 4 =$ the number of quarters. The collection contains 29 coins, so the number of pennies is

$$29 - x - (x - 4) = 33 - 2x.$$

	Number of Coins	Denomination	Value
Pennies	$33 - 2x$	.01	$.01(33 - 2x)$
Nickels	x	.05	$.05x$
Quarters	$x - 4$	.25	$.25(x - 4)$
Total	29		\$2.69

From the last column:

$$\begin{aligned} .01(33 - 2x) + .05x + .25(x - 4) &= 2.69 \\ 1(33 - 2x) + 5x + 25(x - 4) &= 269 \end{aligned}$$

Multiply by 100.

$$\begin{aligned} 33 - 2x + 5x + 25x - 100 &= 269 \\ 28x - 67 &= 269 \\ 28x &= 336 \\ x &= 12 \end{aligned}$$

There are $33 - 2(12) = 9$ pennies, 12 nickels, and $12 - 4 = 8$ quarters.

39. Clark's rule:

$$\frac{\text{Weight of child in pounds}}{150} \times \begin{matrix}\text{adult}\\ \text{dose}\end{matrix} = \begin{matrix}\text{child's}\\ \text{dose}\end{matrix}$$

If the child weighs 55 lb and the adult dosage is 120 mg, then

$$\frac{55}{150} \times 120 = 44.$$

The child's dosage is 44 mg.

40. **(a)** 1990: 1611
1998: 1489

$1611 - 1489 = 122$

The number decreased by 122 newspapers.

(b) $\dfrac{122}{1611} \approx .076$ or 7.6%.

The number decreased by approximately 7.6%.

CHAPTER 3 GRAPHS, LINEAR EQUATIONS, AND FUNCTIONS

Section 3.1

1. **(a)** x represents the year; y represents the percent of women in math or computer science professions.

   (b) The line goes down from 1990 to 2000. Therefore, from 1990–2000, the percent of women in math or computer science professions decreased.

   (c) The dot over the year 1990 appears to be at about 36, so the ordered pair $(x, y) = (1990, 36)$.

   (d) In 2000, the percent of women in math or computer science professions was 30%.

3. The point with coordinates $(0, 0)$ is called the origin of a rectangular coordinate system.

5. The x-intercept is the point where a line crosses the x-axis. To find the x-intercept of a line, we let y equal 0 and solve for x.

   The y-intercept is the point where a line crosses the y-axis. To find the y-intercept of a line, we let x equal 0 and solve for y.

7. To graph a straight line, we must find a minimum of *two* points. A third point is sometimes found to check the accuracy of the first two points.

9. **(a)** The point $(1, 6)$ is located in quadrant I, since the x- and y-coordinates are both positive.

   (b) The point $(-4, -2)$ is located in quadrant III, since the x- and y-coordinates are both negative.

   (c) The point $(-3, 6)$ is located in quadrant II, since the x-coordinate is negative and the y-coordinate is positive.

   (d) The point $(7, -5)$ is located in quadrant IV, since the x-coordinate is positive and the y-coordinate is negative.

   (e) The point $(-3, 0)$ is located on the x-axis, so it does not belong to any quadrant.

11. **(a)** If $xy > 0$, then both x and y have the same sign.
    (x, y) is in quadrant I if x and y are positive.
    (x, y) is in quadrant III if x and y are negative.

    (b) If $xy < 0$, then x and y have different signs.
    (x, y) is in quadrant II if $x < 0$ and $y > 0$.
    (x, y) is in quadrant IV if $x > 0$ and $y < 0$.

    (c) If $\frac{x}{y} < 0$, then x and y have different signs.
    (x, y) is in either quadrant II or IV. (See part (b).)

    (d) If $\frac{x}{y} > 0$, then x and y have the same sign.
    (x, y) is in either quadrant I or III. (See part (a).)

For Exercises 13–21, see the rectangular coordinate system after Exercise 21.

13. To plot $(2, 3)$, go two units from zero to the right along the x-axis, and then go three units up parallel to the y-axis.

15. To plot $(-3, -2)$, go three units from zero to the left along the x-axis, and then go two units down parallel to the y-axis.

17. To plot $(0, 5)$, do not move along the x-axis at all since the x-coordinate is 0. Move five units up along the y-axis.

19. To plot $(-2, 4)$, go two units from zero to the left along the x-axis, and then go four units up parallel to the y-axis.

21. To plot $(-2, 0)$, go two units to the left along the x-axis. Do not move up or down since the y-coordinate is 0.

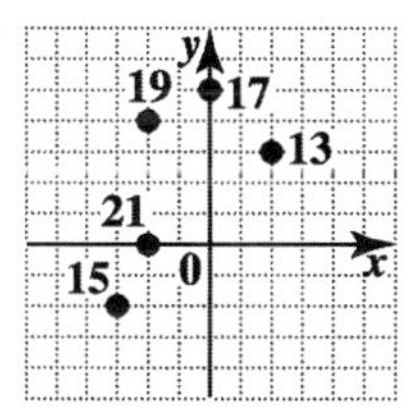

23. $x - y = 3$
    To complete the table, substitute the given values for x and y in the equation.

    For $x = 0$: $x - y = 3$
    $0 - y = 3$
    $y = -3$ $(0, -3)$

    For $y = 0$: $x - y = 3$
    $x - 0 = 3$
    $x = 3$ $(3, 0)$

    For $x = 5$: $x - y = 3$
    $5 - y = 3$
    $-y = -2$
    $y = 2$ $(5, 2)$

    For $x = 2$: $x - y = 3$
    $2 - y = 3$
    $-y - 1$
    $y = -1$ $(2, -1)$

    Plot the ordered pairs and draw the line through them.

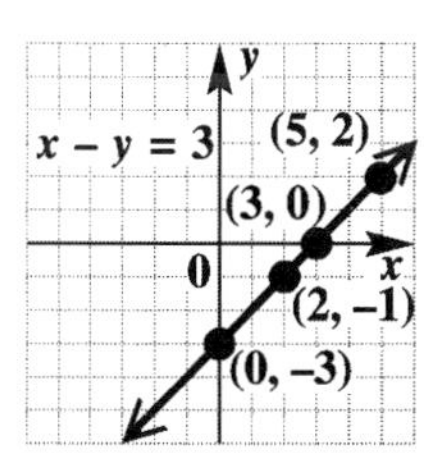

25. $x + 2y = 5$

To complete the table, substitute the given values for x or y in the equation.

For $x = 0$:
$$\begin{aligned} x + 2y &= 5 \\ 0 + 2y &= 5 \\ 2y &= 5 \\ y &= \frac{5}{2} \quad \left(0, \frac{5}{2}\right) \end{aligned}$$

For $y = 0$:
$$\begin{aligned} x + 2y &= 5 \\ x + 2(0) &= 5 \\ x + 0 &= 5 \\ x &= 5 \quad (5, 0) \end{aligned}$$

For $x = 2$:
$$\begin{aligned} x + 2y &= 5 \\ 2 + 2y &= 5 \\ 2y &= 3 \\ y &= \frac{3}{2} \quad \left(2, \frac{3}{2}\right) \end{aligned}$$

For $y = 2$:
$$\begin{aligned} x + 2y &= 5 \\ x + 2(2) &= 5 \\ x + 4 &= 5 \\ x &= 1 \quad (1, 2) \end{aligned}$$

Plot the ordered pairs and draw the line through them.

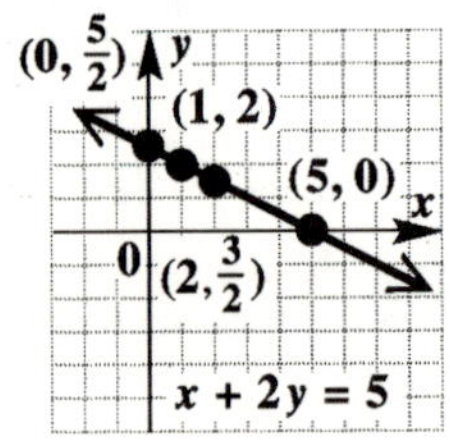

27. $4x - 5y = 20$

For $x = 0$:
$$\begin{aligned} 4x - 5y &= 20 \\ 4(0) - 5y &= 20 \\ -5y &= 20 \\ y &= -4 \quad (0, -4) \end{aligned}$$

For $y = 0$:
$$\begin{aligned} 4x - 5y &= 20 \\ 4x - 5(0) &= 20 \\ 4x &= 20 \\ x &= 5 \quad (5, 0) \end{aligned}$$

For $x = 2$:
$$\begin{aligned} 4x - 5y &= 20 \\ 4(2) - 5y &= 20 \\ 8 - 5y &= 20 \\ -5y &= 12 \\ y &= -\frac{12}{5} \quad \left(2, -\frac{12}{5}\right) \end{aligned}$$

For $y = -3$:
$$\begin{aligned} 4x - 5y &= 20 \\ 4x - 5(-3) &= 20 \\ 4x + 15 &= 20 \\ 4x &= 5 \\ x &= \frac{5}{4} \quad \left(\frac{5}{4}, -3\right) \end{aligned}$$

Plot the ordered pairs and draw the line through them.

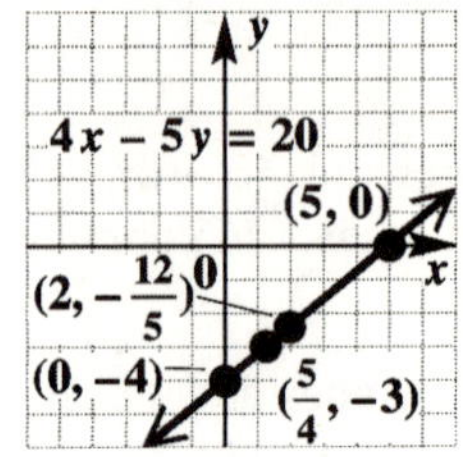

29. In quadrant III, both coordinates of the ordered pairs are negative. If $x + y = k$ and k is positive, then either x or y must be positive, because the sum of two negative numbers is negative.

31. Choose a value *other than* 0 for either x or y and then solve $4x + 5y = 0$ for the other variable. For example, if $x = -5$, then $-20 + 5y = 0$ and $y = 4$. The student should then plot the points $(0, 0)$ and $(-5, 4)$ and draw a line through them.

33. $2x + 3y = 12$

To find the x-intercept, let $y = 0$.
$$\begin{aligned} 2x + 3y &= 12 \\ 2x + 3(0) &= 12 \\ 2x &= 12 \\ x &= 6 \end{aligned}$$

The x-intercept is $(6, 0)$.

To find the y-intercept, let $x = 0$.
$$\begin{aligned} 2x + 3y &= 12 \\ 2(0) + 3y &= 12 \\ 3y &= 12 \\ y &= 4 \end{aligned}$$

The y-intercept is $(0, 4)$.

Plot the intercepts and draw the line through them.

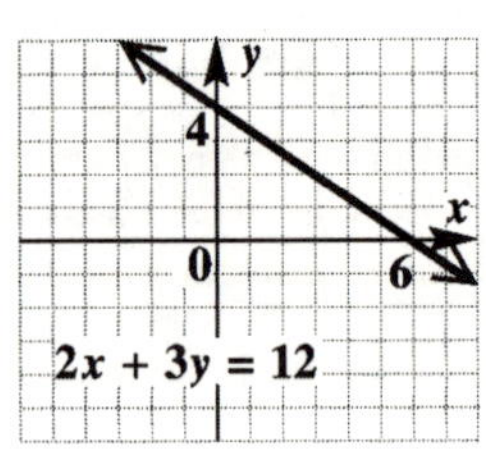

35. $x - 3y = 6$

To find the x-intercept, let $y = 0$.
$$\begin{aligned} x - 3y &= 6 \\ x - 3(0) &= 6 \\ x - 0 &= 6 \\ x &= 6 \end{aligned}$$

The x-intercept is $(6, 0)$.

To find the y-intercept, let $x = 0$.

$$
\begin{aligned}
x - 3y &= 6 \\
0 - 3y &= 6 \\
-3y &= 6 \\
y &= -2
\end{aligned}
$$

The y-intercept is $(0, -2)$.
Plot the intercepts and draw the line through them.

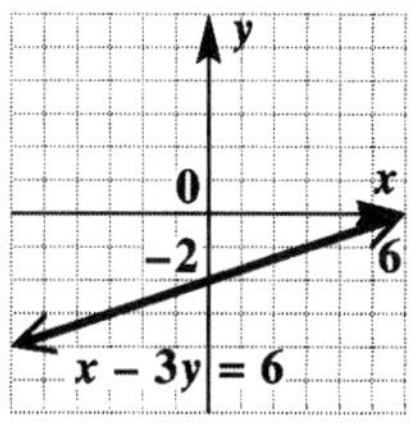

37. $\frac{2}{3}x - 3y = 7$

To find the x-intercept, let $y = 0$.

$$
\begin{aligned}
\frac{2}{3}x - 3(0) &= 7 \\
\frac{2}{3}x &= 7 \\
x &= \frac{3}{2} \cdot 7 = \frac{21}{2}
\end{aligned}
$$

The x-intercept is $\left(\frac{21}{2}, 0\right)$.

To find the y-intercept, let $x = 0$.

$$
\begin{aligned}
\frac{2}{3}(0) - 3y &= 7 \\
-3y &= 7 \\
y &= -\frac{7}{3}
\end{aligned}
$$

The y-intercept is $\left(0, -\frac{7}{3}\right)$.

Plot the intercepts and draw the line through them.

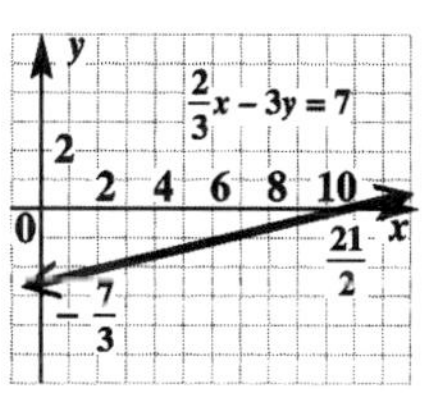

39. $y = 5$

This is a horizontal line. Every point has y-coordinate 5, so no point has y-coordinate 0. There is no x-intercept.
Since every point of the line has y-coordinate 5, the y-intercept is $(0, 5)$. Draw the horizontal line through $(0, 5)$.

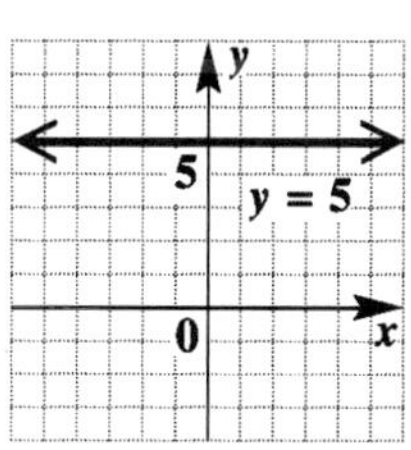

41. $x = 2$

This is a vertical line. Every point has x-coordinate 2, so the x-intercept is $(2, 0)$.
Since every point of the line has x-coordinate 2, no point has x-coordinate 0. There is no y-intercept. Draw the vertical line through $(2, 0)$.

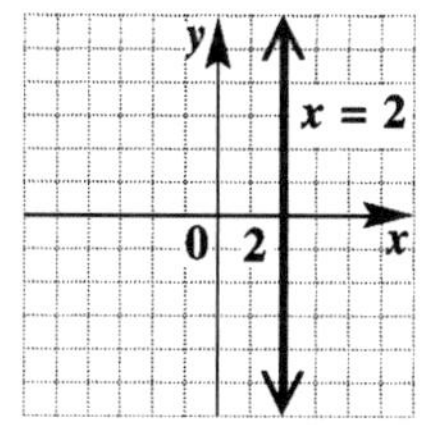

43. $x + 4 = 0 \quad (x = -4)$

This is a vertical line. Every point has x-coordinate -4, so the x-intercept is $(-4, 0)$.
Since every point of the line has x-coordinate -4, no point has x-coordinate 0. There is no y-intercept. Draw the vertical line through $(-4, 0)$.

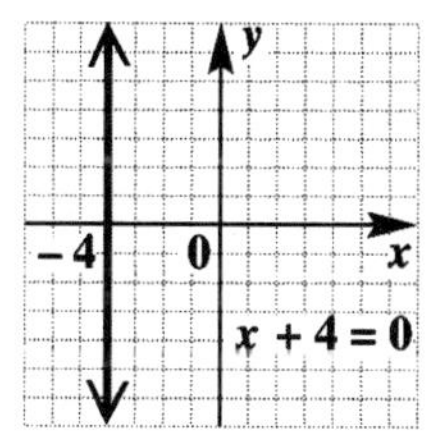

45. $x + 5y = 0$

To find the x-intercept, let $y = 0$.

$$
\begin{aligned}
x + 5y &= 0 \\
x + 5(0) &= 0 \\
x &= 0
\end{aligned}
$$

The x-intercept is $(0, 0)$, and since $x = 0$, this is also the y-intercept. Since the intercepts are the same, another point is needed to graph the line. Choose any number for y, say $y = -1$, and solve the equation for x.

$$
\begin{aligned}
x + 5y &= 0 \\
x + 5(-1) &= 0 \\
x &= 5
\end{aligned}
$$

This gives the ordered pair $(5, -1)$. Plot $(5, -1)$ and $(0, 0)$, and draw the line through them.

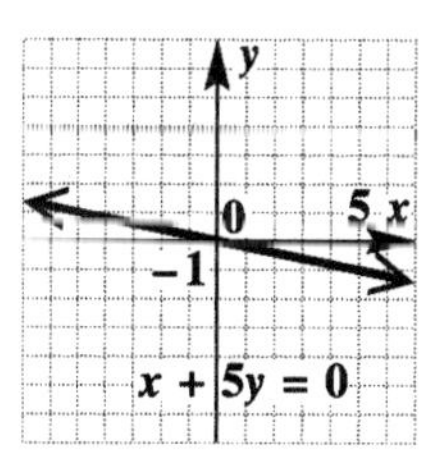

47. $2x = 3y$
If $x = 0$, then $y = 0$, so the x- and y-intercepts are $(0, 0)$. To get another point, let $x = 3$.

$$2(3) = 3y$$
$$2 = y$$

Plot $(3, 2)$ and $(0, 0)$, and draw the line through them.

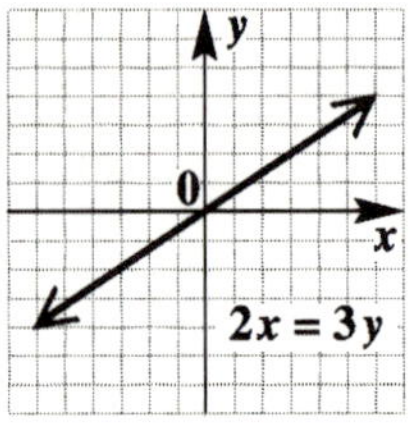

49. $-\frac{2}{3}y = x$
If $x = 0$, then $y = 0$, so the x- and y-intercepts are $(0, 0)$. To get another point, let $y = 3$.

$$-\frac{2}{3}(3) = x$$
$$-2 = x$$

Plot $(-2, 3)$ and $(0, 0)$, and draw the line through them.

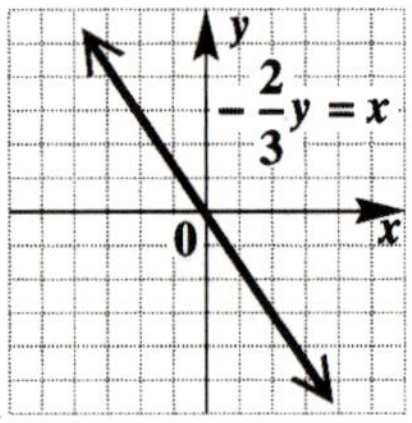

51. $y = 1.22x + 118$
$y = 1.22(30) + 118$ *x=30 corresponds to 1995*
$= 36.6 + 118$
$= 154.6$
From the model, the approximate speed was 154.6 mph.

53. The graph goes through the point $(-2, 0)$ which satisfies only equations **B** and **C**. The graph also goes through the point $(0, 3)$ which satisfies only equations **A** and **B**. Therefore, the correct equation is **B**.

55. The point $(15, 7483.25)$ is on the graph of the equation

$$y = 382.75x + 1742.$$

$x = 15$ corresponds to the year 1995 $(1980 + 15 = 1995)$. $y = 7483.25$ means that in the year 1995, the national average family health care cost was approximately \$7483.25.

57. Usually we need to solve for y.

$$x + 2y = 0$$
$$2y = -x$$
$$y = -\frac{1}{2}x \text{ or } y = -.5x$$

59. We need to solve the given equation for y.

$$5x + 2y = -10$$
$$2y = -5x - 10 \quad \textit{Subtract 5x.}$$
$$y = -\frac{5}{2}x - 5 \quad \textit{Divide by 2.}$$

Graph $Y_1 = -(5/2)X - 5$ in a standard viewing window.

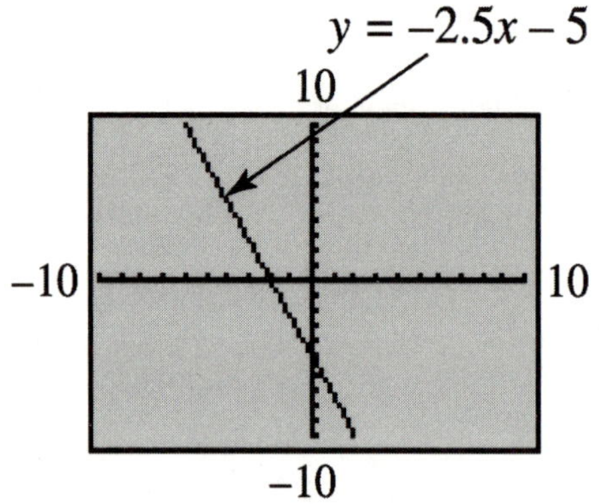

61. $3.6x - y = -5.8$
$-y = -3.6x - 5.8$ *Subtract 3.6x.*
$y = 3.6x + 5.8$ *Multiply by −1.*

Graph $Y_1 = 3.6X + 5.8$ in a standard viewing window.

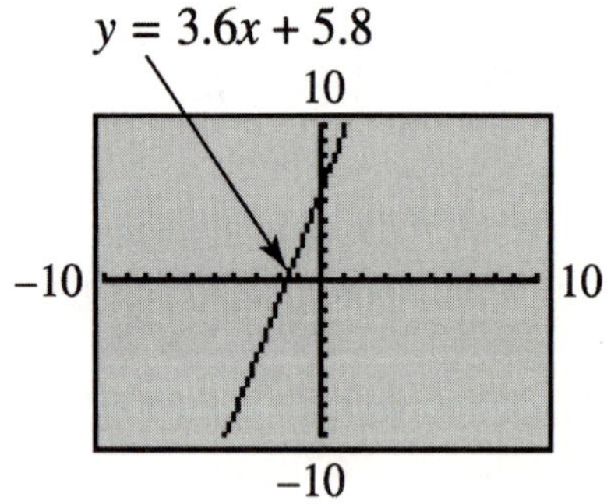

63. R has x-coordinate 6 and y-coordinate -2, so the ordered pair is $(6, -2)$.

64. The midpoint of $\overline{PR}$ has x-coordinate halfway from 4 to 6, which is 5, and y-coordinate -2, so it's the ordered pair $S = (5, -2)$.

65. The midpoint of $\overline{QR}$ has x-coordinate 6 and y-coordinate halfway from -2 to 2, so it's the ordered pair $T = (6, 0)$.

66. $M = (x\text{-coordinate of } S,\ y\text{-coordinate of } T)$
$= (5, 0)$

67. $P = (4, -2)$ and $Q = (6, 2)$. The average of the x-coordinates of P and Q is $\frac{4+6}{2} = \frac{10}{2} = 5$. The average of the y-coordinates of P and Q is $\frac{-2+2}{2} = \frac{0}{2} = 0$.

68. The x-coordinate of M is the average of the x-coordinates of P and Q. The y-coordinate of M is the average of the y-coordinates of P and Q.

69. By the Midpoint Formula, the midpoint of the segment with endpoints $(-8, 4)$ and $(-2, -6)$ is

$$\left(\frac{-8+(-2)}{2}, \frac{4+(-6)}{2}\right) = \left(\frac{-10}{2}, \frac{-2}{2}\right) = (-5, -1).$$

71. By the Midpoint Formula, the midpoint of the segment with endpoints $(3, -6)$ and $(6, 3)$ is

$$\left(\frac{3+6}{2}, \frac{-6+3}{2}\right) = \left(\frac{9}{2}, \frac{-3}{2}\right) = \left(\frac{9}{2}, -\frac{3}{2}\right).$$

73. By the Midpoint Formula, the midpoint of the segment with endpoints $(-9, 3)$ and $(9, 8)$ is

$$\left(\frac{-9+9}{2}, \frac{3+8}{2}\right) = \left(\frac{0}{2}, \frac{11}{2}\right) = \left(0, \frac{11}{2}\right).$$

75. By the Midpoint Formula, the midpoint of the segment with endpoints $(2.5, 3.1)$ and $(1.7, -1.3)$ is

$$\left(\frac{2.5+1.7}{2}, \frac{3.1+(-1.3)}{2}\right) = \left(\frac{4.2}{2}, \frac{1.8}{2}\right) = (2.1, .9).$$

Section 3.2

1.
$$\text{slope} = \frac{\text{change in vertical position}}{\text{change in horizontal position}} = \frac{-30 \text{ feet}}{100 \text{ feet}}$$

Choices **A**, $-.3$, **B**, $-\frac{3}{10}$, and **D**, $-\frac{30}{100}$ are all correct.

3. **(a)** Graph **C** indicates that sales leveled off during the second quarter.

(b) Graph **A** indicates that sales leveled off during the fourth quarter.

(c) Graph **D** indicates that sales rose sharply during the first quarter, and then fell to the original level during the second quarter.

(d) Graph **B** is the only graph that indicates that sales fell during the first two quarters.

5. $\text{slope of } BC = \frac{\text{rise}}{\text{run}} = \frac{0}{-4} = 0$

7. $\text{slope of } DE = \frac{\text{rise}}{\text{run}} = \frac{-1}{3} = -\frac{1}{3}$

9. $m = \frac{6-2}{5-3} = \frac{4}{2} = 2$

11. $m = \frac{4-(-1)}{-3-(-5)} = \frac{4+1}{-3+5} = \frac{5}{2}$

13. $m = \frac{-5-(-5)}{3-2} = \frac{-5+5}{1} = \frac{0}{1} = 0$

15. **B** and **D** are correct. Choice **A** is wrong because the order of subtraction must be the same in the numerator and denominator. Choice **C** is wrong because slope is defined as the change in y divided by the change in x.

17. Let $(x_1, y_1) = (-4, 3)$ and $(x_2, y_2) = (-3, 4)$. Then

$$m = \frac{y_2 - y_1}{x_2 - x_1} = \frac{4-3}{-3-(-4)} = \frac{1}{1} = 1.$$

The slope is 1.

19. Let $(x_1, y_1) = (-3, -3)$ and $(x_2, y_2) = (5, 6)$. Then

$$m = \frac{y_2 - y_1}{x_2 - x_1} = \frac{6-(-3)}{5-(-3)} = \frac{9}{8}.$$

The slope is $\frac{9}{8}$.

21. Let $(x_1, y_1) = (-6, 3)$ and $(x_2, y_2) = (2, 3)$. Then

$$m = \frac{y_2 - y_1}{x_2 - x_1} = \frac{3-3}{2-(-6)} = \frac{0}{8} = 0.$$

The slope is 0.

23. The points shown on the line are $(1, -1)$ and $(4, 5)$. The slope is

$$m = \frac{5-(-1)}{4-1} = \frac{6}{3} = 2.$$

25. "The line has positive slope" means that the line goes up from left to right. This is line B.

27. "The line has slope 0" means that there is no vertical change; that is, the line is horizontal. This is line A.

29. To find the slope of

$$x + 2y = 4,$$

first find the intercepts. Replace y with 0 to find that the x-intercept is $(4, 0)$; replace x with 0 to find that the y-intercept is $(0, 2)$. The slope is then

$$m = \frac{2-0}{0-4} = -\frac{2}{4} = -\frac{1}{2}.$$

To sketch the graph, plot the intercepts and draw the line through them.

continued

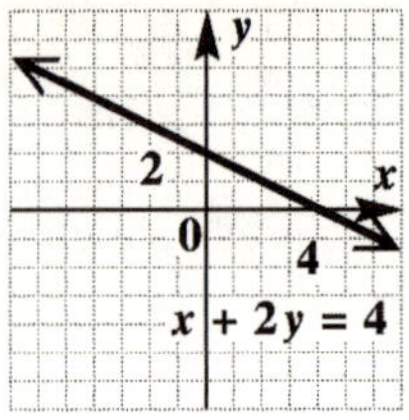

31. To find the slope of

$$5x - 2y = 10,$$

first find the intercepts. Replace y with 0 to find that the x-intercept is $(2, 0)$; replace x with 0 to find that the y-intercept is $(0, -5)$. The slope is then

$$m = \frac{-5 - 0}{0 - 2} = \frac{-5}{-2} = \frac{5}{2}.$$

To sketch the graph, plot the intercepts and draw the line through them.

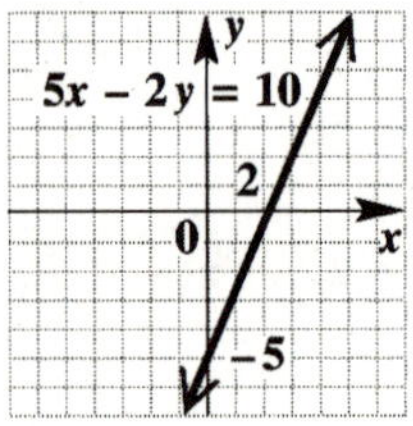

33. In the equation

$$y = 4x,$$

replace x with 0 and then x with 1 to get the ordered pairs $(0, 0)$ and $(1, 4)$, respectively. (There are other possibilities for ordered pairs.) The slope is then

$$m = \frac{4 - 0}{1 - 0} = \frac{4}{1} = 4.$$

To sketch the graph, plot the two points and draw the line through them.

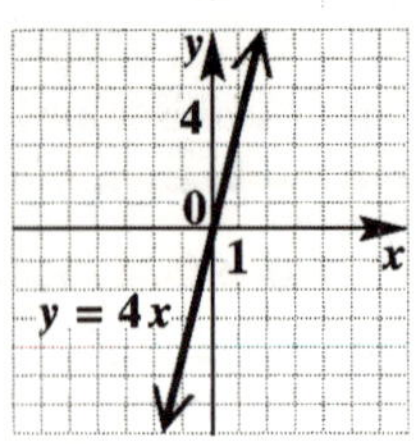

35. $x - 3 = 0 \quad (x = 3)$
The graph of $x = 3$ is the vertical line with x-intercept $(3, 0)$. The slope of a vertical line is undefined.

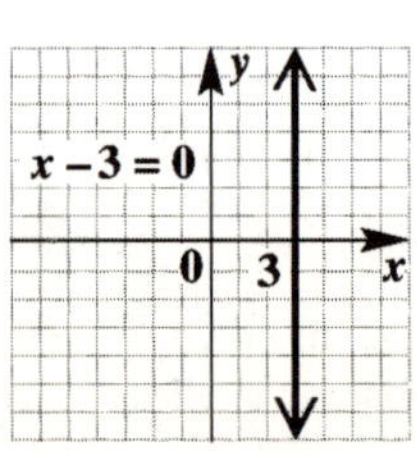

37. $y = -4$
The graph of $y = -4$ is the horizontal line with y-intercept $(0, -4)$. The slope of a horizontal line is 0.

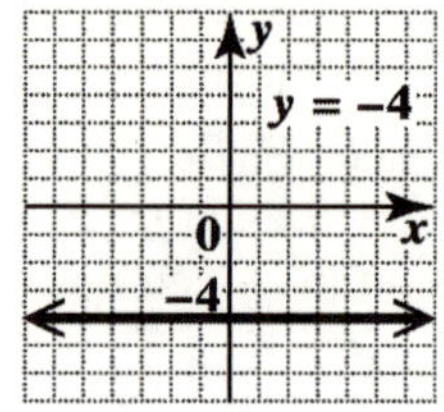

39. To graph the line through $(-2, -3)$ with slope $m = \frac{5}{4}$, locate $(-2, -3)$ on the graph. To find a second point, use the definition of slope.

$$m = \frac{\text{change in } y}{\text{change in } x} = \frac{5}{4}$$

From $(-2, -3)$, go up 5 units. Then go 4 units to the right to get to $(2, 2)$. Draw the line through $(-2, -3)$ and $(2, 2)$.

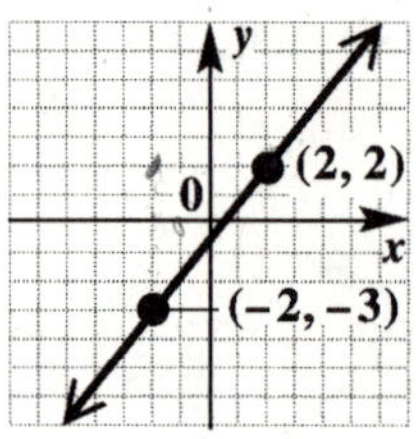

41. To graph the line through $(0, -4)$ with slope $m = -\frac{3}{2}$, locate the point $(0, -4)$ on the graph. To find a second point on the line, use the definition of slope, writing $-\frac{3}{2}$ as $\frac{-3}{2}$.

$$m = \frac{\text{change in } y}{\text{change in } x} = \frac{-3}{2}$$

From $(0, -4)$, move 3 units down and then 2 units to the right. Draw a line through this second point and $(0, -4)$. The slope could also be written as $\frac{3}{-2}$. In this case, move 3 units up and 2 units to the left to get another point on the same line, as shown in the figure.

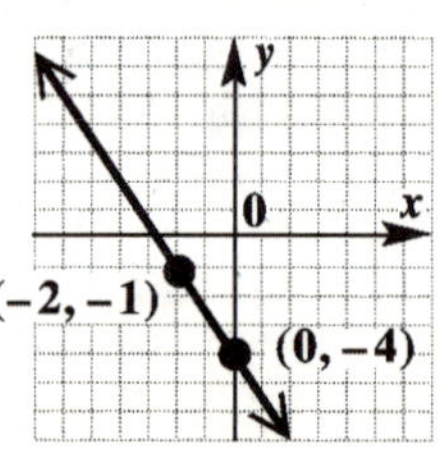

43. Locate $(-2, -4)$. Then use $m = 4 = \frac{4}{1}$ to go 4 units up and 1 unit right to $(-1, 0)$.

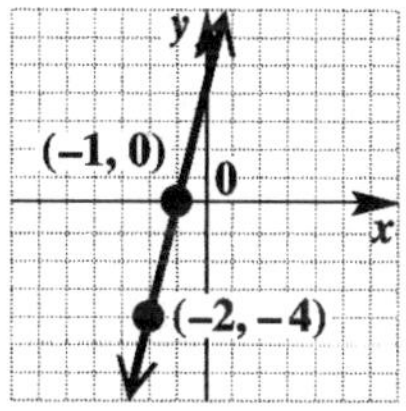

45. Locate $(-3, 1)$. Since the slope is undefined, the line is vertical. The x-value of every point is -3.

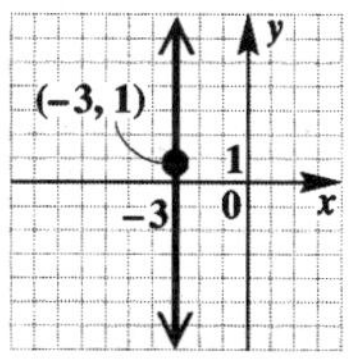

47. Locate $(5, 3)$. A slope of 0 means that the line is horizontal, so $y = 3$ at every point. Draw the horizontal line through $(5, 3)$.

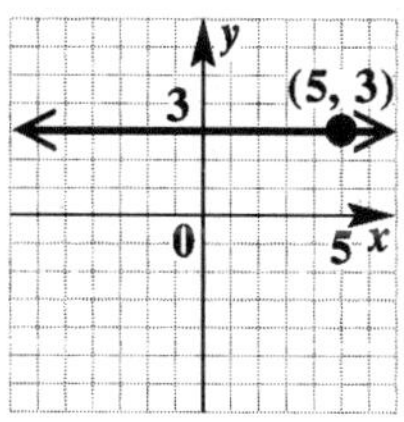

49. The slope of the line through $(4, 6)$ and $(-8, 7)$ is

$$m = \frac{7-6}{-8-4} = \frac{1}{-12} = -\frac{1}{12}.$$

The slope of the line through $(7, 4)$ and $(-5, 5)$ is

$$m = \frac{5-4}{-5-7} = \frac{1}{-12} = -\frac{1}{12}.$$

Since the slopes are equal, the two lines are *parallel.*

51. $2x + 5y = -7$ and $5x - 2y = 1$
Solve the equations for y.

$$5y = -2x - 7 \qquad -2y = -5x + 1$$
$$y = -\frac{2}{5}x - \frac{7}{5} \qquad y = \frac{5}{2}x - \frac{1}{2}$$

The slopes, $-\frac{2}{5}$ and $\frac{5}{2}$, are negative reciprocals of one another, so the lines are *perpendicular.*

53. $2x + y = 6$ and $x - y = 4$
Solve the equations for y.

$$y = -2x + 6 \qquad -y = -x + 4$$
$$y = x - 4$$

The slopes are -2 and 1. The lines are *neither* parallel nor perpendicular.

55. $3x = y$ and $2y - 6x = 5$
The slope of the first line is the coefficient of x, namely 3. Solve the second equation for y.

$$2y = 6x + 5$$
$$y = 3x + \frac{5}{2}$$

The slope of the second line is also 3, so the lines are *parallel.*

57. $2x + 5y = -8$ and $6 + 2x = 5y$
Solve the equations for y.

$$5y = -2x - 8 \qquad 5y = 2x + 6$$
$$y = -\frac{2}{5}x - \frac{8}{5} \qquad y = \frac{2}{5}x + \frac{6}{5}$$

The slopes are $-\frac{2}{5}$ and $\frac{2}{5}$. The lines are *neither* parallel nor perpendicular.

59. $4x - 3y = 8$ and $4y + 3x = 12$
Solve the equations for y.

$$-3y = -4x + 8 \qquad 4y = -3x + 12$$
$$y = \frac{4}{3}x - \frac{8}{3} \qquad y = -\frac{3}{4}x + 3$$

The slopes, $\frac{4}{3}$ and $-\frac{3}{4}$, are negative reciprocals of one another, so the lines are *perpendicular.*

61. The vertical change is 63 ft, and the horizontal change is $250 - 160 = 90$ ft.
The slope is $\frac{63}{90} = \frac{7}{10}$.

63. Use the points $(0, 20)$ and $(4, 4)$.

average rate of change

$$= \frac{\text{change in } y}{\text{change in } x} = \frac{4-20}{4-0} = \frac{-16}{4} = -4$$

The average rate of change is $-\$4000$ per year, that is, the value of the machine is decreasing \$4000 each year during these years.

65. We can see that there is no change in the percent of pay raise. Thus, the average rate of change is 0% per year, that is, the percent of pay raise is not changing—it is 3% each year during these years.

67. **(a)** For 1995–1996:

$$m = \frac{20,000 - 19,000}{1996 - 1995} = \frac{1000}{1} = 1000$$

For 1995–1999:

$$m = \frac{23,000 - 19,000}{1999 - 1995} = \frac{4000}{4} = 1000$$

For 1998–2000:

$$m = \frac{24,000 - 22,000}{2000 - 1998} = \frac{2000}{2} = 1000$$

The answers have units of million per year.

(b) The average rate of change is the same. When graphed, the data points lie on a straight line.

69. **(a)** Use $(1995, 1280)$ and $(1999, 2080)$.

$$m = \frac{2080 - 1280}{1999 - 1995} = \frac{800}{4} = 200$$

The average rate of change is \$200 million per year.

(b) The positive slope means expenditures *increased* an average of \$200 million each year.

71. Use $(1997, 500)$ and $(2002, 155)$.

$$m = \frac{155 - 500}{2002 - 1997} = \frac{-345}{5} = -69$$

The average rate of change in price is $-\$69$ per year, that is, the price decreased an average of \$69 each year from 1997 to 2002.

73. Line A has negative slope and line B has positive slope, so line A must be $y_1 = -2x + 3$ and line B must be $y_2 = 3x - 4$.

75. Label the points as shown in the figure.

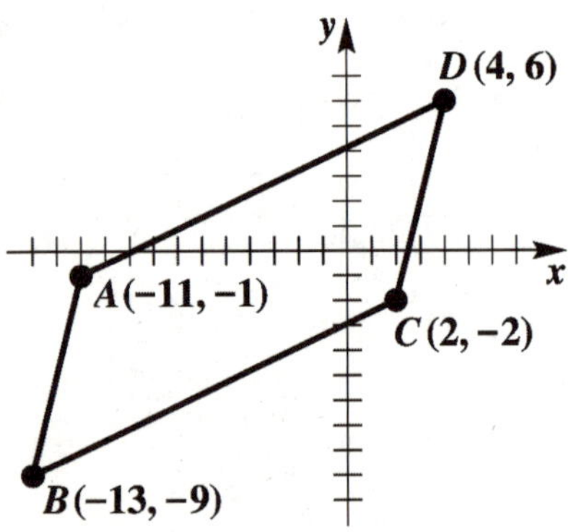

In order to determine whether $ABCD$ is a parallelogram, we need to show that the slope of $\overline{AB}$ equals the slope of $\overline{CD}$ and that the slope of $\overline{AD}$ equals the slope of $\overline{BC}$.

$$\text{Slope of } \overline{AB} = \frac{-9 - (-1)}{-13 - (-11)} = \frac{-8}{-2} = 4$$

$$\text{Slope of } \overline{CD} = \frac{6 - (-2)}{4 - 2} = \frac{8}{2} = 4$$

$$\text{Slope of } \overline{AD} = \frac{6 - (-1)}{4 - (-11)} = \frac{7}{15}$$

$$\text{Slope of } \overline{BC} = \frac{-2 - (-9)}{2 - (-13)} = \frac{7}{15}$$

Thus, the figure is a parallelogram.

77. For $A(3, 1)$ and $B(6, 2)$, the slope of $\overline{AB}$ is

$$m = \frac{1 - 2}{3 - 6} = \frac{-1}{-3} = \frac{1}{3}.$$

78. For $B(6, 2)$ and $C(9, 3)$, the slope of $\overline{BC}$ is

$$m = \frac{2 - 3}{6 - 9} = \frac{-1}{-3} = \frac{1}{3}.$$

79. For $A(3, 1)$ and $C(9, 3)$, the slope of $\overline{AC}$ is

$$m = \frac{1 - 3}{3 - 9} = \frac{-2}{-6} = \frac{1}{3}.$$

80. The slope of $\overline{AB}$ = slope of $\overline{BC}$
= slope of $\overline{AC}$
$= \frac{1}{3}$.

81. For $A(1, -2)$ and $B(3, -1)$, the slope of $\overline{AB}$ is

$$m = \frac{-1 - (-2)}{3 - 1} = \frac{1}{2}.$$

For $B(3, -1)$ and $C(5, 0)$, the slope of $\overline{BC}$ is

$$m = \frac{0 - (-1)}{5 - 3} = \frac{1}{2}.$$

For $A(1, -2)$ and $C(5, 0)$, the slope of $\overline{AC}$ is

$$m = \frac{0 - (-2)}{5 - 1} = \frac{2}{4} = \frac{1}{2}.$$

Since the three slopes are the same, the three points are collinear.

82. For $A(0, 6)$ and $B(4, -5)$, the slope of $\overline{AB}$ is

$$m = \frac{-5 - 6}{4 - 0} = \frac{-11}{4} = -\frac{11}{4}.$$

For $B(4, -5)$ and $C(-2, 12)$, the slope of $\overline{BC}$ is

$$m = \frac{12 - (-5)}{-2 - 4} = \frac{17}{-6} = -\frac{17}{6}.$$

Since these two slopes are not the same, the three points are not collinear.

Section 3.3

1. Choice **A**, $3x - 2y = 5$, is in the form $Ax + By = C$ with $A \geq 0$ and integers A, B, and C having no common factor (except 1).

3. Choice **A**, $y = 6x + 2$, is in the form $y = mx + b$.

5.
$$y = -3x + 10$$
$$3x + y = 10 \qquad \textit{Standard form}$$

7. $y = 2x + 3$
This line is in slope-intercept form with slope $m = 2$ and y-intercept $(0, b) = (0, 3)$. The only graph with positive slope and with a positive y-coordinate of its y-intercept is **A**.

9. $y = -2x - 3$
This line is in slope-intercept form with slope $m = -2$ and y-intercept $(0, b) = (0, -3)$. The only graph with negative slope and with a negative y-coordinate of its y-intercept is **C**.

11. $y = 2x$
This line has slope $m = 2$ and y-intercept $(0, b) = (0, 0)$. The only graph with positive slope and with y-intercept $(0, 0)$ is **H**.

13. $y = 3$
This line is a horizontal line with y-intercept $(0, 3)$. Its y-coordinate is positive. The only graph that has these characteristics is **B**.

15. $m = 5; b = 15$
Substitute these values in the slope-intercept form.

$$\begin{aligned} y &= mx + b \\ y &= 5x + 15 \end{aligned}$$

17. $m = -\frac{2}{3}; b = \frac{4}{5}$
Substitute these values in the slope-intercept form.

$$\begin{aligned} y &= mx + b \\ y &= -\frac{2}{3}x + \frac{4}{5} \end{aligned}$$

19. Slope $\frac{2}{5}$; y-intercept $(0, 5)$

Here, $m = \frac{2}{5}$ and $b = 5$. Substitute these values in the slope-intercept form.

$$\begin{aligned} y &= mx + b \\ y &= \frac{2}{5}x + 5 \end{aligned}$$

21. To get to the point $(3, 3)$ from the y-intercept $(0, 1)$, we must go up 2 units and to the right 3 units, so the slope is $\frac{2}{3}$. The slope-intercept form is

$$y = \frac{2}{3}x + 1.$$

23. $-x + y = 4$
(a) Solve for y to get the equation in slope-intercept form.

$$\begin{aligned} -x + y &= 4 \\ y &= x + 4 \end{aligned}$$

(b) The slope is the coefficient of x, 1.

(c) The y-intercept is the point $(0, b)$, or $(0, 4)$.

(d) 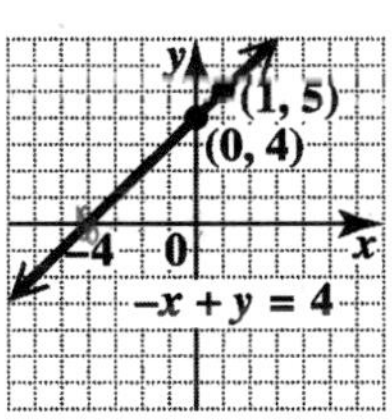

25. $6x + 5y = 30$
(a) Solve for y to get the equation in slope-intercept form.

$$\begin{aligned} 6x + 5y &= 30 \\ 5y &= -6x + 30 \\ y &= -\frac{6}{5}x + 6 \end{aligned}$$

(b) The slope is the coefficient of x, $-\frac{6}{5}$.

(c) The y-intercept is the point $(0, b)$, or $(0, 6)$.

(d) 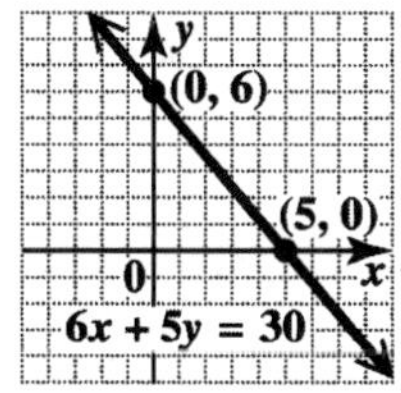

27. $4x - 5y = 20$
(a) Solve for y to get the equation in slope-intercept form.

$$\begin{aligned} 4x - 5y &= 20 \\ -5y &= -4x + 20 \\ y &= \frac{4}{5}x - 4 \end{aligned}$$

(b) The slope is the coefficient of x, $\frac{4}{5}$.

(c) The y-intercept is the point $(0, b)$, or $(0, -4)$.

(d) 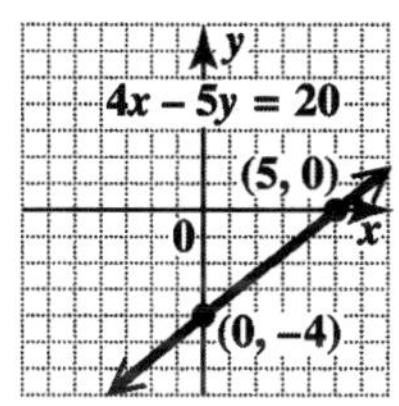

29. $x + 2y = -4$
(a) Solve for y to get the equation in slope-intercept form.

$$\begin{aligned} x + 2y &= -4 \\ 2y &= -x - 4 \\ y &= -\frac{1}{2}x - 2 \end{aligned}$$

(b) The slope is the coefficient of x, $-\frac{1}{2}$.

(c) The y-intercept is the point $(0, b)$, or $(0, -2)$.

(d) 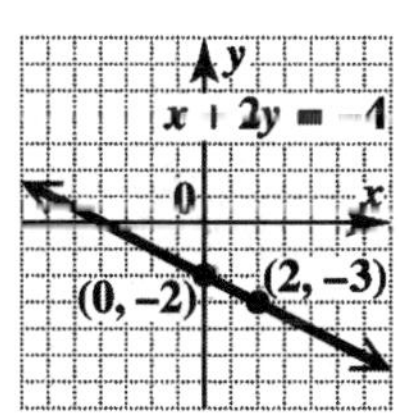

31. Through $(-2, 4)$; slope $= -\frac{3}{4}$
Use the point-slope form with $(x_1, y_1) = (-2, 4)$ and $m = -\frac{3}{4}$.

$$\begin{aligned} y - y_1 &= m(x - x_1) \\ y - 4 &= -\frac{3}{4}[x - (-2)] \\ 4(y - 4) &= -3(x + 2) \\ 4y - 16 &= -3x - 6 \\ 3x + 4y &= 10 \end{aligned}$$

33. Through $(5, 8)$; slope $= -2$
Use the point-slope form with $(x_1, y_1) = (5, 8)$ and $m = -2$.

$$\begin{aligned} y - y_1 &= m(x - x_1) \\ y - 8 &= -2(x - 5) \\ y - 8 &= -2x + 10 \\ 2x + y &= 18 \end{aligned}$$

35. Through $(-5, 4)$; slope $= \frac{1}{2}$
Use the point-slope form with $(x_1, y_1) = (-5, 4)$ and $m = \frac{1}{2}$.

$$\begin{aligned} y - y_1 &= m(x - x_1) \\ y - 4 &= \frac{1}{2}[x - (-5)] \\ 2(y - 4) &= 1(x + 5) \\ 2y - 8 &= x + 5 \\ -x + 2y &= 13 \\ x - 2y &= -13 \end{aligned}$$

37. Through $(3, 0)$; slope $= 4$
Use the point-slope form with $(x_1, y_1) = (3, 0)$ and $m = 4$.

$$\begin{aligned} y - y_1 &= m(x - x_1) \\ y - 0 &= 4(x - 3) \\ y &= 4x - 12 \\ -4x + y &= -12 \\ 4x - y &= 12 \end{aligned}$$

39. The slope-intercept form, $y = mx + b$, is used when the slope and y-intercept are known. The point-slope form, $y - y_1 = m(x - x_1)$, is used when the slope and one point on a line or two points on a line are known. The standard form, $Ax + By = C$, is not useful for writing the equation, but the intercepts can be found quickly and used to graph the equation. The form $y = b$ is used for a horizontal line through the point (a, b). The form $x = a$ is used for a vertical line through the point (a, b).

41. Through $(9, 5)$; slope 0
A line with slope 0 is a horizontal line. A horizontal line through the point (x, k) has equation $y = k$. Here $k = 5$, so an equation is $y = 5$.

43. Through $(9, 10)$; undefined slope
A vertical line has undefined slope and equation $x = c$. Since the x-value in $(9, 10)$ is 9, the equation is $x = 9$.

45. Through $(.5, .2)$; vertical
A vertical line through the point (k, y) has equation $x = k$. Here $k = .5$, so the equation is $x = .5$.

47. Through $(-7, 8)$; horizontal
A horizontal line through the point (x, k) has equation $y = k$, so the equation is $y = 8$.

49. $(3, 4)$ and $(5, 8)$
Find the slope.

$$m = \frac{8 - 4}{5 - 3} = \frac{4}{2} = 2$$

Use the point-slope form with $(x_1, y_1) = (3, 4)$ and $m = 2$.

$$\begin{aligned} y - y_1 &= m(x - x_1) \\ y - 4 &= 2(x - 3) \\ y - 4 &= 2x - 6 \\ -2x + y &= -2 \\ 2x - y &= 2 \end{aligned}$$

51. $(6, 1)$ and $(-2, 5)$
Find the slope.

$$m = \frac{5 - 1}{-2 - 6} = \frac{4}{-8} = -\frac{1}{2}$$

Use the point-slope form with $(x_1, y_1) = (6, 1)$ and $m = -\frac{1}{2}$.

$$\begin{aligned} y - y_1 &= m(x - x_1) \\ y - 1 &= -\frac{1}{2}(x - 6) \\ 2(y - 1) &= -1(x - 6) \\ 2y - 2 &= -x + 6 \\ x + 2y &= 8 \end{aligned}$$

53. $\left(-\frac{2}{5}, \frac{2}{5}\right)$ and $\left(\frac{4}{3}, \frac{2}{3}\right)$
Find the slope.

$$m = \frac{\frac{2}{3} - \frac{2}{5}}{\frac{4}{3} - \left(-\frac{2}{5}\right)} = \frac{\frac{10 - 6}{15}}{\frac{20 + 6}{15}}$$

$$= \frac{\frac{4}{15}}{\frac{26}{15}} = \frac{4}{26} = \frac{2}{13}$$

Use the point-slope form with
$(x_1, y_1) = \left(-\frac{2}{5}, \frac{2}{5}\right)$ and $m = \frac{2}{13}$.

$$\begin{aligned} y - \frac{2}{5} &= \frac{2}{13}\left[x - \left(-\frac{2}{5}\right)\right] \\ 13\left(y - \frac{2}{5}\right) &= 2\left(x + \frac{2}{5}\right) \\ 13y - \frac{26}{5} &= 2x + \frac{4}{5} \\ -2x + 13y &= \frac{30}{5} \\ 2x - 13y &= -6 \end{aligned}$$

55. $(2, 5)$ and $(1, 5)$
Find the slope.

$$m = \frac{5-5}{1-2} = \frac{0}{-1} = 0$$

A line with slope 0 is horizontal. A horizontal line through the point (x, k) has equation $y = k$, so the equation is $y = 5$.

57. $(7, 6)$ and $(7, -8)$
Find the slope.

$$m = \frac{-8-6}{7-7} = \frac{-14}{0} \quad \textit{Undefined}$$

A line with undefined slope is a vertical line. The equation of a vertical line is $x = k$, where k is the common x-value. So the equation is $x = 7$.

59. $(1, -3)$ and $(-1, -3)$
Find the slope.

$$m = \frac{-3-(-3)}{-1-1} = \frac{0}{-2} = 0$$

A line with slope 0 is horizontal. A horizontal line through the point (x, k) has equation $y = k$, so the equation is $y = -3$.

61. Through $(7, 2)$; parallel to $3x - y = 8$
Find the slope of $3x - y = 8$.

$$\begin{aligned} -y &= -3x + 8 \\ y &= 3x - 8 \end{aligned}$$

The slope is 3, so a line parallel to it also has slope 3. Use $m = 3$ and $(x_1, y_1) = (7, 2)$ in the point-slope form.

$$\begin{aligned} y - y_1 &= m(x - x_1) \\ y - 2 &= 3(x - 7) \\ y - 2 &= 3x - 21 \\ y &= 3x - 19 \end{aligned}$$

63. Through $(-2, -2)$; parallel to $-x + 2y = 10$
Find the slope of $-x + 2y = 10$.

$$\begin{aligned} 2y &= x + 10 \\ y &= \frac{1}{2}x + 5 \end{aligned}$$

The slope is $\frac{1}{2}$, so a line parallel to it also has slope $\frac{1}{2}$. Use $m = \frac{1}{2}$ and $(x_1, y_1) = (-2, -2)$ in the point-slope form.

$$\begin{aligned} y - y_1 &= m(x - x_1) \\ y - (-2) &= \frac{1}{2}[x - (-2)] \\ y + 2 &= \frac{1}{2}(x + 2) \\ y + 2 &= \frac{1}{2}x + 1 \\ y &= \frac{1}{2}x - 1 \end{aligned}$$

65. Through $(8, 5)$; perpendicular to $2x - y = 7$
Find the slope of $2x - y = 7$.

$$\begin{aligned} -y &= -2x + 7 \\ y &= 2x - 7 \end{aligned}$$

The slope of the line is 2. Therefore, the slope of the line perpendicular to it is $-\frac{1}{2}$ since $2\left(-\frac{1}{2}\right) = -1$. Use $m = -\frac{1}{2}$ and $(x_1, y_1) = (8, 5)$ in the point-slope form.

$$\begin{aligned} y - y_1 &= m(x - x_1) \\ y - 5 &= -\frac{1}{2}(x - 8) \\ y - 5 &= -\frac{1}{2}x + 4 \\ y &= -\frac{1}{2}x + 9 \end{aligned}$$

67. Through $(-2, 7)$; perpendicular to $x = 9$
$x = 9$ is a vertical line so a line perpendicular to it will be a horizontal line. It goes through $(-2, 7)$ so its equation is

$$y = 7.$$

69. Distance = (rate)(time), so

$$y = 45x.$$

x	$y = 45x$	Ordered Pair
0	$45(0) = 0$	$(0, 0)$
5	$45(5) = 225$	$(5, 225)$
10	$45(10) = 450$	$(10, 450)$

71. Total cost = (cost/gal)(number of gallons), so

$$y = 1.50x.$$

x	$y = 1.50x$	Ordered Pair
0	$1.50(0) = 0$	$(0, 0)$
5	$1.50(5) = 7.50$	$(5, 7.50)$
10	$1.50(10) = 15.00$	$(10, 15.00)$

73. **(a)** The fixed cost is \$99, so that is the value of b. The variable cost is \$39, so

$$y = mx + b = 39x + 99.$$

(b) If $x = 5$, $y = 39(5) + 99 = 294$. The ordered pair is $(5, 294)$. The cost of a 5-month membership is \$294.

(c) If $x = 12$, $y = 39(12) + 99 = 567$. The cost of the first year's membership is \$567.

75. **(a)** The fixed cost is \$25, so that is the value of b. The variable cost is \$50, so

$$y = mx + b = 50x + 25.$$

(b) If $x = 5$, $y = 50(5) + 25 = 275$. The ordered pair is $(5, 275)$. The cost of the plan for 5 months is \$275.

(c) For a 2-year contract, $x = 24$, so $y = 50(24) + 25 = 1225$. The cost of the plan for 2 years is \$1225.

77. **(a)** The fixed cost is \$50, so that is the value of b. The variable cost is \$.20, so

$$y = mx + b = .20x + 50.$$

(b) If $x = 5$, $y = .20(5) + 50 = 51$. The ordered pair is $(5, 51)$. The charge for driving 5 miles is \$51.

(c)
$$\begin{aligned} 84.60 &= .20x + 50 \qquad \textit{Let } y = 84.60. \\ 34.60 &= .20x \\ x &= \frac{34.60}{.20} = 173 \end{aligned}$$

The car was driven 173 miles.

79. **(a)** Use $(0, 9)$ and $(5, 37)$.

$$m = \frac{37 - 9}{5 - 0} = \frac{28}{5} = 5.6$$

The equation is $y = 5.6x + 9$. The slope tells us that the percent of households accessing the Internet by broadband is increasing 5.6% per year.

(b) The year 2006 corresponds to $x = 6$, so $y = 5.6(6) + 9 = 42.6 \approx 43\%$.

81. **(a)** Use $(5, 28{,}392)$ and $(10, 27{,}876)$.

$$m = \frac{27{,}876 - 28{,}392}{10 - 5} = \frac{-516}{5} = -103.2$$

Now use the point-slope form.

$$\begin{aligned} y - 27{,}876 &= -103.2(x - 10) \\ y - 27{,}876 &= -103.2x + 1032 \\ y &= -103.2x + 28{,}908 \end{aligned}$$

(b) The year 1998 corresponds to $x = 8$, so $y = -103.2(8) + 28{,}908 = 28{,}082.4 \approx 28{,}082$. This value is slightly higher than the actual value of 27,952.

83. **(a)**
$$\begin{aligned} 2x + 7 - x &= 4x - 2 \\ x + 7 &= 4x - 2 \\ -3x + 9 &= 0 \\ -3x + 9 &= y \end{aligned}$$

(b) From the screen, we see that $x = 3$ is the solution.

(c)
$$\begin{aligned} 2x + 7 - x &= 4x - 2 \\ x + 7 &= 4x - 2 \\ 9 &= 3x \\ 3 &= x \end{aligned}$$

Solution set: $\{3\}$

85. **(a)**
$$\begin{aligned} 3(2x + 1) - 2(x - 2) &= 5 \\ 6x + 3 - 2x + 4 - 5 &= 0 \\ 4x + 2 &= 0 \\ 4x + 2 &= y \end{aligned}$$

(b) From the screen, we see that $x = -.5$ is the solution.

(c)
$$\begin{aligned} 3(2x + 1) - 2(x - 2) &= 5 \\ 6x + 3 - 2x + 4 &= 5 \\ 4x + 7 &= 5 \\ 4x &= -2 \\ x &= -\frac{1}{2} \quad \text{or} \quad -.5 \end{aligned}$$

Solution set: $\{-.5\}$

87. The solution to the equation $y_1 = 0$ is the x-coordinate of the x-intercept. In this case, the x-intercept is greater than 10, so the correct choice must be **D**.

89. When $C = 0°$, $F = \underline{32}°$, and when $C = 100°$, $F = \underline{212}°$. These are the freezing and boiling temperatures for water.

90. The two points of the form (C, F) would be $(0, 32)$ and $(100, 212)$.

91. $m = \dfrac{212 - 32}{100 - 0} = \dfrac{180}{100} = \dfrac{9}{5}$

92. Let $m = \dfrac{9}{5}$ and $(x_1, y_1) = (0, 32)$.

$$\begin{aligned} y - y_1 &= m(x - x_1) \\ F - 32 &= \frac{9}{5}(C - 0) \\ F - 32 &= \frac{9}{5}C \\ F &= \frac{9}{5}C + 32 \end{aligned}$$

93.
$$\begin{aligned} F &= \frac{9}{5}C + 32 \\ F - 32 &= \frac{9}{5}C \\ \frac{5}{9}(F - 32) &= C \end{aligned}$$

94. A temperature of $50° C$ corresponds to a temperature of $122° F$.

Section 3.4

1. The boundary of the graph of $y \leq -x + 2$ will be a <u>solid</u> line (since the inequality involves $\leq$), and the shading will be <u>below</u> the line (since the inequality sign is $\leq$ or $<$).

3. The boundary of the graph of $y > -x + 2$ will be a <u>dashed</u> line (since the inequality involves $>$), and the shading will be <u>above</u> the line (since the inequality sign is $\geq$ or $>$).

5. The graph of $Ax + By = C$ divides the plane into two regions. In one of these regions, the ordered pairs satisfy $Ax + By < C$; in the other, they satisfy $Ax + By > C$.

7. $x + y \leq 2$
Graph the line $x + y = 2$ by drawing a solid line (since the inequality involves $\leq$) through the intercepts $(2, 0)$ and $(0, 2)$.
Test a point not on this line, such as $(0, 0)$.

$$\begin{aligned} x + y &\leq 2 \\ 0 + 0 &\leq 2 \\ 0 &\leq 2 \quad \textit{True} \end{aligned}$$

Shade that side of the line containing the test point $(0, 0)$.

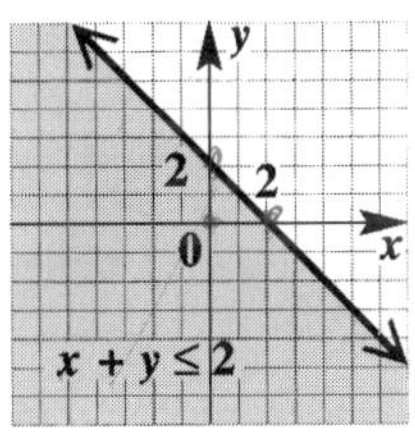

9. $4x - y < 4$
Graph the line $4x - y = 4$ by drawing a dashed line (since the inequality involves $<$) through the intercepts $(1, 0)$ and $(0, -4)$. Instead of using a test point, we will solve the inequality for y.

$$\begin{aligned} -y &< -4x + 4 \\ y &> 4x - 4 \end{aligned}$$

Since we have "$y >$ " in the last inequality, shade the region *above* the boundary line.

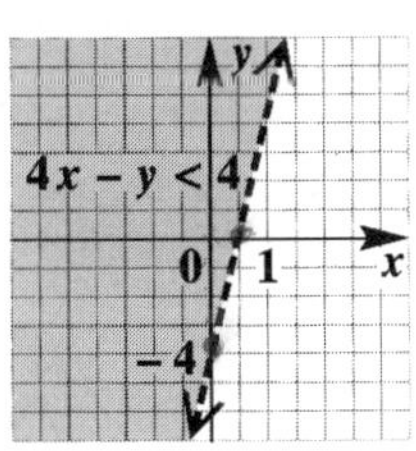

11. $x + 3y \geq -2$
Graph the solid line $x + 3y = -2$ (since the inequality involves $\geq$) through the intercepts $(-2, 0)$ and $\left(0, -\frac{2}{3}\right)$.
Test a point not on this line such as $(0, 0)$.

$$\begin{aligned} 0 + 3(0) &\geq -2 \\ 0 &\geq -2 \quad \textit{True} \end{aligned}$$

Shade that side of the line containing the test point $(0, 0)$.

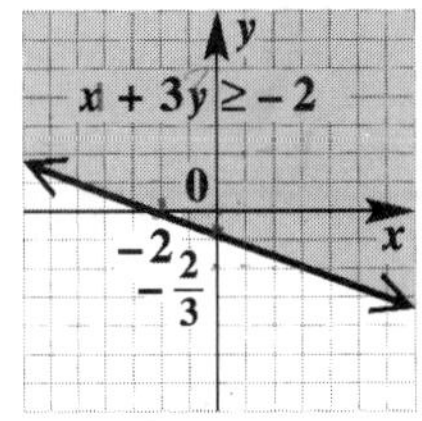

13. $x + y > 0$
Graph the line $x + y = 0$, which includes the points $(0, 0)$ and $(2, -2)$, as a dashed line (since the inequality involves $>$). Solving the inequality for y gives us

$$y > -x,$$

So shade the region above the boundary line.

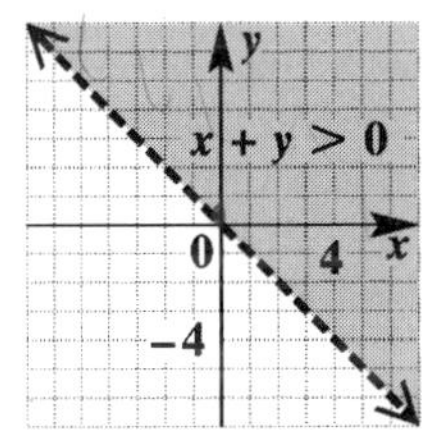

15. $x - 3y \leq 0$
Graph the solid line $x - 3y = 0$ through the points $(0, 0)$ and $(3, 1)$.
Solve the inequality for y.

$$\begin{aligned} -3y &\leq -x \\ y &\geq \frac{1}{3}x \end{aligned}$$

Shade the region above the boundary line.

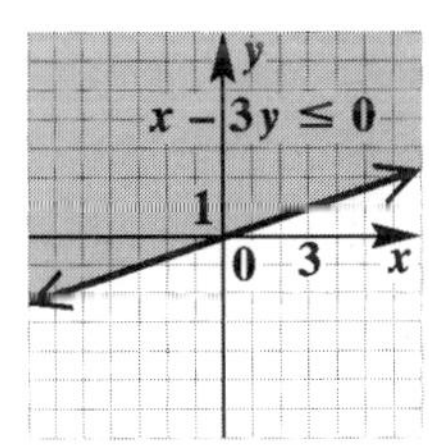

17. $y < x$
Graph the dashed line $y = x$ through $(0, 0)$ and $(2, 2)$. Since we have "$y <$ " in the inequality, shade the region *below* the boundary line.

continued

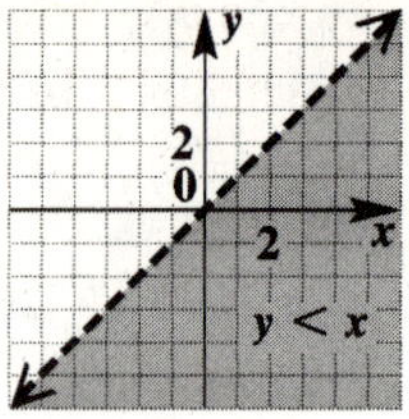

19. $x + y \le 1$ and $x \ge 1$
Graph the solid line $x + y = 1$ through $(0, 1)$ and $(1, 0)$. The inequality $x + y \le 1$ can be written as $y \le -x + 1$, so shade the region below the boundary line.
Graph the solid vertical line $x = 1$ through $(1, 0)$ and shade the region to the right. The required graph is the common shaded area as well as the portions of the lines that bound it.

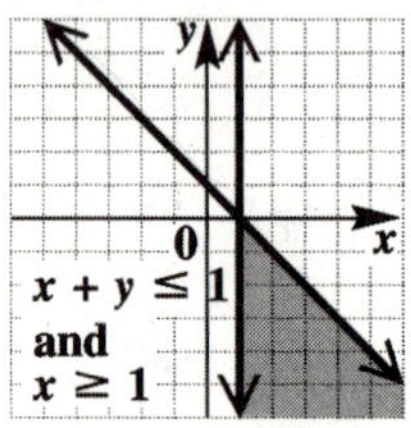

21. $2x - y \ge 2$ and $y < 4$
Graph the solid line $2x - y = 2$ through the intercepts $(1, 0)$ and $(0, -2)$. Test $(0, 0)$ to get $0 \ge 2$, a false statement. Shade that side of the graph not containing $(0, 0)$. To graph $y < 4$ on the same axes, graph the dashed horizontal line through $(0, 4)$. Test $(0, 0)$ to get $0 < 4$, a true statement. Shade that side of the dashed line containing $(0, 0)$.
The word "and" indicates the intersection of the two graphs. The final solution set consists of the region where the two shaded regions overlap.

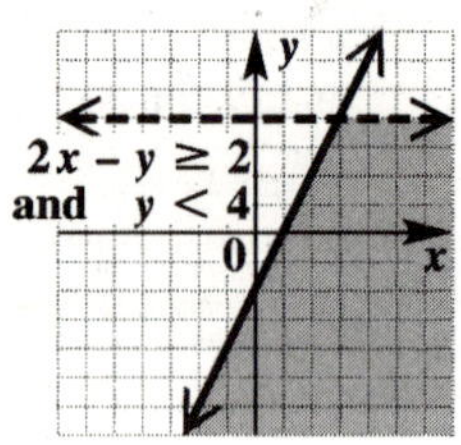

23. $x + y > -5$ and $y < -2$
Graph $x + y = -5$, which has intercepts $(-5, 0)$ and $(0, -5)$, as a dashed line. Test $(0, 0)$, which yields $0 > -5$, a true statement. Shade the region that includes $(0, 0)$.
Graph $y = -2$ as a dashed horizontal line. Shade the region below $y = -2$. The required graph of the intersection is the region common to both graphs.

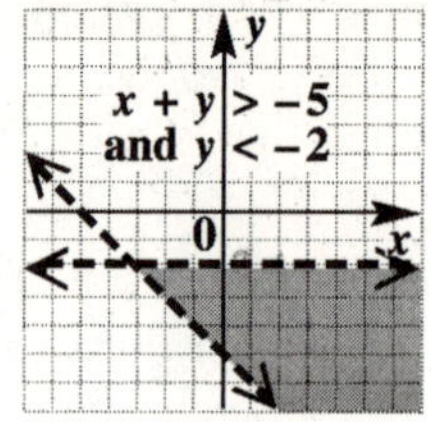

25. $|x| < 3$ can be rewritten as $-3 < x < 3$. The boundaries are the dashed vertical lines $x = -3$ and $x = 3$. Since x is between -3 and 3, the graph includes all points between the lines.

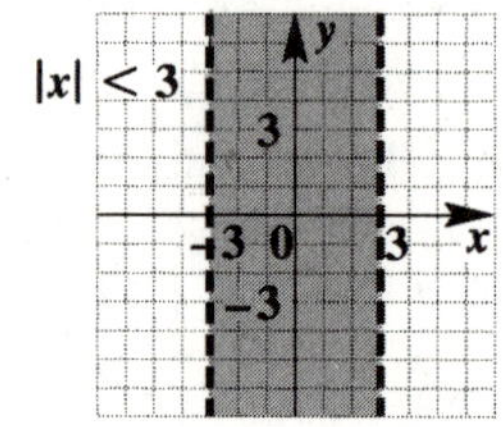

27. $|x + 1| < 2$ can be rewritten as

$$-2 < x + 1 < 2$$
$$-3 < x < 1.$$

The boundaries are the dashed vertical lines $x = -3$ and $x = 1$. Since x is between -3 and 1, the graph includes all points between the lines.

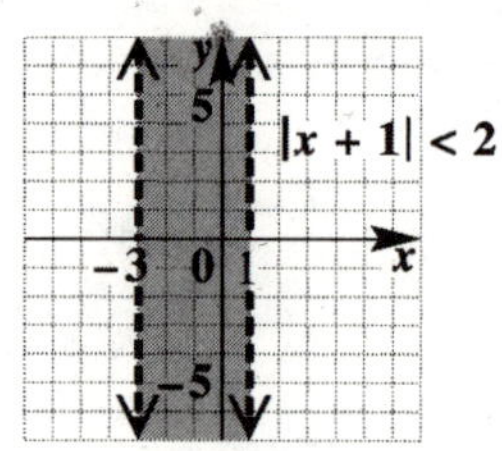

29. $x - y \ge 1$ or $y \ge 2$
Graph the solid line $x - y = 1$, which crosses the y-axis at -1 and the x-axis at 1. Use $(0, 0)$ as a test point, which yields $0 \ge 1$, a false statement. Shade the region that does not include $(0, 0)$.
Now graph the solid line $y = 2$. Since the inequality is $y \ge 2$, shade above this line.
The required graph of the union includes all the shaded regions, that is, all the points that satisfy either inequality.

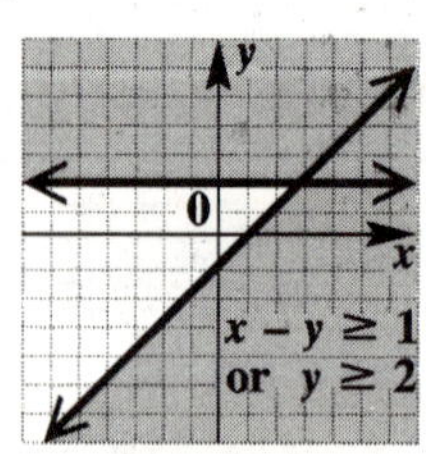

31. $x - 2 > y$ or $x < 1$
Graph $x - 2 = y$, which has intercepts $(2, 0)$ and $(0, -2)$, as a dashed line. Test $(0, 0)$, which yields $-2 > 0$, a false statement. Shade the region that does not include $(0, 0)$.
Graph $x = 1$ as a dashed vertical line. Shade the

region to the left of $x = 1$.
The required graph of the union includes all the shaded regions, that is, all the points that satisfy either inequality.

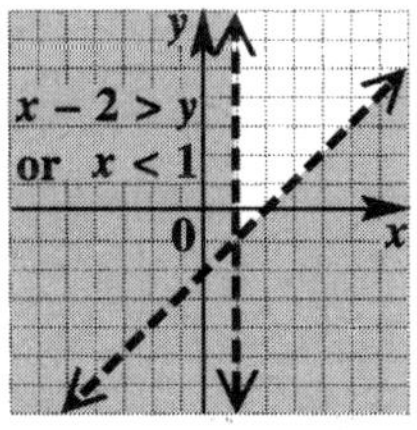

33. $3x + 2y < 6$ or $x - 2y > 2$
Graph $3x + 2y = 6$, which has intercepts $(2, 0)$ and $(0, 3)$, as a dashed line. Test $(0, 0)$, which yields $0 < 6$, a true statement. Shade the region that includes $(0, 0)$.
Graph $x - 2y = 2$, which has intercepts $(2, 0)$ and $(0, -1)$, as a dashed line. Test $(0, 0)$, which yields $0 > 2$, a false statement. Shade the region that does not include $(0, 0)$.
The required graph of the union includes all the shaded regions, that is, all the points that satisfy either inequality.

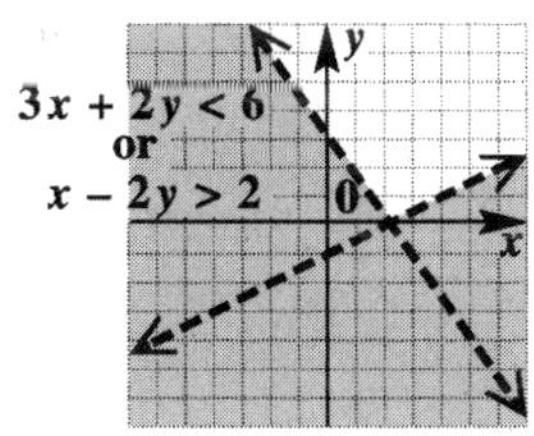

35. $y \le 3x - 6$
The boundary line, $y = 3x - 6$, has slope 3 and y-intercept -6. This would be graph **B** or graph **C**. Since we want the region less than or equal to $3x - 6$, we want the region on or below the boundary line. The answer is graph **C**.

37. $y \le -3x - 6$
The slope of the boundary line $y = -3x - 6$ is -3, and the y-intercept is -6. This would be graph **A** or graph **D**. The inequality sign is $\le$, so we want the region on or below the boundary line. The answer is graph **A**.

39. **(a)** The x-intercept is $(-4, 0)$, so the solution to $y = 0$ is $\{-4\}$.

(b) The solution to $y < 0$ is $(-\infty, -4)$, since the graph is below the x-axis for these values of x.

(c) The solution to $y > 0$ is $(-4, \infty)$, since the graph is above the x-axis for these values of x.

41. **(a)** The x-intercept is $(3.5, 0)$, so the solution to $y = 0$ is $\{3.5\}$.

(b) The solution to $y < 0$ is $(3.5, \infty)$, since the graph is below the x-axis for these values of x.

(c) The solution to $y > 0$ is $(-\infty, 3.5)$, since the graph is above the x-axis for these values of x.

43. **(a)**
$$\begin{aligned} 5x + 3 &= 0 \\ 5x &= -3 \\ x &= -\frac{3}{5} = -.6 \end{aligned}$$
Solution set: $\{-.6\}$

(b)
$$\begin{aligned} 5x + 3 &> 0 \\ 5x &> -3 \\ x &> -\frac{3}{5} \text{ or } -.6 \end{aligned}$$
Solution set: $(-.6, \infty)$

(c)
$$\begin{aligned} 5x + 3 &< 0 \\ 5x &< -3 \\ x &< -\frac{3}{5} \text{ or } -.6 \end{aligned}$$
Solution set: $(-\infty, -.6)$

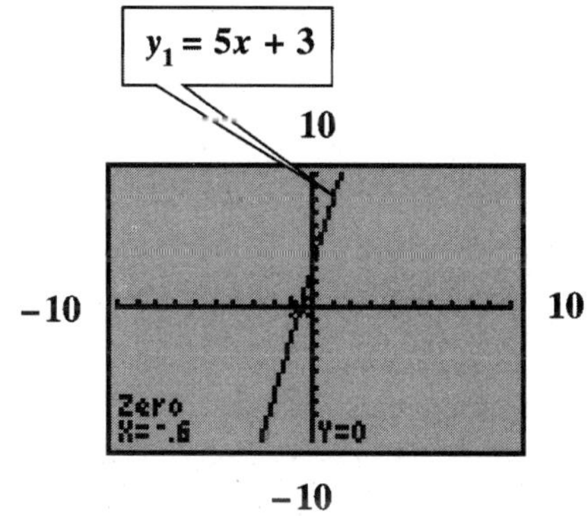

The x-intercept is $(-.6, 0)$, as in part (a). The graph is above the x-axis for $x > -.6$, as in part (b), and below the x-axis for $x < -.6$, as in part (c).

45. **(a)**
$$\begin{aligned} -8x - (2x + 12) &= 0 \\ -8x - 2x - 12 &= 0 \\ -10x - 12 &= 0 \\ -10x &= 12 \\ x &= -1.2 \end{aligned}$$
Solution set: $\{-1.2\}$

(b)
$$\begin{aligned} -8x - (2x + 12) &\ge 0 \\ -8x - 2x - 12 &\ge 0 \\ -10x - 12 &\ge 0 \\ -10x &\ge 12 \\ x &\le -1.2 \end{aligned}$$
Solution set: $(-\infty, -1.2]$

continued

(c) $-8x - (2x + 12) \leq 0$

$$-8x - 2x - 12 \leq 0$$
$$-10x - 12 \leq 0$$
$$-10x \leq 12$$
$$x \geq -1.2$$

Solution set: $[-1.2, \infty)$

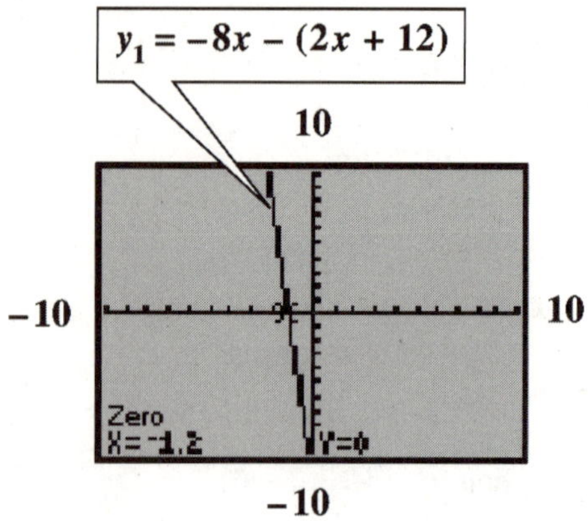

The x-intercept is $(-1.2, 0)$, as in part (a). The graph is on or above the x-axis for $(-\infty, -1.2]$, as in part (b), and on or below the x-axis for $[-1.2, \infty)$, as in part (c).

Section 3.5

1. We give one of many possible answers here. A function is a set of ordered pairs in which each first element corresponds to exactly one second element. For example, $\{(0, 1), (1, 2), (2, 3), (3, 4) \ldots\}$ is a function.

3. In an ordered pair of a relation, the first element is the independent variable.

5. $\{(5, 1), (3, 2), (4, 9), (7, 6)\}$
The relation is a function since for each x-value, there is only one y-value.

7. $\{(2, 4), (0, 2), (2, 5)\}$
The relation is not a function since the x-value 2 has two different y-values associated with it, 4 and 5.

9. $\{(-3, 1), (4, 1), (-2, 7)\}$
The relation is a function since for each x-value, there is only one y-value.

11. $\{(1, 1), (1, -1), (0, 0), (2, 4), (2, -4)\}$
The relation is not a function since the x-value 1 has two different y-values associated with it, 1 and -1. (A similar statement can be made for $x = 2$.)

The domain is the set of x-values: $\{0, 1, 2\}$.
The range is the set of y-values: $\{-4, -1, 0, 1, 4\}$.

13. The relation can be described by the set of ordered pairs

$$\{(2, 1), (5, 1), (11, 7), (17, 20), (3, 20)\}.$$

The relation is a function since for each x-value, there is only one y-value.

The domain is the set of x-values: $\{2, 3, 5, 11, 17\}$.
The range is the set of y-values: $\{1, 7, 20\}$.

15. The relation can be described by the set of ordered pairs

$$\{(1, 5), (1, 2), (1, -1), (1, -4)\}.$$

The relation is not a function since the x-value 1 has four different y-values associated with it, 5, 2, -1, and -4.

The domain is the set of x-values: $\{1\}$.
The range is the set of y-values: $\{5, 2, -1, -4\}$.

17. Using the vertical line test, we find any vertical line will intersect the graph at most once. This indicates that the graph represents a function. This graph extends indefinitely to the left $(-\infty)$ and indefinitely to the right (∞). Therefore, the domain is $(-\infty, \infty)$. This graph extends indefinitely downward $(-\infty)$, and indefinitely upward (∞). Thus, the range is $(-\infty, \infty)$.

19. Using the vertical line test, we find any vertical line will intersect the graph at most once. This indicates that the graph represents a function. This graph extends indefinitely to the left $(-\infty)$ and indefinitely to the right (∞). Therefore, the domain is $(-\infty, \infty)$. This graph extends indefinitely downward $(-\infty)$, and reaches a high point at $y = 4$. Therefore, the range is $(-\infty, 4]$.

21. Since a vertical line can intersect the graph of the relation in more than one point, the relation is not a function. The domain, the x-values of the points on the graph, is $[-4, 4]$. The range, the y-values of the points on the graph, is $[-3, 3]$.

23. $y = x^2$
Each value of x corresponds to one y-value. For example, if $x = 3$, then $y = 3^2 = 9$. Therefore, $y = x^2$ defines y as a function of x.
Since any x-value, positive, negative, or zero, can be squared, the domain is $(-\infty, \infty)$.

25. $x = y^6$
The ordered pairs $(64, 2)$ and $(64, -2)$ both satisfy the equation. Since one value of x, 64, corresponds to two values of y, 2 and -2, the relation does not define a function. Because x is equal to the sixth power of y, the values of x must always be nonnegative. The domain is $[0, \infty)$.

27. $y = 2x - 6$

For any value of x, there is exactly one value of y, so this equation defines a function. The domain is the set of all real numbers, $(-\infty, \infty)$.

29. $x + y < 4$
For a particular x-value, more than one y-value can be selected to satisfy $x + y < 4$. For example, if $x = 2$ and $y = 0$, then

$$2 + 0 < 4. \quad \textit{True}$$

Now, if $x = 2$ and $y = 1$, then

$$2 + 1 < 4. \quad \textit{Also true}$$

Therefore, $x + y < 4$ does not define y as a function of x.
The graph of $x + y < 4$ consists of the shaded region below the dashed line $x + y = 4$, which extends indefinitely from left to right. Therefore, the domain is $(-\infty, \infty)$.

31. $y = \sqrt{x}$
For any value of x, there is exactly one corresponding value for y, so this relation defines a function. Since the radicand must be a nonnegative number, x must always be nonnegative. The domain is $[0, \infty)$.

33. $xy = 1$
Rewrite $xy = 1$ as $y = \frac{1}{x}$. Note that x can never equal 0, otherwise the denominator would equal 0. The domain is $(-\infty, 0) \cup (0, \infty)$.
Each nonzero x-value gives exactly one y-value. Therefore, $xy = 1$ defines y as a function of x.

35. $y = \sqrt{4x + 2}$
To determine the domain of $y = \sqrt{4x + 2}$, recall that the radicand must be nonnegative. Solve the inequality $4x + 2 \geq 0$, which gives us $x \geq -\frac{1}{2}$.
Therefore, the domain is $\left[-\frac{1}{2}, \infty\right)$.
Each x-value from the domain produces exactly one y-value. Therefore, $y = \sqrt{4x + 2}$ defines a function.

37. $y = \dfrac{2}{x - 9}$
Given any value of x, y is found by subtracting 9, then dividing the result into 2. This process produces exactly one value of y for each x-value, so the relation represents a function. The domain includes all real numbers except those that make the denominator 0, namely 9. The domain is $(-\infty, 9) \cup (9, \infty)$.

39. $f(3)$ is the value of the dependent variable when the independent variable is 3—choice **B**.

41.
$$\begin{aligned} f(x) &= -3x + 4 \\ f(0) &= -3(0) + 4 \\ &= 0 + 4 \\ &= 4 \end{aligned}$$

43.
$$\begin{aligned} g(x) &= -x^2 + 4x + 1 \\ g(-2) &= -(-2)^2 + 4(-2) + 1 \\ &= -(4) - 8 + 1 \\ &= -11 \end{aligned}$$

45.
$$\begin{aligned} f(x) &= -3x + 4 \\ f(p) &= -3(p) + 4 \\ &= -3p + 4 \end{aligned}$$

47.
$$\begin{aligned} f(x) &= -3x + 4 \\ f(-x) &= -3(-x) + 4 \\ &= 3x + 4 \end{aligned}$$

49.
$$\begin{aligned} f(x) &= -3x + 4 \\ f(x + 2) &= -3(x + 2) + 4 \\ &= -3x - 6 + 4 \\ &= -3x - 2 \end{aligned}$$

51.
$$\begin{aligned} f(x) &= -3x + 4 \\ f(2m - 3) &= -3(2m - 3) + 4 \\ &= -6m + 9 + 4 \\ &= -6m + 13 \end{aligned}$$

53. **(a)** When $x = 2$, $y = 2$, so $f(2) = 2$.

(b) When $x = -1$, $y = 3$, so $f(-1) = 3$.

55. **(a)** When $x = 2$, $y = 15$, so $f(2) = 15$.

(b) When $x = -1$, $y = 10$, so $f(-1) = 10$.

57. **(a)** The point $(2, 3)$ is on the graph of f, so $f(2) = 3$.

(b) The point $(-1, -3)$ is on the graph of f, so $f(-1) = -3$.

59. **(a)** Solve the equation for y.

$$\begin{aligned} x + 3y &= 12 \\ 3y &= 12 - x \\ y &= \frac{12 - x}{3} \end{aligned}$$

Since $y = f(x)$,

$$f(x) = \frac{12 - x}{3}.$$

(b) $f(3) = \dfrac{12 - 3}{3} = \dfrac{9}{3} = 3$

61. **(a)** Solve the equation for y.

$$\begin{aligned} y + 2x^2 &= 3 \\ y &= 3 - 2x^2 \end{aligned}$$

Since $y = f(x)$,

$$f(x) = 3 - 2x^2.$$

(b)
$$\begin{aligned} f(3) &= 3 - 2(3)^2 \\ &= 3 - 2(9) \\ &= -15 \end{aligned}$$

63. **(a)** Solve the equation for y.

$$4x - 3y = 8$$
$$-3y = 8 - 4x$$
$$y = \frac{8 - 4x}{-3}$$

Since $y = f(x)$,

$$f(x) = \frac{8 - 4x}{-3}.$$

(b) $f(3) = \dfrac{8 - 4(3)}{-3} = \dfrac{8 - 12}{-3}$
$= \dfrac{-4}{-3} = \dfrac{4}{3}$

65. The equation $2x + y = 4$ has a straight line as its graph. To find y in $(3, y)$, let $x = 3$ in the equation.

$$2x + y = 4$$
$$2(3) + y = 4$$
$$6 + y = 4$$
$$y = -2$$

To use functional notation for $2x + y = 4$, solve for y to get

$$y = -2x + 4.$$

Replace y with $f(x)$ to get

$$f(x) = -2x + 4.$$
$$f(3) = -2(3) + 4 = -2$$

Because $y = -2$ when $x = 3$, the point $(3, -2)$ lies on the graph of the function.

67. $f(x) = -2x + 5$
The graph will be a line. The intercepts are $(0, 5)$ and $\left(\frac{5}{2}, 0\right)$.
The domain is $(-\infty, \infty)$. The range is $(-\infty, \infty)$.

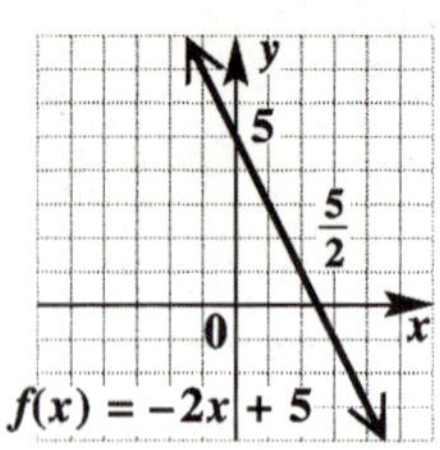

69. $h(x) = \frac{1}{2}x + 2$
The graph will be a line. The intercepts are $(0, 2)$ and $(-4, 0)$.
The domain is $(-\infty, \infty)$. The range is $(-\infty, \infty)$.

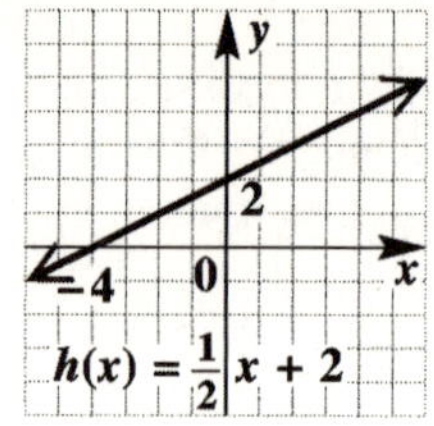

71. $G(x) = 2x$
This line includes the points $(0, 0)$, $(1, 2)$, and $(2, 4)$. The domain is $(-\infty, \infty)$. The range is $(-\infty, \infty)$.

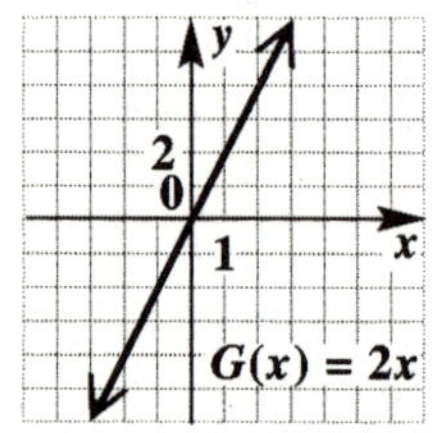

73. $g(x) = -4$
Using a y-intercept of $(0, -4)$ and a slope of $m = 0$, we graph the horizontal line. From the graph we see that the domain is $(-\infty, \infty)$. The range is $\{-4\}$.

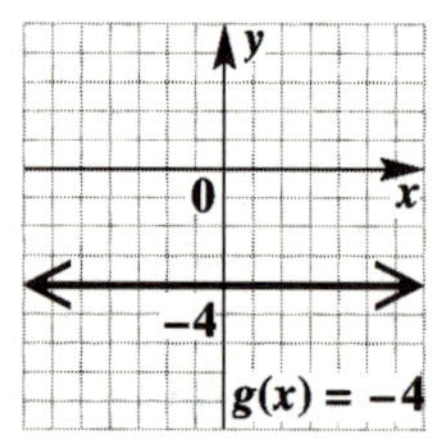

75. **(a)** $f(x) = 2.75x$
$f(3) = 2.75(3)$
$= 8.25$ (dollars)

(b) 3 is the value of the independent variable, which represents a package weight of 3 pounds; $f(3)$ is the value of the dependent variable representing the cost to mail a 3-pound package.

(c) $2.75(5) = \$13.75$, the cost to mail a 5-lb package. Using function notation, we have $f(5) = 13.75$.

77. Since the length of a man's femur is given, use the formula $h(r) = 69.09 + 2.24r$.

$h(56) = 69.09 + 2.24(56)$ *Let $r = 56$.*
$= 194.53$

The man is 194.53 cm tall.

79. Since the length of a woman's femur is given, use the formula $h(r) = 61.41 + 2.32r$.

$h(50) = 61.41 + 2.32(50)$ *Let $r = 50$.*
$= 177.41$

The woman is 177.41 cm tall.

81. $f(x) = .91(3.14)x^2$
$f(.8) = .91(3.14)(.8)^2$
$= 1.828736$
To the nearest hundredth, the volume of the pool is 1.83 m^3.

83. $f(x) = .91(3.14)x^2$
$f(1.2) = .91(3.14)(1.2)^2$
$= 4.114656$
To the nearest hundredth, the volume of the pool is 4.11 m^3.

85. **(a)** The graph passes the vertical line test, so it is the graph of a function.

(b) We start the day at midnight and end the day at midnight. The domain is $[0, 24]$.

(c) At 8 in the morning, it appears that the number of megawatts used is halfway between 1100 and 1300. So 1200 is a good estimate.

(d) The most electricity was used at 17 hours (5 P.M.) and the least electricity was used at 4 hours (4 A.M.).

(e) $f(12) = 2100$; At 12 noon, electricity use is 2100 megawatts.

87. The graph shows $x = 3$ and $y = 7$. In function notation, this is

$$f(3) = 7.$$

Section 3.6

1. This suggests *direct* variation.
As the number of tickets you buy increases, so does the probability that you will win.

3. This suggests *direct* variation.
As the pressure put on the accelerator increases, so does the speed of the car.

5. This suggests *inverse* variation.
As your age gets larger, the probability gets smaller that you believe in Santa Claus.

7. This suggests *inverse* variation.
As the number of days gets smaller, the number of home runs increases.

9. The equation $y = \frac{3}{x}$ represents *inverse* variation.
y varies inversely as x because x is in the denominator.

11. The equation $y = 10x^2$ represents *direct* variation. The number 10 is the constant of variation, and y varies directly as the square of x.

13. The equation $y = 3xz^4$ represents *joint* variation. y varies directly as x and z^4.

15. The equation $y = \frac{4x}{wz}$ represents *combined* variation. In the numerator, 4 is the constant of variation, and y varies directly as x. In the denominator, y varies inversely as w and z.

17. For $k > 0$, if y varies directly as x (then $y = kx$), when x increases, y *increases*, and when x decreases, y *decreases*.

19. "x varies directly as y" means

$$x = ky$$

for some constant k.
Substitute $x = 9$ and $y = 3$ in the equation and solve for k.

$$x = ky$$
$$9 = k(3)$$
$$k = \frac{9}{3} = 3$$

So $x = 3y$.
To find x when $y = 12$, substitute 12 for y in the equation.

$$x = 3y$$
$$x = 3(12)$$
$$x = 36$$

21. "a varies directly as the square of b" means

$$a = kb^2$$

for some constant k.
Substitute $a = 4$ and $b = 3$ in the equation and solve for k.

$$a = kb^2$$
$$4 = k(3)^2$$
$$k = \frac{4}{9}$$

So $a = \frac{4}{9}b^2$.
To find a when $b = 2$, substitute 2 for b in the equation.

$$a = \frac{4}{9}b^2$$
$$a = \frac{4}{9}(2)^2$$
$$a = \frac{4 \cdot 4}{9} = \frac{16}{9}$$

23. "z varies inversely as w" means

$$z = \frac{k}{w}$$

for some constant k. Since $z = 10$ when $w = .5$, substitute these values in the equation and solve for k.

continued

$$z = \frac{k}{w}$$
$$10 = \frac{k}{.5}$$
$$k = 10(.5) = 5$$

So $z = \frac{5}{w}$.

To find z when $w = 8$, substitute 8 for w in the equation.

$$z = \frac{5}{w}$$
$$z = \frac{5}{8} \text{ or } .625$$

25. "m varies inversely as p^2" means

$$m = \frac{k}{p^2}$$

for some constant k. Since $m = 20$ when $p = 2$, substitute these values in the equation and solve for k.

$$m = \frac{k}{p^2}$$
$$20 = \frac{k}{2^2}$$
$$k = 20(4) = 80$$

So $m = \frac{80}{p^2}$. Now let $p = 5$.

$$m = \frac{80}{p^2}$$
$$m = \frac{80}{5^2} = \frac{16}{5}$$

27. "p varies jointly as q and r^2" means

$$p = kqr^2$$

for some constant k. Given that $p = 200$ when $q = 2$ and $r = 3$, solve for k.

$$p = kqr^2$$
$$200 = k(2)(3)^2$$
$$200 = 18k$$
$$k = \frac{200}{18} = \frac{100}{9}$$

So $p = \frac{100}{9}qr^2$. Now let $q = 5$ and $r = 2$.

$$p = \frac{100}{9}qr^2$$
$$p = \frac{100}{9}(5)(2)^2$$
$$= \frac{100}{9}(20)$$
$$= \frac{2000}{9} \text{ or } 222\frac{2}{9}$$

29. In a direct variation, both variables either increase or decrease.

In an inverse variation, one variable increases while the other decreases.

31. Let $x =$ the number of gallons he bought and let $C =$ the cost.
C varies directly as x, so

$$C = kx.$$

Since $C = 13.59$ when $x = 8$,

$$13.59 = k(8)$$
$$k = \frac{13.59}{8} = 1.69875 \approx 1.699.$$

The price per gallon is $\$1.69\frac{9}{10}$.

33. Let $V =$ the volume of the can and let $h =$ the height of the can.
V varies directly as h, so

$$V = kh.$$

Since $V = 300$ when $h = 10.62$,

$$300 = k(10.62)$$
$$k = \frac{300}{10.62} \approx 28.25.$$

So $V = 28.25h$. Now let $h = 15.92$.

$$V = 28.25h$$
$$V = 28.25(15.92) = 449.74$$

The volume is about 450 cm^3.

35. Let $y =$ the weight of an object on earth and $x =$ the weight of the object on the moon.
y varies directly as x, so

$$y = kx$$

for some constant k. Since $y = 200$ when $x = 32$, substitute these values in the equation and solve for k.

$$y = kx$$
$$200 = k(32)$$
$$k = \frac{200}{32} = 6.25$$

So $y = 6.25x$.
To find x when $y = 50$, substitute 50 for y in the equation

$$y = 6.25x.$$
$$50 = 6.25x$$
$$x = \frac{50}{6.25} = 8$$

The dog would weigh 8 lb on the moon.

37. Let $d =$ the distance and $t =$ the time.
d varies directly as the square of t, so $d = kt^2$.
Let $d = -576$ and $t = 6$. (You could also use $d = 576$, but the negative sign indicates the direction of the body.)

$$
\begin{aligned}
-576 &= k(6)^2 \\
-576 &= 36k \\
-16 &= k
\end{aligned}
$$

So $d = -16t^2$. Now let $t = 4$.

$$d = -16(4)^2 = -256$$

The object fell 256 ft in the first 4 seconds.

39. Let $s =$ the speed and $t =$ the time.
The speed varies inversely with time, so there is a constant k such that $s = k/t$. Find the value of k by replacing s with 160 and t with $\frac{1}{2}$.

$$
\begin{aligned}
s &= \frac{k}{t} \\
160 &= \frac{k}{\frac{1}{2}} \\
k &= 160\left(\tfrac{1}{2}\right) = 80
\end{aligned}
$$

So $s = \dfrac{80}{t}$. Now let $t = \frac{3}{4}$.

$$
\begin{aligned}
s &= \frac{80}{\frac{3}{4}} \\
s &= \frac{80}{1} \cdot \frac{4}{3} = \frac{320}{3} \quad \text{or} \quad 106\frac{2}{3}
\end{aligned}
$$

A speed $106\frac{2}{3}$ miles per hour is needed to go the same distance in $\frac{3}{4}$ minute.

41. Let $f =$ the frequency of a string in cycles per second and $s =$ the length in feet.
f varies inversely as s, so

$$f = \frac{k}{s}$$

for some constant k. Since $f = 250$ when $s = 2$, substitute these values in the equation and solve for k.

$$
\begin{aligned}
f &= \frac{k}{s} \\
250 &= \frac{k}{2} \\
k &= 250(2) = 500
\end{aligned}
$$

So $f = \dfrac{500}{s}$. Now let $s = 5$.

$$f = \frac{500}{5} = 100.$$

The string would have a frequency of 100 cycles per second.

43. Let $I =$ the illumination produced by a light source and $d =$ the distance from the source.
I varies inversely as d^2, so

$$I = \frac{k}{d^2}$$

for some constant k. Since $I = 768$ when $d = 1$, substitute these values in the equation and solve for k.

$$
\begin{aligned}
I &= \frac{k}{d^2} \\
768 &= \frac{k}{1^2} \\
768 &= k
\end{aligned}
$$

So $I = \dfrac{768}{d^2}$. Now let $d = 6$.

$$
\begin{aligned}
I &= \frac{768}{d^2} \\
I &= \frac{768}{6^2} = \frac{768}{36} = \frac{64}{3} \quad \text{or} \quad 21\frac{1}{3}
\end{aligned}
$$

The illumination produced by the light source is $21\dfrac{1}{3}$ foot-candles.

45. Let $I =$ the simple interest, P the principal, and t the time.
Since I varies jointly as the principal and time, there is a constant k such that $I = kPt$. Find k by replacing I with 280, P with 2000, and t with 4.

$$
\begin{aligned}
I &= kPt \\
280 &= k(2000)(4) \\
k &= \frac{280}{8000} = .035
\end{aligned}
$$

So $I = .035Pt$. Now let $t = 6$.

$$
\begin{aligned}
I &= .035(2000)(6) \\
&= 420
\end{aligned}
$$

The interest would be \$420.

47. Let $F =$ the force, $w =$ the weight of the car, $s =$ the speed, and $r =$ the radius.
The force varies inversely as the radius and jointly as the weight and the square of the speed, so

$$F = \frac{kws^2}{r}.$$

Let $F = 242$, $w = 2000$, $r = 500$, and $s = 30$.

$$
\begin{aligned}
242 &= \frac{k(2000)(30)^2}{500} \\
k &= \frac{242(500)}{2000(900)} = \frac{121}{1800}
\end{aligned}
$$

So $F = \dfrac{121ws^2}{1800r}$.
Let $r = 750$, $s = 50$, and $w = 2000$.

$$F = \frac{121(2000)(50)^2}{1800(750)} \approx 448.1$$

Approximately 448.1 lb of force would be needed.

49. Let N = the number of long distance calls,
p_1 = the population of City 1,
p_2 = the population of City 2,
and d = the distance between them.

$$N = \frac{kp_1p_2}{d}$$

Let $N = 80,000$, $p_1 = 70,000$, $p_2 = 100,000$, and $d = 400$.

$$\begin{aligned} 80,000 &= \frac{k(70,000)(100,000)}{400} \\ 80,000 &= 17,500,000k \\ k &= \frac{80,000}{17,500,000} = \frac{4}{875} \\ N &= \frac{4}{875}\left(\frac{p_1p_2}{d}\right) \end{aligned}$$

Let $p_1 = 50,000$, $p_2 = 75,000$, and $d = 250$.

$$\begin{aligned} N &= \frac{4}{875}\left(\frac{50,000 \cdot 75,000}{250}\right) \\ &= \frac{480,000}{7} \text{ or } 68,571\frac{3}{7} \end{aligned}$$

Rounded to the nearest hundred, there are approximately 68,600 calls.

51. Let V = the volume of gas, p = the pressure, and t = the temperature.

Then $V = \dfrac{kt}{p}$.

$V = 1.3$ when $t = 300$ and $p = 18$.

$$\begin{aligned} 1.3 &= \frac{k(300)}{18} \\ k &= \frac{1.3(18)}{300} = .078 \end{aligned}$$

So $V = \dfrac{.078t}{p}$.

Let $t = 340$ and $p = 24$.

$$V = \frac{.078(340)}{24} = 1.105$$

The volume is 1.105 L.

53. The weight W of a bass varies jointly as its girth G and the square of its length L, so

$$W = kGL^2$$

for some constant k. Substitute 22.7 for W, 21 for G, and 36 for L.

$$\begin{aligned} 22.7 &= k(21)(36)^2 \\ k &= \frac{22.7}{27{,}216} \approx .000834 \end{aligned}$$

So $W = .000834GL^2$. Now let $G = 18$ and $L = 28$.

$$\begin{aligned} W &= .000834GL^2 \\ &= .000834(18)(28)^2 \\ &\approx 11.8 \end{aligned}$$

The bass would weigh about 11.8 pounds.

55. The ordered pairs in the form of (gallons, price) are $(0, 0)$ and $(1, 1.25)$.

56. Let $(x_1, y_1) = (0, 0)$ and $(x_2, y_2) = (1, 1.25)$. Then

$$m = \frac{\text{change in } y}{\text{change in } x} = \frac{1.25 - 0}{1 - 0} = 1.25.$$

The slope is 1.25.

57. Since $m = 1.25$ and $b = 0$, the equation is

$$y = 1.25x + 0$$
$$\text{or} \quad y = 1.25x.$$

58. If $f(x) = ax + b$, then $a = 1.25$ and $b = 0$.

59. The value of a, 1.25, is the price in dollars per gallon of gasoline. It is also the slope of the line.

60. Since $f(x) = 1.25x$, it fits the form for a direct variation, that is, $y = kx$. The value of a, 1.25, is the constant of variation $(k = a)$.

61. The point $(4.6, 5.75)$ means that 4.6 gal of gasoline cost \$5.75.

62. X = 12 and Y_1 = 15 means that 12 gal of gas cost \$15.00.

Chapter 3 Review Exercises

1. $3x + 2y = 10$

For $x = 0$:

$$\begin{aligned} 3(0) + 2y &= 10 \\ 2y &= 10 \\ y &= 5 \quad (0, 5) \end{aligned}$$

For $y = 0$:

$$\begin{aligned} 3x + 2(0) &= 10 \\ 3x &= 10 \\ x &= \frac{10}{3} \quad \left(\frac{10}{3}, 0\right) \end{aligned}$$

For $x = 2$:

$$\begin{aligned} 3(2) + 2y &= 10 \\ 6 + 2y &= 10 \\ 2y &= 4 \\ y &= 2 \quad (2, 2) \end{aligned}$$

For $y = -2$:

$$\begin{aligned} 3x + 2(-2) &= 10 \\ 3x - 4 &= 10 \\ 3x &= 14 \end{aligned}$$

$$x = \frac{14}{3} \quad \left(\frac{14}{3}, -2\right)$$

Plot the ordered pairs, and draw the line through them.

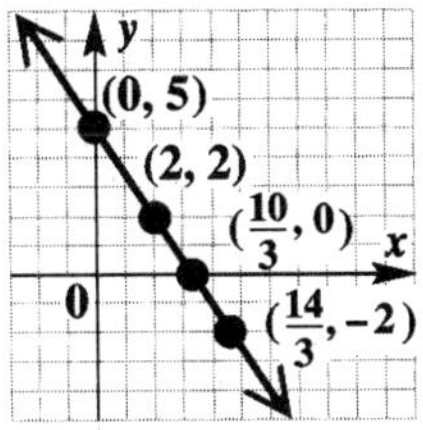

2. $x - y = 8$
For $x = 2$:

$$\begin{aligned} 2 - y &= 8 \\ -y &= 6 \\ y &= -6 \quad (2, -6) \end{aligned}$$

For $y = -3$:

$$\begin{aligned} x - (-3) &= 8 \\ x + 3 &= 8 \\ x &= 5 \quad (5, -3) \end{aligned}$$

For $x = 3$:

$$\begin{aligned} 3 - y &= 8 \\ -y &= 5 \\ y &= -5 \quad (3, -5) \end{aligned}$$

For $y = -2$:

$$\begin{aligned} x - (-2) &= 8 \\ x + 2 &= 8 \\ x &= 6 \quad (6, -2) \end{aligned}$$

Plot the ordered pairs, and draw the line through them.

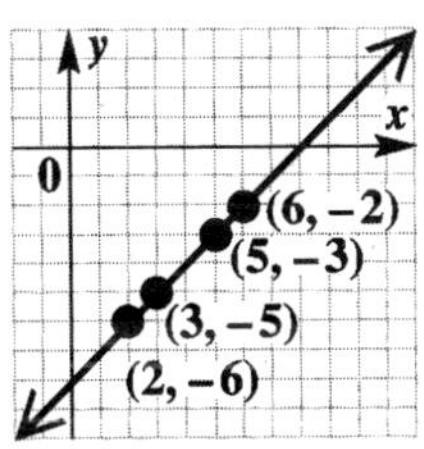

3. $4x - 3y = 12$
To find the x-intercept, let $y = 0$.

$$\begin{aligned} 4x - 3y &= 12 \\ 4x - 3(0) &= 12 \\ 4x &= 12 \\ x &= 3 \end{aligned}$$

The x-intercept is $(3, 0)$.
To find the y-intercept, let $x = 0$.

$$\begin{aligned} 4x - 3y &= 12 \\ 4(0) - 3y &= 12 \\ -3y &= 12 \\ y &= -4 \end{aligned}$$

The y-intercept is $(0, -4)$.
Plot the intercepts and draw the line through them.

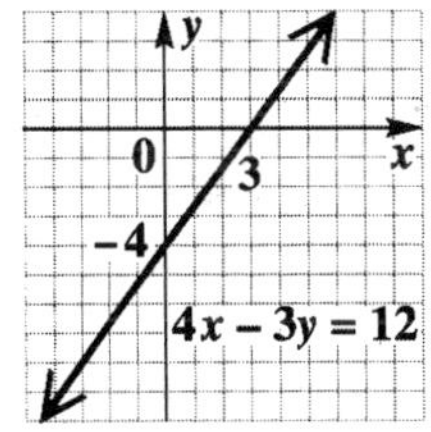

4. $5x + 7y = 28$
To find the x-intercept, let $y = 0$.

$$\begin{aligned} 5x + 7y &= 28 \\ 5x + 7(0) &= 28 \\ 5x &= 28 \\ x &= \frac{28}{5} \end{aligned}$$

The x-intercept is $\left(\frac{28}{5}, 0\right)$.

To find the y-intercept, let $x = 0$.

$$\begin{aligned} 5x + 7y &= 28 \\ 5(0) + 7y &= 28 \\ 7y &= 28 \\ y &= 4 \end{aligned}$$

The y-intercept is $(0, 4)$.
Plot the intercepts and draw the line through them.

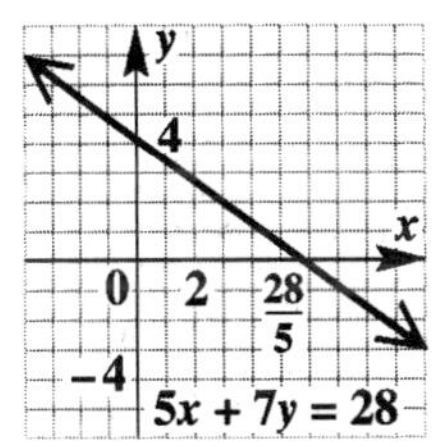

5. $2x + 5y = 20$
To find the x-intercept, let $y = 0$.

$$\begin{aligned} 2x + 5y &= 20 \\ 2x + 5(0) &= 20 \\ 2x &= 20 \\ x &= 10 \end{aligned}$$

The x-intercept is $(10, 0)$.
To find the y-intercept, let $x = 0$.

$$\begin{aligned} 2x + 5y &= 20 \\ 2(0) + 5y &= 20 \\ 5y &= 20 \\ y &= 4 \end{aligned}$$

The y-intercept is $(0, 4)$.
Plot the intercepts and draw the line through them.

continued

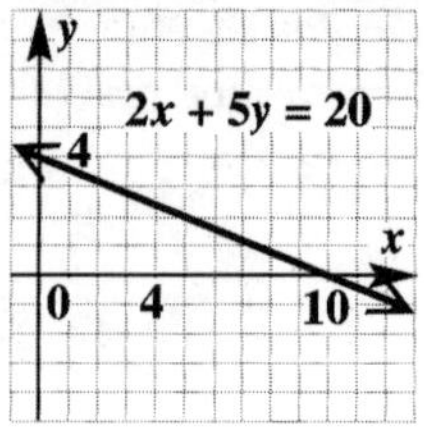

6. $x - 4y = 8$
To find the x-intercept, let $y = 0$.

$$\begin{aligned} x - 4y &= 8 \\ x - 4(0) &= 8 \\ x &= 8 \end{aligned}$$

The x-intercept is $(8, 0)$.
To find the y-intercept, let $x = 0$.

$$\begin{aligned} 0 - 4y &= 8 \\ -4y &= 8 \\ y &= -2 \end{aligned}$$

The y-intercept is $(0, -2)$.
Plot the intercepts and draw the line through them.

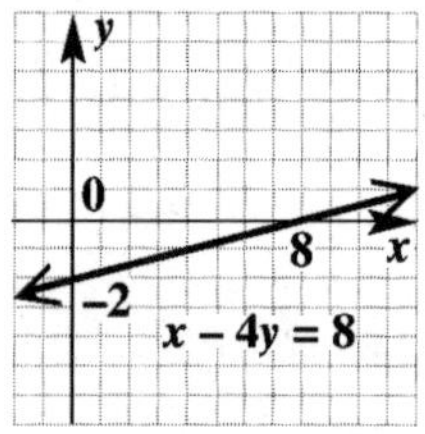

7. If both coordinates are positive, the point lies in quadrant I. If the first coordinate is negative and the second is positive, the point lies in quadrant II. To lie in quadrant III, the point must have both coordinates negative. To lie in quadrant IV, the first coordinate must be positive and the second must be negative.

8. Through $(-1, 2)$ and $(4, -5)$

$$m = \frac{\text{change in } y}{\text{change in } x} = \frac{2 - (-5)}{-1 - 4} = \frac{7}{-5} = -\frac{7}{5}$$

9. Through $(0, 3)$ and $(-2, 4)$
Let $(x_1, y_1) = (0, 3)$ and $(x_2, y_2) = (-2, 4)$.

$$m = \frac{y_2 - y_1}{x_2 - x_1} = \frac{4 - 3}{-2 - 0} = \frac{1}{-2} = -\frac{1}{2}$$

10. The slope of $y = 2x + 3$ is 2, the coefficient of x.

11. $3x - 4y = 5$
Write the equation in slope-intercept form.

$$\begin{aligned} -4y &= -3x + 5 \\ y &= \frac{3}{4}x - \frac{5}{4} \end{aligned}$$

The slope is $\frac{3}{4}$.

12. $x = 5$ is a vertical line and has *undefined* slope.

13. Parallel to $3y = 2x + 5$
Write the equation in slope-intercept form.

$$\begin{aligned} 3y &= 2x + 5 \\ y &= \frac{2}{3}x + \frac{5}{3} \end{aligned}$$

The slope of $3y = 2x + 5$ is $\frac{2}{3}$; all lines parallel to it will also have a slope of $\frac{2}{3}$.

14. Perpendicular to $3x - y = 4$
Solve for y.

$$y = 3x - 4$$

The slope is 3; the slope of a line perpendicular to it is $-\frac{1}{3}$ since

$$3\left(-\frac{1}{3}\right) = -1.$$

15. Through $(-1, 5)$ and $(-1, -4)$

$$m = \frac{5 - (-4)}{-1 - (-1)} = \frac{9}{0} \quad \textit{Undefined}$$

This is a vertical line; it has undefined slope.

16. $y + 6 = 0$ is the horizontal line with y-intercept $(0, -6)$. Its slope is 0.

17. The x-intercept is $(2, 0)$ and the y-intercept is $(0, 2)$. The slope is

$$m = \frac{\text{change in } y}{\text{change in } x} = \frac{2 - 0}{0 - 2} = \frac{2}{-2} = -1.$$

18. The line goes up from left to right, so it has positive slope.

19. The line goes down from left to right, so it has negative slope.

20. The line is vertical, so it has *undefined* slope.

21. The line is horizontal, so it has 0 slope.

22. The slope is $\frac{2}{10}$ which can be written as

$$.2,\ 20\%,\ \frac{20}{100},\ \text{or}\ \frac{1}{5}.$$

The correct responses are **A**, **B**, **C**, **D**, and **F**.

23. To rise 1 foot, we must move 4 feet in the horizontal direction. To rise 3 feet, we must move $3(4) = 12$ feet in the horizontal direction.

24. Let $(x_1, y_1) = (1970,\ 10{,}000)$ and $(x_2, y_2) = (1999,\ 49{,}000)$. Then

$$m = \frac{49,000 - 10,000}{1999 - 1970} = \frac{39,000}{29} \approx 1345.$$

The average rate of change is \$1345 per year.

25. Slope $-\frac{1}{3}$, y-intercept $(0, -1)$

Use the slope-intercept form with $m = -\frac{1}{3}$ and $b = -1$.

$$y = mx + b$$
$$y = -\frac{1}{3}x - 1$$

26. Slope 0, y-intercept $(0, -2)$
Use the slope-intercept form with $m = 0$ and $b = -2$.

$$y = mx + b$$
$$y = (0)x - 2$$
$$y = -2$$

27. Slope $-\frac{4}{3}$, through $(2, 7)$

Use the point-slope form with $m = -\frac{4}{3}$ and $(x_1, y_1) = (2, 7)$.

$$y - y_1 = m(x - x_1)$$
$$y - 7 = -\frac{4}{3}(x - 2)$$
$$y - 7 = -\frac{4}{3}x + \frac{8}{3}$$
$$y = -\frac{4}{3}x + \frac{29}{3}$$

28. Slope 3, through $(-1, 4)$
Use the point-slope form with $m = 3$ and $(x_1, y_1) = (-1, 4)$.

$$y - y_1 = m(x - x_1)$$
$$y - 4 = 3[x - (-1)]$$
$$y - 4 = 3(x + 1)$$
$$y - 4 = 3x + 3$$
$$y = 3x + 7$$

29. Vertical, through $(2, 5)$
The equation of any vertical line is in the form $x = k$. Since the line goes through $(2, 5)$, the equation is $x = 2$. (Slope-intercept form is not possible.)

30. Through $(2, -5)$ and $(1, 4)$
Find the slope.

$$m = \frac{4 - (-5)}{1 - 2} = \frac{9}{-1} = -9$$

Use the point-slope form with $m = -9$ and $(x_1, y_1) = (2, -5)$.

$$y - y_1 = m(x - x_1)$$
$$y - (-5) = -9(x - 2)$$
$$y + 5 = -9x + 18$$
$$y = -9x + 13$$

31. Through $(-3, -1)$ and $(2, 6)$
Find the slope.

$$m = \frac{6 - (-1)}{2 - (-3)} = \frac{7}{5}$$

Use the point-slope form with $m = \frac{7}{5}$ and $(x_1, y_1) = (2, 6)$.

$$y - y_1 = m(x - x_1)$$
$$y - 6 = \frac{7}{5}(x - 2)$$
$$y - 6 = \frac{7}{5}x - \frac{14}{5}$$
$$y = \frac{7}{5}x + \frac{16}{5}$$

32. From Exercise 17, we have $m = -1$ and a y-intercept of $(0, 2)$. The slope-intercept form is

$$y = -1x + 2 \quad \text{or} \quad y = -x + 2.$$

33. Parallel to $4x - y = 3$ and through $(7, -1)$
Writing $4x - y = 3$ in slope-intercept form gives us $y = 4x - 3$, which has slope 4. Lines parallel to it will also have slope 4. The line with slope 4 through $(7, -1)$ is :

$$y - y_1 = m(x - x_1)$$
$$y - (-1) = 4(x - 7)$$
$$y + 1 = 4x - 28$$
$$y = 4x - 29$$

34. Perpendicular to $2x - 5y = 7$ and through $(4, 3)$
Write the equation in slope-intercept form.

$$2x - 5y = 7$$
$$-5y = -2x + 7$$
$$y = \frac{2}{5}x - \frac{7}{5}$$

$y = \frac{2}{5}x - \frac{7}{5}$ has slope $\frac{2}{5}$ and is perpendicular to lines with slope $-\frac{5}{2}$.

The line with slope $-\frac{5}{2}$ through $(4, 3)$ is

$$y - y_1 = m(x - x_1)$$
$$y - 3 = -\frac{5}{2}(x - 4)$$
$$y - 3 = -\frac{5}{2}x + 10$$
$$y = -\frac{5}{2}x + 13$$

35. **(a)** The fixed cost is \$159, so that is the value of b. The variable cost is \$57, so

$$y = mx + b = 57x + 159.$$

The cost of a 1-year membership can be found by substituting 12 for x.

$$\begin{aligned} y &= 57(12) + 159 \\ &= 684 + 159 = 843 \end{aligned}$$

The cost is \$843.

(b) As in part (a),

$$\begin{aligned} y &= 47x + 159. \\ y &= 47(12) + 159 \\ &= 564 + 159 = 723 \end{aligned}$$

The cost is \$723.

36. **(a)** Use $(6, 12.6)$ and $(11, 30.7)$.

$$m = \frac{30.7 - 12.6}{11 - 6} = \frac{18.1}{5} = 3.62$$

Use the point-slope form of a line.

$$\begin{aligned} y - y_1 &= m(x - x_1) \\ y - 12.6 &= 3.62(x - 6) \\ y - 12.6 &= 3.62x - 21.72 \\ y &= 3.62x - 9.12 \end{aligned}$$

The slope, 3.62, indicates that the number of e-filing taxpayers increased by 3.62% each year from 1996 to 2001.

(b) The year 2005 corresponds to $x = 15$.

$$\begin{aligned} y &= 3.62(15) - 9.12 \\ &= 54.3 - 9.12 = 45.18 \end{aligned}$$

According to the equation from part (a), 45.18% of tax returns will be filed electronically in 2005.

37. $3x - 2y \le 12$
Graph $3x - 2y = 12$ as a solid line through $(0, -6)$ and $(4, 0)$. Use $(0, 0)$ as a test point. Since $(0, 0)$ satisfies the inequality, shade the region on the side of the line containing $(0, 0)$.

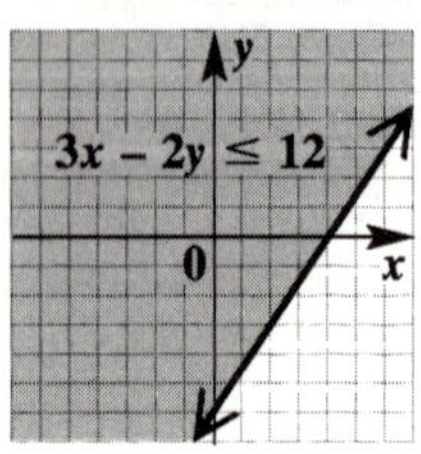

38. $5x - y > 6$
Graph $5x - y = 6$ as a dashed line through $(0, -6)$ and $\left(\frac{6}{5}, 0\right)$. Use $(0, 0)$ as a test point. Since $(0, 0)$ does not satisfy the inequality, shade the region on the side of the line that does not contain $(0, 0)$.

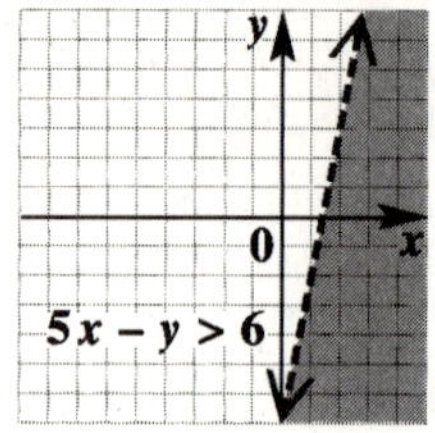

39. $x \ge 2$
Graph $x = 2$ as a solid vertical line, crossing the x-axis at 2. Since $(0, 0)$ does not satisfy the inequality, shade the region to the right of the line.

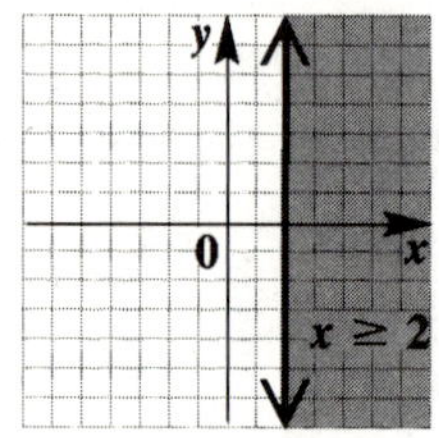

40. $2x + y \le 1$ and $x \ge 2y$
Graph $2x + y = 1$ as a solid line through $\left(\frac{1}{2}, 0\right)$ and $(0, 1)$, and shade the region on the side containing $(0, 0)$ since it satisfies the inequality. Next, graph $x = 2y$ as a solid line through $(0, 0)$ and $(2, 1)$, and shade the region on the side containing $(2, 0)$ since $2 > 2(0)$ or $2 > 0$ is true. The intersection is the region where the graphs overlap.

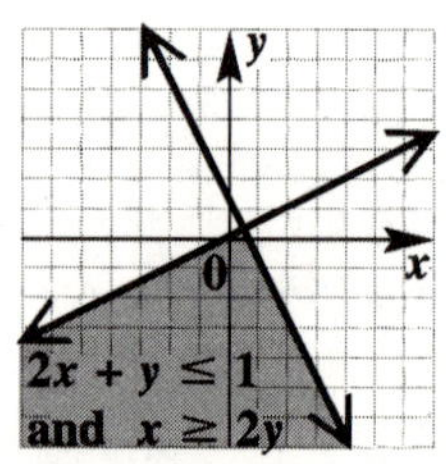

41. $x \ge 2$
Graph $x = 2$ as a solid vertical line through $(2, 0)$. Shade the region to the right of $x = 2$.

$y \ge 2$
Graph $y = 2$ as a solid horizontal line through $(0, 2)$. Shade the region above $y = 2$. The graph of

$$x \ge 2 \quad \text{or} \quad y \ge 2$$

includes all the shaded regions.

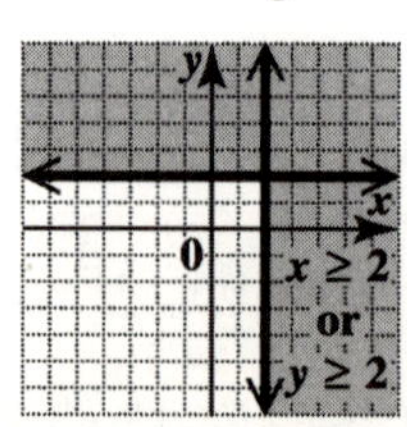

42. $\{(-4,2),(-4,-2),(1,5),(1,-5)\}$
The domain, the set of x-values, is $\{-4,1\}$.
The range, the set of y-values, is $\{2,-2,5,-5\}$.
Since each x-value has more than one y-value, the relation is not a function.

43. The relation can be described by the set of ordered pairs

$$\{(9,32),(11,47),(4,47),(17,69),(25,14)\}.$$

The relation is a function since for each x-value, there is only one y-value.

The domain is the set of x-values:
$\{9,11,4,17,25\}$.
The range is the set of y-values: $\{32,47,69,14\}$.

44. The domain, the x-values of the points on the graph, is $[-4,4]$. The range, the y-values of the points on the graph, is $[0,2]$. Since a vertical line intersects the graph of the relation in at most one point, the relation is a function.

45. The x-values are negative or zero, so the domain is $(-\infty,0]$. The y-values can be any real number, so the range is $(-\infty,\infty)$. A vertical line, such as $x=-3$, will intersect the graph twice, so by the vertical line test, the relation is not a function.

46. $y=3x-3$
For any value of x, there is exactly one value of y, so the equation defines a function, actually a linear function. The domain is the set of all real numbers, $(-\infty,\infty)$.

47. $y<x+2$
For any value of x, there are many values of y. For example, $(1,0)$ and $(1,1)$ are both solutions of the inequality that have the same x-value but different y-values. The inequality does not define a function. The domain is the set of all real numbers, $(-\infty,\infty)$.

48. $y=|x|$
For any value of x, there is exactly one value of y, so the equation defines a function. The domain is the set of all real numbers, $(-\infty,\infty)$.

49. $y=\sqrt{4x+7}$
Given any value of x, y is found by multiplying x by 4, adding 7, and taking the square root of the result. This process produces exactly one value of y for each x-value, so the equation defines a function. Since the radicand must be nonnegative,

$$\begin{aligned}4x+7&\geq 0\\4x&\geq -7\\x&\geq -\frac{7}{4}.\end{aligned}$$

The domain is $\left[-\frac{7}{4},\infty\right)$.

50. $x=y^2$
The ordered pairs $(4,2)$ and $(4,-2)$ both satisfy the equation. Since one value of x, 4, corresponds to two values of y, 2 and -2, the equation does not define a function. Because x is equal to the square of y, the values of x must always be nonnegative. The domain is $[0,\infty)$.

51. $y=\dfrac{7}{x-6}$
Given any value of x, y is found by subtracting 6, then dividing the result into 7. This process produces exactly one value of y for each x-value, so the equation defines a function. The domain includes all real numbers except those that make the denominator 0, namely 6. The domain is $(-\infty,6)\cup(6,\infty)$.

52. If no vertical line intersects the graph in more than one point, then it is the graph of a function.

In Exercises 53–56, use

$$f(x)=-2x^2+3x-6.$$

53. $f(0)=-2(0)^2+3(0)-6=-6$

54. $$\begin{aligned}f(2.1)&=-2(2.1)^2+3(2.1)-6\\&=-8.82+6.3-6=-8.52\end{aligned}$$

55. $$\begin{aligned}f\left(-\frac{1}{2}\right)&=-2\left(-\frac{1}{2}\right)^2+3\left(-\frac{1}{2}\right)-6\\&=-\frac{1}{2}-\frac{3}{2}-6=-8\end{aligned}$$

56. $f(k)=-2k^2+3k-6$

57. **(a)** For each year, there is exactly one profit associated with the year, so the table defines a function.

(b) The domain is the set of years, that is, $\{1994,1995,1996,1997,1998,1999\}$.
The range is the set of profits, that is, $\{40,60,80,130,180,200\}$.

(c) Answers will vary. Two possible answers are $(1994,40)$ and $(1995,60)$.

(d) $f(1994)=40$. In 1994, CNBC profits were \$40 million.

(e) Since $f(1998)=180$, $x=1998$.

58. $$\begin{aligned}2x^2-y&=0\\-y&=-2x^2\\y&=2x^2\end{aligned}$$
Since $y=f(x)$,
$f(x)=2x^2$,
and $f(3)=2(3)^2=2(9)=18$.

59. Solve for y in terms of x.

$$\begin{aligned} 2x - 5y &= 7 \\ 2x - 7 &= 5y \\ \frac{2x-7}{5} &= y \end{aligned}$$

This is the same as choice **C**,

$$f(x) = \frac{-7+2x}{5}.$$

60. No, because the equation of a line with undefined slope is $x = a$, so the ordered pairs have the form (a, y), where a is a constant and y is a variable. Thus, the number a corresponds to an infinite number of values of y.

61. The slope is negative since the line falls from left to right.

62. Use the points $(-1, 5)$ and $(3, -1)$.

$$m = \frac{5-(-1)}{-1-3} = \frac{5+1}{-1-3} = \frac{6}{-4} = -\frac{3}{2}$$

63. Since $m = -\frac{3}{2}$, the slope of any line parallel to this line is also $-\frac{3}{2}$, whereas the slope of any line perpendicular to this line is $\frac{2}{3}$ since $\frac{2}{3}$ is the negative reciprocal of $-\frac{3}{2}$.

64. $2y = -3x + 7$
To find the x-intercept, let $y = 0$.

$$\begin{aligned} 2(0) &= -3x + 7 \\ 3x &= 7 \\ x &= \frac{7}{3} \end{aligned}$$

The x-intercept is $\left(\frac{7}{3}, 0\right)$.

65. $2y = -3x + 7$
To find the y-intercept, let $x = 0$.

$$\begin{aligned} 2y &= -3(0) + 7 \\ 2y &= 7 \\ y &= \frac{7}{2} \end{aligned}$$

The y-intercept is $\left(0, \frac{7}{2}\right)$.

66. Solve $2y = -3x + 7$ for y.

$$y = -\frac{3}{2}x + \frac{7}{2}$$

Since $y = f(x)$,

$$f(x) = -\frac{3}{2}x + \frac{7}{2}.$$

67.
$$\begin{aligned} f(x) &= -\frac{3}{2}x + \frac{7}{2} \\ f(8) &= -\frac{3}{2}(8) + \frac{7}{2} \\ &= -\frac{24}{2} + \frac{7}{2} = -\frac{17}{2} \end{aligned}$$

68.
$$\begin{aligned} f(x) &= -\frac{3}{2}x + \frac{7}{2} \\ -8 &= -\frac{3}{2}x + \frac{7}{2} && \textit{Let } f(x) = -8. \\ -16 &= -3x + 7 && \textit{Multiply by 2.} \\ -23 &= -3x && \textit{Subtract 7.} \\ x &= \frac{23}{3} && \textit{Divide by } -3. \end{aligned}$$

69.
$$\begin{aligned} f(x) &\geq 0 \\ -\frac{3}{2}x + \frac{7}{2} &\geq 0 \\ -\frac{3}{2}x &\geq -\frac{7}{2} \\ x &\leq \left(-\frac{7}{2}\right)\left(-\frac{2}{3}\right) \\ x &\leq \frac{7}{3} \end{aligned}$$

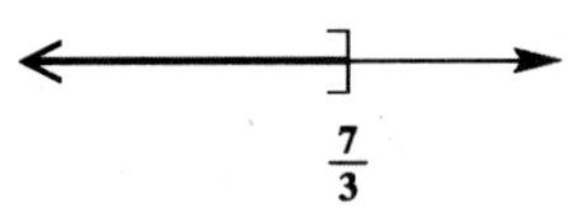

70. $f(x) = 0$ is equivalent to $y = 0$, which is the equation we solved in Exercise 64 to find the x-intercept.

Solution set: $\left\{\frac{7}{3}\right\}$

71. The graph is below the x-axis for $x > \frac{7}{3}$, so the solution set of $f(x) < 0$ is $\left(\frac{7}{3}, \infty\right)$.

72. The graph is above the x-axis for $x < \frac{7}{3}$, so the solution set of $f(x) > 0$ is $\left(-\infty, \frac{7}{3}\right)$.

73. If y varies inversely as x, then $y = \frac{k}{x}$, for some constant k. This form fits choice **C**.

74. Let v = the viewing distance and e = the amount of enlargement.
v varies directly as e, so

$$v = ke$$

for some constant k. Since $v = 250$ when $e = 5$, substitute these values in the equation and solve for k.

$$v = ke$$
$$250 = k(5)$$
$$50 = k$$

So $v = 50e$. Now let $e = 8.6$.

$$v = 50(8.6) = 430$$

It should be viewed from 430 mm.

75. The frequency f of a vibrating guitar string varies inversely as its length L, so

$$f = \frac{k}{L}$$

for some constant k. Substitute .65 for L and 4.3 for f.

$$4.3 = \frac{k}{6.5}$$
$$k = 4.3(.65) = 2.795$$

So $f = \dfrac{2.795}{L}$. Now let $L = .5$.

$$f = \frac{2.795}{.5} = 5.59$$

The frequency would be 5.59 vibrations per second.

76. The volume V of a rectangular box of a given height is proportional to its width W and length L, so

$$V = kWL$$

for some constant k. Substitute 2 for W, 4 for L, and 12 for V.

$$12 = k(2)(4)$$
$$k = \frac{12}{8} = \frac{3}{2}$$

So $V = \frac{3}{2}WL$. Now let $W = 3$ and $L = 5$.

$$V = \frac{3}{2}(3)(5) = \frac{45}{2}$$

The volume is 22.5 ft^3.

Chapter 3 Test

1. $2x - 3y = 12$

For $x = 1$:

$$2(1) - 3y = 12$$
$$2 - 3y = 12$$
$$-3y = 10$$
$$y = -\frac{10}{3} \quad \left(1, -\frac{10}{3}\right)$$

For $x = 3$:

$$2(3) - 3y = 12$$
$$6 - 3y = 12$$
$$-3y = 6$$
$$y = -2 \quad (3, -2)$$

For $y = -4$:

$$2x - 3(-4) = 12$$
$$2x + 12 = 12$$
$$2x = 0$$
$$x = 0 \quad (0, -4)$$

2. $3x - 2y = 20$

To find the x-intercept, let $y = 0$.

$$3x - 2(0) = 20$$
$$3x = 20$$
$$x = \frac{20}{3}$$

The x-intercept is $\left(\frac{20}{3}, 0\right)$.

To find the y-intercept, let $x = 0$.

$$3(0) - 2y = 20$$
$$-2y = 20$$
$$y = -10$$

The y-intercept is $(0, -10)$.

Draw the line through these two points.

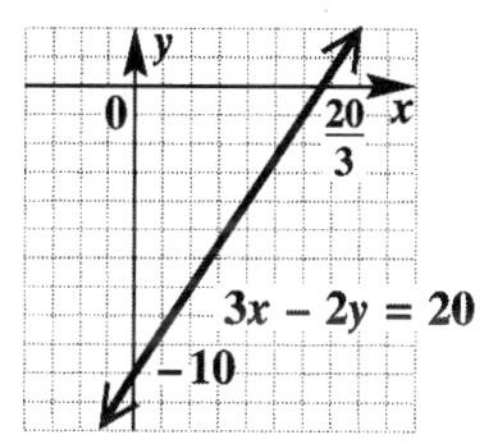

3. The graph of $y = 5$ is the horizontal line with slope 0 and y-intercept $(0, 5)$. There is no x-intercept.

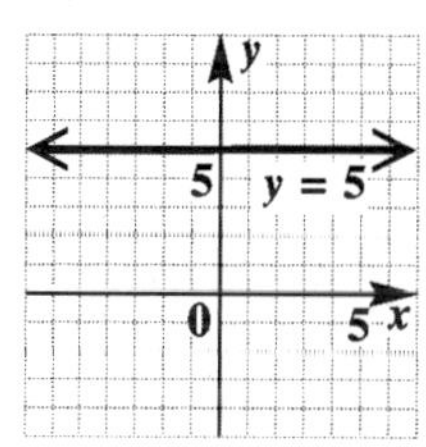

4. The graph of $x = 2$ is the vertical line with x-intercept at $(2, 0)$. There is no y-intercept.

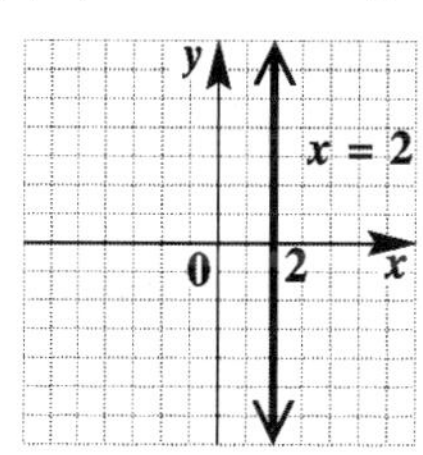

5. Through $(6, 4)$ and $(-4, -1)$

$$m = \frac{4-(-1)}{6-(-4)} = \frac{4+1}{6+4} = \frac{5}{10} = \frac{1}{2}$$

The slope of the line is $\frac{1}{2}$.

6. The graph of a line with undefined slope is the graph of a vertical line.

7. Find the slope of each line.

$$\begin{aligned} 5x - y &= 8 \\ -y &= -5x + 8 \\ y &= 5x - 8 \end{aligned}$$

The slope is 5.

$$\begin{aligned} 5y &= -x + 3 \\ y &= -\frac{1}{5}x + \frac{3}{5} \end{aligned}$$

The slope is $-\frac{1}{5}$.

Since $5\left(-\frac{1}{5}\right) = -1$, the two slopes are negative reciprocals and the lines are perpendicular.

8. Find the slope of each line.

$$\begin{aligned} 2y &= 3x + 12 \\ y &= \frac{3}{2}x + 6 \end{aligned}$$

The slope is $\frac{3}{2}$.

$$\begin{aligned} 3y &= 2x - 5 \\ y &= \frac{2}{3}x - \frac{5}{3} \end{aligned}$$

The slope is $\frac{2}{3}$.
The lines are neither parallel nor perpendicular.

9. Use the points $(1980, 119{,}000)$ and $(2001, 93{,}500)$.

average rate of change

$$\begin{aligned} &= \frac{\text{change in } y}{\text{change in } x} = \frac{93{,}500 - 119{,}000}{2001 - 1980} \\ &= \frac{-25{,}500}{21} \approx -1214 \end{aligned}$$

The average rate of change is about -1214 farms per year, that is, the number of farms decreased by about 1214 each year from 1980 to 2001.

10. Through $(4, -1)$; $m = -5$
Let $m = -5$ and $(x_1, y_1) = (4, -1)$ in the point-slope form.

$$\begin{aligned} y - y_1 &= m(x - x_1) \\ y - (-1) &= -5(x - 4) \\ y + 1 &= -5x + 20 \\ y &= -5x + 19 \end{aligned}$$

11. Through $(-3, 14)$; horizontal
A horizontal line has equation $y = k$. Here $k = 14$, so the line has equation $y = 14$.

12. Through $(-7, 2)$ and parallel to $3x + 5y = 6$
To find the slope of $3x + 5y = 6$, write the equation in slope-intercept form by solving for y.

$$\begin{aligned} 3x + 5y &= 6 \\ 5y &= -3x + 6 \\ y &= -\frac{3}{5}x + \frac{6}{5} \end{aligned}$$

The slope is $-\frac{3}{5}$, so a line parallel to it also has slope $-\frac{3}{5}$. Let $m = -\frac{3}{5}$ and $(x_1, y_1) = (-7, 2)$ in the point-slope form.

$$\begin{aligned} y - y_1 &= m(x - x_1) \\ y - 2 &= -\frac{3}{5}[x - (-7)] \\ y - 2 &= -\frac{3}{5}(x + 7) \\ y - 2 &= -\frac{3}{5}x - \frac{21}{5} \\ y &= -\frac{3}{5}x - \frac{11}{5} \end{aligned}$$

13. Through $(-7, 2)$ and perpendicular to $y = 2x$
Since $y = 2x$ is in slope-intercept form $(b = 0)$, the slope, m, of $y = 2x$ is 2. A line perpendicular to it has a slope that is the negative reciprocal of 2, that is, $-\frac{1}{2}$. Let $m = -\frac{1}{2}$ and $(x_1, y_1) = (-7, 2)$ in the point-slope form.

$$\begin{aligned} y - y_1 &= m(x - x_1) \\ y - 2 &= -\frac{1}{2}(x + 7) \\ y - 2 &= -\frac{1}{2}x - \frac{7}{2} \\ y &= -\frac{1}{2}x - \frac{3}{2} \end{aligned}$$

14. Through $(-2, 3)$ and $(6, -1)$
First find the slope.

$$m = \frac{3-(-1)}{-2-6} = \frac{3+1}{-8} = \frac{4}{-8} = -\frac{1}{2}$$

Use $m = -\frac{1}{2}$ and $(x_1, y_1) = (-2, 3)$ in the point-slope form.

$$\begin{aligned} y - y_1 &= m(x - x_1) \\ y - 3 &= -\frac{1}{2}[x - (-2)] \\ y - 3 &= -\frac{1}{2}(x + 2) \end{aligned}$$

$$y - 3 = -\frac{1}{2}x - 1$$
$$y = -\frac{1}{2}x + 2$$

15. Positive slope means that the line goes up from left to right. The only line that has positive slope and a negative y-coordinate for its y-intercept is choice **B**.

16. **(a)** Use the points $(5, 22{,}860)$ and $(9, 30{,}735)$.

$$m = \frac{30{,}735 - 22{,}860}{9 - 5} = \frac{7875}{4} = 1968.75$$

Use the point-slope form with $m = 1968.75$ and $(x_1, y_1) = (5, 22{,}860)$.

$$y - 22{,}860 = 1968.75(x - 5)$$
$$y - 22{,}860 = 1968.75x - 9843.75$$
$$y = 1968.75x + 13{,}016.25$$

(b) $y = 1968.75(7) + 13{,}016.25$ *Let $x = 7$.*
$= 13{,}781.25 + 13{,}016.25$
$= 26{,}797.50 \approx \$26{,}798$,
which is close to the actual value.

17. $3x - 2y > 6$
Graph the line $3x - 2y = 6$, which has intercepts $(2, 0)$ and $(0, -3)$, as a dashed line since the inequality involves $>$. Test $(0, 0)$, which yields $0 > 6$, a false statement. Shade the region that does not include $(0, 0)$.

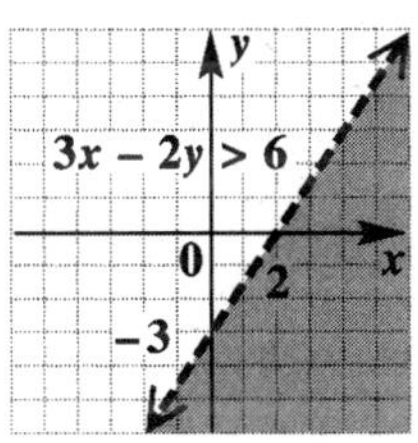

18. $y < 2x - 1$ and $x - y < 3$
First graph $y = 2x - 1$ as a dashed line through $(2, 3)$ and $(0, -1)$. Test $(0, 0)$, which yields $0 < -1$, a false statement. Shade the side of the line not containing $(0, 0)$.
Next, graph $x - y = 3$ as a dashed line through $(3, 0)$ and $(0, -3)$. Test $(0, 0)$, which yields $0 < 3$, a true statement. Shade the side of the line containing $(0, 0)$. The intersection is the region where the graphs overlap.

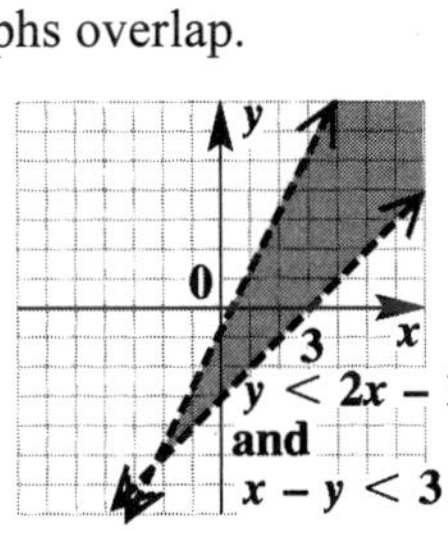

19. Choice **D** is the only graph that passes the vertical line test.

20. Choice **D** does not define a function, since its domain (input) element 0 is paired with two different range (output) elements, 1 and 2.

21. **(a)** The x-values are greater than or equal to zero, so the domain is $[0, \infty)$. Since y can be any value, the range is $(-\infty, \infty)$.

(b) The domain is the set of x-values: $\{0, -2, 4\}$. The range is the set of y-values: $\{1, 3, 8\}$.

22. $f(x) = -x^2 + 2x - 1$
$f(1) = -(1)^2 + 2(1) - 1$
$= -1 + 2 - 1$
$= 0$
$f(a) = -a^2 + 2a - 1$

23. $f(x) = \frac{2}{3}x - 1$
This function represents a line with y-intercept $(0, -1)$ and x-intercept $\left(\frac{3}{2}, 0\right)$.
Draw the line through these two points.
The domain is $(-\infty, \infty)$, and the range is $(-\infty, \infty)$.

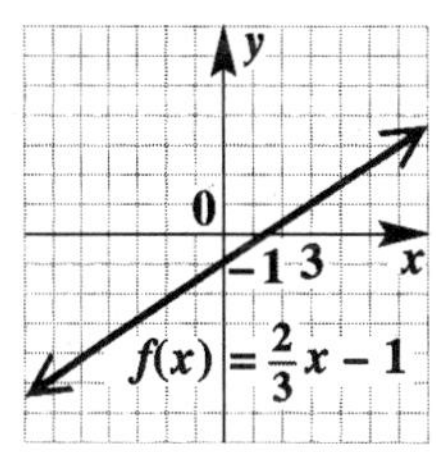

24. The current I is inversely proportional to the resistance R, so

$$I = \frac{k}{R}$$

for some constant k. Let $I = 80$ and $R = 30$. Find k.

$$80 = \frac{k}{30}$$
$$k = 80(30) = 2400$$

So $I = \frac{2400}{R}$. Now let $R = 12$.

$$I = \frac{2400}{12} = 200$$

The current is 200 amps.

25. The force F of the wind blowing on a vertical surface varies jointly as the area A of the surface and the square of the velocity V, so

$$F = kAV^2$$

for some constant k. Let $F = 50$, $A = 500$, and $V = 40$. Find k.

continued

$$50 = k(500)(40)^2$$
$$k = \frac{50}{500(1600)} = \frac{1}{16{,}000}$$

So $F = \frac{1}{16{,}000}AV^2$. Now let $A = 2$ and $V = 80$.

$$F = \frac{1}{16{,}000}(2)(80)^2 = .8$$

The force of the wind is .8 lb.

Cumulative Review Exercises Chapters 1–3

1. The absolute value of a negative number is a positive number and the additive inverse of the same negative number is the same positive number. For example, suppose the negative number is -5:

$$|-5| = -(-5) = 5$$
$$\text{and} \quad -(-5) = 5$$

The statement is *always true*.

2. The statement is *always true*; in fact, it is the definition of a rational number.

3. The sum of two negative numbers is another negative number, so the statement is *never true*.

4. The statement is *sometimes true*. For example,

$$3 + (-3) = 0,$$
$$\text{but} \quad 3 + (-1) = 2 \neq 0.$$

5. $$\begin{aligned} -|-2| - 4 + |-3| + 7 &= -2 - 4 + 3 + 7 \\ &= -6 + 3 + 7 \\ &= -3 + 7 \\ &= 4 \end{aligned}$$

6. $(-.8)^2 = (-.8)(-.8) = .64$

7. $\sqrt{-64}$ is not a real number.

8. The product of two numbers that have the same sign is positive.

$$-\frac{2}{3}\left(-\frac{12}{5}\right) = \frac{2 \cdot 12}{3 \cdot 5} = \frac{2 \cdot 4}{5} = \frac{8}{5}$$

9. $$\begin{aligned} -(-4m + 3) &= -(-4m) - 3 \\ &= 4m - 3 \end{aligned}$$

10. $$\begin{aligned} &3x^2 - 4x + 4 + 9x - x^2 \\ &= 3x^2 - x^2 - 4x + 9x + 4 \\ &= 2x^2 + 5x + 4 \end{aligned}$$

11. $$\begin{aligned} \frac{3\left(\sqrt{16}\right) - (-1)(7)}{4 + (-6)} &= \frac{3(4) - (-7)}{-2} \\ &= \frac{12 + 7}{-2} = -\frac{19}{2} \end{aligned}$$

12. $-3 < x \leq 5$
This is the set of numbers between -3 and 5, not including -3 (use a parenthesis), but including 5 (use a bracket). In interval notation, the set is $(-3, 5]$.

13. $$\sqrt{\frac{-2+4}{-5}} = \sqrt{\frac{2}{-5}} = \sqrt{-\frac{2}{5}}$$

This is not a real number since the number under the radical sign is negative.

For Exercises 14–16, let $p = -4$, $q = -2$, and $r = 5$.

14. $$\begin{aligned} -3(2q - 3p) &= -3[2(-2) - 3(-4)] \\ &= -3(-4 + 12) \\ &= -3(8) \\ &= -24 \end{aligned}$$

15. $$\begin{aligned} |p|^3 - |q^3| &= |-4|^3 - |(-2)^3| \\ &= 4^3 - |-8| \\ &= 64 - 8 \\ &= 56 \end{aligned}$$

16. $$\begin{aligned} \frac{\sqrt{r}}{-p + 2q} &= \frac{\sqrt{5}}{-(-4) + 2(-2)} \\ &= \frac{\sqrt{5}}{4 - 4} \\ &= \frac{\sqrt{5}}{0} \end{aligned}$$

This expression is undefined since the denominator is zero.

17. $$\begin{aligned} 2z - 5 + 3z &= 4 - (z + 2) \\ 5z - 5 &= 4 - z - 2 \\ 5z - 5 &= 2 - z \\ 6z &= 7 \\ z &= \frac{7}{6} \end{aligned}$$

Solution set: $\left\{\frac{7}{6}\right\}$

18. $$\frac{3a - 1}{5} + \frac{a + 2}{2} = -\frac{3}{10}$$

Multiply both sides by the LCD, 10.

$$\begin{aligned} 10\left(\frac{3a - 1}{5} + \frac{a + 2}{2}\right) &= 10\left(-\frac{3}{10}\right) \\ 2(3a - 1) + 5(a + 2) &= -3 \\ 6a - 2 + 5a + 10 &= -3 \\ 11a + 8 &= -3 \\ 11a &= -11 \\ a &= -1 \end{aligned}$$

Solution set: $\{-1\}$

19. Solve $V = \frac{1}{3}\pi r^2 h$ for h.

$3V = \pi r^2 h$ *Multiply by 3.*

$\frac{3V}{\pi r^2} = h$ *Divide by πr^2.*

20. $F = \frac{9}{5}C + 32$

$$F = \frac{9}{5}(-55) + 32 \qquad \textit{Let } C = -55.$$
$$= -99 + 32 = -67$$

$-55°C$ is equivalent to $-67°F$.

21. Let x denote the side of the original square and $4x$ the perimeter. Now $x + 4$ is the side of the new square and $4(x + 4)$ is its perimeter.
"The perimeter would be 8 inches less than twice the perimeter of the original square " translates as

$$\begin{aligned} 4(x + 4) &= 2(4x) - 8. \\ 4x + 16 &= 8x - 8 \\ 24 &= 4x \\ 6 &= x \end{aligned}$$

The length of a side of the original square is 6 inches.

22. Let $x =$ the time it takes for the planes to be 2100 miles apart.
Make a table. Use the formula $d = rt$.

	r	*t*	*d*
Eastbound Plane	550	x	$550x$
Westbound Plane	500	x	$500x$

The total distance is 2100 miles.

$$\begin{aligned} 550x + 500x &= 2100 \\ 1050x &= 2100 \\ x &= 2 \end{aligned}$$

It will take 2 hr for the planes to be 2100 mi apart.

23. $-4 < 3 - 2k < 9$
$-7 < -2k < 6$
Divide by -2; reverse the inequality symbols.
$\frac{7}{2} > k > -3$ or $-3 < k < \frac{7}{2}$

Solution set: $\left(-3, \frac{7}{2}\right)$

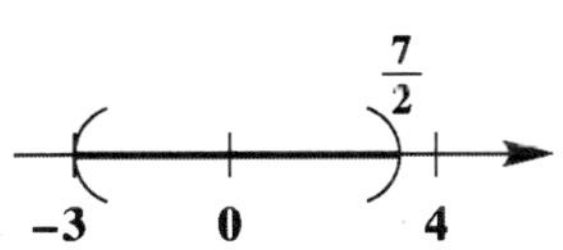

24. $-.3x + 2.1(x - 4) \le -6.6$
$3x + 21(x - 4) \le -66$
Multiply by 10.
$$\begin{aligned} -3x + 21x - 84 &\le -66 \\ 18x - 84 &\le -66 \\ 18x &\le 18 \\ x &\le 1 \end{aligned}$$

Solution set: $(-\infty, 1]$

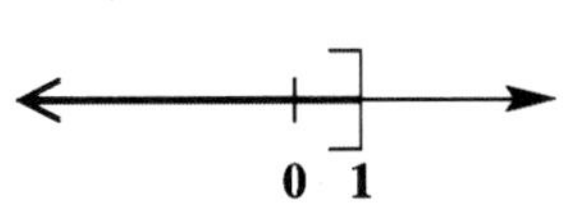

25. $x > 6$ and $x < 8$
The graph of the solution set is all numbers both greater than 6 *and* less than 8. This is the intersection. The elements common to both sets are the numbers between 6 and 8, not including the endpoints. The solution set is $(6, 8)$.

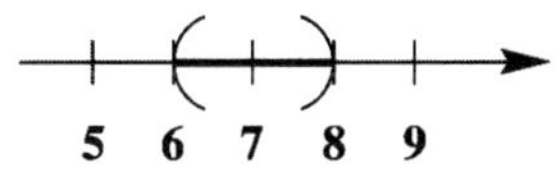

26. $-5x + 1 \ge 11$ or $3x + 5 > 26$
$-5x \ge 10$ $\qquad$ $3x > 21$
$x \le -2$ or $x > 7$
The graph of the solution set is all numbers either less than or equal to -2 *or* greater than 7. This is the union. The solution set is $(-\infty, -2] \cup (7, \infty)$.

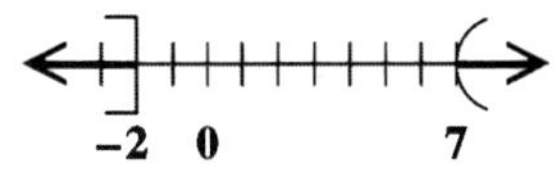

27. $|2k - 7| + 4 = 11$
$|2k - 7| = 7$
$2k - 7 = 7$ or $2k - 7 = -7$
$2k = 14$ $\qquad$ $2k = 0$
$k = 7$ or $k = 0$
Solution set: $\{0, 7\}$

28. $|3m + 6| \ge 0$
The absolute value of an expression is always nonnegative, so the inequality is true for any real number m.
Solution set: $(-\infty, \infty)$

29. The union of the three solution sets is $(-\infty, \infty)$; that is, the set of all real numbers.

30. To complete the table of ordered pairs, substitute the given values for x or y in the equation $3x - 4y = 12$.

For $x = 0$:
$$\begin{aligned} 3x - 4y &= 12 \\ 3(0) - 4y &= 12 \\ -4y &= 12 \\ y &= -3 \end{aligned}$$

The ordered pair is $(0, -3)$.

For $y = 0$:
$$\begin{aligned} 3x - 4y &= 12 \\ 3x - 4(0) &= 12 \\ 3x &= 12 \\ x &= 4 \end{aligned}$$

The ordered pair is $(4, 0)$.

For $x = 2$:
$$\begin{aligned} 3x - 4y &= 12 \\ 3(2) - 4y &= 12 \\ 6 - 4y &= 12 \\ -4y &= 6 \\ y &= \frac{6}{-4} = -\frac{3}{2} \end{aligned}$$

The ordered pair is $\left(2, -\frac{3}{2}\right)$.

31. $3x + 5y = 12$
To find the x-intercept, let $y = 0$.

$$\begin{aligned} 3x + 5(0) &= 12 \\ 3x &= 12 \\ x &= 4 \end{aligned}$$

The x-intercept is $(4, 0)$.
To find the y-intercept, let $x = 0$.

$$\begin{aligned} 3(0) + 5y &= 12 \\ 5y &= 12 \\ y &= \frac{12}{5} \end{aligned}$$

The y-intercept is $\left(0, \frac{12}{5}\right)$.

Plot the intercepts and draw the line through them.

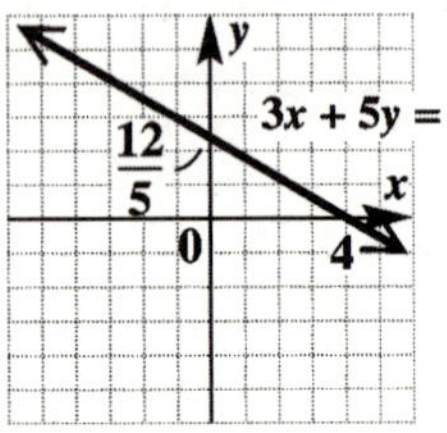

32. $A(-2, 1)$ and $B(3, -5)$
(a) The slope of line *AB* is

$$m = \frac{1-(-5)}{-2-3} = \frac{6}{-5} = -\frac{6}{5}.$$

(b) The slope of a line perpendicular to line *AB* is the negative reciprocal of $-\frac{6}{5}$, which is $\frac{5}{6}$.

33. $-2x + y < -6$
Graph the line $-2x + y = -6$, which has intercepts $(3, 0)$ and $(0, -6)$, as a dashed line since the inequality involves $<$. Test $(0, 0)$, which yields $0 < -6$, a false statement. Shade the region that does not include $(0, 0)$.

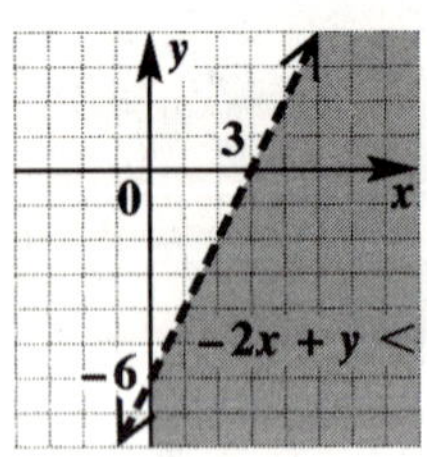

34. Slope $-\frac{3}{4}$; y-intercept $(0, -1)$

To write an equation of this line, let $m = -\frac{3}{4}$ and $b = -1$ in the slope-intercept form.

$$\begin{aligned} y &= mx + b \\ y &= -\frac{3}{4}x - 1 \end{aligned}$$

35. Horizontal; through $(2, -2)$
A horizontal line through the point (c, d) has equation $y = d$. Here $d = -2$, so the equation of the line is $y = -2$.

36. Through $(4, -3)$ and $(1, 1)$
First find the slope of the line.

$$m = \frac{1-(-3)}{1-4} = \frac{4}{-3} = -\frac{4}{3}$$

Now substitute $(x_1, y_1) = (4, -3)$ and $m = -\frac{4}{3}$ in the point-slope form. Then solve for y.

$$\begin{aligned} y - y_1 &= m(x - x_1) \\ y - (-3) &= -\frac{4}{3}(x - 4) \\ y + 3 &= -\frac{4}{3}x + \frac{16}{3} \\ y &= -\frac{4}{3}x + \frac{7}{3} \end{aligned}$$

37. The domain of the relation consists of the elements in the leftmost figure; that is, $\{14, 91, 75, 23\}$.

The range of the relation consists of the elements in the rightmost figure; that is, $\{9, 70, 56, 5\}$.

Since the element 75 in the domain is paired with two different values, 70 and 56, in the range, the relation is not a function.

38. $f(x) = -4x + 10$

(a) The variable x can be any real number, so the domain is $(-\infty, \infty)$. The function is a non-constant linear function, so its range is $(-\infty, \infty)$.

(b) $f(-3) = -4(-3) + 10 = 12 + 10 = 22$

39. Use $(1992, 11)$ and $(2000, 95)$.

$$m = \frac{\text{change in } y}{\text{change in } x} = \frac{95 - 11}{2000 - 1992} = \frac{84}{8} = 10.5$$

So the average rate of change is 10.5 million per year; that is, the number of U.S. cell phone subscribers increased by 10.5 million per year from 1992 to 2000.

40. Let $C =$ the cost of a pizza
and $r =$ the radius of the pizza.
C varies directly as r^2, so

$$C = kr^2$$

for some constant k. Since $C = 6$ when $r = 7$, substitute these values in the equation and solve for k.

$$\begin{aligned} C &= kr^2 \\ 6 &= k(7)^2 \\ 6 &= 49k \\ \frac{6}{49} &= k \end{aligned}$$

So $C = \frac{6}{49}r^2$. Now let $r = 9$.

$$C = \frac{6}{49}(9)^2 = \frac{6}{49}(81) = \frac{486}{49} \approx 9.92.$$

A pizza with a 9-inch radius should cost \$9.92.

CHAPTER 4 SYSTEMS OF LINEAR EQUATIONS

Section 4.1

1. If $(3, -6)$ is a solution of a linear system in two variables, then substituting $\underline{3}$ for x and $\underline{-6}$ for y leads to true statements in *both* equations.

3. If the solution process leads to a false statement such as $0 = 5$ when solving a system, the solution set is $\underline{\emptyset}$.

5. If the two lines forming a system have the same slope and different y-intercepts, the system has $\underline{0}$ solutions. (The lines are parallel.)

7. **D**; The ordered pair solution must be in quadrant IV, since that is where the graphs of the equations intersect.

9. **(a)** $x - y = 0$ implies that $x = y$, so the coordinates of the solution must be equal—this limits our choice to **B** or **C**. Since $x + y = 6$, graph **B** is correct.

(b) As in part (a), we must have **B** or **C**. Since $x + y = -6$, graph **C** is correct.

(c) $x + y = 0$ implies that $x = -y$, so the coordinates of the solution must be opposites—this limits our choice to **A** or **D**. Since $x - y = -6$, graph **A** is correct.

(d) As in part (c), we must have **A** or **D**. Since $x - y = 6$, graph **D** is correct.

11. $x - y = 17$
$x + y = -1$
To decide if $(8, -9)$ is a solution, replace x with 8 and y with -9 in each equation of the system.

$$\begin{aligned} x - y &= 17 \\ 8 - (-9) &= 17 \\ 17 &= 17 \quad \textit{True} \end{aligned}$$

$$\begin{aligned} x + y &= -1 \\ 8 + (-9) &= -1 \\ -1 &= -1 \quad \textit{True} \end{aligned}$$

Since $(8, -9)$ makes both equations true, $(8, -9)$ is a solution of the system.

13. $3x - 5y = -12$
$x - y = 1$
Replace x with -1 and y with 2.

$$\begin{aligned} 3(-1) - 5(2) &= -12 \\ -3 - 10 &= -12 \\ -13 &= -12 \quad \textit{False} \end{aligned}$$

It is not necessary to check if $(-1, 2)$ satisfies the second equation since it does not satisfy the first equation. Therefore, $(-1, 2)$ is *not* a solution of the system.

15. $x + y = -5$
$-2x + y = 1$
Graph the line $x + y = -5$ through its intercepts, $(-5, 0)$ and $(0, -5)$, and the line $-2x + y = 1$ through its intercepts, $\left(-\frac{1}{2}, 0\right)$ and $(0, 1)$. The lines appear to intersect at $(-2, -3)$.

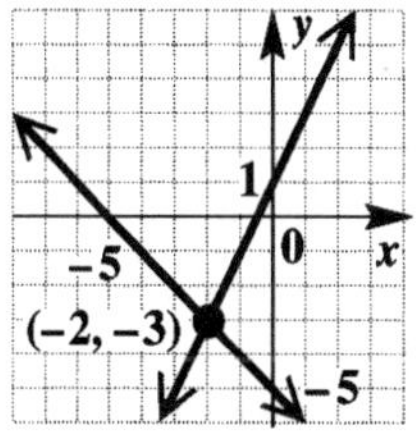

Check this ordered pair in the system.

$$\begin{aligned} (-2) + (-3) &= -5 \\ -5 &= -5 \quad \textit{True} \end{aligned}$$

$$\begin{aligned} -2(-2) + (-3) &= 1 \\ 4 + (-3) &= 1 \quad \textit{True} \end{aligned}$$

Solution set: $\{(-2, -3)\}$

17. $2x - 5y = 11 \quad (1)$
$3x + y = 8 \quad (2)$
To eliminate y, multiply equation (2) by 5 and add the result to equation (1).

$$\begin{array}{rcrcrl} 2x & - & 5y & = & 11 & (1) \\ 15x & + & 5y & = & 40 & (3)\quad 5 \times (2) \\ \hline 17x & & & = & 51 & \\ x & & & = & 3 & \end{array}$$

To find y, substitute 3 for x in equation (2).

$$\begin{aligned} 3x + y &= 8 \quad (2) \\ 3(3) + y &= 8 \\ 9 + y &= 8 \\ y &= -1 \end{aligned}$$

The ordered pair $(3, -1)$ satisfies both equations, so it checks.

Solution set: $\{(3, -1)\}$

19. $3x + 4y = -6 \quad (1)$
$5x + 3y = 1 \quad (2)$
To eliminate x, multiply equation (1) by 5 and equation (2) by -3. Then add the results.

$$\begin{array}{rcrcrl} 15x & + & 20y & = & -30 & (3)\quad 5 \times (1) \\ -15x & - & 9y & = & -3 & (4)\quad -3 \times (2) \\ \hline & & 11y & = & -33 & \\ & & y & = & -3 & \end{array}$$

To find x, substitute -3 for y in equation (2).

continued

$$\begin{aligned} 5x + 3y &= 1 \quad (2) \\ 5x + 3(-3) &= 1 \\ 5x - 9 &= 1 \\ 5x &= 10 \\ x &= 2 \end{aligned}$$

The ordered pair $(2, -3)$ satisfies both equations, so it checks.

Solution set: $\{(2, -3)\}$

21. $3x + 3y = 0 \quad (1)$
$4x + 2y = 3 \quad (2)$
To eliminate y, multiply equation (1) by 2 and equation (2) by -3. Then add the results.

$$\begin{array}{rcrcrl} 6x & + & 6y & = & 0 & (3) \quad 2 \times (1) \\ -12x & - & 6y & = & -9 & (4) \quad -3 \times (2) \\ \hline -6x & & & = & -9 & \end{array}$$

$$x = \frac{-9}{-6} = \frac{3}{2}$$

To find y, substitute $\frac{3}{2}$ for x in equation (1).

$$\begin{aligned} 3x + 3y &= 0 \quad (1) \\ 3\left(\frac{3}{2}\right) + 3y &= 0 \\ \frac{9}{2} + 3y &= 0 \\ 3y &= -\frac{9}{2} \\ y &= \frac{1}{3}\left(-\frac{9}{2}\right) = -\frac{3}{2} \end{aligned}$$

The solution $\left(\frac{3}{2}, -\frac{3}{2}\right)$ checks.

Solution set: $\left\{\left(\frac{3}{2}, -\frac{3}{2}\right)\right\}$

When you get a solution that has non-integer components, it is sometimes more difficult to check the problem than it was to solve it. A graphing calculator can be very helpful in this case. Just store the values for x and y in their respective memory locations, and then type the expressions as shown in the following screen. The results 0 and 3 (the right sides of the equations) indicate that we have found the correct solution.

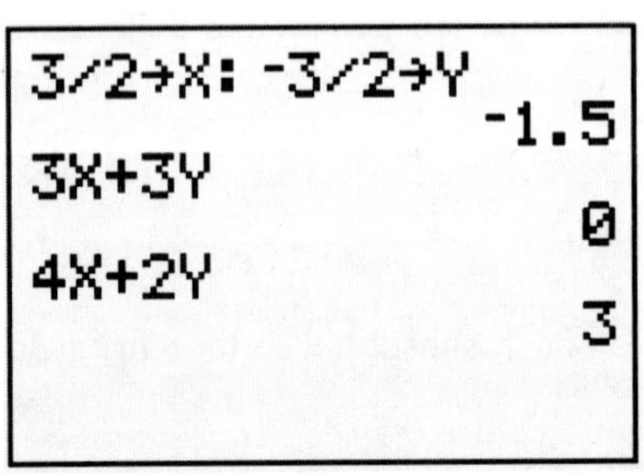

23. $7x + 2y = 6 \quad (1)$
$-14x - 4y = -12 \quad (2)$
To eliminate y, multiply equation (1) by 2 and add the result to equation (2).

$$\begin{array}{rcrcrl} 14x & + & 4y & = & 12 & (3) \quad 2 \times (1) \\ -14x & - & 4y & = & -12 & (2) \\ \hline & & 0 & = & 0 & \textit{True} \end{array}$$

Multiplying equation (1) by -2 gives equation (2). The equations are dependent, and the solution is the set of all points on the line.

Solution set: $\{(x, y) | 7x + 2y = 6\}$

25. $5x - 5y = 3 \quad (1)$
$x - y = 12 \quad (2)$
To eliminate x, multiply equation (2) by -5 and add the result to equation (1).

$$\begin{array}{rcrcrl} 5x & - & 5y & = & 3 & (1) \\ -5x & + & 5y & = & -60 & (3) \quad -5 \times (2) \\ \hline & & 0 & = & -57 & \textit{False} \end{array}$$

The system is *inconsistent*. Since the graphs of the equations are parallel lines, there are no ordered pairs that satisfy both equations.

Solution set: $\emptyset$

27. $x - \frac{1}{2}y = 2 \quad (1)$
$-x + \frac{2}{5}y = -\frac{8}{5} \quad (2)$

To eliminate x, add equations (1) and (2) to get

$$-\frac{1}{2}y + \frac{2}{5}y = 2 - \frac{8}{5}.$$

Multiply by the LCD, 10.

$$\begin{aligned} -5y + 4y &= 20 - 16 \\ -y &= 4 \\ y &= -4 \end{aligned}$$

Substitute -4 for y in equation (1) to find x.

$$\begin{aligned} x - \frac{1}{2}(-4) &= 2 \\ x + 2 &= 2 \\ x &= 0 \end{aligned}$$

The solution $(0, -4)$ checks.

Solution set: $\{(0, -4)\}$

29. $x + y = 0 \quad (1)$
$2x - 2y = 0 \quad (2)$

To eliminate y, multiply equation (1) by 2 and add the result to equation (2).

$$\begin{array}{rcrcrl} 2x & + & 2y & = & 0 & (3) \quad 2 \times (1) \\ 2x & - & 2y & = & 0 & (2) \\ \hline 4x & & & = & 0 & \\ & & x & = & 0 & \end{array}$$

Substitute 0 for x in (1) to get $y = 0$. The solution $(0, 0)$ checks.

Solution set: $\{(0, 0)\}$

31. $\frac{1}{2}x + \frac{1}{3}y = -\frac{1}{3}$ (1)

$\frac{1}{2}x + 2y = -7$ (2)

Eliminate the fractions by multiplying equation (1) by -6 and equation (2) by 6. Then add the results to eliminate x.

$$\begin{array}{rcrl} -3x - 2y &=& 2 & (3) \quad -6 \times (1) \\ 3x + 12y &=& -42 & (4) \quad 6 \times (2) \\ \hline 10y &=& -40 & \\ y &=& -4 & \end{array}$$

To find x, substitute -4 for y in equation (3).

$$\begin{aligned} -3x - 2y &= 2 \quad (3) \\ -3x - 2(-4) &= 2 \\ -3x + 8 &= 2 \\ -3x &= -6 \\ x &= 2 \end{aligned}$$

The solution $(2, -4)$ checks.

Solution set: $\{(2, -4)\}$

33. $3x + 7y = 4$ (1)

$6x + 14y = 3$ (2)

Write each equation in slope-intercept form by solving for y.

$$\begin{aligned} 3x + 7y &= 4 \quad (1) \\ 7y &= -3x + 4 \\ y &= -\frac{3}{7}x + \frac{4}{7} \end{aligned}$$

$$\begin{aligned} 6x + 14y &= 3 \quad (2) \\ 14y &= -6x + 3 \\ y &= -\frac{6}{14}x + \frac{3}{14} \\ y &= -\frac{3}{7}x + \frac{3}{14} \end{aligned}$$

Since the equations have the same slope, $-\frac{3}{7}$, but different y-intercepts, $\frac{4}{7}$ and $\frac{3}{14}$, the lines when graphed are parallel. The system is inconsistent and has no solution.

35. $2x = -3y + 1$ (1)

$6x = -9y + 3$ (2)

Write each equation in slope-intercept form by solving for y.

$$\begin{aligned} 2x &= -3y + 1 \quad (1) \\ 3y &= -2x + 1 \\ y &= -\frac{2}{3}x + \frac{1}{3} \end{aligned}$$

$$\begin{aligned} 6x &= -9y + 3 \quad (2) \\ 9y &= -6x + 3 \\ y &= -\frac{6}{9}x + \frac{3}{9} \\ y &= -\frac{2}{3}x + \frac{1}{3} \end{aligned}$$

Since both equations are the same, the solution set is all points on the line $y = -\frac{2}{3}x + \frac{1}{3}$. The system has infinitely many solutions.

37. **(a)** The graphs of these two linear equations will intersect once.

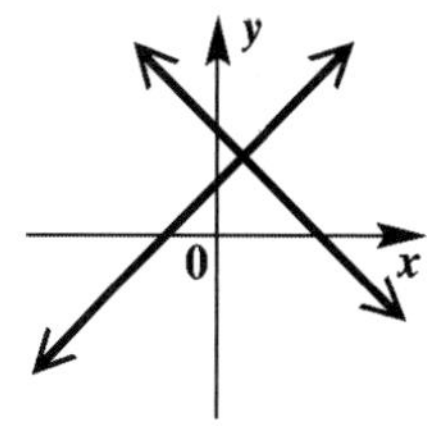

(b) The graphs of these two linear equations will not intersect. They are parallel lines.

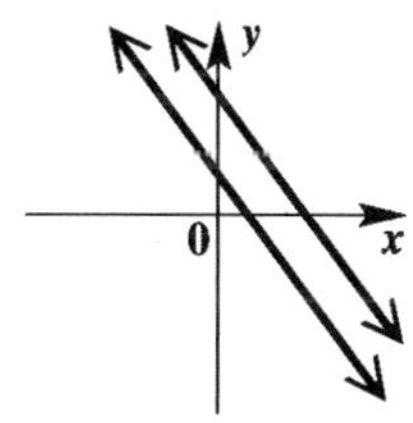

(c) The graphs of these two linear equations will be the same line.

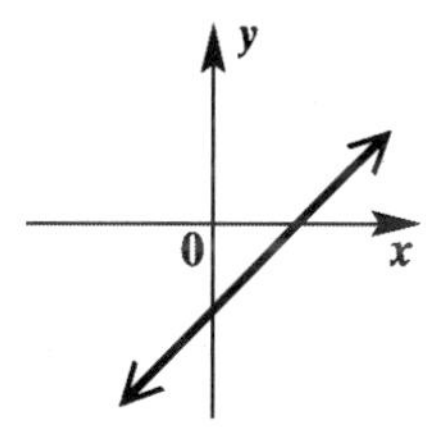

39. $4x + y = 6$ (1)

$y = 2x$ (2)

Since equation (2) is already solved for y, substitute $2x$ for y in equation (1).

$$\begin{aligned} 4x + y &= 6 \quad (1) \\ 4x + 2x &= 6 \quad \textit{Let } y=2x. \\ 6x &= 6 \\ x &= 1 \end{aligned}$$

Substitute 1 for x in (2).

$$y = 2(1) = 2$$

The solution $(1, 2)$ checks.

Solution set: $\{(1, 2)\}$

41. $3x - 4y = -22$ (1)
$-3x + y = 0$ (2)
Solve equation (2) for y to get

$$y = 3x.$$

Substitute $3x$ for y in equation (1).

$$\begin{aligned} 3x - 4y &= -22 && (1) \\ 3x - 4(3x) &= -22 && \textit{Let } y=3x. \\ 3x - 12x &= -22 \\ -9x &= -22 \\ x &= \frac{-22}{-9} = \frac{22}{9} \end{aligned}$$

Substitute $\frac{22}{9}$ for x in $y = 3x$ to get

$$y = 3\left(\frac{22}{9}\right) = \frac{22}{3}.$$

The solution $\left(\frac{22}{9}, \frac{22}{3}\right)$ checks.

Solution set: $\left\{\left(\frac{22}{9}, \frac{22}{3}\right)\right\}$

43. $-x - 4y = -14$ (1)
$2x - y = 1$ (2)
Solve equation (2) for y.

$$\begin{aligned} 2x &= y + 1 \\ 2x - 1 &= y && (3) \end{aligned}$$

Substitute $2x - 1$ for y in equation (1) and solve for x.

$$\begin{aligned} -x - 4y &= -14 && (1) \\ -x - 4(2x - 1) &= -14 \\ -x - 8x + 4 &= -14 \\ -9x &= -18 \\ x &= 2 \end{aligned}$$

Substitute 2 for x in (3).

$$y = 2(2) - 1 = 3$$

The solution $(2, 3)$ checks.

Solution set: $\{(2, 3)\}$

45. $5x - 4y = 9$ (1)
$3 - 2y = -x$ (2)
Solve equation (2) for x.

$$\begin{aligned} 3 - 2y &= -x && (2) \\ 2y - 3 &= x && (3) \quad -1 \times (2) \end{aligned}$$

Substitute $2y - 3$ for x in equation (1).

$$\begin{aligned} 5x - 4y &= 9 && (1) \\ 5(2y - 3) - 4y &= 9 && \textit{Let } x=2y-3. \\ 10y - 15 - 4y &= 9 \\ 6y - 15 &= 9 \\ 6y &= 24 \\ y &= 4 \end{aligned}$$

Substitute 4 for y in $x = 2y - 3$.

$$x = 2(4) - 3 = 5$$

The solution $(5, 4)$ checks.
Solution set: $\{(5, 4)\}$

47. $x = 3y + 5$ (1)
$x = \frac{3}{2}y$ (2)
Both equations are given in terms of x. Choose equation (2), and substitute $\frac{3}{2}y$ for x in equation (1).

$$\begin{aligned} x &= 3y + 5 && (1) \\ \frac{3}{2}y &= 3y + 5 && \textit{Let } x = \tfrac{3}{2}y. \\ 3y &= 6y + 10 && \textit{Multiply by 2.} \\ -3y &= 10 \\ y &= -\frac{10}{3} \end{aligned}$$

Since $x = \frac{3}{2}y$ and $y = -\frac{10}{3}$,

$$x = \frac{3}{2}\left(-\frac{10}{3}\right) = -\frac{10}{2} = -5.$$

The solution $\left(-5, -\frac{10}{3}\right)$ checks.

Solution set: $\left\{\left(-5, -\frac{10}{3}\right)\right\}$

49. $\frac{1}{2}x + \frac{1}{3}y = 3$ (1)
$-3x + y = 0$ (2)
Multiply (1) by its LCD, 6, to eliminate fractions

$$3x + 2y = 18 \qquad (3)$$

From (2), we have $y = 3x$, so substitute $3x$ for y in (3).

$$\begin{aligned} 3x + 2(3x) &= 18 \\ 3x + 6x &= 18 \\ 9x &= 18 \\ x &= 2 \end{aligned}$$

Substitute 2 for x in $y = 3x$.

$$y = 3(2) = 6$$

The solution $(2, 6)$ checks.

Solution set: $\{(2, 6)\}$

51.
$$\begin{aligned} y &= 2x \quad &(1)\\ 4x - 2y &= 0 \quad &(2)\end{aligned}$$

From equation (1), substitute $2x$ for y in equation (2).

$$\begin{aligned} 4x - 2y &= 0 \quad (2)\\ 4x - 2(2x) &= 0\\ 4x - 4x &= 0\\ 0 &= 0 \quad \textit{True}\end{aligned}$$

The equations are dependent, and the solution is the set of all points on the line.

Solution set: $\{(x, y) | \, y = 2x\}$

53.
$$\begin{aligned} x &= 5y \quad &(1)\\ 5x - 25y &= 5 \quad &(2)\end{aligned}$$

From equation (1), substitute $5y$ for x in equation (2).

$$\begin{aligned} 5x - 25y &= 5 \quad (2)\\ 5(5y) - 25y &= 5\\ 25y - 25y &= 5\\ 0 &= 5 \quad \textit{False}\end{aligned}$$

The system is *inconsistent*. Since the graphs of the equations are parallel lines, there are no ordered pairs that satisfy both equations.

Solution set: $\emptyset$

55. The table shows that when $\text{X} = 3$, $\text{Y}_1 = -4$ and $\text{Y}_2 = -4$. Since no other values of X in the table give the same values for Y_1 and Y_2, and since the functions are linear, the point $(3, -4)$ is the only point of intersection for the two graphs.

57. $y_1 = 3x - 5$

$y_2 = -4x + 2$

Both of the graphs in **B** have negative y-intercepts. But the y-intercept of y_2 is positive, so the graph in **B** is not acceptable. The slope of y_1 is positive, and y_1 has a negative y-intercept. The slope of y_2 is negative, and y_2 has a positive y-intercept. This fits graph **A**.

59. **(a)**
$$\begin{array}{rcrcrl} x & + & y & = & 10 & (1)\\ 2x & - & y & = & 5 & (2)\\ \hline 3x & & & = & 15 & \\ & & x & = & 5 & \end{array}$$

Substitute 5 for x in equation (1).

$$\begin{aligned} x + y &= 10 \quad (1)\\ 5 + y &= 10\\ y &= 5\end{aligned}$$

Solution set: $\{(5, 5)\}$

(b) Solve (1) and (2) for y.

$$\begin{aligned} x + y &= 10 \quad &(1)\\ y &= -x + 10 \quad &(3)\end{aligned}$$

$$\begin{aligned} 2x - y &= 5 \quad &(2)\\ 2x - 5 &= y \quad &(4)\end{aligned}$$

Now graph (3) and (4).

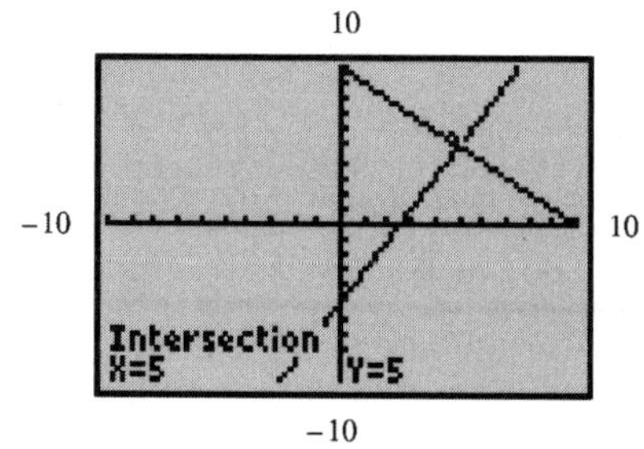

61. **(a)**
$$\begin{array}{rcrcrl} 3x & - & 2y & = & 4 & (1)\\ 3x & + & y & = & -2 & (2)\end{array}$$

Multiply equation (2) by -1 and add the result to equation (1).

$$\begin{array}{rcrcrll} 3x & - & 2y & = & 4 & (1) & \\ -3x & - & y & = & 2 & (3) & -1 \times (2)\\ \hline & & 3y & = & 6 & & \\ & & y & = & -2 & & \end{array}$$

Substitute -2 for y in equation (1).

$$\begin{aligned} 3x - 2y &= 4 \quad (1)\\ 3x - 2(-2) &= 4\\ 3x + 4 &= 4\\ 3x &= 0\\ x &= 0\end{aligned}$$

Solution set: $\{(0, -2)\}$

(b) Solving (1) and (2) for y gives us

$$y = \frac{3}{2}x - 2 \quad (3)$$

and $\quad y = -3x - 2. \quad (4)$

Now graph (3) and (4).

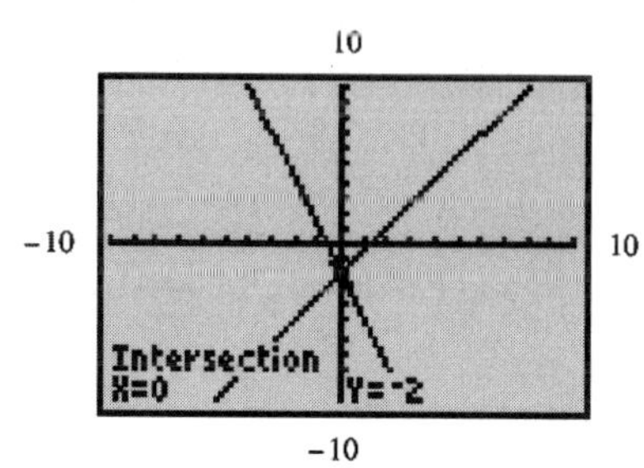

63. **(a)** The monthly payment for the fixed-rate mortgage is about \$650, which is more than the monthly payment for the variable-rate mortgage for years 0 to 6.

(b) The graphs intersect at year 6, so the monthly payments would be the same in year 6, and they are about \$650.

65. **(a)** The graph for ABC is above the others for years 89–97. Therefore, ABC dominated between 1989 and 1997.

(b) ABC and NBC intersect at $(1997, 17)$. Therefore, ABC's dominance ended in 1997. NBC equaled ABC's share of 17%.

(c) The graphs of ABC and CBS intersect at $(1989, 20)$ and $(1998, 16)$. In 1989, with a share of 20% and in 1998, with a share of 16%, ABC and CBS had equal shares.

(d) The graphs of NBC and ABC intersect at $(2000, 16)$, so they had equal shares of 16% in 2000.

(e) Viewership has generally declined during these years.

67. The graph for Hewlett Packard is below that of Packard Bell, NEC for the years 1995–1997.

69.
$$\begin{aligned} y &= -.5x + 7 \quad (1) \\ y &= x + 3.5 \quad (2) \end{aligned}$$

From (1), substitute $-.5x + 7$ for y in equation (2).

$$\begin{aligned} -.5x + 7 &= x + 3.5 \\ -x + 14 &= 2x + 7 \quad \textit{Multiply by 2.} \\ 7 &= 3x \\ x &= \frac{7}{3} \end{aligned}$$

Substitute $\frac{7}{3}$ for x in equation (2).

$$y = \frac{7}{3} + \frac{7}{2} = \frac{14 + 21}{6} = \frac{35}{6}$$

Rounded to the nearest tenth, we have $x = 2.3$ and $y = 5.8$.

Solution set: $\{(2.3, 5.8)\}$

71.
$$\begin{aligned} \frac{3}{x} + \frac{4}{y} &= \frac{5}{2} \quad (1) \\ \frac{5}{x} - \frac{3}{y} &= \frac{7}{4} \quad (2) \end{aligned}$$

If $p = \frac{1}{x}$ and $q = \frac{1}{y}$, equations (1) and (2) can be written as

$$\begin{aligned} 3p + 4q &= \frac{5}{2} \quad (3) \\ 5p - 3q &= \frac{7}{4}. \quad (4) \end{aligned}$$

To eliminate q, multiply equation (3) by 3 and equation (4) by 4. Then add the results.

$$\begin{aligned} 9p + 12q &= \frac{15}{2} \quad (5)\ 3 \times (3) \\ 20p - 12q &= 7 \quad (6)\ 4 \times (4) \\ \hline 29p &= \frac{29}{2} \\ p &= \frac{1}{2} \end{aligned}$$

To find q, substitute $\frac{1}{2}$ for p in equation (3).

$$\begin{aligned} 3p + 4q &= \frac{5}{2} \quad (3) \\ 3\left(\frac{1}{2}\right) + 4q &= \frac{5}{2} \quad \textit{Let } p = \frac{1}{2}. \\ \frac{3}{2} + 4q &= \frac{5}{2} \\ 4q &= 1 \\ q &= \frac{1}{4} \end{aligned}$$

Since $p = \frac{1}{x}$ and $p = \frac{1}{2}$, $\frac{1}{2} = \frac{1}{x}$ and $x = 2$.

Since $q = \frac{1}{y}$ and $q = \frac{1}{4}$, $\frac{1}{4} = \frac{1}{y}$ and $y = 4$.

The solution $(2, 4)$ checks.

Solution set: $\{(2, 4)\}$

73.
$$\begin{aligned} \frac{2}{x} - \frac{5}{y} &= \frac{3}{2} \quad (1) \\ \frac{4}{x} + \frac{1}{y} &= \frac{4}{5} \quad (2) \end{aligned}$$

Let $p = \frac{1}{x}$ and $q = \frac{1}{y}$. Rewrite equations (1) and (2).

$$\begin{aligned} 2p - 5q &= \frac{3}{2} \quad (3) \\ 4p + q &= \frac{4}{5} \quad (4) \end{aligned}$$

To clear the fractions multiply equation (3) by 2 and equation (4) by 5.

$$\begin{aligned} 4p - 10q &= 3 \quad (5)\ 2 \times (3) \\ 20p + 5q &= 4 \quad (6)\ 5 \times (4) \end{aligned}$$

To eliminate q, multiply equation (6) by 2. Then add the result to equation (5).

$$\begin{aligned} 4p - 10q &= 3 \quad (5) \\ 40p + 10q &= 8 \quad (7)\ 2 \times (6) \\ \hline 44p &= 11 \\ p &= \frac{1}{4} \end{aligned}$$

Substitute $\frac{1}{4}$ for p in equation (4).

$$\begin{aligned} 4p + q &= \frac{4}{5} \quad (4) \\ 4\left(\frac{1}{4}\right) + q &= \frac{4}{5} \quad \textit{Let } p = \frac{1}{4}. \end{aligned}$$

$$1+q=\frac{4}{5}$$
$$q=-\frac{1}{5}$$

Since $p=\frac{1}{x}$ and $p=\frac{1}{4}$, $\frac{1}{4}=\frac{1}{x}$ and $x=4$.
Since $q=\frac{1}{y}$ and $q=-\frac{1}{5}$, $-\frac{1}{5}=\frac{1}{y}$ and $y=-5$.
The solution $(4,-5)$ checks.
Solution set: $\{(4,-5)\}$

75. Since the coefficients of x, a and $-a$, are opposites, we can simply add the two equations to eliminate x.

$$\begin{array}{rcll} ax + by &=& 2 & (1)\\ -ax + 2by &=& 1 & (2)\\ \hline 3by &=& 3 & \\ y &=& \frac{3}{3b} = \frac{1}{b} & \end{array}$$

Substitute $\frac{1}{b}$ for y in equation (1).

$$ax+by=2 \quad (1)$$
$$ax+b\left(\frac{1}{b}\right)=2$$
$$ax+1=2$$
$$ax=1$$
$$x=\frac{1}{a}$$

The solution $\left(\frac{1}{a},\frac{1}{b}\right)$ checks.

Solution set: $\left\{\left(\frac{1}{a},\frac{1}{b}\right)\right\}$

77. $3ax+2y=1 \quad (1)$
$-ax+\ y=2 \quad (2)$
To eliminate x, multiply equation (2) by 3 and add the result to equation (1).

$$\begin{array}{rcll} 3ax + 2y &=& 1 & (1)\\ -3ax + 3y &=& 6 & (3)\quad 3\times(2)\\ \hline 5y &=& 7 & \\ y &=& \frac{7}{5} & \end{array}$$

Substitute $\frac{7}{5}$ for y in equation (2).

$$-ax+\frac{7}{5}=2$$
$$-ax=\frac{3}{5}$$
$$x=-\frac{3}{5a}$$

The solution $\left(-\frac{3}{5a},\frac{7}{5}\right)$ checks.

Solution set: $\left\{\left(-\frac{3}{5a},\frac{7}{5}\right)\right\}$

79. $3x+\ y=6 \quad (1)$
$-2x+3y=7 \quad (2)$
Multiply equation (1) by -3 and add the result to equation (2).

$$\begin{array}{rcll} -9x - 3y &=& -18 & (3)\quad -3\times(1)\\ -2x + 3y &=& 7 & (2)\\ \hline -11x &=& -11 & \\ x &=& 1 & \end{array}$$

To find y, substitute 1 for x in equation (1).

$$3x+y=6 \quad (1)$$
$$3(1)+y=6$$
$$y=3$$

The solution $(1,3)$ checks.

Solution set: $\{(1,3)\}$

80. $3x+y=6$
$y=-3x+6$
Replace y with $f(x)$.

$$f(x)=-3x+6$$

Since f is in the form $f(x)=mx+b$, it is a linear function.

81. $-2x+3y=7$
$3y=2x+7$
$y=\frac{2}{3}x+\frac{7}{3}$
Replace y with $g(x)$.

$$g(x)=\frac{2}{3}x+\frac{7}{3}$$

Since g is in the form $g(x)=mx+b$, it is a linear function.

82. Because the graphs of f and g are straight lines that are neither parallel nor coincide, they intersect in exactly <u>one</u> point. The coordinates of the point are (<u>1</u>, <u>3</u>). Using functional notation, this is given by $f(\underline{1})=\underline{3}$ and $g(\underline{1})=\underline{3}$.

83. Begin by multiplying one or both equations by a number that will lead to opposite coefficients for one variable in the system. Then add the two equations to eliminate that variable. Solve for the remaining variable. Once that value is found, find the other variable value by substituting back into one of the original equations. For example:

$$2x-y=9$$
$$x+2y=7.$$

Multiply the first equation by 2 and add to get $5x=25$, or $x=5$. Let $x=5$ in either equation to get $y=1$. The solution set is $\{(5,1)\}$.

Section 4.2

1. The statement means that when -1 is substituted for x, 2 is substituted for y, and 3 is substituted for z in the three equations, the resulting three statements are true.

3.
$$\begin{aligned} 2x - 5y + 3z &= -1 \quad (1)\\ x + 4y - 2z &= 9 \quad (2)\\ x - 2y - 4z &= -5 \quad (3)\end{aligned}$$

Eliminate x by adding equation (1) to -2 times equation (2).

$$\begin{aligned} 2x - 5y + 3z &= -1 \quad (1)\\ -2x - 8y + 4z &= -18 \quad (4) \quad -2\times(2)\\ \hline -13y + 7z &= -19 \quad (5)\end{aligned}$$

Now eliminate x by adding equation (2) to -1 times equation (3).

$$\begin{aligned} x + 4y - 2z &= 9 \quad (2)\\ -x + 2y + 4z &= 5 \quad (6) \quad -1\times(3)\\ \hline 6y + 2z &= 14 \quad (7)\end{aligned}$$

Use equations (5) and (7) to eliminate z. Multiply equation (5) by -2 and add the result to 7 times equation (7).

$$\begin{aligned} 26y - 14z &= 38 \quad -2\times(5)\\ 42y + 14z &= 98 \quad 7\times(7)\\ \hline 68y &= 136\\ y &= 2\end{aligned}$$

Substitute 2 for y in equation (7) to find z.

$$\begin{aligned} 6y + 2z &= 14 \quad (7)\\ 6(2) + 2z &= 14\\ 12 + 2z &= 14\\ 2z &= 2\\ z &= 1\end{aligned}$$

Substitute 2 for y and 1 for z in equation (3) to find x.

$$\begin{aligned} x - 2y - 4z &= -5 \quad (3)\\ x - 2(2) - 4(1) &= -5\\ x - 4 - 4 &= -5\\ x - 8 &= -5\\ x &= 3\end{aligned}$$

The solution $(3, 2, 1)$ checks in all three of the original equations.
Solution set: $\{(3, 2, 1)\}$

5.
$$\begin{aligned} 3x + 2y + z &= 8 \quad (1)\\ 2x - 3y + 2z &= -16 \quad (2)\\ x + 4y - z &= 20 \quad (3)\end{aligned}$$

Eliminate z by adding equations (1) and (3).

$$\begin{aligned} 3x + 2y + z &= 8 \quad (1)\\ x + 4y - z &= 20 \quad (3)\\ \hline 4x + 6y &= 28 \quad (4)\end{aligned}$$

To get another equation without z, multiply equation (3) by 2 and add the result to equation (2).

$$\begin{aligned} 2x + 8y - 2z &= 40 \quad 2\times(3)\\ 2x - 3y + 2z &= -16 \quad (2)\\ \hline 4x + 5y &= 24 \quad (5)\end{aligned}$$

Use equations (4) and (5) to eliminate x. Multiply equation (4) by -1 and add the result to equation (5).

$$\begin{aligned} -4x - 6y &= -28 \quad -1\times(4)\\ 4x + 5y &= 24 \quad (5)\\ \hline -y &= -4\\ y &= 4\end{aligned}$$

Substitute 4 for y in equation (5) to find x.

$$\begin{aligned} 4x + 5y &= 24 \quad (5)\\ 4x + 5(4) &= 24\\ 4x + 20 &= 24\\ 4x &= 4\\ x &= 1\end{aligned}$$

Substitute 1 for x and 4 for y in equation (3) to find z.

$$\begin{aligned} x + 4y - z &= 20 \quad (3)\\ 1 + 4(4) - z &= 20\\ 1 + 16 - z &= 20\\ 17 - z &= 20\\ -z &= 3\\ z &= -3\end{aligned}$$

The solution $(1, 4, -3)$ checks in all three of the original equations.
Solution set: $\{(1, 4, -3)\}$

7.
$$\begin{aligned} 2x + 5y + 2z &= 0 \quad (1)\\ 4x - 7y - 3z &= 1 \quad (2)\\ 3x - 8y - 2z &= -6 \quad (3)\end{aligned}$$

Add equations (1) and (3) to eliminate z.

$$\begin{aligned} 2x + 5y + 2z &= 0 \quad (1)\\ 3x - 8y - 2z &= -6 \quad (3)\\ \hline 5x - 3y &= -6 \quad (4)\end{aligned}$$

Multiply equation (1) by 3 and equation (2) by 2. Then add the results to eliminate z again.

$$\begin{aligned} 6x + 15y + 6z &= 0 \quad 3\times(1)\\ 8x - 14y - 6z &= 2 \quad 2\times(2)\\ \hline 14x + y &= 2 \quad (5)\end{aligned}$$

Solve the system

$$\begin{aligned} 5x - 3y &= -6 \quad (4)\\ 14x + y &= 2. \quad (5)\end{aligned}$$

Multiply equation (5) by 3 then add this result to (4).

$$\begin{array}{rcrl} 5x - 3y &=& -6 & (4) \\ 42x + 3y &=& 6 & 3 \times (5) \\ \hline 47x &=& 0 & \\ x &=& 0 & \end{array}$$

To find y, substitute $x = 0$ into equation (4).

$$\begin{aligned} 5x - 3y &= -6 \quad (4) \\ 5(0) - 3y &= -6 \\ y &= 2 \end{aligned}$$

To find z, substitute $x = 0$ and $y = 2$ in equation (1).

$$\begin{aligned} 2x + 5y + 2z &= 0 \quad (1) \\ 2(0) + 5(2) + 2z &= 0 \\ 10 + 2z &= 0 \\ 2z &= -10 \\ z &= -5 \end{aligned}$$

Solution set: $\{(0, 2, -5)\}$

9.

$$\begin{array}{rcrl} x + 2y + z &=& 4 & (1) \\ 2x + y - z &=& -1 & (2) \\ x - y - z &=& -2 & (3) \end{array}$$

Add (1) and (2) to eliminate z.

$$\begin{array}{rcrl} x + 2y + z &=& 4 & (1) \\ 2x + y - z &=& -1 & (2) \\ \hline 3x + 3y &=& 3 & (4) \end{array}$$

Add (1) and (3) to eliminate z.

$$\begin{array}{rcrl} x + 2y + z &=& 4 & (1) \\ x - y - z &=& -2 & (3) \\ \hline 2x + y &=& 2 & (5) \end{array}$$

Multiply equation (5) by -3.

$$\begin{array}{rcrl} 3x + 3y &=& 3 & (4) \; -3 \times (5) \\ -6x - 3y &=& -6 & (6) \\ \hline -3x &=& -3 & \\ x &=& 1 & \end{array}$$

Substitute 1 for x in equation (5) and solve for y.

$$\begin{aligned} 2(1) + y &= 2 \\ y &= 0 \end{aligned}$$

Substitute 1 for x and 0 for y in (1).

$$\begin{aligned} (1) + 2(0) + z &= 4 \\ z &= 3 \end{aligned}$$

Solution set: $\{(1, 0, 3)\}$

11.

$$\begin{array}{rcrl} \frac{1}{3}x + \frac{1}{6}y - \frac{2}{3}z &=& -1 & (1) \\ -\frac{3}{4}x - \frac{1}{3}y - \frac{1}{4}z &=& 3 & (2) \\ \frac{1}{2}x + \frac{3}{2}y + \frac{3}{4}z &=& 21 & (3) \end{array}$$

Eliminate all fractions.

$$\begin{array}{rcrll} 2x + y - 4z &=& -6 & (4) & 6 \times (1) \\ -9x - 4y - 3z &=& 36 & (5) & 12 \times (2) \\ 2x + 6y + 3z &=& 84 & (6) & 4 \times (3) \end{array}$$

Multiply (4) by 4.

$$\begin{array}{rcrl} 8x + 4y - 16z &=& -24 & \\ -9x - 4y - 3z &=& 36 & (5) \\ \hline -x - 19z &=& 12 & (7) \end{array}$$

Multiply (4) by -6.

$$\begin{array}{rcrl} -12x - 6y + 24z &=& 36 & \\ 2x + 6y + 3z &=& 84 & (6) \\ \hline -10x + 27z &=& 120 & (8) \end{array}$$

Multiply (7) by -10.

$$\begin{array}{rcr} 10x + 190z &=& -120 \\ -10x + 27z &=& 120 \\ \hline 217z &=& 0 \\ z &=& 0 \end{array}$$

Substituting 0 for z in (7) gives us $x = -12$. Substitute -12 for x and 0 for z in (4).

$$\begin{aligned} 2(-12) + y - 4(0) &= -6 \\ -24 + y &= -6 \\ y &= 18 \end{aligned}$$

Solution set: $\{(-12, 18, 0)\}$

13.

$$\begin{array}{rcrl} -x + 2y + 6z &=& 2 & (1) \\ 3x + 2y + 6z &=& 6 & (2) \\ x + 4y - 3z &=& 1 & (3) \end{array}$$

Eliminate y and z by adding equation (1) to -1 times equation (2).

$$\begin{array}{rcrl} -x + 2y + 6z &=& 2 & (1) \\ -3x - 2y - 6z &=& -6 & -1 \times (2) \\ \hline -4x &=& -4 & \\ x &=& 1 & \end{array}$$

Eliminate z by adding equation (2) to 2 times equation (3).

$$\begin{array}{rcrl} 3x + 2y + 6z &=& 6 & (2) \\ 2x + 8y - 6z &=& 2 & 2 \times (3) \\ \hline 5x + 10y &=& 8 & (4) \end{array}$$

Substitute 1 for x in equation (4).

continued

$$\begin{aligned} 5x+10y&=8 \qquad (4)\\ 5(1)+10y&=8\\ 10y&=3\\ y&=\frac{3}{10} \end{aligned}$$

Substitute 1 for x and $\frac{3}{10}$ for y in equation (1).

$$\begin{aligned} -x+2y+6z&=2 \qquad (1)\\ -1+2\left(\frac{3}{10}\right)+6z&=2\\ \frac{3}{5}+6z&=3\\ 6z&=\frac{12}{5}\\ z&=\frac{2}{5} \end{aligned}$$

Solution set: $\left\{\left(1,\frac{3}{10},\frac{2}{5}\right)\right\}$

15.
$$\begin{aligned} x+y-z&=-2 \qquad (1)\\ 2x-y+z&=-5 \qquad (2)\\ -x+2y-3z&=-4 \qquad (3) \end{aligned}$$

Eliminate y and z by adding equations (1) and (2).

$$\begin{aligned} x+y-z&=-2 \quad (1)\\ 2x-y+z&=-5 \quad (2)\\ \hline 3x&=-7\\ x&=-\frac{7}{3} \end{aligned}$$

To get another equation without y, multiply equation (2) by 2 and add the result to equation (3).

$$\begin{aligned} 4x-2y+2z&=-10 \qquad 2\times(2)\\ -x+2y-3z&=-4 \quad (3)\\ \hline 3x-z&=-14 \quad (4) \end{aligned}$$

Substitute $-\frac{7}{3}$ for x in equation (4) to find z.

$$\begin{aligned} 3x-z&=-14 \quad (4)\\ 3\left(-\frac{7}{3}\right)-z&=-14\\ -7-z&=-14\\ -z&=-7\\ z&=7 \end{aligned}$$

Substitute $-\frac{7}{3}$ for x and 7 for z in equation (1) to find y.

$$\begin{aligned} x+y-z&=-2 \quad (1)\\ -\frac{7}{3}+y-7&=-2\\ -7+3y-21&=-6 \quad \textit{Multiply by 3.}\\ 3y-28&=-6\\ 3y&=22\\ y&=\frac{22}{3} \end{aligned}$$

Solution set: $\left\{\left(-\frac{7}{3},\frac{22}{3},7\right)\right\}$

A calculator check reduces the probability of making any arithmetic errors and is highly recommended. The following screen shows the substitution of the solution for x, y, and z along with the left sides of the three original equations. The evaluation of the three expressions, -2, -5, and -4 (the right sides of the three equations), indicates that we have found the correct solution.

```
-7/3→X:22/3→Y:7→
Z:X+Y-Z
                 -2
2X-Y+Z
                 -5
-X+2Y-3Z
                 -4
```

17.
$$\begin{aligned} 2x-3y+2z&=-1 \qquad (1)\\ x+2y+z&=17 \qquad (2)\\ 2y-z&=7 \qquad (3) \end{aligned}$$

Multiply equation (2) by -2, and add the result to equation (1).

$$\begin{aligned} 2x-3y+2z&=-1 \quad (1)\\ -2x-4y-2z&=-34 \qquad -2\times(2)\\ \hline -7y&=-35\\ y&=5 \end{aligned}$$

To find z, substitute 5 for y in equation (3).

$$\begin{aligned} 2y-z&=7 \quad (3)\\ 2(5)-z&=7\\ 10-z&=7\\ -z&=-3\\ z&=3 \end{aligned}$$

To find x, substitute $y=5$ and $z=3$ into equation (1).

$$\begin{aligned} 2x-3y+2z&=-1 \quad (1)\\ 2x-3(5)+2(3)&=-1\\ 2x-9&=-1\\ 2x&=8\\ x&=4 \end{aligned}$$

Solution set: $\{(4,5,3)\}$

19.

$$\begin{aligned} 4x + 2y - 3z &= 6 \quad (1)\\ x - 4y + z &= -4 \quad (2)\\ -x + 2z &= 2 \quad (3)\end{aligned}$$

Equation (3) is missing y. Eliminate y again by multiplying equation (1) by 2 and adding the result to equation (2).

$$\begin{aligned} 8x + 4y - 6z &= 12 \quad 2\times(1)\\ x - 4y + z &= -4 \quad (2)\\ \hline 9x - 5z &= 8 \quad (4)\end{aligned}$$

Use equations (3) and (4) to eliminate x. Multiply equation (3) by 9 and add the result to equation (4).

$$\begin{aligned} -9x + 18z &= 18 \quad 9\times(3)\\ 9x - 5z &= 8 \quad (4)\\ \hline 13z &= 26\\ z &= 2\end{aligned}$$

Substitute 2 for z in equation (3) to find x.

$$\begin{aligned} -x + 2z &= 2 \quad (3)\\ -x + 2(2) &= 2\\ -x + 4 &= 2\\ -x &= -2\\ x &= 2\end{aligned}$$

Substitute 2 for x and 2 for z in equation (2) to find y.

$$\begin{aligned} x - 4y + z &= -4 \quad (2)\\ 2 - 4y + 2 &= -4\\ -4y + 4 &= -4\\ -4y &= -8\\ y &= 2\end{aligned}$$

Solution set: $\{(2, 2, 2)\}$

21.

$$\begin{aligned} 2x + y &= 6 \quad (1)\\ 3y - 2z &= -4 \quad (2)\\ 3x - 5z &= -7 \quad (3)\end{aligned}$$

To eliminate y, multiply equation (1) by -3 and add the result to equation (2).

$$\begin{aligned} -6x - 3y &= -18 \quad -3\times(1)\\ 3y - 2z &= -4 \quad (2)\\ \hline -6x - 2z &= -22 \quad (4)\end{aligned}$$

Since equation (3) does not have a y-term, we can multiply equation (3) by 2 and add the result to equation (4) to eliminate x and solve for z.

$$\begin{aligned} 6x - 10z &= -14 \quad 2\times(3)\\ -6x - 2z &= -22 \quad (4)\\ \hline -12z &= -36\\ z &= 3\end{aligned}$$

To find x, substitute 3 for z into equation (3).

$$\begin{aligned} 3x - 5z &= -7 \quad (3)\\ 3x - 5(3) &= -7\\ 3x &= 8\\ x &= \frac{8}{3}\end{aligned}$$

To find y, substitute 3 for z into equation (2).

$$\begin{aligned} 3y - 2z &= -4 \quad (2)\\ 3y - 2(3) &= -4\\ 3y &= 2\\ y &= \frac{2}{3}\end{aligned}$$

Solution set: $\left\{\left(\frac{8}{3}, \frac{2}{3}, 3\right)\right\}$

23.

$$\begin{aligned} -5x + 2y + z &= 5 \quad (1)\\ -3x - 2y - z &= 3 \quad (2)\\ -x + 6y &= 1 \quad (3)\end{aligned}$$

Add (1) and (2) to eliminate y and z.

$$\begin{aligned} -8x &= 8\\ x &= -1\end{aligned}$$

Substitute -1 for x in equation (3).

$$\begin{aligned} -x + 6y &= 1 \quad (3)\\ -(-1) + 6y &= 1\\ 6y &= 0\\ y &= 0\end{aligned}$$

Substitute -1 for x and 0 for y in (1).

$$\begin{aligned} -5x + 2y + z &= 5 \quad (1)\\ -5(-1) + 2(0) + z &= 5\\ 5 + z &= 5\\ z &= 0\end{aligned}$$

Solution set: $\{(-1, 0, 0)\}$

25.

$$\begin{aligned} 4x - z &= -6 \quad (1)\\ \frac{3}{5}y + \frac{1}{2}z &= 0 \quad (2)\\ \frac{1}{3}x + \frac{2}{3}z &= -5 \quad (3)\end{aligned}$$

Eliminate fractions first.

$$\begin{aligned} 6y + 5z &= 0 \quad (4) \quad 10\times(2)\\ x + 2z &= -15 \quad (5) \quad 3\times(3)\end{aligned}$$

Eliminate z by adding (5) to 2 times (1).

$$\begin{aligned} 8x - 2z &= -12 \quad 2\times(1)\\ x + 2z &= -15 \quad (5)\\ \hline 9x &= -27\\ x &= -3\end{aligned}$$

Substitute -3 for x in (5).

continued

$$\begin{aligned} x + 2z &= -15 \quad (5) \\ -3 + 2z &= -15 \\ 2z &= -12 \\ z &= -6 \end{aligned}$$

Substitute -6 for z in (4).

$$\begin{aligned} 6y + 5z &= 0 \quad (4) \\ 6y + 5(-6) &= 0 \\ 6y &= 30 \\ y &= 5 \end{aligned}$$

Solution set: $\{(-3, 5, -6)\}$

27. Answers will vary.

(a) One example is two connecting walls and the ceiling (or floor) of a room, which meet at one point.

(b) One example is two opposite walls and the floor, which have no points in common. Another is the floors of three different levels of an office building.

(c) One example is the plane through the ceiling of a house, the plane through an outside wall, and the plane through one side of a slanted roof, which all meet in one line; therefore, they intersect in infinitely many points. Another example is three pages of this book, which intersect in the spine.

29.

$$\begin{aligned} 2x + 2y - 6z &= 5 \quad (1) \\ -3x + y - z &= -2 \quad (2) \\ -x - y + 3z &= 4 \quad (3) \end{aligned}$$

Multiply equation (3) by 2 and add the result to equation (1).

$$\begin{aligned} 2x + 2y - 6z &= 5 \quad (1) \\ -2x - 2y + 6z &= 8 \quad 2 \times (3) \\ \hline 0 &= 13 \quad \textit{False} \end{aligned}$$

Solution set: $\emptyset$; inconsistent system

31.

$$\begin{aligned} -5x + 5y - 20z &= -40 \quad (1) \\ x - y + 4z &= 8 \quad (2) \\ 3x - 3y + 12z &= 24 \quad (3) \end{aligned}$$

Dividing equation (1) by -5 gives equation (2). Dividing equation (3) by 3 also gives equation (2). The resulting equations are the same, so the three equations are dependent.
Solution set: $\{(x, y, z) | x - y + 4z = 8\}$

33.

$$\begin{aligned} 2x + y - z &= 6 \quad (1) \\ 4x + 2y - 2z &= 12 \quad (2) \\ -x - \frac{1}{2}y + \frac{1}{2}z &= -3 \quad (3) \end{aligned}$$

Multiplying equation (1) by 2 gives equation (2). Multiplying equation (3) by -4 also gives equation (2). The resulting equations are the same, so the three equations are dependent.
Solution set: $\{(x, y, z) | 2x + y - z = 6\}$

35.

$$\begin{aligned} x + y - 2z &= 0 \quad (1) \\ 3x - y + z &= 0 \quad (2) \\ 4x + 2y - z &= 0 \quad (3) \end{aligned}$$

Eliminate z by adding equations (2) and (3).

$$\begin{aligned} 3x - y + z &= 0 \quad (2) \\ 4x + 2y - z &= 0 \quad (3) \\ \hline 7x + y &= 0 \quad (4) \end{aligned}$$

To get another equation without z, multiply equation (2) by 2 and add the result to equation (1).

$$\begin{aligned} 6x - 2y + 2z &= 0 \quad 2 \times (2) \\ x + y - 2z &= 0 \quad (1) \\ \hline 7x - y &= 0 \quad (5) \end{aligned}$$

Add equations (4) and (5) to find x.

$$\begin{aligned} 7x + y &= 0 \quad (4) \\ 7x - y &= 0 \quad (5) \\ \hline 14x &= 0 \\ x &= 0 \end{aligned}$$

Substitute 0 for x in equation (4) to find y.

$$\begin{aligned} 7x + y &= 0 \quad (4) \\ 7(0) + y &= 0 \\ 0 + y &= 0 \\ y &= 0 \end{aligned}$$

Substitute 0 for x and 0 for y in equation (1) to find z.

$$\begin{aligned} x + y - 2z &= 0 \quad (1) \\ 0 + 0 - 2z &= 0 \\ -2z &= 0 \\ z &= 0 \end{aligned}$$

Solution set: $\{(0, 0, 0)\}$

37.

$$\begin{aligned} x + y + z - w &= 5 \quad (1) \\ 2x + y - z + w &= 3 \quad (2) \\ x - 2y + 3z + w &= 18 \quad (3) \\ -x - y + z + 2w &= 8 \quad (4) \end{aligned}$$

Eliminate w. Add equations (1) and (2).

$$\begin{aligned} x + y + z - w &= 5 \quad (1) \\ 2x + y - z + w &= 3 \quad (2) \\ \hline 3x + 2y &= 8 \quad (5) \end{aligned}$$

Eliminate w again. Add equations (1) and (3).

$$\begin{aligned} x + y + z - w &= 5 \quad (1) \\ x - 2y + 3z + w &= 18 \quad (3) \\ \hline 2x - y + 4z &= 23 \quad (6) \end{aligned}$$

Eliminate w again. Multiply equation (2) by -2. Add the result to equation (4).

$$\begin{array}{rcll} -4x - 2y + 2z - 2w &=& -6 & -2\times(2) \\ -x - y + z + 2w &=& 8 & (4) \\ \hline -5x - 3y + 3z &=& 2 & (7) \end{array}$$

Equations (5), (6), and (7) do not contain a w-term. Since (5) does not have a z-term, we will find another equation without a z-term.

Eliminate z. Multiply equation (6) by 3 and equation (7) by -4. Then add the results.

$$\begin{array}{rcll} 6x - 3y + 12z &=& 69 & 3\times(6) \\ 20x + 12y - 12z &=& -8 & -4\times(7) \\ \hline 26x + 9y &=& 61 & (8) \end{array}$$

Eliminate z again. Multiply equation (5) by 9 and equation (8) by -2. Then add the results.

$$\begin{array}{rcll} 27x + 18y &=& 72 & 9\times(5) \\ -52x - 18y &=& -122 & -2\times(8) \\ \hline -25x &=& -50 & \\ x &=& 2 & \end{array}$$

To find y, substitute $x = 2$ into equation (5).

$$\begin{aligned} 3x + 2y &= 8 \quad (5) \\ 3(2) + 2y &= 8 \\ 2y &= 2 \\ y &= 1 \end{aligned}$$

To find z, substitute $x = 2$ and $y = 1$ into equation (6).

$$\begin{aligned} 2x - y + 4z &= 23 \quad (6) \\ 2(2) - 1 + 4z &= 23 \\ 4z &= 20 \\ z &= 5 \end{aligned}$$

To find w, substitute $x = 2$, $y = 1$, and $z = 5$ into equation (1).

$$\begin{aligned} x + y + z - w &= 5 \quad (1) \\ 2 + 1 + 5 - w &= 5 \\ -w &= -3 \\ w &= 3 \end{aligned}$$

Solution set: $\{(2, 1, 5, 3)\}$

39.
$$\begin{array}{rcll} 3x + y - z + w &=& -3 & (1) \\ 2x + 4y + z - w &=& -7 & (2) \\ -2x + 3y - 5z + w &=& 3 & (3) \\ 5x + 4y - 5z + 2w &=& -7 & (4) \end{array}$$

Eliminate w. Add equations (1) and (2).

$$\begin{array}{rcll} 3x + y - z + w &=& -3 & (1) \\ 2x + 4y + z - w &=& -7 & (2) \\ \hline 5x + 5y &=& -10 & (5) \end{array}$$

Eliminate w again. Add equations (2) and (3).

$$\begin{array}{rcll} 2x + 4y + z - w &=& -7 & (2) \\ -2x + 3y - 5z + w &=& 3 & (3) \\ \hline 7y - 4z &=& -4 & (6) \end{array}$$

Eliminate w again. Multiply equation (2) by 2. Add the result to equation (4).

$$\begin{array}{rcll} 4x + 8y + 2z - 2w &=& -14 & 2\times(2) \\ 5x + 4y - 5z + 2w &=& -7 & (4) \\ \hline 9x + 12y - 3z &=& -21 & (7) \end{array}$$

Equations (5), (6), and (7) do not contain a w-term. Since (5) does not have a z-term, we will find another equation without a z-term.

Eliminate z. Multiply equation (6) by 3 and equation (7) by -4. Then add the results.

$$\begin{array}{rcll} 21y - 12z &=& -12 & 3\times(6) \\ -36x - 48y + 12z &=& 84 & -4\times(7) \\ \hline -36x - 27y &=& 72 & (8) \end{array}$$

Eliminate y using (5) and (8).

$$\begin{array}{rcll} 3x + 3y &=& -6 & \frac{3}{5}\times(5) \\ -4x - 3y &=& 8 & (8)\div 9 \\ \hline -x &=& 2 & \\ x &=& -2 & \end{array}$$

Substitute -2 for x in (5).

$$\begin{aligned} 5(-2) + 5y &= -10 \\ 5y &= 0 \\ y &= 0 \end{aligned}$$

Substitute 0 for y in (6).

$$\begin{aligned} 7(0) - 4z &= -4 \\ -4z &= -4 \\ z &= 1 \end{aligned}$$

Substitute -2 for x, 0 for y, and 1 for z in (1).

$$\begin{aligned} 3(-2) + (0) - (1) + w &= -3 \\ -6 - 1 + w &= -3 \\ w &= 4 \end{aligned}$$

Solution set: $\{(-2, 0, 1, 4)\}$

For Exercises 41–48,

$$f(x) = ax^2 + bx + c \quad (a \neq 0).$$

41. $f(1) = a(1)^2 + b(1) + c$
$= a + b + c$
Since $f(1) = 128$, the first equation is

$$a + b + c = 128.$$

42. $f(1.5) = a(1.5)^2 + b(1.5) + c$
$= 2.25a + 1.5b + c$
Since $f(1.5) = 140$, the second equation is

$$2.25a + 1.5b + c = 140.$$

43. $f(3) = a(3)^2 + b(3) + c$
$= 9a + 3b + c$
Since $f(3) = 80$, the third equation is

$$9a + 3b + c = 80.$$

44. Using the three equations from Exercises 41–43, the system is

$$\begin{aligned} a + b + c &= 128 \quad (1) \\ 2.25a + 1.5b + c &= 140 \quad (2) \\ 9a + 3b + c &= 80. \quad (3) \end{aligned}$$

Multiply equation (2) by -1 and add the result to equation (1).

$$\begin{array}{rcrl} a + b + c &=& 128 & (1) \\ -2.25a - 1.5b - c &=& -140 & (4) \\ \hline -1.25a - .5b &=& -12 & (5) \end{array}$$

Multiply equation (3) by -1 and add the result to equation (1).

$$\begin{array}{rcrl} a + b + c &=& 128 & (1) \\ -9a - 3b - c &=& -80 & (6) \\ \hline -8a - 2b &=& 48 & (7) \end{array}$$

Use equations (5) and (7) to eliminate b. Multiply equation (5) by -4 and add the result to equation (7).

$$\begin{array}{rcrl} 5a + 2b &=& 48 & -4 \times (5) \\ -8a - 2b &=& 48 & (7) \\ \hline -3a &=& 96 & \\ a &=& -32 & \end{array}$$

To find b, substitute $a = -32$ into equation (7).

$$\begin{aligned} -8a - 2b &= 48 \quad (7) \\ -8(-32) - 2b &= 48 \\ 256 - 2b &= 48 \\ -2b &= -208 \\ b &= 104 \end{aligned}$$

To find c, substitute $a = -32$ and $b = 104$ into equation (1).

$$\begin{aligned} a + b + c &= 128 \quad (1) \\ -32 + 104 + c &= 128 \\ 72 + c &= 128 \\ c &= 56 \end{aligned}$$

Solution set: $\{(-32, 104, 56)\}$

45. If $(a, b, c) = (-32, 104, 56)$, then
$f(x) = -32x^2 + 104x + 56.$

46. In the function f written in Exercise 45, the *height* of the projectile is a function of the *time* elapsed since it was projected.

47. $f(x) = -32x^2 + 104x + 56$
$f(0) = -32(0)^2 + 104(0) + 56$
$= 56$
The initial height is 56 ft.

48. $f(1.625) = -32(1.625)^2 + 104(1.625) + 56$
$= -84.5 + 169 + 56$
$= 140.5$
The maximum height is 140.5 ft.

49. $f(x) = ax^2 + bx + c$
$f(1) = a(1)^2 + b(1) + c$
$= a + b + c$
So, $a + b + c = 2.$ (1)

$f(-1) = a(-1)^2 + b(-1) + c$
$= a - b + c$
So, $a - b + c = 0.$ (2)

$f(-2) = a(-2)^2 + b(-2) + c$
$= 4a - 2b + c$
So, $4a - 2b + c = 8.$ (3)

Add equations (1) and (2).

$$\begin{array}{rcrl} a + b + c &=& 2 & (1) \\ a - b + c &=& 0 & (2) \\ \hline 2a + 2c &=& 2 & (4) \end{array}$$

Multiply equation (1) by 2 and add the result to equation (3).

$$\begin{array}{rcrl} 2a + 2b + 2c &=& 4 & 2 \times (1) \\ 4a - 2b + c &=& 8 & (3) \\ \hline 6a + 3c &=& 12 & (5) \end{array}$$

Multiply equation (4) by -3 and add the result to equation (5).

$$\begin{array}{rcrl} -6a - 6c &=& -6 & -3 \times (4) \\ 6a + 3c &=& 12 & (5) \\ \hline -3c &=& 6 & \\ c &=& -2 & \end{array}$$

To find a, substitute $c = -2$ into equation (4).

$$\begin{aligned} 2a + 2c &= 2 \quad (4) \\ 2a + 2(-2) &= 2 \\ 2a - 4 &= 2 \\ 2a &= 6 \\ a &= 3 \end{aligned}$$

To find b, substitute $a = 3$ and $c = -2$ into equation (1).

$$\begin{aligned} a + b + c &= 2 \quad (1) \\ 3 + b - 2 &= 2 \\ b + 1 &= 2 \\ b &= 1 \end{aligned}$$

So, $a = 3$, $b = 1$, and $c = -2$ and the equation is

$$f(x) = ax^2 + bx + c$$
$$f(x) = 3x^2 + x - 2.$$

50. $Y_1 = aX^2 + bX + c$
Using $(1, 8)$ gives
$$a(1)^2 + b(1) + c = 8$$
$$a + b + c = 8. \quad (1)$$
Using $(2, 15)$ gives
$$a(2)^2 + b(2) + c = 15$$
$$4a + 2b + c = 15. \quad (2)$$
Using $(3, 24)$ gives
$$a(3)^2 + b(3) + c = 24$$
$$9a + 3b + c = 24. \quad (3)$$
The system is
$$\begin{aligned} a + b + c &= 8 & (1) \\ 4a + 2b + c &= 15 & (2) \\ 9a + 3b + c &= 24. & (3) \end{aligned}$$

Multiply equation (2) by -1 and add the result to equation (1).
$$\begin{array}{rcll} a + b + c &=& 8 & (1) \\ -4a - 2b - c &=& -15 & -1 \times (2) \\ \hline -3a - b &=& -7 & (4) \end{array}$$

Multiply equation (3) by -1 and add the result to equation (1).
$$\begin{array}{rcll} a + b + c &=& 8 & (1) \\ -9a - 3b - c &=& -24 & -1 \times (3) \\ \hline -8a - 2b &=& -16 & \end{array}$$

Divide by -2.
$$4a + b = 8 \quad (5)$$

Add equations (4) and (5).
$$\begin{array}{rcll} -3a - b &=& -7 & (4) \\ 4a + b &=& 8 & (5) \\ \hline a &=& 1 & \end{array}$$

To find b, substitute $a = 1$ into equation (4).
$$\begin{aligned} -3a - b &= -7 \quad (4) \\ -3(1) - b &= -7 \\ -3 - b &= -7 \\ -b &= -4 \\ b &= 4 \end{aligned}$$

To find c, substitute $a = 1$ and $b = 4$ into equation (1).
$$\begin{aligned} a + b + c &= 8 \quad (1) \\ 1 + 4 + c &= 8 \\ c &= 3 \end{aligned}$$

So, $a = 1$, $b = 4$, and $c = 3$, and the equation is
$$\begin{aligned} Y_1 &= aX^2 + bX + c \\ Y_1 &= X^2 + 4X + 3. \end{aligned}$$

51. If one were to eliminate *different* variables in the first two steps, the result would be two equations in three variables, and it would not be possible to solve for a single variable in the next step.

Section 4.3

1. *Step 2*
Let $x =$ the number of games that the Cardinals won and let $y =$ the number of games that they lost.

Step 3
They played 162 games, so
$$x + y = 162. \quad (1)$$
They won 28 more games than they lost, so
$$x = 28 + y. \quad (2)$$

Step 4
Substitute $28 + y$ for x in (1).
$$\begin{aligned} (28 + y) + y &= 162 \\ 28 + 2y &= 162 \\ 2y &= 134 \\ y &= 67 \end{aligned}$$
Substitute 67 for y in (2),
$$x = 28 + 67 = 95$$

Step 5
The Cardinals win-loss record was 95 wins and 67 losses.

Step 6
95 is 28 more than 67 and the sum of 95 and 67 is 162.

3. Let $W =$ the width of the tennis court and $L =$ the length of the court.

Since the length is 42 ft more than the width,
$$L = W + 42. \quad (1)$$
The perimeter of a rectangle is given by
$$2W + 2L = P.$$
With perimeter $P = 228$ ft,
$$2W + 2L = 228. \quad (2)$$
Substitute $W + 42$ for L in equation (2).
$$\begin{aligned} 2W + 2(W + 42) &= 228 \\ 2W + 2W + 84 &= 228 \\ 4W &= 144 \\ W &= 36 \end{aligned}$$
Substitute $W = 36$ into equation (1).
$$\begin{aligned} L &= W + 42 \quad (1) \\ L &= 36 + 42 \\ &= 78 \end{aligned}$$
The length is 78 ft; the width is 36 ft.

5. Let $x =$ the revenue, (in billions of dollars) for ExxonMobil, and $y =$ the revenue (in billions of dollars) for General Motors.

The total revenue was \$399 billion.

$$x + y = 399 \quad (1)$$

ExxonMobil's revenue was \$29 billion more than General Motors.

$$x = y + 29 \quad (2)$$

Substitute $y + 29$ for x in equation (1).

$$\begin{aligned}(y + 29) + y &= 399\\ 2y &= 370\\ y &= 185\end{aligned}$$

Substitute 185 for y in equation (2).

$$x = 185 + 29 = 214$$

ExxonMobil's revenue was \$214 billion and General Motors' revenue was \$185 billion.

7. From the figure in the text, the angles marked y and $3x + 10$ are supplementary, so

$$(3x + 10) + y = 180. \quad (1)$$

Also, the angles x and y are complementary, so

$$x + y = 90. \quad (2)$$

Solve equation (2) for y to get

$$y = 90 - x. \quad (3)$$

Substitute $90 - x$ for y in equation (1).

$$\begin{aligned}(3x + 10) + (90 - x) &= 180\\ 2x + 100 &= 180\\ 2x &= 80\\ x &= 40\end{aligned}$$

Substitute $x = 40$ into equation (3) to get

$$y = 90 - x = 90 - 40 = 50.$$

The angles measure 40° and 50°.

9. Let $x =$ the hockey FCI and $y =$ the basketball FCI.

The sum is \$423.12, so

$$x + y = 423.12. \quad (1)$$

The hockey FCI was \$16.36 more than the basketball FCI, so

$$x = y + 16.36. \quad (2)$$

From (2), substitute $y + 16.36$ for x in (1).

$$\begin{aligned}(y + 16.36) + y &= 423.12\\ 2y + 16.36 &= 423.12\\ 2y &= 406.76\\ y &= 203.38\end{aligned}$$

From (2),
$x = y + 16.36 = 203.38 + 16.36 = 219.74.$

The hockey FCI was \$219.74 and the basketball FCI was \$203.38.

11. Let $x =$ the cost of a single hamburger, and $y =$ the cost of a double hamburger.

15 singles and 10 doubles cost \$63.25.

$$15x + 10y = 63.25 \quad (1)$$

30 singles and 5 doubles cost \$78.65.

$$30x + 5y = 78.65 \quad (2)$$

Multiply equation (1) by -2 and add to equation (2).

$$\begin{array}{rcrcrl} -30x & - & 20y & = & -126.50 & -2 \times (1)\\ 30x & + & 5y & = & 78.65 & (2)\\ \hline & - & 15y & = & -47.85 & \\ & & y & = & 3.19 & \end{array}$$

Substitute 3.19 for y in equation (1).

$$\begin{aligned}15x + 10y &= 63.25 \quad (1)\\ 15x + 10(3.19) &= 63.25\\ 15x &= 31.35\\ x &= 2.09\end{aligned}$$

A single hamburger costs \$2.09 and a double hamburger costs \$3.19.

13. Use the formula (rate of percent) • (base amount) = amount (percentage) of pure acid to compute parts (a) – (d).

(a) $.10(60) = 6$ oz

(b) $.25(60) = 15$ oz

(c) $.40(60) = 24$ oz

(d) $.50(60) = 30$ oz

15. The cost is the price per pound, \$.99, times the number of pounds, x, or $\$.99x$.

17. Let $x =$ the amount of 25% alcohol solution, and $y =$ the amount of 35% alcohol solution.

Make a table. The percent times the amount of solution gives the amount of pure alcohol in the third column.

Gallons of Solution	Percent (as a decimal)	Gallons of Pure Alcohol
x	$25\% = .25$	$.25x$
y	$35\% = .35$	$.35y$
20	$32\% = .32$	$.32(20) = 6.4$

The third row gives the total amounts of solution and pure alcohol. From the columns in the table, write a system of equations.

$$\begin{aligned} x + \quad y &= 20 & (1) \\ .25x + .35y &= 6.4 & (2) \end{aligned}$$

Solve the system. Multiply equation (1) by -25 and equation (2) by 100. Then add the results.

$$\begin{array}{rcrl} -25x - 25y &=& -500 & -25 \times (1) \\ 25x + 35y &=& 640 & 100 \times (2) \\ \hline 10y &=& 140 & \\ y &=& 14 & \end{array}$$

Substitute $y = 14$ into equation (1).

$$\begin{aligned} x + y &= 20 \quad (1) \\ x + 14 &= 20 \\ x &= 6 \end{aligned}$$

Mix 6 gal of 25% solution and 14 gal of 35% solution.

19. Let $x =$ the amount of pure acid and $y =$ the amount of 10% acid.

Make a table.

Liters of Solution	Percent (as a decimal)	Liters of Pure Acid
x	$100\% = 1$	$1.00x = x$
y	$10\% = .10$	$.10y$
54	$20\% = .20$	$.20(54) = 10.8$

Solve the following system.

$$\begin{aligned} x + \quad y &= 54 & (1) \\ x + .10y &= 10.8 & (2) \end{aligned}$$

Multiply equation (2) by 10 to clear the decimals.

$$10x + y = 108 \qquad (3)$$

To eliminate y, multiply equation (1) by -1 and add the result to equation (3).

$$\begin{array}{rcrl} -x - y &=& -54 & -1 \times (1) \\ 10x + y &=& 108 & (3) \\ \hline 9x &=& 54 & \\ x &=& 6 & \end{array}$$

Since $x = 6$,

$$\begin{aligned} x + y &= 54 \quad (1) \\ 6 + y &= 54 \\ y &= 48. \end{aligned}$$

Use 6 L of pure acid and 48 L of 10% acid.

21. Complete the table.

	Number of Kilograms	Price per Kilogram	Value
Nuts	x	2.50	$2.50x$
Cereal	y	1.00	$1.00y$
Mixture	30	1.70	$1.70(30)$

From the "Number of Kilograms" column,

$$x + y = 30. \quad (1)$$

From the "Value" column,

$$2.50x + 1.00y = 1.70(30) = 51. \quad (2)$$

Solve the system.

$$\begin{array}{rcrl} -10x - 10y &=& -300 & -10 \times (1) \\ 25x + 10y &=& 510 & 10 \times (2) \\ \hline 15x &=& 210 & \\ x &=& 14 & \end{array}$$

From (1), $14 + y = 30$, so $y = 16$.
The party mix should be made from 14 kg of nuts and 16 kg of cereal.

23. From the "Principal" column in the text,

$$x + y = 3000. \quad (1)$$

From the "Interest" column in the text,

$$.02x + .04y = 100. \quad (2)$$

Multiply equation (2) by 100 to clear the decimals.

$$2x + 4y = 10{,}000 \quad (3)$$

To eliminate x, multiply equation (1) by -2 and add the result to equation (3).

$$\begin{array}{rcrl} -2x - 2y &=& -6000 & -2 \times (1) \\ 2x + 4y &=& 10{,}000 & (3) \\ \hline 2y &=& 4000 & \\ y &=& 2000 & \end{array}$$

From (1), $x + 2000 = 3000$, so $x = 1000$.
\$1000 is invested at 2%, and \$2000 is invested at 4%.

25. In the formula $d = rt$, substitute 25 for r (rate or speed) and y for t (time in hours) to get

$$d = 25y \text{ mi}.$$

27. Let $x =$ the speed of the train and $y =$ the speed of the plane.

	r	t	d
Train	x	$\frac{150}{x}$	150
Plane	y	$\frac{400}{y}$	400

The times are equal, so

$$\frac{150}{x} = \frac{400}{y}. \quad (1)$$

The speed of the plane is 20 km per hour less than 3 times the speed of the train, so

$$3x - 20 = y. \quad (2)$$

Multiply (1) by xy (or use cross products).

$$150y = 400x \quad (3)$$

continued

From (2), substitute $3x - 20$ for y in (3).

$$\begin{aligned} 150(3x - 20) &= 400x \\ 450x - 3000 &= 400x \\ 50x &= 3000 \\ x &= 60 \end{aligned}$$

From (2), $y = 3(60) - 20 = 160$.
The speed of the train is 60 km/hr, and the speed of the plane is 160 km/hr.

29. Let $x =$ the speed of the boat in still water and $y =$ the speed of the current.

Furthermore,

$$\begin{aligned} \text{rate upstream} &= x - y \\ \text{and} \quad \text{rate downstream} &= x + y. \end{aligned}$$

Use these rates and the information in the problem to make a table.

	r	t	d
Upstream	$x - y$	2	36
Downstream	$x + y$	1.5	36

From the table, use the formula $d = rt$ to write a system of equations.

$$\begin{aligned} 36 &= 2(x - y) \\ 36 &= 1.5(x + y) \end{aligned}$$

Remove the parentheses and move the variables to the left side.

$$\begin{aligned} 2x - 2y &= 36 \quad (1) \\ 1.5x + 1.5y &= 36 \quad (2) \end{aligned}$$

Solve the system. Multiply equation (1) by -3 and equation (2) by 4. Then add the results.

$$\begin{array}{rcrl} -6x + 6y &=& -108 & -3 \times (1) \\ 6x + 6y &=& 144 & 4 \times (2) \\ \hline 12y &=& 36 & \\ y &=& 3 & \end{array}$$

Substitute $y = 3$ into equation (1).

$$\begin{aligned} 2x - 2y &= 36 \quad (1) \\ 2x - 2(3) &= 36 \\ 2x &= 42 \\ x &= 21 \end{aligned}$$

The speed of the boat is 21 mph, and the speed of the current is 3 mph.

31. Let $x =$ revenue from Turner (in millions of dollars) and $y =$ revenue from 'N Sync (in millions of dollars).

The total is \$157 million, so

$$x + y = 157. \quad (1)$$

'N Sync took in \$3.8 million less than Turner, so

$$y = x - 3.8. \quad (2)$$

Substitute $x - 3.8$ for y in equation (1).

$$\begin{aligned} x + y &= 157 \quad (1) \\ x + (x - 3.8) &= 157 \\ 2x &= 160.8 \\ x &= 80.4 \end{aligned}$$

From (2), $y = 80.4 - 3.8 = 76.6$.
Turner's revenue was \$80.4 million and 'N Sync's revenue was \$76.6 million.

33. Let $x =$ the number of general admission tickets and $y =$ the number of student tickets.

Make a table.

Ticket	Number	Value of Tickets
General	x	$5 \cdot x = 5x$
Student	y	$4 \cdot y = 4y$
Total	184	812

Solve the system.

$$\begin{aligned} x + y &= 184 \quad (1) \\ 5x + 4y &= 812 \quad (2) \end{aligned}$$

To eliminate y, multiply equation (1) by -4 and add the result to equation (2).

$$\begin{array}{rcrl} -4x - 4y &=& -736 & -4 \times (1) \\ 5x + 4y &=& 812 & (2) \\ \hline x &=& 76 & \end{array}$$

From (1), $76 + y = 184$, so $y = 108$.

76 general admission tickets and 108 student tickets were sold.

35. Let $x =$ the price for a citron and let $y =$ the price for a wood apple.

"9 citrons and 7 fragrant wood apples is 107" gives us

$$9x + 7y = 107. \quad (1)$$

"7 citrons and 9 fragrant wood apples is 101" gives us

$$7x + 9y = 101. \quad (2)$$

Multiply equation (1) by -7 and equation (2) by 9. Then add.

$$\begin{array}{rcrl} -63x - 49y &=& -749 & -7 \times (1) \\ 63x + 81y &=& 909 & 9 \times (2) \\ \hline 32y &=& 160 & \\ y &=& 5 & \end{array}$$

Substitute 5 for y in equation (1).

$$\begin{aligned} 9x + 7(5) &= 107 \\ 9x + 35 &= 107 \\ 9x &= 72 \\ x &= 8 \end{aligned}$$

The prices are 8 for a citron and 5 for a wood apple.

37. Let $x =$ the measure of one angle,
$y =$ the measure of another angle,
and $z =$ the measure of the last angle.

Two equations are given, so

$$\begin{aligned} z &= x + 10 \\ \text{or} \quad -x + z &= 10 \quad (1) \\ \text{and} \quad x + y &= 100. \quad (2) \end{aligned}$$

Since the sum of the measures of the angles of a triangle is 180°, the third equation of the system is

$$x + y + z = 180. \quad (3)$$

Equation (1) is missing y. To eliminate y again, multiply equation (2) by -1 and add the result to equation (3).

$$\begin{array}{rcrl} -x - y & = & -100 & -1 \times (2) \\ x + y + z & = & 180 & (3) \\ \hline z & = & 80 & \end{array}$$

Since $z = 80$,

$$\begin{aligned} -x + z &= 10 \quad (1) \\ -x + 80 &= 10 \\ -x &= -70 \\ x &= 70. \end{aligned}$$

From (2), $70 + y = 100$, so $y = 30$.
The measures of the angles are 70°, 30°, and 80°.

39. Let $x =$ the measure of the first angle,
$y =$ the measure of the second angle, and
$z =$ the measure of the third angle.

The sum of the angles in a triangle equals 180°, so

$$x + y + z = 180. \quad (1)$$

The measure of the second angle is 10° more than 3 times that of the first angle, so

$$y = 3x + 10. \quad (2)$$

The third angle is equal to the sum of the other two, so

$$z = x + y. \quad (3)$$

Solve the system. Substitute z for $x + y$ in equation (1).

$$\begin{aligned} (x + y) + z &= 180 \quad (1) \\ z + z &= 180 \\ 2z &= 180 \\ z &= 90 \end{aligned}$$

Substitute $z = 90$ and $3x + 10$ for y in equation (3).

$$\begin{aligned} z &= x + y \quad (3) \\ 90 &= x + (3x + 10) \\ 80 &= 4x \\ 20 &= x \end{aligned}$$

Substitute $x = 20$ and $z = 90$ into equation (3).

$$\begin{aligned} z &= x + y \quad (3) \\ 90 &= 20 + y \\ 70 &= y \end{aligned}$$

The three angles have measures of 20°, 70°, and 90°.

41. Let $x =$ the length of the longest side,
$y =$ the length of the middle side,
and $z =$ the length of the shortest side.

Perimeter is the sum of the measures of the sides, so

$$x + y + z = 70. \quad (1)$$

The longest side is 4 cm less than the sum of the other sides, so

$$\begin{aligned} x &= y + z - 4 \\ \text{or} \quad x - y - z &= -4. \quad (2) \end{aligned}$$

Twice the shortest side is 9 cm less than the longest side, so

$$\begin{aligned} 2z &= x - 9 \\ \text{or} \quad -x + 2z &= -9 \quad (3) \end{aligned}$$

Add equations (1) and (2) to eliminate y and z.

$$\begin{array}{rcrl} x + y + z & = & 70 & (1) \\ x - y - z & = & -4 & (2) \\ \hline 2x & = & 66 & \\ x & = & 33 & \end{array}$$

Since $x = 33$,

$$\begin{aligned} -x + 2z &= -9 \quad (3) \\ -33 + 2z &= -9 \\ 2z &= 24 \\ z &= 12. \end{aligned}$$

Since $x = 33$ and $z = 12$,

$$\begin{aligned} x + y + z &= 70 \quad (1) \\ 33 + y + 12 &= 70 \\ y + 45 &= 70 \\ y &= 25. \end{aligned}$$

The shortest side is 12 cm long, the middle side is 25 cm long, and the longest side is 33 cm long.

43. Let $x =$ the number of Republicans, $y =$ the number of Democrats, and $z =$ the number of Independents. There are 100 Americans in the sample, so

$$x + y + z = 100. \quad (1)$$

10 more Americans identify themselves as Independents than Republicans, so

$$z = x + 10. \quad (2)$$

continued

Six fewer Americans identify themselves as Republicans than Democrats, so

$$y - 6 = x. \quad (3)$$

Substitute $x + 6$ for y and $x + 10$ for z in (1).

$$\begin{aligned} x + (x + 6) + (x + 10) &= 100 \\ 3x + 16 &= 100 \\ 3x &= 84 \\ x &= 28 \end{aligned}$$

From (3), $y - 6 = 28$, so $y = 34$.
From (2), $z = 28 + 10 = 38$.

There are 28 Republicans, 34 Democrats, and 38 Independents.

45. Let $x =$ the number of \$10 tickets, $y =$ the number of \$18 tickets, and $z =$ the number of VIP \$30 tickets.

Five times as many \$18 tickets have been sold as VIP tickets, so

$$y = 5z. \quad (1)$$

The number of \$10 tickets equals the number of \$18 tickets plus twice the number of VIP tickets, so

$$x = y + 2z. \quad (2)$$

From (1) and (2),

$$\begin{aligned} x &= y + 2z \\ &= 5z + 2z \\ &= 7z. \quad (3) \end{aligned}$$

Sales of these tickets totaled \$9500.

$$\begin{aligned} 10x + 18y + 30z &= 9500 \\ 10(7z) + 18(5z) + 30z &= 9500 \\ 190z &= 9500 \\ z &= 50 \end{aligned}$$

From (1), $y = 5(50) = 250$.
From (3), $x = 7(50) = 350$.

There were 350 \$10 tickets, 250 \$18 tickets, and 50 \$30 tickets sold.

47. Let $x =$ the number of type A units,
$y =$ the number of type B units,
and $z =$ the number of type C units.

They must make 10 units more of type C than the total of the other two.

$$z = x + y + 10 \quad (1)$$

They must make twice as many of type B as type A.

$$y = 2x \quad (2)$$

They must produce a total of 490 units per day.

$$x + y + z = 490 \quad (3)$$

Rewrite (1) in standard form.

$$\begin{array}{rcrcrcrl} -x & - & y & + & z & = & 10 & (4) \\ x & + & y & + & z & = & 490 & (3) \\ \hline & & & & 2z & = & 500 & \\ & & & & z & = & 250 & \end{array}$$

Substitute 250 for z and $2x$ for y in (3).

$$\begin{aligned} x + 2x + 250 &= 490 \\ 3x &= 240 \\ x &= 80 \end{aligned}$$

From (2), $y = 2x = 2(80) = 160$.
They can make 80 units of type A clamps, 160 units of type B, and 250 units of type C.

49. Let x, y, and z denote the number of kilograms of the first, second, and third chemicals, respectively. The mix must include 60% of the first and second chemicals, so

$$x + y = .60(750) = 450. \quad (1)$$

The second and third chemicals must be in a ratio of 4 to 3 by weight, so

$$y = \tfrac{4}{3}z. \quad (2)$$

From (1),

$$x = 450 - y. \quad (3)$$

From (2),

$$z = \tfrac{3}{4}y. \quad (4)$$

The total is 750, so

$$\begin{aligned} x + y + z &= 750. \\ (450 - y) + y + \frac{3}{4}y &= 750 \\ \frac{3}{4}y &= 300 \\ y &= 400 \end{aligned}$$

From (3), $x = 450 - 400 = 50$.
From (4), $z = \tfrac{3}{4}(400) = 300$.

Use 50 kg of the first chemical, 400 kg of the second chemical, and 300 kg of the third chemical to make the plant food.

51. *Step 2*
Let $x =$ the number of wins,
$y =$ the number of losses,
and $z =$ the number of ties.

Step 3
They played 82 games, so

$$x + y + z = 82. \quad (1)$$

Their wins and losses totaled 74, so

$$x + y = 74. \quad (2)$$

They tied 18 fewer games than they lost, so

$$z = y - 18. \qquad (3)$$

Step 4
Multiply (2) by -1 and add to (1).

$$\begin{array}{rrrrrl} x & + y & + z & = & 82 & (1) \\ -x & - y & & = & -74 & -1 \times (2) \\ \hline & & z & = & 8 & \end{array}$$

Substitute 8 for z in (3).

$$8 = y - 18$$
$$26 = y$$

Substitute 26 for y in (2).

$$x + 26 = 74$$
$$x = 48$$

Step 5
The Stars won 48 games, lost 26 games, and tied 8 games.

Step 6
Adding 48, 26, and 8 gives 82 total games. The wins and losses add up to 74, and there were 18 fewer ties than losses. The solution is correct.

Section 4.4

1. $\begin{bmatrix} -2 & 3 & 1 \\ 0 & 5 & -3 \\ 1 & 4 & 8 \end{bmatrix}$

(a) The elements of the second row are 0, 5, and -3.

(b) The elements of the third column are 1, -3, and 8.

(c) The matrix is square since the number of rows (three) is the same as the number of columns.

(d) The matrix obtained by interchanging the first and third rows is

$$\begin{bmatrix} 1 & 4 & 8 \\ 0 & 5 & -3 \\ -2 & 3 & 1 \end{bmatrix}$$

(e) The matrix obtained by multiplying the first row by $-\frac{1}{2}$ is

$$\begin{bmatrix} -2(-\frac{1}{2}) & 3(-\frac{1}{2}) & 1(-\frac{1}{2}) \\ 0 & 5 & -3 \\ 1 & 4 & 8 \end{bmatrix} = \begin{bmatrix} 1 & -\frac{3}{2} & -\frac{1}{2} \\ 0 & 5 & -3 \\ 1 & 4 & 8 \end{bmatrix}.$$

(f) The matrix obtained by multiplying the third row by 3 and adding to the first row is

$$\begin{bmatrix} -2+3(1) & 3+3(4) & 1+3(8) \\ 0 & 5 & -3 \\ 1 & 4 & 8 \end{bmatrix} = \begin{bmatrix} 1 & 15 & 25 \\ 0 & 5 & -3 \\ 1 & 4 & 8 \end{bmatrix}.$$

3. $4x + 8y = 44$
$2x - y = -3$

$$\left[\begin{array}{rr|r} 4 & 8 & 44 \\ 2 & -1 & -3 \end{array}\right]$$

$$\left[\begin{array}{rr|r} 1 & \underline{2} & \underline{11} \\ 2 & -1 & -3 \end{array}\right] \quad \tfrac{1}{4}R_1$$

$$\left[\begin{array}{rr|r} 1 & 2 & 11 \\ 0 & \underline{-5} & \underline{-25} \end{array}\right] \quad -2R_1 + R_2$$

Note: $\begin{cases} -2(2) + (-1) = -5 \\ -2(11) + (-3) = -25 \end{cases}$

$$\left[\begin{array}{rr|r} 1 & 2 & 11 \\ 0 & 1 & \underline{5} \end{array}\right] \quad -\tfrac{1}{5}R_1$$

This represents the system

$$x + 2y = 11$$
$$y = 5.$$

Substitute $y = 5$ in the first equation.

$$x + 2y = 11$$
$$x + 2(5) = 11$$
$$x + 10 = 11$$
$$x = 1$$

Solution set: $\{(1, 5)\}$

5. $x + y = 5$
$x - y = 3$

Write the augmented matrix for this system.

$$\left[\begin{array}{rr|r} 1 & 1 & 5 \\ 1 & -1 & 3 \end{array}\right]$$

$$\left[\begin{array}{rr|r} 1 & 1 & 5 \\ 0 & -2 & -2 \end{array}\right] \quad -1R_1 + R_2$$

$$\left[\begin{array}{rr|r} 1 & 1 & 5 \\ 0 & 1 & 1 \end{array}\right] \quad -\tfrac{1}{2}R_2$$

This matrix gives the system

$$x + y = 5$$
$$y = 1.$$

Substitute $y = 1$ in the first equation.

$$x + y = 5$$
$$x + 1 = 5$$
$$x = 4$$

Solution set: $\{(4, 1)\}$

7. $2x + 4y = 6$
$3x - y = 2$

Write the augmented matrix.

$$\left[\begin{array}{rr|r} 2 & 4 & 6 \\ 3 & -1 & 2 \end{array}\right]$$

continued

The easiest way to get a 1 in the first row, first column position is to multiply the elements in the first row by $\frac{1}{2}$.

$$\left[\begin{array}{rr|r} 1 & 2 & 3 \\ 3 & -1 & 2 \end{array}\right] \quad \tfrac{1}{2}R_1$$

To get a 0 in row two, column 1, we need to subtract 3 from the 3 that is in that position. To do this we will multiply row 1 by -3 and add the result to row 2.

$$\left[\begin{array}{rr|r} 1 & 2 & 3 \\ 0 & -7 & -7 \end{array}\right] \quad -3R_1 + R_2$$

$$\left[\begin{array}{rr|r} 1 & 2 & 3 \\ 0 & 1 & 1 \end{array}\right] \quad -\tfrac{1}{7}R_2$$

This matrix gives the system

$$\begin{aligned} x + 2y &= 3 \\ y &= 1. \end{aligned}$$

Substitute $y = 1$ in the first equation.

$$\begin{aligned} x + 2y &= 3 \\ x + 2(1) &= 3 \\ x &= 1 \end{aligned}$$

Solution set: $\{(1, 1)\}$

9. $3x + 4y = 13$
$2x - 3y = -14$

Write the augmented matrix.

$$\left[\begin{array}{rr|r} 3 & 4 & 13 \\ 2 & -3 & -14 \end{array}\right]$$

$$\left[\begin{array}{rr|r} 1 & 7 & 27 \\ 2 & -3 & -14 \end{array}\right] \quad -1R_2 + R_1$$

$$\left[\begin{array}{rr|r} 1 & 7 & 27 \\ 0 & -17 & -68 \end{array}\right] \quad -2R_1 + R_2$$

$$\left[\begin{array}{rr|r} 1 & 7 & 27 \\ 0 & 1 & 4 \end{array}\right] \quad -\tfrac{1}{17}R_2$$

This matrix gives the system

$$\begin{aligned} x + 7y &= 27 \\ y &= 4. \end{aligned}$$

Substitute $y = 4$ in the first equation.

$$\begin{aligned} x + 7y &= 27 \\ x + 7(4) &= 27 \\ x + 28 &= 27 \\ x &= -1 \end{aligned}$$

Solution set: $\{(-1, 4)\}$

11. $-4x + 12y = 36$
$x - 3y = 9$

Write the augmented matrix.

$$\left[\begin{array}{rr|r} -4 & 12 & 36 \\ 1 & -3 & 9 \end{array}\right]$$

$$\left[\begin{array}{rr|r} -1 & 3 & 9 \\ 1 & -3 & 9 \end{array}\right] \quad \tfrac{1}{4}R_1$$

$$\left[\begin{array}{rr|r} -1 & 3 & 9 \\ 0 & 0 & 18 \end{array}\right] \quad R_1 + R_2$$

The corresponding system is

$$\begin{aligned} -x + 3y &= 9 \\ 0 &= 18 \quad \textit{False} \end{aligned}$$

which is inconsistent and has no solution.

Solution set: $\emptyset$

13. $2x + y = 4$
$4x + 2y = 8$

Write the augmented matrix.

$$\left[\begin{array}{rr|r} 2 & 1 & 4 \\ 4 & 2 & 8 \end{array}\right]$$

$$\left[\begin{array}{rr|r} 1 & \frac{1}{2} & 2 \\ 4 & 2 & 8 \end{array}\right] \quad \tfrac{1}{2}R_1$$

$$\left[\begin{array}{rr|r} 1 & \frac{1}{2} & 2 \\ 0 & 0 & 0 \end{array}\right] \quad -4R_1 + R_2$$

Row 2, $0 = 0$, indicates that the system has dependent equations.

Solution set: $\{(x, y) \mid 2x + y = 4\}$

15.
$$\begin{aligned} x + y - z &= -3 \\ 2x + y + z &= 4 \\ 5x - y + 2z &= 23 \end{aligned}$$

Write the augmented matrix.

$$\left[\begin{array}{rrr|r} 1 & 1 & -1 & -3 \\ 2 & 1 & 1 & 4 \\ 5 & -1 & 2 & 23 \end{array}\right]$$

$$\left[\begin{array}{rrr|r} 1 & 1 & -1 & -3 \\ 0 & \underline{-1} & \underline{3} & \underline{10} \\ 0 & \underline{-6} & \underline{7} & \underline{38} \end{array}\right] \quad \begin{array}{l} \\ -2R_1 + R_2 \\ -5R_1 + R_3 \end{array}$$

$$\left[\begin{array}{rrr|r} 1 & 1 & -1 & -3 \\ 0 & 1 & \underline{-3} & \underline{-10} \\ 0 & -6 & 7 & 38 \end{array}\right] \quad -1R_2$$

$$\left[\begin{array}{rrr|r} 1 & 1 & -1 & -3 \\ 0 & 1 & -3 & -10 \\ 0 & 0 & \underline{-11} & \underline{-22} \end{array}\right] \quad 6R_2 + R_3$$

$$\left[\begin{array}{rrr|r} 1 & 1 & -1 & -3 \\ 0 & 1 & -3 & -10 \\ 0 & 0 & 1 & \underline{2} \end{array}\right] \quad -\tfrac{1}{11}R_3$$

This matrix gives the system

$$\begin{aligned} x + y - z &= -3 \\ y - 3z &= -10 \\ z &= 2. \end{aligned}$$

Substitute $z = 2$ in the second equation.

$$\begin{aligned} y - 3z &= -10 \\ y - 3(2) &= -10 \\ y - 6 &= -10 \\ y &= -4 \end{aligned}$$

Substitute $y = -4$ and $z = 2$ in the first equation.

$$\begin{aligned} x + y - z &= -3 \\ x - 4 - 2 &= -3 \\ x - 6 &= -3 \\ x &= 3 \end{aligned}$$

Solution set: $\{(3, -4, 2)\}$

17. $\begin{aligned} x + y - 3z &= 1 \\ 2x - y + z &= 9 \\ 3x + y - 4z &= 8 \end{aligned}$

Write the augmented matrix.

$$\left[\begin{array}{rrr|r} 1 & 1 & -3 & 1 \\ 2 & -1 & 1 & 9 \\ 3 & 1 & -4 & 8 \end{array}\right]$$

$$\left[\begin{array}{rrr|r} 1 & 1 & -3 & 1 \\ 0 & -3 & 7 & 7 \\ 0 & -2 & 5 & 5 \end{array}\right] \quad \begin{array}{l} -2R_1 + R_2 \\ -3R_1 + R_3 \end{array}$$

$$\left[\begin{array}{rrr|r} 1 & 1 & -3 & 1 \\ 0 & 1 & -\frac{7}{3} & -\frac{7}{3} \\ 0 & -2 & 5 & 5 \end{array}\right] \quad -\tfrac{1}{3}R_2$$

$$\left[\begin{array}{rrr|r} 1 & 1 & -3 & 1 \\ 0 & 1 & -\frac{7}{3} & -\frac{7}{3} \\ 0 & 0 & \frac{1}{3} & \frac{1}{3} \end{array}\right] \quad 2R_2 + R_3$$

$$\left[\begin{array}{rrr|r} 1 & 1 & -3 & 1 \\ 0 & 1 & -\frac{7}{3} & -\frac{7}{3} \\ 0 & 0 & 1 & 1 \end{array}\right] \quad 3R_3$$

This matrix gives the system

$$\begin{aligned} x + y - 3z &= 1 \\ y - \frac{7}{3}z &= -\frac{7}{3} \\ z &= 1. \end{aligned}$$

Substitute $z = 1$ in the second equation.

$$\begin{aligned} y - \frac{7}{3}z &= -\frac{7}{3} \\ y - \frac{7}{3}(1) &= -\frac{7}{3} \\ y &= 0 \end{aligned}$$

Substitute $y = 0$ and $z = 1$ in the first equation.

$$\begin{aligned} x + y - 3z &= 1 \\ x + 0 - 3(1) &= 1 \\ x - 3 &= 1 \\ x &= 4 \end{aligned}$$

Solution set: $\{(4, 0, 1)\}$

19. $\begin{aligned} x + y - z &= 6 \\ 2x - y + z &= -9 \\ x - 2y + 3z &= 1 \end{aligned}$

Write the augmented matrix.

$$\left[\begin{array}{rrr|r} 1 & 1 & -1 & 6 \\ 2 & -1 & 1 & -9 \\ 1 & -2 & 3 & 1 \end{array}\right]$$

$$\left[\begin{array}{rrr|r} 1 & 1 & -1 & 6 \\ 0 & -3 & 3 & -21 \\ 0 & -3 & 4 & -5 \end{array}\right] \quad \begin{array}{l} -2R_1 + R_2 \\ -1R_1 + R_3 \end{array}$$

$$\left[\begin{array}{rrr|r} 1 & 1 & -1 & 6 \\ 0 & 1 & -1 & 7 \\ 0 & -3 & 4 & -5 \end{array}\right] \quad -\tfrac{1}{3}R_2$$

$$\left[\begin{array}{rrr|r} 1 & 1 & -1 & 6 \\ 0 & 1 & -1 & 7 \\ 0 & 0 & 1 & 16 \end{array}\right] \quad 3R_2 + R_3$$

This matrix gives the system

$$\begin{aligned} x + y - z &= 6 \\ y - z &= 7 \\ z &= 16. \end{aligned}$$

Substitute $z = 16$ in the second equation.

$$\begin{aligned} y - 16 &= 7 \\ y &= 23 \end{aligned}$$

Substitute $y = 23$ and $z = 16$ in the first equation.

$$\begin{aligned} x + 23 - 16 &= 6 \\ x + 7 &= 6 \\ x &= -1 \end{aligned}$$

Solution set: $\{(-1, 23, 16)\}$

21. $\begin{aligned} x - y \phantom{{}- z} &= 1 \\ y - z &= 6 \\ x + \phantom{y - {}} z &= -1 \end{aligned}$

Write the augmented matrix.

$$\left[\begin{array}{rrr|r} 1 & -1 & 0 & 1 \\ 0 & 1 & -1 & 6 \\ 1 & 0 & 1 & -1 \end{array}\right]$$

$$\left[\begin{array}{rrr|r} 1 & -1 & 0 & 1 \\ 0 & 1 & -1 & 6 \\ 0 & 1 & 1 & -2 \end{array}\right] \quad -1R_1 + R_3$$

$$\left[\begin{array}{rrr|r} 1 & -1 & 0 & 1 \\ 0 & 1 & -1 & 6 \\ 0 & 0 & 2 & -8 \end{array}\right] \quad -1R_2 + R_3$$

$$\left[\begin{array}{rrr|r} 1 & -1 & 0 & 1 \\ 0 & 1 & -1 & 6 \\ 0 & 0 & 1 & -4 \end{array}\right] \quad \tfrac{1}{2}R_3$$

This matrix gives the system

$$\begin{aligned} x - y \phantom{{}- z} &= 1 \\ y - z &= 6 \\ z &= -4. \end{aligned}$$

continued

Substitute $z = -4$ in the second equation.

$$\begin{aligned} y - z &= 6 \\ y - (-4) &= 6 \\ y &= 2 \end{aligned}$$

Substitute $y = 2$ in the first equation.

$$\begin{aligned} x - y &= 1 \\ x - 2 &= 1 \\ x &= 3 \end{aligned}$$

Solution set: $\{(3, 2, -4)\}$

23.

$$\begin{aligned} x - 2y + z &= 4 \\ 3x - 6y + 3z &= 12 \\ -2x + 4y - 2z &= -8 \end{aligned}$$

Write the augmented matrix.

$$\left[\begin{array}{rrr|r} 1 & -2 & 1 & 4 \\ 3 & -6 & 3 & 12 \\ -2 & 4 & -2 & -8 \end{array}\right]$$

$$\left[\begin{array}{rrr|r} 1 & -2 & 1 & 4 \\ 1 & -2 & 1 & 4 \\ -1 & 2 & -1 & -4 \end{array}\right] \quad \begin{array}{l} \\ \frac{1}{3}R_2 \\ \frac{1}{2}R_3 \end{array}$$

$$\left[\begin{array}{rrr|r} 1 & -2 & 1 & 4 \\ 0 & 0 & 0 & 0 \\ 0 & 0 & 0 & 0 \end{array}\right] \quad \begin{array}{r} \\ -1R_1 + R_2 \\ R_1 + R_3 \end{array}$$

This augmented matrix represents a system of dependent equations.

Solution set: $\{(x, y, z) | x - 2y + z = 4\}$

25.

$$\begin{aligned} x + 2y + 3z &= -2 \\ 2x + 4y + 6z &= -5 \\ x - y + 2z &= 6 \end{aligned}$$

Write the augmented matrix.

$$\left[\begin{array}{rrr|r} 1 & 2 & 3 & -2 \\ 2 & 4 & 6 & -5 \\ 1 & -1 & 2 & 6 \end{array}\right]$$

$$\left[\begin{array}{rrr|r} 1 & 2 & 3 & -2 \\ 0 & 0 & 0 & -1 \\ 0 & -3 & -1 & 8 \end{array}\right] \quad \begin{array}{r} \\ -2R_1 + R_2 \\ -R_1 + R_3 \end{array}$$

From the second row, $0 = -1$, we see that the system is inconsistent.

Solution set: $\emptyset$

27. Examples will vary.

(a) A matrix is a rectangular array of numbers.

(b) A horizontal arrangement of elements in a matrix is a row of a matrix.

(c) A vertical arrangement of elements in a matrix is a column of a matrix.

(d) A square matrix contains the same number of rows as columns.

(e) A matrix formed by the coefficients and constants of a linear system is an augmented matrix for the system.

(f) Row operations on a matrix allow it to be transformed into another matrix in which the solution of the associated system can be found more easily.

29.

$$\begin{aligned} 4x + y &= 5 \\ 2x + y &= 3 \end{aligned}$$

Enter the augmented matrix as [A].

$$\begin{bmatrix} 4 & 1 & 5 \\ 2 & 1 & 3 \end{bmatrix}$$

The TI-83 screen for A follows. (Use MATRX EDIT.)

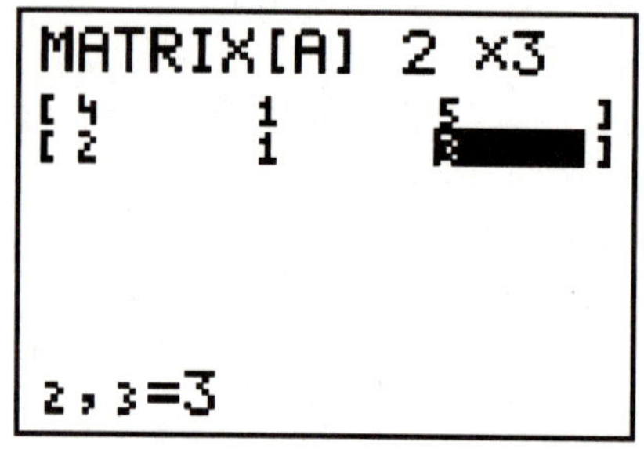

Now use the reduced row echelon form (rref) command to simplify the system. (Use MATRX MATH ALPHA B for rref and MATRX 1 for [A].)

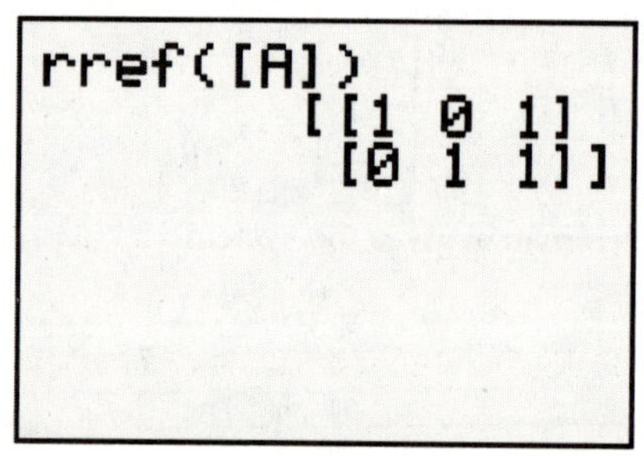

This matrix gives the system

$$\begin{aligned} 1x + 0y &= 1 \\ 0x + 1y &= 1, \end{aligned}$$

or simply, $x = 1$ and $y = 1$.

Solution set: $\{(1, 1)\}$

31.

$$\begin{aligned} 5x + y - 3z &= -6 \\ 2x + 3y + z &= 5 \\ -3x - 2y + 4z &= 3 \end{aligned}$$

Enter the augmented matrix as [C].

$$\begin{bmatrix} 5 & 1 & -3 & -6 \\ 2 & 3 & 1 & 5 \\ -3 & -2 & 4 & 3 \end{bmatrix}$$

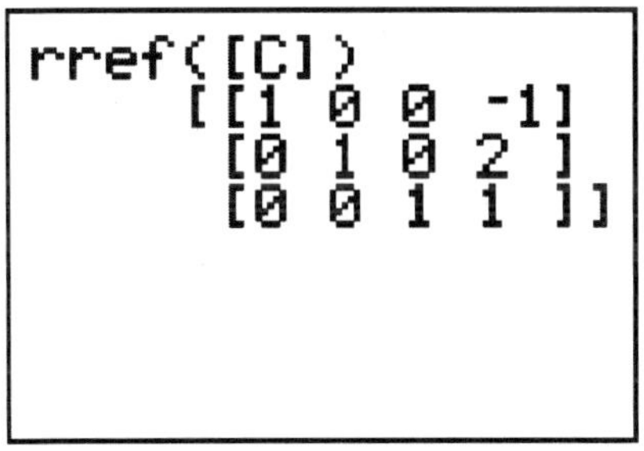

Solution set: $\{(-1, 2, 1)\}$

33. $x + z = -3$
$y + z = 3$
$x + y = 8$

Enter the augmented matrix as [E].

$$\begin{bmatrix} 1 & 0 & 1 & -3 \\ 0 & 1 & 1 & 3 \\ 1 & 1 & 0 & 8 \end{bmatrix}$$

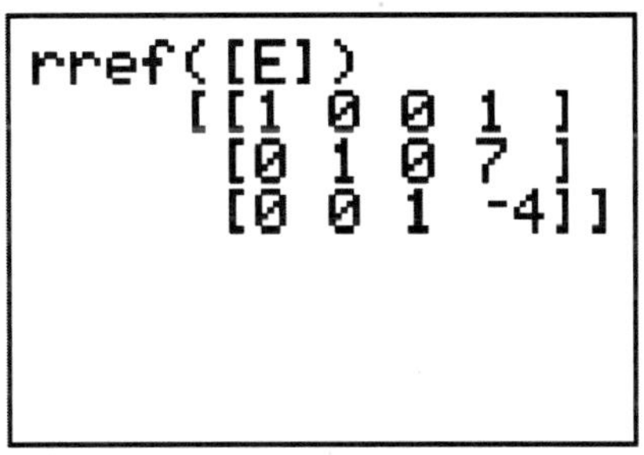

Solution set: $\{(1, 7, -4)\}$

Chapter 4 Review Exercises

1. **(a)** The graphs meet between the years of 1978 and 1982. Therefore, the number of men and women reached equal numbers between 1978 and 1982.

(b) The number was just less than 500,000.

2. $x + 3y = 8$ (1)
$2x - y = 2$ (2)

Graph the two lines. They appear to intersect at the point $(2, 2)$. Check $(2, 2)$ in both equations.

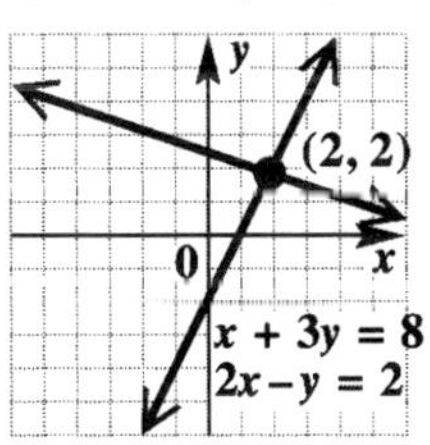

$$x + 3y = 8 \quad (1)$$
$$2 + 3(2) = 8$$
$$8 = 8 \quad \textit{True}$$

$$2x - y = 2 \quad (2)$$
$$2(2) - 2 = 2$$
$$2 = 2 \quad \textit{True}$$

Solution set: $\{(2, 2)\}$

3. Checking the ordered pairs in choices **A**, **B**, and **C** yields true statements.

For choice **D**, we have:

$$3x + 2y = 6$$
$$3(3) + 2(-2) = 6 \quad ?$$
$$9 - 4 = 6 \quad ?$$
$$5 = 6 \quad \textit{False}$$

So **D** is the answer.

4. $5x + y = 12$ (1)
$2x - 2y = 0$ (2)
To eliminate y, multiply equation (1) by 2 and add the result to equation (2).

$$\begin{array}{rcl} 10x + 2y & = & 24 \quad 2 \times (1) \\ 2x - 2y & = & 0 \quad (2) \\ \hline 12x & = & 24 \\ x & = & 2 \end{array}$$

To find y, substitute 2 for x in equation (2).

$$2x - 2y = 0 \quad (2)$$
$$2(2) - 2y = 0$$
$$4 = 2y$$
$$2 = y$$

The ordered pair $(2, 2)$ satisfies both equations, so it checks.
Solution set: $\{(2, 2)\}$

5. $x - 4y = -4$ (1)
$3x + y = 1$ (2)

To eliminate y, multiply equation (2) by 4 and add the result to equation (1).

$$\begin{array}{rcl} x - 4y & = & -4 \quad (1) \\ 12x + 4y & = & 4 \quad 4 \times (2) \\ \hline 13x & = & 0 \\ x & = & 0 \end{array}$$

Substitute 0 for x in equation (1) to find y.

$$x - 4y = -4 \quad (1)$$
$$0 - 4y = -4$$
$$y = 1$$

Solution set: $\{(0, 1)\}$

6. $6x + 5y = 4$ (1)
$-4x + 2y = 8$ (2)

To eliminate x, multiply equation (1) by 2 and equation (2) by 3. Then add the results.

continued

$$\begin{array}{rcrl} 12x + 10y &=& 8 & 2\times(1) \\ -12x + 6y &=& 24 & 3\times(2) \\ \hline 16y &=& 32 & \\ y &=& 2 & \end{array}$$

Since $y = 2$,

$$\begin{aligned} -4x + 2y &= 8 \quad (2) \\ -4x + 2(2) &= 8 \\ -4x + 4 &= 8 \\ -4x &= 4 \\ x &= -1. \end{aligned}$$

Solution set: $\{(-1, 2)\}$

7. $\frac{1}{6}x + \frac{1}{6}y = -\frac{1}{2}$ (1)
$x - y = -9$ (2)

Multiply equation (1) by 6 to clear the fractions. Add the result to equation (2) to eliminate y.

$$\begin{array}{rcrl} x + y &=& -3 & 6\times(1) \\ x - y &=& -9 & (2) \\ \hline 2x &=& -12 & \\ x &=& -6 & \end{array}$$

Since $x = -6$,

$$\begin{aligned} x - y &= -9 \quad (2) \\ -6 - y &= -9 \\ -y &= -3 \\ y &= 3. \end{aligned}$$

Solution set: $\{(-6, 3)\}$

8. $-3x + y = 6$ (1)
$y = 6 + 3x$ (2)

Since equation (2) can be rewritten as $-3x + y = 6$, the two equations are the same, and hence, dependent.
Solution set: $\{(x, y) \mid -3x + y = 6\}$

9. $5x - 4y = 2$ (1)
$-10x + 8y = 7$ (2)

Multiply equation (1) by 2 and add the result to equation (2).

$$\begin{array}{rcrl} 10x - 8y &=& 4 & 2\times(1) \\ -10x + 8y &=& 7 & (2) \\ \hline 0 &=& 11 & \textit{False} \end{array}$$

Since a false statement results, the system is *inconsistent*. The solution set is $\emptyset$.

10. $3x + y = -4$ (1)
$x = \frac{2}{3}y$ (2)

Substitute $\frac{2}{3}y$ for x in equation (1) and solve for y.

$$\begin{aligned} 3x + y &= -4 \quad (1) \\ 3\left(\frac{2}{3}y\right) + y &= 4 \\ 2y + y &= -4 \\ 3y &= -4 \\ y &= -\frac{4}{3} \end{aligned}$$

Since $x = \frac{2}{3}y$ and $y = -\frac{4}{3}$,

$$x = \frac{2}{3}\left(-\frac{4}{3}\right) = -\frac{8}{9}.$$

Solution set: $\left\{\left(-\frac{8}{9}, -\frac{4}{3}\right)\right\}$

11. $9x - y = -4$ (1)
$y = x + 4$ (2)

Substitute $x + 4$ for y in equation (1) and solve for x.

$$\begin{aligned} 9x - y &= -4 \quad (1) \\ 9x - (x + 4) &= -4 \\ 9x - x - 4 &= -4 \\ 8x &= 0 \\ x &= 0 \end{aligned}$$

Since $x = 0$, $y = x + 4 = 0 + 4 = 4$.
Solution set: $\{(0, 4)\}$

12. $-5x + 2y = -2$ (1)
$x + 6y = 26$ (2)

Solve equation (2) for x.

$$x = 26 - 6y$$

Substitute $26 - 6y$ for x in equation (1).

$$\begin{aligned} -5x + 2y &= -2 \quad (1) \\ -5(26 - 6y) + 2y &= -2 \\ -130 + 30y + 2y &= -2 \\ -130 + 32y &= -2 \\ 32y &= 128 \\ y &= 4 \end{aligned}$$

Since $x = 26 - 6y$ and $y = 4$,

$$x = 26 - 6(4) = 26 - 24 = 2.$$

Solution set: $\{(2, 4)\}$

13. **(a)** The system is inconsistent. Parallel lines do not intersect, so there is no ordered pair that would satisfy both equations.

(b) The equations are equivalent and have the same line for their graph. The equations are dependent.

14. $y = 3x + 2$
$y = 3x - 4$

Since both equations are in slope-intercept form, their slopes and y-intercepts can be easily determined. Since they have the same slope but different y-intercepts, they are parallel lines. Thus, there is no point common to both lines, and the system has $\emptyset$ as its solution set.

15.
$$\begin{array}{rcrcrcrl} 2x & + & 3y & - & z & = & -16 & (1) \\ x & + & 2y & + & 2z & = & -3 & (2) \\ -3x & + & y & + & z & = & -5 & (3) \end{array}$$

To eliminate z, add equations (1) and (3).

$$\begin{array}{rcrcrcrl} 2x & + & 3y & - & z & = & -16 & (1) \\ -3x & + & y & + & z & = & -5 & (3) \\ \hline -x & + & 4y & & & = & -21 & (4) \end{array}$$

To eliminate z again, multiply equation (1) by 2 and add the result to equation (2).

$$\begin{array}{rcrcrcrl} 4x & + & 6y & - & 2z & = & -32 & 2 \times (1) \\ x & + & 2y & + & 2z & = & -3 & (2) \\ \hline 5x & + & 8y & & & = & -35 & (5) \end{array}$$

Use equations (4) and (5) to eliminate x. Multiply equation (4) by 5 and add the result to equation (5).

$$\begin{array}{rcrcrl} -5x & + & 20y & = & -105 & 5 \times (4) \\ 5x & + & 8y & = & -35 & (5) \\ \hline & & 28y & = & -140 & \\ & & y & = & -5 & \end{array}$$

Substitute -5 for y in equation (4) to find x.

$$\begin{aligned} -x + 4y &= -21 \quad (4) \\ -x + 4(-5) &= -21 \\ -x - 20 &= -21 \\ -x &= -1 \\ x &= 1 \end{aligned}$$

Substitute 1 for x and -5 for y in equation (2) to find z.

$$\begin{aligned} x + 2y + 2z &= -3 \quad (2) \\ 1 + 2(-5) + 2z &= -3 \\ 1 - 10 + 2z &= -3 \\ 2z &= 6 \\ z &= 3 \end{aligned}$$

The solution set is $\{(1, -5, 3)\}$.

16.
$$\begin{array}{rcrcrcrl} 4x & - & y & & & = & 2 & (1) \\ & & 3y & + & z & = & 9 & (2) \\ x & & & + & 2z & = & 7 & (3) \end{array}$$

To eliminate y, multiply equation (1) by 3 and add the result to equation (2).

$$\begin{array}{rcrcrcrl} 12x & - & 3y & & & = & 6 & 3 \times (1) \\ & & 3y & + & z & = & 9 & (2) \\ \hline 12x & & & + & z & = & 15 & (4) \end{array}$$

To eliminate z, multiply equation (4) by -2 and add the result to equation (3).

$$\begin{array}{rcrcrl} -24x & - & 2z & = & -30 & -2 \times (4) \\ x & + & 2z & = & 7 & (3) \\ \hline -23x & & & = & -23 & \\ x & & & = & 1 & \end{array}$$

Substitute 1 for x in equation (3) to find z.

$$\begin{aligned} x + 2z &= 7 \quad (3) \\ 1 + 2z &= 7 \\ 2z &= 6 \\ z &= 3 \end{aligned}$$

Substitute 1 for x in equation (1) to find y.

$$\begin{aligned} 4x - y &= 2 \quad (1) \\ 4(1) - y &= 2 \\ 4 - y &= 2 \\ -y &= -2 \\ y &= 2 \end{aligned}$$

The solution set is $\{(1, 2, 3)\}$.

17.
$$\begin{array}{rcrcrcrl} 3x & - & y & - & z & = & -8 & (1) \\ 4x & + & 2y & + & 3z & = & 15 & (2) \\ -6x & + & 2y & + & 2z & = & 10 & (3) \end{array}$$

To eliminate y, multiply equation (1) by 2 and add the result to equation (3).

$$\begin{array}{rcrcrcrl} 6x & - & 2y & - & 2z & = & -16 & 2 \times (1) \\ -6x & + & 2y & + & 2z & = & 10 & (3) \\ \hline & & & & 0 & = & -6 & \textit{False} \end{array}$$

Since a false statement results, equations (1) and (3) have no common solution. The system is *inconsistent*. The solution set is $\emptyset$.

18. Let $x =$ the width of the rink, and $y =$ the length of the rink.

The length is 30 ft longer than twice the width.

$$y = 2x + 30 \quad (1)$$

The perimeter is 570 ft.

$$2x + 2y = 570 \quad (2)$$

Substitute $2x + 30$ for y in equation (2).

$$\begin{aligned} 2x + 2y &= 570 \quad (2) \\ 2x + 2(2x + 30) &= 570 \\ 2x + 4x + 60 &= 570 \\ 6x &= 510 \\ x &= 85 \end{aligned}$$

From (1), $y = 2(85) + 30 = 200$.

The width is 85 ft and the length is 200 ft.

19. Let $x =$ the number of days he rented the car at the weekday rate, and $y =$ the number of days he rented the car at the weekend rate.

Since the business trip lasted 6 days,

$$x + y = 6. \quad (1)$$

At \$53/day at weekday rates and \$35/day at weekend rates, his total bill was \$264.

$$53x + 35y = 264 \quad (2)$$

Solve the equation (1) for y.

$$y = 6 - x \quad (3)$$

Substitute $6 - x$ for y in equation (2).

$$\begin{aligned} 53x + 35(6 - x) &= 264 \\ 53x + 210 - 35x &= 264 \\ 18x &= 54 \\ x &= 3 \end{aligned}$$

From (3), $y = 6 - 3 = 3$.

He rented the car for 3 weekdays and 3 weekend days.

20. Let $x =$ the speed of the plane and $y =$ the speed of the wind.

Complete the chart.

	r	t	d
With Wind	$x+y$	1.75	$1.75(x+y)$
Against Wind	$x-y$	2	$2(x-y)$

The distance each way is 560 miles. From the chart,

$$1.75(x + y) = 560.$$

Divide by 1.75.

$$x + y = 320 \quad (1)$$

From the chart,

$$\begin{aligned} 2(x - y) &= 560 \\ x - y &= 280. \quad (2) \end{aligned}$$

Solve the system by adding equations (1) and (2) to eliminate y.

$$\begin{array}{rcl} x + y &=& 320 \quad (1) \\ x - y &=& 280 \quad (2) \\ \hline 2x &=& 600 \\ x &=& 300 \end{array}$$

From (1), $300 + y = 320$, so $y = 20$.
The speed of the plane was 300 mph, and the speed of the wind was 20 mph.

21. Let $x =$ the amount of \$2-per-pound nuts and $y =$ the amount of \$1-per-pound chocolate candy.

Complete the chart.

	Number of Pounds	Price per Pound	Value
Nuts	x	2	$2x$
Chocolate	y	1	$1y = y$
Mixture	100	1.30	$1.30(100) = 130$

Solve the system formed from the first and third columns.

$$\begin{aligned} x + y &= 100 \quad (1) \\ 2x + y &= 130 \quad (2) \end{aligned}$$

Solve equation (1) for y.

$$y = 100 - x \quad (3)$$

Substitute $100 - x$ for y in equation (2).

$$\begin{aligned} 2x + (100 - x) &= 130 \\ x &= 30 \end{aligned}$$

From (3), $y = 100 - 30 = 70$.

She should use 30 lb of \$2-per-pound nuts and 70 lb of \$1-per-pound candy.

22. Let $x =$ the measure of the largest angle, $y =$ the measure of the middle-sized angle, and $z =$ the measure of the smallest angle.

Since the sum of the measures of the angles of a triangle is 180°,

$$x + y + z = 180. \quad (1)$$

Since the largest angle measures 10° less than the sum of the other two,

$$\begin{aligned} x &= y + z - 10 \\ \text{or} \quad x - y - z &= -10. \quad (2) \end{aligned}$$

Since the measure of the middle-sized angle is the average of the other two,

$$\begin{aligned} y &= \frac{x + z}{2} \\ 2y &= x + z \\ -x + 2y - z &= 0. \quad (3) \end{aligned}$$

Solve the system.

$$\begin{array}{rcl} x + y + z &=& 180 \quad (1) \\ x - y - z &=& -10 \quad (2) \\ -x + 2y - z &=& 0 \quad (3) \end{array}$$

Add equations (1) and (2) to find x.

$$\begin{array}{rcl} x + y + z &=& 180 \quad (1) \\ x - y - z &=& -10 \quad (2) \\ \hline 2x &=& 170 \\ x &=& 85 \end{array}$$

Add equations (1) and (3), to find y.

$$\begin{array}{rcrcrcrl} x & + & y & + & z & = & 180 & (1) \\ -x & + & 2y & - & z & = & 0 & (3) \\ \hline & & 3y & & & = & 180 & \\ & & & & y & = & 60 & \end{array}$$

Substitute 85 for x and 60 for y in equation (1) to find z.

$$\begin{aligned} x+y+z &= 180 \quad (1) \\ 85+60+z &= 180 \\ 145+z &= 180 \\ z &= 35 \end{aligned}$$

The three angles are 85°, 60°, and 35°.

23. Let $x =$ the value of sales at 10%,
$y =$ the value of sales at 6%,
and $z =$ the value of sales at 5%.

Since her total sales were $\$280,000$,

$$x+y+z = 280,000 \quad (1)$$

Since her commissions on the sales totaled $\$17,000$,

$$.10x+.06y+.05z = 17,000.$$

Multiply by 100 to clear the decimals, so

$$10x+6y+5z = 1,700,000. \quad (2)$$

Since the 5% sale amounted to the sum of the other two sales,

$$z = x+y. \quad (3)$$

Solve the system.

$$\begin{array}{rcrcrcll} x & + & y & + & z & = & 280,000 & (1) \\ 10x & + & 6y & + & 5z & = & 1,700,000 & (2) \\ & & & & z & = & x+y & (3) \end{array}$$

Since equation (3) is given in terms of z, substitute $x+y$ for z in equations (1) and (2).

$$\begin{aligned} x+y+z &= 280,000 \quad (1) \\ x+y+(x+y) &= 280,000 \\ 2x+2y &= 280,000 \\ x+y &= 140,000 \quad (4) \end{aligned}$$

$$\begin{aligned} 10x+6y+5z &= 1,700,000 \quad (2) \\ 10x+6y+5(x+y) &= 1,700,000 \\ 10x+6y+5x+5y &= 1,700,000 \\ 15x+11y &= 1,700,000 \quad (5) \end{aligned}$$

To eliminate x, multiply equation (4) by -11 and add the result to equation (5).

$$\begin{array}{rcrcrll} -11x & - & 11y & = & -1,540,000 & & -11\times(4) \\ 15x & + & 11y & = & 1,700,000 & (5) & \\ \hline 4x & & & = & 160,000 & & \\ & & x & = & 40,000 & & \end{array}$$

From (4), $y = 100,000$.
From (3), $z = 40,000+100,000 = 140,000$.

He sold $\$40,000$ at 10%, $\$100,000$ at 6%, and $\$140,000$ at 5%.

24. Let $x =$ the number of liters of 8% solution,
$y =$ the number of liters of 10% solution,
and $z =$ the number of liters of 20% solution.

Since the amount of the mixture will be 8 L,

$$x+y+z = 8. \quad (1)$$

Since the final solution will be 12.5% hydrogen peroxide,

$$.08x+.10y+.20z = .125(8).$$

Multiply by 100 to clear the decimals.

$$8x+10y+20z = 100 \quad (2)$$

Since the amount of 8% solution used must be 2 L more than the amount of 20% solution,

$$x = z+2. \quad (3)$$

Solve the system.

$$\begin{array}{rcrcrcll} x & + & y & + & z & = & 8 & (1) \\ 8x & + & 10y & + & 20z & = & 100 & (2) \\ & & & & x & = & z+2 & (3) \end{array}$$

Since equation (3) is given in terms of x, substitute $z+2$ for x in equations (1) and (2).

$$\begin{aligned} x+y+z &= 8 \quad (1) \\ (z+2)+y+z &= 8 \\ y+2z &= 6 \quad (4) \end{aligned}$$

$$\begin{aligned} 8x+10y+20z &= 100 \quad (2) \\ 8(z+2)+10y+20z &= 100 \\ 8z+16+10y+20z &= 100 \\ 10y+28z &= 84 \quad (5) \end{aligned}$$

To eliminate y, multiply equation (4) by -10 and add the result to equation (5).

$$\begin{array}{rcrcrll} -10y & - & 20z & = & -60 & & -10\times(4) \\ 10y & + & 28z & = & 84 & (5) & \\ \hline & & 8z & = & 24 & & \\ & & z & = & 3 & & \end{array}$$

From (3), $x = z+2 = 3+2 = 5$.
From (4), $y = 6-2z = 6-2(3) = 0$.
Mix 5 L of 8% solution, none of 10% solution, and 3 L of 20% solution.

25. Let $x =$ the number of home runs hit by Mantle,
$y =$ the number of home runs hit by Maris,
and $z =$ the number of home runs hit by Blanchard.

They combined for 136 home runs, so

$$x+y+z = 136. \quad (1)$$

Mantle hit 7 fewer than Maris, so

continued

$$x = y - 7. \quad (2)$$

Maris hit 40 more than Blanchard, so

$$y = z + 40 \quad \text{or} \quad z = y - 40. \quad (3)$$

Substitute $y - 7$ for x and $y - 40$ for z in (1).

$$\begin{aligned} x + y + z &= 136 \quad (1) \\ (y-7) + y + (y-40) &= 136 \\ 3y - 47 &= 136 \\ 3y &= 183 \\ y &= 61 \end{aligned}$$

From (2), $x = y - 7 = 61 - 7 = 54$.
From (3), $z = y - 40 = 61 - 40 = 21$.
Mantle hit 54 home runs, Maris hit 61 home runs, and Blanchard hit 21 home runs.

26. $\begin{aligned} 2x + 5y &= -4 \\ 4x - y &= 14 \end{aligned}$

Write the augmented matrix.

$$\left[\begin{array}{cc|c} 2 & 5 & -4 \\ 4 & -1 & 14 \end{array}\right]$$

$$\left[\begin{array}{cc|c} 2 & 5 & -4 \\ 0 & -11 & 22 \end{array}\right] \quad -2R_1 + R_2$$

$$\left[\begin{array}{cc|c} 2 & 5 & -4 \\ 0 & 1 & -2 \end{array}\right] \quad -\tfrac{1}{11}R_2$$

This matrix gives the system

$$\begin{aligned} 2x + 5y &= -4 \\ y &= -2. \end{aligned}$$

Substitute $y = -2$ in the first equation.

$$\begin{aligned} 2x + 5y &= -4 \\ 2x + 5(-2) &= -4 \\ 2x - 10 &= -4 \\ 2x &= 6 \\ x &= 3 \end{aligned}$$

Solution set: $\{(3, -2)\}$

27. $\begin{aligned} 6x + 3y &= 9 \\ -7x + 2y &= 17 \end{aligned}$

Write the augmented matrix.

$$\left[\begin{array}{cc|c} 6 & 3 & 9 \\ -7 & 2 & 17 \end{array}\right]$$

$$\left[\begin{array}{cc|c} 1 & -5 & -26 \\ -7 & 2 & 17 \end{array}\right] \quad -R_1 - R_2$$

$$\left[\begin{array}{cc|c} 1 & -5 & -26 \\ 0 & -33 & -165 \end{array}\right] \quad 7R_1 + R_2$$

$$\left[\begin{array}{cc|c} 1 & -5 & -26 \\ 0 & 1 & 5 \end{array}\right] \quad -\tfrac{1}{33}R_2$$

This matrix gives the system

$$\begin{aligned} x - 5y &= -26 \\ y &= 5. \end{aligned}$$

Substitute $y = 5$ in the first equation.

$$\begin{aligned} x - 5y &= -26 \\ x - 5(5) &= -26 \\ x - 25 &= -26 \\ x &= -1 \end{aligned}$$

Solution set: $\{(-1, 5)\}$

28. $\begin{aligned} x + 2y - z &= 1 \\ 3x + 4y + 2z &= -2 \\ -2x - y + z &= -1 \end{aligned}$

$$\left[\begin{array}{ccc|c} 1 & 2 & -1 & 1 \\ 3 & 4 & 2 & -2 \\ -2 & -1 & 1 & -1 \end{array}\right]$$

$$\left[\begin{array}{ccc|c} 1 & 2 & -1 & 1 \\ 0 & -2 & 5 & -5 \\ 0 & 3 & -1 & 1 \end{array}\right] \quad \begin{array}{l} -3R_1 + R_2 \\ 2R_1 + R_3 \end{array}$$

$$\left[\begin{array}{ccc|c} 1 & 2 & -1 & 1 \\ 0 & 1 & 4 & -4 \\ 0 & 3 & -1 & 1 \end{array}\right] \quad R_3 + R_2$$

$$\left[\begin{array}{ccc|c} 1 & 2 & -1 & 1 \\ 0 & 1 & 4 & -4 \\ 0 & 0 & -13 & 13 \end{array}\right] \quad -3R_2 + R_3$$

$$\left[\begin{array}{ccc|c} 1 & 2 & -1 & 1 \\ 0 & 1 & 4 & -4 \\ 0 & 0 & 1 & -1 \end{array}\right] \quad -\tfrac{1}{13}R_3$$

This matrix gives the system

$$\begin{aligned} x + 2y - z &= 1 \\ y + 4z &= -4 \\ z &= -1. \end{aligned}$$

Substitute $z = -1$ in the second equation.

$$\begin{aligned} y + 4z &= -4 \\ y + 4(-1) &= -4 \\ y &= 0 \end{aligned}$$

Substitute $y = 0$ and $z = -1$ in the first equation.

$$\begin{aligned} x + 2y - z &= 1 \\ x + 2(0) - (-1) &= 1 \\ x + 1 &= 1 \\ x &= 0 \end{aligned}$$

Solution set: $\{(0, 0, -1)\}$

29. $\begin{aligned} x + 3y \phantom{{}+z} &= 7 \\ 3x \phantom{{}+3y} + z &= 2 \\ y - 2z &= 4 \end{aligned}$

$$\left[\begin{array}{ccc|c} 1 & 3 & 0 & 7 \\ 3 & 0 & 1 & 2 \\ 0 & 1 & -2 & 4 \end{array}\right]$$

$$\left[\begin{array}{ccc|c} 1 & 3 & 0 & 7 \\ 0 & -9 & 1 & -19 \\ 0 & 1 & -2 & 4 \end{array}\right] \quad -3R_1 + R_2$$

$$\left[\begin{array}{ccc|c} 1 & 3 & 0 & 7 \\ 0 & 1 & -2 & 4 \\ 0 & -9 & 1 & -19 \end{array}\right] \quad R_2 \leftrightarrow R_3$$

We use $\leftrightarrow$ to represent the interchanging of 2 rows.

$$\left[\begin{array}{ccc|c} 1 & 3 & 0 & 7 \\ 0 & 1 & -2 & 4 \\ 0 & 0 & -17 & 17 \end{array}\right] \quad 9R_2 + R_3$$

$$\left[\begin{array}{ccc|c} 1 & 3 & 0 & 7 \\ 0 & 1 & -2 & 4 \\ 0 & 0 & 1 & -1 \end{array}\right] \quad -\tfrac{1}{17}R_3$$

This matrix gives the system

$$\begin{aligned} x + 3y &= 7 \\ y - 2z &= 4 \\ z &= -1. \end{aligned}$$

Substitute $z = -1$ in the second equation.

$$\begin{aligned} y - 2z &= 4 \\ y - 2(-1) &= 4 \\ y + 2 &= 4 \\ y &= 2 \end{aligned}$$

Substitute $y = 2$ in the first equation.

$$\begin{aligned} x + 3y &= 7 \\ x + 3(2) &= 7 \\ x + 6 &= 7 \\ x &= 1 \end{aligned}$$

Solution set: $\{(1, 2, -1)\}$

30. $\frac{2}{3}x + \frac{1}{6}y = \frac{19}{2}$ (1)

$\frac{1}{3}x - \frac{2}{9}y = 2$ (2)

Multiply equation (1) by 6 and equation (2) by 9 to clear the fractions.

$$\begin{aligned} 4x + y &= 57 \quad (3)\ 6 \times (1) \\ 3x - 2y &= 18 \quad (4)\ 9 \times (2) \end{aligned}$$

To eliminate y, multiply equation (3) by 2 and add the result to equation (4).

$$\begin{array}{rcll} 8x + 2y &=& 114 & 2 \times (3) \\ 3x - 2y &=& 18 & (4) \\ \hline 11x &=& 132 & \\ x &=& 12 & \end{array}$$

Substitute 12 for x in equation (3) to find y.

$$\begin{aligned} 4x + y &= 57 \quad (3) \\ 4(12) + y &= 57 \\ 48 + y &= 57 \\ y &= 9 \end{aligned}$$

Solution set: $\{(12, 9)\}$

31.
$$\begin{array}{rcll} 2x + 5y - z &=& 12 & (1) \\ -x + y - 4z &=& -10 & (2) \\ -8x - 20y + 4z &=& 31 & (3) \end{array}$$

Multiply equation (1) by 4 and add the result to equation (3).

$$\begin{array}{rcll} 8x + 20y - 4z &=& 48 & 4 \times (1) \\ -8x - 20y + 4z &=& 31 & (3) \\ \hline 0 &=& 79 & \textit{False} \end{array}$$

Since a false statement results, the system is *inconsistent*. The solution set is $\emptyset$.

32. $x = 7y + 10$ (1)

$2x + 3y = 3$ (2)

Since equation (1) is given in terms of x, substitute $7y + 10$ for x in equation (2) and solve for y.

$$\begin{aligned} 2(7y + 10) + 3y &= 3 \\ 14y + 20 + 3y &= 3 \\ 17y &= -17 \\ y &= -1 \end{aligned}$$

From (1), $x = 7(-1) + 10 = 3$.

Solution set: $\{(3, -1)\}$

33.
$$\begin{array}{rcll} x + 4y &=& 17 & (1) \\ -3x + 2y &=& -9 & (2) \end{array}$$

To eliminate x, multiply equation (1) by 3 and add the result to equation (2).

$$\begin{array}{rcll} 3x + 12y &=& 51 & 3 \times (1) \\ -3x + 2y &=& -9 & (2) \\ \hline 14y &=& 42 & \\ y &=& 3 & \end{array}$$

Substitute 3 for y in equation (1) to find x.

$$\begin{aligned} x + 4y &= 17 \quad (1) \\ x + 4(3) &= 17 \\ x + 12 &= 17 \\ x &= 5 \end{aligned}$$

Solution set: $\{(5, 3)\}$

34.
$$\begin{array}{rcll} -7x + 3y &=& 12 & (1) \\ 5x + 2y &=& 8 & (2) \end{array}$$

To eliminate y, multiply equation (1) by 2 and equation (2) by -3. Then add the results.

continued

$$\begin{array}{rcrl} -14x + 6y &=& 24 & 2 \times (1) \\ -15x - 6y &=& -24 & -3 \times (2) \\ \hline -29x &=& 0 & \\ x &=& 0 & \end{array}$$

Substitute 0 for x in equation (1) to find y.

$$\begin{aligned} -7x + 3y &= 12 \quad (1) \\ -7(0) + 3y &= 12 \\ 3y &= 12 \\ y &= 4 \end{aligned}$$

Solution set: $\{(0, 4)\}$

35. $$\begin{array}{rcrl} 2x - 5y &=& 8 & (1) \\ 3x + 4y &=& 10 & (2) \end{array}$$

To eliminate y, multiply equation (1) by 4 and equation (2) by 5 and add the results.

$$\begin{array}{rcrl} 8x - 20y &=& 32 & 4 \times (1) \\ 15x + 20y &=& 50 & 5 \times (2) \\ \hline 23x &=& 82 & \\ x &=& \frac{82}{23} & \end{array}$$

Instead of substituting to find y, we'll choose different multipliers and eliminate x from the original system.

$$\begin{array}{rcrl} 6x - 15y &=& 24 & 3 \times (1) \\ -6x - 8y &=& -20 & -2 \times (2) \\ \hline -23y &=& 4 & \\ y &=& -\frac{4}{23} & \end{array}$$

Solution set: $\left\{\left(\frac{82}{23}, -\frac{4}{23}\right)\right\}$

36. Let $x =$ the number of liters of 5% solution and $y =$ the number of liters of 10% solution.

Liters of Solution	Percent (as a decimal)	Liters of Pure Acid
x	.05	$.05x$
10	.20	$.20(10) = 2$
y	.10	$.10y$

Solve the system formed from the first and third columns.

$$\begin{aligned} x + 10 &= y \quad (1) \\ .05x + 2 &= .10y \quad (2) \end{aligned}$$

Multiply equation (2) by 100 to clear the decimals.

$$5x + 200 = 10y \quad (3)$$

Substitute $x + 10$ for y in equation (3) and solve for x.

$$\begin{aligned} 5x + 200 &= 10y \quad (3) \\ 5x + 200 &= 10(x + 10) \\ 5x + 200 &= 10x + 100 \\ 100 &= 5x \\ 20 &= x \end{aligned}$$

He should use 20 L of 5% solution.

37. Let x, y, and z denote the number of medals won by the United States, Russia, and China, respectively.

The total number of medals won was 244, so

$$x + y + z = 244. \quad (1)$$

The U.S. won 9 more medals than Russia, so

$$x = y + 9. \quad (2)$$

China won 29 fewer medals than Russia, so

$$z = y - 29. \quad (3)$$

Substitute $y + 9$ for x and $y - 29$ for z in (1).

$$\begin{aligned} (y + 9) + y + (y - 29) &= 244 \\ 3y - 20 &= 244 \\ 3y &= 264 \\ y &= 88 \end{aligned}$$

From (2), $x = 88 + 9 = 97$.
From (3), $z = 88 - 29 = 59$.

The United States won 97 medals, Russia won 88 medals, and China won 59 medals.

38. $$x^2 + y^2 + ax + by + c = 0$$

Let $x = 2$ and $y = 1$.

$$\begin{aligned} 2^2 + 1^2 + a(2) + b(1) + c &= 0 \\ 4 + 1 + 2a + b + c &= 0 \\ 2a + b + c &= -5 \end{aligned}$$

39. $$x^2 + y^2 + ax + by + c = 0$$

Let $x = -1$ and $y = 0$.

$$\begin{aligned} (-1)^2 + 0^2 + a(-1) + b(0) + c &= 0 \\ 1 - a + c &= 0 \\ -a + c &= -1 \\ a - c &= 1 \end{aligned}$$

40. $$x^2 + y^2 + ax + by + c = 0$$

Let $x = 3$ and $y = 3$.

$$\begin{aligned} 3^2 + 3^2 + a(3) + b(3) + c &= 0 \\ 9 + 9 + 3a + 3b + c &= 0 \\ 3a + 3b + c &= -18 \end{aligned}$$

41. Use the equations from Exercises 38–40 to form a system of equations.

$$\begin{array}{rcrl} 2a + b + c &=& -5 & (1) \\ a - c &=& 1 & (2) \\ 3a + 3b + c &=& -18 & (3) \end{array}$$

Add equations (1) and (2).

$$\begin{array}{rcrcrcrl} 2a & + & b & + & c & = & -5 & (1) \\ a & & & - & c & = & 1 & (2) \\ \hline 3a & + & b & & & = & -4 & (4) \end{array}$$

Add equations (2) and (3).

$$\begin{array}{rcrcrcrl} a & & & - & c & = & 1 & (2) \\ 3a & + & 3b & + & c & = & -18 & (3) \\ \hline 4a & + & 3b & & & = & -17 & (5) \end{array}$$

Multiply equation (4) by -3 and add the result to equation (5).

$$\begin{array}{rcrcrl} -9a & - & 3b & = & 12 & -3 \times (4) \\ 4a & + & 3b & = & -17 & (5) \\ \hline -5a & & & = & -5 & \\ & & a & = & 1 & \end{array}$$

Substitute $a = 1$ into equation (4).

$$\begin{aligned} 3a + b &= -4 \quad (4) \\ 3(1) + b &= -4 \\ b &= -7 \end{aligned}$$

Substitute $a = 1$ into equation (2).

$$\begin{aligned} a - c &= 1 \quad (2) \\ 1 - c &= 1 \\ c &= 0 \end{aligned}$$

Since $a = 1$, $b = -7$, and $c = 0$, the equation of the circle is

$$\begin{aligned} x^2 + y^2 + ax + by + c &= 0 \\ x^2 + y^2 + x - 7y &= 0. \end{aligned}$$

42. The graph of the circle is not the graph of a function because it fails the vertical line test.

Chapter 4 Test

1. No; The graph for Babe Ruth lies completely below the graph for Aaron, indicating that Ruth's total was always lower than Aaron's.

2. Aaron had more and Ruth had fewer.

3. When each equation of the system

$$\begin{aligned} x + y &= 7 \\ x - y &= 5 \end{aligned}$$

is graphed, the point of intersection appears to be $(6, 1)$. To check, substitute 6 for x and 1 for y in each of the equations. Since $(6, 1)$ makes both equations true, the solution set of the system is $\{(6, 1)\}$.

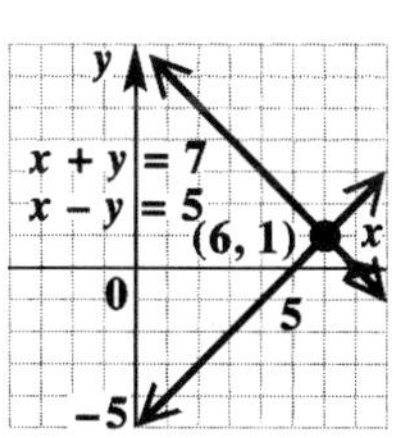

4. $$\begin{array}{rcrcrl} 3x & + & y & = & 12 & (1) \\ 2x & - & y & = & 3 & (2) \end{array}$$

To eliminate y, add equations (1) and (2).

$$\begin{array}{rcrcrl} 3x & + & y & = & 12 & (1) \\ 2x & - & y & = & 3 & (2) \\ \hline 5x & & & = & 15 & \\ & & x & = & 3 & \end{array}$$

Substitute 3 for x in equation (1) to find y.

$$\begin{aligned} 3x + y &= 12 \quad (1) \\ 3(3) + y &= 12 \\ 9 + y &= 12 \\ y &= 3 \end{aligned}$$

Solution set: $\{(3, 3)\}$

5. $$\begin{array}{rcrcrl} -5x & + & 2y & = & -4 & (1) \\ 6x & + & 3y & = & -6 & (2) \end{array}$$

To eliminate x, multiply equation (1) by 6 and equation (2) by 5. Then add the results.

$$\begin{array}{rcrcrl} -30x & + & 12y & = & -24 & 6 \times (1) \\ 30x & + & 15y & = & -30 & 5 \times (2) \\ \hline & & 27y & = & -54 & \\ & & y & = & -2 & \end{array}$$

Substitute -2 for y in equation (1) to find x.

$$\begin{aligned} -5x + 2y &= -4 \quad (1) \\ -5x + 2(-2) &= -4 \\ -5x - 4 &= -4 \\ -5x &= 0 \\ x &= 0 \end{aligned}$$

Solution set: $\{(0, -2)\}$

6. $$\begin{array}{lrcrcrl} & 3x & + & 4y & = & 8 & (1) \\ & & & 8y & = & 7 - 6x & \\ \text{or} & 6x & + & 8y & = & 7 & (2) \end{array}$$

Multiply equation (1) by -2 and add the result to equation (2).

$$\begin{array}{rcrcrl} -6x & - & 8y & = & -16 & -2 \times (1) \\ 6x & + & 8y & = & 7 & \\ \hline & & 0 & = & -9 & \textit{False} \end{array}$$

Since a false statement results, the system is *inconsistent*. The solution set is $\emptyset$.

7. $$\begin{array}{rcrcrcrl} 3x & + & 5y & + & 3z & = & 2 & (1) \\ 6x & + & 5y & + & z & = & 0 & (2) \\ 3x & + & 10y & - & 2z & = & 6 & (3) \end{array}$$

To eliminate x, multiply equation (1) by -1 and add the result to equation (3).

$$\begin{array}{rcrcrcrl} -3x & - & 5y & - & 3z & = & -2 & -1 \times (1) \\ 3x & + & 10y & - & 2z & = & 6 & (3) \\ \hline & & 5y & - & 5z & = & 4 & (4) \end{array}$$

continued

To eliminate x again, multiply equation (1) by -2 and add the result to equation (2).

$$\begin{array}{rrrrl} -6x & -\ 10y & -\ 6z & = -4 & -2\times(1) \\ 6x & +\ 5y & +\ z & = 0 & (2) \\ \hline & -5y & -\ 5z & = -4 & (5) \end{array}$$

To eliminate y, add equations (4) and (5).

$$\begin{array}{rrrl} 5y & -\ 5z & = 4 & (4) \\ -5y & -\ 5z & = -4 & (5) \\ \hline & -10z & = 0 & \\ & z & = 0 & \end{array}$$

Substitute 0 for z in equation (4) to find y.

$$\begin{aligned} 5y - 5z &= 4 \quad (4) \\ 5y - 5(0) &= 4 \\ 5y - 0 &= 4 \\ 5y &= 4 \\ y &= \frac{4}{5} \end{aligned}$$

Substitute $\frac{4}{5}$ for y and 0 for z in equation (1) to find x.

$$\begin{aligned} 3x + 5y + 3z &= 2 \quad (1) \\ 3x + 5\left(\frac{4}{5}\right) + 3(0) &= 2 \\ 3x + 4 + 0 &= 2 \\ 3x &= -2 \\ x &= -\frac{2}{3} \end{aligned}$$

Solution set: $\left\{\left(-\frac{2}{3}, \frac{4}{5}, 0\right)\right\}$

8. $$\begin{array}{rrrrl} 4x & +\ y & +\ z & = 11 & (1) \\ x & -\ y & -\ z & = 4 & (2) \\ & y & +\ 2z & = 0 & (3) \end{array}$$

To eliminate x, multiply equation (2) by -4 and add the result to equation (1).

$$\begin{array}{rrrrl} 4x & +\ y & +\ z & = 11 & (1) \\ -4x & +\ 4y & +\ 4z & = -16 & -4\times(2) \\ \hline & 5y & +\ 5z & = -5 & (4) \end{array}$$

To eliminate y, multiply equation (3) by -5 and add the result to equation (4).

$$\begin{array}{rrrl} -5y & -\ 10z & = 0 & -5\times(3) \\ 5y & +\ 5z & = -5 & (4) \\ \hline & -5z & = -5 & \\ & z & = 1 & \end{array}$$

From (3), $y + 2(1) = 0$, so $y = -2$.
From (2), $x - (-2) - 1 = 4$, so $x = 3$.

Solution set: $\{(3, -2, 1)\}$

9. $$\begin{aligned} 2x - 3y &= 24 \quad (1) \\ y &= -\frac{2}{3}x \quad (2) \end{aligned}$$

Since equation (2) is solved for y, substitute $-\frac{2}{3}x$ for y in equation (1) and solve for x.

$$\begin{aligned} 2x - 3y &= 24 \quad (1) \\ 2x - 3\left(-\frac{2}{3}x\right) &= 24 \\ 2x + 2x &= 24 \\ 4x &= 24 \\ x &= 6 \end{aligned}$$

From (2), $y = -\frac{2}{3}(6) = -4$.

Solution set: $\{(6, -4)\}$

10. $$\begin{aligned} 12x - 5y &= 8 \quad (1) \\ 3x &= \frac{5}{4}y + 2 \\ \text{or} \quad x &= \frac{5}{12}y + \frac{2}{3} \quad (2) \end{aligned}$$

Substitute $\frac{5}{12}y + \frac{2}{3}$ for x in equation (1) and solve for y.

$$\begin{aligned} 12x - 5y &= 8 \quad (1) \\ 12\left(\frac{5}{12}y + \frac{2}{3}\right) - 5y &= 8 \\ 5y + 8 - 5y &= 8 \\ 8 &= 8 \quad \textit{True} \end{aligned}$$

Equations (1) and (2) are dependent.
Solution set: $\{(x, y) \mid 12x - 5y = 8\}$

11. Let x = the gross (in millions of dollars) for *Pretty Woman*, and y = the gross (in millions of dollars) for *Runaway Bride*.

Together the movies grossed \$330.7 million, so

$$x + y = 330.7. \quad (1)$$

Runaway Bride grossed \$26.1 million less than *Pretty Woman*, so

$$y = x - 26.10. \quad (2)$$

Substitute $x - 26.1$ for y in equation (1).

$$\begin{aligned} x + y &= 330.7 \quad (1) \\ x + (x - 26.1) &= 330.7 \\ 2x &= 356.8 \\ x &= 178.4 \end{aligned}$$

From (2), $y = 178.4 - 26.1 = 152.3$.

Pretty Woman grossed \$178.4 million and *Runaway Bride* grossed \$152.3 million.

12. Let x = the speed of the faster car and y = the speed of the slower car.

Make a table.

	r	t	d
Faster Car	x	3.5	$3.5x$
Slower Car	y	3.5	$3.5y$

Since the fast car travels 30 mph faster than the slow car,

$$x - y = 30. \quad (1)$$

Since the cars travel a total of 420 miles,

$$3.5x + 3.5y = 420.$$

Multiply by 10 to clear the decimals.

$$35x + 35y = 4200 \quad (2)$$

To eliminate y, multiply equation (1) by 35 and add the result to equation (2).

$$\begin{array}{rcrcrl} 35x & - & 35y & = & 1050 & 35 \times (1) \\ 35x & + & 35y & = & 4200 & (2) \\ \hline 70x & & & = & 5250 & \\ & & x & = & 75 & \end{array}$$

Substitute 75 for x in equation (1) to find y.

$$\begin{aligned} x - y &= 30 \quad (1) \\ 75 - y &= 30 \\ -y &= -45 \\ y &= 45 \end{aligned}$$

The faster car is traveling at 75 mph, and the slower car is traveling at 45 mph.

13. Let x = the number of liters of 20% solution and y = the number of liters of 50% solution.

Make a table.

Liters of Solution	**Percent (as a decimal)**	**Liters of Pure Alcohol**
x	.20	$.20x$
y	.50	$.50y$
12	.40	$.40(12) = 4.8$

Since 12 L of the mixture are needed,

$$x + y = 12. \quad (1)$$

Since the amount of pure alcohol in the 20% solution plus the amount of pure alcohol in the 50% solution must equal the amount of alcohol in the mixture,

$$.2x + .5y = 4.8.$$

Multiply by 10 to clear the decimals.

$$2x + 5y = 48 \quad (2)$$

Multiply equation (1) by -2 and add the result to equation (2).

$$\begin{array}{rcrcrl} -2x & - & 2y & = & -24 & -2 \times (1) \\ 2x & + & 5y & = & 48 & (2) \\ \hline & & 3y & = & 24 & \\ & & y & = & 8 & \end{array}$$

From (1), $x + 8 = 12$, so $x = 4$.
4 L of 20% solution and 8 L of 50% solution are needed.

14. Let x = the price of an AC adaptor and y = the price of a rechargeable flashlight.

Since 7 AC adaptors and 2 rechargeable flashlights cost \$86,

$$7x + 2y = 86. \quad (1)$$

Since 3 AC adaptors and 4 rechargeable flashlights cost \$84,

$$3x + 4y = 84. \quad (2)$$

Solve the system.

$$\begin{array}{rcrcrl} 7x & + & 2y & = & 86 & (1) \\ 3x & + & 4y & = & 84 & (2) \end{array}$$

To eliminate y, multiply equation (1) by -2 and add the result to equation (2).

$$\begin{array}{rcrcrl} -14x & - & 4y & = & -172 & -2 \times (1) \\ 3x & + & 4y & = & 84 & (2) \\ \hline -11x & & & = & -88 & \\ & & x & = & 8 & \end{array}$$

Substitute 8 for x in equation (1) to find y.

$$\begin{aligned} 7x + 2y &= 86 \quad (1) \\ 7(8) + 2y &= 86 \\ 56 + 2y &= 86 \\ 2y &= 30 \\ y &= 15 \end{aligned}$$

An AC adaptor costs \$8, and a rechargeable flashlight costs \$15.

15. Let x = the amount of Orange Pekoe, y = the amount of Irish Breakfast, and z = the amount of Earl Grey.

The owner wants 100 oz of tea, so

$$x + y + z = 100. \quad (1)$$

An equation which relates the prices of the tea is

$$.80x + .85y + .95z = .83(100).$$

Multiply by 100 to clear the decimals.

$$80x + 85y + 95z = 8300 \quad (2)$$

The mixture must use twice as much Orange Pekoe as Irish Breakfast, so

$$x = 2y. \quad (3)$$

continued

To eliminate z, multiply equation (1) by -95 and add the result to equation (2).

$$\begin{array}{rcrcrcrl} -95x & - & 95y & - & 95z & = & -9500 & \quad -95 \times (1) \\ 80x & + & 85y & + & 95z & = & 8300 & \quad (2) \\ \hline -15x & - & 10y & & & = & -1200 & \quad (4) \end{array}$$

Divide equation (4) by -5.

$$3x + 2y = 240 \qquad (5)$$

Substitute $2y$ for x in equation (5) to find y.

$$\begin{aligned} 3x + 2y &= 240 \qquad (5) \\ 3(2y) + 2y &= 240 \\ 8y &= 240 \\ y &= 30 \end{aligned}$$

From (3), $x = 2(30) = 60$.
Substitute 60 for x and 30 for y in equation (1) to find z.

$$\begin{aligned} x + y + z &= 100 \qquad (1) \\ 60 + 30 + z &= 100 \\ z &= 10 \end{aligned}$$

He should use 60 oz of Orange Pekoe, 30 oz of Irish Breakfast, and 10 oz of Earl Grey.

16. $$\begin{aligned} 3x + 2y &= 4 \\ 5x + 5y &= 9 \end{aligned}$$

Write the augmented matrix.

$$\left[\begin{array}{rr|r} 3 & 2 & 4 \\ 5 & 5 & 9 \end{array}\right]$$

$$\left[\begin{array}{rr|r} 1 & -1 & -1 \\ 5 & 5 & 9 \end{array}\right] \quad 2R_1 - R_2$$

$$\left[\begin{array}{rr|r} 1 & -1 & -1 \\ 0 & 10 & 14 \end{array}\right] \quad -5R_1 + R_2$$

$$\left[\begin{array}{rr|r} 1 & -1 & -1 \\ 0 & 1 & \frac{7}{5} \end{array}\right] \quad \tfrac{1}{10}R_2$$

This matrix gives the system

$$\begin{aligned} x - y &= -1 \\ y &= \frac{7}{5}. \end{aligned}$$

Substitute $y = \frac{7}{5}$ in the first equation.

$$\begin{aligned} x - \frac{7}{5} &= -1 \\ x &= -1 + \frac{7}{5} \\ &= -\frac{5}{5} + \frac{7}{5} = \frac{2}{5} \end{aligned}$$

Solution set: $\left\{\left(\frac{2}{5}, \frac{7}{5}\right)\right\}$

17. $$\begin{aligned} x + 3y + 2z &= 11 \\ 3x + 7y + 4z &= 23 \\ 5x + 3y - 5z &= -14 \end{aligned}$$

Write the augmented matrix.

$$\left[\begin{array}{rrr|r} 1 & 3 & 2 & 11 \\ 3 & 7 & 4 & 23 \\ 5 & 3 & -5 & -14 \end{array}\right]$$

$$\left[\begin{array}{rrr|r} 1 & 3 & 2 & 11 \\ 0 & -2 & -2 & -10 \\ 0 & -12 & -15 & -69 \end{array}\right] \quad \begin{array}{l} -3R_1 + R_2 \\ -5R_1 + R_3 \end{array}$$

$$\left[\begin{array}{rrr|r} 1 & 3 & 2 & 11 \\ 0 & 1 & 1 & 5 \\ 0 & -12 & -15 & -69 \end{array}\right] \quad -\tfrac{1}{2}R_2$$

$$\left[\begin{array}{rrr|r} 1 & 3 & 2 & 11 \\ 0 & 1 & 1 & 5 \\ 0 & 0 & -3 & -9 \end{array}\right] \quad 12R_2 + R_3$$

$$\left[\begin{array}{rrr|r} 1 & 3 & 2 & 11 \\ 0 & 1 & 1 & 5 \\ 0 & 0 & 1 & 3 \end{array}\right] \quad -\tfrac{1}{3}R_3$$

This matrix gives the system

$$\begin{aligned} x + 3y + 2z &= 11 \\ y + z &= 5 \\ z &= 3. \end{aligned}$$

Substitute $z = 3$ in the second equation.

$$\begin{aligned} y + z &= 5 \\ y + 3 &= 5 \\ y &= 2 \end{aligned}$$

Substitute $y = 2$ and $z = 3$ in the first equation.

$$\begin{aligned} x + 3y + 2z &= 11 \\ x + 3(2) + 2(3) &= 11 \\ x + 6 + 6 &= 11 \\ x &= -1 \end{aligned}$$

Solution set: $\{(-1, 2, 3)\}$

18. $$\begin{aligned} 4x - 2y \qquad &= -8 \\ 3y - 5z &= 14 \\ 2x \qquad + z &= -10 \end{aligned}$$

Write the augmented matrix.

$$\left[\begin{array}{rrr|r} 4 & -2 & 0 & -8 \\ 0 & 3 & -5 & 14 \\ 2 & 0 & 1 & -10 \end{array}\right]$$

$$\left[\begin{array}{rrr|r} 1 & -\frac{1}{2} & 0 & -2 \\ 0 & 3 & -5 & 14 \\ 2 & 0 & 1 & -10 \end{array}\right] \quad \tfrac{1}{4}R_1$$

$$\left[\begin{array}{rrr|r} 1 & -\frac{1}{2} & 0 & -2 \\ 0 & 3 & -5 & 14 \\ 0 & 1 & 1 & -6 \end{array}\right] \quad -2R_1 + R_3$$

$$\left[\begin{array}{ccc|c} 1 & -\frac{1}{2} & 0 & -2 \\ 0 & 1 & -\frac{5}{3} & \frac{14}{3} \\ 0 & 1 & 1 & -6 \end{array}\right] \quad \tfrac{1}{3}R_2$$

$$\left[\begin{array}{ccc|c} 1 & -\frac{1}{2} & 0 & -2 \\ 0 & 1 & -\frac{5}{3} & \frac{14}{3} \\ 0 & 0 & \frac{8}{3} & -\frac{32}{3} \end{array}\right] \quad -1R_2 + R_3$$

$$\left[\begin{array}{ccc|c} 1 & -\frac{1}{2} & 0 & -2 \\ 0 & 1 & -\frac{5}{3} & \frac{14}{3} \\ 0 & 0 & 1 & -4 \end{array}\right] \quad \tfrac{3}{8}R_3$$

This matrix gives the system

$$\begin{aligned} x - \frac{1}{2}y &= -2 \\ y - \frac{5}{3}z &= \frac{14}{3} \\ z &= -4. \end{aligned}$$

Substitute $z = -4$ in the second equation.

$$\begin{aligned} y - \frac{5}{3}z &= \frac{14}{3} \\ y - \frac{5}{3}(-4) &= \frac{14}{3} \\ y + \frac{20}{3} &= \frac{14}{3} \\ y &= -2 \end{aligned}$$

Substitute $y = -2$ in the first equation.

$$\begin{aligned} x - \frac{1}{2}y &= -2 \\ x - \frac{1}{2}(-2) &= -2 \\ x + 1 &= -2 \\ x &= -3 \end{aligned}$$

Solution set: $\{(-3, -2, -4)\}$

Cumulative Review Exercises Chapters 1–4

1. $(-3)^4 = (-3)(-3)(-3)(-3) = 81$

2. $-3^4 = -(3)(3)(3)(3) = -81$

3. $-(-3)^4 = -(-3)(-3)(-3)(-3) = -81$

4. $\sqrt{.49} = .7$, since $.7$ is positive and $(.7)^2 = .49$.

5. $-\sqrt{.49} = -.7$, since $(.7)^2 = .49$ and the negative sign in front of the radical must be applied.

6. $\sqrt{-.49}$ is not a real number because of the negative sign under the radical. No real number squared is negative.

7. Using the same concept as with $\sqrt{\ }$, $\sqrt[3]{64} = 4$, since $4^3 = 64$.

8. $\sqrt[3]{-64} = -4$, since $(-4)^3 = -64$.

In Exercises 9–11, let $x = -4$, $y = 3$, and $z = 6$.

9.
$$\begin{aligned} &|2x| + 3y - z^3 \\ &= |(2)(-4)| + 3(3) - (6)^3 \\ &= |-8| + 9 - 216 \\ &= 8 + 9 - 216 \\ &= -199 \end{aligned}$$

10.
$$\begin{aligned} -5(x^3 - y^3) &= -5\left[(-4)^3 - (3)^3\right] \\ &= -5(-64 - 27) \\ &= -5(-91) \\ &= 455 \end{aligned}$$

11. The *commutative property* says that $3 \cdot 6 = 6 \cdot 3$, so that is the property that justifies the given statement.

12.
$$\begin{aligned} 7(2x + 3) - 4(2x + 1) &= 2(x + 1) \\ 14x + 21 - 8x - 4 &= 2x + 2 \\ 6x + 17 &= 2x + 2 \\ 4x &= -15 \\ x &= -\frac{15}{4} \end{aligned}$$

Solution set: $\left\{-\frac{15}{4}\right\}$

13. $|6x - 8| - 4$

$$\begin{aligned} 6x - 8 &= 4 \quad \text{or} \quad 6x - 8 = -4 \\ 6x &= 12 \qquad\qquad 6x = 4 \\ x &= 2 \quad \text{or} \quad x = \frac{4}{6} = \frac{2}{3} \end{aligned}$$

Solution set: $\left\{\frac{2}{3}, 2\right\}$

14. $ax + by = cx + d$

To solve for x, get all terms with x alone on one side of the equals sign.

$$\begin{aligned} ax - cx &= d - by \\ x(a - c) &= d - by \\ x &= \frac{d - by}{a - c} \\ \text{or} \qquad x &= \frac{by - d}{c - a} \end{aligned}$$

if the x-terms were put on the right side of the equals sign.

15. $.04x + .06(x - 1) = 1.04$

Multiply both sides by 100 to clear the decimals.

$$\begin{aligned} 4x + 6(x - 1) &= 104 \\ 4x + 6x - 6 &= 104 \\ 10x - 6 &= 104 \\ 10x &= 110 \\ x &= 11 \end{aligned}$$

Solution set: $\{11\}$

16. $\frac{2}{3}x + \frac{5}{12}x \le 20$

Multiply both sides by 12.

$$\begin{aligned} 12\left(\frac{2}{3}x + \frac{5}{12}x\right) &\le 12(20) \\ 8x + 5x &\le 240 \\ 13x &\le 240 \\ x &\le \frac{240}{13} \end{aligned}$$

Solution set: $\left(-\infty, \frac{240}{13}\right]$

17. $|3x+2| \le 4$

$-4 \le 3x + 2 \le 4$

$-6 \le 3x \le 2$ *Subtract 2.*

$-2 \le x \le \frac{2}{3}$ *Divide by 3.*

Solution set: $\left[-2, \frac{2}{3}\right]$

18. $|12t+7| \ge 0$

The solution set is $(-\infty, \infty)$ since the absolute value of any number is greater than or equal to 0.

19. 80.4% of 2500 is $.804(2500) = 2010$

72.5% of 2500 is $.725(2500) = 1812.5 \approx 1813$

$\frac{1570}{2500} = .628$ or 62.8%

$\frac{1430}{2500} = .572$ or 57.2%

Product or Company	Percent	Actual Number
Charmin	80.4%	2010
Wheaties	72.5%	1813
Budweiser	62.8%	1570
State Farm	57.2%	1430

20. Let h = the height of the triangle.

Use the formula $A = \frac{1}{2}bh$. Here, $A = 42$ and $b = 14$, so substitute these values in the formula and solve for h.

$$\begin{aligned} A &= \frac{1}{2}bh \\ 42 &= \frac{1}{2}(14)h \\ 42 &= 7h \\ 6 &= h \end{aligned}$$

The height is 6 m.

21. Let x = the number of nickels,
$x+1$ = the number of dimes, and
$x+6$ = the number of pennies.

The total value is \$4.80, so

$$.05x + .10(x+1) + .01(x+6) = 4.80.$$

Multiply both sides by 100 to clear the decimals.

$$\begin{aligned} 5x + 10(x+1) + 1(x+6) &= 480 \\ 5x + 10x + 10 + x + 6 &= 480 \\ 16x + 16 &= 480 \\ 16x &= 464 \\ x &= 29 \end{aligned}$$

Then, $x + 1 = 29 + 1 = 30$,
and $x + 6 = 29 + 6 = 35$.
There are 35 pennies, 29 nickels, and 30 dimes.

22. Let x = the measure of the equal angles and
$2x - 4$ = the measure of the third angle.

The sum of the measures of the angles in a triangle is 180, so

$$\begin{aligned} x + x + (2x - 4) &= 180 \\ 4x - 4 &= 180 \\ 4x &= 184 \\ x &= 46. \end{aligned}$$

So, $2x - 4 = 2(46) - 4 = 92 - 4 = 88$.
The measures of the angles are 46°, 46°, and 88°.

23. A horizontal line through the point (x, k) has equation $y = k$. Since point A has coordinates $(-2, 6)$, $k = 6$. The equation of the horizontal line through A is $y = 6$.

24. A vertical line through the point (k, y) has equation $x = k$. Since point B has coordinates $(4, -2)$, $k = 4$. The equation of the vertical line through B is $x = 4$.

25. Let $(x_1, y_1) = (-2, 6)$ and $(x_2, y_2) = (4, -2)$. Then,

$$m = \frac{y_2 - y_1}{x_2 - x_1} = \frac{-2-6}{4-(-2)} = \frac{-8}{6} = -\frac{4}{3}.$$

The slope is $-\frac{4}{3}$.

26. Perpendicular lines have slopes that are negative reciprocals of each other. The slope of line *AB* is $-\frac{4}{3}$ (from Exercise 25). The negative reciprocal of $-\frac{4}{3}$ is $\frac{3}{4}$, so the slope of a line perpendicular to line *AB* is $\frac{3}{4}$.

27. Let $m = -\frac{4}{3}$ and $(x_1, y_1) = (4, -2)$ in the point-slope form.

$$\begin{aligned} y - y_1 &= m(x - x_1) \\ y - (-2) &= -\frac{4}{3}(x - 4) \end{aligned}$$

$$y + 2 = -\frac{4}{3}x + \frac{16}{3}$$

Multiply by 3 to clear the fractions, and then write the equation in standard form, $Ax + By = C$.

$$\begin{aligned} 3y + 6 &= -4x + 16 \\ 4x + 3y &= 10 \end{aligned}$$

28. First locate the point $(-1, -3)$ on a graph. Then use the definition of slope to find a second point on the line.

$$m = \frac{\text{change in } y}{\text{change in } x} = \frac{2}{3}$$

From $(-1, -3)$, move 2 units up and 3 units to the right. The line through $(-1, -3)$ and the new point, $(2, -1)$, is the graph.

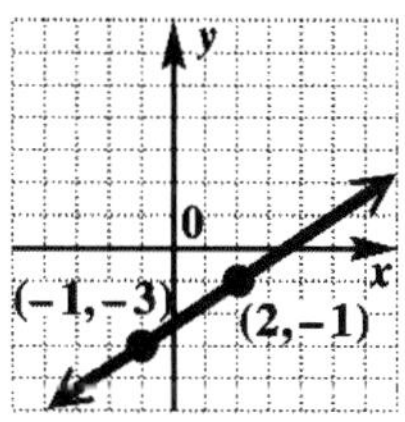

29. $-3x - 2y \le 6$
Graph the line $-3x - 2y = 6$ through its intercepts, $(-2, 0)$ and $(0, -3)$, as a solid line, since the inequality involves $\le$.
To determine the region that belongs to the graph, test $(0, 0)$.

$$\begin{aligned} -3x - 2y &\le 6 \\ -3(0) - 2(0) &\le 6 \\ 0 &\le 6 \quad \textit{True} \end{aligned}$$

Since the result is true, shade the region that includes $(0, 0)$.

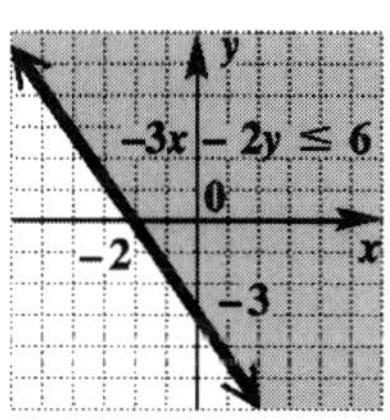

30. $f(x) = x^2 + 3x - 6$

(a) $f(-3) = (-3)^2 + 3(-3) - 6$
$= 9 - 9 - 6 = -6$

(b) $f(a) - (a)^2 + 3(a) - 6$
$= a^2 + 3a - 6$

31. If y varies directly as x, then

$$y = kx$$

for some constant k. If $y = 5$ when $x = 12$, then

$$5 = k(12),$$

so $\quad k = \dfrac{5}{12}$

and $\quad y = \dfrac{5}{12}x.$

When $x = 42$,

$$\begin{aligned} y &= \frac{5}{12}(42) \\ &= \frac{5 \cdot 6 \cdot 7}{2 \cdot 6} = \frac{35}{2} \quad \text{or} \quad 17.5. \end{aligned}$$

32.
$$\begin{aligned} -2x + 3y &= -15 \quad (1) \\ 4x - y &= 15 \quad (2) \end{aligned}$$

To eliminate x, multiply equation (1) by 2 and add the result to equation (2).

$$\begin{aligned} -4x + 6y &= -30 \quad 2 \times (1) \\ 4x - y &= 15 \quad (2) \\ \hline 5y &= -15 \\ y &= -3 \end{aligned}$$

Substitute -3 for y in equation (2) to find x.

$$\begin{aligned} 4x - y &= 15 \quad (2) \\ 4x - (-3) &= 15 \\ 4x + 3 &= 15 \\ 4x &= 12 \\ x &= 3 \end{aligned}$$

Solution set: $\{(3, -3)\}$

33.
$$\begin{aligned} x - 3y &= 7 \quad (1) \\ 2x - 6y &= 14 \quad (2) \end{aligned}$$

Multiplying equation (1) by 2 gives equation (2). Since those equations are the same, the two equations are dependent.

Solution set: $\{(x, y) \mid x - 3y = 7\}$

34.
$$\begin{aligned} x + y + z &= 10 \quad (1) \\ x - y - z &= 0 \quad (2) \\ -x + y - z &= -4 \quad (3) \end{aligned}$$

Add equations (1) and (2) to eliminate y and z. The result is

$$\begin{aligned} 2x - 10 \\ x = 5. \end{aligned}$$

Add equations (2) and (3) to eliminate x and y. The result is

$$\begin{aligned} -2z &= -4 \\ z &= 2. \end{aligned}$$

Substitute 5 for x and 2 for z in equation (1) to find y.

$$\begin{aligned} x + y + z &= 10 \quad (1) \\ 5 + y + 2 &= 10 \\ y + 7 &= 10 \\ y &= 3 \end{aligned}$$

Solution set: $\{(5, 3, 2)\}$

35. Let $x =$ the number of pounds of oranges and $y =$ the number of pounds of apples.

Since she bought 6 lb of fruit,

$$x + y = 6. \quad (1)$$

The total cost of x lb of oranges at \$.90/lb and y lb of apples at \$.70/lb is \$5.20, so

$$.90x + .70y = 5.20.$$

Multiply by 10 to clear the decimals.

$$9x + 7y = 52 \quad (2)$$

To solve the system, solve equation (1) for x.

$$\begin{aligned} x + y &= 6 \qquad (1) \\ x &= 6 - y \end{aligned}$$

Substitute $6 - y$ for x in equation (2) and solve for y.

$$\begin{aligned} 9x + 7y &= 52 \quad (2) \\ 9(6 - y) + 7y &= 52 \\ 54 - 9y + 7y &= 52 \\ -2y &= -2 \\ y &= 1 \end{aligned}$$

Since $x = 6 - y$ and $y = 1$,

$$x = 6 - 1 = 5.$$

She bought 5 lb of oranges and 1 lb of apples.

36. Let $x =$ the average retail price of Elmo and $y =$ the average retail price of Kid.

Elmo cost \$8.63 less than Kid, so

$$x = y - 8.63. \quad (1)$$

Together they cost \$63.89, so

$$x + y = 63.89. \quad (2)$$

From (1), substitute $y - 8.63$ for x in equation (2).

$$\begin{aligned} (y - 8.63) + y &= 63.89 \\ 2y - 8.63 &= 63.89 \\ 2y &= 72.52 \\ y &= 36.26 \end{aligned}$$

From (1), $x = y - 8.63 = 36.26 - 8.63 = 27.63$. The average retail price of Elmo was \$27.63 and of Kid was \$36.26.

37. Let $x =$ the cost of a small box and $y =$ the cost of a large box.

The cost of 10 small and 20 large boxes is \$65, so

$$10x + 20y = 65. \quad (1)$$

The cost of 6 small and 10 large boxes is \$34, so

$$6x + 10y = 34. \quad (2)$$

To eliminate y, multiply equation (2) by -2 and add the result to equation (1).

$$\begin{array}{rcrcrl} 10x & + & 20y & = & 65 & (1) \\ -12x & - & 20y & = & -68 & \quad -2 \times (2) \\ \hline -2x & & & = & -3 & \\ & & x & = & \frac{3}{2} & \text{or } 1.5 \end{array}$$

Substitute $\frac{3}{2}$ for x in equation (2) to find y.

$$\begin{aligned} 6x + 10y &= 34 \qquad (2) \\ 6\left(\frac{3}{2}\right) + 10y &= 34 \\ 9 + 10y &= 34 \\ 10y &= 25 \\ y &= \frac{25}{10} \text{ or } \frac{5}{2} \text{ or } 2.5 \end{aligned}$$

A small box costs \$1.50, and a large box costs \$2.50.

38. Let $x =$ the cost of a pound of peanuts and $y =$ the cost of a pound of cashews.

The cost of 6 lb of peanuts and 12 lb of cashews is \$60, so

$$6x + 12y = 60. \quad (1)$$

The cost of 3 lb of peanuts and 4 lb of cashews is \$22, so

$$3x + 4y = 22. \quad (2)$$

To eliminate x, multiply equation (2) by -2 and add the result to equation (1).

$$\begin{array}{rcrcrl} 6x & + & 12y & = & 60 & (1) \\ -6x & - & 8y & = & -44 & \quad -2 \times (2) \\ \hline & & 4y & = & 16 & \\ & & y & = & 4 & \end{array}$$

Substitute 4 for y in equation (2) to find x.

$$\begin{aligned} 3x + 4y &= 22 \quad (2) \\ 3x + 4(4) &= 22 \\ 3x + 16 &= 22 \\ 3x &= 6 \\ x &= 2 \end{aligned}$$

Peanuts cost \$2/lb, and cashews cost \$4/lb.

39. The lines intersect at $(8, 3000)$, so the cost equals the revenue at $x = 8$ (which is 800 items). The revenue is \$3000.

40. On the sale of 1100 parts ($x = 11$), the revenue is about \$4100 and the cost is about \$3700.

$$\begin{aligned} \text{Profit} &= \text{Revenue} - \text{Cost} \\ &\approx 4100 - 3700 \\ &= 400 \end{aligned}$$

The profit is about \$400.

CHAPTER 5 EXPONENTS, POLYNOMIALS, AND POLYNOMIAL FUNCTIONS

Section 5.1

1. $(ab)^2 = a^2b^2$ by a power rule. Since $a^2b^2 \neq ab^2$, the expression $(ab)^2 = ab^2$ has been simplified incorrectly. The exponent should apply to both a and b.

3. $\left(\frac{4}{a}\right)^3 = \frac{4^3}{a^3}$

   Since $\frac{4^3}{a^3} \neq \frac{4^3}{a}$, the expression
   $$\left(\frac{4}{a}\right)^3 = \frac{4^3}{a}$$
   has been simplified incorrectly.

5. The product rule says that when like bases are multiplied, the base stays the same and the exponents are added. For example, $x^5 \cdot x^6 = x^{11}$.

7. $13^4 \cdot 13^8 = 13^{4+8} = 13^{12}$

9. $x^3 \cdot x^5 \cdot x^9 = x^{3+5+9} = x^{17}$

11. $(-3w^5)(9w^3) = (-3)(9)w^{5+3} = -27w^8$

13. $(2x^2y^5)(9xy^3) = (2)(9)x^{2+1}y^{5+3} = 18x^3y^8$

15. $r^2 \cdot s^4$ cannot be simplified because the product rule does not apply.

17. **(a)** $9^0 = 1$ **(B)**

    (b) $-9^0 = -(9^0) = -(1) = -1$ **(C)**

    (c) $(-9)^0 = 1$ **(B)**

    (d) $-(-9)^0 = -1$ **(C)**

19. $25^0 = 1$, since $a^0 = 1$ for any nonzero base a.

21. $-7^0 = -(7^0) = -(1) = -1$

23. $(-15)^0 = 1$ since -15 is in parentheses.

25. $3^0 + (-3)^0 = 1 + 1 = 2$

27. $-3^0 + 3^0 = -1 + 1 = 0$

29. $-4^0 - m^0 = -1 - 1 = -2$

31. **(a)** $4^{-2} = \frac{1}{4^2} = \frac{1}{16}$ **(B)**

    (b) $-4^{-2} = -\frac{1}{4^2} = -\frac{1}{16}$ **(D)**

    (c) $(-4)^{-2} = \frac{1}{(-4)^2} = \frac{1}{16}$ **(B)**

    (d) $-(-4)^{-2} = -\frac{1}{(-4)^2} = -\frac{1}{16}$ **(D)**

33. $5^{-4} = \frac{1}{5^4} = \frac{1}{625}$

35. $8^{-1} = \frac{1}{8^1} = \frac{1}{8}$

37. $(4x)^{-2} = \frac{1}{(4x)^2} = \frac{1}{4^2x^2} = \frac{1}{16x^2}$

39. $4x^{-2} = \frac{4}{x^2}$

41. $-a^{-3} = -\frac{1}{a^3}$

43. $(-a)^{-4} = \frac{1}{(-a)^4} = \frac{1}{a^4}$

45. $5^{-1} + 6^{-1} = \frac{1}{5} + \frac{1}{6} = \frac{6}{30} + \frac{5}{30} = \frac{11}{30}$

47. $8^{-1} - 3^{-1} = \frac{1}{8} - \frac{1}{3} = \frac{3}{24} - \frac{8}{24} = -\frac{5}{24}$

49. Consider $-a^n$ and $(-a)^n$. Let $a = 2$.

    If $n = 2$, then
    $-a^n = -2^2 = -4$ and $(-a)^n = (-2)^2 = 4$.
    If $n = 3$, then
    $-a^n = -2^3 = -8$ and $(-a)^n = (-2)^3 = -8$.
    If $n = 4$, then
    $-a^n = -2^4 = -16$ and $(-a)^n = (-2)^4 = 16$.
    If $n = 5$, then
    $-a^n = -2^5 = -32$ and $(-a)^n = (-2)^5 = -32$.
    If $n = 6$, then
    $-a^n = -2^6 = -64$ and $(-a)^n = (-2)^6 = 64$.

    Based on these cases, when n is even, the expressions are opposites. When n is odd, the expressions are equal.

51. $\frac{1}{4^{-2}} = 4^2 = 16$

53. $\frac{2^{-2}}{3^{-3}} = \frac{3^3}{2^2} = \frac{27}{4}$

55. $\left(\frac{2}{3}\right)^{-3} = \left(\frac{3}{2}\right)^3 = \frac{3^3}{2^3} = \frac{27}{8}$

57. $\left(\frac{4}{5}\right)^{-2} = \left(\frac{5}{4}\right)^2 = \frac{5^2}{4^2} = \frac{25}{16}$

59. **(a)** $\left(\frac{1}{3}\right)^{-1} = \left(\frac{3}{1}\right)^1 = 3$ **(B)**

    (b) $\left(-\frac{1}{3}\right)^{-1} = \left(-\frac{3}{1}\right)^1 = -3$ **(D)**

    (c) $-\left(\frac{1}{3}\right)^{-1} = -\left(\frac{3}{1}\right)^1 = -3$ **(D)**

(d) $-\left(-\frac{1}{3}\right)^{-1} = -\left(-\frac{3}{1}\right)^{1} = -(-3) = 3$ **(B)**

61. The quotient rule says that when like bases are divided, the base stays the same and the exponents are subtracted. For example, $\frac{x^8}{x^5} = x^3$.

63. $\frac{4^8}{4^6} = 4^{8-6} = 4^2$ or 16

65. $\frac{x^{12}}{x^8} = x^{12-8} = x^4$

67. $\frac{r^7}{r^{10}} = r^{7-10} = r^{-3} = \frac{1}{r^3}$

69. $\frac{6^4}{6^{-2}} = 6^{4-(-2)} = 6^{4+2} = 6^6$

71. $\frac{6^{-3}}{6^7} = 6^{-3-7} = 6^{-10} = \frac{1}{6^{10}}$

73. $\frac{7}{7^{-1}} = 7^{1-(-1)} = 7^2$ or 49

75. $\frac{r^{-3}}{r^{-6}} = r^{-3-(-6)} = r^{-3+6} = r^3$

77. $\frac{x^3}{y^2}$ cannot be simplified because the quotient rule does not apply.

79. $(x^3)^6 = x^{3 \cdot 6} = x^{18}$

81. $\left(\frac{3}{5}\right)^3 = \frac{3^3}{5^3} = \frac{27}{125}$

83. $(4t)^3 = 4^3t^3 = 64t^3$

85. $(-6x^2)^3 = (-6)^3x^{2 \cdot 3} = -216x^6$

87. $\left(\frac{-4m^2}{t}\right)^3 = \frac{(-4)^3m^{2\cdot 3}}{t^3} = \frac{-64m^6}{t^3} = -\frac{64m^6}{t^3}$

89. $3^5 \cdot 3^{-6} = 3^{5+(-6)} = 3^{-1} = \frac{1}{3^1} = \frac{1}{3}$

91. $a^{-3}a^2a^{-4} = a^{-3+2+(-4)} = a^{-5} = \frac{1}{a^5}$

93.
$$\begin{aligned}(k^2)^{-3}k^4 &= k^{2(-3)}k^4\\ &= k^{-6}k^4\\ &= k^{-6+4}\\ &= k^{-2}\\ &= \frac{1}{k^2}\end{aligned}$$

95.
$$\begin{aligned}-4r^{-2}(r^4)^2 &= -4r^{-2}(r^8)\\ &= -4r^{-2+8}\\ &= -4r^6\end{aligned}$$

97.
$$\begin{aligned}(5a^{-1})^4(a^2)^{-3} &= 5^4a^{-1\cdot 4}a^{2(-3)}\\ &= 5^4a^{-4}a^{-6}\\ &= 5^4a^{-4-6}\\ &= 5^4a^{-10}\\ &= \frac{5^4}{a^{10}} \text{ or } \frac{625}{a^{10}}\end{aligned}$$

99.
$$\begin{aligned}(z^{-4}x^3)^{-1} &= z^{-4(-1)}x^{3(-1)}\\ &= z^4x^{-3}\\ &= \frac{z^4}{x^3}\end{aligned}$$

101.
$$\begin{aligned}7k^2(-2k)(4k^{-5})^0 &= 7(-2)k^2k \cdot 1\\ &= -14k^{2+1}\\ &= -14k^3\end{aligned}$$

103. $\frac{(p^{-2})^0}{5p^{-4}} = \frac{1 \cdot p^4}{5} = \frac{p^4}{5}$

105.
$$\begin{aligned}\frac{(3pq)q^2}{6p^2q^4} &= \frac{3p^1q^{1+2}}{6p^2q^4} = \frac{3pq^3}{6p^2q^4}\\ &= \frac{1}{2}p^{1-2}q^{3-4}\\ &= \frac{1}{2}p^{-1}q^{-1} = \frac{1}{2} \cdot \frac{1}{p^1} \cdot \frac{1}{q^1}\\ &= \frac{1}{2pq}\end{aligned}$$

107.
$$\begin{aligned}\frac{4a^5(a^{-1})^3}{(a^{-2})^{-2}} &= \frac{4a^5a^{-1\cdot 3}}{a^{-2(-2)}} = \frac{4a^5a^{-3}}{a^4}\\ &= 4a^{5-3-4}\\ &= 4a^{-2}\\ &= \frac{4}{a^2}\end{aligned}$$

109. The first step may be the most confusing.

$$\begin{aligned}(-y^{-4})^2 &= (-1 \cdot y^{-4})^2\\ &= (-1)^2(y^{-4})^2\\ &= 1 \cdot y^{-4(2)} = y^{-8}\end{aligned}$$

It is not necessary to include these steps once the concept of squaring a negative is committed to memory.

$$\begin{aligned}\frac{(-y^{-4})^2}{6(y^{-5})^{-1}} &= \frac{y^{-4(2)}}{6y^{-5(-1)}}\\ &= \frac{y^{-8}}{6y^5}\\ &= \frac{1}{6y^5y^8}\\ &= \frac{1}{6y^{13}}\end{aligned}$$

111. $\dfrac{(2k)^2m^{-5}}{(km)^{-3}} = \dfrac{2^2k^2m^{-5}}{k^{-3}m^{-3}}$

$= 2^2k^{2-(-3)}m^{-5-(-3)}$

$= 2^2k^5m^{-2}$

$= \dfrac{2^2k^5}{m^2}$ or $\dfrac{4k^5}{m^2}$

113. $\dfrac{(2k)^2k^3}{k^{-1}k^{-5}}\left(5k^{-2}\right)^{-3} = \dfrac{2^2k^2k^3}{k^{-1}k^{-5}}\left(5^{-3}k^{-2(-3)}\right)$

$= \dfrac{2^2k^5}{k^{-6}}\left(5^{-3}k^6\right)$

$= 2^2k^{11}\left(5^{-3}k^6\right)$

$= 2^2 \cdot 5^{-3}k^{11+6}$

$= \dfrac{2^2k^{17}}{5^3}$ or $\dfrac{4k^{17}}{125}$

115. $\left(\dfrac{3k^{-2}}{k^4}\right)^{-1} \cdot \dfrac{2}{k} = \dfrac{\left(3k^{-2}\right)^{-1}}{\left(k^4\right)^{-1}} \cdot \dfrac{2}{k}$

$= \dfrac{3^{-1}k^2 \cdot 2}{k^{-4}k^1}$

$= \dfrac{3^{-1} \cdot 2k^2}{k^{-3}}$

$= \dfrac{2k^2k^3}{3}$

$= \dfrac{2k^5}{3}$

117. $\left(\dfrac{2p}{q^2}\right)^3\left(\dfrac{3p^4}{q^{-4}}\right)^{-1} = \dfrac{(2p)^3}{\left(q^2\right)^3} \cdot \dfrac{\left(3p^4\right)^{-1}}{\left(q^{-4}\right)^{-1}}$

$= \dfrac{2^3p^3}{q^6} \cdot \dfrac{3^{-1}p^{-4}}{q^4}$

$= \dfrac{2^3p^{3-4}}{3^1q^{6+4}}$

$= \dfrac{8p^{-1}}{3q^{10}} = \dfrac{8}{3pq^{10}}$

119. $\dfrac{2^2y^4(y^{-3})^{-1}}{2^5y^{-2}} = \dfrac{y^4y^3y^2}{2^{5-2}}$

$= \dfrac{y^{4+3+2}}{2^3}$

$= \dfrac{y^9}{8}$

121. $\dfrac{\left(2m^2p^3\right)^2\left(4m^2p\right)^{-2}}{\left(-3mp^4\right)^{-1}\left(2m^3p^4\right)^3}$

$= \dfrac{2^2m^4p^64^{-2}m^{-4}p^{-2}}{(-3)^{-1}m^{-1}p^{-4}2^3m^9p^{12}}$ *Power rule*

$= \dfrac{4(-3)^1\left(m^{4-4}\right)\left(p^{6-2}\right)}{4^2 \cdot 2^3\left(m^{-1+9}\right)\left(p^{-4+12}\right)}$ *Product rule*

$= \dfrac{-3m^0p^4}{4 \cdot 2^3m^8p^8}$

$= \dfrac{-3\left(p^{4-8}\right)}{32m^8}$ *Quotient rule*

$= \dfrac{-3p^{-4}}{32m^8}$

$= -\dfrac{3}{32m^8p^4}$

123. $\dfrac{\left(-3y^3x^3\right)\left(-4y^4x^2\right)\left(x^2\right)^{-4}}{18x^3y^2\left(y^3\right)^3\left(x^3\right)^{-2}}$

$= \dfrac{\left(-3y^3x^3\right)\left(-4y^4x^2\right)\left(x^{-8}\right)}{18x^3y^2y^9x^{-6}}$ *Power rule*

$= \dfrac{12x^{3+2-8}y^{3+4}}{18x^{3-6}y^{2+9}}$ *Product rule*

$= \dfrac{2x^{-3}y^7}{3x^{-3}y^{11}}$

$= \dfrac{2x^{-3-(-3)}y^{7-11}}{3}$ *Quotient rule*

$= \dfrac{2y^{-4}}{3}$

$= \dfrac{2}{3y^4}$

125. $\left(\dfrac{p^2q^{-1}}{2p^{-2}}\right)^2 \cdot \left(\dfrac{p^3 \cdot 4q^{-2}}{3q^{-5}}\right)^{-1} \cdot \left(\dfrac{pq^{-5}}{q^{-2}}\right)^3$

$= \dfrac{p^4q^{-2}p^{-3}4^{-1}q^2p^3q^{-15}}{2^2p^{-4}3^{-1}q^5q^{-6}}$ *Power rule*

$= \dfrac{4^{-1}p^{4-3+3}q^{-2+2-15}}{2^23^{-1}p^{-4}q^{5-6}}$ *Product rule*

$= \dfrac{3p^4q^{-15}}{2^2 \cdot 4p^{-4}q^{-1}}$

$= \dfrac{3p^{4-(-4)}q^{-15-(-1)}}{4 \cdot 4}$ *Quotient rule*

$= \dfrac{3p^8q^{-14}}{16}$

$= \dfrac{3p^8}{16q^{14}}$

127. $530 = 5_{\wedge}30.$ ← Decimal point

Count 2 places.

Since the number 5.3 is to be made larger, the exponent on 10 is positive.

$$530 = 5.3 \times 10^2$$

129. $.830 = .8_{\wedge}30$

Count 1 place.

Since the number 8.3 is to be made smaller, the exponent on 10 is negative.

$$.830 = 8.3 \times 10^{-1}$$

131. $.00000692 = .000006_\wedge 92$

Count 6 places.

Since the number 6.92 is to be made smaller, the exponent on 10 is negative.

$$.00000692 = 6.92 \times 10^{-6}$$

133. $-38,500 = -3_\wedge 8500.$

Count 4 places.

Since the number 3.85 is to be made larger, the exponent on 10 is positive. Also, affix a negative sign in front of the number.

$$-38,500 = -3.85 \times 10^4$$

135. $7.2 \times 10^4 = 72,000$
Move the decimal point 4 places to the *right* because of the *positive* exponent. Attach extra zeros.

137. $2.54 \times 10^{-3} = .00254$
Since the exponent is *negative*, move the decimal point 3 places to the *left*.

139. $-6 \times 10^4 = -60,000$
Move the decimal point 4 places to the *right* because of the *positive* exponent. Attach extra zeros.

141. $1.2 \times 10^{-5} = .000012$
Since the exponent is *negative*, move the decimal point 5 places to the *left*.

143.
$$\begin{aligned}\frac{12 \times 10^4}{2 \times 10^6} &= \frac{12}{2} \times \frac{10^4}{10^6}\\ &= 6 \times 10^{4-6}\\ &= 6 \times 10^{-2}\\ &= .06\end{aligned}$$

145.
$$\begin{aligned}\frac{3 \times 10^{-2}}{12 \times 10^3} &= \frac{3 \times 10^{-2}}{1.2 \times 10^4}\\ &= \frac{3}{1.2} \times \frac{10^{-2}}{10^4}\\ &= 2.5 \times 10^{-6}\\ &= .0000025\end{aligned}$$

147.
$$\begin{aligned}\frac{.05 \times 1600}{.0004} &= \frac{5 \times 10^{-2} \times 1.6 \times 10^3}{4 \times 10^{-4}}\\ &= \frac{5(1.6)}{4} \times \frac{10^{-2} \times 10^3}{10^{-4}}\\ &= 2 \times 10^{-2+3-(-4)}\\ &= 2 \times 10^5\\ &= 200,000\end{aligned}$$

149.
$$\begin{aligned}\frac{20,000 \times .018}{300 \times .0004} &= \frac{2 \times 10^4 \times 1.8 \times 10^{-2}}{3 \times 10^2 \times 4 \times 10^{-4}}\\ &= \frac{2 \times 1.8}{3 \times 4} \times \frac{10^4 \times 10^{-2}}{10^2 \times 10^{-4}}\\ &= .3 \times 10^{4-2-2-(-4)}\\ &= .3 \times 10^4\\ &= 3000\end{aligned}$$

151.
$$\begin{aligned}\$1{,}000{,}000{,}000 &= \$1 \times 10^9\\ \$1{,}000{,}000{,}000{,}000 &= \$1 \times 10^{12}\\ \$2{,}128{,}000{,}000{,}000 &= \$2.128 \times 10^{12}\\ 144{,}419 &= 1.44419 \times 10^5\end{aligned}$$

153. If x represents the number of operations per second that the normal desktop computer is capable of, then

$$\begin{aligned}15,000x &= 3.9 \times 10^8.\\ x &= \frac{3.9 \times 10^8}{1.5 \times 10^4}\\ &= \frac{3.9}{1.5} \times \frac{10^8}{10^4}\\ &= 2.6 \times 10^4\\ &= 26,000\end{aligned}$$

155. Since $d = rt$, $t = \frac{d}{r}$. Divide the distance traveled by the speed of light.

$$\frac{9 \times 10^{12}}{3 \times 10^{10}} = \frac{9}{3} \times 10^{12-10} = 3 \times 10^2$$

It will take 300 seconds.

157. First find the number of seconds in a year.

$$\begin{aligned}1 \text{ year} &= 365 \text{ days}\\ &= 365(24) \text{ hr}\\ &= 8760 \text{ hr}\\ &= 8760(60) \text{ min}\\ &= 525,600 \text{ min}\\ &= 525,600(60) \text{ sec}\\ &= 31,536,000 \text{ sec}\\ &= 3.1536 \times 10^7 \text{ sec}\end{aligned}$$

Now use $d = rt$ and multiply the rate light travels by the number of seconds in a year.

$$\begin{aligned}1 \text{ light year} &= (1.86 \times 10^5 \text{ mi/sec})\\ &\quad \cdot (3.1536 \times 10^7 \text{ sec})\\ &= 1.86(3.1536) \times 10^{5+7}\\ &\approx 5.87 \times 10^{12} \text{ mi}\end{aligned}$$

There are about 5.87×10^{12} miles in a light year.

159. **(a)** The distance from Mercury to the sun is 3.6×10^7 mi and the distance from Venus to the sun is 6.7×10^7 mi, so the distance between Mercury and Venus in miles is

$$\begin{aligned}&(6.7\times10^7)-(3.6\times10^7)\\&=(6.7-3.6)\times10^7\\&=3.1\times10^7.\end{aligned}$$

Use $d=rt$, or $\frac{d}{r}=t$, where $d=3.1\times10^7$ and $r=1.55\times10^3$.

$$\begin{aligned}\frac{3.1\times10^7}{1.55\times10^3}&=\frac{3.1}{1.55}\times10^{7-3}\\&=2\times10^4\\&=20,000\end{aligned}$$

It would take $20,000$ hr.

(b) From part (a), it takes $20,000$ hr for a spacecraft to travel from Venus to Mercury. Convert this to days (24 hours = 1 day).

$$\begin{aligned}20,000\text{ hr}&=\frac{20,000}{24}\text{ days}\\&\approx 833\text{ days}\end{aligned}$$

It would take about 833 days.

161. $(1.5\text{E}12)*(5\text{E}^{-}3)$
$$\begin{aligned}&=(1.5\times10^{12})(5\times10^{-3})\\&=(1.5\times5)(10^{12-3})\\&=7.5\times10^9\end{aligned}$$

163. $(8.4\text{E}14)/(2.1\text{E}^{-}3)$
$$\begin{aligned}&=\frac{8.4\times10^{14}}{2.1\times10^{-3}}\\&=\frac{8.4}{2.1}\times\frac{10^{14}}{10^{-3}}\\&=4\times10^{14-(-3)}\\&=4\times10^{17}\end{aligned}$$

Section 5.2

1. $2x^3+x-3x^2+4$
The polynomial is written in descending powers of the variable if the exponents on the terms of the polynomial decrease from left to right.

$$2x^3-3x^2+x+4$$

3. $4p^3-8p^5+p^7=p^7-8p^5+4p^3$

5. $-m^3+5m^2+3m^4+10$
$$=3m^4-m^3+5m^2+10$$

7. In $7z$, the coefficient is 7 and, since $7z=7z^1$, the degree is 1.

9. In $-15p^2$, the coefficient is -15 and the degree is 2.

11. In x^4, since $x^4=1x^4$, the coefficient is 1 and the degree is 4.

13. In $\frac{t}{6}=\frac{1}{6}t$, the coefficient is $\frac{1}{6}$ and since $\frac{t}{6}=\frac{1}{6}t^1$, the degree is 1.

15. In $-mn^5$, the coefficient is -1, since $-mn^5=-1mn^5$, and the degree is 6, since the sum of the exponents on m and n is $1+5=6$.

17. 25 is one term, so it's a *monomial.* 25 is a nonzero constant, so it has degree zero.

19. $7m-22$ has two terms, so it's a *binomial.* The exponent on m is 1, so $7m-22$ has degree 1.

21. $-7y^6+11y^8$ is a *binomial* of degree 8.

23. $-5m^3+6m-9m^2$ has three terms, so it's a *trinomial.* The greatest exponent is 3, so the degree is 3.

25. $-6p^4q-3p^3q^2+2pq^3-q^4$ has four terms, so it is classified as *none of these.* The greatest sum of exponents on any term is 5, so the polynomial has degree 5.

27. Only choice **A** is a trinomial (it has three terms) with descending powers and having degree 6.

29. $5z^4+3z^4=(5+3)z^4=8z^4$

31. $-m^3+2m^3+6m^3=(-1+2+6)m^3=7m^3$

33. $x+x+x+x+x$
$$\begin{aligned}&=(1+1+1+1+1)x\\&=5x\end{aligned}$$

35. m^4-3m^2+m is *already simplified* since there are no like terms to be combined.

37. $$\begin{aligned}y^2+7y-4y^2&=(y^2-4y^2)+7y\\&=(1-4)y^2+7y\\&=-3y^2+7y\end{aligned}$$

39. $2k+3k^2+5k^2-7$
$$\begin{aligned}&=(3k^2+5k^2)+2k-7\\&=(3+5)k^2+2k-7\\&=8k^2+2k-7\end{aligned}$$

41. $n^4-2n^3+n^2-3n^4+n^3$
$$\begin{aligned}&=n^4-3n^4-2n^3+n^3+n^2\\&=(1-3)n^4+(-2+1)n^3+n^2\\&=-2n^4-n^3+n^2\end{aligned}$$

43. $3ab^2+7a^2b-5ab^2+13a^2b$
$$\begin{aligned}&=(3-5)ab^2+(7+13)a^2b\\&=-2ab^2+20a^2b\end{aligned}$$

45. $4-(2+3m)+6m+9$
$$\begin{aligned}&=4-2-3m+6m+9\\&=(-3+6)m+(4-2+9)\\&=3m+11\end{aligned}$$

47. $(6+3p)-(2p+1)-(2p+9)$
$=6+3p-2p-1-2p-9$
$=(3-2-2)p+(6-1-9)$
$=-p-4$

49. A *monomial* (or *term*) is a numeral, a variable, or a product of numerals and variables raised to positive integer powers. Some examples of monomials are 6, x, and $-4x^2y^3$. A *binomial* is a sum or difference of exactly two terms such as x^2+y^2 or x^2-y^2. A *trinomial* consists of exactly three terms, such as x^2-3x+8. These are all examples of *polynomials*.

51. $(5x^2+7x-4)+(3x^2-6x+2)$
$=5x^2+3x^2+7x-6x-4+2$
$=8x^2+x-2$

53. $(6t^2-4t^4-t)+(3t^4-4t^2+5)$
$=-4t^4+3t^4+6t^2-4t^2-t+5$
$=(-4+3)t^4+(6-4)t^2-t+5$
$=-t^4+2t^2-t+5$

55. $(y^3+3y+2)+(4y^3-3y^2+2y-1)$
$=y^3+4y^3-3y^2+3y+2y+2-1$
$=(1+4)y^3-3y^2+(3+2)y+(2-1)$
$=5y^3-3y^2+5y+1$

57. $(3r+8)-(2r-5)$

Change all signs in the second polynomial and add.

$=(3r+8)+(-2r+5)$
$=3r+8-2r+5$
$=3r-2r+8+5$
$=r+13$

59. $(2a^2+3a-1)-(4a^2+5a+6)$
$=(2a^2+3a-1)+(-4a^2-5a-6)$
$=2a^2-4a^2+3a-5a-1-6$
$=-2a^2-2a-7$

61. $(z^5+3z^2+2z)-(4z^5+2z^2-5z)$
$=z^5+3z^2+2z-4z^5-2z^2+5z$
$=z^5-4z^5+3z^2-2z^2+2z+5z$
$=-3z^5+z^2+7z$

63.
$$\begin{array}{r} 21p-8 \\ -9p+4 \\ \hline 12p-4 \end{array}$$
Add vertically.

65.
$$\begin{array}{r} -12p^2+4p-1 \\ 3p^2+7p-8 \\ \hline -9p^2+11p-9 \end{array}$$
Add vertically.

67.
$$\begin{array}{r} 12a+15 \\ 7a-3 \\ \hline \end{array}$$

To subtract, change all the signs in the second polynomial, and add.

$$\begin{array}{r} 12a+15 \\ -7a+3 \\ \hline 5a+18 \end{array}$$

69.
$$\begin{array}{r} 6m^2-11m+5 \\ -8m^2+2m-1 \\ \hline \end{array}$$

To subtract, change all the signs in the second polynomial, and add.

$$\begin{array}{r} 6m^2-11m+5 \\ 8m^2-2m+1 \\ \hline 14m^2-13m+6 \end{array}$$

71. Add column by column to obtain the result on the bottom line.

$$\begin{array}{r} 12z^2-11z+8 \\ 5z^2+16z-2 \\ -4z^2+5z-9 \\ \hline 13z^2+10z-3 \end{array}$$

73.
$$\begin{array}{r} 6y^3-9y^2\qquad+8 \\ 4y^3+2y^2+5y\qquad \\ \hline 10y^3-7y^2+5y+8 \end{array}$$

75.
$$\begin{array}{r} -5a^4\qquad+8a^2-9 \\ 6a^3-a^2+2 \\ \hline \end{array}$$

To subtract, change all the signs in the second polynomial, and add.

$$\begin{array}{r} -5a^4\qquad+8a^2-9 \\ -6a^3+a^2-2 \\ \hline -5a^4-6a^3+9a^2-11 \end{array}$$

77. $7y^2-6y+5-(4y^2-2y+3)$
$=7y^2-6y+5-4y^2+2y-3$
$=7y^2-4y^2-6y+2y+5-3$
$=3y^2-4y+2$

79. Simplify the expression in brackets first.

$(3m^2-5n^2+2n)+(-3m^2)+4n^2$
$=3m^2-5n^2+2n-3m^2+4n^2$
$=3m^2-3m^2-5n^2+4n^2+2n$
$=-n^2+2n$

Now perform the subtraction.

$(-4m^2+3n^2-5n)-(-n^2+2n)$
$=-4m^2+3n^2-5n+n^2-2n$
$=-4m^2+3n^2+n^2-5n-2n$
$=-4m^2+4n^2-7n$

81. $\left[-\left(y^4 - y^2 + 1\right) - \left(y^4 + 2y^2 + 1\right)\right]$
$+ \left(3y^4 - 3y^2 - 2\right)$
$= -y^4 + y^2 - 1 - y^4 - 2y^2 - 1$
$+ 3y^4 - 3y^2 - 2$
$= -y^4 - y^4 + 3y^4 + y^2 - 2y^2 - 3y^2$
$- 1 - 1 - 2$
$= y^4 - 4y^2 - 4$

83. $-\left[3z^2 + 5z - \left(2z^2 - 6z\right)\right]$
$+ \left[\left(8z^2 - \left[5z - z^2\right]\right) + 2z^2\right]$
$= -\left(3z^2 + 5z - 2z^2 + 6z\right)$
$+ \left(8z^2 - 5z + z^2 + 2z^2\right)$
$= -\left(3z^2 - 2z^2 + 5z + 6z\right)$
$+ \left(8z^2 + z^2 + 2z^2 - 5z\right)$
$= -\left(z^2 + 11z\right) + \left(11z^2 - 5z\right)$
$= -z^2 - 11z + 11z^2 - 5z$
$= -z^2 + 11z^2 - 11z - 5z$
$= 10z^2 - 16z$

Section 5.3

1. $f(x) = 6x - 4$

(a) $f(-1) = 6(-1) - 4$
$= -6 - 4 = -10$

(b) $f(2) = 6(2) - 4$
$= 12 - 4 = 8$

3. $f(x) = x^2 - 3x + 4$

(a) $f(-1) = (-1)^2 - 3(-1) + 4$
$= 1 + 3 + 4 = 8$

(b) $f(2) = (2)^2 - 3(2) + 4$
$= 4 - 6 + 4 = 2$

5. $f(x) = 5x^4 - 3x^2 + 6$

(a) $f(-1) = 5(-1)^4 - 3(-1)^2 + 6$
$= 5 \cdot 1 - 3 \cdot 1 + 6$
$= 5 - 3 + 6 = 8$

(b) $f(2) = 5(2)^4 - 3(2)^2 + 6$
$= 5 \cdot 16 - 3 \cdot 4 + 6$
$= 80 - 12 + 6 = 74$

7. $f(x) = -x^2 + 2x^3 - 8$

(a) $f(-1) = -(-1)^2 + 2(-1)^3 - 8$
$= -(1) + 2(-1) - 8$
$= -1 - 2 - 8 = -11$

(b) $f(2) = -(2)^2 + 2(2)^3 - 8$
$= -(4) + 2 \cdot 8 - 8$
$= -4 + 16 - 8 = 4$

9. $f(x) = -6.77x^2 + 445.34x + 11{,}279.82$

(a) $x = 1970 - 1970 = 0$

$f(0) = -6.77(0)^2 + 445.34(0) + 11{,}279.82$
$\approx 11{,}280$

(b) $x = 1985 - 1970 = 15$

$f(15) = -6.77(15)^2 + 445.34(15) + 11{,}279.82$
$\approx 16{,}437$

(c) $x = 1997 - 1970 = 27$

$f(27) = -6.77(27)^2 + 445.34(27) + 11{,}279.82$
$\approx 18{,}369$

11. $P(x) = -.31x^3 + 5.8x^2 - 15x + 9$

(a) $x = 1990 - 1990 = 0$

$P(0) = -.31(0)^3 + 5.8(0)^2 - 15(0) + 9$
$= 9$ million

(b) $x = 1996 - 1990 = 6$

$P(6) = -.31(6)^3 + 5.8(6)^2 - 15(6) + 9$
$= 60.84 \approx 61$ million

(c) $x = 1999 - 1990 = 9$

$P(9) = -.31(9)^3 + 5.8(9)^2 - 15(9) + 9$
$= 117.81 \approx 118$ million

13. **(a)** $(f + g)(x) = f(x) + g(x)$
$= (5x - 10) + (3x + 7)$
$= 8x - 3$

(b) $(f - g)(x) = f(x) - g(x)$
$= (5x - 10) - (3x + 7)$
$= (5x - 10) + (-3x - 7)$
$= 2x - 17$

15. **(a)**

$(f + g)(x) = f(x) + g(x)$
$= (4x^2 + 8x - 3) + (-5x^2 + 4x - 9)$
$= -x^2 + 12x - 12$

(b)

$(f - g)(x) = f(x) - g(x)$
$= (4x^2 + 8x - 3) - (-5x^2 + 4x - 9)$
$= (4x^2 + 8x - 3) + (5x^2 - 4x + 9)$
$= 9x^2 + 4x + 6$

For Exercises 17–27, let $f(x) = x^2 - 9$, $g(x) = 2x$, and $h(x) = x - 3$.

17. $(f + g)(x) = f(x) + g(x)$
$= \left(x^2 - 9\right) + (2x)$
$= x^2 + 2x - 9$

19. $(f + g)(3) = f(3) + g(3)$
$= \left(3^2 - 9\right) + 2(3)$
$= 0 + 6 = 6$

Alternatively, we could evaluate the polynomial in Exercise 17, $x^2 + 2x - 9$, using $x = 3$.

21. $$\begin{aligned}(f-h)(x) &= f(x)-h(x)\\ &= (x^2-9)-(x-3)\\ &= x^2-9-x+3\\ &= x^2-x-6\end{aligned}$$

23. $$\begin{aligned}(f-h)(-3) &= f(-3)-h(-3)\\ &= \left[(-3)^2-9\right]-[(-3)-3]\\ &= (9-9)-(-6)\\ &= 0+6=6\end{aligned}$$

25. $$\begin{aligned}(g+h)(-10) &= g(-10)+h(-10)\\ &= 2(-10)+[(-10)-3]\\ &= -20+(-13)\\ &= -33\end{aligned}$$

27. $$\begin{aligned}(g-h)(-3) &= g(-3)-h(-3)\\ &= 2(-3)-[(-3)-3]\\ &= -6-(-6)\\ &= -6+6=0\end{aligned}$$

29. Answers will vary. Let $f(x)=x^3$ and $g(x)=x^4$.

$$\begin{aligned}(f-g)(x) &= f(x)-g(x)\\ &= x^3-x^4\\ (g-f)(x) &= g(x)-f(x)\\ &= x^4-x^3\end{aligned}$$

Because the two differences are not equal, subtraction of polynomial functions is not commutative.

31.

x	$f(x)=-2x+1$
-2	$-2(-2)+1=5$
-1	$-2(-1)+1=3$
0	$-2(0)+1=1$
1	$-2(1)+1=-1$
2	$-2(2)+1=-3$

This is a linear function, so plot the points and draw a line through them.

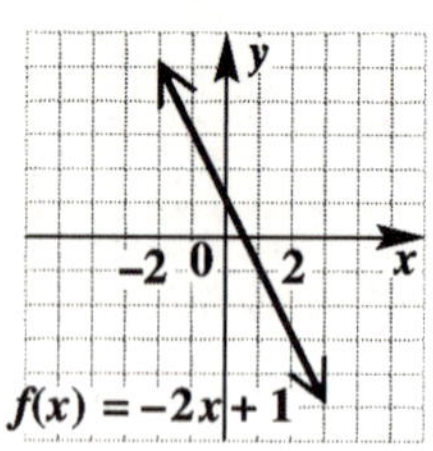

Any x-value can be used, so the domain is $(-\infty,\infty)$. From the graph, we see that any y-value can be obtained from the function, so the range is $(-\infty,\infty)$.

33.

x	$f(x)=-3x^2$
-2	$-3(-2)^2=-12$
-1	$-3(-1)^2=-3$
0	$-3(0)^2=0$
1	$-3(1)^2=-3$
2	$-3(2)^2=-12$

Since the greatest exponent is 2, the graph of f is a parabola.

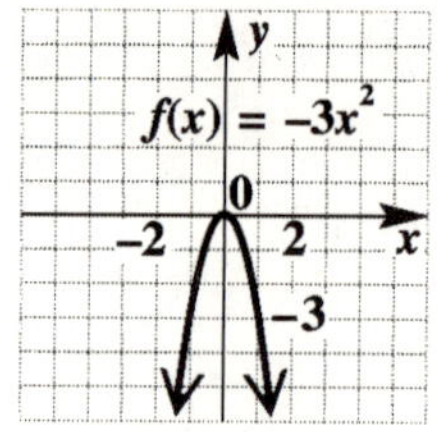

Any x-value can be used, so the domain is $(-\infty,\infty)$. From the graph, we see that the y-values are at most 0, so the range is $(-\infty,0]$.

35.

x	$f(x)=x^3+1$
-2	$(-2)^3+1=-7$
-1	$(-1)^3+1=0$
0	$(0)^3+1=1$
1	$(1)^3+1=2$
2	$(2)^3+1=9$

The greatest exponent is 3, so the graph of f is s-shaped.

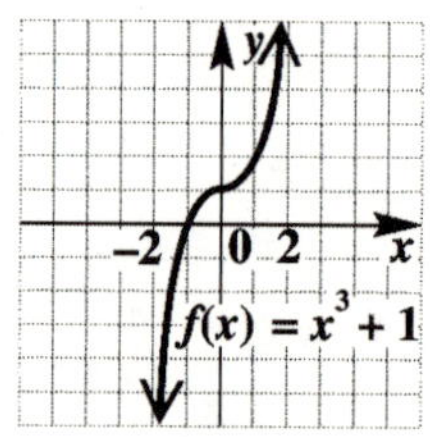

Any x-value can be used, so the domain is $(-\infty,\infty)$. From the graph, we see that any y-value can be obtained from the function, so the range is $(-\infty,\infty)$.

Section 5.4

1. $(2x-5)(3x+4)$

$$\begin{aligned}&\quad\ \ \textbf{F}\qquad\quad \textbf{O}\qquad\quad \textbf{I}\qquad\qquad \textbf{L}\\ &= (2x)(3x)+(2x)(4)+(-5)(3x)+(-5)(4)\\ &= 6x^2+8x-15x-20\\ &= 6x^2-7x-20\quad \text{(Choice }\textbf{C}\text{)}\end{aligned}$$

3. $(2x-5)(3x-4)$

$$\begin{aligned}&\quad\ \ \textbf{F}\qquad\quad \textbf{O}\qquad\quad \textbf{I}\qquad\qquad \textbf{L}\\ &= (2x)(3x)+(2x)(-4)+(-5)(3x)+(-5)(-4)\\ &= 6x^2-8x-15x+20\\ &= 6x^2-23x+20\quad \text{(Choice }\textbf{D}\text{)}\end{aligned}$$

5. $-8m^3(3m^2)=-8(3)m^{3+2}=-24m^5$

7. $$\begin{aligned}\left(14x^2y^3\right)\left(-2x^5y\right) &= 14(-2)x^{2+5}y^{3+1}\\ &= -28x^7y^4\end{aligned}$$

9. $3x(-2x+5)=3x(-2x)+3x(5)=-6x^2+15x$

11. $-q^3(2+3q)=-q^3(2)-q^3(3q)=-2q^3-3q^4$

13. $6k^2(3k^2+2k+1)$
$= 6k^2(3k^2) + 6k^2(2k) + 6k^2(1)$
$= 18k^4 + 12k^3 + 6k^2$

15. $(2m+3)(3m^2-4m-1)$
$= 2m(3m^2-4m-1) + 3(3m^2-4m-1)$
$= 2m(3m^2) + 2m(-4m) + 2m(-1) + 3(3m^2) + 3(-4m) + 3(-1)$
$= 6m^3 - 8m^2 - 2m + 9m^2 - 12m - 3$
$= 6m^3 - 8m^2 + 9m^2 - 2m - 12m - 3$
$= 6m^3 + m^2 - 14m - 3$

17. $m(m+5)(m-8)$
$= m(m^2 - 8m + 5m - 40)$
$= m(m^2 - 3m - 40)$
$= m(m^2) + m(-3m) + m(-40)$
$= m^3 - 3m^2 - 40m$

19. $4z(2z+1)(3z-4)$
$= 4z(6z^2 - 8z + 3z - 4)$
$= 4z(6z^2 - 5z - 4)$
$= 4z(6z^2) + 4z(-5z) + 4z(-4)$
$= 24z^3 - 20z^2 - 16z$

21. $4x^3(x-3)(x+2)$
$= 4x^3(x^2 + 2x - 3x - 6)$
$= 4x^3(x^2 - x - 6)$
$= 4x^3(x^2) + 4x^3(-x) + 4x^3(-6)$
$= 4x^5 - 4x^4 - 24x^3$

23.
$$\begin{array}{rrrl} & 2y & +\ 3 & \\ & 3y & -\ 4 & \\ \hline & -\ 8y & -\ 12 & \leftarrow -4(2y+3) \\ 6y^2 & +\ 9y & & \leftarrow 3y(2y+3) \\ \hline 6y^2 & +\ y & -\ 12 & \textit{Combine like terms.} \end{array}$$

25.
$$\begin{array}{rrrr} & -b^2 & +\ 3b & +\ 3 \\ & & 2b & +\ 4 \\ \hline & -\ 4b^2 & +\ 12b & +\ 12 \\ -2b^3 & +\ 6b^2 & +\ 6b & \\ \hline -2b^3 & +\ 2b^2 & +\ 18b & +\ 12 \end{array}$$

27.
$$\begin{array}{rrr} & 5m & -\ 3n \\ & 5m & +\ 3n \\ \hline & 15mn & -\ 9n^2 \\ 25m^2 & -\ 15mn & \\ \hline 25m^2 & & -\ 9n^2 \end{array}$$

29.
$$\begin{array}{rrrrr} & 2z^3 & -\ 5z^2 & +\ 8z & -\ 1 \\ & & & 4z & +\ 3 \\ \hline & 6z^3 & -\ 15z^2 & +\ 24z & -\ 3 \\ 8z^4 & -\ 20z^3 & +\ 32z^2 & -\ 4z & \\ \hline 8z^4 & -\ 14z^3 & +\ 17z^2 & +\ 20z & -\ 3 \end{array}$$

31.
$$\begin{array}{rrrrr} & & 2p^2 & +\ 3p & +\ 6 \\ & & 3p^2 & -\ 4p & -\ 1 \\ \hline & & -\ 2p^2 & -\ 3p & -\ 6 \\ & -\ 8p^3 & -\ 12p^2 & -\ 24p & \\ 6p^4 & +\ 9p^3 & +\ 18p^2 & & \\ \hline 6p^4 & +\ p^3 & +\ 4p^2 & -\ 27p & -\ 6 \end{array}$$

33. $(m+5)(m-8)$
F O I L
$= m^2 - 8m + 5m - 40$
$= m^2 - 3m - 40$

35. $(4k+3)(3k-2)$
F O I L
$= 12k^2 - 8k + 9k - 6$
$= 12k^2 + k - 6$

37. $(z-w)(3z+4w)$
F O I L
$= 3z^2 + 4zw - 3zw - 4w^2$
$= 3z^2 + zw - 4w^2$

39. $(6c-d)(2c+3d)$
F O I L
$= 12c^2 + 18cd - 2cd - 3d^2$
$= 12c^2 + 16cd - 3d^2$

41. $(.2x+1.3)(.5x-.1)$
F O I L
$= .1x^2 - .02x + .65x - .13$
$= .1x^2 + .63x - .13$

43. $\left(3w + \frac{1}{4}z\right)(w - 2z)$
F O I L
$= 3w^2 - 6wz + \frac{1}{4}wz - \frac{1}{2}z^2$
$= 3w^2 - \frac{23}{4}wz - \frac{1}{2}z^2$

45. The product of two binomials is the sum of the product of the first terms, the product of the outer terms, the product of the inner terms, and the product of the last terms.

47. Use the formula for the product of the sum and difference of two terms.

$(2p-3)(2p+3) = (2p)^2 - (3)^2$
$= 4p^2 - 9$

49. $(5m-1)(5m+1) = (5m)^2 - (1)^2$
$= 25m^2 - 1$

51. $(3a+2c)(3a-2c) = (3a)^2 - (2c)^2$
$= 9a^2 - 4c^2$

53. $\left(4x-\frac{2}{3}\right)\left(4x+\frac{2}{3}\right)=(4x)^2-\left(\frac{2}{3}\right)^2$
$=16x^2-\frac{4}{9}$

55. $(4m+7n^2)(4m-7n^2)=(4m)^2-(7n^2)^2$
$=16m^2-49n^4$

57. $(5y^3+2)(5y^3-2)=(5y^3)^2-(2)^2$
$=25y^6-4$

59. Use the formula for the square of a binomial.

$$\begin{aligned}(y-5)^2&=y^2-2(y)(5)+5^2\\&=y^2-10y+25\end{aligned}$$

61. $(2p+7)^2=(2p)^2+2(2p)(7)+7^2$
$=4p^2+28p+49$

63. $(4n+3m)^2$
$=(4n)^2+2(4n)(3m)+(3m)^2$
$=16n^2+24nm+9m^2$

65. $\left(k-\frac{5}{7}p\right)^2=k^2-2(k)\left(\frac{5}{7}p\right)+\left(\frac{5}{7}p\right)^2$
$=k^2-\frac{10}{7}kp+\frac{25}{49}p^2$

67. $(x+y)^2=x^2+2xy+y^2$. The expression x^2+y^2 is missing the $2xy$ term, so $(x+y)^2\neq x^2+y^2$ in general.

69. $[(5x+1)+6y]^2$
$=(5x+1)^2+2(5x+1)(6y)+(6y)^2$ *Square of a binomial*
$=(5x+1)^2+12y(5x+1)+36y^2$
$=\left[(5x)^2+2(5x)(1)+1^2\right]$
$+60xy+12y+36y^2$ *Square of a binomial*
$=25x^2+10x+1+60xy+12y+36y^2$

71. $[(2a+b)-3]^2$
$=(2a+b)^2-2(2a+b)(3)+3^2$
$=\left[(2a)^2+2(2a)(b)+b^2\right]$
$-6(2a+b)+3^2$
$=4a^2+4ab+b^2-12a-6b+9$

73. $[(2a+b)-3][(2a+b)+3]$
$=(2a+b)^2-3^2$ *Product of the sum and difference of two terms*
$=\left[(2a)^2+2(2a)(b)+b^2\right]-9$ *Square of a binomial*
$=4a^2+4ab+b^2-9$

75. $[(2h-k)+j][(2h-k)-j]$
$=(2h-k)^2-j^2$
$=(2h)^2-2(2h)(k)+k^2-j^2$
$=4h^2-4hk+k^2-j^2$

77. $(y+2)^3=(y+2)^2(y+2)$
$=\left[y^2+2(y)(2)+2^2\right](y+2)$
$=(y^2+4y+4)(y+2)$

$$\begin{array}{rrrr} & y^2 & +\ 4y & +\ 4\\ & & y & +\ 2\\ \hline & 2y^2 & +\ 8y & +\ 8\\ y^3 & +\ 4y^2 & +\ 4y & \\ \hline y^3 & +\ 6y^2 & +\ 12y & +\ 8\end{array}$$

79. $(5r-s)^3$
$=(5r-s)^2(5r-s)$
$=(25r^2-10rs+s^2)(5r-s)$
$=125r^3-50r^2s+5rs^2-25r^2s+10rs^2-s^3$
$=125r^3-75r^2s+15rs^2-s^3$

81. $(q-2)^4=(q-2)^2(q-2)^2$
$=(q^2-4q+4)(q^2-4q+4)$

$$\begin{array}{rrrrr} & & q^2 & -\ 4q & +\ 4\\ & & q^2 & -\ 4q & +\ 4\\ \hline & & 4q^2 & -\ 16q & +\ 16\\ & -\ 4q^3 & +\ 16q^2 & -\ 16q & \\ q^4 & -\ 4q^3 & +\ 4q^2 & & \\ \hline q^4 & -\ 8q^3 & +\ 24q^2 & -\ 32q & +\ 16\end{array}$$

83. $(2a+b)(3a^2+2ab+b^2)$
Rewrite vertically and multiply.

$$\begin{array}{rrrr} & 3a^2 & +\ 2ab & +\ b^2\\ & & 2a & +\ b\\ \hline & 3a^2b & +\ 2ab^2 & +\ b^3\\ 6a^3 & +\ 4a^2b & +\ 2ab^2 & \\ \hline 6a^3 & +\ 7a^2b & +\ 4ab^2 & +\ b^3\end{array}$$

85.

$$\begin{array}{rrrrr} & z^3 & -\ 4z^2x & +\ 2zx^2 & -\ x^3\\ & & & 4z & -\ x\\ \hline & -\ z^3x & +\ 4z^2x^2 & -\ 2zx^3 & +\ x^4\\ 4z^4 & -\ 16z^3x & +\ 8z^2x^2 & -\ 4zx^3 & \\ \hline 4z^4 & -\ 17z^3x & +\ 12z^2x^2 & -\ 6zx^3 & +\ x^4\end{array}$$

87.

$$\begin{array}{rrrrr} & & m^2 & -\ 2mp & +\ p^2\\ & & m^2 & +\ 2mp & -\ p^2\\ \hline & & -\ m^2p^2 & +\ 2mp^3 & -\ p^4\\ & 2m^3p & -\ 4m^2p^2 & +\ 2mp^3 & \\ m^4 & -\ 2m^3p & +\ m^2p^2 & & \\ \hline m^4 & & -\ 4m^2p^2 & +\ 4mp^3 & -\ p^4\end{array}$$

89. $(a+b)(a+2b)(a-3b)$
$=[(a+b)(a+2b)](a-3b)$
$=\left[a^2+3ab+2b^2\right](a-3b)$

$$\begin{array}{rrrr} & a^2 + 3ab & + 2b^2 & \\ & & a & - 3b \\ \hline & -3a^2b & -9ab^2 & -6b^3 \\ a^3 + 3a^2b & + 2ab^2 & & \\ \hline a^3 & & -7ab^2 & -6b^3 \end{array}$$

Now multiply the last polynomial times ab.

$$\begin{aligned} &ab(a^3 - 7ab^2 - 6b^3) \\ &= a^4b - 7a^2b^3 - 6ab^4 \end{aligned}$$

In Exercises 91–93, substitute 3 for x and 4 for y.

91. $(x+y)^2 = (3+4)^2 = 7^2 = 49$
$x^2 + y^2 = 3^2 + 4^2 = 9 + 6 = 25$
Since $49 \neq 25$, $(x+y)^2 \neq x^2 + y^2$.

93. $(x+y)^4 = (3+4)^4 = 7^4 = 2401$
$x^4 + y^4 = 3^4 + 4^4 = 81 + 256 = 337$
Since $2401 \neq 337$, $(x+y)^4 \neq x^4 + y^4$.

95. The formula for the area of a triangle is $A = \frac{1}{2}bh$. Use $b = 3x + 2y$ and $h = 3x - 2y$.

$$\begin{aligned} A &= \frac{1}{2}(3x+2y)(3x-2y) \\ &= \frac{1}{2}(9x^2 - 4y^2) \\ &= \frac{9}{2}x^2 - 2y^2 \end{aligned}$$

97. The formula for the area of a parallelogram is $A = bh$. Use $b = 5x + 6$ and $h = 3x - 4$.

$$\begin{aligned} A &= (5x+6)(3x-4) \\ &= 15x^2 - 20x + 18x - 24 \\ &= 15x^2 - 2x - 24 \end{aligned}$$

99. The length of each side of the entire square is a. To find the length of each side of the blue square, subtract b, the length of a side of the green rectangle, that is, $a - b$.

100. The formula for the area A of a square with side s is $A = s^2$. Since $s = a - b$, the formula for the area of the blue square would be $A = (a-b)^2$.

101. The formula for the area of a rectangle is $A = LW$. The green rectangle has length $a - b$ and width b, so each green rectangle has an area of $\underline{(a-b)b}$ or $\underline{ab - b^2}$.

Since there are two green rectangles, the total area in green is $\underline{2(ab - b^2)}$ or $\underline{2ab - 2b^2}$.

102. The length of each side of the yellow square is b, so the yellow square has an area of $\underline{b^2}$.

103. The area of the entire colored region is $\underline{a^2}$, because each side of the entire colored region has length $\underline{a}$.

104. Using the results from Exercises 99–103, the area of the blue square equals

$$a^2 - (2ab - 2b^2) - b^2 = a^2 - 2ab + b^2.$$

105. **(a)** Both expressions for the area of the blue square must be equal to each other; that is, $(a-b)^2$ from Exercise 100 and $a^2 - 2ab + b^2$ from Exercise 104.

(b) From Exercise 100, the area of the blue square is $(a-b)^2$. From Exercise 104, the area of the blue square is $a^2 - 2ab + b^2$. Since these expressions must be equal

$$(a-b)^2 = a^2 - 2ab + b^2.$$

This reinforces the special product for the square of a binomial difference.

106.

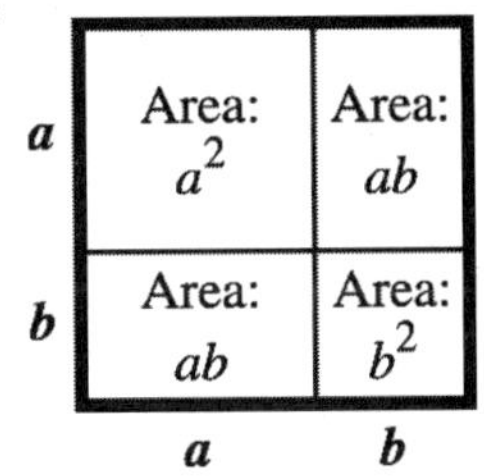

The large square is made up of two smaller squares and two congruent rectangles. The sum of the areas is

$$a^2 + 2ab + b^2.$$

Since these expressions represent the same quantity, they must be equal. Thus,

$$(a+b)^2 = a^2 + 2ab + b^2.$$

107. $f(x) = 2x$, $g(x) = 5x - 1$

$$\begin{aligned} (fg)(x) &= f(x) \cdot g(x) \\ &= 2x(5x-1) \\ &= 10x^2 - 2x \end{aligned}$$

109. $f(x) = x + 1$, $g(x) = 2x - 3$

$$\begin{aligned} (fg)(x) &= f(x) \cdot g(x) \\ &= (x+1)(2x-3) \\ &= 2x^2 - 3x + 2x - 3 \\ &= 2x^2 - x - 3 \end{aligned}$$

111. $f(x) = 2x - 3$, $g(x) = 4x^2 + 6x + 9$

$$\begin{aligned} (fg)(x) &= f(x) \cdot g(x) \\ &= (2x-3)(4x^2 + 6x + 9) \end{aligned}$$

Multiply vertically.

$$\begin{array}{rrrr} & 4x^2 + & 6x + & 9 \\ & & 2x - & 3 \\ \hline & -12x^2 & -18x & -27 \\ 8x^3 + & 12x^2 + & 18x & \\ \hline 8x^3 & & & -27 \end{array}$$

For Exercises 113–119, let $f(x) = x^2 - 9$, $g(x) = 2x$, and $h(x) = x - 3$.

113. $(fg)(x) = f(x) \cdot g(x)$
$= (x^2 - 9)(2x)$
$= 2x^3 - 18x$

115. $(fg)(2) = f(2) \cdot g(2)$
$= (2^2 - 9)[2(2)]$
$= -5 \cdot 4 = -20$

117. $(gh)(x) = g(x) \cdot h(x)$
$= (2x)(x - 3)$
$= 2x^2 - 6x$

119. $(gh)(-3) = g(-3) \cdot h(-3)$
$= [2(-3)] \cdot [(-3) - 3]$
$= -6(-6) = 36$

Section 5.5

1. We find the quotient of two monomials by using the quotient rule for exponents.

3. When dividing polynomials that are not monomials, first write them in descending powers.

5. $$\frac{15x^3 - 10x^2 + 5}{5} = \frac{15x^3}{5} - \frac{10x^2}{5} + \frac{5}{5}$$
$$= 3x^3 - 2x^2 + 1$$

7. $$\frac{9y^2 + 12y - 15}{3y} = \frac{9y^2}{3y} + \frac{12y}{3y} - \frac{15}{3y}$$
$$= 3y + 4 - \frac{5}{y}$$

9. $$\frac{15m^3 + 25m^2 + 30m}{5m^2}$$
$$= \frac{15m^3}{5m^2} + \frac{25m^2}{5m^2} + \frac{30m}{5m^2}$$
$$= 3m + 5 + \frac{6}{m}$$

11. $$\frac{14m^2n^2 - 21mn^3 + 28m^2n}{14m^2n}$$
$$= \frac{14m^2n^2}{14m^2n} - \frac{21mn^3}{14m^2n} + \frac{28m^2n}{14m^2n}$$
$$= n - \frac{3n^2}{2m} + 2$$

13. $$\frac{8wxy^2 + 3wx^2y + 12w^2xy}{4wx^2y}$$
$$= \frac{8wxy^2}{4wx^2y} + \frac{3wx^2y}{4wx^2y} + \frac{12w^2xy}{4wx^2y}$$
$$= \frac{2y}{x} + \frac{3}{4} + \frac{3w}{x}$$

15.
$$\begin{array}{r}
r^2 - 7r + 6 \\
3r - 1\,\overline{)\,3r^3 - 22r^2 + 25r - 6} \\
\underline{3r^3 - r^2 \qquad\qquad} \\
-21r^2 + 25r \qquad \\
\underline{-21r^2 + 7r \qquad} \\
18r - 6 \\
\underline{18r - 6} \\
0
\end{array}$$

Answer: $r^2 - 7r + 6$

17.
$$\begin{array}{r}
y - 3 \\
y + 6\,\overline{)\,y^2 + 3y - 18} \\
\underline{y^2 + 6y \qquad} \\
-3y - 18 \\
\underline{-3y - 18} \\
0
\end{array}$$

Answer: $y - 3$

19.
$$\begin{array}{r}
t + 5 \\
3t + 2\,\overline{)\,3t^2 + 17t + 10} \\
\underline{3t^2 + 2t \qquad} \\
15t + 10 \\
\underline{15t + 10} \\
0
\end{array}$$

Answer: $t + 5$

21.
$$\begin{array}{r}
p - 4 \\
p + 6\,\overline{)\,p^2 + 2p + 20} \\
\underline{p^2 + 6p \qquad} \\
-4p + 20 \\
\underline{-4p - 24} \\
44
\end{array}$$

Remainder

Answer: $p - 4 + \dfrac{44}{p + 6}$

23.
$$\begin{array}{r}
m^2 + 2m - 1 \\
3m - 1\,\overline{)\,3m^3 + 5m^2 - 5m + 1} \\
\underline{3m^3 - m^2 \qquad\qquad} \\
6m^2 - 5m \qquad \\
\underline{6m^2 - 2m \qquad} \\
-3m + 1 \\
\underline{-3m + 1} \\
0
\end{array}$$

Answer: $m^2 + 2m - 1$

25.
$$\begin{array}{r}
z^2 \qquad + 3 \\
2z - 5\,\overline{)\,2z^3 - 5z^2 + 6z - 15} \\
\underline{2z^3 - 5z^2 \qquad\qquad} \\
6z - 15 \\
\underline{6z - 15} \\
0
\end{array}$$

Answer: $z^2 + 3$

27.

$$\begin{array}{r}
x^2 + 2x - 3 \\
4x + 1 \overline{\smash{)}\, 4x^3 + 9x^2 - 10x + 3} \\
\underline{4x^3 + x^2} \qquad\qquad \\
8x^2 - 10x \qquad \\
\underline{8x^2 + 2x} \qquad \\
-12x + 3 \\
\underline{-12x - 3} \\
6
\end{array}$$

Remainder

Answer: $x^2 + 2x - 3 + \dfrac{6}{4x+1}$

29.

$$\begin{array}{r}
2x - 5 \\
3x^2 - 2x + 4 \overline{\smash{)}\, 6x^3 - 19x^2 + 14x - 15} \\
\underline{6x^3 - 4x^2 + 8x} \qquad \\
-15x^2 + 6x - 15 \\
\underline{-15x^2 + 10x - 20} \\
-4x + 5
\end{array}$$

Remainder

Answer: $2x - 5 + \dfrac{-4x+5}{3x^2 - 2x + 4}$

31.

$$\begin{array}{r}
x^2 + x + 3 \\
x - 1 \overline{\smash{)}\, x^3 + 0x^2 + 2x - 3} \\
\underline{x^3 - x^2} \qquad\qquad \\
x^2 + 2x \qquad \\
\underline{x^2 - x} \qquad \\
3x - 3 \\
\underline{3x - 3} \\
0
\end{array}$$

Answer: $x^2 + x + 3$

33.

$$\begin{array}{r}
3x^2 + 6x + 11 \\
x - 2 \overline{\smash{)}\, 3x^3 + 0x^2 - x + 4} \\
\underline{3x^3 - 6x^2} \qquad\qquad \\
6x^2 - x \qquad \\
\underline{6x^2 - 12x} \qquad \\
11x + 4 \\
\underline{11x - 22} \\
26
\end{array}$$

Answer: $3x^2 + 6x + 11 + \dfrac{26}{x-2}$

35.

$$\begin{array}{r}
2k^2 + 3k - 1 \\
2k^2 + 1 \overline{\smash{)}\, 4k^4 + 6k^3 + 0k^2 + 3k - 1} \\
\underline{4k^4 \qquad + 2k^2} \qquad\qquad \\
6k^3 - 2k^2 + 3k \qquad \\
\underline{6k^3 \qquad + 3k} \qquad \\
-2k^2 \qquad - 1 \\
\underline{-2k^2 \qquad - 1} \\
0
\end{array}$$

Answer: $2k^2 + 3k - 1$

37.

$$\begin{array}{r}
x^2 - 4x + 2 \\
x^2 + 3 \overline{\smash{)}\, x^4 - 4x^3 + 5x^2 - 3x + 2} \\
\underline{x^4 \qquad + 3x^2} \qquad\qquad \\
-4x^3 + 2x^2 - 3x \qquad \\
\underline{-4x^3 \qquad - 12x} \qquad \\
2x^2 + 9x + 2 \\
\underline{2x^2 \qquad + 6} \\
9x - 4
\end{array}$$

Answer: $x^2 - 4x + 2 + \dfrac{9x-4}{x^2+3}$

39.

$$\begin{array}{r}
p^2 + \frac{5}{2}p + 2 \\
2p + 2 \overline{\smash{)}\, 2p^3 + 7p^2 + 9p + 2} \\
\underline{2p^3 + 2p^2} \qquad\qquad \\
5p^2 + 9p \qquad \\
\underline{5p^2 + 5p} \qquad \\
4p + 2 \\
\underline{4p + 4} \\
-2
\end{array}$$

Answer: $p^2 + \dfrac{5}{2}p + 2 + \dfrac{-2}{2p+2}$

41.

$$\begin{array}{r}
p^2 + p + 1 \\
p - 1 \overline{\smash{)}\, p^3 + 0p^2 + 0p - 1} \\
\underline{p^3 - p^2} \qquad\qquad \\
p^2 + 0p \qquad \\
\underline{p^2 - p} \qquad \\
p - 1 \\
\underline{p - 1} \\
0
\end{array}$$

Answer: $p^2 + p + 1$

43. To start: $\dfrac{2x^2}{3x} = \dfrac{2}{3}x$

$$\begin{array}{r}
\frac{2}{3}x - 1 \\
3x + 1 \overline{\smash{)}\, 2x^2 - \frac{7}{3}x - 1} \\
\underline{2x^2 + \frac{2}{3}x} \qquad \\
-3x - 1 \\
\underline{-3x - 1} \\
0
\end{array}$$

Answer: $\dfrac{2}{3}x - 1$

45.

$$\begin{array}{r|llll} & \frac{3}{4}a & - & 2 & \\ \hline 4a+3 & 3a^2 & - & \frac{23}{4}a & - 5 \\ & 3a^2 & + & \frac{9}{4}a & \\ \hline & & & -8a & - 5 \\ & & & -8a & - 6 \\ \hline & & & & 1 \end{array}$$

Remainder

Answer: $\frac{3}{4}a - 2 + \frac{1}{4a+3}$

47. The volume of a box is the product of the height, length, and width. Use the formula $V = LWH$.

$$V = LWH$$
$$\frac{V}{LH} = W$$

Here,
$L \bullet H = (p+4)p = p^2 + 4p$, so
$W = \frac{V}{LH} = \frac{2p^3 + 15p^2 + 28p}{p^2 + 4p}$.

$$\begin{array}{r|l} & 2p + 7 \\ \hline p^2 + 4p & 2p^3 + 15p^2 + 28p \\ & 2p^3 + 8p^2 \\ \hline & \qquad 7p^2 + 28p \\ & \qquad 7p^2 + 28p \\ \hline & \qquad\qquad 0 \end{array}$$

The width is $2p + 7$.

49. $P(x) = x^3 - 4x^2 + 3x - 5$
$P(-1) = (-1)^3 - 4(-1)^2 + 3(-1) - 5$
$= -1 - 4 - 3 - 5 = -13$

Now divide the given polynomial by $x + 1$.

$$\begin{array}{r|l} & x^2 - 5x + 8 \\ \hline x+1 & x^3 - 4x^2 + 3x - 5 \\ & x^3 + x^2 \\ \hline & \quad -5x^2 + 3x \\ & \quad -5x^2 - 5x \\ \hline & \qquad\quad 8x - 5 \\ & \qquad\quad 8x + 8 \\ \hline & \qquad\qquad -13 \end{array}$$

Remainder

The remainder in the division is the same as $P(-1)$, -13. This suggests that if a polynomial is divided by $x - r$, in this case $x - (-1)$ or $x + 1$, then the remainder is equal to $P(r)$, in this case $P(-1)$.

51. $\left(\frac{f}{g}\right)(x) = \frac{f(x)}{g(x)} = \frac{10x^2 - 2x}{2x}$
$= \frac{10x^2}{2x} - \frac{2x}{2x}$
$= 5x - 1$

The x-values that are not in the domain of the quotient function g are found by solving $g(x) = 0$.

$$2x = 0$$
$$x = 0$$

53.

$$\begin{array}{r|l} & 2x - 3 \\ \hline x+1 & 2x^2 - x - 3 \\ & 2x^2 + 2x \\ \hline & \quad -3x - 3 \\ & \quad -3x - 3 \\ \hline & \qquad\quad 0 \end{array}$$

Quotient: $2x - 3$

$$g(x) = 0$$
$$x + 1 = 0$$
$$x = -1$$

55.

$$\begin{array}{r|l} & 4x^2 + 6x + 9 \\ \hline 2x-3 & 8x^3 + 0x^2 + 0x - 27 \\ & 8x^3 - 12x^2 \\ \hline & \quad 12x^2 \\ & \quad 12x^2 - 18x \\ \hline & \qquad\quad 18x - 27 \\ & \qquad\quad 18x - 27 \\ \hline & \qquad\qquad 0 \end{array}$$

Quotient: $4x^2 + 6x + 9$

$$g(x) = 0$$
$$2x - 3 = 0$$
$$2x = 3$$
$$x = \frac{3}{2}$$

For Exercises 57–63, let $f(x) = x^2 - 9$, $g(x) = 2x$, and $h(x) = x - 3$.

57. $\left(\frac{f}{g}\right)(x) = \frac{f(x)}{g(x)} = \frac{x^2 - 9}{2x}$

We must exclude any values of x that make the denominator equal to zero, that is, $x \neq 0$.

59. $\left(\frac{f}{g}\right)(2) = \frac{f(2)}{g(2)} = \frac{2^2 - 9}{2(2)} = \frac{-5}{4} = -\frac{5}{4}$

61. $\left(\frac{h}{g}\right)(x) = \frac{h(x)}{g(x)} = \frac{x - 3}{2x}, x \neq 0$

63. $\left(\frac{h}{g}\right)(3) = \frac{h(3)}{g(3)} = \frac{(3) - 3}{2(3)} = \frac{0}{6} = 0$

Chapter 5 Review Exercises

1. $4^3 = 4 \cdot 4 \cdot 4 = 64$

2. $\left(\frac{1}{3}\right)^4 = \frac{1}{3} \cdot \frac{1}{3} \cdot \frac{1}{3} \cdot \frac{1}{3} = \frac{1}{81}$

3. $(-5)^3 = (-5)(-5)(-5) = -125$

4. $$\begin{aligned} \frac{2}{(-3)^{-2}} &= \frac{2}{\frac{1}{(-3)^2}} \\ &= 2 \cdot (-3)^2 \\ &= 2 \cdot (-3)(-3) \\ &= 18 \end{aligned}$$

5. $$\begin{aligned} \left(\frac{2}{3}\right)^{-4} &= \left(\frac{3}{2}\right)^4 \\ &= \frac{3}{2} \cdot \frac{3}{2} \cdot \frac{3}{2} \cdot \frac{3}{2} \\ &= \frac{81}{16} \end{aligned}$$

6. $\left(\frac{5}{4}\right)^{-2} = \left(\frac{4}{5}\right)^2 = \frac{4}{5} \cdot \frac{4}{5} = \frac{16}{25}$

7. $5^{-1} + 6^{-1} = \frac{1}{5} + \frac{1}{6} = \frac{6}{30} + \frac{5}{30} = \frac{11}{30}$

8. $(5+6)^{-1} = 11^{-1} = \frac{1}{11}$

9. $-3^0 + 3^0 = -1 + 1 = 0$

10. $\left(3^{-4}\right)^2 = 3^{(-4) \cdot 2} = 3^{-8} = \frac{1}{3^8}$

11. $\left(x^{-4}\right)^{-2} = x^{-4(-2)} = x^8$

12. $$\begin{aligned} \left(xy^{-3}\right)^{-2} &= x^{1(-2)}y^{(-3)(-2)} \\ &= x^{-2}y^6 \\ &= \frac{1}{x^2} \cdot y^6 = \frac{y^6}{x^2} \end{aligned}$$

13. $$\begin{aligned} \left(z^{-3}\right)^3 z^{-6} &= z^{-9}z^{-6} \\ &= z^{-9+(-6)} \\ &= z^{-15} = \frac{1}{z^{15}} \end{aligned}$$

14. $$\begin{aligned} \left(5m^{-3}\right)^2 \left(m^4\right)^{-3} &= 5^2 m^{-6} m^{-12} \\ &= 25m^{-6-12} \\ &= 25m^{-18} = \frac{25}{m^{18}} \end{aligned}$$

15. $$\begin{aligned} \frac{(3r)^2 r^4}{r^{-2}r^{-3}}\left(9r^{-3}\right)^{-2} &= \frac{3^2r^2r^4 9^{-2}r^6}{r^{-2}r^{-3}} \\ &= \frac{9^1 9^{-2} r^{2+4+6}}{r^{-2-3}} \\ &= \frac{9^{-1}r^{12}}{r^{-5}} \\ &= \frac{r^{12-(-5)}}{9} = \frac{r^{17}}{9} \end{aligned}$$

16. $$\begin{aligned} \left(\frac{5z^{-3}}{z^{-1}}\right)\frac{5}{z^2} &= \frac{25z^{-3}}{z^{-1+2}} \\ &= \frac{25}{z^3 z^1} \\ &= \frac{25}{z^{3+1}} = \frac{25}{z^4} \end{aligned}$$

17. $$\begin{aligned} \left(\frac{6m^{-4}}{m^{-9}}\right)^{-1}\left(\frac{m^{-2}}{16}\right) &= \frac{6^{-1}m^4}{m^9} \cdot \frac{m^{-2}}{16} \\ &= \frac{1}{6 \cdot 16} m^{4+(-2)-9} \\ &= \frac{1}{96}m^{-7} = \frac{1}{96m^7} \end{aligned}$$

18. $$\begin{aligned} &\left(\frac{3r^5}{5r^{-3}}\right)^{-2}\left(\frac{9r^{-1}}{2r^{-5}}\right)^3 \\ &= \frac{3^{-2}r^{-10}}{5^{-2}r^6} \cdot \frac{9^3r^{-3}}{2^3r^{-15}} \\ &= \frac{5^2}{3^2}r^{-10-6} \cdot \frac{9^3}{2^3}r^{-3-(-15)} \\ &= \frac{25}{9}r^{-16} \cdot \frac{729}{8}r^{12} \\ &= \frac{(25)(729)}{(9)(8)}r^{-16+12} \\ &= \frac{2025}{8}r^{-4} = \frac{2025}{8r^4} \end{aligned}$$

19. $$\begin{aligned} \left(-3x^4y^3\right)\left(4x^{-2}y^5\right) &= -3(4)x^4x^{-2}y^3y^5 \\ &= -12x^{4-2}y^{3+5} \\ &= -12x^2y^8 \end{aligned}$$

20. $$\begin{aligned} \frac{6m^{-4}n^3}{-3mn^2} &= -2m^{-4-1}n^{3-2} \\ &= -2m^{-5}n^1 \\ &= -\frac{2n}{m^5} \text{ or } \frac{-2n}{m^5} \end{aligned}$$

21. $$\begin{aligned} \frac{(5p^{-2}q)(4p^5q^{-3})}{2p^{-5}q^5} &= \frac{20p^{-2+5}q^{1-3}}{2p^{-5}q^5} \\ &= \frac{10p^3q^{-2}}{p^{-5}q^5} \\ &= 10p^{3-(-5)}q^{-2-5} \\ &= 10p^8q^{-7} \\ &= \frac{10p^8}{q^7} \end{aligned}$$

22. $$\begin{aligned} &\left(\frac{a^{-2}b^{-1}}{3a^2}\right)^{-2}\left(\frac{b^{-2} \cdot 3a^4}{2b^{-3}}\right)^{-2}\left(\frac{a^{-4}b^5}{a^3}\right)^{-2} \\ &= \frac{a^4b^2b^4 3^{-2}a^{-8}a^8b^{-10}}{3^{-2}a^{-4}2^{-2}b^6a^{-6}} \\ &= \frac{4a^{4-8+8}b^{2+4-10}}{a^{-4-6}b^6} \\ &= \frac{4a^4b^{-4}}{a^{-10}b^6} \end{aligned}$$

continued

$= 4a^{4-(-10)}b^{-4-6}$

$= 4a^{14}b^{-10} = \dfrac{4a^{14}}{b^{10}}$

23. In $(-6)^0$, the base is -6 and the expression simplifies to 1. In -6^0, the base is 6 and the expression simplifies to -1.

24. For example, if $a = 4$,

$(2a)^{-3} = (2 \cdot 4)^{-3} = 8^{-3} = \dfrac{1}{512}$, while

$\dfrac{2}{a^3} = \dfrac{2}{4^3} = \dfrac{2}{64} = \dfrac{1}{32}$. $\dfrac{1}{512} \neq \dfrac{1}{32}$.

25. Yes, $\left(\dfrac{a}{b}\right)^{-1} = \dfrac{a^{-1}}{b^{-1}}$ for all $a, b \neq 0$.

26. No, $(ab)^{-1} \neq ab^{-1}$ for all $a, b \neq 0$. For example, let $a = 3$ and $b = 4$.

Then $(ab)^{-1} = (3 \cdot 4)^{-1} = 12^{-1} = \dfrac{1}{12}$,

while $ab^{-1} = 3 \cdot 4^{-1} = 3 \cdot \dfrac{1}{4} = \dfrac{3}{4}$. $\dfrac{1}{12} \neq \dfrac{3}{4}$.

27. Let $x = 2$ and $y = 3$. Then

$$\begin{aligned} (x^2 + y^2)^2 &= (2^2 + 3^2)^2 \\ &= (4 + 9)^2 \\ &= 13^2 = 169, \end{aligned}$$

and

$$\begin{aligned} x^4 + y^4 &= 2^4 + 3^4 \\ &= 16 + 81 = 97. \end{aligned}$$

Since $169 \neq 97$,

$$(x^2 + y^2)^2 \neq x^4 + y^4.$$

28. $13,450 = 1_{\wedge}3450.$

Place a caret to the right of the first nonzero digit. Count 4 places.

Since the number 1.345 is to be made larger, the exponent on 10 is positive.

$$13,450 = 1.345 \times 10^4$$

29. $.0000000765 = .00000007_{\wedge}65$

Count 8 places

Since the number 7.65 is to be made smaller, the exponent on 10 is negative.

$$.0000000765 = 7.65 \times 10^{-8}$$

30. $.138 = .1_{\wedge}38$

Count 1 place.

Since the number 1.38 is to be made smaller, the exponent on 10 is negative.

$$.138 = 1.38 \times 10^{-1}$$

31. $281,400,000 = 2.814 \times 10^8$

$50,454 = 5.0454 \times 10^4$

$100 = 1 \times 10^2$

32. $1.21 \times 10^6 = 1,210,000$

Move the decimal point 6 places to the right because the exponent is positive. Attach extra zeros.

33. $5.8 \times 10^{-3} = .0058$

Move the decimal point 3 places to the left because the exponent is negative.

34.
$$\begin{aligned} \frac{16 \times 10^4}{8 \times 10^8} &= \frac{16}{8} \times 10^{4-8} \\ &= 2 \times 10^{-4} \text{ or } .0002 \end{aligned}$$

35.
$$\begin{aligned} \frac{6 \times 10^{-2}}{4 \times 10^{-5}} &= \frac{6}{4} \times 10^{-2-(-5)} \\ &= 1.5 \times 10^3 \text{ or } 1500 \end{aligned}$$

36.
$$\begin{aligned} \frac{.0000000164}{.0004} &= \frac{1.64 \times 10^{-8}}{4 \times 10^{-4}} \\ &= \frac{1.64}{4} \times 10^{-8-(-4)} \\ &= .41 \times 10^{-4} \\ &= 4.1 \times 10^{-5} \text{ or } .000041 \end{aligned}$$

37.
$$\begin{aligned} &\frac{.0009 \times 12,000,000}{400,000} \\ &= \frac{9 \times 10^{-4} \times 1.2 \times 10^7}{4 \times 10^5} \\ &= \frac{9 \times 1.2}{4} \times \frac{10^{-4} \times 10^7}{10^5} \\ &= \frac{10.8}{4} \times 10^{-4+7-5} \\ &= 2.7 \times 10^{-2} \text{ or } .027 \end{aligned}$$

38. The density D is the population P divided by the area A.

$$D = \frac{P}{A}$$

We want to find the area, so solve the formula for A.

$$\begin{aligned} DA &= P \\ A &= \frac{P}{D} \\ &= \frac{3.92 \times 10^5}{400} \\ &= \frac{3.92 \times 10^5}{4 \times 10^2} \\ &= .98 \times 10^{5-2} \\ &= .98 \times 10^3 = 980 \end{aligned}$$

The area is 980 mi^2.

39. **(a)** $5449 = 5_{\wedge}449.$

Count 3 places.

$= 5.449 \times 10^3$

(b) As in Exercise 38, use $A = \dfrac{P}{D}$.

$$\begin{aligned}\frac{3.45 \times 10^5}{5.449 \times 10^3} &= \frac{3.45}{5.449} \times \frac{10^5}{10^3} \\ &\approx .63 \times 10^{5-3} \\ &= .63 \times 10^2 \\ &= 63\end{aligned}$$

The area is approximately 63 mi^2.

40. The coefficient of $14p^5$ is 14.

41. The coefficient of $-z$ is -1.

42. The coefficient of $\frac{x}{10} = \frac{1}{10}x$ is $\frac{1}{10}$.

43. The coefficient of $504p^3r^5$ is 504.

44. $9k + 11k^3 - 3k^2$

(a) In descending powers of k, the polynomial is

$$11k^3 - 3k^2 + 9k.$$

(b) The polynomial is a trinomial since it has three terms.

(c) The degree of the polynomial is 3 since the highest power of k is 3.

45. $14m^6 + 9m^7$

(a) In descending powers of m, the polynomial is

$$9m^7 + 14m^6.$$

(b) The polynomial is a binomial since it has two terms.

(c) The degree of the polynomial is 7 since the highest power of m is 7.

46. $-5y^4 + 3y^3 + 7y^2 - 2y$

(a) The polynomial is already written in descending powers of y.

(b) The polynomial has four terms, so it is none of these choices.

(c) The degree of the polynomial is 4 since the highest power of y is 4.

47. $-7q^5r^3$

(a) The polynomial is already written in descending powers.

(b) The polynomial is a monomial since it has just one term.

(c) The degree is $5 + 3 = 8$, the sum of the exponents of this term.

48. One example of a polynomial in the variable x that has degree 5, is lacking a third-degree term, and is written in descending powers of the variable is

$$x^5 + 2x^4 - x^2 + x + 2.$$

49. Add by columns.

$$\begin{array}{rrrl} 3x^2 & -\ 5x & +\ 6 & \\ -4x^2 & +\ 2x & -\ 5 & \\ \hline -1x^2 & -\ 3x & +\ 1 & \text{or } -x^2 - 3x + 1 \end{array}$$

50. Subtract.

$$\begin{array}{rrrr} -5y^3 & & +\ 8y & -\ 3 \\ & 4y^2 & +\ 2y & +\ 9 \\ \hline \end{array}$$

Change the signs in the second polynomial and add.

$$\begin{array}{rrrr} -5y^3 & & +\ 8y & -\ 3 \\ & -\ 4y^2 & -\ 2y & -\ 9 \\ \hline -5y^3 & -\ 4y^2 & +\ 6y & -\ 12 \end{array}$$

51. $$\begin{aligned}&(4a^3 - 9a + 15) - (-2a^3 + 4a^2 + 7a) \\ &= 4a^3 - 9a + 15 + 2a^3 - 4a^2 - 7a \\ &= 4a^3 + 2a^3 - 4a^2 - 9a - 7a + 15 \\ &= 6a^3 - 4a^2 - 16a + 15\end{aligned}$$

52. $$\begin{aligned}&(3y^2 + 2y - 1) + (5y^2 - 11y + 6) \\ &= 3y^2 + 5y^2 + 2y - 11y - 1 + 6 \\ &= 8y^2 - 9y + 5\end{aligned}$$

53. To find the perimeter, add the measures of the three sides.

$$\begin{aligned}&(4x^2 + 2) + (6x^2 + 5x + 2) + (2x^2 + 3x + 1) \\ &= 4x^2 + 6x^2 + 2x^2 + 5x + 3x + 2 + 2 + 1 \\ &= 12x^2 + 8x + 5\end{aligned}$$

The perimeter is $12x^2 + 8x + 5$.

54. $f(x) = -2x^2 + 5x + 7$

(a) $$\begin{aligned}f(-2) &= -2(-2)^2 + 5(-2) + 7 \\ &= -2(4) - 10 + 7 \\ &= -8 - 3 = -11\end{aligned}$$

(b) $$\begin{aligned}f(3) &= -2(3)^2 + 5(3) + 7 \\ &= -2(9) + 15 + 7 \\ &= -18 + 22 = 4\end{aligned}$$

55. **(a)** $$\begin{aligned}(f + g)(x) &= f(x) + g(x) \\ &= (2x + 3) + (5x^2 - 3x + 2) \\ &= 5x^2 + 2x - 3x + 3 + 2 \\ &= 5x^2 - x + 5\end{aligned}$$

(b) $$\begin{aligned}(f - g)(x) &= f(x) - g(x) \\ &= (2x + 3) - (5x^2 - 3x + 2) \\ &= 2x + 3 - 5x^2 + 3x - 2 \\ &= -5x^2 + 5x + 1\end{aligned}$$

(c) $(f+g)(-1)$
$= f(-1)+g(-1)$
$= [2(-1)+3]+\left[5(-1)^2-3(-1)+2\right]$
$= [1]+[10]$
$= 11$

(d) $(f-g)(-1) = f(-1)-g(-1)$
$= 1-10$ *from part (c)*
$= -9$

56. $f(x) = .241x^2 + 3.26x + 30.0$

(a) $x = 1990 - 1990 = 0$

$f(0) = .241(0)^2 + 3.26(0) + 30.0$
$= 30$ million

(b) $x = 1995 - 1990 = 5$

$f(5) = .241(5)^2 + 3.26(5) + 30.0$
$= 52.325$ million

(c) $x = 2000 - 1990 = 10$

$f(10) = .241(10)^2 + 3.26(10) + 30.0$
$= 86.7$ million

57.

x	$f(x) = -2x+5$
-2	$-2(-2)+5=9$
-1	$-2(-1)+5=7$
0	$-2(0)+5=5$
1	$-2(1)+5=3$
2	$-2(2)+5=1$

This is a linear function, so plot the points and draw a line through them.

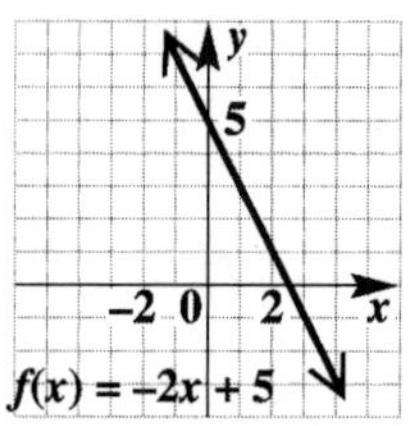

Any x-value can be used, so the domain is $(-\infty,\infty)$. From the graph, we see that any y-value can be obtained from the function, so the range is $(-\infty,\infty)$.

58.

x	$f(x) = x^2-6$
-2	$(-2)^2-6=-2$
-1	$(-1)^2-6=-5$
0	$(0)^2-6=-6$
1	$(1)^2-6=-5$
2	$(2)^2-6=-2$

Since the greatest exponent is 2, the graph of f is a parabola.

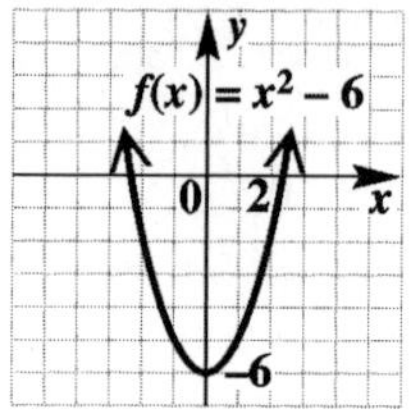

Any x-value can be used, so the domain is $(-\infty,\infty)$. From the graph, we see that the y-values are at least -6, so the range is $[-6,\infty)$.

59.

x	$f(x) = -x^3+1$
-2	$-(-2)^3+1=9$
-1	$-(-1)^3+1=2$
0	$-(0)^3+1=1$
1	$-(1)^3+1=0$
2	$-(2)^3+1=-7$

The greatest exponent is 3, so the graph of f is s-shaped.

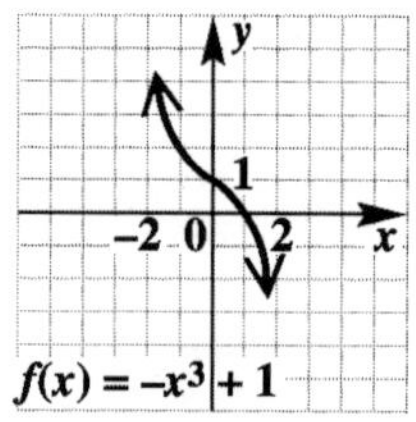

Any x-value can be used, so the domain is $(-\infty,\infty)$. From the graph, we see that any y-value can be obtained from the function, so the range is $(-\infty,\infty)$.

60. $-6k(2k^2+7) = -6k(2k^2) - 6k(7)$
$= -12k^3 - 42k$

61. $(3m-2)(5m+1)$
F O I L
$= 15m^2 + 3m - 10m - 2$
$= 15m^2 - 7m - 2$

62. $(7y-8)(2y+3)$
F O I L
$= 14y^2 + 21y - 16y - 24$
$= 14y^2 + 5y - 24$

63. $(3w-2t)(2w-3t)$
F O I L
$= 6w^2 - 9wt - 4wt + 6t^2$
$= 6w^2 - 13wt + 6t^2$

64. $(2p^2+6p)(5p^2-4)$
F O I L
$= 10p^4 - 8p^2 + 30p^3 - 24p$
$= 10p^4 + 30p^3 - 8p^2 - 24p$

65. $(3q^2+2q-4)(q-5)$

$$\begin{aligned}&= (3q^2+2q-4)(q)+(3q^2+2q-4)(-5)\\&= 3q^3+2q^2-4q-15q^2-10q+20\\&= 3q^3+2q^2-15q^2-4q-10q+20\\&= 3q^3-13q^2-14q+20\end{aligned}$$

66. $(3z^3-2z^2+4z-1)(3z-2)$

$$\begin{aligned}&= (3z^3-2z^2+4z-1)(3z)\\&\quad+(3z^3-2z^2+4z-1)(-2)\\&= 9z^4-6z^3+12z^2-3z\\&\quad-6z^3+4z^2-8z+2\\&= 9z^4-6z^3-6z^3+12z^2+4z^2\\&\quad-3z-8z+2\\&= 9z^4-12z^3+16z^2-11z+2\end{aligned}$$

67. $(6r^2-1)(6r^2+1)=(6r^2)^2-1^2$
$$=36r^4-1$$

68. $\left(z+\frac{3}{5}\right)\left(z-\frac{3}{5}\right)=z^2-\left(\frac{3}{5}\right)^2$
$$=z^2-\frac{9}{25}$$

69. $(4m+3)^2=(4m)^2+2(4m)(3)+3^2$
$$=16m^2+24m+9$$

70. $(2n-10)^2$

$$\begin{aligned}&=(2n)^2-2(2n)(10)+10^2\\&=4n^2-40n+100\end{aligned}$$

71. $\dfrac{4y^3-12y^2+5y}{4y}=\dfrac{4y^3}{4y}-\dfrac{12y^2}{4y}+\dfrac{5y}{4y}$
$$=y^2-3y+\frac{5}{4}$$

72. $\dfrac{2p^3+9p^2+27}{2p-3}$

$$\begin{array}{r|rrrr}
 & p^2 & +\ 6p & +\ 9 & \\ \hline
2p-3 & 2p^3 & +\ 9p^2 & +\ 0p & +\ 27\\
 & 2p^3 & -\ 3p^2 & & \\ \hline
 & & 12p^2 & +\ 0p & \\
 & & 12p^2 & -\ 18p & \\ \hline
 & & & 18p & +\ 27\\
 & & & 18p & -\ 27\\ \hline
 & & & & 54
\end{array}$$

Remainder

Answer: $p^2+6p+9+\dfrac{54}{2p-3}$

73. $\dfrac{5p^4+15p^3-33p^2-9p+18}{5p^2-3}$

$$\begin{array}{r|rrrrr}
 & & & p^2 & +\ 3p & -\ 6\\ \hline
5p^2-3 & 5p^4 & +\ 15p^3 & -\ 33p^2 & -\ 9p & +\ 18\\
 & 5p^4 & & -\ 3p^2 & & \\ \hline
 & & 15p^3 & -\ 30p^2 & -\ 9p & \\
 & & 15p^3 & & -\ 9p & \\ \hline
 & & & -30p^2 & & +\ 18\\
 & & & -30p^2 & & +\ 18\\ \hline
 & & & & & 0
\end{array}$$

Answer: p^2+3p-6

74. **(a)** $4^{-2}=\dfrac{1}{4^2}=\dfrac{1}{16}$ **(A)**

(b) $-4^2=-(4^2)=-16$ **(G)**

(c) $4^0=1$ **(C)**

(d) $(-4)^0=1$ **(C)**

(e) $(-4)^{-2}=\dfrac{1}{(-4)^2}=\dfrac{1}{16}$ **(A)**

(f) $-4^0=-(4^0)=-1$ **(E)**

(g) $-4^0+4^0=-1+1=0$ **(B)**

(h) $-4^0-4^0=(-1)-1=-2$ **(H)**

(i) $4^{-2}+4^{-1}=\dfrac{1}{4^2}+\dfrac{1}{4}=\dfrac{1}{16}+\dfrac{4}{16}=\dfrac{5}{16}$ **(F)**

75. $(4x+1)(2x-3)$

F O I L

$$\begin{aligned}&=8x^2-12x+2x-3\\&=8x^2-10x-3\end{aligned}$$

76. $\dfrac{6^{-1}y^3(y^2)^{-2}}{6y^{-4}(y^{-1})}=\dfrac{y^3y^{-4}}{6^1\cdot 6y^{-4}y^{-1}}$

$$\begin{aligned}&=\frac{y^{3-4-(-4-1)}}{36}\\&=\frac{y^{-1+5}}{36}=\frac{y^4}{36}\end{aligned}$$

77. $5^{-3}=\dfrac{1}{5^3}=\dfrac{1}{5\cdot 5\cdot 5}=\dfrac{1}{125}$

78. $(y^6)^{-5}(2y^{-3})^{-4}$

$$\begin{aligned}&=y^{-30}(2)^{-4}y^{12}\\&=\frac{y^{-30+12}}{2^4}\\&=\frac{y^{-18}}{2^4}=\frac{1}{16y^{18}}\end{aligned}$$

79. $7p^5(3p^4+p^3+2p^2)$

$$\begin{aligned}&=7p^5(3p^4)+7p^5(p^3)+7p^5(2p^2)\\&=21p^9+7p^8+14p^7\end{aligned}$$

80. $(2x-9)^2 = (2x)^2 - 2(2x)(9) + 9^2$
$= 4x^2 - 36x + 81$

81. $\dfrac{(-z^{-2})^3}{5(z^{-3})^{-1}} = \dfrac{(-1)^3 z^{-2(3)}}{5z^{-3(-1)}}$
$= \dfrac{-z^{-6}}{5z^3}$
$= \dfrac{-z^{-6-3}}{5}$
$= \dfrac{-z^{-9}}{5} = -\dfrac{1}{5z^9}$

82. $-(-3)^2 = -(9) = -9$

83.
$$\begin{array}{r|l} & 8x + 1 \\ x - 3 & 8x^2 - 23x + 2 \\ & 8x^2 - 24x \\ & \quad\quad x + 2 \\ & \quad\quad x - 3 \\ & \quad\quad\quad 5 \end{array}$$

Answer: $8x + 1 + \dfrac{5}{x-3}$

84. $\dfrac{(5z^2x^3)^2(2zx^2)^{-1}}{(-10zx^{-3})^{-2}(3z^{-1}x^{-4})^2}$
$= \dfrac{5^2z^4x^6 2^{-1}z^{-1}x^{-2}}{(-10)^{-2}z^{-2}x^6 3^2 z^{-2}x^{-8}}$
$= \dfrac{25(-10)^2z^3x^4}{2 \cdot 9z^{-4}x^{-2}}$
$= \dfrac{25(100)z^7x^6}{2 \cdot 9}$
$= \dfrac{1250z^7x^6}{9}$

85. $[(3m-5n)+p][(3m-5n)-p]$
$= (3m-5n)^2 - (p)^2$
$= (3m)^2 - 2(3m)(5n) + (5n)^2 - p^2$
$= 9m^2 - 30mn + 25n^2 - p^2$

86. $\dfrac{20y^3x^3 + 15y^4x + 25yx^4}{10yx^2}$
$= \dfrac{20y^3x^3}{10yx^2} + \dfrac{15y^4x}{10yx^2} + \dfrac{25yx^4}{10yx^2}$
$= 2y^2x + \dfrac{3y^3}{2x} + \dfrac{5x^2}{2}$

87. $(2k-1) - (3k^2 - 2k + 6)$
$= 2k - 1 - 3k^2 + 2k - 6$
$= -3k^2 + 2k + 2k - 1 - 6$
$= -3k^2 + 4k - 7$

88. $(-5 + 11w) + (6 + 5w) + (-15 - 8w^2)$
$= -5 + 6 - 15 + 11w + 5w - 8w^2$
$= -14 + 16w - 8w^2$

Chapter 5 Test

1. **(a)** $7^{-2} = \dfrac{1}{7^2} = \dfrac{1}{49}$ **(C)**

(b) $7^0 = 1$ **(A)**

(c) $-7^0 = -(1) = -1$ **(D)**

(d) $(-7)^0 = 1$ **(A)**

(e) $-7^2 = -49$ **(E)**

(f) $7^{-1} + 2^{-1} = \dfrac{1}{7} + \dfrac{1}{2}$
$= \dfrac{2}{14} + \dfrac{7}{14} = \dfrac{9}{14}$ **(F)**

(g) $(7+2)^{-1} = 9^{-1} = \dfrac{1}{9}$ **(B)**

(h) $\dfrac{7^{-1}}{2^{-1}} = \dfrac{2^1}{7^1} = \dfrac{2}{7}$ **(G)**

(i) $(-7)^{-2} = \dfrac{1}{(-7)^2} = \dfrac{1}{49}$ **(C)**

2. $(3x^{-2}y^3)^{-2}(4x^3y^{-4})$
$= 3^{-2}x^{-2(-2)}y^{3(-2)}4x^3y^{-4}$
$= 3^{-2}x^4y^{-6}4x^3y^{-4}$
$= \dfrac{4x^{4+3}y^{-6-4}}{3^2}$
$= \dfrac{4x^7y^{-10}}{9} = \dfrac{4x^7}{9y^{10}}$

3. $\dfrac{36r^{-4}(r^2)^{-3}}{6r^4} = \dfrac{36r^{-4}r^{2(-3)}}{6r^4}$
$= \dfrac{6r^{-4}r^{-6}}{r^4}$
$= \dfrac{6r^{-10}}{r^4} = \dfrac{6}{r^{14}}$

4. $\left(\dfrac{4p^2}{q^4}\right)^3\left(\dfrac{6p^8}{q^{-8}}\right)^{-2}$
$= \dfrac{4^3p^6}{q^{12}} \cdot \dfrac{6^{-2}p^{-16}}{q^{16}}$
$= \dfrac{4^3p^{-10}}{6^2q^{28}}$
$= \dfrac{64}{36p^{10}q^{28}} = \dfrac{16}{9p^{10}q^{28}}$

5. $(-2x^4y^{-3})^0(-4x^{-3}y^{-8})^2$
$= 1(-4)^2x^{-6}y^{-16}$
$= \dfrac{16}{x^6y^{16}}$

6. $9.1 \times 10^{-7} = .00000091$
Move the decimal point 7 places to the left because the exponent is negative.

7. $\dfrac{(2,500,000)(.00003)}{(.05)(5,000,000)}$

$= \dfrac{(2.5 \times 10^6)(3 \times 10^{-5})}{(5 \times 10^{-2})(5 \times 10^6)}$

$= \dfrac{7.5 \times 10^1}{25 \times 10^4}$

$= .3 \times 10^{1-4}$

$= .3 \times 10^{-3}$

$= 3 \times 10^{-4}$ or $.0003$

8. **(a)** $f(x) = -2x^2 + 5x - 6$

$f(4) = -2(4)^2 + 5(4) - 6$

$= -2 \cdot 16 + 20 - 6$

$= -32 + 14 = -18$

(b) $(f+g)(x) = f(x) + g(x)$

$= (-2x^2 + 5x - 6) + (7x - 3)$

$= -2x^2 + 12x - 9$

(c) $(f-g)(x) = f(x) - g(x)$

$= (-2x^2 + 5x - 6) - (7x - 3)$

$= -2x^2 + 5x - 6 - 7x + 3$

$= -2x^2 - 2x - 3$

(d) Using the answer in part (c), we have

$(f-g)(-2) = -2(-2)^2 - 2(-2) - 3$

$= -8 + 4 - 3 = -7.$

9.

x	$f(x) = -2x^2 + 3$
-2	$-2(-2)^2 + 3 = -5$
-1	$-2(-1)^2 + 3 = 1$
0	$-2(0)^2 + 3 = 3$
1	$-2(1)^2 + 3 = 1$
2	$-2(2)^2 + 3 = -5$

Since the greatest exponent is 2, the graph of f is a parabola.

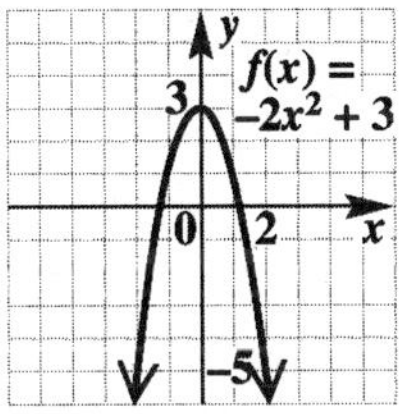

10. $f(x) = .178x^2 + 19.1x + 614$

$f(x)$ represents the number of medical doctors, in thousands, in the United States.

	Year	x	$f(x)$
(a)	1990	0	614
(b)	1996	6	$735.008 \approx 735$
(c)	1999	9	$800.318 \approx 800$

11. $(4x^3 - 3x^2 + 2x - 5)$

$- (3x^3 + 11x + 8) + (x^2 - x)$

$= 4x^3 - 3x^2 + 2x - 5 - 3x^3 - 11x$

$- 8 + x^2 - x$

$= x^3 - 2x^2 - 10x - 13$

12. $(5x - 3)(2x + 1)$

F O I L

$= 10x^2 + 5x - 6x - 3$

$= 10x^2 - x - 3$

13. $(2m - 5)(3m^2 + 4m - 5)$

$= 2m(3m^2 + 4m - 5)$

$+ (-5)(3m^2 + 4m - 5)$

$= 6m^3 + 8m^2 - 10m$

$- 15m^2 - 20m + 25$

$= 6m^3 - 7m^2 - 30m + 25$

14. $(6x + y)(6x - y) = (6x)^2 - y^2$

$= 36x^2 - y^2$

15. $(3k + q)^2 = (3k)^2 + 2(3k)(q) + q^2$

$= 9k^2 + 6kq + q^2$

16. $[2y + (3z - x)][2y - (3z - x)]$

$= (2y)^2 - (3z - x)^2$

$= 4y^2 - (9z^2 - 6zx + x^2)$

$= 4y^2 - 9z^2 + 6zx - x^2$

17. $\dfrac{16p^3 - 32p^2 + 24p}{4p^2}$

$= \dfrac{16p^3}{4p^2} - \dfrac{32p^2}{4p^2} + \dfrac{24p}{4p^2}$

$= 4p - 8 + \dfrac{6}{p}$

18. $(x^3 + 3x^2 - 4) \div (x - 1)$

Insert $0x$ for the missing x-term.

$$\begin{array}{r|l} & x^2 + 4x + 4 \\ x - 1 & x^3 + 3x^2 + 0x - 4 \\ & \underline{x^3 - x^2} \\ & \quad 4x^2 + 0x \\ & \quad \underline{4x^2 - 4x} \\ & \qquad 4x - 4 \\ & \qquad \underline{4x - 4} \\ & \qquad\qquad 0 \end{array}$$

Answer: $x^2 + 4x + 4$

19. **(a)** $(fg)(x) = f(x) \cdot g(x)$

$= (x^2 + 3x + 2)(x + 1)$

$= (x^2 + 3x + 2)(x)$

$+ (x^2 + 3x + 2)(1)$

$= x^3 + 3x^2 + 2x + x^2 + 3x + 2$

$= x^3 + 4x^2 + 5x + 2$

(b) $(fg)(-2) = f(-2)\cdot g(-2)$
$$= [(-2)^2 + 3(-2) + 2]\cdot[(-2) + 1]$$
$$= [4 - 6 + 2]\cdot[-1]$$
$$= 0(-1) = 0$$

Alternatively, we could have substituted -2 for x into our answer from part (a).

20. (a) $\left(\frac{f}{g}\right)(x) = \frac{f(x)}{g(x)} = \frac{x^2 + 3x + 2}{x + 1}$

$$\begin{array}{r|l} & x + 2 \\ \hline x + 1 & x^2 + 3x + 2 \\ & x^2 + x \\ \hline & 2x + 2 \\ & 2x + 2 \\ \hline & 0 \end{array}$$

Thus, $\left(\frac{f}{g}\right)(x) = x + 2$ if $x + 1 \neq 0$, that is, $x \neq -1$.

(b) Using our answer from part (a),

$$\left(\frac{f}{g}\right)(-2) = (-2) + 2 = 0.$$

Cumulative Review Exercises Chapters 1–5

1. 34 is a natural number, so it is also a whole number, an integer, a rational number, and a real number. **A, B, C, D, F**

2. 0 is a whole number, so it is also an integer, a rational number, and a real number. **B, C, D, F**

3. 2.16 is a rational number, so it is also a real number. **D, F**

4. $-\sqrt{36} = -6$ is an integer, so it is also a rational number and a real number. **C, D, F**

5. $\sqrt{13}$ is an irrational number, so it is also a real number. **E, F**

6. $-\frac{4}{5}$ is a rational number, so it is also a real number. **D, F**

7. $9\cdot 4 - 16 \div 4 = (9\cdot 4) - (16 \div 4) = 36 - 4 = 32$

8. $-|8 - 13| - |-4| + |-9| = -|-5| - 4 + 9$
$$= -5 - 4 + 9$$
$$= -9 + 9 = 0$$

9. $-5(8 - 2z) + 4(7 - z) = 7(8 + z) - 3$
$$-40 + 10z + 28 - 4z = 56 + 7z - 3 \quad \textit{Distributive property}$$
$$6z - 12 = 7z + 53 \quad \textit{Combine like terms}$$
$$-65 = z \quad \textit{Subtract 6z; 53}$$

Thus, the solution set is $\{-65\}$.

10. $3(x + 2) - 5(x + 2) = -2x - 4$
$$3x + 6 - 5x - 10 = -2x - 4$$
$$-2x - 4 = -2x - 4$$

The last statement is true for all real numbers, so the solution set is $(-\infty, \infty)$.

11. Solve $A = p + prt$ for t.
$$A = p + prt$$
$$A - p = prt \quad \textit{Subtract p}$$
$$\frac{A - p}{pr} = t \quad \textit{Divide by pr}$$

12. $2(m + 5) - 3m + 1 > 5$
$$2m + 10 - 3m + 1 > 5$$
$$-m + 11 > 5$$
$$-m > -6$$
$$m < 6$$

Solution set: $(-\infty, 6)$

13. $|3x - 1| = 2$

$3x - 1 = 2$ or $3x - 1 = -2$
$3x = 3$ $\quad$ $3x = -1$
$x = 1$ or $x = -\frac{1}{3}$

Solution set: $\left\{-\frac{1}{3}, 1\right\}$

14. $|3z + 1| \geq 7$

$3z + 1 \geq 7$ or $3z + 1 \leq -7$
$3z \geq 6$ $\quad$ $3z \leq -8$
$z \geq 2$ or $z \leq -\frac{8}{3}$

Solution set: $\left(-\infty, -\frac{8}{3}\right] \cup [2, \infty)$

15. Personal computer: $\frac{480}{1500} = 32\%$

Pacemaker:

$$26\% \text{ of } 1500 = .26(1500) = 390$$

Wireless communication:

$$18\% \text{ of } 1500 = .18(1500) = 270$$

Television: $\frac{150}{1500} = 10\%$

16. The sum of the measures of the angles of any triangle is 180°, so

$$(x + 15) + (6x + 10) + (x - 5) = 180.$$

Solve this equation.

$$8x + 20 = 180$$
$$8x = 160$$
$$x = 20$$

Substitute 20 for x to find the measures of the angles.

$$x-5=20-5=15$$
$$x+15=20+15=35$$
$$6x+10=6(20)+10=130$$

The measures of the angles of the triangle are 15°, 35°, and 130°.

17. Through $(-4,5)$ and $(2,-3)$

Use the definition of slope with $x_1=-4$, $y_1=5$, $x_2=2$, and $y_2=-3$.

$$m=\frac{y_2-y_1}{x_2-x_1}=\frac{-3-5}{2-(-4)}=\frac{-8}{6}=-\frac{4}{3}$$

18. Horizontal, through $(4,5)$

The slope of every horizontal line is 0.

19. Through $(4,-1)$, $m=-4$

Use the point-slope form with $x_1=4$, $y_1=-1$, and $m=-4$.

$$y-y_1=m(x-x_1)$$
$$y-(-1)=-4(x-4)$$
$$y+1=-4x+16$$
$$y=-4x+15$$

20. Through $(0,0)$ and $(1,4)$

Find the slope.

$$m=\frac{4-0}{1-0}=4$$

Because the slope is 4 and the y-intercept is 0, the equation of the line in slope-intercept form is

$$y=4x.$$

21. $-3x+4y=12$

If $y=0$, $x=-4$, so the x-intercept is $(-4,0)$.

If $x=0$, $y=3$, so the y-intercept is $(0,3)$.

Draw a line through these intercepts. A third point may be used as a check.

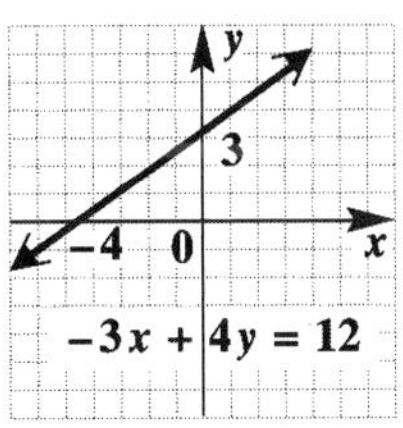

22. $y\le 2x-6$

Graph the boundary, $y=2x-6$, as a solid line through the intercepts $(3,0)$ and $(0,-6)$. A third point such as $(1,-4)$ can be used as a check. Using $(0,0)$ as a test point results in the false inequality $0\le -6$, so shade the region *not* containing the origin. This is the region below the line. The solid line shows that the boundary is part of the graph.

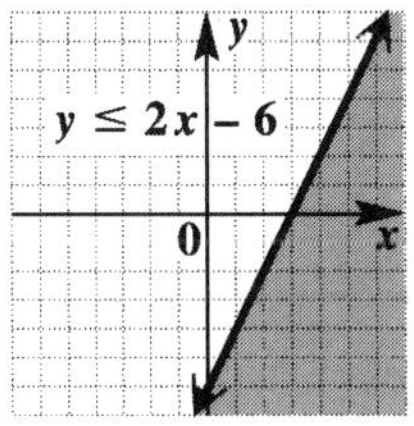

23. $3x+2y<0$

Graph the boundary, $3x+2y=0$, as a dashed line through $(0,0)$, $(-2,3)$, and $(2,-3)$. Choose a test point not on the line. Using $(1,1)$ results in the false statement $5<0$, so shade the region *not* containing $(1,1)$. This is the region below the line. The dashed line shows that the boundary is not part of the graph.

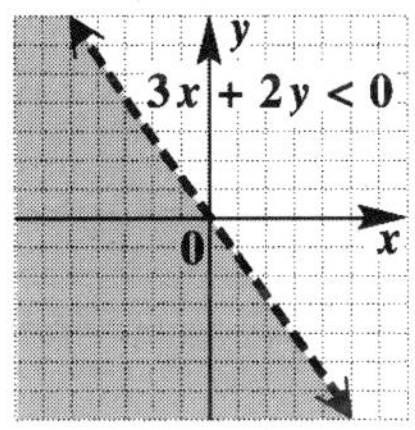

24. **(a)** $m=\dfrac{118{,}916-93{,}865}{10-0}=\dfrac{25{,}051}{10}=2505.1$

The average rate of change is 2505.1 twin births per year, that is, the number of twin births increased an average of 2505.1 per year.

(b) $y=mx+b$
$y=2505.1x+93{,}865$

(c) If $x=2002-1990=12$, then

$$y=2505.1(12)+93{,}865$$
$$=123{,}926.2$$

The model predicts about 123,926 twin births in 2002.

25. $\{(-4,-2),(-1,0),(2,0),(5,2)\}$

The domain is the set of first components, that is, $\{-4,-1,2,5\}$.
The range is the set of second components, that is, $\{-2,0,2\}$.
The relation is a function since each first component is paired with a unique second component.

26. Let $F =$ the force of the wind, $A =$ the area, and $v =$ the velocity.
Then $F = kAv^2$.

Let $F = 50$, $A = \frac{1}{2}$, and $v = 40$.

$$\begin{aligned} 50 &= k\left(\frac{1}{2}\right)(40)^2 \\ 50 &= k\left(\frac{1}{2}\right)(1600) \\ 50 &= 800k \\ k &= \frac{50}{800} = \frac{1}{16} \end{aligned}$$

So $F = \frac{1}{16}Av^2$.
Let $A = 2$ and $v = 80$.

$$\begin{aligned} F &= \frac{1}{16}(2)(80)^2 \\ &= \frac{1}{16}(2)(6400) = 800 \end{aligned}$$

The force is 800 lb.

27. $3x - 4y = 1$ (1)
$2x + 3y = 12$ (2)

To eliminate y, multiply equation (1) by 3 and equation (2) by 4. Then add the results.

$$\begin{array}{rcrcrl} 9x & - & 12y & = & 3 & 3 \times (1) \\ 8x & + & 12y & = & 48 & 4 \times (2) \\ \hline 17x & & & = & 51 & \\ & & x & = & 3 & \end{array}$$

Since $x = 3$,

$$\begin{aligned} 3x - 4y &= 1 \quad (1) \\ 3(3) - 4y &= 1 \\ 9 - 4y &= 1 \\ -4y &= -8 \\ y &= 2 \end{aligned}$$

Solution set: $\{(3, 2)\}$

28. $3x - 2y = 4$ (1)
$-6x + 4y = 7$ (2)

Multiply equation (1) by 2 and add the result to equation (2).

$$\begin{array}{rcrcrl} 6x & - & 4y & = & 8 & 2 \times (1) \\ -6x & + & 4y & = & 7 & (2) \\ \hline & & 0 & = & 15 & \textit{False} \end{array}$$

Since a false statement results, the system is *inconsistent*. The solution set is $\emptyset$.

29.
$$\begin{aligned} x + 3y - 6z &= 7 \quad (1) \\ 2x - y + z &= 1 \quad (2) \\ x + 2y + 2z &= -1 \quad (3) \end{aligned}$$

To eliminate x, multiply equation (1) by -2 and add the result to equation (2).

$$\begin{array}{rcrcrcrl} -2x & - & 6y & + & 12z & = & -14 & -2 \times (1) \\ 2x & - & y & + & z & = & 1 & (2) \\ \hline & - & 7y & + & 13z & = & -13 & (4) \end{array}$$

To eliminate x again, multiply equation (3) by -2 and add the result to equation (2).

$$\begin{array}{rcrcrcrl} 2x & - & y & + & z & = & 1 & (2) \\ -2x & - & 4y & - & 4z & = & 2 & -2 \times (3) \\ \hline & - & 5y & - & 3z & = & 3 & (5) \end{array}$$

Use equations (4) and (5) to eliminate z. Multiply equation (4) by 3 and add the result to 13 times equation (5).

$$\begin{array}{rcrcrl} -21y & + & 39z & = & -39 & 3 \times (4) \\ -65y & - & 39z & = & 39 & 13 \times (5) \\ \hline -86y & & & = & 0 & \\ & & y & = & 0 & \end{array}$$

From (5), $-3z = 3$, so $z = -1$.
From (3), $x - 2 = -1$, so $x = 1$.

Solution set: $\{(1, 0, -1)\}$

30. The length L of the rectangular flag measured 12 feet more than its width W, so

$$L = W + 12. \quad (1)$$

The perimeter is 144 feet.

$$P = 2L + 2W \quad (2)$$

Substitute $W + 12$ for L into equation (2).

$$\begin{aligned} 144 &= 2(W + 12) + 2W \\ 144 &= 2W + 24 + 2W \\ 120 &= 4W \\ 30 &= W \end{aligned}$$

From (1), $L = 30 + 12 = 42$.

The length is 42 feet and the width is 30 feet.

31. Make a chart.

Number of Liters	Percent (as a decimal)	Pure Liters of Alcohol
x	.15	$.15x$
y	.30	$.30y$
9	.20	$.20(9) = 1.8$

From the first and third columns, we have the following system:

$$\begin{aligned} x + y &= 9 \quad (1) \\ .15x + .30y &= 1.8 \quad (2) \end{aligned}$$

To eliminate x, multiply equation (1) by -15 and add the result to 100 times equation (2).

$$\begin{array}{rcl} -15x - 15y &=& -135 \\ 15x + 30y &=& 180 \\ \hline 15y &=& 45 \\ y &=& 3 \end{array}$$

From (1), $x + 3 = 9$, so $x = 6$.

She should use 6 L of 15% solution and 3 L of 30% solution.

32. $$\left(\frac{2m^3n}{p^2}\right)^3 = \frac{2^3(m^3)^3n^3}{(p^2)^3} = \frac{8m^9n^3}{p^6}$$

33. $$\frac{x^{-6}y^3z^{-1}}{x^7y^{-4}z} = \frac{y^4y^3}{x^6x^7z^1z} = \frac{y^7}{x^{13}z^2}$$

34. $(2m^{-2}n^3)^{-3}$

$$= 2^{-3}(m^{-2})^{-3}(n^3)^{-3}$$
$$= 2^{-3}m^{(-2)(-3)}n^{3(-3)}$$
$$= 2^{-3}m^6n^{-9}$$
$$= \frac{m^6}{2^3n^9} = \frac{m^6}{8n^9}$$

35. $2(3x^2 - 8x + 1) - 4(x^2 - 3x - 9)$

$$= 6x^2 - 16x + 2 - 4x^2 + 12x + 36$$

Distributive property

$$= (6x^2 - 4x^2) + (-16x + 12x) + (2 + 36)$$

Combine like terms

$$= 2x^2 - 4x + 38$$

36. $(3x + 2y)(5x - y)$

F O I L

$$= 3x(5x) + 3x(-y) + 2y(5x) + 2y(-y)$$
$$= 15x^2 - 3xy + 10xy - 2y^2$$
$$= 15x^2 + 7xy - 2y^2$$

37. $(8m + 5n)(8m - 5n) = (8m)^2 - (5n)^2$
$$= 64m^2 - 25n^2$$

38. $(x + 2y)(x^2 - 2xy + 4y^2)$

Multiply vertically.

$$\begin{array}{rrrr} & x^2 & -\ 2xy & +\ 4y^2 \\ & & x & +\ 2y \\ \hline & 2x^2y & -\ 4xy^2 & +\ 8y^3 \\ x^3 & -\ 2x^2y & +\ 4xy^2 & \\ \hline x^3 & & & +\ 8y^3 \end{array}$$

Thus,

$$(x + 2y)(x^2 - 2xy + 4y^2) = x^3 + 8y^3.$$

39. $$\frac{m^3 - 3m^2 + 5m - 3}{m - 1}$$

$$\begin{array}{r|l} & m^2 - 2m + 3 \\ m - 1 & m^3 - 3m^2 + 5m - 3 \\ & \underline{m^3 - m^2} \\ & \quad -2m^2 + 5m \\ & \quad \underline{-2m^2 + 2m} \\ & \qquad\quad 3m - 3 \\ & \qquad\quad \underline{3m - 3} \\ & \qquad\qquad\quad 0 \end{array}$$

The remainder is 0. The answer is the quotient,

$$m^2 - 2m + 3.$$

40. **(a)** 281.4 million
$$= 281{,}400{,}000$$
$$= 2.814 \times 10^8$$

(b) $\$1{,}000{,}000{,}000{,}000 = \$1 \times 10^{12}$

(c) Divide the amount by the number of people to determine how much each person would have to contribute.

$$\frac{\$1 \times 10^{12}}{2.814 \times 10^8} \approx \$.3554 \times 10^4$$
$$= \$3554$$

CHAPTER 6 FACTORING

Section 6.1

1. $12m+60=12\cdot m+12\cdot 5$
$=12(m+5)$

3. $8k^3+24k=8k\cdot k^2+8k\cdot 3$
$=8k(k^2+3)$

5. $xy-5xy^2=xy\cdot 1-xy\cdot 5y$
$=xy(1-5y)$

7. $-4p^3q^4-2p^2q^5=-2p^2q^4\cdot 2p-2p^2q^4\cdot q$
$=-2p^2q^4(2p+q)$

9. $21x^5+35x^4+14x^3$
$=7x^3(3x^2+5x+2)$

11. $10t^5-8t^4-2t^3$
$=2t^3(5t^2-4t-1)$

13. $15a^2c^3-25ac^2+5ac$
$=5ac(3ac^2-5c+1)$

15. $16z^2n^6+64zn^7-32z^3n^3$
$=16zn^3(zn^3+4n^4-2z^2)$

17. $-27m^3p^5+36m^4p^3-72m^5p^4$
$=-9m^3p^3(3p^2-4m+8m^2p)$

19. $14a^3b^2+7a^2b-21a^5b^3+42ab^4$
$=7ab(2a^2b+a-3a^4b^2+6b^3)$

21. $(m-4)(m+2)+(m-4)(m+3)$
The GCF is $(m-4)$.
$=(m-4)[(m+2)+(m+3)]$
$=(m-4)(m+2+m+3)$
$=(m-4)(2m+5)$

23. $(2z-1)(z+6)-(2z-1)(z-5)$
$=(2z-1)[(z+6)-(z-5)]$
$=(2z-1)[z+6-z+5]$
$=(2z-1)(11)$
$=11(2z-1)$

25. $5(2-x)^2-(2-x)^3+4(2-x)$
$=(2-x)[5(2-x)-(2-x)^2+4]$
$=(2-x)[10-5x-(4-4x+x^2)+4]$
$=(2-x)[10-5x-4+4x-x^2+4]$
$=(2-x)(10-x-x^2)$

27. $4(3-x)^2-(3-x)^3+3(3-x)$
$=(3-x)[4(3-x)-(3-x)^2+3]$
$=(3-x)[12-4x-(9-6x+x^2)+3]$
$=(3-x)[12-4x-9+6x-x^2+3]$
$=(3-x)(6+2x-x^2)$

29. $15(2z+1)^3+10(2z+1)^2-25(2z+1)$
The GCF is $5(2z+1)$.
$=5(2z+1)$
$\cdot[3(2z+1)^2+2(2z+1)-5]$
$=5(2z+1)$
$\cdot[3(4z^2+4z+1)+4z+2-5]$
$=5(2z+1)[12z^2+12z+3+4z+2-5]$
$=5(2z+1)[12z^2+16z]$
$=5(2z+1)[4z(3z+4)]$
$=20z(2z+1)(3z+4)$

31. $5(m+p)^3-10(m+p)^2-15(m+p)^4$
The GCF is $5(m+p)^2$.
$=5(m+p)^2[(m+p)-2-3(m+p)^2]$
$=5(m+p)^2$
$\cdot[m+p-2-3(m^2+2mp+p^2)]$
$=5(m+p)^2$
$\cdot(m+p-2-3m^2-6mp-3p^2)$

33. $-r^3+3r^2+5r$
Factor out r.
$=r(-r^2+3r+5)$
Factor out $-r$.
$=-r(r^2-3r-5)$

35. $-12s^5+48s^4$
Factor out $12s^4$.
$=12s^4(-s+4)$
Factor out $-12s^4$.
$=-12s^4(s-4)$

37. $-2x^5+6x^3+4x^2$
Factor out $2x^2$.
$=2x^2(-x^3+3x+2)$
Factor out $-2x^2$.
$=-2x^2(x^3-3x-2)$

39. $mx+3qx+my+3qy$
$=(mx+3qx)+(my+3qy)$
$=x(m+3q)+y(m+3q)$
$=(m+3q)(x+y)$

41. $10m+2n+5mk+nk$
$=(10m+2n)+(5mk+nk)$
$=2(5m+n)+k(5m+n)$
$=(5m+n)(2+k)$

43. $4-2q-6p+3pq$
$=(4-2q)+(-6p+3pq)$
$=2(2-q)-3p(2-q)$
$=(2-q)(2-3p)$

45. $p^2 - 4zq + pq - 4pz$
$= (p^2 + pq) + (-4zq - 4pz)$
$= p(p+q) - 4z(q+p)$
$= (p+q)(p-4z)$

47. $2xy - 8y + 3x - 12$
$= (2xy - 8y) + (3x - 12)$
$= 2y(x-4) + 3(x-4)$
$= (x-4)(2y+3)$

49. $m^3 + 4m^2 - 6m - 24$
$= (m^3 + 4m^2) + (-6m - 24)$
$= m^2(m+4) - 6(m+4)$
$= (m+4)(m^2 - 6)$

51. $-3a^3 - 3ab^2 + 2a^2b + 2b^3$
$= (-3a^3 - 3ab^2) + (2a^2b + 2b^3)$
$= -3a(a^2 + b^2) + 2b(a^2 + b^2)$
$= (a^2 + b^2)(-3a + 2b)$

53. $4 + xy - 2y - 2x$
$= xy - 2x - 2y + 4$
$= (xy - 2x) + (-2y + 4)$
$= x(y-2) - 2(y-2)$
$= (y-2)(x-2)$

55. $8 + 9y^4 - 6y^3 - 12y$
$= 9y^4 - 6y^3 - 12y + 8$
$= (9y^4 - 6y^3) + (-12y + 8)$
$= 3y^3(3y-2) - 4(3y-2)$
$= (3y-2)(3y^3 - 4)$

57. $1 - a + ab - b$
$= 1 - a - b + ab$
$= (1-a) + (-b + ab)$
$= 1(1-a) - b(1-a)$
$= (1-a)(1-b)$

59. $3m^{-5} + m^{-3}$

Factor out m^{-5} since -5 is the smaller exponent.

$= m^{-5}(3) + m^{-5}(m^{-3-(-5)})$

$= m^{-5}(3 + m^2)$ or $\dfrac{3+m^2}{m^5}$

61. $3p^{-3} + 2p^{-2}$

Factor out p^{-3} since -3 is the smaller exponent.

$= p^{-3}(3) + p^{-3}(2p^{-2-(-3)})$

$= p^{-3}(3 + 2p)$ or $\dfrac{3+2p}{p^3}$

63. The directions indicated to factor the polynomial *completely*. The student's response is not the complete factored form, which is $4xy^3(xy^2 - 2)$. The teacher was justified because the directions were not fulfilled.

65. Factoring a polynomial involves writing the polynomial as the product of two or more simpler polynomials. Here, choice **A** is the sum of two polynomials; choices **B** and **D** are differences of two polynomials. Only choice **C** involves the product of two polynomials and is an example of a polynomial in factored form. The correct answer is choice **C**.

Section 6.2

1. **D** is not valid.

$(8x)(4x) = 32x^2 \neq 12x^2$

3. **B** is not a factored form.

$$(-x-10)(x+6) = -x^2 - 16x - 60$$
$$\neq -x^2 + 16x - 60$$

5. To factor $y^2 + 7y - 30$, we need two integer factors whose sum is 7 (coefficient of the middle term) and whose product is -30 (the last term). Since $-3 + 10 = 7$ and $-3 \cdot 10 = -30$, we have

$$y^2 + 7y - 30 = (y-3)(y+10).$$

7. $p^2 + 15p + 56$
Two integer factors whose product is 56 and whose sum is 15 are 8 and 7.
$= (p+8)(p+7)$

9. $-m^2 + 11m - 60 = -1(m^2 - 11m + 60)$

To factor $m^2 - 11m + 60$, we need two integer factors whose product is 60 (the last term) and whose sum is -11 (coefficient of the middle term).

Factors	Sum
$-1, -60$	-61
$-2, -30$	-32
$-3, -20$	-23
$-4, -15$	-19
$-5, -12$	-17
$-6, -10$	-16

No sum is -11, so the trinomial is prime.

11. $a^2 - 2ab - 35b^2$
Two integer factors whose product is -35 and whose sum is -2 are 5 and -7.
$= (a + 5b)(a - 7b)$

13. $y^2 - 3yq - 15q^2$
There are no integer factors of -15 that add up to -3, so this trinomial is prime.

15. $x^2y^2 + 11xy + 18$
Two integer factors whose product is 18 and whose sum is 11 are 9 and 2.
$= (xy + 9)(xy + 2)$

17. $-6m^2 - 13m + 15 = -1(6m^2 + 13m - 15)$
We'll try to factor $6m^2 + 13m - 15$.
Multiply the first and last coefficients to get $6(-15) = -90$.
Two integer factors whose product is -90 and whose sum is 13 are -5 and 18.
Rewrite the trinomial in a form that can be factored by grouping.

$$\begin{aligned}&6m^2 + (+13m) - 15\\ &= 6m^2 + (-5m + 18m) - 15\\ &= (6m^2 - 5m) + (18m - 15)\\ &= m(6m - 5) + 3(6m - 5)\\ &= (6m - 5)(m + 3)\end{aligned}$$

Thus, the final factored form is

$$-1(6m - 5)(m + 3).$$

Note: These exercises can be worked using the alternative method of repeated combinations and FOIL or the grouping method.

19. $10x^2 + 3x - 18$
Two integer factors whose product is $(10)(-18) = -180$ and whose sum is 3 are 15 and -12.
Rewrite the trinomial in a form that can be factored by grouping.

$$\begin{aligned}&10x^2 + 3x - 18\\ &= 10x^2 + 15x - 12x - 18\\ &= 5x(2x + 3) - 6(2x + 3)\\ &= (2x + 3)(5x - 6)\end{aligned}$$

21. $20k^2 + 47k + 24$
Two integer factors whose product is $(20)(24) = 480$ and whose sum is 47 are 15 and 32.
Rewrite the trinomial in a form that can be factored by grouping.

$$\begin{aligned}&= 20k^2 + 15k + 32k + 24\\ &= 5k(4k + 3) + 8(4k + 3)\\ &= (4k + 3)(5k + 8)\end{aligned}$$

23. $15a^2 - 22ab + 8b^2$
Two integer factors whose product is $(15)(8) = 120$ and whose sum is -22 are -10 and -12.
Rewrite the trinomial in a form that can be factored by grouping.

$$\begin{aligned}&= 15a^2 - 10ab - 12ab + 8b^2\\ &= 5a(3a - 2b) - 4b(3a - 2b)\\ &= (3a - 2b)(5a - 4b)\end{aligned}$$

25. $36m^2 - 60m + 25$
Use the alternative method and write $36m^2$ as $6m \cdot 6m$ and 25 as $5 \cdot 5$. Use these factors in the binomial factors to obtain

$$\begin{aligned}36m^2 - 60m + 25 &= (6m - 5)(6m - 5)\\ &= (6m - 5)^2.\end{aligned}$$

27. $40x^2 + xy + 6y^2$
There are no integer factors of $(40)(6) = 240$ that add up to 1, so this trinomial is prime.

29. $6x^2z^2 + 5xz - 4$
Two integer factors whose product is $(6)(-4) = -24$ and whose sum is 5 are 8 and -3.
Rewrite the trinomial in a form that can be factored by grouping.

$$\begin{aligned}&= 6x^2z^2 + 8xz - 3xz - 4\\ &= 2xz(3xz + 4) - 1(3xz + 4)\\ &= (3xz + 4)(2xz - 1)\end{aligned}$$

31. $24x^2 + 42x + 15$
Always factor out the GCF first.

$$= 3(8x^2 + 14x + 5)$$

Now factor $8x^2 + 14x + 5$ by the alternative method.

$$8x^2 + 14x + 5 = (4x + 5)(2x + 1)$$

The final factored form is

$$3(4x + 5)(2x + 1).$$

33. $-15a^2 - 70a + 120$

$$\begin{aligned}&= -5(3a^2 + 14a - 24)\\ &= -5(a + 6)(3a - 4)\end{aligned}$$

35. $-11x^3 + 110x^2 - 264x$

$$\begin{aligned}&= -11x(x^2 - 10x + 24)\\ &= -11x(x - 6)(x - 4)\end{aligned}$$

37. $2x^3y^3 - 48x^2y^4 + 288xy^5$

$$\begin{aligned}&= 2xy^3(x^2 - 24xy + 144y^2)\\ &= 2xy^3(x - 12y)(x - 12y)\\ \text{or}\quad &2xy^3(x - 12y)^2\end{aligned}$$

39. $6a^3 + 12a^2 - 90a$

$$\begin{aligned}&= 6a(a^2 + 2a - 15)\\ &= 6a(a - 3)(a + 5)\end{aligned}$$

41. $13y^3 + 39y^2 - 52y$

$$\begin{aligned}&= 13y(y^2 + 3y - 4)\\ &= 13y(y + 4)(y - 1)\end{aligned}$$

43. $12p^3 - 12p^2 + 3p$

$$\begin{aligned}&= 3p(4p^2 - 4p + 1)\\ &= 3p(2p - 1)(2p - 1)\\ \text{or}\quad &3p(2p - 1)^2\end{aligned}$$

45. There is a GCF of 2. She did not factor the polynomial *completely*. The factor $(4x + 10)$ can be factored further as $2(2x + 5)$, giving the final form as $2(2x + 5)(x - 2)$.

47. In $12p^6 - 32p^3r + 5r^2$, let $x = p^3$ to obtain

$$12x^2 - 32xr + 5r^2 = (6x - r)(2x - 5r).$$

Replace x with p^3.

$$12p^6 - 32p^3r + 5r^2 = (6p^3 - r)(2p^3 - 5r)$$

49. $10(k+1)^2 - 7(k+1) + 1$
Let $x = k + 1$ to obtain
$10x^2 - 7x + 1 = (5x - 1)(2x - 1)$.
Replace x with $k + 1$.
$10(k+1)^2 - 7(k+1) + 1$
$= [5(k+1) - 1][2(k+1) - 1]$
$= (5k + 5 - 1)(2k + 2 - 1)$
$= (5k + 4)(2k + 1)$

51. $3(m+p)^2 - 7(m+p) - 20$
Let $x = m + p$ to obtain
$3x^2 - 7x - 20 = (3x + 5)(x - 4)$.
Replace x with $m + p$.
$3(m+p)^2 - 7(m+p) - 20$
$= [3(m+p) + 5][(m+p) - 4]$
$= (3m + 3p + 5)(m + p - 4)$

53. $a^2(a+b)^2 - ab(a+b)^2 - 6b^2(a+b)^2$
Factor out the GCF, $(a+b)^2$.
$= (a+b)^2(a^2 - ab - 6b^2)$
Factor the trinomial.
$= (a+b)^2(a - 3b)(a + 2b)$

55. $p^2(p+q) + 4pq(p+q) + 3q^2(p+q)$
Factor out the GCF, $p + q$.
$= (p+q)(p^2 + 4pq + 3q^2)$

Factor the trinomial.
$= (p+q)(p+q)(p+3q)$
$= (p+q)^2(p+3q)$

57. $z^2(z-x) - zx(x-z) - 2x^2(z-x)$
Factor out -1 from the middle term:
$x - z = -1(z - x)$.
$= z^2(z-x) + zx(z-x) - 2x^2(z-x)$
Factor out the GCF, $z - x$.
$= (z-x)(z^2 + zx - 2x^2)$
Factor the trinomial.
$= (z-x)(z+2x)(z-x)$
$= (z-x)^2(z+2x)$

59. In $p^4 - 10p^2 + 16$, let $x = p^2$ to obtain

$$x^2 - 10x + 16 = (x - 8)(x - 2).$$

Replace x with p^2.

$$p^4 - 10p^2 + 16 = (p^2 - 8)(p^2 - 2)$$

61. In $2x^4 - 9x^2 - 18$, let $y = x^2$ to obtain

$$2y^2 - 9y - 18 = (2y + 3)(y - 6).$$

Replace y with x^2.

$$2x^4 - 9x^2 - 18 = (2x^2 + 3)(x^2 - 6)$$

63. In $16x^4 + 16x^2 + 3$, let $m = x^2$ to obtain

$$16m^2 + 16m + 3 = (4m + 3)(4m + 1).$$

Replace m with x^2.

$$16x^4 + 16x^2 + 3 = (4x^2 + 3)(4x^2 + 1)$$

Section 6.3

1. **A.** Yes, 64 and m^2 are squares.

B. No, $2x^2$ is not a square.

C. No, $k^2 + 9$ is a *sum* of squares.

D. Yes, $4z^2$ and 49 are squares.

So, the binomials that are differences of squares are **A** and **D**.

3. **A.** Since $x^2 - 8x - 16$ has a negative third term, it is not a perfect square trinomial.

B. $4m^2 + 20m + 25 = (2m)^2 + 20m + 5^2$

$2(2m)(5) = 20m$, the middle term. Therefore, this trinomial is a perfect square.

C. $9z^4 + 30z^2 + 25 = (3z^2)^2 + 30z^2 + 5^2$

$2(3z^2)(5) = 30z^2$, the middle term. Therefore, this trinomial is a perfect square.

D. $25a^2 - 45a + 81 = (5a)^2 - 45a + (-9)^2$
$2(5a)(-9) = -90a$

This is not the middle term, so the trinomial is not a perfect square.

So, the perfect square trinomials are **B** and **C**.

5. The sum of two squares can be factored only if the binomial has a common factor. For example,

$$9x^2 + 81 = 9(x^2 + 9).$$

7. $p^2 - 16 = p^2 - 4^2$
$= (p + 4)(p - 4)$

9. $25x^2 - 4 = (5x)^2 - 2^2$
$= (5x + 2)(5x - 2)$

11. $18a^2 - 98b^2 = 2(9a^2 - 49b^2)$
$= 2[(3a)^2 - (7b)^2]$
$= 2[(3a + 7b)(3a - 7b)]$
$= 2(3a + 7b)(3a - 7b)$

13. $64m^4 - 4y^4$
$= 4(16m^4 - y^4)$
$= 4\left[(4m^2)^2 - (y^2)^2\right]$
Factor the difference of squares.
$= 4(4m^2 + y^2)(4m^2 - y^2)$
$= 4(4m^2 + y^2)\left[(2m)^2 - y^2\right]$
Factor the difference of squares again.
$= 4(4m^2 + y^2)(2m + y)(2m - y)$

15. $(y+z)^2 - 81$
$= (y+z)^2 - 9^2$
$= [(y+z)+9][(y+z)-9]$
$= (y+z+9)(y+z-9)$

17. $16 - (x+3y)^2$
$= 4^2 - z^2$ *Let* $z = (x+3y)$.
$= (4+z)(4-z)$
Substitute $x + 3y$ for z.
$= [4 + (x+3y)][4 - (x+3y)]$
$= (4+x+3y)(4-x-3y)$

19. $(p+q)^2 - (p-q)^2$
$= [(p+q)+(p-q)]$
$\cdot [(p+q)-(p-q)]$
$= (p+q+p-q)(p+q-p+q)$
$= 2p(2q)$
$= 4pq$

21. $k^2 - 6k + 9 = (k)^2 - 2(k)(3) + 3^2$
$= (k-3)^2$

23. $4z^2 + 4zw + w^2$
$= (2z)^2 + 2(2z)(w) + w^2$
$= (2z+w)^2$

25. $16m^2 - 8m + 1 - n^2$
Group the first three terms.
$= (16m^2 - 8m + 1) - n^2$
$= \left[(4m)^2 - 2(4m)(1) + 1^2\right] - n^2$
$= (4m-1)^2 - n^2$
$= [(4m-1)+n][(4m-1)-n]$
$= (4m-1+n)(4m-1-n)$

27. $4r^2 - 12r + 9 - s^2$
Group the first three terms.
$= (4r^2 - 12r + 9) - s^2$
$= \left[(2r)^2 - 2(2r)(3) + 3^2\right] - s^2$
$= (2r-3)^2 - s^2$
$= [(2r-3)+s][(2r-3)-s]$
$= (2r-3+s)(2r-3-s)$

29. $x^2 - y^2 + 2y - 1$
Group the last three terms.
$= x^2 - (y^2 - 2y + 1)$
$= x^2 - (y-1)^2$
$= [x + (y-1)][x - (y-1)]$
$= (x+y-1)(x-y+1)$

31. $98m^2 + 84mn + 18n^2$
$= 2(49m^2 + 42mn + 9n^2)$
$= 2\left[(7m)^2 + 2(7m)(3n) + (3n)^2\right]$
$= 2(7m+3n)^2$

33. $(p+q)^2 + 2(p+q) + 1$
$= x^2 + 2x + 1$ *Let* $x = p + q$.
$= (x+1)^2$
$= (p+q+1)^2$ *Resubstitute.*

35. $(a-b)^2 + 8(a-b) + 16$
$= (a-b)^2 + 2(a-b)(4) + 4^2$
$= [(a-b)+4]^2$
$= (a-b+4)^2$

37. $8x^3 - y^3$
$= (2x)^3 - y^3$
$= [2x - y]\left[(2x)^2 + (2x)(y) + y^2\right]$
$= (2x-y)(4x^2 + 2xy + y^2)$

39. $64g^3 - 27h^3$
$= (4g)^3 - (3h)^3$
$= (4g - 3h)\left[(4g)^2 + (4g)(3h) + (3h)^2\right]$
$= (4g-3h)(16g^2 + 12gh + 9h^2)$

41. $24n^3 + 81p^3$
$= 3(8n^3 + 27p^3)$
$= 3\left[(2n)^3 + (3p)^3\right]$
$= 3[2n+3p]\left[(2n)^2 - (2n)(3p) + (3p)^2\right]$
$= 3(2n+3p)(4n^2 - 6np + 9p^2)$

43. $(y+z)^3 + 64$
$= (y+z)^3 + 4^3$
$= [(y+z)+4]\left[(y+z)^2 - (y+z)(4) + 4^2\right]$
$= (y+z+4)(y^2 + 2yz + z^2 - 4y - 4z + 16)$

45. $(a+b)^3 - (a-b)^3$
$= [(a+b) - (a-b)]$
$\cdot\left[(a+b)^2 + (a+b)(a-b) + (a-b)^2\right]$
$= (a+b-a+b)$
$\cdot\left[(a^2 + 2ab + b^2) + (a^2 - b^2) + (a^2 - 2ab + b^2)\right]$
$= 2b(3a^2 + b^2)$

47. $m^6 - 125 = (m^2)^3 - (5)^3$
$= \left[m^2 - 5\right]\left[(m^2)^2 + (m^2)(5) + 5^2\right]$
$= (m^2 - 5)(m^4 + 5m^2 + 25)$

49. $1000x^9 - 27$
$= (10x^3)^3 - 3^3$
$= (10x^3 - 3)\left[(10x^3)^2 + (10x^3)(3) + 3^2\right]$
$= (10x^3 - 3)(100x^6 + 30x^3 + 9)$

51. $125y^6 + z^6$
$= (5y^2)^3 + (z^2)^3$
$= (5y^2 + z^2)\left[(5y^2)^2 - (5y^2)(z^2) + (z^2)^2\right]$
$= (5y^2 + z^2)(25y^4 - 5y^2z^2 + z^4)$

52. $x^6 - y^6$
$= (x^3)^2 - (y^3)^2$
$= (x^3 + y^3)(x^3 - y^3)$
$= [(x+y)(x^2 - xy + y^2)]$
$\cdot [(x-y)(x^2 + xy + y^2)]$
$= (x+y)(x^2 - xy + y^2)(x - y)$
$\cdot (x^2 + xy + y^2)$

53. $x^6 - y^6 = (x-y)(x+y)$
$\cdot \underline{(x^2 + xy + y^2)(x^2 - xy + y^2)}$

54. $x^6 - y^6$
$= (x^2)^3 - (y^2)^3$
$= (x^2 - y^2)(x^4 + x^2y^2 + y^4)$
$= (x+y)(x-y)(x^4 + x^2y^2 + y^4)$

55. $x^6 - y^6 = (x-y)(x+y)$
$\cdot \underline{(x^4 + x^2y^2 + y^4)}$

56. The product written on the blank in Exercise 53 must equal the product written on the blank in Exercise 55. To verify this, multiply the two factors written in Exercise 53.

$(x^2 + xy + y^2)(x^2 - xy + y^2)$
$= x^2(x^2 - xy + y^2)$
$+ xy(x^2 - xy + y^2)$
$+ y^2(x^2 - xy + y^2)$
$= x^4 - x^3y + x^2y^2 + x^3y - x^2y^2$
$+ xy^3 + x^2y^2 - xy^3 + y^4$
$= x^4 + x^2y^2 + y^4$

They are equal.

57. Start by factoring as the difference of squares since doing so resulted in the complete factorization more directly.

59. $125p^3 + 25p^2 + 8q^3 - 4q^2$
$= (125p^3 + 8q^3) + (25p^2 - 4q^2)$
$= [(5p)^3 + (2q)^3] + [(5p)^2 - (2q)^2]$
Factor within groups.
$= [(5p + 2q)(25p^2 - 10pq + 4q^2)]$
$+ [(5p + 2q)(5p - 2q)]$
Factor out the GCF, $5p + 2q$.
$= (5p + 2q)$
$\cdot [(25p^2 - 10pq + 4q^2) + (5p - 2q)]$
$= (5p + 2q)$
$\cdot (25p^2 - 10pq + 4q^2 + 5p - 2q)$

61. $27a^3 + 15a - 64b^3 - 20b$
$= (27a^3 - 64b^3) + (15a - 20b)$
$= [(3a)^3 - (4b)^3] + (15a - 20b)$
Factor within groups.
$= (3a - 4b)(9a^2 + 12ab + 16b^2)$
$+ 5(3a - 4b)$
Factor out the GCF, $3a - 4b$.
$= (3a - 4b)[(9a^2 + 12ab + 16b^2) + 5]$
$= (3a - 4b)(9a^2 + 12ab + 16b^2 + 5)$

63. $8t^4 - 24t^3 + t - 3$
$= (8t^4 - 24t^3) + (t - 3)$
$= 8t^3(t - 3) + 1(t - 3)$
Factor out the GCF, $t - 3$.
$= (t - 3)(8t^3 + 1)$
Factor the sum of cubes.
$= (t - 3)(2t + 1)(4t^2 - 2t + 1)$

65. $64m^2 - 512m^3 - 81n^2 + 729n^3$
$= (64m^2 - 81n^2) - (512m^3 - 729n^3)$
$= [(8m)^2 - (9n)^2] - [(8m)^3 - (9n)^3]$
Factor within groups.
$= (8m + 9n)(8m - 9n)$
$- (8m - 9n)(64m^2 + 72mn + 81n^2)$
Factor out the GCF, $8m - 9n$.
$= (8m - 9n)$
$\cdot [(8m + 9n) - (64m^2 + 72mn + 81n^2)]$
$= (8m - 9n)$
$\cdot (8m + 9n - 64m^2 - 72mn - 81n^2)$

Section 6.4

1. $100a^2 - 9b^2$
$= (10a)^2 - (3b)^2$ *Difference of squares*
$= (10a + 3b)(10a - 3b)$

3. $3p^4 - 3p^3 - 90p^2$
$= 3p^2(p^2 - p - 30)$
$= 3p^2(p - 6)(p + 5)$

5. $3a^2pq + 3abpq - 90b^2pq$
$= 3pq(a^2 + ab - 30b^2)$
$= 3pq(a + 6b)(a - 5b)$

7. $225p^2 + 256$ is the *sum* of squares and cannot be factored. The binomial is prime.

9. $6b^2 - 17b - 3$
Two integer factors whose product is $(6)(-3) = -18$ and whose sum is -17 are -18 and 1.
$= 6b^2 - 18b + b - 3$
$= 6b(b-3) + 1(b-3)$
$= (b-3)(6b+1)$

11. $x^3 - 1000$
$= (x)^3 - 10^3$ *Difference of cubes*
$= (x-10)(x^2 + 10x + 100)$

13. $4(p+2) + m(p+2) = (p+2)(4+m)$

15. $9m^2 - 45m + 18m^3$
Factor out the GCF, $9m$.
$= 9m(m - 5 + 2m^2)$ or $9m(2m^2 + m - 5)$
There is no pair of integers with a product of -10 and a sum of 1, so this cannot be factored further.

17. $54m^3 - 2000$
Factor out the GCF, 2.
$2(27m^3 - 1000)$
$= 2[(3m)^3 - 10^3]$ *Difference of cubes*
$= 2(3m - 10)$
$\cdot [(3m)^2 + (3m)(10) + 10^2]$
$= 2(3m-10)(9m^2 + 30m + 100)$

19. $9m^2 - 30mn + 25n^2$
$= (3m)^2 - 2(3m)(5n) + (5n)^2$
Perfect square trinomial
$= (3m - 5n)^2$

21. $kq - 9q + kr - 9r$
$= q(k-9) + r(k-9)$
$= (k-9)(q+r)$

23. $16z^3x^2 - 32z^2x = 16z^2x(zx - 2)$

25. $x^2 + 2x - 35$

The numbers 7 and -5 have a product of -35 and a sum of 2.

$= (x+7)(x-5)$

27. $x^4 - 625$
$= (x^2)^2 - 25^2$ *Difference of squares*
$= (x^2 + 25)(x^2 - 25)$
$= (x^2 + 25)(x+5)(x-5)$ *Difference of squares again*

29. $p^3 + 64$
$= p^3 + 4^3$ *Sum of cubes*
$= (p+4)(p^2 - 4p + 4^2)$
$= (p+4)(p^2 - 4p + 16)$

31. $64m^2 - 625$
$= (8m)^2 - 25^2$ *Difference of squares*
$= (8m + 25)(8m - 25)$

33. $12z^3 - 6z^2 + 18z$
Factor out the GCF, $6z$.

$$6z(2z^2 - z + 3)$$

Further factoring is not possible. There is no pair of integers whose product is 6 and whose sum is -1.

35. $256b^2 - 400c^2$
$= 16(16b^2 - 25c^2)$
$= 16[(4b)^2 - (5c)^2]$ *Difference of squares*
$= 16(4b + 5c)(4b - 5c)$

37. $1000z^3 + 512$
$= 8(125z^3 + 64)$
$= 8[(5z)^3 + 4^3]$ *Sum of cubes*
$= 8[5z + 4][(5z)^2 - (5z)(4) + 4^2]$
$= 8(5z+4)(25z^2 - 20z + 16)$

39. $10r^2 + 23rs - 5s^2$
Two integer factors whose product is $(10)(-5) = -50$ and whose sum is 23 are 25 and -2.
$= 10r^2 + 25rs - 2rs - 5s^2$
$= 5r(2r + 5s) - s(2r + 5s)$
$= (2r + 5s)(5r - s)$

41. $24p^3q + 52p^2q^2 + 20pq^3$
$= 4pq(6p^2 + 13pq + 5q^2)$
Two integer factors whose product is $(6)(5) = 30$ and whose sum is 13 are 10 and 3.
$= 4pq(6p^2 + 10pq + 3pq + 5q^2)$
$= 4pq[2p(3p + 5q) + q(3p + 5q)]$
$= 4pq(3p + 5q)(2p + q)$

43. $48k^4 - 243$
$= 3(16k^4 - 81)$
$= 3[(4k^2)^2 - 9^2]$
$= 3(4k^2 + 9)(4k^2 - 9)$
$= 3(4k^2 + 9)[(2k)^2 - 3^2]$
$= 3(4k^2 + 9)(2k + 3)(2k - 3)$

45. $m^3 + m^2 - n^3 - n^2$
$= (m^3 - n^3) + (m^2 - n^2)$
Difference of cubes; Difference of squares
$= (m-n)(m^2 + mn + n^2)$
$+ (m+n)(m-n)$
$= (m-n)[(m^2 + mn + n^2) + (m+n)]$
$= (m-n)(m^2 + mn + n^2 + m + n)$

47. $x^2 - 4m^2 - 4mn - n^2$
$= x^2 - (4m^2 + 4mn + n^2)$
$= x^2 - [(2m)^2 + 2(2m)n + n^2]$ *Perfect square trinomial*
$= x^2 - (2m + n)^2$ *Difference of squares*
$= [x + (2m + n)][x - (2m + n)]$
$= (x + 2m + n)(x - 2m - n)$

49. $18p^5 - 24p^3 + 12p^6$

Factor out the GCF, $6p^3$.

$$6p^3(3p^2 - 4 + 2p^3)$$

Further factoring is not possible.

51. $2x^2 - 2x - 40 = 2(x^2 - x - 20)$
$= 2(x + 4)(x - 5)$

53. $(2m + n)^2 - (2m - n)^2$ *Difference of squares*
$= [(2m + n) + (2m - n)]$
$\cdot [(2m + n) - (2m - n)]$
$= (2m + n + 2m - n)(2m + n - 2m + n)$
$= 4m(2n)$
$= 8mn$

55. $50p^2 - 162$
$= 2(25p^2 - 81)$
$= 2[(5p)^2 - 9^2]$
$= 2(5p + 9)(5p - 9)$

57. $12m^2rx + 4mnrx + 40n^2rx$
Factor out the GCF, $4rx$.
$= 4rx(3m^2 + mn + 10n^2)$

59. $21a^2 - 5ab - 4b^2$
Two integer factors whose product is $(21)(-4) = -84$ and whose sum is -5 are -12 and 7.
$= 21a^2 - 12ab + 7ab - 4b^2$
$= 3a(7a - 4b) + b(7a - 4b)$
$= (7a - 4b)(3a + b)$

61. $x^2 - y^2 - 4$ cannot be factored. The polynomial is *prime*.

63. $(p + 8q)^2 - 10(p + 8q) + 25$
$= x^2 - 10x + 25$ Let $x = p + 8q$.
$= (x - 5)^2$
$= [(p + 8q) - 5]^2$ *Resubstitute.*
$= (p + 8q - 5)^2$

65. $21m^4 - 32m^2 - 5$
$= 21x^2 - 32x - 5$ Let $x = m^2$.
Two integer factors whose product is $(21)(-5) = -105$ and whose sum is -32 are -35 and 3.
$= 21x^2 - 35x + 3x - 5$
$= 7x(3x - 5) + 1(3x - 5)$
$= (3x - 5)(7x + 1)$
$= (3m^2 - 5)(7m^2 + 1)$ *Resubstitute.*

67. $(r + 2t)^3 + (r - 3t)^3$
Let $x = r + 2t$ and $y = r - 3t$.
$= x^3 + y^3$ *Sum of cubes*
$= (x + y)(x^2 - xy + y^2)$
Substitute $r + 2t$ for x and $r - 3t$ for y.
$= [(r + 2t) + (r - 3t)]$
$\cdot [(r + 2t)^2 - (r + 2t)(r - 3t) + (r - 3t)^2]$
$= (2r - t)$
$\cdot (r^2 + 4rt + 4t^2 - r^2 + rt + 6t^2 + r^2 - 6rt + 9t^2)$
$= (2r - t)(r^2 - rt + 19t^2)$

69. $x^5 + 3x^4 - x - 3$
$= x^4(x + 3) - 1(x + 3)$
$= (x + 3)(x^4 - 1)$
$= (x + 3)\left[(x^2)^2 - 1^2\right]$
$= (x + 3)(x^2 + 1)(x^2 - 1)$
$= (x + 3)(x^2 + 1)[x^2 - 1^2]$
$= (x + 3)(x^2 + 1)(x + 1)(x - 1)$

71. $m^2 - 4m + 4 - n^2 + 6n - 9$
$= (m^2 - 4m + 4) - (n^2 - 6n + 9)$
$= (m^2 - 2(2)m + 2^2)$
$- (n^2 - 2(3)n + 3^2)$ *Perfect square trinomials*
$= (m - 2)^2 - (n - 3)^2$
$= [(m - 2) + (n - 3)]$
$\cdot [(m - 2) - (n - 3)]$ *Difference of two squares*
$= (m - 2 + n - 3)(m - 2 - n + 3)$
$= (m + n - 5)(m - n + 1)$

Section 6.5

1. First rewrite the equation so one side is 0. Factor the other side and set each factor equal to 0. The solutions of these linear equations are solutions of the quadratic equation.

In the exercises in this section, check all solutions to the equations by substituting them back in the original equations.

3. $(x - 5)(x + 10) = 0$

$x - 5 = 0$ or $x + 10 = 0$
$x = 5$ or $x = -10$
Check $x = 5$: $(0)(15) = 0$ *True*

Check $x = -10: \ (-15)(0) = 0$ *True*

Solution set: $\{5, -10\}$

5. $(2k - 5)(3k + 8) = 0$

$2k - 5 = 0$ or $3k + 8 = 0$

$2k = 5$ $\quad 3k = -8$

$k = \frac{5}{2}$ or $k = -\frac{8}{3}$

Solution set: $\left\{\frac{5}{2}, -\frac{8}{3}\right\}$

7. $m^2 - 3m - 10 = 0$

$(m + 2)(m - 5) = 0$

Use the zero-factor property.

$m + 2 = 0$ or $m - 5 = 0$

$m = -2$ or $m = 5$

Solution set: $\{-2, 5\}$

9. $z^2 + 9z + 18 = 0$

$(z + 6)(z + 3) = 0$

$z + 6 = 0$ or $z + 3 = 0$

$z = -6$ or $z = -3$

Solution set: $\{-6, -3\}$

11. $2x^2 = 7x + 4$

Get 0 on one side.

$2x^2 - 7x - 4 = 0$

$(2x + 1)(x - 4) = 0$

$2x + 1 = 0$ or $x - 4 = 0$

$2x = -1$ $\quad x = 4$

$x = -\frac{1}{2}$

Solution set: $\left\{-\frac{1}{2}, 4\right\}$

13. $15k^2 - 7k = 4$

$15k^2 - 7k - 4 = 0$

$(3k + 1)(5k - 4) = 0$

$3k + 1 = 0$ or $5k - 4 = 0$

$3k = -1$ $\quad 5k = 4$

$k = -\frac{1}{3}$ or $k = \frac{4}{5}$

Solution set: $\left\{-\frac{1}{3}, \frac{4}{5}\right\}$

15. $2x^2 - 12 - 4x = x^2 - 3x$

$x^2 - x - 12 = 0$

$(x + 3)(x - 4) = 0$

$x + 3 = 0$ or $x - 4 = 0$

$x = -3$ or $x = 4$

Solution set: $\{-3, 4\}$

17. $(5z + 1)(z + 3) = -2(5z + 1)$

$(5z + 1)(z + 3) + 2(5z + 1) = 0$

$(5z + 1)[(z + 3) + 2] = 0$

$(5z + 1)(z + 5) = 0$

$5z + 1 = 0$ or $z + 5 = 0$

$5z = -1$ $\quad z = -5$

$z = -\frac{1}{5}$

Solution set: $\left\{-5, -\frac{1}{5}\right\}$

19. $4p^2 + 16p = 0$

$4p(p + 4) = 0$

$4p = 0$ or $p + 4 = 0$

$p = 0$ $\quad p = -4$

Solution set: $\{-4, 0\}$

21. $6m^2 - 36m = 0$

$6m(m - 6) = 0$

$6m = 0$ or $m - 6 = 0$

$m = 0$ or $m = 6$

Solution set: $\{0, 6\}$

23. $-3m^2 + 27 = 0$

$-3(m^2 - 9) = 0$

$-3(m + 3)(m - 3) = 0$

Note that the leading -3 does not affect the solution set of the equation.

$m + 3 = 0$ or $m - 3 = 0$

$m = -3$ or $m = 3$

Solution set: $\{-3, 3\}$

25. $4p^2 - 16 = 0$

$4(p^2 - 4) = 0$

$4(p + 2)(p - 2) = 0$

$p + 2 = 0$ or $p - 2 = 0$

$p = -2$ or $p = 2$

Solution set: $\{-2, 2\}$

27. $-x^2 = 9 - 6x$

$0 = x^2 - 6x + 9$

$0 = (x - 3)(x - 3)$

$0 = x - 3$

$3 = x$

Solution set: $\{3\}$

29. $9k^2 + 24k + 16 = 0$

$(3k + 4)(3k + 4) = 0$

$3k + 4 = 0$

$3k = -4$

$k = -\frac{4}{3}$

Solution set: $\left\{-\frac{4}{3}\right\}$

31. $(x-3)(x+5)=-7$
Multiply the factors, and then add 7 on both sides of the equation to get 0 on the right.
$$x^2+5x-3x-15=-7$$
$$x^2+2x-8=0$$
Now factor the polynomial.
$$(x+4)(x-2)=0$$
$x+4=0$ or $x-2=0$
$x=-4$ or $x=2$
Solution set: $\{-4,2\}$

33. $$(2x+1)(x-3)=6x+3$$
$$2x^2-6x+x-3=6x+3$$
$$2x^2-5x-3=6x+3$$
$$2x^2-11x-6=0$$
$$(2x+1)(x-6)=0$$
$2x+1=0$ or $x-6=0$
$2x=-1$ $\quad x=6$
$x=-\frac{1}{2}$
Solution set: $\left\{-\frac{1}{2},6\right\}$

35. $(x+3)(x-6)=(2x+2)(x-6)$
$$x^2-3x-18=2x^2-10x-12$$
$$0=x^2-7x+6$$
$$0=(x-1)(x-6)$$
$x-1=0$ or $x-6=0$
$x=1$ or $x=6$
Solution set: $\{1,6\}$

37. $$2x^3-9x^2-5x=0$$
$$x(2x^2-9x-5)=0$$
$$x(2x+1)(x-5)=0$$
$x=0$ or $2x+1=0$ or $x-5=0$
$2x=-1$ $\quad x=5$
$x=-\frac{1}{2}$
Solution set: $\left\{-\frac{1}{2},0,5\right\}$

39. $$9t^3=16t$$
$$9t^3-16t=0$$
$$t(9t^2-16)=0$$
$$t(3t+4)(3t-4)=0$$
$t=0$ or $3t+4=0$ or $3t-4=0$
$3t=-4$ $\quad 3t=4$
$t=-\frac{4}{3}$ or $t=\frac{4}{3}$
Solution set: $\left\{-\frac{4}{3},0,\frac{4}{3}\right\}$

41. $$2r^3+5r^2-2r-5=0$$
Factor by grouping.
$$(2r^3-2r)+(5r^2-5)=0$$
$$2r(r^2-1)+5(r^2-1)=0$$
$$(r^2-1)(2r+5)=0$$
$$(r+1)(r-1)(2r+5)=0$$
$r+1=0$ or $r-1=0$ or $2r+5=0$
$r=-1$ or $r=1$ or $r=-\frac{5}{2}$
Solution set: $\left\{-\frac{5}{2},-1,1\right\}$

43. $$-x^3+6x^2+9x-54=0$$
Factor by grouping.
$$(-x^3+6x^2)+(9x-54)=0$$
$$-x^2(x-6)+9(x-6)=0$$
$$(x-6)(-x^2+9)=0$$
$$(x-6)(9-x^2)=0$$
$$(x-6)(3+x)(3-x)=0$$
$x-6=0$ or $3+x=0$ or $3-x=0$
$x=6$ $\quad x=-3$ $\quad 3=x$
Solution set: $\{-3,3,6\}$

45. $$x^3-3x^2-4x+12=0$$
Factor by grouping.
$$(x^3-3x^2)+(-4x+12)=0$$
$$x^2(x-3)-4(x-3)=0$$
$$(x-3)(x^2-4)=0$$
$$(x-3)(x+2)(x-2)=0$$
$x-3=0$ or $x+2=0$ or $x-2=0$
$x=3$ or $x=-2$ or $x=2$
Solution set: $\{-2,2,3\}$

47. An equation will have 0 in its solution set if the factor x appears in each term. Thus, the correct choice is **D**.

49. $4(2k+3)^2-(2k+3)-3=0$
Let $x=2k+3$.
$$4x^2-x-3=0$$
$$(4x+3)(x-1)=0$$
$4x+3=0$ or $x-1=0$
$4x=-3$ $\quad x=1$
$x=-\frac{3}{4}$
Substitute $2k+3$ for x.

$$2k + 3 = -\frac{3}{4} \quad \text{or} \quad 2k + 3 = 1$$
$$2k = -\frac{15}{4} \qquad 2k = -2$$
$$k = -\frac{15}{8} \quad \text{or} \quad k = -1$$

Solution set: $\left\{\frac{15}{8}, -1\right\}$

51. $2(m+3)^2 = 5(m+3) - 2$

Let $x = m + 3$.

$$2x^2 = 5x - 2$$
$$2x^2 - 5x + 2 = 0$$
$$(2x - 1)(x - 2) = 0$$

$$2x - 1 = 0 \quad \text{or} \quad x - 2 = 0$$
$$x = \frac{1}{2} \qquad x = 2$$

Substitute $m + 3$ for x.

$$m + 3 = \frac{1}{2} \quad \text{or} \quad m + 3 = 2$$
$$m = -\frac{5}{2} \qquad m = -1$$

Solution set: $\left\{-\frac{5}{2}, -1\right\}$

53. $9p^2 = (5p+2)^2$

$$9p^2 = 25p^2 + 20p + 4$$
$$0 = 16p^2 + 20p + 4$$
$$0 = 4p^2 + 5p + 1 \qquad \textit{Divide by 4.}$$
$$0 = (p+1)(4p+1)$$

$$p + 1 = 0 \quad \text{or} \quad 4p + 1 = 0$$
$$p = -1 \qquad 4p = -1$$
$$p = -\frac{1}{4}$$

Solution set: $\left\{-1, -\frac{1}{4}\right\}$

55. Let $x =$ the length of the side of the square painting. Then $x - 2 =$ the length of the side of the square mirror.

The formula for the area of a square is $A = s^2$, so the area of the painting is x^2 and the area of the mirror is $(x-2)^2$. The difference between their areas is 32 square feet.

$$x^2 - (x-2)^2 = 32$$
$$x^2 - \left(x^2 - 4x + 4\right) = 32$$
$$x^2 - x^2 + 4x - 4 = 32$$
$$4x - 4 = 32$$
$$4x = 36$$
$$x = 9$$

The length of the side of the painting is 9 feet. The length of the side of the mirror is $9 - 2 = 7$ feet.

57. Let $x =$ the height of the parallelogram. Then $x + 7 =$ the base of the parallelogram.

The area is 60 ft^2, so use the formula $A = bh$ and substitute 60 for A, $x + 7$ for b, and x for h.

$$A = bh$$
$$60 = (x+7)x$$
$$60 = x^2 + 7x$$
$$0 = x^2 + 7x - 60$$
$$0 = (x+12)(x-5)$$

$$x + 12 = 0 \quad \text{or} \quad x - 5 = 0$$
$$x = -12 \quad \text{or} \quad x = 5$$

A parallelogram cannot have a height that is negative, so reject -12 as a solution. The only possible solution is 5.
The height of the parallelogram is 5 ft and the base is $5 + 7 = 12$ ft.

59. Let $x =$ the width. Then $x + 200 =$ the length.

The area is 30,000 ft^2, so use the formula $A = LW$ and substitute 30,000 for A, $x + 200$ for L, and x for W.

$$LW = A$$
$$(x+200)x = 30{,}000$$
$$x^2 + 200x = 30{,}000$$
$$x^2 + 200x - 30{,}000 = 0$$
$$(x - 100)(x + 300) = 0$$

$$x - 100 = 0 \quad \text{or} \quad x + 300 = 0$$
$$x = 100 \quad \text{or} \quad x = -300$$

The width cannot be a negative measure, so reject -300 as a solution. The only possible solution is 100.
The width is 100 ft and the length is $100 + 200 = 300$ ft.

61. Let x and $x + 1$ denote the two consecutive integers.

Their product is 72, so

$$x(x+1) = 72.$$
$$x^2 + x = 72$$
$$x^2 + x - 72 = 0$$
$$(x+9)(x-8) = 0$$

$$x + 9 = 0 \quad \text{or} \quad x - 8 = 0$$
$$x = -9 \quad \text{or} \quad x = 8$$

The two possible pairs of consecutive integers are -9 and -8 or 8 and 9.

63. Let $x =$ the length of the side of the base of the box.

The surface area is 161.

$$\begin{aligned} S(x) &= x^2 + 16x \\ 161 &= x^2 + 16x \\ 0 &= x^2 + 16x - 161 \\ 0 &= (x+23)(x-7) \end{aligned}$$

$x + 23 = 0$ or $x - 7 = 0$
$x = -23$ or $x = 7$

The box cannot have a length that is a negative measure, so reject -23 as a solution. The only possible solution is 7.
Each side of the base of the box measures 7 in.

65. When the object hits the ground, its height is given by $f(t) = 0$. Let $f(t) = 0$ in the given equation, and solve for t.

$$\begin{aligned} -16t^2 + 64t + 80 &= f(t) \\ -16t^2 + 64t + 80 &= 0 \\ t^2 - 4t - 5 &= 0 \quad \textit{Divide by } -16. \\ (t-5)(t+1) &= 0 \end{aligned}$$

$t - 5 = 0$ or $t + 1 = 0$
$t = 5$ or $t = -1$

Time cannot be negative, so reject -1 as a solution. The only possible solution is 5.
The object will hit the ground 5 seconds after it is thrown.

67. Use $f(t) = -16t^2 + 576$ with $f(t) = 0$.

$$\begin{aligned} 0 &= -16t^2 + 576 \\ 0 &= t^2 - 36 \quad \textit{Divide by } -16. \\ 0 &= (t+6)(t-6) \end{aligned}$$

$t + 6 = 0$ or $t - 6 = 0$
$t = -6$ or $t = 6$

Time cannot be negative, so reject -6 as a solution. The only possible solution is 6.
The rock will hit the ground after 6 seconds.

69. $2x^2 + 7x - 15 = 0$
$(x+5)(2x-3) = 0$

$x + 5 = 0$ or $2x - 3 = 0$
$x = -5$ $\quad 2x = 3$
$x = \dfrac{3}{2}$

These values correspond to the zeros shown on the screens.
Solution set: $\{-5, 1.5\}$

71. $-x^2 + x = -12$

$$\begin{aligned} 0 &= x^2 - x - 12 \\ 0 &= (x+3)(x-4) \end{aligned}$$

$x + 3 = 0$ or $x - 4 = 0$
$x = -3$ $\quad x = 4$

These values correspond to the zeros shown on the screens.
Solution set: $\{-3, 4\}$

Chapter 6 Review Exercises

1. $12p^2 - 6p = 6p(2p-1)$

2. $21x^2 + 35x = 7x(3x+5)$

3. $12q^2b + 8qb^2 - 20q^3b^2$
$= 4qb(3q + 2b - 5q^2b)$

4. $6r^3t - 30r^2t^2 + 18rt^3$
$= 6rt(r^2 - 5rt + 3t^2)$

5. $(x+3)(4x-1) - (x+3)(3x+2)$

The GCF is $(x+3)$.

$$\begin{aligned} &= (x+3)[(4x-1) - (3x+2)] \\ &= (x+3)(4x - 1 - 3x - 2) \\ &= (x+3)(x-3) \end{aligned}$$

6. $(z+1)(z-4) + (z+1)(2z+3)$
$= (z+1)[(z-4) + (2z+3)]$
$= (z+1)(3z-1)$

7. $4m + nq + mn + 4q$

Rearrange the terms.

$$\begin{aligned} &= 4m + 4q + mn + nq \\ &= 4(m+q) + n(m+q) \\ &= (m+q)(4+n) \end{aligned}$$

8. $x^2 + 5y + 5x + xy$
$= x^2 + xy + 5x + 5y$
$= x(x+y) + 5(x+y)$
$= (x+y)(x+5)$

9. $2m + 6 - am - 3a$
$= 2(m+3) - a(m+3)$
$= (m+3)(2-a)$

10. $x^2 + 3x - 3y - xy$
$= (x^2 + 3x) + (-3y - xy)$
$= x(x+3) - y(3+x)$
$= x(x+3) - y(x+3)$
$= (x+3)(x-y)$

11. $3p^2 - p - 4$
Two integer factors whose product is $(3)(-4) = -12$ and whose sum is -1 are -4 and 3.
$= 3p^2 - 4p + 3p - 4$
$= p(3p-4) + 1(3p-4)$
$= (3p-4)(p+1)$

12. $6k^2 + 11k - 10$
Two integer factors whose product is $(6)(-10) = -60$ and whose sum is 11 are 15 and -4.
$= 6k^2 + 15k - 4k - 10$
$= 3k(2k + 5) - 2(2k + 5)$
$= (2k + 5)(3k - 2)$

13. $12r^2 - 5r - 3$
Two integer factors whose product is $(12)(-3) = -36$ and whose sum is -5 are -9 and 4.
$= 12r^2 - 9r + 4r - 3$
$= 3r(4r - 3) + 1(4r - 3)$
$= (4r - 3)(3r + 1)$

14. $10m^2 + 37m + 30$
Two integer factors whose product is $(10)(30) = 300$ and whose sum is 37 are 12 and 25.
$= 10m^2 + 12m + 25m + 30$
$= 2m(5m + 6) + 5(5m + 6)$
$= (5m + 6)(2m + 5)$

15. $10k^2 - 11kh + 3h^2$
Two integer factors whose product is $(10)(3) = 30$ and whose sum is -11 are -6 and -5.
$= 10k^2 - 6kh - 5kh + 3h^2$
$= 2k(5k - 3h) - h(5k - 3h)$
$= (5k - 3h)(2k - h)$

16. $9x^2 + 4xy - 2y^2$
There are no integers that have a product of $9(-2) = -18$ and a sum of 4. Therefore, the trinomial cannot be factored and is *prime*.

17. $24x - 2x^2 - 2x^3$
$= 2x(12 - x - x^2)$
$= 2x(4 + x)(3 - x)$

18. $6b^3 - 9b^2 - 15b$
$= 3b(2b^2 - 3b - 5)$
$= 3b(2b - 5)(b + 1)$

19. $y^4 + 2y^2 - 8$
$= (y^2)^2 + 2y^2 - 8$
$= (y^2 + 4)(y^2 - 2)$

20. $2k^4 - 5k^2 - 3$
$= 2(k^2)^2 - 5k^2 - 3$
$= (2k^2 + 1)(k^2 - 3)$

21. $p^2(p + 2)^2 + p(p + 2)^2 - 6(p + 2)^2$
Factor out $(p + 2)^2$.
$= (p + 2)^2(p^2 + p - 6)$
$= (p + 2)^2(p + 3)(p - 2)$

22. $3(r + 5)^2 - 11(r + 5) - 4$
$= 3x^2 - 11x - 4$ *Let x=r+5.*
$= (3x + 1)(x - 4)$
$= [3(r + 5) + 1][(r + 5) - 4]$ *Resubstitute.*
$= (3r + 15 + 1)(r + 1)$
$= (3r + 16)(r + 1)$

23. The student's answer is

$$x^2y^2 - 6x^2 + 5y^2 - 30 = x^2(y^2 - 6) + 5(y^2 - 6).$$

This is incorrect because the polynomial still has two terms, so it is not factored. The correct answer is

$$(y^2 - 6)(x^2 + 5).$$

24. Since area equals length times width, factor the polynomial given for the area.

$$4p^2 + 3p - 1 = (4p - 1)(p + 1)$$

Since the length is given as $4p - 1$, the width is $p + 1$.

25. $16x^2 - 25 = (4x)^2 - 5^2$ *Difference of squares*
$= (4x + 5)(4x - 5)$

26. $9t^2 - 49 = (3t)^2 - 7^2$
$= (3t + 7)(3t - 7)$

27. $36m^2 - 25n^2 = (6m)^2 - (5n)^2$
$= (6m + 5n)(6m - 5n)$

28. $x^2 + 14x + 49 = x^2 + 2(x)(7) + 7^2$
Perfect square trinomial
$= (x + 7)^2$

29. $9k^2 - 12k + 4 = (3k)^2 - 2(3k)(2) + 2^2$
$= (3k - 2)^2$

30. $r^3 + 27 = r^3 + 3^3$ *Sum of cubes*
$= (r + 3)(r^2 - 3r + 9)$

31. $125x^3 - 1 = (5x)^3 - 1^3$ *Difference of cubes*
$= (5x - 1)(25x^2 + 5x + 1)$

32. $m^6 - 1 = (m^3)^2 - 1^2$
Difference of squares
$= (m^3 + 1)(m^3 - 1)$
$= (m^3 + 1^3)(m^3 - 1^3)$
Sum of cubes; Difference of cubes
$= (m + 1)(m^2 - m + 1)$
$\cdot (m - 1)(m^2 + m + 1)$

33. $x^8 - 1 = (x^4)^2 - 1^2$

Difference of squares

$= (x^4 + 1)(x^4 - 1)$

$= (x^4 + 1)\left[(x^2)^2 - 1^2\right]$

Difference of squares again

$= (x^4 + 1)(x^2 + 1)(x^2 - 1)$

Difference of squares again

$= (x^4 + 1)(x^2 + 1)(x + 1)(x - 1)$

34. $x^2 + 6x + 9 - 25y^2$

$= (x^2 + 6x + 9) - 25y^2$

$= [x^2 + 2(x)(3) + 3^2] - 25y^2$

Perfect square trinomial

$= (x + 3)^2 - (5y)^2$

Difference of squares

$= [(x + 3) + 5y][(x + 3) - 5y]$

$= (x + 3 + 5y)(x + 3 - 5y)$

35. $(a + b)^3 - (a - b)^3$

Difference of cubes

$= [(a + b) - (a - b)]$

$\cdot\left[(a + b)^2 + (a + b)(a - b)\right.$

$\left.+ (a - b)^2\right]$

$= [2b](a^2 + 2ab + b^2 + a^2 - b^2$

$+ a^2 - 2ab + b^2)$

$= 2b(3a^2 + b^2)$

36. $x^5 - x^3 - 8x^2 + 8$

$= x^3(x^2 - 1) - 8(x^2 - 1)$

$= (x^2 - 1)(x^3 - 8)$

$= (x^2 - 1^2)(x^3 - 2^3)$

Difference of squares; *Difference of cubes*

$= (x + 1)(x - 1)(x - 2)(x^2 + 2x + 4)$

37. $x^2 - 8x + 16 = 0$

$(x - 4)(x - 4) = 0$

$x - 4 = 0$

$x = 4$

Solution set: $\{4\}$

38. $(5x + 2)(x + 1) = 0$

$5x + 2 = 0$ or $x + 1 = 0$

$5x = -2$ $\qquad x = -1$

$x = -\dfrac{2}{5}$

Solution set: $\left\{-1, -\dfrac{2}{5}\right\}$

39. $p^2 - 5p + 6 = 0$

$(p - 2)(p - 3) = 0$

$p - 2 = 0$ or $p - 3 = 0$

$p = 2$ or $p = 3$

Solution set: $\{2, 3\}$

40. $q^2 + 2q = 8$

$q^2 + 2q - 8 = 0$

$(q + 4)(q - 2) = 0$

$q + 4 = 0$ or $q - 2 = 0$

$q = -4$ or $q = 2$

Solution set: $\{-4, 2\}$

41. $6z^2 = 5z + 50$

$6z^2 - 5z - 50 = 0$

$(3z - 10)(2z + 5) = 0$

$3z - 10 = 0$ or $2z + 5 = 0$

$3z = 10$ $\qquad 2z = -5$

$z = \dfrac{10}{3}$ or $z = -\dfrac{5}{2}$

Solution set: $\left\{-\dfrac{5}{2}, \dfrac{10}{3}\right\}$

42. $6r^2 + 7r = 3$

$6r^2 + 7r - 3 = 0$

$(2r + 3)(3r - 1) = 0$

$2r + 3 = 0$ or $3r - 1 = 0$

$2r = -3$ $\qquad 3r = 1$

$r = -\dfrac{3}{2}$ or $r = \dfrac{1}{3}$

Solution set: $\left\{-\dfrac{3}{2}, \dfrac{1}{3}\right\}$

43. $8k^2 + 14k + 3 = 0$

$(2k + 3)(4k + 1) = 0$

$2k + 3 = 0$ or $4k + 1 = 0$

$2k = -3$ $\qquad 4k = -1$

$k = -\dfrac{3}{2}$ or $k = -\dfrac{1}{4}$

Solution set: $\left\{-\dfrac{3}{2}, -\dfrac{1}{4}\right\}$

44. $-4m^2 + 36 = 0$

$m^2 - 9 = 0$ *Divide by –4.*

$(m + 3)(m - 3) = 0$

$m + 3 = 0$ or $m - 3 = 0$

$m = -3$ or $m = 3$

Solution set: $\{-3, 3\}$

45. $6x^2 + 9x = 0$

$3x(2x + 3) = 0$

$3x = 0$ or $2x + 3 = 0$

$x = 0$ $\qquad 2x = -3$

$x = -\dfrac{3}{2}$

Solution set: $\left\{-\frac{3}{2}, 0\right\}$

46.
$$\begin{aligned}(2x+1)(x-2) &= -3\\ 2x^2-3x-2 &= -3\\ 2x^2-3x+1 &= 0\\ (2x-1)(x-1) &= 0\end{aligned}$$

$$2x-1=0 \quad \text{or} \quad x-1=0$$
$$2x=1 \qquad\qquad x=1$$
$$x=\frac{1}{2}$$

Solution set: $\left\{\frac{1}{2}, 1\right\}$

47.
$$\begin{aligned}(r+2)(r-2) &= (r-2)(r+3)-2\\ r^2-4 &= r^2+r-6-2\\ -4 &= r-8\\ 4 &= r\end{aligned}$$

Solution set: $\{4\}$

48.
$$\begin{aligned}2x^3-x^2-28x &= 0\\ x\left(2x^2-x-28\right) &= 0\\ x(2x+7)(x-4) &= 0\end{aligned}$$

$$x=0 \quad \text{or} \quad 2x+7=0 \quad \text{or} \quad x-4=0$$
$$2x=-7 \qquad\qquad x=4$$
$$x=-\frac{7}{2}$$

Solution set: $\left\{-\frac{7}{2}, 0, 4\right\}$

49.
$$\begin{aligned}-t^3-3t^2+4t+12 &= 0\\ t^3+3t^2-4t-12 &= 0 \quad \textit{Multiply by } -1.\\ t^2(t+3)-4(t+3) &= 0\\ (t+3)\left(t^2-4\right) &= 0\\ (t+3)(t+2)(t-2) &= 0\end{aligned}$$

$$t+3=0 \quad \text{or} \quad t+2=0 \quad \text{or} \quad t-2=0$$
$$t=-3 \quad \text{or} \quad t=-2 \quad \text{or} \quad t=2$$

Solution set: $\{-3, -2, 2\}$

50.
$$\begin{aligned}(r+2)\left(5r^2-9r-18\right) &= 0\\ (r+2)(5r+6)(r-3) &= 0\end{aligned}$$

$$r+2=0 \quad \text{or} \quad 5r+6=0 \quad \text{or} \quad r-3=0$$
$$r=-2 \qquad 5r=-6 \qquad r=3$$
$$r=-\frac{6}{5}$$

Solution set: $\left\{-2, -\frac{6}{5}, 3\right\}$

51. Let x be the length of the shorter side. Then the length of the longer side will be $2x+1$. The area is 10.5 ft^2. Use the formula for area of a triangle, $A=\frac{1}{2}bh$.

$$\begin{aligned}\frac{1}{2}(2x+1)(x) &= 10.5\\ x(2x+1) &= 21 \quad \textit{Multiply by 2.}\\ 2x^2+x &= 21\\ 2x^2+x-21 &= 0\\ (2x+7)(x-3) &= 0\end{aligned}$$

$$2x+7=0 \quad \text{or} \quad x-3=0$$
$$2x=-7 \qquad\qquad x=3$$
$$x=-\frac{7}{2}$$

The side cannot have a negative length, so reject $x=-\frac{7}{2}$.

The length of the shorter side is 3 ft.

52. Let w be the width of the lot. Then $w+20$ will be the length of the lot. The area is 2400 ft^2. Use the formula $LW=A$.

$$\begin{aligned}(w+20)w &= 2400\\ w^2+20w &= 2400\\ w^2+20w-2400 &= 0\\ (w-40)(w+60) &= 0\end{aligned}$$

$$w-40=0 \quad \text{or} \quad w+60=0$$
$$w=40 \quad \text{or} \quad w=\ \ 60$$

The lot cannot have a negative width, so reject $w=-60$.

The width of the lot is 40 ft, and the length is $40+20=60$ ft.

53. The height is 0 when the rock returns to the ground.

$$\begin{aligned}f(t) &= -16t^2+256t\\ 0 &= -16t^2+256t\\ 0 &= -16t(t-16)\end{aligned}$$

$$-16t=0 \quad \text{or} \quad t-16=0$$
$$t=0 \quad \text{or} \quad t=16$$

The rock is on the ground when $t=0$. It will return to the ground again after 16 seconds.

54.
$$\begin{aligned}f(t) &= -16t^2+256t\\ 240 &= -16t^2+256t \quad \textit{Let } f(t)=240.\\ 0 &= -16t^2+256t-240\\ 0 &= -16\left(t^2-16t+15\right)\\ 0 &= -16(t-15)(t-1)\end{aligned}$$

$$t-15=0 \quad \text{or} \quad t-1=0$$
$$t=15 \quad \text{or} \quad t=1$$

The rock will be 240 ft above the ground after 1 second and again after 15 seconds.

55. The question in Exercise 54 has two answers because the rock will be 240 ft above the ground after 1 second on the way up and again after 15 seconds on the way back down.

56. $30a + am - am^2$
$= a(30 + m - m^2)$
$= a(6 - m)(5 + m)$

57. $16 - 81k^2 = 4^2 - (9k)^2$
$= (4 + 9k)(4 - 9k)$

58. $8 - a^3$
$= 2^3 - a^3$
$= (2 - a)(2^2 + 2a + a^2)$
$= (2 - a)(4 + 2a + a^2)$

59. $9x^2 + 13xy - 3y^2$ is *prime* since it cannot be factored further.

60. $15y^3 + 20y^2 = 5y^2(3y + 4)$

61. $25z^2 - 30zm + 9m^2$
$= (5z)^2 - 2(5z)(3m) + (3m)^2$
$= (5z - 3m)^2$

62. $5x^2 - 17x = 12$
$5x^2 - 17x - 12 = 0$
$(5x + 3)(x - 4) = 0$

$5x + 3 = 0$ or $x - 4 = 0$
$5x = -3$ $\quad x = 4$
$x = -\frac{3}{5}$

Solution set: $\left\{-\frac{3}{5}, 4\right\}$

63. $3m^2 - 9m = 0$
$3m(m - 3) = 0$

$3m = 0$ or $m - 3 = 0$
$m = 0$ $\quad m = 3$

Solution set: $\{0, 3\}$

64. $x^3 - x = 0$
$x(x^2 - 1) = 0$
$x(x + 1)(x - 1) = 0$

$x = 0$ or $x + 1 = 0$ or $x - 1 = 0$
$x = -1$ $\quad x = 1$

Solution set: $\{-1, 0, 1\}$

65. Let x be the width of the frame. Then $x + 2$ will be the length of the frame. The area is 48 in^2. Use the formula $LW = A$.

$$(x + 2)x = 48$$
$$x^2 + 2x = 48$$
$$x^2 + 2x - 48 = 0$$
$$(x + 8)(x - 6) = 0$$

$x + 8 = 0$ or $x - 6 = 0$
$x = -8$ or $x = 6$

The frame cannot have a negative width, so reject $x = -8$.
The width of the frame is 6 in.

66. Let x be the width of the floor. Then $x + 85$ will be the length of the floor. The area is 2750 ft^2. Use the formula $LW = A$.

$$(x + 85)x = 2750$$
$$x^2 + 85x = 2750$$
$$x^2 + 85x - 2750 = 0$$
$$(x + 110)(x - 25) = 0$$

$x + 110 = 0$ or $x - 25 = 0$
$x = -110$ or $x = 25$

The floor cannot have a negative width, so reject $x = -110$.
The width of the floor was 25 feet and the length was $25 + 85 = 110$ feet.

Chapter 6 Test

1. $11z^2 - 44z = 11z(z - 4)$

2. $10x^2y^5 - 5x^2y^3 - 25x^5y^3$
Factor out the GCF, $5x^2y^3$.
$= 5x^2y^3(2y^2 - 1 - 5x^3)$

3. $3x + by + bx + 3y$
$= 3x + 3y + bx + by$
$= 3(x + y) + b(x + y)$
$= (x + y)(3 + b)$

4. $-2x^2 - x + 36 = -1(2x^2 + x - 36)$

Two integer factors whose product is $(2)(-36) = -72$ and whose sum is 1 are 9 and -8.

$2x^2 + x - 36 = 2x^2 + 9x - 8x - 36$
$= x(2x + 9) - 4(2x + 9)$
$= (2x + 9)(x - 4)$

Thus, the final factored form is

$$-(2x + 9)(x - 4).$$

5. $6x^2 + 11x - 35$

Two integer factors whose product is $(6)(-35) = -210$ and whose sum is 11 are 21 and -10.

$= 6x^2 + 21x - 10x - 35$
$= 3x(2x + 7) - 5(2x + 7)$
$= (2x + 7)(3x - 5)$

6. $4p^2 + 3pq - q^2$
Two integer factors whose product is $(4)(-1) = -4$ and whose sum is 3 are 4 and -1.

$$= 4p^2 + 4pq - pq - q^2$$
$$= 4p(p+q) - q(p+q)$$
$$= (p+q)(4p-q)$$

7. $16a^2 + 40ab + 25b^2$
$$= (4a)^2 + 2(4a)(5b) + (5b)^2$$
$$= (4a+5b)^2$$

8. $x^2 + 2x + 1 - 4z^2$
$$= \left(x^2 + 2x + 1\right) - 4z^2$$
$$= (x+1)^2 - (2z)^2$$
$$= [(x+1) + 2z][(x+1) - 2z]$$
$$= (x+1+2z)(x+1-2z)$$

9. $a^3 + 2a^2 - ab^2 - 2b^2$
$$= a^2(a+2) - b^2(a+2)$$
$$= (a+2)\left(a^2 - b^2\right)$$
$$= (a+2)(a+b)(a-b)$$

10. $9k^2 - 121j^2 = (3k)^2 - (11j)^2$
$$= (3k+11j)(3k-11j)$$

11. $y^3 - 216 = y^3 - 6^3$
$$= (y-6)\left(y^2 + 6y + 6^2\right)$$
$$= (y-6)\left(y^2 + 6y + 36\right)$$

12. $6k^4 - k^2 - 35 = 6\left(k^2\right)^2 - k^2 - 35$

Two integer factors whose product is $(6)(-35) = -210$ and whose sum is -1 are -15 and 14.
$$= 6\left(k^2\right)^2 - 15k^2 + 14k^2 - 35$$
$$= 3k^2\left(2k^2 - 5\right) + 7\left(2k^2 - 5\right)$$
$$= \left(2k^2 - 5\right)\left(3k^2 + 7\right)$$

13. $27x^6 + 1$
$$= \left(3x^2\right)^3 + (1)^3$$
$$= \left(3x^2 + 1\right)\left[\left(3x^2\right)^2 - \left(3x^2\right)(1) + 1^2\right]$$
$$= \left(3x^2 + 1\right)\left(9x^4 - 3x^2 + 1\right)$$

14. It is not in factored form because there are two terms: $(x^2 + 2y)p$ and $3(x^2 + 2y)$. The common factor is $x^2 + 2y$, and the factored form is $(x^2 + 2y)(p + 3)$.

15. **A** $(3-x)(x+4)$
$$= 3x + 12 - x^2 - 4x$$
$$= -x^2 - x + 12$$

B $-(x-3)(x+4)$
$$= -\left(x^2 + 4x - 3x - 12\right)$$
$$= -\left(x^2 + x - 12\right)$$
$$= -x^2 - x + 12$$

C $(-x+3)(x+4)$
$$= -x^2 - 4x + 3x + 12$$
$$= -x^2 - x + 12$$

D $(x-3)(-x+4)$
$$= -x^2 + 4x + 3x - 12$$
$$= -x^2 + 7x - 12$$

Therefore, only **D** is *not* a factored form of $-x^2 - x + 12$.

16.
$$3x^2 + 8x = -4$$
$$3x^2 + 8x + 4 = 0$$
$$(x+2)(3x+2) = 0$$

$x + 2 = 0$ or $3x + 2 = 0$

$x = -2$ $\quad 3x = -2$

$x = -\frac{2}{3}$

Solution set: $\left\{-2, -\frac{2}{3}\right\}$

17.
$$3x^2 - 5x = 0$$
$$x(3x-5) = 0$$

$x = 0$ or $3x - 5 = 0$

$3x = 5$

$x = \frac{5}{3}$

Solution set: $\left\{0, \frac{5}{3}\right\}$

18.
$$5m(m-1) = 2(1-m)$$
$$5m^2 - 5m = 2 - 2m$$
$$5m^2 - 3m - 2 = 0$$
$$(5m+2)(m-1) = 0$$

$5m + 2 = 0$ or $m - 1 = 0$

$5m = -2$ $\quad m = 1$

$m = -\frac{2}{5}$

Solution set: $\left\{-\frac{2}{5}, 1\right\}$

19. Using $A = LW$, substitute 40 for A, $x + 7$ for L, and $2x + 3$ for W.
$$A = LW$$
$$40 = (x+7)(2x+3)$$
$$40 = 2x^2 + 3x + 14x + 21$$
$$40 = 2x^2 + 17x + 21$$
$$0 = 2x^2 + 17x - 19$$
$$0 = (2x+19)(x-1)$$

$2x + 19 = 0$ or $x - 1 = 0$

$2x = -19$ $\quad x = 1$

$x = -\frac{19}{2}$

Length and width will be negative if $x = -\frac{19}{2}$, so reject it as a possible solution. If $x = 1$, then

continued

$$x + 7 = 1 + 7 = 8$$

and

$$2x + 3 = 2(1) + 3 = 5.$$

The length is 8 inches, and the width is 5 inches.

20. Substitute 128 for $f(t)$ in the equation.

$$f(t) = -16t^2 + 96t$$
$$128 = -16t^2 + 96t$$
$$16t^2 - 96t + 128 = 0$$
$$16(t^2 - 6t + 8) = 0$$
$$16(t-4)(t-2) = 0$$

$$t - 4 = 0 \quad \text{or} \quad t - 2 = 0$$
$$t = 4 \quad \text{or} \quad t = 2$$

The ball is 128 ft high at 2 seconds (on the way up) and again at 4 seconds (on the way down).

Cumulative Review Exercises Chapters 1–6

1. $-2(m-3) = -2(m) - 2(-3) = -2m + 6$

2. $-(-4m+3) = -(-4m) - (3) = 4m - 3$

3. $3x^2 - 4x + 4 + 9x - x^2$
$= 3x^2 - x^2 - 4x + 9x + 4$
$= 2x^2 + 5x + 4$

For Exercises 4–7, let $p = -4, q = -2$, and $r = 5$.

4. $-3(2q - 3p) = -3[2(-2) - 3(-4)]$
$= -3(-4 + 12)$
$= -3(8) = -24$

5. $8r^2 + q^2 = 8(5)^2 + (-2)^2$
$= 8(25) + 4$
$= 200 + 4 = 204$

6. $$\frac{\sqrt{r}}{-p+2q} = \frac{\sqrt{5}}{-(-4)+2(-2)} = \frac{\sqrt{5}}{4-4} = \frac{\sqrt{5}}{0}$$

This is undefined since the denominator is zero.

7. $$\frac{5p+6r^2}{p^2+q-1} = \frac{5(-4)+6(5)^2}{(-4)^2+(-2)-1} = \frac{-20+6(25)}{16-2-1} = \frac{-20+150}{13} = \frac{130}{13} = 10$$

8. $2z - 5 + 3z = 4 - (z+2)$
$5z - 5 = 4 - z - 2$
$5z - 5 = 2 - z$
$6z = 7$
$z = \frac{7}{6}$

Solution set: $\left\{\frac{7}{6}\right\}$

9. $$\frac{3a-1}{5} + \frac{a+2}{2} = -\frac{3}{10}$$
$2(3a-1) + 5(a+2) = -3$ *Multiply by 10.*
$6a - 2 + 5a + 10 = -3$
$11a + 8 = -3$
$11a = -11$
$a = -1$

Solution set: $\{-1\}$

10. $-\frac{4}{3}d \geq -5$
$-\frac{3}{4}\left(-\frac{4}{3}d\right) \leq -\frac{3}{4}(-5)$
$d \leq \frac{15}{4}$

Solution set: $\left(-\infty, \frac{15}{4}\right]$

11. $3 - 2(m+3) < 4m$
$3 - 2m - 6 < 4m$
$-2m - 3 < 4m$
$-6m < 3$
$m > -\frac{3}{6}$ or $-\frac{1}{2}$

Solution set: $\left(-\frac{1}{2}, \infty\right)$

12. $2k + 4 < 10$ and $3k - 1 > 5$
$2k < 6$ $\quad$ $3k > 6$
$k < 3$ and $k > 2$

The overlap of these inequalities is the set of all numbers between 2 and 3.
Solution set: $(2, 3)$

13. $2k + 4 > 10$ or $3k - 1 < 5$
$2k > 6$ $\quad$ $3k < 6$
$k > 3$ or $k < 2$

The solution set is the set of numbers that are either greater than 3 or less than 2.

Solution set: $(-\infty, 2) \cup (3, \infty)$

14. $|5x + 3| - 10 = 3$
$|5x + 3| = 13$

$5x + 3 = 13$ or $5x + 3 = -13$
$5x = 10$ $\quad$ $5x = -16$
$x = 2$ or $x = -\frac{16}{5}$

Solution set: $\left\{-\frac{16}{5}, 2\right\}$

15. $|x + 2| < 9$
$-9 < x + 2 < 9$
$-11 < x < 7$

Solution set: $(-11, 7)$

16. $|2x - 5| \geq 9$

$$\begin{aligned} 2x - 5 &\geq 9 \quad \text{or} \quad & 2x - 5 &\leq -9 \\ 2x &\geq 14 & 2x &\leq -4 \\ x &\geq 7 \quad \text{or} & x &\leq -2 \end{aligned}$$

Solution set: $(-\infty, -2] \cup [7, \infty)$

17. Solve $V = lwh$ for h.

$$\frac{V}{lw} = \frac{lwh}{lw} \qquad \textit{Divide by lw.}$$

$$\frac{V}{lw} = h$$

18. Let x be the time it takes for the planes to be 2100 mi apart. Use the formula $d = rt$ to complete the table.

Plane	r	t	d
Eastbound	550	x	$550x$
Westbound	500	x	$500x$

The total distance is 2100 mi.

$$\begin{aligned} 550x + 500x &= 2100 \\ 1050x &= 2100 \\ x &= 2 \end{aligned}$$

It will take 2 hours for the planes to be 2100 miles apart.

19. $4x + 2y = -8$
Draw a line through the x- and y-intercepts, $(-2, 0)$ and $(0, -4)$, respectively.

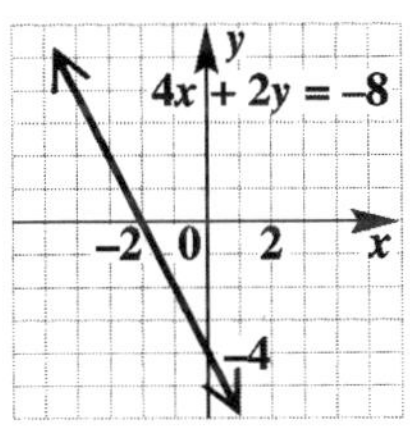

20. The slope m of the line through $(-4, 8)$ and $(-2, 6)$ is

$$m = \frac{6 - 8}{-2 - (-4)} = \frac{-2}{2} = -1.$$

21. $y = -3$ is an equation of a horizontal line. Its slope is 0.

22.

$$\begin{aligned} f(x) &= 2x + 7 \\ f(-4) &= 2(-4) + 7 \\ &= -8 + 7 = -1 \end{aligned}$$

23. To find the x-intercept of the graph of $f(x) = 2x + 7$, let $f(x) = 0$ (which is the same as letting $y = 0$) and solve for x.

$$\begin{aligned} 0 &= 2x + 7 \\ -7 &= 2x \\ -\frac{7}{2} &= x \end{aligned}$$

The x-intercept is $\left(-\frac{7}{2}, 0\right)$.

24. $f(0) = 7$, so the y-intercept is $(0, 7)$.

25.

$$\begin{aligned} 3x - 2y &= -7 \quad (1) \\ 2x + 3y &= 17 \quad (2) \end{aligned}$$

To eliminate y, multiply (1) by 3 and (2) by 2, and then add the resulting equations.

$$\begin{aligned} 9x - 6y &= -21 \quad 3 \times (1) \\ 4x + 6y &= 34 \quad 2 \times (2) \\ \hline 13x &= 13 \\ x &= 1 \end{aligned}$$

Substitute 1 for x in (1).

$$\begin{aligned} 3x - 2y &= -7 \quad (1) \\ 3(1) - 2y &= -7 \\ -2y &= -10 \\ y &= 5 \end{aligned}$$

Solution set: $\{(1, 5)\}$

26.

$$\begin{aligned} 2x + 3y - 6z &= 5 \quad (1) \\ 8x - y + 3z &= 7 \quad (2) \\ 3x + 4y - 3z &= 7 \quad (3) \end{aligned}$$

To eliminate z, add (2) and (3).

$$\begin{aligned} 8x - y + 3z &= 7 \quad (2) \\ 3x + 4y \quad 3z &= 7 \quad (3) \\ \hline 11x + 3y &= 14 \quad (4) \end{aligned}$$

To eliminate z again, multiply (2) by 2 and add the result to (1).

$$\begin{aligned} 2x + 3y - 6z &= 5 \quad (1) \\ 16x - 2y + 6z &= 14 \quad 2 \times (2) \\ \hline 18x + y &= 19 \quad (5) \end{aligned}$$

To eliminate y, multiply (5) by -3 and add the result to (4).

$$\begin{aligned} 11x + 3y &= 14 \quad (4) \\ -54x - 3y &= -57 \quad -3 \times (5) \\ \hline -43x &= -43 \\ x &= 1 \end{aligned}$$

From (5), $18(1) + y - 19$, so $y = 1$.
To find z, let $x = 1$ and $y = 1$ in (1).

$$\begin{aligned} 2x + 3y - 6z &= 5 \quad (1) \\ 2(1) + 3(1) - 6z &= 5 \\ 5 - 6z &= 5 \\ -6z &= 0 \\ z &= 0 \end{aligned}$$

Solution set: $\{(1, 1, 0)\}$

27.

$$\begin{aligned} &(3x^2y^{-1})^{-2}(2x^{-3}y)^{-1} \\ &= 3^{-2}x^{-4}y^2 \cdot 2^{-1}x^3y^{-1} \\ &= 3^{-2}2^{-1}x^{-1}y \\ &= \frac{y}{3^2 \cdot 2x} = \frac{y}{18x} \end{aligned}$$

28. $$\frac{5m^{-2}y^3}{3m^{-3}y^{-1}} = \frac{5}{3} \cdot \frac{m^{-2}}{m^{-3}} \cdot \frac{y^3}{y^{-1}} = \frac{5}{3}m^{-2-(-3)}y^{3-(-1)} = \frac{5}{3}m^1y^4 \text{ or } \frac{5my^4}{3}$$

29. $$(3x^3 + 4x^2 - 7) - (2x^3 - 8x^2 + 3x) = 3x^3 + 4x^2 - 7 - 2x^3 + 8x^2 - 3x = x^3 + 12x^2 - 3x - 7$$

30. $$(7x + 3y)^2 = (7x)^2 + 2(7x)(3y) + (3y)^2 = 49x^2 + 42xy + 9y^2$$

31. $$(2p + 3)(5p^2 - 4p - 8) = 10p^3 - 8p^2 - 16p + 15p^2 - 12p - 24 = 10p^3 + 7p^2 - 28p - 24$$

32. $16w^2 + 50wz - 21z^2$

Two integer factors whose product is $(16)(-21) = -336$ and whose sum is 50 are 56 and -6.

$$= 16w^2 + 56wz - 6wz - 21z^2 = 8w(2w + 7z) - 3z(2w + 7z) = (2w + 7z)(8w - 3z)$$

33. $4x^2 - 4x + 1 - y^2$

Group the first three terms.

$$= (4x^2 - 4x + 1) - y^2 = (2x - 1)^2 - y^2 = [(2x - 1) + y][(2x - 1) - y] = (2x - 1 + y)(2x - 1 - y)$$

34. $$4y^2 - 36y + 81 = (2y)^2 - 2(2y)(9) + 9^2 = (2y - 9)^2$$

35. $$100x^4 - 81 = (10x^2)^2 - 9^2 = (10x^2 + 9)(10x^2 - 9)$$

36. $$8p^3 + 27 = (2p)^3 + 3^3 = (2p + 3)[(2p)^2 - (2p)(3) + 3^2] = (2p + 3)(4p^2 - 6p + 9)$$

37. $(p - 1)(2p + 3)(p + 4) = 0$

$p - 1 = 0$ or $2p + 3 = 0$ or $p + 4 = 0$

$p = 1$ $\quad 2p = -3$ $\quad p = -4$

$p = -\frac{3}{2}$

Solution set: $\left\{-4, -\frac{3}{2}, 1\right\}$

38. $$9q^2 = 6q - 1$$
$$9q^2 - 6q + 1 = 0$$
$$(3q - 1)^2 = 0$$
$$3q - 1 = 0$$
$$3q = 1$$
$$q = \frac{1}{3}$$

Solution set: $\left\{\frac{1}{3}\right\}$

39. Let x be the length of the base. Then $x + 3$ will be the height. The area is 14 square feet. Use the formula $A = \frac{1}{2}bh$, and substitute 14 for A, x for b, and $x + 3$ for h.

$$\frac{1}{2}bh = A$$
$$\frac{1}{2}(x)(x + 3) = 14$$
$$x(x + 3) = 28 \quad \textit{Multiply by 2.}$$
$$x^2 + 3x = 28$$
$$x^2 + 3x - 28 = 0$$
$$(x + 7)(x - 4) = 0$$

$x + 7 = 0$ or $x - 4 = 0$

$x = -7$ or $x = 4$

The length cannot be negative, so reject -7 as a solution. The only possible solution is 4. The base is 4 feet long.

40. Let x be the distance between the longer sides. (This is actually the width.) Then $x + 2$ will be the length of the longer side. The area of the rectangle is 288 in^2. Use the formula $LW = A$. Substitute 288 for A, $x + 2$ for L, and x for W.

$$(x + 2)x = 288$$
$$x^2 + 2x = 288$$
$$x^2 + 2x - 288 = 0$$
$$(x + 18)(x - 16) = 0$$

$x + 18 = 0$ or $x - 16 = 0$

$x = -18$ or $x = 16$

The distance cannot be negative, so reject -18 as a solution. The only possible solution is 16. The distance between the longer sides is 16 inches, and the length of the longer sides is $16 + 2 = 18$ inches.

CHAPTER 7 RATIONAL EXPRESSIONS AND FUNCTIONS

Section 7.1

1. $$\frac{x-3}{x+4}=\frac{(-1)(x-3)}{(-1)(x+4)}=\frac{-x+3}{-x-4}\text{ or }\frac{3-x}{-x-4}\quad\textbf{(C)}$$

3. $$\frac{x-3}{x-4}=\frac{(-1)(x-3)}{(-1)(x-4)}=\frac{-x+3}{-x+4}\quad\textbf{(D)}$$

5. $$\frac{3-x}{x+4}=\frac{(-1)(3-x)}{(-1)(x+4)}=\frac{-3+x}{-x-4}\text{ or }\frac{x-3}{-x-4}\quad\textbf{(E)}$$

7. Replacing x with 2 makes the denominator 0 and the value of the expression undefined. To find the values excluded from the domain, set the denominator equal to 0 and solve the equation. All solutions of the equation are excluded from the domain.

9. $f(x)=\frac{x}{x-7}$
Set the denominator equal to zero, and solve the equation.
$$x-7=0$$
$$x=7$$
The number 7 makes the rational expression undefined, so 7 is not in the domain of the function.

11. $f(x)=\frac{6x-5}{7x+1}$
Set the denominator equal to zero, and solve the equation.
$$7x+1=0$$
$$7x=-1$$
$$x=-\frac{1}{7}$$
The number $-\frac{1}{7}$ makes the rational expression undefined, so $-\frac{1}{7}$ is not in the domain of the function.

13. $f(x)=\frac{12x+3}{x}$
Set the denominator equal to zero and solve.
$$x=0$$
The number 0 makes the rational expression undefined, so 0 is not in the domain of the function.

15. $f(x)=\frac{3x+1}{2x^2+x-6}$
Set the denominator equal to zero and solve.
$$2x^2+x-6=0$$
$$(x+2)(2x-3)=0$$
$$x+2=0\quad\text{or}\quad 2x-3=0$$
$$x=-2\qquad\qquad 2x=3$$
$$x=\frac{3}{2}$$
The numbers -2 and $\frac{3}{2}$ are not in the domain of the function.

17. $f(x)=\frac{x+2}{14}$
The denominator is never zero, so all numbers are in the domain of the function.

19. $f(x)=\frac{2x^2-3x+4}{3x^2+8}$
Set the denominator equal to zero and solve.
$$3x^2+8=0$$
$$3x^2=-8$$
$$x^2=-\frac{8}{3}$$
The square of any real number x is positive or zero. There are no real numbers which make this rational expression undefined, so all numbers are in the domain of the function.

21. **(a)** $\frac{x^2+4x}{x+4}$
The two terms in the numerator are x^2 and $4x$. The two terms in the denominator are x and 4.

(b) To express the rational expression in lowest terms, factor the numerator and denominator and replace the quotient of common factors with 1.
$$\frac{x^2+4x}{x+4}=\frac{x(x+4)}{x+4}=x\cdot 1=x$$

23. **A.** $\frac{3-x}{x-4}=\frac{-1(x-3)}{-1(4-x)}=\frac{x-3}{4-x}$

B. $\frac{x+3}{4+x}$ cannot be transformed to equal $\frac{x-3}{4-x}$.

C. $-\frac{3-x}{4-x}=\frac{-(3-x)}{4-x}=\frac{x-3}{4-x}$

D. $-\frac{x-3}{x-4}=\frac{x-3}{-(x-4)}=\frac{x-3}{4-x}$

Only the expression in **B** is *not* equivalent to $\frac{x-3}{4-x}$.

25. $\dfrac{x^2(x+1)}{x(x+1)} = \dfrac{x}{1} \cdot \dfrac{x(x+1)}{x(x+1)} = \dfrac{x}{1} \cdot 1 = x$

27. $\dfrac{(x+4)(x-3)}{(x+5)(x+4)} = \dfrac{x-3}{x+5} \cdot \dfrac{x+4}{x+4} = \dfrac{x-3}{x+5}$

29. $\dfrac{4x(x+3)}{8x^2(x-3)} = \dfrac{(x+3) \cdot 4x}{2x(x-3) \cdot 4x}$
$= \dfrac{x+3}{2x(x-3)}$

31. $\dfrac{3x+7}{3}$ Since the numerator and denominator have no common factors, the expression is already in lowest terms.

33. $\dfrac{6m+18}{7m+21} = \dfrac{6(m+3)}{7(m+3)} = \dfrac{6}{7}$

35. $\dfrac{3z^2+z}{18z+6} = \dfrac{z(3z+1)}{6(3z+1)} = \dfrac{z}{6}$

37. $\dfrac{2t+6}{t^2-9} = \dfrac{2(t+3)}{(t-3)(t+3)}$
$= \dfrac{2}{t-3}$

39. $\dfrac{x^2+2x-15}{x^2+6x+5} = \dfrac{(x+5)(x-3)}{(x+5)(x+1)}$
$= \dfrac{x-3}{x+1}$

41. $\dfrac{8x^2-10x-3}{8x^2-6x-9} = \dfrac{(4x+1)(2x-3)}{(4x+3)(2x-3)}$
$= \dfrac{4x+1}{4x+3}$

43. $\dfrac{a^3+b^3}{a+b} = \dfrac{(a+b)(a^2-ab+b^2)}{a+b}$
$= a^2-ab+b^2$

45. $\dfrac{2c^2+2cd-60d^2}{2c^2-12cd+10d^2}$
$= \dfrac{2(c^2+cd-30d^2)}{2(c^2-6cd+5d^2)}$
$= \dfrac{2(c+6d)(c-5d)}{2(c-d)(c-5d)}$
$= \dfrac{c+6d}{c-d}$

47. $\dfrac{ac-ad+bc-bd}{ac-ad-bc+bd}$
$= \dfrac{a(c-d)+b(c-d)}{a(c-d)-b(c-d)}$
$= \dfrac{(c-d)(a+b)}{(c-d)(a-b)}$ *Factor by grouping*
$= \dfrac{a+b}{a-b}$

49. $\dfrac{7-b}{b-7} = \dfrac{-1(b-7)}{b-7} = -1$

In Exercises 51–53, there are other acceptable ways to express each answer.

51. $\dfrac{x^2-y^2}{y-x} = \dfrac{(x-y)(x+y)}{y-x}$
$= \dfrac{-1(y-x)(x+y)}{y-x}$
$= -(x+y)$

53. $\dfrac{(a-3)(x+y)}{(3-a)(x-y)} = \dfrac{(a-3)(x+y)}{-1(a-3)(x-y)}$
$= \dfrac{x+y}{-1(x-y)}$
$= -\dfrac{x+y}{x-y}$

55. $\dfrac{5k-10}{20-10k} = \dfrac{-5(2-k)}{10(2-k)} = \dfrac{-5}{10} = -\dfrac{1}{2}$

57. $\dfrac{a^2-b^2}{a^2+b^2} = \dfrac{(a+b)(a-b)}{a^2+b^2}$
The numerator and denominator have no common factors except 1, so the original expression is already in lowest terms.

59. Multiply the numerators, multiply the denominators, and factor each numerator and denominator. (Factoring can be performed first.) Divide the numerator and denominator by any common factors to write in lowest terms. For example,

$\dfrac{6r-5s}{3r+2s} \cdot \dfrac{6r+4s}{5s-6r}$
$= \dfrac{(6r-5s)(6r+4s)}{(3r+2s)(5s-6r)}$
$= \dfrac{(6r-5s)2(3r+2s)}{(3r+2s)(-1)(6r-5s)} = \dfrac{2}{-1} = -2.$

61. $\dfrac{x^3}{3y} \cdot \dfrac{9y^2}{x^5} = \dfrac{3y \cdot 3x^3y}{x^2 \cdot 3x^3y} = \dfrac{3y}{x^2} \cdot 1 = \dfrac{3y}{x^2}$

63. $\dfrac{5a^4b^2}{16a^2b} \div \dfrac{25a^2b}{60a^3b^2}$
Multiply by the reciprocal of the divisor.
$= \dfrac{5a^4b^2}{16a^2b} \cdot \dfrac{60a^3b^2}{25a^2b}$
$= \dfrac{300a^7b^4}{400a^4b^2}$
$= \dfrac{100a^4b^2 \cdot 3a^3b^2}{100a^4b^2 \cdot 4}$
$= \dfrac{3a^3b^2}{4}$

65. $\dfrac{(-3mn)^2 \cdot (4m^2n)^3}{16m^2n^4(mn^2)^3} \div \dfrac{24(m^2n^2)^4}{(3m^2n^3)^2}$
$= \dfrac{9m^2n^2 \cdot 64m^6n^3}{16m^2n^4 \cdot m^3n^6} \cdot \dfrac{9m^4n^6}{24m^8n^8}$

$$= \frac{9 \cdot 4 \cdot 9m^{12}n^{11}}{3 \cdot 4 \cdot 2m^{13}n^{18}}$$
$$= \frac{3 \cdot 9}{2m^1n^7} = \frac{27}{2mn^7}$$

67. $\frac{(x+2)(x+1)}{(x+3)(x-2)} \cdot \frac{(x+3)(x+4)}{(x+2)(x+1)}$
$$= \frac{(x+1)(x+2)(x+3)}{(x+1)(x+2)(x+3)} \cdot \frac{x+4}{x-2} = \frac{x+4}{x-2}$$

69. $\frac{(2x+3)(x-4)}{(x+8)(x-4)} \div \frac{(x-4)(x+2)}{(x-4)(x+8)}$
$$= \frac{2x+3}{x+8} \div \frac{x+2}{x+8}$$
$$= \frac{2x+3}{x+8} \cdot \frac{x+8}{x+2}$$
$$= \frac{2x+3}{x+2}$$

71. $\frac{4x}{8x+4} \cdot \frac{14x+7}{6} = \frac{4x \cdot 7(2x+1)}{4(2x+1) \cdot 6}$
$$= \frac{7x}{6}$$

For Exercise 73, there are several other ways to express the answer.

73. $\frac{p^2-25}{4p} \cdot \frac{2}{5-p} = \frac{(p+5)(p-5)2}{2 \cdot 2p(-1)(p-5)}$
$$= \frac{p+5}{(2p)(-1)} = -\frac{p+5}{2p}$$

75. $(7k+7) \div \frac{4k+4}{5}$
$$= \frac{7(k+1)}{1} \cdot \frac{5}{4(k+1)}$$
$$= \frac{7 \cdot 5}{4} = \frac{35}{4}$$

77. $(z^2-1) \cdot \frac{1}{1-z}$
$$= \frac{(z+1)(z-1)}{1} \cdot \frac{1}{-1(z-1)}$$
$$= \frac{z+1}{-1} = -(z+1) \quad \text{or} \quad -z-1$$

For Exercise 79, there are several other ways to express the answer.

79. $\frac{m^2-49}{m+1} \div \frac{7-m}{m}$
$$= \frac{(m-7)(m+7)}{m+1} \cdot \frac{m}{7-m}$$
$$= \frac{(-1)(7-m)(m+7)}{m+1} \cdot \frac{m}{7-m}$$
$$= \frac{-m(m+7)}{m+1}$$

81. $\frac{12x-10y}{3x+2y} \cdot \frac{6x+4y}{10y-12x}$
$$= \frac{2(6x-5y) \cdot 2(3x+2y)}{(3x+2y) \cdot 2(5y-6x)}$$
$$= \frac{2(-1)(5y-6x)}{(5y-6x)} = -2$$

83. $\frac{x^2-25}{x^2+x-20} \cdot \frac{x^2+7x+12}{x^2-2x-15}$
$$= \frac{(x-5)(x+5)}{(x+5)(x-4)} \cdot \frac{(x+3)(x+4)}{(x-5)(x+3)}$$
$$= \frac{x+4}{x-4}$$

85. $\frac{6x^2+5xy-6y^2}{12x^2-11xy+2y^2} \div \frac{4x^2-12xy+9y^2}{8x^2-14xy+3y^2}$
$$= \frac{(3x-2y)(2x+3y)}{(3x-2y)(4x-y)} \cdot \frac{(4x-y)(2x-3y)}{(2x-3y)(2x-3y)}$$
$$= \frac{2x+3y}{2x-3y}$$

87. $\frac{3k^2+17kp+10p^2}{6k^2+13kp-5p^2} \div \frac{6k^2+kp-2p^2}{6k^2-5kp+p^2}$
$$= \frac{(3k+2p)(k+5p)}{(3k-p)(2k+5p)} \cdot \frac{(3k-p)(2k-p)}{(3k+2p)(2k-p)}$$
$$= \frac{k+5p}{2k+5p}$$

89. $\left(\frac{6k^2-13k-5}{k^2+7k} \div \frac{2k-5}{k^3+6k^2-7k}\right) \cdot \frac{k^2-5k+6}{3k^2-8k-3}$

Factor k from the denominator of the divisor; multiply by the reciprocal.
$$= \left[\frac{6k^2-13k-5}{k^2+7k} \cdot \frac{k(k^2+6k-7)}{2k-5}\right] \cdot \frac{k^2-5k+6}{3k^2-8k-3}$$
$$= \left[\frac{(3k+1)(2k-5)}{k(k+7)} \cdot \frac{k(k+7)(k-1)}{2k-5}\right] \cdot \frac{(k-2)(k-3)}{(3k+1)(k-3)}$$
$$= (k-1)(k-2)$$

91. $\frac{a^2(2a+b)+6a(2a+b)+5(2a+b)}{3a^2(a+2b)-2a(a+2b)-(a+2b)} \div \frac{a+1}{a-1}$
$$= \frac{(2a+b)(a^2+6a+5)}{(a+2b)(3a^2-2a-1)} \cdot \frac{a-1}{a+1}$$
$$= \frac{(2a+b)(a+5)(a+1)(a-1)}{(a+2b)(3a+1)(a-1)(a+1)}$$
$$= \frac{(a+5)(2a+b)}{(3a+1)(a+2b)}$$

Section 7.2

1. $\frac{1}{x}+\frac{1}{y}=\frac{1}{4}+\frac{1}{2}$ *Let x=4 and y=2.*

$=\frac{1}{4}+\frac{2}{4}=\frac{3}{4}$

2. $\frac{1}{x+y}=\frac{1}{4+2}$ *Let x=4 and y=2.*

$=\frac{1}{6}$

3. The answers for Exercises 1 and 2 are not the same. We conclude that we cannot find the sum $\frac{1}{x}+\frac{1}{y}$ by adding the denominators and keeping the common numerator.

4. $\frac{1}{x}-\frac{1}{y}=\frac{1}{3}-\frac{1}{5}$ *Let x=3 and y=5.*

$=\frac{5}{15}-\frac{3}{15}=\frac{2}{15}$

5. $\frac{1}{x-y}=\frac{1}{3-5}$ *Let x=3 and y=5.*

$=\frac{1}{-2}=-\frac{1}{2}$

6. The answers for Exercises 4 and 5 are not the same. We conclude that we cannot find the difference $\frac{1}{x}-\frac{1}{y}$ by subtracting the denominators and keeping the common numerator.

7. $\frac{7}{t}+\frac{2}{t}=\frac{7+2}{t}=\frac{9}{t}$

9. $\frac{11}{5x}-\frac{1}{5x}=\frac{11-1}{5x}=\frac{10}{5x}=\frac{2}{x}$

11. $\frac{5x+4}{6x+5}+\frac{x+1}{6x+5}$

$=\frac{5x+4+x+1}{6x+5}$

$=\frac{6x+5}{6x+5}=1$

13. $\frac{x^2}{x+5}-\frac{25}{x+5}=\frac{x^2-25}{x+5}$

$=\frac{(x+5)(x-5)}{x+5}$

$=x-5$

15. $\frac{-3p+7}{p^2+7p+12}+\frac{8p+13}{p^2+7p+12}$

$=\frac{-3p+7+8p+13}{p^2+7p+12}$

$=\frac{5p+20}{(p+3)(p+4)}$

$=\frac{5(p+4)}{(p+3)(p+4)}=\frac{5}{p+3}$

17. $\frac{a^3}{a^2+ab+b^2}-\frac{b^3}{a^2+ab+b^2}$

$=\frac{a^3-b^3}{a^2+ab+b^2}$

$=\frac{(a-b)(a^2+ab+b^2)}{a^2+ab+b^2}$

$=a-b$

19. First add or subtract the numerators. Then place the result over the common denominator. Write the answer in lowest terms. We give one example:

$$\frac{5}{x}-\frac{3x+1}{x}=\frac{5-(3x+1)}{x}$$
$$=\frac{5-3x-1}{x}=\frac{4-3x}{x}.$$

21. $18x^2y^3$, $24x^4y^5$

Factor each denominator.

$$18x^2y^3=2\cdot3\cdot3\cdot x^2\cdot y^3$$
$$=2\cdot3^2\cdot x^2\cdot y^3$$
$$24x^4y^5=2\cdot2\cdot2\cdot3\cdot x^4\cdot y^5$$
$$=2^3\cdot3\cdot x^4\cdot y^5$$

The least common denominator (LCD) is the product of all the different factors, with each factor raised to the greatest power in any denominator.

$$\text{LCD}=2^3\cdot3^2\cdot x^4\cdot y^5$$
$$=8\cdot9\cdot x^4\cdot y^5$$
$$=72x^4y^5$$

The LCD is $72x^4y^5$.

23. $z-2$, z

Both $z-2$ and z have only 1 and themselves for factors.

LCD $=z(z-2)$

25. $2y+8$, $y+4$

Factor each denominator.

$$2y+8=2(y+4)$$

The second denominator, $y+4$, is already factored. The LCD is

$$2(y+4).$$

27. x^2-81, $x^2+18x+81$

Factor each denominator.

$$x^2-81=(x+9)(x-9)$$
$$x^2+18x+81=(x+9)(x+9)$$

LCD $=(x+9)^2(x-9)$

29. $m+n,\ m-n,\ m^2-n^2$

Both $m+n$ and $m-n$ have only 1 and themselves for factors, while $m^2-n^2=(m+n)(m-n)$.

LCD $=(m+n)(m-n)$

31. $x^2-3x-4,\ x+x^2$

Factor each denominator.

$$x^2-3x-4=(x-4)(x+1)$$
$$x+x^2=x(1+x)=x(x+1)$$

The LCD is $x(x-4)(x+1)$.

33. $2t^2+7t-15,\ t^2+3t-10$

Factor each denominator.

$$2t^2+7t-15=(2t-3)(t+5)$$
$$t^2+3t-10=(t+5)(t-2)$$

The LCD is $(2t-3)(t+5)(t-2)$.

35. $2y+6,\ y^2-9,\ y$

Factor each denominator.

$$2y+6=2(y+3)$$
$$y^2-9=(y+3)(y-3)$$

Remember the factor y from the third denominator. The LCD is

$$2y(y+3)(y-3).$$

37. Yes, they could both be correct because the expressions are equivalent. Multiplying $\frac{3}{5-y}$ by 1 in the form $\frac{-1}{-1}$ gives $\frac{-3}{y-5}$.

39. $\frac{8}{t}+\frac{7}{3t}$ The LCD is $3t$.

$$\frac{8}{t}+\frac{7}{3t}=\frac{8\cdot 3}{t\cdot 3}+\frac{7}{3t}$$
$$=\frac{24+7}{3t}=\frac{31}{3t}$$

41. $\frac{5}{12x^2y}-\frac{11}{6xy}$ The LCD is $12x^2y$.

$$\frac{5}{12x^2y}-\frac{11}{6xy}=\frac{5}{12x^2y}-\frac{11\cdot 2x}{6xy\cdot 2x}$$
$$=\frac{5}{12x^2y}-\frac{22x}{12x^2y}$$
$$=\frac{5-22x}{12x^2y}$$

43. $\frac{1}{x-1}-\frac{1}{x}$ LCD $=x(x-1)$

$$\frac{1}{x-1}-\frac{1}{x}=\frac{1\cdot x}{(x-1)x}-\frac{1\cdot(x-1)}{x(x-1)}$$
$$=\frac{x-(x-1)}{x(x-1)}$$
$$=\frac{x-x+1}{x(x-1)}$$
$$=\frac{1}{x(x-1)}$$

45. $\frac{3a}{a+1}+\frac{2a}{a-3}$ LCD $=(a+1)(a-3)$

$$=\frac{3a(a-3)}{(a+1)(a-3)}+\frac{2a(a+1)}{(a-3)(a+1)}$$
$$=\frac{3a(a-3)+2a(a+1)}{(a+1)(a-3)}$$
$$=\frac{3a^2-9a+2a^2+2a}{(a+1)(a-3)}$$
$$=\frac{5a^2-7a}{(a+1)(a-3)}$$

47. $\frac{17y+3}{9y+7}-\frac{-10y-18}{9y+7}$

$$=\frac{17y+3-(-10y-18)}{9y+7}$$
$$=\frac{17y+3+10y+18}{9y+7}$$
$$=\frac{27y+21}{9y+7}$$
$$=\frac{3(9y+7)}{9y+7}$$
$$=3$$

49. $\frac{2}{4-x}+\frac{5}{x-4}$

To get a common denominator of $x-4$, multiply both the numerator and denominator of the first expression by -1.

$$=\frac{(2)(-1)}{(4-x)(-1)}+\frac{5}{x-4}$$
$$=\frac{-2}{x-4}+\frac{5}{x-4}$$
$$=\frac{-2+5}{x-4}$$
$$=\frac{3}{x-4}$$

If you chose $4-x$ for the LCD, then you should have obtained the equivalent answer, $\frac{-3}{4-x}$.

51. $\frac{w}{w-z}-\frac{z}{z-w}$

$w-z$ and $z-w$ are opposites, so factor out -1 from $z-w$ to get a common denominator.

$$=\frac{w}{w-z}-\frac{z}{-1(w-z)}$$
$$=\frac{w}{w-z}+\frac{z}{w-z}$$
$$=\frac{w+z}{w-z}\quad\text{or}\quad\frac{-w-z}{z-w}$$

53. $\dfrac{5}{12+4x}-\dfrac{7}{9+3x}=\dfrac{5}{4(3+x)}-\dfrac{7}{3(3+x)}$

The LCD is $12(3+x)$.

$$=\frac{5\cdot 3}{4(3+x)\cdot 3}-\frac{7\cdot 4}{3(3+x)\cdot 4}$$
$$=\frac{15-28}{12(3+x)}$$
$$=\frac{-13}{12(3+x)}$$

55. $\dfrac{4x}{x-1}-\dfrac{2}{x+1}-\dfrac{4}{x^2-1}$

$x^2-1=(x+1)(x-1)$, the LCD.

$$\frac{4x}{x-1}-\frac{2}{x+1}-\frac{4}{x^2-1}$$
$$=\frac{4x(x+1)}{(x-1)(x+1)}-\frac{2(x-1)}{(x+1)(x-1)}$$
$$-\frac{4}{(x+1)(x-1)}$$
$$=\frac{4x(x+1)-2(x-1)-4}{(x+1)(x-1)}$$
$$=\frac{4x^2+4x-2x+2-4}{(x-1)(x+1)}$$
$$=\frac{4x^2+2x-2}{(x-1)(x+1)}$$
$$=\frac{2(2x^2+x-1)}{(x-1)(x+1)}$$
$$=\frac{2(2x-1)(x+1)}{(x-1)(x+1)}$$
$$=\frac{2(2x-1)}{x-1}$$

57. $\dfrac{15}{y^2+3y}+\dfrac{2}{y}+\dfrac{5}{y+3}$

$y^2+3y=y(y+3)$, the LCD.

$$=\frac{15}{y(y+3)}+\frac{2(y+3)}{y(y+3)}+\frac{5y}{(y+3)y}$$
$$=\frac{15+2(y+3)+5y}{y(y+3)}$$
$$=\frac{15+2y+6+5y}{y(y+3)}$$
$$=\frac{7y+21}{y(y+3)}$$
$$=\frac{7(y+3)}{y(y+3)}=\frac{7}{y}$$

59. $\dfrac{5}{x-2}+\dfrac{1}{x}+\dfrac{2}{x^2-2x}$

$x^2-2x=x(x-2)$, the LCD.

$$\frac{5}{x-2}+\frac{1}{x}+\frac{2}{x^2-2x}$$
$$=\frac{5x}{(x-2)x}+\frac{1(x-2)}{x(x-2)}+\frac{2}{x(x-2)}$$
$$=\frac{5x+x-2+2}{x(x-2)}$$
$$=\frac{6x}{x(x-2)}=\frac{6}{x-2}$$

61. $\dfrac{3x}{x+1}+\dfrac{4}{x-1}-\dfrac{6}{x^2-1}$

$x^2-1=(x+1)(x-1)$, the LCD.

$$=\frac{3x(x-1)}{(x+1)(x-1)}+\frac{4(x+1)}{(x-1)(x+1)}$$
$$-\frac{6}{(x+1)(x-1)}$$
$$=\frac{3x(x-1)+4(x+1)-6}{(x+1)(x-1)}$$
$$=\frac{3x^2-3x+4x+4-6}{(x+1)(x-1)}$$
$$=\frac{3x^2+x-2}{(x+1)(x-1)}$$
$$=\frac{(3x-2)(x+1)}{(x+1)(x-1)}$$
$$=\frac{3x-2}{x-1}$$

63. $\dfrac{4}{x+1}+\dfrac{1}{x^2-x+1}-\dfrac{12}{x^3+1}$

$x^3+1=(x+1)(x^2-x+1)$, the LCD.

$$=\frac{4(x^2-x+1)}{(x+1)(x^2-x+1)}$$
$$+\frac{1\cdot(x+1)}{(x^2-x+1)(x+1)}$$
$$-\frac{12}{(x+1)(x^2-x+1)}$$
$$=\frac{4(x^2-x+1)+(x+1)-12}{(x+1)(x^2-x+1)}$$
$$=\frac{4x^2-4x+4+x+1-12}{(x+1)(x^2-x+1)}$$
$$=\frac{4x^2-3x-7}{(x+1)(x^2-x+1)}$$
$$=\frac{(4x-7)(x+1)}{(x+1)(x^2-x+1)}$$
$$=\frac{4x-7}{x^2-x+1}$$

65. $\dfrac{2x+4}{x+3}+\dfrac{3}{x}-\dfrac{6}{x^2+3x}$

$x^2+3x=x(x+3)$, the LCD.

$$=\frac{(2x+4)x}{(x+3)x}+\frac{3(x+3)}{x(x+3)}-\frac{6}{x(x+3)}$$
$$=\frac{(2x+4)x+3(x+3)-6}{x(x+3)}$$
$$=\frac{2x^2+4x+3x+9-6}{x(x+3)}$$

$$= \frac{2x^2+7x+3}{x(x+3)}$$
$$= \frac{(2x+1)(x+3)}{x(x+3)} = \frac{2x+1}{x}$$

67. $\dfrac{3}{(p-2)^2} - \dfrac{5}{p-2} + 4$

$$= \frac{3}{(p-2)^2} - \frac{5(p-2)}{(p-2)^2} + \frac{4(p-2)^2}{(p-2)^2}$$

LCD=(p–2)²

$$= \frac{3-5(p-2)+4(p^2-4p+4)}{(p-2)^2}$$
$$= \frac{3-5p+10+4p^2-16p+16}{(p-2)^2}$$
$$= \frac{4p^2-21p+29}{(p-2)^2}$$

69. $\dfrac{3}{x^2-5x+6} - \dfrac{2}{x^2-4x+4}$

$$= \frac{3}{(x-2)(x-3)} - \frac{2}{(x-2)(x-2)}$$
$$= \frac{3(x-2)}{(x-2)(x-3)(x-2)}$$
$$- \frac{2(x-3)}{(x-2)(x-2)(x-3)}$$

LCD=(x–2)²(x–3)

$$= \frac{3x-6-2x+6}{(x-2)^2(x-3)}$$
$$= \frac{x}{(x-2)^2(x-3)}$$

71. $\dfrac{5x}{x^2+xy-2y^2} - \dfrac{3x}{x^2+5xy-6y^2}$

Factor each denominator.

$$x^2+xy-2y^2 = (x+2y)(x-y)$$
$$x^2+5xy-6y^2 = (x+6y)(x-y)$$

The LCD is $(x+2y)(x-y)(x+6y)$.

$$\frac{5x}{(x+2y)(x-y)} - \frac{3x}{(x+6y)(x-y)}$$
$$= \frac{(5x)(x+6y)}{(x+2y)(x-y)(x+6y)}$$
$$- \frac{(3x)(x+2y)}{(x+6y)(x-y)(x+2y)}$$
$$= \frac{(5x)(x+6y)-(3x)(x+2y)}{(x+6y)(x-y)(x+2y)}$$
$$= \frac{5x^2+30xy-(3x^2+6xy)}{(x+2y)(x-y)(x+6y)}$$
$$= \frac{2x^2+24xy}{(x+2y)(x-y)(x+6y)}$$
$$= \frac{2x(x+12y)}{(x+2y)(x-y)(x+6y)}$$

73. $\dfrac{5x-y}{x^2+xy-2y^2} - \dfrac{3x+2y}{x^2+5xy-6y^2}$

Factor each denominator.

$$x^2+xy-2y^2 = (x+2y)(x-y)$$
$$x^2+5xy-6y^2 = (x+6y)(x-y)$$

The LCD is $(x+2y)(x-y)(x+6y)$.

$$\frac{5x-y}{(x+2y)(x-y)} - \frac{3x+2y}{(x+6y)(x-y)}$$
$$= \frac{(5x-y)(x+6y)}{(x+2y)(x-y)(x+6y)}$$
$$- \frac{(3x+2y)(x+2y)}{(x+6y)(x-y)(x+2y)}$$
$$= \frac{(5x-y)(x+6y)-(3x+2y)(x+2y)}{(x+6y)(x-y)(x+2y)}$$
$$= \frac{5x^2+29xy-6y^2-(3x^2+8xy+4y^2)}{(x+2y)(x-y)(x+6y)}$$
$$= \frac{2x^2+21xy-10y^2}{(x+2y)(x-y)(x+6y)}$$

75. $\dfrac{r+s}{3r^2+2rs-s^2} - \dfrac{s-r}{6r^2-5rs+s^2}$

Factor each denominator.

$$3r^2+2rs-s^2 = (3r-s)(r+s)$$
$$6r^2-5rs+s^2 = (3r-s)(2r-s)$$

The LCD is $(3r-s)(r+s)(2r-s)$.

$$\frac{r+s}{3r^2+2rs-s^2} - \frac{s-r}{6r^2-5rs+s^2}$$
$$= \frac{r+s}{(3r-s)(r+s)} - \frac{s-r}{(3r-s)(2r-s)}$$
$$= \frac{(r+s)(2r-s)}{(3r-s)(r+s)(2r-s)}$$
$$- \frac{(s-r)(r+s)}{(3r-s)(2r-s)(r+s)}$$
$$= \frac{(r+s)(2r-s)-(s-r)(s+r)}{(3r-s)(r+s)(2r-s)}$$
$$= \frac{2r^2+rs-s^2-(s^2-r^2)}{(3r-s)(r+s)(2r-s)}$$
$$= \frac{2r^2+rs-s^2-s^2+r^2}{(3r-s)(r+s)(2r-s)}$$
$$= \frac{3r^2+rs-2s^2}{(3r-s)(r+s)(2r-s)}$$
$$= \frac{(3r-2s)(r+s)}{(3r-s)(r+s)(2r-s)}$$
$$= \frac{3r-2s}{(3r-s)(2r-s)}$$

77. $\dfrac{3}{x^2+4x+4}+\dfrac{7}{x^2+5x+6}$

$=\dfrac{3}{(x+2)^2}+\dfrac{7}{(x+2)(x+3)}$

$LCD=(x+2)^2(x+3)$

$=\dfrac{3(x+3)}{(x+2)^2(x+3)}+\dfrac{7(x+2)}{(x+2)^2(x+3)}$

$=\dfrac{3x+9+7x+14}{(x+2)^2(x+3)}$

$=\dfrac{10x+23}{(x+2)^2(x+3)}$

79. **(a)** $c(x)=\dfrac{1010}{49(101-x)}-\dfrac{10}{49}$

$=\dfrac{1010}{49(101-x)}-\dfrac{10(101-x)}{49(101-x)}$

$=\dfrac{1010-1010+10x}{49(101-x)}$

$=\dfrac{10x}{49(101-x)}$

(b) $c(95)=\dfrac{10(95)}{49(101-95)}$

$=\dfrac{950}{294}\approx 3.23$

It would cost approximately 3.23 thousand dollars to win 95 points.

81. $\dfrac{3}{7}+\dfrac{5}{9}-\dfrac{6}{63}$

The LCD is $7(9)=63$.

$=\dfrac{3\cdot 9}{7\cdot 9}+\dfrac{5\cdot 7}{9\cdot 7}-\dfrac{6}{63}$

$=\dfrac{27+35-6}{63}$

$=\dfrac{56}{63}=\dfrac{8}{9}$

82. From Example 6,

$$\frac{3}{x-2}+\frac{5}{x}-\frac{6}{x^2-2x}.$$

Substitute 9 for x.

$$\frac{3}{9-2}+\frac{5}{9}-\frac{6}{9^2-2(9)}=\frac{3}{7}+\frac{5}{9}-\frac{6}{63}$$

The problems in Exercises 81 and 82 are the same.

83. From Exercise 81,

$$\frac{3}{7}+\frac{5}{9}-\frac{6}{63}=\frac{8}{9}.$$

From Example 6, the answer is $\frac{8}{x}$. If we substitute 9 for x, the answer becomes $\frac{8}{9}$. The answers agree.

84. Answers will vary. For example, suppose the last name is Bush so that $x=4$. The problem in Example 6 becomes

$$\frac{3}{4-2}+\frac{5}{4}-\frac{6}{4^2-2(4)}=\frac{3}{2}+\frac{5}{4}-\frac{6}{8}.$$

The predicted answer is

$$\frac{8}{x},\text{ or }\frac{8}{4}=2.$$

Perform the operations to verify our prediction.

$\dfrac{3}{2}+\dfrac{5}{4}-\dfrac{6}{8}=\dfrac{3\cdot 4}{2\cdot 4}+\dfrac{5\cdot 2}{4\cdot 2}-\dfrac{6}{8}$

$=\dfrac{12+10-6}{8}$

$=\dfrac{16}{8}=2$

The prediction is correct.

85. If $x=2$, then the problem from Example 6,

$$\frac{3}{x-2}+\frac{5}{x}-\frac{6}{x^2-2x},$$

becomes

$$\frac{3}{2-2}+\frac{5}{2}-\frac{6}{2^2-2(2)}=\frac{3}{0}+\frac{5}{2}-\frac{6}{0}.$$

Thus, if $x=2$, then

$$\frac{3}{x-2}\text{ and }\frac{6}{x^2-2x}$$

are undefined.

86. If $x=0$, then

$$\frac{5}{x}=\frac{5}{0}\text{ and }\frac{6}{x^2-2x}=\frac{6}{0},$$

which are undefined. Therefore, 0 is not allowed as a value of x.

Section 7.3

1. *Method 1*: Begin by simplifying the numerator to a single fraction. Then simplify the denominator to a single fraction. Write as a division problem, and multiply by the reciprocal of the denominator. Simplify the result, if possible.
Method 2: Find the LCD of all fractions in the complex fraction. Multiply the numerator and denominator of the complex fraction by this LCD. Simplify the result, if possible.

3. $\dfrac{\frac{12}{x-1}}{\frac{6}{x}}=\dfrac{12}{x-1}\div\dfrac{6}{x}$

Multiply by the reciprocal of the divisor.

$=\dfrac{12}{x-1}\cdot\dfrac{x}{6}$

$=\dfrac{2x}{x-1}$

5. $$\frac{\frac{k+1}{2k}}{\frac{3k-1}{4k}} = \frac{k+1}{2k} \cdot \frac{4k}{3k-1} = \frac{4k(k+1)}{2k(3k-1)} = \frac{2(k+1)}{3k-1}$$

7. $$\frac{\frac{4z^2x^4}{9}}{\frac{12x^2z^5}{15}} = \frac{\frac{4z^2x^4}{9}}{\frac{4x^2z^5}{5}} = \frac{4z^2x^4}{9} \div \frac{4x^2z^5}{5} = \frac{4z^2x^4}{9} \cdot \frac{5}{4x^2z^5} = \frac{5z^2x^4}{9x^2z^5} = \frac{5x^2}{9z^3}$$

9. $$\frac{\frac{1}{x}+1}{-\frac{1}{x}+1}$$

Multiply the numerator and denominator by x, the LCD of all the fractions.

$$= \frac{x\left(\frac{1}{x}+1\right)}{x\left(-\frac{1}{x}+1\right)} = \frac{x \cdot \frac{1}{x} + x \cdot 1}{x\left(-\frac{1}{x}\right) + x \cdot 1} = \frac{1+x}{-1+x}$$

11. $$\frac{\frac{3}{x}+\frac{3}{y}}{\frac{3}{x}-\frac{3}{y}}$$

Multiply the numerator and denominator by xy, the LCD of all the fractions.

$$= \frac{\left(\frac{3}{x}+\frac{3}{y}\right)xy}{\left(\frac{3}{x}-\frac{3}{y}\right)xy} = \frac{\frac{3}{x} \cdot xy + \frac{3}{y} \cdot xy}{\frac{3}{x} \cdot xy - \frac{3}{y} \cdot xy} = \frac{3y+3x}{3y-3x} = \frac{3(y+x)}{3(y-x)} = \frac{y+x}{y-x}$$

13. $$\frac{\frac{8x-24y}{10}}{\frac{x-3y}{5x}} = \frac{8x-24y}{10} \cdot \frac{5x}{x-3y} = \frac{8(x-3y)5x}{10(x-3y)} = \frac{40x}{10} = 4x$$

15. $$\frac{\frac{x^2-16y^2}{xy}}{\frac{1}{y}-\frac{4}{x}}$$

Multiply the numerator and denominator by xy, the LCD of all the fractions.

$$= \frac{\left(\frac{x^2-16y^2}{xy}\right)xy}{\left(\frac{1}{y}-\frac{4}{x}\right)xy} = \frac{x^2-16y^2}{\frac{1}{y} \cdot xy - \frac{4}{x} \cdot xy} = \frac{x^2-16y^2}{x-4y} = \frac{(x+4y)(x-4y)}{x-4y} = x+4y$$

17. $$\frac{y - \frac{y-3}{3}}{\frac{4}{9}+\frac{2}{3y}}$$

Multiply the numerator and denominator by $9y$, the LCD of all the fractions.

$$= \frac{9y\left(y-\frac{y-3}{3}\right)}{9y\left(\frac{4}{9}+\frac{2}{3y}\right)} = \frac{9y^2-3y(y-3)}{4y+6} = \frac{9y^2-3y^2+9y}{4y+6} = \frac{6y^2+9y}{4y+6} = \frac{3y(2y+3)}{2(2y+3)} = \frac{3y}{2}$$

19. $$\frac{\frac{x+2}{x}+\frac{1}{x+2}}{\frac{5}{x}+\frac{x}{x+2}}$$

Multiply the numerator and denominator by $x(x+2)$, the LCD of all the fractions.

continued

$$= \frac{x(x+2)\left(\frac{x+2}{x} + \frac{1}{x+2}\right)}{x(x+2)\left(\frac{5}{x} + \frac{x}{x+2}\right)}$$

$$= \frac{x(x+2)\left(\frac{x+2}{x}\right) + x(x+2)\left(\frac{1}{x+2}\right)}{x(x+2)\left(\frac{5}{x}\right) + x(x+2)\left(\frac{x}{x+2}\right)}$$

$$= \frac{(x+2)(x+2)+x}{5(x+2)+x^2}$$

$$= \frac{x^2+4x+4+x}{5x+10+x^2}$$

$$= \frac{x^2+5x+4}{x^2+5x+10}$$

21. To add the fractions in the numerator, use the LCD $m(m-1)$.

$$\frac{4}{m} + \frac{m+2}{m-1} = \frac{4(m-1)}{m(m-1)} + \frac{m(m+2)}{m(m-1)}$$

$$= \frac{4m-4+m^2+2m}{m(m-1)}$$

$$= \frac{m^2+6m-4}{m(m-1)}$$

22. To subtract the fractions in the denominator, use the same LCD, $m(m-1)$.

$$\frac{m+2}{m} - \frac{2}{m-1}$$

$$= \frac{(m+2)(m-1)}{m(m-1)} - \frac{m \cdot 2}{m(m-1)}$$

$$= \frac{m^2+m-2-2m}{m(m-1)}$$

$$= \frac{m^2-m-2}{m(m-1)}$$

23. Exercise 21 answer ↓ Exercise 22 answer ↓

$$\frac{m^2+6m-4}{m(m-1)} \div \frac{m^2-m-2}{m(m-1)}$$

Multiply by the reciprocal.

$$= \frac{m^2+6m-4}{m(m-1)} \cdot \frac{m(m-1)}{m^2-m-2}$$

$$= \frac{m^2+6m-4}{m^2-m-2}$$

24. The LCD of all the denominators in the complex fraction is $m(m-1)$.

25.
$$\frac{\left(\frac{4}{m} + \frac{m+2}{m-1}\right) \cdot m(m-1)}{\left(\frac{m+2}{m} - \frac{2}{m-1}\right) \cdot m(m-1)}$$

$$= \frac{4(m-1)+m(m+2)}{(m+2)(m-1)-2m}$$

$$= \frac{4m-4+m^2+2m}{m^2+m-2-2m}$$

$$= \frac{m^2+6m-4}{m^2-m-2}$$

26. Answers will vary. Because of the complicated nature of the numerator and denominator of the complex fraction, using Method 1 takes much longer to simplify the complex fraction. Method 2 is a simpler, more direct means of simplifying and is most likely the preferred method.

27.
$$\frac{1}{x^{-2}+y^{-2}} = \frac{1}{\frac{1}{x^2} + \frac{1}{y^2}}$$

$$= \frac{x^2y^2(1)}{x^2y^2\left(\frac{1}{x^2} + \frac{1}{y^2}\right)} \qquad LCD = x^2y^2$$

$$= \frac{x^2y^2}{y^2+x^2}$$

29.
$$\frac{x^{-2}+y^{-2}}{x^{-1}+y^{-1}}$$

$$= \frac{\frac{1}{x^2} + \frac{1}{y^2}}{\frac{1}{x} + \frac{1}{y}}$$

Multiply the numerator and denominator by x^2y^2, the LCD of all the fractions.

$$= \frac{x^2y^2\left(\frac{1}{x^2} + \frac{1}{y^2}\right)}{x^2y^2\left(\frac{1}{x} + \frac{1}{y}\right)}$$

$$= \frac{x^2y^2 \cdot \frac{1}{x^2} + x^2y^2 \cdot \frac{1}{y^2}}{x^2y^2 \cdot \frac{1}{x} + x^2y^2 \cdot \frac{1}{y}}$$

$$= \frac{y^2+x^2}{xy^2+x^2y} \text{ or } \frac{y^2+x^2}{xy(y+x)}$$

31.
$$\frac{x^{-1}+2y^{-1}}{2y+4x} = \frac{\frac{1}{x} + \frac{2}{y}}{2y+4x}$$

Multiply the numerator and denominator by xy, the LCD of all the fractions.

$$= \frac{xy\left(\frac{1}{x} + \frac{2}{y}\right)}{xy(2y+4x)}$$

$$= \frac{y+2x}{2xy(y+2x)}$$

$$= \frac{1}{2xy}$$

33. **(a)** $\dfrac{\dfrac{3}{mp}-\dfrac{4}{p}+\dfrac{8}{m}}{2m^{-1}-3p^{-1}}$

$=\dfrac{\dfrac{3}{mp}-\dfrac{4}{p}+\dfrac{8}{m}}{\dfrac{2}{m}-\dfrac{3}{p}}$

(b) $2m^{-1}=\dfrac{2}{m}$, not $\dfrac{1}{2m}$, since the exponent applies only to m, not to 2. Likewise, $3p^{-1}=\dfrac{3}{p}$, not $\dfrac{1}{3p}$.

(c) $\dfrac{\dfrac{3}{mp}-\dfrac{4}{p}+\dfrac{8}{m}}{2m^{-1}-3p^{-1}}=\dfrac{\dfrac{3}{mp}-\dfrac{4}{p}+\dfrac{8}{m}}{\dfrac{2}{m}-\dfrac{3}{p}}$

Multiply the numerator and denominator by mp, the LCD of all the fractions.

$$=\frac{mp\left(\frac{3}{mp}-\frac{4}{p}+\frac{8}{m}\right)}{mp\left(\frac{2}{m}-\frac{3}{p}\right)}$$

$$=\frac{mp\cdot\frac{3}{mp}-mp\cdot\frac{4}{p}+mp\cdot\frac{8}{m}}{mp\cdot\frac{2}{m}-mp\cdot\frac{3}{p}}$$

$$=\frac{3-4m+8p}{2p-3m}$$

Section 7.4

1. **(a)** $\dfrac{1}{x+1}-\dfrac{1}{x-2}=0$

Set each denominator equal to 0 and solve.

$$x+1=0 \quad\text{or}\quad x-2=0$$
$$x=-1 \quad\text{or}\quad x=2$$

Solutions of -1 and 2 would be rejected since these values would make a denominator of the original equation equal to 0.

(b) The domain is $\{x|x\neq-1,2\}$.

3. **(a)** $\dfrac{5}{3x+5}-\dfrac{1}{x}=\dfrac{1}{2x+3}$

Set each denominator equal to 0 and solve.

$$3x+5=0 \quad\text{or}\quad x=0 \quad\text{or}\quad 2x+3=0$$
$$3x=-5 \qquad\qquad 2x=-3$$
$$x=-\frac{5}{3} \qquad\qquad x=-\frac{3}{2}$$

So $-\dfrac{5}{3}$, 0, and $-\dfrac{3}{2}$ would have to be rejected as potential solutions.

(b) The domain is $\left\{x|x\neq-\dfrac{5}{3},0,-\dfrac{3}{2}\right\}$.

5. **(a)** $\dfrac{1}{3x}+\dfrac{1}{2x}=\dfrac{x}{3}$

Only 0 would make any of the denominators equal to 0, so 0 would have to be rejected as a potential solution.

(b) The domain is $\{x|x\neq0\}$.

7. **(a)** $\dfrac{3x+1}{x-4}=\dfrac{6x+5}{2x-7}$

Set each denominator equal to 0 and solve.

$$x-4=0 \quad\text{or}\quad 2x-7=0$$
$$x=4 \qquad\qquad 2x=7$$
$$x=\frac{7}{2}$$

So 4 and $\dfrac{7}{2}$ would have to be rejected as potential solutions.

(b) The domain is $\left\{x|x\neq4,\dfrac{7}{2}\right\}$.

9. **(a)** $\dfrac{2}{x^2-x}+\dfrac{1}{x+3}=\dfrac{4}{x-2}$

$x^2-x=x(x-1)$ is 0 if $x=0$ or $x=1$.
$x+3$ is 0 if $x=-3$.
$x-2$ is 0 if $x=2$.

So 0, 1, -3, and 2 would have to be rejected as potential solutions.

(b) The domain is $\{x|x\neq0,1,-3,2\}$.

In Exercises 11–39, check each potential solution in the original equation.

11. $\dfrac{-5}{2x}+\dfrac{3}{4x}=\dfrac{-7}{4}$

Multiply by the LCD, $4x$. $(x\neq0)$

$$4x\left(\frac{-5}{2x}+\frac{3}{4x}\right)=4x\left(\frac{-7}{4}\right)$$
$$2(-5)+1(3)=x(-7)$$
$$-10+3=-7x$$
$$-7=-7x$$
$$1=x$$

Check $x=1$: $-\dfrac{5}{2}+\dfrac{3}{4}=-\dfrac{7}{4}$ *True*

Solution set: $\{1\}$

13. $$x - \frac{24}{x} = -2$$

Multiply by the LCD, x. $(x \neq 0)$

$$x\left(x - \frac{24}{x}\right) = -2 \cdot x$$
$$x^2 - 24 = -2x$$
$$x^2 + 2x - 24 = 0$$
$$(x + 6)(x - 4) = 0$$
$$x + 6 = 0 \quad \text{or} \quad x - 4 = 0$$
$$x = -6 \quad \text{or} \quad x = 4$$

Check $x = -6$: $-6 + 4 = -2$ *True*
Check $x = 4$: $4 - 6 = -2$ *True*
Solution set: $\{-6, 4\}$

15. $$\frac{x-4}{x+6} = \frac{2x+3}{2x-1}$$

Multiply by the LCD, $(x + 6)(2x - 1)$.
Note that $x \neq -6$ and $x \neq \frac{1}{2}$.

$$(x+6)(2x-1)\left(\frac{x-4}{x+6}\right) = (x+6)(2x-1)\left(\frac{2x+3}{2x-1}\right)$$
$$(2x - 1)(x - 4) = (x + 6)(2x + 3)$$
$$2x^2 - 9x + 4 = 2x^2 + 15x + 18$$
$$-24x = 14$$
$$x = \frac{14}{-24} = -\frac{7}{12}$$

A calculator check is suggested.

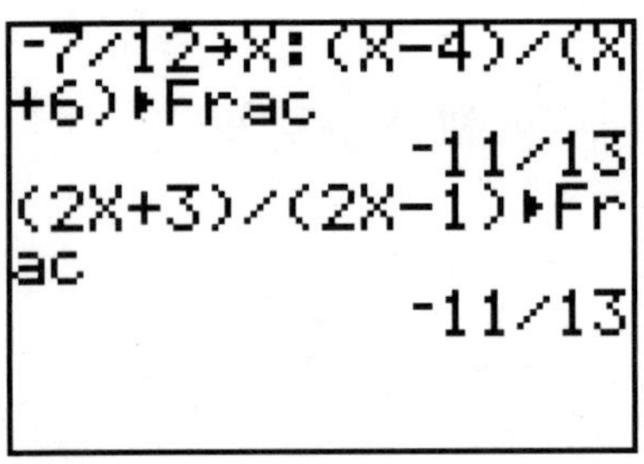

Check $x = -\frac{7}{12}$: $-\frac{11}{13} = -\frac{11}{13}$ *True*

Solution set: $\left\{-\frac{7}{12}\right\}$

17. $$\frac{3x+1}{x-4} = \frac{6x+5}{2x-7}$$

Multiply by the LCD, $(x - 4)(2x - 7)$.

Note that $x \neq 4$ and $x \neq \frac{7}{2}$.

$$(x-4)(2x-7)\left(\frac{3x+1}{x-4}\right) = (x-4)(2x-7)\left(\frac{6x+5}{2x-7}\right)$$
$$(2x - 7)(3x + 1) = (x - 4)(6x + 5)$$
$$6x^2 - 19x - 7 = 6x^2 - 19x - 20$$
$$-7 = -20 \quad \textit{False}$$

The false statement indicates that the original equation has no solution.
Solution set: $\emptyset$

19. $$\frac{1}{y-1} + \frac{5}{12} = \frac{-2}{3y-3}$$
$$\frac{1}{y-1} + \frac{5}{12} = \frac{-2}{3(y-1)}$$

Multiply by the LCD, $12(y - 1)$. $(y \neq 1)$

$$12(y-1)\left(\frac{1}{y-1} + \frac{5}{12}\right) = 12(y-1)\left(\frac{-2}{3(y-1)}\right)$$
$$12 + 5(y - 1) = -8$$
$$12 + 5y - 5 = -8$$
$$5y + 7 = -8$$
$$5y = -15$$
$$y = -3$$

Check $y = -3$: $-\frac{1}{4} + \frac{5}{12} = \frac{1}{6}$ *True*
Solution set: $\{-3\}$

21. $$\frac{-2}{3t-6} - \frac{1}{36} = \frac{-3}{4t-8}$$
$$\frac{-2}{3(t-2)} - \frac{1}{36} = \frac{-3}{4(t-2)}$$

Multiply by the LCD, $36(t - 2)$. $(t \neq 2)$

$$36(t-2)\left(\frac{-2}{3(t-2)} - \frac{1}{36}\right) = 36(t-2)\left(\frac{-3}{4(t-2)}\right)$$
$$12(-2) - 1(t - 2) = 9(-3)$$
$$-24 - t + 2 = -27$$
$$-t - 22 = -27$$
$$-t = -5$$
$$t = 5$$

Check $t = 5$: $-\frac{2}{9} - \frac{1}{36} = \frac{-3}{12}$ *True*

Solution set: $\{5\}$

23. $$\frac{3}{k+2} - \frac{2}{k^2-4} = \frac{1}{k-2}$$
$$\frac{3}{k+2} - \frac{2}{(k+2)(k-2)} = \frac{1}{k-2}$$

Multiply by the LCD, $(k + 2)(k - 2)$.
$(k \neq -2, 2)$

$$(k+2)(k-2)\left(\frac{3}{k+2} - \frac{2}{(k+2)(k-2)}\right)$$
$$= (k+2)(k-2)\left(\frac{1}{k-2}\right)$$
$$3(k - 2) - 2 = k + 2$$
$$3k - 6 - 2 = k + 2$$
$$3k - 8 = k + 2$$
$$2k = 10$$
$$k = 5$$

Check $k = 5$: $\frac{3}{7} - \frac{2}{21} = \frac{1}{3}$ *True*

Solution set: $\{5\}$

25. $$\frac{1}{y+2}+\frac{3}{y+7}=\frac{5}{y^2+9y+14}$$
$$\frac{1}{y+2}+\frac{3}{y+7}=\frac{5}{(y+2)(y+7)}$$
Multiply by the LCD, $(y+2)(y+7)$. $(y \neq -2, -7)$
$$(y+2)(y+7)\left(\frac{1}{y+2}+\frac{3}{y+7}\right)=(y+2)(y+7)\left(\frac{5}{(y+2)(y+7)}\right)$$
$$\begin{aligned}(y+7)+3(y+2)&=5\\ y+7+3y+6&=5\\ 4y+13&=5\\ 4y&=-8\\ y&=-2\end{aligned}$$
But y cannot equal -2 because that would make the denominator $y+2$ equal to 0. Since division by 0 is undefined, the equation has no solution.

Solution set: $\emptyset$

27. $$\frac{9}{x}+\frac{4}{6x-3}=\frac{2}{6x-3}$$
Multiply by the LCD, $x(6x-3)$. $\left(x \neq 0, \frac{1}{2}\right)$
$$x(6x-3)\left(\frac{9}{x}+\frac{4}{6x-3}\right)=x(6x-3)\left(\frac{2}{6x-3}\right)$$
$$\begin{aligned}9(6x-3)+4x&=2x\\ 54x-27+4x&=2x\\ 56x&=27\\ x&=\frac{27}{56}\end{aligned}$$
Check $x=\frac{27}{56}$: $\frac{56}{3}+\left(-\frac{112}{3}\right)=-\frac{56}{3}$ *True*

Solution set: $\left\{\frac{27}{56}\right\}$

29. $$\frac{6}{w+3}+\frac{-7}{w-5}=\frac{-48}{w^2-2w-15}$$
$$\frac{6}{w+3}+\frac{-7}{w-5}=\frac{-48}{(w+3)(w-5)}$$
Multiply by the LCD, $(w+3)(w-5)$. $(w \neq -3, 5)$
$$(w+3)(w-5)\left(\frac{6}{w+3}+\frac{-7}{w-5}\right)=(w+3)(w-5)\left[\frac{-48}{(w+3)(w-5)}\right]$$
$$\begin{aligned}6(w-5)-7(w+3)&=-48\\ 6w-30-7w-21&=-48\\ -w-51&=-48\\ -w&=3\\ w&=-3\end{aligned}$$
But w cannot equal -3 because that would make the denominator $w+3$ equal to 0. Since division by 0 is undefined, the equation has no solution.

Solution set: $\emptyset$

31. $$\frac{x}{x-3}+\frac{4}{x+3}=\frac{18}{x^2-9}$$
$$\frac{x}{x-3}+\frac{4}{x+3}=\frac{18}{(x-3)(x+3)}$$
Multiply by the LCD, $(x-3)(x+3)$. $(x \neq 3, -3)$
$$(x-3)(x+3)\left(\frac{x}{x-3}+\frac{4}{x+3}\right)=(x-3)(x+3)\left(\frac{18}{(x-3)(x+3)}\right)$$
$$\begin{aligned}x(x+3)+4(x-3)&=18\\ x^2+3x+4x-12&=18\\ x^2+7x-30&=0\\ (x-3)(x+10)&=0\end{aligned}$$
$$x-3=0 \quad \text{or} \quad x+10=0$$
$$x=3 \quad \text{or} \quad x=-10$$
But $x \neq 3$ since a denominator of 0 results. The only solution to check is -10.

Check $x=-10$: $\frac{10}{13}+\left(-\frac{4}{7}\right)=\frac{18}{91}$ *True*

Solution set: $\{-10\}$

33. $$\frac{1}{x+4}+\frac{x}{x-4}=\frac{-8}{x^2-16}$$
$$\frac{1}{x+4}+\frac{x}{x-4}=\frac{-8}{(x+4)(x-4)}$$
Multiply by the LCD, $(x+4)(x-4)$. $(x \neq -4, 4)$
$$(x+4)(x-4)\left(\frac{1}{x+4}+\frac{x}{x-4}\right)=(x+4)(x-4)\left(\frac{-8}{x^2-16}\right)$$
$$\begin{aligned}(x-4)+x(x+4)&=-8\\ x-4+x^2+4x&=-8\\ x^2+5x+4&=0\\ (x+4)(x+1)&=0\end{aligned}$$
$$x+4=0 \quad \text{or} \quad x+1=0$$
$$x=-4 \qquad\qquad x=-1$$
But x cannot equal -4, so we only need to check -1.

Check $x=-1$: $\frac{1}{3}+\frac{1}{5}=\frac{8}{15}$ *True*

Solution set: $\{-1\}$

35. $$\frac{2}{4x+7}+\frac{x}{3}=\frac{6}{12x+21}$$
$$\frac{2}{4x+7}+\frac{x}{3}=\frac{6}{3(4x+7)}$$
Multiply by the LCD, $3(4x+7)$. $\left(x \neq -\frac{7}{4}\right)$
$$\begin{aligned}2(3)+x(4x+7)&=6\\ 6+4x^2+7x&=6\\ 4x^2+7x&=0\\ x(4x+7)&=0\end{aligned}$$

continued

$$x = 0 \quad \text{or} \quad 4x + 7 = 0$$
$$4x = -7$$
$$x = -\frac{7}{4}$$

But $x \neq -\frac{7}{4}$, so the only solution is 0.

Solution set: $\{0\}$

37. $$\frac{5}{p^2 + 3p + 2} - \frac{3}{p^2 - 4} = \frac{1}{p^2 - p - 2}$$
$$\frac{5}{(p+1)(p+2)} - \frac{3}{(p+2)(p-2)} = \frac{1}{(p-2)(p+1)}$$
Multiply by the LCD, $(p+1)(p+2)(p-2)$.
$(p \neq -1, -2, 2)$
$$5(p-2) - 3(p+1) = 1(p+2)$$
$$5p - 10 - 3p - 3 = p + 2$$
$$2p - 13 = p + 2$$
$$p = 15$$

Check $p = 15$: $\frac{5}{272} - \frac{3}{221} = \frac{1}{208}$ *True*

Solution set: $\{15\}$

39. $$\frac{4x - 7}{4x^2 - 9} = \frac{-2x^2 + 5x - 4}{4x^2 - 9} + \frac{x + 1}{2x + 3}$$
$$\frac{4x - 7}{(2x+3)(2x-3)} = \frac{-2x^2 + 5x - 4}{(2x+3)(2x-3)} + \frac{x+1}{2x+3}$$
Multiply by the LCD, $(2x+3)(2x-3)$.
$\left(x \neq -\frac{3}{2}, \frac{3}{2}\right)$
$$4x - 7 = -2x^2 + 5x - 4 + (2x-3)(x+1)$$
$$4x - 7 = -2x^2 + 5x - 4 + 2x^2 - x - 3$$
$$4x - 7 = 4x - 7 \quad \textit{True}$$
This equation is true for every real number value of x, but we have already determined that $x \neq -\frac{3}{2}$ or $x \neq \frac{3}{2}$. So every real number except $-\frac{3}{2}$ and $\frac{3}{2}$ is a solution.

Solution set: $\left\{x \middle| x \neq -\frac{3}{2}, \frac{3}{2}\right\}$ or

$$\left(-\infty, -\frac{3}{2}\right) \cup \left(-\frac{3}{2}, \frac{3}{2}\right) \cup \left(\frac{3}{2}, \infty\right)$$

41. $f(x) = \frac{2}{x}$ is not defined when $x = 0$, so an equation of the vertical asymptote is $x = 0$.

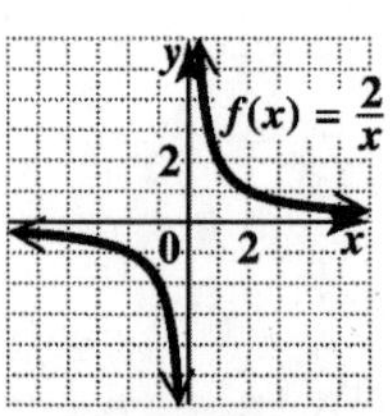

43. $f(x) = \frac{1}{x - 2}$ is not defined when $x = 2$, so an equation of the vertical asymptote is $x = 2$.

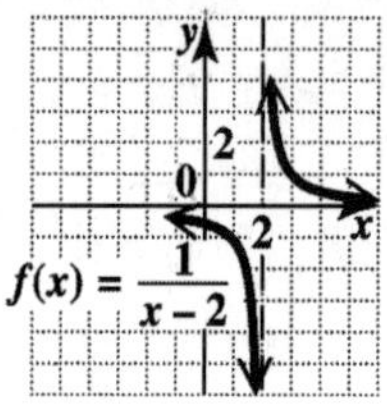

45. $$w(x) = \frac{x^2}{2(1 - x)}$$

(a) $$w(.1) = \frac{(.1)^2}{2(1 - .1)} = \frac{.01}{2(.9)} \approx .006$$

To the nearest tenth, $w(.1)$ is 0.

(b) $$w(.8) = \frac{(.8)^2}{2(1 - .8)} = \frac{.64}{2(.2)} = 1.6$$

(c) $$w(.9) = \frac{(.9)^2}{2(1 - .9)} = \frac{.81}{2(.1)} = 4.05 \approx 4.1$$

(d) Based on the answers in (a), (b), and (c), we see that as the traffic intensity increases, the waiting time also increases.

47. **(a)** $$F(r) = \frac{225,000}{r}$$
$$450 = \frac{225,000}{r} \quad \textit{Let } F(r)=450.$$
$$450r = 225,000$$
$$r = \frac{225,000}{450} = 500$$
The radius must be 500 feet.

(b) As r increases, the fraction $\frac{225,000}{r}$ gets smaller, so the force, $f(r)$, decreases.

49. The number of solutions of the equation $f(x) = 0$ is the same as the number of x-intercepts, four.

51. The x-intercepts are -2, 0, and 3, so the solution set is $\{-2, 0, 3\}$.

Summary Exercises on Operations and Equations with Rational Expressions

1. $$\frac{x}{2} - \frac{x}{4} = 5$$
There is an equals sign, so this is an *equation.*
$$4\left(\frac{x}{2} - \frac{x}{4}\right) = 4(5) \quad \textit{Multiply by 4.}$$

$$2x - x = 20$$
$$x = 20$$

Check $x = 20$: $10 - 5 = 5$ *True*

Solution set: $\{20\}$

3. No equals sign appears so this is an *operation.*

$$\frac{6}{7x} - \frac{4}{x}$$
$$= \frac{6}{7x} - \frac{4 \cdot 7}{x \cdot 7} \qquad \textit{LCD=7x}$$
$$= \frac{6 - 28}{7x} = \frac{-22}{7x}$$

5. $$\frac{5}{7t} = \frac{52}{7} - \frac{3}{t}$$

There is an equals sign, so this is an *equation.* Multiply by the LCD, $7t$. $(t \neq 0)$

$$7t\left(\frac{5}{7t}\right) = 7t\left(\frac{52}{7} - \frac{3}{t}\right)$$
$$5 = 52t - 3(7)$$
$$5 = 52t - 21$$
$$26 = 52t$$
$$t = \frac{26}{52} = \frac{1}{2}$$

Check $t = \frac{1}{2}$: $\frac{10}{7} = \frac{52}{7} - 6$ *True*

Solution set: $\left\{\frac{1}{2}\right\}$

7. No equals sign appears so this is an *operation.*

$$\frac{7}{6x} + \frac{5}{8x}$$
$$= \frac{7}{3 \cdot 2x} + \frac{5}{4 \cdot 2x}$$
$$= \frac{7(4)}{4 \cdot 3 \cdot 2x} + \frac{5(3)}{3 \cdot 4 \cdot 2x} \qquad \textit{LCD=24x}$$
$$= \frac{28 + 15}{3 \cdot 4 \cdot 2x} = \frac{43}{24x}$$

9. No equals sign appears so this is an *operation.*

$$\frac{\frac{6}{x+1} - \frac{1}{x}}{\frac{2}{x} - \frac{4}{x+1}}$$

Multiply the numerator and denominator by the LCD of all the fractions, $x(x + 1)$.

$$= \frac{6(x) - 1(x+1)}{2(x+1) - 4(x)}$$
$$= \frac{6x - x - 1}{2x + 2 - 4x}$$
$$= \frac{5x - 1}{-2x + 2} \text{ or } \frac{5x - 1}{-2(x - 1)}$$

11. No equals sign appears so this is an *operation.*

$$\frac{x}{x+y} + \frac{2y}{x-y}$$
$$\frac{x(x-y)}{(x+y)(x-y)} + \frac{2y(x+y)}{(x+y)(x-y)}$$
$$\textit{LCD=(x+y)(x–y)}$$
$$= \frac{x^2 - xy + 2xy + 2y^2}{(x+y)(x-y)}$$
$$= \frac{x^2 + xy + 2y^2}{(x+y)(x-y)}$$

13. No equals sign appears so this is an *operation.*

$$\frac{x-2}{9} \cdot \frac{5}{8 - 4x}$$
$$= \frac{x-2}{9} \cdot \frac{5}{-4(x-2)}$$
$$= \frac{5}{-36} = -\frac{5}{36}$$

15. No equals sign appears so this is an *operation.*

$$\frac{b^2 + b - 6}{b^2 + 2b - 8} \cdot \frac{b^2 + 8b + 16}{3b + 12}$$
$$= \frac{(b+3)(b-2)}{(b+4)(b-2)} \cdot \frac{(b+4)(b+4)}{3(b+4)}$$
$$= \frac{b+3}{3}$$

17. No equals sign appears so this is an *operation.*

$$\frac{5}{x^2 - 2x} - \frac{3}{x^2 - 4}$$
$$= \frac{5}{x(x-2)} - \frac{3}{(x+2)(x-2)}$$
$$= \frac{5(x+2)}{x(x-2)(x+2)} - \frac{3x}{x(x-2)(x+2)}$$
$$= \frac{5x + 10 - 3x}{x(x-2)(x+2)}$$
$$= \frac{2x + 10}{x(x-2)(x+2)}$$

19. No equals sign appears so this is an *operation.*

$$\frac{\frac{5}{x} - \frac{3}{y}}{\frac{9x^2 - 25y^2}{x^2y}}$$

Multiply the numerator and denominator by the LCD of all the fractions, x^2y.

$$= \frac{x^2y\left(\frac{5}{x} - \frac{3}{y}\right)}{x^2y\left(\frac{9x^2 - 25y^2}{x^2y}\right)}$$
$$= \frac{5xy - 3x^2}{9x^2 - 25y^2}$$

continued

$$= \frac{-x(3x-5y)}{(3x+5y)(3x-5y)}$$
$$= \frac{-x}{(3x+5y)}$$

21. No equals sign appears so this is an *operation.*

$$\frac{4y^2-13y+3}{2y^2-9y+9} \div \frac{4y^2+11y-3}{6y^2-5y-6}$$
$$= \frac{(4y-1)(y-3)}{(2y-3)(y-3)} \cdot \frac{(2y-3)(3y+2)}{(4y-1)(y+3)}$$
$$= \frac{3y+2}{y+3}$$

23. $\frac{3r}{r-2} = 1 + \frac{6}{r-2}$

There is an equals sign, so this is an *equation.*
Multiply by the LCD, $r-2$. $(r \neq 2)$

$$3r = r - 2 + 6$$
$$2r = 4$$
$$r = 2$$

But $r \neq 2$.
Solution set: $\emptyset$

25. No equals sign appears so this is an *operation.*

$$\frac{-1}{3-x} - \frac{2}{x-3} = \frac{-1}{-(x-3)} - \frac{2}{x-3}$$
$$= \frac{1}{x-3} - \frac{2}{x-3}$$
$$= \frac{-1}{x-3}$$
$$\text{or } \frac{-1}{-(3-x)} = \frac{1}{3-x}$$

27. $\frac{2}{y+1} - \frac{3}{y^2-y-2} = \frac{3}{y-2}$

$$\frac{2}{y+1} - \frac{3}{(y-2)(y+1)} = \frac{3}{y-2}$$

There is an equals sign, so this is an *equation.*
Multiply by the LCD, $(y-2)(y+1)$.
$(y \neq -1, 2)$

$$2(y-2) - 3 = 3(y+1)$$
$$2y - 4 - 3 = 3y + 3$$
$$2y - 7 = 3y + 3$$
$$-10 = y$$

Check $y = -10$: $-\frac{2}{9} - \frac{1}{36} = -\frac{1}{4}$ *True*

Solution set: $\{-10\}$

29. $\frac{3}{y-3} - \frac{3}{y^2-5y+6} = \frac{2}{y-2}$

$$\frac{3}{y-3} - \frac{3}{(y-3)(y-2)} = \frac{2}{y-2}$$

There is an equals sign, so this is an *equation.*
Multiply by the LCD, $(y-3)(y-2)$. $(y \neq 2, 3)$

$$3(y-2) - 3 = 2(y-3)$$
$$3y - 6 - 3 = 2y - 6$$
$$3y - 9 = 2y - 6$$
$$y = 3$$

But $y \neq 3$.
Solution set: $\emptyset$

Section 7.5

1. **A.** $b = \frac{p}{r}$ is the same as $p = br$.

B. $r = \frac{b}{p}$ is the same as $b = pr$.

C. $b = \frac{r}{p}$ is the same as $r = bp$.

D. $p = \frac{r}{b}$ is the same as $r = bp$.

Choice **A** is correct.

3. **A.** $a = mF$ is the same as $m = \frac{a}{F}$.

B. $F = \frac{m}{a}$ is the same as $m = Fa$.

C. $F = \frac{a}{m}$ is the same as $Fm = a$, which is the same as $m = \frac{a}{F}$.

D. $F = ma$ is the same as $m = \frac{F}{a}$.

Choice **D** is correct.

5. $\frac{PV}{T} = \frac{pv}{t}$

Suppose that $T = 300$, $t = 350$, $V = 9$, $P = 50$, $v = 8$.

$$\frac{50 \cdot 9}{300} = \frac{p \cdot 8}{350}$$
$$p = \frac{50 \cdot 9 \cdot 350}{300 \cdot 8}$$
$$= 65.625$$

7. $c = \frac{100b}{L}$

$80 = \frac{100 \cdot 5}{L}$ *Let c=80, b=5.*

$80L = 500$ *Multiply by L.*

$L = \frac{500}{80} = \frac{25}{4}$ or 6.25

9. Solve $F = \frac{GMm}{d^2}$ for G.

$Fd^2 = GMm$ *Multiply by d^2.*

$\frac{Fd^2}{Mm} = G$ *Divide by Mm.*

11. Solve $\frac{1}{a} = \frac{1}{b} + \frac{1}{c}$ for a.

Multiply by the LCD, abc.

$$abc\left(\frac{1}{a}\right) = abc\left(\frac{1}{b} + \frac{1}{c}\right)$$
$$bc = ac + ab$$
$$bc = a(c + b) \quad \textit{Factor out a.}$$
$$\frac{bc}{c + b} = a \quad \textit{Divide by } c + b.$$

13. Solve $\frac{PV}{T} = \frac{pv}{t}$ for v.

$$\frac{PVt}{T} = pv \quad \textit{Multiply by t.}$$
$$\frac{PVt}{pT} = v \quad \textit{Divide by p.}$$

15. Solve $I = \frac{nE}{R + nr}$ for r.

$$I(R + nr) = nE$$
$$IR + Inr = nE$$
$$Inr = nE - IR$$
$$r = \frac{nE - IR}{In}$$

17. Solve $A = \frac{1}{2}h(B + b)$ for b.

$$\frac{2}{h}(A) = \frac{2}{h}\left[\frac{1}{2}h(B + b)\right] \quad \textit{Multiply by } \frac{2}{h}.$$
$$\frac{2A}{h} = B + b$$
$$\frac{2A}{h} - B = b \text{ or } b = \frac{2A - Bh}{h}$$

19. $\frac{E}{e} = \frac{R + r}{r}$ for r

$$Er = e(R + r) \quad \textit{Multiply by er.}$$
$$Er = eR + er$$
$$Er - er = eR \quad \textit{Subtract er.}$$
$$r(E - e) = eR$$
$$r = \frac{eR}{E - e} \quad \textit{Divide by } E - e.$$

21. To solve the equation $m = \frac{ab}{a - b}$ for a, the first step is to multiply both sides of the equation by the LCD, $a - b$.

23. Let $x =$ the number of girls in the class. Write and solve a proportion.

$$\frac{3}{4} = \frac{x}{20}$$

Multiply by the LCD, 20.

$$20\left(\frac{3}{4}\right) = 20\left(\frac{x}{20}\right)$$
$$15 = x$$

There are 15 girls and $20 - 15 = 5$ boys in the class.

25.
$$\text{Marin's rate} = \frac{1 \text{ job}}{\text{time to complete 1 job}} = \frac{1 \text{ job}}{2 \text{ hours}} = \frac{1}{2} \text{ job per hour}$$

27. 1996: $\frac{4.8}{13.3} \approx .361$ (least)

1997: $\frac{7.7}{16.0} \approx .481$

1998: $\frac{7.3}{13.8} \approx .529$

1999: $\frac{6.2}{11.6} \approx .534$ (greatest)

The ratio of truck accidents to car accidents was smallest in 1996.

29. 1996: $\frac{13.3}{4.8} \approx 2.77$

1997: $\frac{16.0}{7.7} \approx 2.08$

1998: $\frac{13.8}{7.3} \approx 1.89$

1999: $\frac{11.6}{6.2} \approx 1.87$

The ratio of car accidents to truck accidents was closest to 3 to 1 in 1996.

31. Let $x =$ the number of teachers.

Write a proportion.

$$\frac{1}{24} = \frac{x}{554}$$

Multiply each side by $554 \cdot 24$.

$$554 \cdot 24 \cdot \frac{1}{24} = 554 \cdot 24 \cdot \frac{x}{554}$$
$$554 = 24x$$
$$x = \frac{554}{24} \approx 23$$

There would be 23 teachers.

33. Let $x =$ the number of fish in the lake. Write and solve a proportion.

$$\frac{\text{total in lake}}{\text{tagged in lake}} = \frac{\text{total in sample}}{\text{tagged in sample}}$$
$$\frac{x}{500} = \frac{400}{8}$$
$$x = 500(50)$$
$$= 25,000$$

There are approximately $25,000$ fish in the lake.

35. $\dfrac{100 \text{ mg}}{2 \text{ mL}} = \dfrac{120 \text{ mg}}{x \text{ mL}}$

Multiply by the LCD, $2x$.

$$2x\left(\frac{100}{2}\right) = 2x\left(\frac{120}{x}\right)$$
$$100x = 240$$
$$x = 2.4$$

The correct dose is 2.4 mL.

37. Since $\frac{2}{3} = \frac{4}{6} = \frac{6}{9}$, use the proportion

$$\frac{2}{3} = \frac{2x+1}{2x+5}.$$
$$2(2x+5) = 3(2x+1)$$
$$4x + 10 = 6x + 3$$
$$7 = 2x$$
$$\frac{7}{2} = x$$

If $x = \frac{7}{2}$, then

$$AC = 2x + 1 = 2\left(\frac{7}{2}\right) + 1 = 8$$

and

$$DF = 2x + 5 = 2\left(\frac{7}{2}\right) + 5 = 12.$$

39. *Step 2*
Let x represent the speed of the current of the river. The boat goes 12 mph, so the downstream speed is $12 + x$ and the upstream speed is $12 - x$.

Use $t = \frac{d}{r}$ and make a table.

	d	r	t
Downstream	10	$12 + x$	$\frac{10}{12+x}$
Upstream	6	$12 - x$	$\frac{6}{12-x}$

Step 3
Because the time upstream equals the time downstream,

$$\frac{6}{12-x} = \frac{10}{12+x}$$

Step 4
Multiply by the LCD, $(12 - x)(12 + x)$.

$$(12-x)(12+x)\left(\frac{6}{12-x}\right) = (12-x)(12+x)\left(\frac{10}{12+x}\right)$$
$$6(12+x) = 10(12-x)$$
$$72 + 6x = 120 - 10x$$
$$16x = 48$$
$$x = 3$$

Step 5
The speed of the current of the river is 3 mph.

Step 6
Check: The rate downstream is $12 + 3 = 15$ mph, so she can go 10 miles in $\frac{10}{15} = \frac{2}{3}$ hour. The rate upstream is $12 - 3 = 9$ mph, so she can go 6 miles in $\frac{6}{9} = \frac{2}{3}$ hour. The times are the same, as required.

41. *Step 2*
Find the distance from Tulsa to Detroit.
Let x represent that distance.

Complete the table.

	d	r	t
Actual trip	x	50	$\frac{x}{50}$
Alternative trip	x	60	$\frac{x}{60}$

Step 3
At 60 mph, his time at 50 mph would be decreased 3 hr.

$$\frac{x}{60} = \frac{x}{50} - 3$$

Step 4
Multiply by 300.

$$5x = 6x - 900$$
$$900 = x$$

Step 5
The distance from Tulsa to Detroit is 900 miles.

Step 6
Check: 900 miles at 50 mph takes $\frac{900}{50}$ or 18 hours; 900 miles at 60 mph takes $\frac{900}{60}$ or 15 hours; $15 = 18 - 3$ as required.

43. *Step 2*
Let $x =$ the distance from San Francisco to the secret rendezvous.

Make a table.

	d	r	t
First trip	x	200	$\frac{x}{200}$
Return trip	x	300	$\frac{x}{300}$

Step 3

Time there	plus	time back	equals	4 hr.
$\frac{x}{200}$	$+$	$\frac{x}{300}$	$=$	4

Step 4
Multiply by the LCD, 600.

$$600\left(\frac{x}{200}+\frac{x}{300}\right)=600(4)$$
$$3x+2x=2400$$
$$5x=2400$$
$$x=480$$

Step 5
The distance is 480 miles.

Step 6

Check: 480 miles at 200 mph takes $\frac{480}{200}$ or 2.4 hours; 480 miles at 300 mph takes $\frac{480}{300}$ or 1.6 hours. The total time is 4 hours, as required.

45. *Step 2*
Let $x=$ the distance on the first part of the trip.

Make a table.

	d	r	t
First part	x	60	$\frac{x}{60}$
Second part	$x+10$	50	$\frac{x+10}{50}$

Step 3
From the problem, the equation is stated in words: (Note that 30 min $=\frac{1}{2}$ hr.) Time for the second part $=$ time for the first part $+\frac{1}{2}$. Use the times given in the table to write the equation.

$$\frac{x+10}{50}=\frac{x}{60}+\frac{1}{2}$$

Step 4
Multiply by the LCD, 300.

$$300\left(\frac{x+10}{50}\right)=300\left(\frac{x}{60}+\frac{1}{2}\right)$$
$$6(x+10)=5x+150$$
$$6x+60=5x+150$$
$$x-90$$

Step 5
The distance for both parts of the trip is given by

$$x+(x+10)=90+(90+10)=190.$$

The distance is 190 miles.

Step 6

Check: 90 miles at 60 mph takes $\frac{90}{60}$ or $1\frac{1}{2}$ hours; 100 miles at 50 mph takes 2 hours. The second part of the trip takes $\frac{1}{2}$ hour more than the first part, as required.

47. Let $x=$ the time it would take them working together.

Complete the table.

Worker	Rate	Time Working Together	Fractional Part of the Job Done
Butch	$\frac{1}{15}$	x	$\frac{1}{15}x$
Peggy	$\frac{1}{12}$	x	$\frac{1}{12}x$

Part done by Butch	$+$	part done by Peggy	$=$	1 whole job.
$\frac{1}{15}x$	$+$	$\frac{1}{12}x$	$=$	1

Multiply by the LCD, 60.

$$60\left(\frac{1}{15}x+\frac{1}{12}x\right)=60\cdot 1$$
$$4x+5x=60$$
$$9x=60$$
$$x=\frac{60}{9}=\frac{20}{3}\text{ or }6\frac{2}{3}$$

Together they could do the job in $\frac{20}{3}$ or $6\frac{2}{3}$ minutes.

49. Let $x=$ the time it would take Jason working alone.

Worker	Rate	Time Working Together	Fractional Part of the Job Done
Ron	$\frac{1}{6}$	4	$\frac{1}{6}(4)=\frac{2}{3}$
Jason	$\frac{1}{x}$	4	$\frac{1}{x}(4)=\frac{4}{x}$

Part done by Ron	$+$	part done by Jason	$=$	1 whole job.
$\frac{2}{3}$	$+$	$\frac{4}{x}$	$=$	1

Multiply by the LCD, $3x$.

$$2x+12=3x$$
$$12=x$$

It would take Jason 12 hours to do the job alone.

51. Let $x=$ the time it will take to fill the vat if both pipes are open.

	Rate	Time to Fill the Vat	Fractional Part of the Job Done
Inlet Pipe	$\frac{1}{10}$	x	$\frac{1}{10}x$
Outlet Pipe	$-\frac{1}{20}$	x	$-\frac{1}{20}x$

continued

Notice that the rate of the outlet pipe is negative because it will empty the vat, not fill it.

Part done with the inlet pipe open	+	Part done with the outlet pipe open	=	1 whole job.
$\frac{1}{10}x$	+	$\left(-\frac{1}{20}x\right)$	=	1

Multiply by the LCD, 20.

$$\begin{aligned} 2x - x &= 20 \\ x &= 20 \end{aligned}$$

It will take 20 hours to fill the vat.

53. Let $x =$ the time from Mimi's arrival home to the time the place is a shambles.

	Rate	Time to Mess up House	Fractional Part of the Job Done
Hortense and Mort	$-\frac{1}{7}$	x	$-\frac{1}{7}x$
Mimi	$\frac{1}{2}$	x	$\frac{1}{2}x$

Notice that Hortense and Mort's rate is negative since they are opposing the messing up by cleaning the house.

Part done by Hortense and Mort	+	Part done by Mimi	=	1 whole job of messing up.
$-\frac{1}{7}x$	+	$\frac{1}{2}x$	=	1

Multiply by the LCD, 14.

$$\begin{aligned} -2x + 7x &= 14 \\ 5x &= 14 \\ x &= \frac{14}{5} \text{ or } 2\frac{4}{5} \end{aligned}$$

It would take $\frac{14}{5}$ or $2\frac{4}{5}$ hours after Mimi got home for the house to be a shambles.

Chapter 7 Review Exercises

1. **(a)** $f(x) = \dfrac{-7}{3x+18}$

Set the denominator equal to zero and solve.

$$\begin{aligned} 3x + 18 &= 0 \\ 3x &= -18 \\ x &= -6 \end{aligned}$$

The number -6 makes the expression undefined, so it is excluded from the domain.

(b) The domain is $\{x|x \neq -6\}$.

2. **(a)** $f(x) = \dfrac{5x+17}{x^2-7x+10}$

Set the denominator equal to zero and solve.

$$\begin{aligned} x^2 - 7x + 10 &= 0 \\ (x-5)(x-2) &= 0 \end{aligned}$$

$$\begin{aligned} x - 5 = 0 \quad &\text{or} \quad x - 2 = 0 \\ x = 5 \quad &\text{or} \quad x = 2 \end{aligned}$$

The numbers 2 and 5 make the expression undefined, so they are excluded from the domain.

(b) The domain is $\{x|x \neq 2, 5\}$.

3. **(a)** $f(x) = \dfrac{9}{x^2-18x+81}$

Set the denominator equal to zero and solve.

$$\begin{aligned} x^2 - 18x + 81 &= 0 \\ (x-9)^2 &= 0 \\ x - 9 &= 0 \\ x &= 9 \end{aligned}$$

The number 9 makes the expression undefined, so it is excluded from the domain.

(b) The domain is $\{x|x \neq 9\}$.

4. $\dfrac{12x^2+6x}{24x+12} = \dfrac{6x(2x+1)}{12(2x+1)} = \dfrac{x}{2}$

5.
$$\begin{aligned} \frac{25m^2-n^2}{25m^2-10mn+n^2} &= \frac{(5m+n)(5m-n)}{(5m-n)(5m-n)} \\ &= \frac{5m+n}{5m-n} \end{aligned}$$

6.
$$\begin{aligned} \frac{r-2}{4-r^2} &= \frac{r-2}{(2+r)(2-r)} \\ &= \frac{(-1)(2-r)}{(2+r)(2-r)} \\ &= \frac{-1}{2+r} \end{aligned}$$

7. The reciprocal of a rational expression is another rational expression such that the two rational expressions have a product of 1.

8.
$$\begin{aligned} &\frac{(2y+3)^2}{5y} \cdot \frac{15y^3}{4y^2-9} \\ &= \frac{15y^3(2y+3)^2}{5y(2y+3)(2y-3)} \\ &= \frac{3y^2(2y+3)}{2y-3} \end{aligned}$$

9.
$$\begin{aligned} &\frac{w^2-16}{w} \cdot \frac{3}{4-w} \\ &= \frac{(w-4)(w+4)}{w} \cdot \frac{3}{4-w} \\ &= \frac{(-1)(4-w)(w+4)}{w} \cdot \frac{3}{4-w} \\ &= \frac{-3(w+4)}{w} \end{aligned}$$

10. $\dfrac{z^2-z-6}{z-6}\cdot\dfrac{z^2-6z}{z^2+2z-15}$

$=\dfrac{(z-3)(z+2)}{z-6}\cdot\dfrac{z(z-6)}{(z-3)(z+5)}$

$=\dfrac{z(z+2)}{z+5}$

11. $\dfrac{m^3-n^3}{m^2-n^2}\div\dfrac{m^2+mn+n^2}{m+n}$

Multiply by the reciprocal.

$=\dfrac{m^3-n^3}{m^2-n^2}\cdot\dfrac{m+n}{m^2+mn+n^2}$

$=\dfrac{(m-n)(m^2+mn+n^2)}{(m-n)(m+n)}\cdot\dfrac{m+n}{m^2+mn+n^2}$

$=1$

12. $32b^3, 24b^5$

Factor each denominator.

$32b^3=2\cdot2\cdot2\cdot2\cdot2\cdot b^3=2^5\cdot b^3$

$24b^5=2\cdot2\cdot2\cdot3\cdot b^5=2^3\cdot3\cdot b^5$

$\text{LCD}=2^5\cdot3\cdot b^5-96b^5$

13. $9r^2, 3r+1, 9$

Factor each denominator.

$$9r^2=3^2\cdot r^2$$

The second denominator is already in factored form. The third denominator is $9=3^2$. The LCD is

$$3^2\cdot r^2\cdot(3r+1)\ \text{ or }\ 9r^2(3r+1).$$

14. $6x^2+13x-5, 9x^2+9x-4$

Factor each denominator.

$$6x^2+13x-5=(3x-1)(2x+5)$$
$$9x^2+9x-4=(3x-1)(3x+4)$$

The LCD is

$$(3x-1)(2x+5)(3x+4).$$

15. $\dfrac{8}{z}-\dfrac{3}{2z^2}$

The LCD is $2z^2$.

$=\dfrac{8\cdot2z}{z\cdot2z}-\dfrac{3}{2z^2}$

$=\dfrac{16z}{2z^2}-\dfrac{3}{2z^2}$

$=\dfrac{16z-3}{2z^2}$

16. $\dfrac{5y+13}{y+1}-\dfrac{1-7y}{y+1}$

$=\dfrac{5y+13-(1-7y)}{y+1}$

$=\dfrac{5y+13-1+7y}{y+1}$

$=\dfrac{12y+12}{y+1}$

$=\dfrac{12(y+1)}{y+1}=12$

17. $\dfrac{6}{5a+10}+\dfrac{7}{6a+12}$

$=\dfrac{6}{5(a+2)}+\dfrac{7}{6(a+2)}$

The LCD is $30(a+2)$.

$=\dfrac{6\cdot6}{5(a+2)\cdot6}+\dfrac{7\cdot5}{6(a+2)\cdot5}$

$=\dfrac{36}{30(a+2)}+\dfrac{35}{30(a+2)}$

$=\dfrac{36+35}{30(a+2)}=\dfrac{71}{30(a+2)}$

18. $\dfrac{3r}{10r^2-3rs-s^2}+\dfrac{2r}{2r^2+rs-s^2}$

$=\dfrac{3r}{(5r+s)(2r-s)}+\dfrac{2r}{(2r-s)(r+s)}$

The LCD is $(5r+s)(2r-s)(r+s)$.

$=\dfrac{3r(r+s)}{(5r+s)(2r-s)(r+s)}$

$+\dfrac{2r(5r+s)}{(2r-s)(r+s)(5r+s)}$

$=\dfrac{3r^2+3rs+10r^2+2rs}{(5r+s)(2r-s)(r+s)}$

$=\dfrac{13r^2+5rs}{(5r+s)(2r-s)(r+s)}$

19. $\dfrac{\frac{3}{t}+2}{\frac{4}{t}-7}$

Multiply the numerator and denominator by the LCD of all the fractions, t.

$$\frac{t\left(\frac{3}{t}+2\right)}{t\left(\frac{4}{t}-7\right)}=\frac{3+2t}{4-7t}$$

20. $\dfrac{\frac{2}{m-3n}}{\frac{1}{3n-m}}=\dfrac{2}{m-3n}\div\dfrac{1}{3n-m}$

$=\dfrac{2}{m-3n}\cdot\dfrac{3n-m}{1}$

$=\dfrac{2}{m-3n}\cdot\dfrac{-1(m-3n)}{1}$

$=\dfrac{2\cdot(-1)}{1}=-2$

21. $\dfrac{\dfrac{3}{p}-\dfrac{2}{q}}{\dfrac{9q^2-4p^2}{qp}}$

Multiply the numerator and denominator by the LCD of all the fractions, qp.

$$=\frac{qp\left(\frac{3}{p}-\frac{2}{q}\right)}{qp\left(\frac{9q^2-4p^2}{qp}\right)}$$
$$=\frac{3q-2p}{9q^2-4p^2}$$
$$=\frac{3q-2p}{(3q+2p)(3q-2p)}$$
$$=\frac{1}{3q+2p}$$

22. $\dfrac{x^{-2}-y^{-2}}{x^{-1}-y^{-1}}=\dfrac{\dfrac{1}{x^2}-\dfrac{1}{y^2}}{\dfrac{1}{x}-\dfrac{1}{y}}$

Multiply the numerator and denominator by the LCD of all the fractions, x^2y^2.

$$=\frac{x^2y^2\left(\frac{1}{x^2}-\frac{1}{y^2}\right)}{x^2y^2\left(\frac{1}{x}-\frac{1}{y}\right)}$$
$$=\frac{y^2-x^2}{xy^2-x^2y}$$
$$=\frac{(y+x)(y-x)}{xy(y-x)}$$
$$=\frac{y+x}{xy}$$

23. $\dfrac{1}{t+4}+\dfrac{1}{2}=\dfrac{3}{2t+8}$

$$\frac{1}{t+4}+\frac{1}{2}=\frac{3}{2(t+4)}$$

Multiply by the LCD, $2(t+4)$. $(t\neq -4)$

$$2(t+4)\left(\frac{1}{t+4}+\frac{1}{2}\right)=2(t+4)\left(\frac{3}{2(t+4)}\right)$$
$$2+(t+4)=3$$
$$t+6=3$$
$$t=-3$$

Check $t=-3$: $1+\frac{1}{2}=\frac{3}{2}$ *True*

Solution set: $\{-3\}$

24. $\dfrac{-5m}{m+1}+\dfrac{m}{3m+3}=\dfrac{56}{6m+6}$

$$\frac{-5m}{m+1}+\frac{m}{3(m+1)}=\frac{56}{6(m+1)}$$
$$\frac{-5m}{m+1}+\frac{m}{3(m+1)}=\frac{28}{3(m+1)}$$

Multiply by the LCD, $3(m+1)$. $(m\neq -1)$

$$3(m+1)\left(\frac{-5m}{m+1}+\frac{m}{3(m+1)}\right)$$
$$=3(m+1)\left(\frac{28}{3(m+1)}\right)$$
$$-15m+m=28$$
$$-14m=28$$
$$m=-2$$

Check $m=-2$: $-10+\frac{2}{3}=-\frac{56}{6}$ *True*

Solution set: $\{-2\}$

25. $\dfrac{2}{k-1}-\dfrac{4k+1}{k^2-1}=\dfrac{-1}{k+1}$

$$\frac{2}{k-1}-\frac{4k+1}{(k+1)(k-1)}=\frac{-1}{k+1}$$

Multiply by the LCD, $(k+1)(k-1)$.

$$(k+1)(k-1)\left(\frac{2}{k-1}-\frac{4k+1}{(k+1)(k-1)}\right)$$
$$=(k+1)(k-1)\left(\frac{-1}{k+1}\right)$$
$$2(k+1)-(4k+1)=-1(k-1)$$
$$2k+2-4k-1=-k+1$$
$$-2k+1=-k+1$$
$$0=k$$

Check $k=0$: $-2+1=-1$ *True*

Solution set: $\{0\}$

26. $\dfrac{5}{x+2}+\dfrac{3}{x+3}=\dfrac{x}{x^2+5x+6}$

$$\frac{5}{x+2}+\frac{3}{x+3}=\frac{x}{(x+2)(x+3)}$$

Multiply by the LCD, $(x+2)(x+3)$. $(x\neq -3,-2)$

$$(x+2)(x+3)\left(\frac{5}{x+2}+\frac{3}{x+3}\right)$$
$$=(x+2)(x+3)\left(\frac{x}{(x+2)(x+3)}\right)$$
$$5(x+3)+3(x+2)=x$$
$$5x+15+3x+6=x$$
$$8x+21=x$$
$$7x=-21$$
$$x=-3$$

Substituting -3 in the original equation results in division by 0, so -3 is not a solution.

Solution set: $\emptyset$

27. Although her algebra was correct, 3 is not a solution because it is not in the domain of the equation. Thus, $\emptyset$ is correct.

28. In simplifying the expression, we are combining terms to get a single fraction with a denominator of $6x$. In solving the equation, we are finding a value for x that makes the equation true.

29. The graph in choice **C** has a vertical asymptote. Its equation is $x=0$.

30. $\frac{1}{A}=\frac{1}{B}+\frac{1}{C}$

Let $B=30$ and $C=10$.

$\frac{1}{A}=\frac{1}{30}+\frac{1}{10}$

To solve for A, multiply both sides by the LCD, $30A$.

$$30A\left(\frac{1}{A}\right)=30A\left(\frac{1}{30}+\frac{1}{10}\right)$$
$$30=A+3A$$
$$30=4A$$
$$A=\frac{30}{4}=\frac{15}{2}$$

31. Solve $F=\frac{GMm}{d^2}$ for m.

$Fd^2=GMm$ *Multiply by d^2.*

$\frac{Fd^2}{GM}=m$ *Divide by GM.*

32. Solve $\mu=\frac{Mv}{M+m}$ for M.

$\mu(M+m)=Mv$ *Multiply by $M+m$.*

$$\mu M+\mu m=Mv$$
$$\mu m=Mv-\mu M$$
$$m\mu=M(v-\mu)$$
$$M=\frac{m\mu}{v-\mu}$$

33. Let $x=$ the number of passenger-kilometers per day provided by high-speed trains.

Write a proportion.

$$\frac{x}{15,000}=\frac{23,200}{58,000}$$
$$x=\frac{23,200(15,000)}{58,000}$$
$$=6000$$

34. Let $x=$ the speed of the boat in still water.

Use $d=rt$, or $t=\frac{d}{r}$, to make a table.

	Distance	Rate	Time
Downstream	40	$x+4$	$\frac{40}{x+4}$
Upstream	24	$x-4$	$\frac{24}{x-4}$

Because the times are equal,

$$\frac{40}{x+4}=\frac{24}{x-4}.$$

Multiply by the LCD, $(x+4)(x-4)$. $(x\neq -4, 4)$

$$(x+4)(x-4)\left(\frac{40}{x+4}\right)=(x+4)(x-4)\left(\frac{24}{x-4}\right)$$
$$40(x-4)=24(x+4)$$
$$40x-160=24x+96$$
$$16x=256$$
$$x=16$$

The speed of the boat in still water is 16 km/hr.

35. Let $x=$ the time it takes to fill the sink with both taps open.

Make a table.

	Rate	Time Working Together	Fractional Part of the Job Done
Cold	$\frac{1}{8}$	x	$\frac{x}{8}$
Hot	$\frac{1}{12}$	x	$\frac{x}{12}$

Part done by cold	plus	part done by hot	equals	1 whole job.
$\frac{x}{8}$	$+$	$\frac{x}{12}$	$=$	1

Multiply by the LCD, 24.

$$24\left(\frac{x}{8}+\frac{x}{12}\right)=24\cdot 1$$
$$3x+2x=24$$
$$5x=24$$
$$x=\frac{24}{5}\text{ or }4\frac{4}{5}$$

The sink will be filled in $\frac{24}{5}$ or $4\frac{4}{5}$ minutes.

36. Let $x=$ the time to do the job working together.

Make a table.

Worker	Rate	Time Working Together	Fractional Part of the Job Done
Amber	$\frac{1}{9}$	x	$\frac{x}{9}$
Steven	$\frac{1}{6}$	x	$\frac{x}{6}$

Part done by Amber	plus	part done by Steven	equals	1 whole job.
$\frac{x}{9}$	$+$	$\frac{x}{6}$	$=$	1

Multiply by the LCD, 36.

$$36\left(\frac{x}{9}+\frac{x}{6}\right)=36\cdot 1$$
$$4x+6x=36$$
$$10x=36$$
$$x=\frac{36}{10}$$
$$x=\frac{18}{5}\text{ or }3\frac{3}{5}$$

Working together, they can do the job in $\frac{18}{5}$ or $3\frac{3}{5}$ hours.

37. $\dfrac{x+2y}{x^2-4y^2} = \dfrac{x+2y}{(x+2y)(x-2y)}$

$= \dfrac{1}{x-2y}$

38. $\dfrac{x^2+2x-15}{x^2-x-6} = \dfrac{(x+5)(x-3)}{(x-3)(x+2)}$

$= \dfrac{x+5}{x+2}$

39. $\dfrac{2}{m} + \dfrac{5}{3m^2}$

The LCD is $3m^2$.

$= \dfrac{2 \cdot 3m}{m \cdot 3m} + \dfrac{5}{3m^2}$

$= \dfrac{6m}{3m^2} + \dfrac{5}{3m^2} = \dfrac{6m+5}{3m^2}$

40. $\dfrac{k^2-6k+9}{1-216k^3} \cdot \dfrac{6k^2+17k-3}{9-k^2}$

Factor $1-216k^3$ as the difference of cubes, $1^3-(6k)^3$.

$= \dfrac{(k-3)(k-3)}{(1-6k)(1+6k+36k^2)} \cdot \dfrac{(6k-1)(k+3)}{(3-k)(3+k)}$

$= \dfrac{(k-3)(k-3)}{(-1)(6k-1)(1+6k+36k^2)} \cdot \dfrac{(6k-1)(k+3)}{(-1)(k-3)(k+3)}$

$= \dfrac{k-3}{1+6k+36k^2}$ or $\dfrac{k-3}{36k^2+6k+1}$

41. $\dfrac{\dfrac{-3}{x}+\dfrac{x}{2}}{1+\dfrac{x+1}{x}}$

Multiply the numerator and denominator by the LCD of all the fractions, $2x$.

$= \dfrac{2x\left(\dfrac{-3}{x}+\dfrac{x}{2}\right)}{2x\left(1+\dfrac{x+1}{x}\right)}$

$= \dfrac{-6+x^2}{2x+2(x+1)}$

$= \dfrac{x^2-6}{2x+2x+2}$

$= \dfrac{x^2-6}{4x+2} = \dfrac{x^2-6}{2(2x+1)}$

42. $\dfrac{9x^2+46x+5}{3x^2-2x-1} \div \dfrac{x^2+11x+30}{x^3+5x^2-6x}$

Multiply by the reciprocal.

$= \dfrac{9x^2+46x+5}{3x^2-2x-1} \cdot \dfrac{x(x^2+5x-6)}{x^2+11x+30}$

$= \dfrac{(9x+1)(x+5)}{(3x+1)(x-1)} \cdot \dfrac{x(x+6)(x-1)}{(x+6)(x+5)}$

$= \dfrac{x(9x+1)}{3x+1}$

43. $\dfrac{\dfrac{3}{x}-5}{6+\dfrac{1}{x}}$

Multiply the numerator and denominator by the LCD of all the fractions, x.

$= \dfrac{x\left(\dfrac{3}{x}-5\right)}{x\left(6+\dfrac{1}{x}\right)}$

$= \dfrac{3-5x}{6x+1}$

44. $\dfrac{9}{3-x} - \dfrac{2}{x-3}$

$= \dfrac{9}{3-x} - \dfrac{2(-1)}{(x-3)(-1)}$

$= \dfrac{9}{3-x} - \dfrac{-2}{3-x}$

$= \dfrac{9-(-2)}{3-x}$

$= \dfrac{11}{3-x}$ or $\dfrac{-11}{x-3}$

45. $\dfrac{4y+16}{30} \div \dfrac{2y+8}{5}$

Multiply by the reciprocal.

$= \dfrac{4y+16}{30} \cdot \dfrac{5}{2y+8}$

$= \dfrac{4(y+4)}{30} \cdot \dfrac{5}{2(y+4)}$

$= \dfrac{4 \cdot 5}{2 \cdot 30} = \dfrac{2}{6} = \dfrac{1}{3}$

46. $\dfrac{t^{-2}+s^{-2}}{t^{-1}-s^{-1}} = \dfrac{\dfrac{1}{t^2}+\dfrac{1}{s^2}}{\dfrac{1}{t}-\dfrac{1}{s}}$

Multiply the numerator and denominator by the LCD of all the fractions, t^2s^2.

$= \dfrac{t^2s^2\left(\dfrac{1}{t^2}+\dfrac{1}{s^2}\right)}{t^2s^2\left(\dfrac{1}{t}-\dfrac{1}{s}\right)}$

$= \dfrac{s^2+t^2}{ts^2-t^2s}$

$= \dfrac{s^2+t^2}{st(s-t)}$

47. $\dfrac{4a}{a^2-ab-2b^2}-\dfrac{6b-a}{a^2+4ab+3b^2}$

$=\dfrac{4a}{(a-2b)(a+b)}-\dfrac{6b-a}{(a+3b)(a+b)}$

The LCD is $(a+3b)(a-2b)(a+b)$.

$=\dfrac{4a(a+3b)}{(a-2b)(a+b)(a+3b)}$

$-\dfrac{(6b-a)(a-2b)}{(a+3b)(a+b)(a-2b)}$

$=\dfrac{4a(a+3b)-(6b-a)(a-2b)}{(a+3b)(a+b)(a-2b)}$

$=\dfrac{4a^2+12ab-(6ab-12b^2-a^2+2ab)}{(a+3b)(a+b)(a-2b)}$

$=\dfrac{4a^2+12ab-6ab+12b^2+a^2-2ab}{(a+3b)(a+b)(a-2b)}$

$=\dfrac{5a^2+4ab+12b^2}{(a+3b)(a+b)(a-2b)}$

48. $\dfrac{a}{b}+\dfrac{b}{c}+\dfrac{c}{d}$

The LCD is bcd.

$=\dfrac{a\cdot cd}{b\cdot cd}+\dfrac{b\cdot bd}{c\cdot bd}+\dfrac{c\cdot bc}{d\cdot bc}$

$=\dfrac{acd+b^2d+bc^2}{bcd}$

49. $\dfrac{x+3}{x^2-5x+4}-\dfrac{1}{x}=\dfrac{2}{x^2-4x}$

$\dfrac{x+3}{(x-4)(x-1)}-\dfrac{1}{x}=\dfrac{2}{x(x-4)}$

Multiply by the LCD, $x(x-4)(x-1)$. $(x\neq 0,1,4)$

$x(x-4)(x-1)\left(\dfrac{x+3}{(x-4)(x-1)}-\dfrac{1}{x}\right)$

$=x(x-4)(x-1)\cdot\left(\dfrac{2}{x(x-4)}\right)$

$x(x+3)-(x-4)(x-1)=2(x-1)$

$x^2+3x-\left(x^2-5x+4\right)=2x-2$

$x^2+3x-x^2+5x-4=2x-2$

$8x-4=2x-2$

$6x=2$

$x=\dfrac{1}{3}$

Check $x=\dfrac{1}{3}$: $\dfrac{15}{11}-3=-\dfrac{18}{11}$ *True*

Solution set: $\left\{\dfrac{1}{3}\right\}$

50. Solve $A=\dfrac{Rr}{R+r}$ for r.

$A(R+r)=Rr$

$AR+Ar=Rr$

$AR=Rr-Ar$

$AR=(R-A)r$

$\dfrac{AR}{R-A}=r$ or $r=\dfrac{-AR}{A-R}$

51. $1-\dfrac{5}{r}=\dfrac{-4}{r^2}$

Multiply by the LCD, r^2. $(r\neq 0)$

$r^2\left(1-\dfrac{5}{r}\right)=r^2\left(\dfrac{-4}{r^2}\right)$

$r^2-5r=-4$

$r^2-5r+4=0$

$(r-4)(r-1)=0$

$r-4=0$ or $r-1=0$

$r=4$ or $r=1$

Check $r=1$: $1-5=-4$ *True*

Check $r=4$: $1-\dfrac{5}{4}=-\dfrac{1}{4}$ *True*

Solution set: $\{1,4\}$

52. $\dfrac{3x}{x-4}+\dfrac{2}{x}=\dfrac{48}{x^2-4x}$

$\dfrac{3x}{x-4}+\dfrac{2}{x}=\dfrac{48}{x(x-4)}$

Multiply by the LCD, $x(x-4)$. $(x\neq 0,4)$

$x(x-4)\left(\dfrac{3x}{x-4}+\dfrac{2}{x}\right)=x(x-4)\left(\dfrac{48}{x(x-4)}\right)$

$3x^2+2(x-4)=48$

$3x^2+2x-8=48$

$3x^2+2x-56=0$

$(3x+14)(x-4)=0$

$3x+14=0$ or $x-4=0$

$3x=-14$

$x=-\dfrac{14}{3}$ or $x=4$

The number 4 is not allowed as a solution because substituting it in the original equation results in division by 0.

Check $x=-\dfrac{14}{3}$: $\dfrac{21}{13}-\dfrac{3}{7}=\dfrac{108}{91}$ *True*

Solution set: $\left\{-\dfrac{14}{3}\right\}$

53. $a=\dfrac{337}{d}$

(a) $a=\dfrac{337}{40.5}$ *Let d=40.5*

≈ 8.32

(b) $7.51=\dfrac{337}{d}$ *Let a=7.51*

$7.51d=337$

$d=\dfrac{337}{7.51}\approx 44.9$

54. Let $x =$ the time to fill the tub working together. Make a table.

	Rate	Time Working Together	Fractional Part of the Job Done
Hot	$\frac{1}{20}$	x	$\frac{x}{20}$
Cold	$\frac{1}{15}$	x	$\frac{x}{15}$

Part done by hot	plus	part done by cold	equals	1 whole job.
$\frac{x}{20}$	$+$	$\frac{x}{15}$	$=$	1

Multiply by the LCD, 60.

$$60\left(\frac{x}{20}+\frac{x}{15}\right) = 60 \cdot 1$$
$$3x + 4x = 60$$
$$7x = 60$$
$$x = \frac{60}{7} \text{ or } 8\frac{4}{7}$$

Working together, the tub can be filled in $\frac{60}{7}$ or $8\frac{4}{7}$ minutes.

55. Let $x =$ the cost of 13 gallons.
Write a proportion.

$$\frac{x}{13} = \frac{4.86}{3}$$
$$x = \frac{13(4.86)}{3}$$
$$= 21.06$$

The cost of 13 gallons of unleaded gasoline is \$21.06.

56. Let $x =$ the number.

"Three-fourths of a number is subtracted from seven-sixths of the number, giving 10" translates to

$$\frac{7}{6}x - \frac{3}{4}x = 10.$$

Multiply each term by the LCD, 12.

$$12\left(\frac{7}{6}x\right) - 12\left(\frac{3}{4}x\right) = 12(10)$$
$$14x - 9x = 120$$
$$5x = 120$$
$$x = 24$$

The number is 24.

Chapter 7 Test

1. $f(x) = \dfrac{x+3}{3x^2+2x-8}$

Set the denominator equal to zero and solve.

$$3x^2 + 2x - 8 = 0$$
$$(3x-4)(x+2) = 0$$
$$3x - 4 = 0 \quad \text{or} \quad x + 2 = 0$$
$$3x = 4$$
$$x = \frac{4}{3} \quad \text{or} \quad x = -2$$

The numbers -2 and $\frac{4}{3}$ make the rational expression undefined and are excluded from the domain of f, which can be written in set notation as $\{x|x \neq -2, \frac{4}{3}\}$.

2.
$$\frac{6x^2-13x-5}{9x^3-x} = \frac{(3x+1)(2x-5)}{x(9x^2-1)}$$
$$= \frac{(3x+1)(2x-5)}{x(3x+1)(3x-1)}$$
$$= \frac{2x-5}{x(3x-1)}$$

3.
$$\frac{(x+3)^2}{4} \cdot \frac{6}{2x+6} = \frac{(x+3)^2}{4} \cdot \frac{2 \cdot 3}{2(x+3)}$$
$$= \frac{3(x+3)}{4}$$

4.
$$\frac{y^2-16}{y^2-25} \cdot \frac{y^2+2y-15}{y^2-7y+12}$$
$$= \frac{(y+4)(y-4)}{(y+5)(y-5)} \cdot \frac{(y+5)(y-3)}{(y-4)(y-3)}$$
$$= \frac{y+4}{y-5}$$

5.
$$\frac{x^2-9}{x^3+3x^2} \div \frac{x^2+x-12}{x^3+9x^2+20x}$$

Multiply by the reciprocal.

$$= \frac{x^2-9}{x^3+3x^2} \cdot \frac{x(x^2+9x+20)}{x^2+x-12}$$
$$= \frac{(x+3)(x-3)}{x^2(x+3)} \cdot \frac{x(x+5)(x+4)}{(x+4)(x-3)}$$
$$= \frac{x+5}{x}$$

6. $t^2 + t - 6, t^2 + 3t, t^2$

Factor each denominator.

$$t^2 + t - 6 = (t+3)(t-2)$$
$$t^2 + 3t = t(t+3)$$

The third denominator is already in factored form. The LCD is

$$t^2(t+3)(t-2).$$

7. $\dfrac{7}{6t^2} - \dfrac{1}{3t}$

The LCD is $6t^2$.

$$= \frac{7}{6t^2} - \frac{1 \cdot 2t}{3t \cdot 2t}$$
$$= \frac{7}{6t^2} - \frac{2t}{6t^2} = \frac{7 - 2t}{6t^2}$$

8. $\dfrac{9}{(x-3)^2} + \dfrac{2}{(x-3)(x+3)}$

The LCD is $(x-3)^2(x+3)$.

$$= \frac{9(x+3)}{(x-3)^2(x+3)} + \frac{2(x-3)}{(x-3)^2(x+3)}$$
$$= \frac{9x + 27 + 2x - 6}{(x-3)^2(x+3)}$$
$$= \frac{11x + 21}{(x-3)^2(x+3)}$$

9. $\dfrac{6}{x+4} + \dfrac{1}{x+2} - \dfrac{3x}{x^2 + 6x + 8}$

$$= \frac{6}{x+4} + \frac{1}{x+2} - \frac{3x}{(x+4)(x+2)}$$

The LCD is $(x+4)(x+2)$.

$$= \frac{6(x+2)}{(x+4)(x+2)} + \frac{1(x+4)}{(x+2)(x+4)} - \frac{3x}{(x+4)(x+2)}$$
$$= \frac{6(x+2) + x + 4 - 3x}{(x+4)(x+2)}$$
$$= \frac{6x + 12 + x + 4 - 3x}{(x+4)(x+2)}$$
$$= \frac{4x + 16}{(x+4)(x+2)}$$
$$= \frac{4(x+4)}{(x+4)(x+2)} = \frac{4}{x+2}$$

10. $\dfrac{\dfrac{12}{r+4}}{\dfrac{11}{6r+24}} = \dfrac{12}{r+4} \div \dfrac{11}{6r+24}$

Multiply by the reciprocal.

$$= \frac{12}{r+4} \cdot \frac{6r+24}{11}$$
$$= \frac{12}{r+4} \cdot \frac{6(r+4)}{11} = \frac{72}{11}$$

11. $\dfrac{\dfrac{1}{a} - \dfrac{1}{b}}{\dfrac{a}{b} - \dfrac{b}{a}}$

Multiply the numerator and the denominator by the LCD of all the fractions, ab.

$$= \frac{ab\left(\dfrac{1}{a} - \dfrac{1}{b}\right)}{ab\left(\dfrac{a}{b} - \dfrac{b}{a}\right)} = \frac{b - a}{a^2 - b^2}$$
$$= \frac{b - a}{(a-b)(a+b)} = \frac{(-1)(a-b)}{(a-b)(a+b)}$$
$$= \frac{-1}{a+b} \text{ or } -\frac{1}{a+b}$$

12. $\dfrac{2x^{-2} + y^{-2}}{x^{-1} - y^{-1}} = \dfrac{\dfrac{2}{x^2} + \dfrac{1}{y^2}}{\dfrac{1}{x} - \dfrac{1}{y}}$

Multiply the numerator and denominator by the LCD of all the fractions, x^2y^2.

$$= \frac{x^2y^2\left(\dfrac{2}{x^2} + \dfrac{1}{y^2}\right)}{x^2y^2\left(\dfrac{1}{x} - \dfrac{1}{y}\right)}$$
$$= \frac{2y^2 + x^2}{xy^2 - x^2y}$$
$$= \frac{2y^2 + x^2}{xy(y - x)}$$

13. (a) Simplify this expression.

$$\frac{2x}{3} + \frac{x}{4} - \frac{11}{2}$$

The LCD is 12.

$$= \frac{2x \cdot 4}{3 \cdot 4} + \frac{x \cdot 3}{4 \cdot 3} - \frac{11 \cdot 6}{2 \cdot 6}$$
$$= \frac{8x}{12} + \frac{3x}{12} - \frac{66}{12}$$
$$= \frac{8x + 3x - 66}{12}$$
$$= \frac{11x - 66}{12} = \frac{11(x-6)}{12}$$

(b) Solve this equation.

$$\frac{2x}{3} + \frac{x}{4} = \frac{11}{2}$$

Multiply by the LCD, 12.

$$12\left(\frac{2x}{3} + \frac{x}{4}\right) - 12\left(\frac{11}{2}\right)$$
$$8x + 3x = 66$$
$$11x = 66$$
$$x = 6$$

Check $x = 6$: $4 + \dfrac{3}{2} = \dfrac{11}{2}$ *True*

Solution set: $\{6\}$

14. $$\frac{1}{x} - \frac{4}{3x} = \frac{1}{x-2}$$

Multiply by the LCD, $3x(x-2)$. $(x \neq 0, 2)$

$$3x(x-2)\left(\frac{1}{x} - \frac{4}{3x}\right) = 3x(x-2)\left(\frac{1}{x-2}\right)$$
$$3(x-2) - 4(x-2) = 3x$$
$$3x - 6 - 4x + 8 = 3x$$
$$-x + 2 = 3x$$
$$-4x = -2$$
$$x = \frac{1}{2}$$

Check $x = \frac{1}{2}$: $2 - \frac{8}{3} = -\frac{2}{3}$ *True*

Solution set: $\left\{\frac{1}{2}\right\}$

15. $$\frac{y}{y+2} - \frac{1}{y-2} = \frac{8}{y^2-4}$$
$$\frac{y}{y+2} - \frac{1}{y-2} = \frac{8}{(y+2)(y-2)}$$

Multiply by the LCD, $(y+2)(y-2)$. $(y \neq -2, 2)$

$$(y+2)(y-2)\left(\frac{y}{y+2} - \frac{1}{y-2}\right)$$
$$= (y+2)(y-2)\left(\frac{8}{(y+2)(y-2)}\right)$$
$$y(y-2) - 1(y+2) = 8$$
$$y^2 - 2y - y - 2 = 8$$
$$y^2 - 3y - 10 = 0$$
$$(y-5)(y+2) = 0$$

$$y - 5 = 0 \quad \text{or} \quad y + 2 = 0$$
$$y = 5 \quad \text{or} \quad y = -2$$

The number -2 is not allowed as a solution because substituting it in the original equation results in division by 0.

Check $y = 5$: $\frac{5}{7} - \frac{1}{3} = \frac{8}{21}$ *True*

Solution set: $\{5\}$

16. A solution cannot make a denominator 0. This is illustrated in Exercise 15 where -2 is not allowed as a solution for this reason.

17. Solve $S = \frac{n}{2}(a + \ell)$ for ℓ.

To solve for ℓ, multiply both sides by $\frac{2}{n}$.

$$\frac{2}{n} \cdot S = \frac{2}{n} \cdot \frac{n}{2}(a + \ell)$$
$$\frac{2S}{n} = a + \ell$$
$$\frac{2S}{n} - a = \ell \quad \text{or} \quad \ell = \frac{2S - na}{n}$$

18. $f(x) = \frac{-2}{x+1}$ is not defined when $x = -1$, so an equation of the vertical asymptote is $x = -1$.

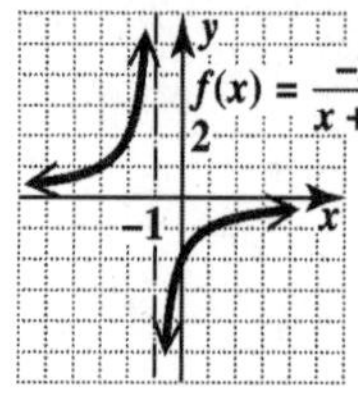

19. Let $x =$ the time to do the job working together. Make a table.

Worker	Rate	Time Working Together	Fractional Part of the Job Done
Wayne	$\frac{1}{9}$	x	$\frac{x}{9}$
Susan	$\frac{1}{5}$	x	$\frac{x}{5}$

Part done by Wayne	plus	part done by Susan	equals	1 whole job.
$\frac{x}{9}$	$+$	$\frac{x}{5}$	$=$	1

Multiply by the LCD, 45.

$$45\left(\frac{x}{9} + \frac{x}{5}\right) = 45 \cdot 1$$
$$5x + 9x = 45$$
$$14x = 45$$
$$x = \frac{45}{14} \quad \text{or} \quad 3\frac{3}{14}$$

Working together, they can do the job in $\frac{45}{14}$ or $3\frac{3}{14}$ hours.

20. Let $x =$ the speed of the boat in still water. Use $d = rt$, or $t = \frac{d}{r}$, to make a table.

	d	r	t
Downstream	36	$x+3$	$\frac{36}{x+3}$
Upstream	24	$x-3$	$\frac{24}{x-3}$

Because the times are equal,

$$\frac{36}{x+3} = \frac{24}{x-3}.$$

Multiply by the LCD, $(x+3)(x-3)$. $(x \neq -3, 3)$

$$(x+3)(x-3)\left(\frac{36}{x+3}\right) = (x+3)(x-3)\left(\frac{24}{x-3}\right)$$
$$36(x-3) = 24(x+3)$$
$$36x - 108 = 24x + 72$$
$$12x = 180$$
$$x = 15$$

The speed of the boat in still water is 15 mph.

21. Let $x =$ the number of fish in Lake Linda. Write a proportion.

$$\frac{x}{600} = \frac{800}{10}$$

Multiply by the LCD, 600.

$$600\left(\frac{x}{600}\right) = 600\left(\frac{800}{10}\right)$$
$$x = 60 \cdot 800$$
$$x = 48,000$$

There are about $48,000$ fish in the lake.

22. $g(x) = \dfrac{5x}{2+x}$

(a)
$$3 = \frac{5x}{2+x} \qquad \textit{Let } g(x)=3$$
$$3(2+x) = 5x$$
$$6 + 3x = 5x$$
$$6 = 2x$$
$$x = 3 \text{ units}$$

(b) If no food is available, then $x = 0$.

$$g(0) = \frac{5(0)}{2+0} = \frac{0}{2} = 0$$

The growth rate is 0.

Cumulative Review Exercises Chapters 1–7

In Exercises 1–2, $x = -4$, $y = 3$, and $z = 6$.

1.
$$|2x| + 3y - z^3$$
$$= |(2)(-4)| + 3(3) - (6)^3$$
$$= |-8| + 9 - 216$$
$$= 8 - 207 = -199$$

2.
$$\frac{x(2x-1)}{3y-z} = \frac{-4[2(-4)-1]}{3(3)-6}$$
$$= \frac{-4[-8-1]}{9-6}$$
$$= \frac{-4(-9)}{3} = 12$$

3.
$$7(2x+3) - 4(2x+1) = 2(x+1)$$
$$14x + 21 - 8x - 4 = 2x + 2$$
$$6x + 17 = 2x + 2$$
$$4x = -15$$
$$x = -\frac{15}{4}$$

Solution set: $\left\{-\frac{15}{4}\right\}$

4.
$$|6x-8| - 4 = 0$$
$$|6x-8| = 4$$

$$6x - 8 = 4 \quad \text{or} \quad 6x - 8 = -4$$
$$6x = 12 \qquad\qquad 6x = 4$$
$$x = 2 \quad \text{or} \quad x = \frac{4}{6} = \frac{2}{3}$$

Solution set: $\left\{\frac{2}{3}, 2\right\}$

5. Solve $ax + by = cx + d$ for x. Get the x-terms on one side.

$$ax - cx = d - by$$
$$x(a-c) = d - by$$
$$x = \frac{d-by}{a-c}$$
$$\text{or} \qquad x = \frac{by-d}{c-a}$$

6.
$$\frac{2}{3}y + \frac{5}{12}y \le 20$$

Multiply both sides by 12.

$$12\left(\frac{2}{3}y + \frac{5}{12}y\right) \le 12(20)$$
$$8y + 5y \le 240$$
$$13y \le 240$$
$$y \le \frac{240}{13}$$

Solution set: $\left(-\infty, \frac{240}{13}\right]$

7.
$$|3x+2| \ge 4$$

$$3x + 2 \ge 4 \quad \text{or} \quad 3x + 2 \le -4$$
$$3x \ge 2 \quad \text{or} \quad 3x \le -6$$
$$x \ge \frac{2}{3} \quad \text{or} \quad x \le -2$$

Solution set: $(-\infty, -2] \cup \left[\frac{2}{3}, \infty\right)$

8. Let $x =$ the amount of money invested at 4% and $2x =$ the amount of money invested at 3%.

Use $I = prt$ with the time, t, equal to 1 yr.

$$.04x + .03(2x) = 400$$
$$.04x + .06x = 400$$
$$.10x = 400$$

Multiply by 10 to clear the decimal.

$$x = 4000$$

Then $2x = 2(4000) = 8000$.

He invested \$4000 at 4% and \$8000 at 3%.

9. Let $h =$ the height of the triangle.
Use the formula $A = \frac{1}{2}bh$. Here, $A = 42$ and $b = 14$.

$$\begin{aligned} A &= \frac{1}{2}bh \\ 42 &= \frac{1}{2}(14)h \\ 42 &= 7h \\ 6 &= h \end{aligned}$$

The height is 6 meters.

10. $-4x + 2y = 8$
To find the x-intercept, let $y = 0$.

$$\begin{aligned} -4x + 2(0) &= 8 \\ -4x &= 8 \\ x &= -2 \end{aligned}$$

The x-intercept is $(-2, 0)$.
To find the y-intercept, let $x = 0$.

$$\begin{aligned} -4(0) + 2y &= 8 \\ 2y &= 8 \\ y &= 4 \end{aligned}$$

The y-intercept is $(0, 4)$. Plot the intercepts, and draw the line through them.

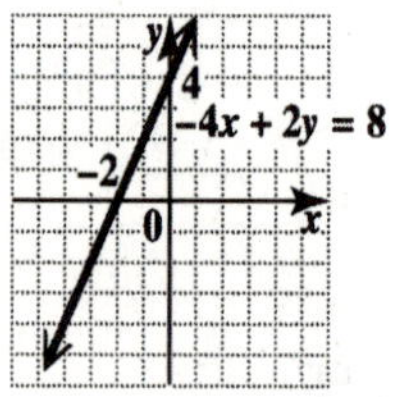

11. Through $(-5, 8)$ and $(-1, 2)$
Let $(x_1, y_1) = (-5, 8)$ and $(x_2, y_2) = (-1, 2)$. Then,

$$m = \frac{y_2 - y_1}{x_2 - x_1} = \frac{2 - 8}{-1 - (-5)} = \frac{-6}{4} = -\frac{3}{2}.$$

The slope is $-\frac{3}{2}$.

12. Perpendicular to $4x - 3y = 12$

Solve for y to write the equation in slope-intercept form and find the slope.

$$\begin{aligned} 4x - 3y &= 12 \\ -3y &= -4x + 12 \\ y &= \frac{4}{3}x - 4 \end{aligned}$$

The slope is $\frac{4}{3}$. Perpendicular lines have slopes that are negative reciprocals of each other. The negative reciprocal of $\frac{4}{3}$ is $-\frac{3}{4}$. The slope of a line perpendicular to the given line is $-\frac{3}{4}$.

13. Use $(x_1, y_1) = (-5, 8)$ and $m = -\frac{3}{2}$ in the point-slope form.

$$\begin{aligned} y - y_1 &= m(x - x_1) \\ y - 8 &= -\frac{3}{2}[x - (-5)] \\ y - 8 &= -\frac{3}{2}(x + 5) \\ y - 8 &= -\frac{3}{2}x - \frac{15}{2} \\ y &= -\frac{3}{2}x + \frac{1}{2} \end{aligned}$$

14. $2x + 5y > 10$
Graph the line $2x + 5y = 10$ by drawing a dashed line (since the inequality involves $>$) through the intercepts $(5, 0)$ and $(0, 2)$.
Test a point not on this line, such as $(0, 0)$.

$$\begin{aligned} 2x + 5y &> 10 \\ 2(0) + 5(0) &> 10 \\ 0 &> 10 \quad \textit{False} \end{aligned}$$

Shade the side of the line not containing $(0, 0)$.

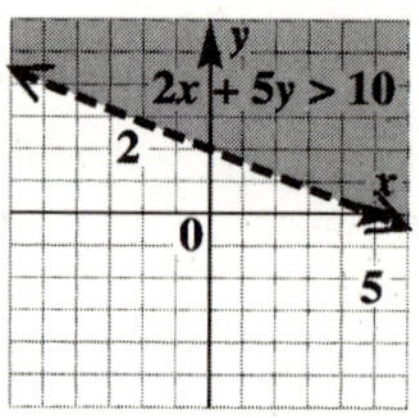

15. $x - y \geq 3$ and $3x + 4y \leq 12$
Graph the solid line $x - y = 3$ through $(3, 0)$ and $(0, -3)$. The inequality $x - y \geq 3$ can be written as $y \leq x - 3$, so shade the region below the boundary line.
Graph the solid line $3x + 4y = 12$ through $(4, 0)$ and $(0, 3)$. The inequality $3x + 4y \leq 12$ can be written as $y \leq -\frac{3}{4}x + 3$, so shade the region below the boundary line.
The required graph is the common shaded area as well as the portions of the lines that bound it.

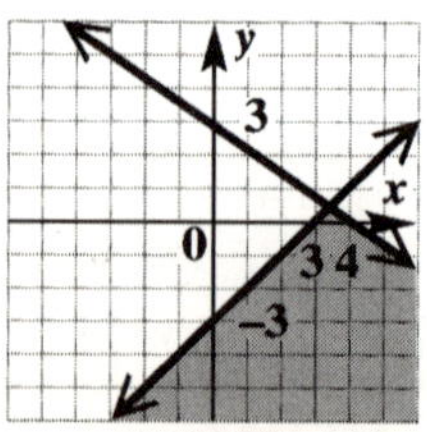

$x - y \geq 3$ and
$3x + 4y \leq 12$

16. Since no year is repeated, the relation defines a function.
The domain is
$\{1990, 1992, 1994, 1996, 1998, 2000\}$.
The range is $\{1.25, 1.61, 1.80, 1.21, 1.94, 2.26\}$.

17. The relation is not a function since a vertical line may intersect the graph in more than one point. The domain is the set of x-values, $[-2, \infty)$. The range is the set of y-values, $(-\infty, \infty)$.

18. $y = -\sqrt{x+2}$
The relation is a function since each input value corresponds to exactly one output value. The radicand must be nonnegative; that is, $x + 2 \geq 0$, or $x \geq -2$. The domain is $[-2, \infty)$. The values of $\sqrt{x+2}$ are nonnegative, so the values of $-\sqrt{x+2}$ are nonpositive. Thus, the range is $(-\infty, 0]$.

19. **(a)** Solve the equation for y.

$$\begin{aligned} 5x - 3y &= 8 \\ 5x - 8 &= 3y \\ \frac{5x - 8}{3} &= y \end{aligned}$$

So $f(x) = \dfrac{5x-8}{3}$ or $f(x) = \dfrac{5}{3}x - \dfrac{8}{3}$.

(b) $f(1) = \dfrac{5(1) - 8}{3} = \dfrac{-3}{3} = -1$

20.
$$\begin{aligned} f(x) &= 3x + 6 \\ f(x+3) &= 3(x+3) + 6 \\ &= 3x + 9 + 6 \\ &= 3x + 15 \end{aligned}$$

21.
$$\begin{aligned} 4x - y &= -7 \quad (1) \\ 5x + 2y &= 1 \quad (2) \end{aligned}$$

To eliminate y, multiply equation (1) by 2 and add the result to (2).

$$\begin{array}{rcrl} 8x - 2y &=& -14 & 2 \times (1) \\ 5x + 2y &=& 1 & (2) \\ \hline 13x &=& -13 & \\ x &=& -1 & \end{array}$$

To find y, substitute -1 for x in (1).

$$\begin{aligned} 4x - y &= -7 \quad (1) \\ 4(-1) - y &= -7 \\ -4 - y &= -7 \\ -y &= -3 \\ y &= 3 \end{aligned}$$

Solution set: $\{(-1, 3)\}$

22.
$$\begin{aligned} x + y - 2z &= -1 \quad (1) \\ 2x - y + z &= -6 \quad (2) \\ 3x + 2y - 3z &= -3 \quad (3) \end{aligned}$$

Add (1) and (2) to eliminate y.

$$\begin{array}{rcrl} x + y - 2z &=& -1 & (1) \\ 2x - y + z &=& -6 & (2) \\ \hline 3x - z &=& -7 & (4) \end{array}$$

Multiply (2) by 2 and add it to (3) to eliminate y.

$$\begin{array}{rcrl} 4x - 2y + 2z &=& -12 & 2 \times (2) \\ 3x + 2y - 3z &=& -3 & (3) \\ \hline 7x - z &=& -15 & (5) \end{array}$$

Multiply (4) by -1 and add to (5) to eliminate z.

$$\begin{array}{rcrl} -3x + z &=& 7 & -1 \times (4) \\ 7x - z &=& -15 & (5) \\ \hline 4x &=& -8 & \\ x &=& -2 & \end{array}$$

Substitute -2 for x into (4) and solve for z.

$$\begin{aligned} 3(-2) - z &= -7 \\ -6 - z &= -7 \\ -z &= -1 \\ z &= 1 \end{aligned}$$

Substitute -2 for x and 1 for z into (1) and solve for y.

$$\begin{aligned} -2 + y - 2(1) &= -1 \\ -2 + y - 2 &= -1 \\ y - 4 &= -1 \\ y &= 3 \end{aligned}$$

Solution set: $\{(-2, 3, 1)\}$

23.
$$\begin{aligned} x + 2y + z &= 5 \quad (1) \\ x - y + z &= 3 \quad (2) \\ 2x + 4y + 2z &= 11 \quad (3) \end{aligned}$$

Multiply (1) by -1 and add to (2).

$$\begin{array}{rcrl} -x - 2y - z &=& -5 & -1 \times (1) \\ x - y + z &=& 3 & (2) \\ \hline -3y &=& -2 & \\ y &=& \frac{2}{3} & \end{array}$$

Substitute $\frac{2}{3}$ for y in (1) and (3), then add (1) and (3).

$$\begin{aligned} x + 2\left(\frac{2}{3}\right) + z &= 5 \qquad \text{Let } y = \tfrac{2}{3} \text{ in (1).} \\ x + \frac{4}{3} + z &= 5 \\ x + z &= \frac{11}{3} \qquad (4) \\ 2x + 4\left(\frac{2}{3}\right) + 2z &= 11 \qquad \text{Let } y = \tfrac{2}{3} \text{ in (3).} \\ 2x + \frac{8}{3} + 2z &= 11 \\ 2x + 2z &= \frac{25}{3} \qquad (5) \end{aligned}$$

continued

Multiply (4) by -2 and add to (5).

$$\begin{aligned} -2x - 2z &= -\frac{22}{3} \quad -2\times(4)\\ 2x + 2z &= \frac{25}{3} \quad (5)\\ \hline 0 &= 1 \quad \textit{False} \end{aligned}$$

Since this statement is false, the solution is $\emptyset$.

24. Let $x =$ the average speed of the automobile. Then $x + 558 =$ the average speed of the airplane. Write a proportion and solve for x.

$$\frac{7}{100} = \frac{x}{x+558}$$

Multiply each side by the LCD, $100(x+558)$.

$$100(x+558)\left(\frac{7}{100}\right) = 100(x+558)\left(\frac{x}{x+558}\right)$$
$$\begin{aligned} 7(x+558) &= 100x\\ 7x + 3906 &= 100x\\ 3906 &= 93x\\ 42 &= x \end{aligned}$$

Since $x = 42$, $x + 558 = 600$.

The average speed of the automobile is 42 km per hr and that of the airplane is 600 km per hr.

25. $\left(\dfrac{a^{-3}b^4}{a^2b^{-1}}\right)^{-2} = \left(\dfrac{b^4b^1}{a^2a^3}\right)^{-2} = \left(\dfrac{b^5}{a^5}\right)^{-2}$
$= \left(\dfrac{a^5}{b^5}\right)^2 = \dfrac{a^{10}}{b^{10}}$

26. $\left(\dfrac{m^{-4}n^2}{m^2n^{-3}}\right)\cdot\left(\dfrac{m^5n^{-1}}{m^{-2}n^5}\right)$
$= \left(\dfrac{n^2n^3}{m^2m^4}\right)\cdot\left(\dfrac{m^5m^2}{n^1n^5}\right)$
$= \left(\dfrac{n^5}{m^6}\right)\cdot\left(\dfrac{m^7}{n^6}\right)$
$= \dfrac{n^5m^7}{n^6m^6} = \dfrac{m}{n}$

27. $(3y^2 - 2y + 6) - (-y^2 + 5y + 12)$
$= 3y^2 - 2y + 6 + y^2 - 5y - 12$
$= 4y^2 - 7y - 6$

28. $-6x^4(x^2 - 3x + 2)$
$= -6x^6 + 18x^5 - 12x^4$

29. $(4f+3)(3f-1) = 12f^2 - 4f + 9f - 3$
$= 12f^2 + 5f - 3$

30. $(7t^3+8)(7t^3-8)$

This is the product of the sum and difference of two terms.

$= (7t^3)^2 - 8^2$
$= 49t^6 - 64$

31. $\left(\dfrac{1}{4}x + 5\right)^2$

Use the formula for the square of a binomial, $(a+b)^2 = a^2 + 2ab + b^2$.

$= \left(\dfrac{1}{4}x\right)^2 + 2\left(\dfrac{1}{4}x\right)(5) + 5^2$
$= \dfrac{1}{16}x^2 + \dfrac{5}{2}x + 25$

32. $(3x^3 + 13x^2 - 17x - 7) \div (3x + 1)$

$$\begin{array}{r|l} & x^2 + 4x - 7 \\ \hline 3x+1 & 3x^3 + 13x^2 - 17x - 7 \\ & 3x^3 + x^2 \\ \hline & 12x^2 - 17x \\ & 12x^2 + 4x \\ \hline & -21x - 7 \\ & -21x - 7 \\ \hline & 0 \end{array}$$

Answer: $x^2 + 4x - 7$

33. **(a)** $(f+g)(x) = f(x) + g(x)$
$= (x^2 + 2x - 3)$
$+ (2x^3 - 3x^2 + 4x - 1)$
$= 2x^3 - 2x^2 + 6x - 4$

(b) $(g-f)(x) = g(x) - f(x)$
$= (2x^3 - 3x^2 + 4x - 1)$
$- (x^2 + 2x - 3)$
$= 2x^3 - 3x^2 + 4x - 1$
$- x^2 - 2x + 3$
$= 2x^3 - 4x^2 + 2x + 2$

(c) Using part (a),

$(f+g)(-1) = 2(-1) - 2(1) + 6(-1) - 4$
$= -2 - 2 - 6 - 4$
$= -14$

34. $2x^2 - 13x - 45 = (2x+5)(x-9)$

35. $100t^4 - 25 = 25(4t^4 - 1)$
$= 25\left[(2t^2)^2 - 1^2\right]$
$= 25(2t^2+1)(2t^2-1)$

36. Use the sum of cubes formula,

$x^3 + y^3 = (x+y)(x^2 - xy + y^2)$.
$8p^3 + 125 = (2p)^3 + 5^3$
$= (2p+5)\left[(2p)^2 - (2p)(5) + 5^2\right]$
$= (2p+5)(4p^2 - 10p + 25)$

37.
$$\begin{aligned} 3x^2+4x&=7\\ 3x^2+4x-7&=0\\ (3x+7)(x-1)&=0 \end{aligned}$$

$3x+7=0$ or $x-1=0$

$3x=-7$

$x=-\frac{7}{3}$ or $x=1$

Solution set: $\left\{-\frac{7}{3},1\right\}$

38.
$$\frac{y^2-16}{y^2-8y+16}=\frac{(y+4)(y-4)}{(y-4)(y-4)}=\frac{y+4}{y-4}$$

39.
$$\frac{8x^2-18}{8x^2+4x-12}=\frac{2(4x^2-9)}{4(2x^2+x-3)}=\frac{2(2x+3)(2x-3)}{4(2x+3)(x-1)}=\frac{2x-3}{2(x-1)}$$

40.
$$\frac{2a^2}{a+b}\cdot\frac{a-b}{4a}=\frac{2a^2(a-b)}{4a(a+b)}=\frac{a(a-b)}{2(a+b)}$$

41.
$$\frac{x+4}{x-2}+\frac{2x-10}{x-2}=\frac{x+4+2x-10}{x-2}=\frac{3x-6}{x-2}=\frac{3(x-2)}{x-2}=3$$

42.
$$\frac{2x}{2x-1}+\frac{4}{2x+1}+\frac{8}{4x^2-1}=\frac{2x}{2x-1}+\frac{4}{2x+1}+\frac{8}{(2x+1)(2x-1)}$$

The LCD is $(2x+1)(2x-1)$.

$$=\frac{2x(2x+1)}{(2x-1)(2x+1)}+\frac{4(2x-1)}{(2x+1)(2x-1)}+\frac{8}{(2x+1)(2x-1)}$$
$$=\frac{2x(2x+1)+4(2x-1)+8}{(2x+1)(2x-1)}$$
$$=\frac{4x^2+2x+8x-4+8}{(2x+1)(2x-1)}$$
$$=\frac{4x^2+10x+4}{(2x+1)(2x-1)}$$
$$=\frac{2(2x^2+5x+2)}{(2x+1)(2x-1)}$$
$$=\frac{2(2x+1)(x+2)}{(2x+1)(2x-1)}$$
$$=\frac{2(x+2)}{2x-1}$$

43.
$$\frac{-3x}{x+1}+\frac{4x+1}{x}=\frac{-3}{x^2+x}$$
$$\frac{-3x}{x+1}+\frac{4x+1}{x}=\frac{-3}{x(x+1)}$$

Multiply by the LCD, $x(x+1)$. $(x\neq-1,0)$

$$x(x+1)\left(\frac{-3x}{x+1}+\frac{4x+1}{x}\right)=x(x+1)\left(\frac{-3}{x(x+1)}\right)$$
$$\begin{aligned} x(-3x)+(x+1)(4x+1)&=-3\\ -3x^2+4x^2+x+4x+1&=-3\\ x^2+5x+4&=0\\ (x+4)(x+1)&=0 \end{aligned}$$

$x+4=0$ or $x+1=0$

$x=-4$ or $x=-1$

The number -1 is not allowed as a solution because substituting it in the original equation results in division by 0.

Check $x=-4$: $-4+\frac{15}{4}=-\frac{1}{4}$ *True*

Solution set: $\{-4\}$

44. Solve $\frac{1}{f}=\frac{1}{p}+\frac{1}{q}$ for q.

Multiply by the LCD, fpq.

$$fpq\left(\frac{1}{f}\right)=fpq\left(\frac{1}{p}+\frac{1}{q}\right)$$
$$pq=fq+fp$$

Get all terms with q on one side.

$$\begin{aligned} pq-fq&=fp\\ q(p-f)&=fp\\ q&=\frac{fp}{p-f} \end{aligned}$$

or $q=\frac{-fp}{f-p}$

45. Let $x =$ the speed of the plane in still air.
Use $d = rt$, or $t = \frac{d}{r}$, to make a table.

	d	r	t
Against the Wind	200	$x - 30$	$\frac{200}{x-30}$
With the Wind	300	$x + 30$	$\frac{300}{x+30}$

The times are the same, so

$$\frac{200}{x-30} = \frac{300}{x+30}.$$

Multiply by the LCD, $(x-30)(x+30)$.

$$(x-30)(x+30)\left(\frac{200}{x-30}\right) = (x-30)(x+30)\left(\frac{300}{x+30}\right)$$

$$\begin{aligned} 200(x+30) &= 300(x-30) \\ 200x + 6000 &= 300x - 9000 \\ -100x &= -15,000 \\ x &= 150 \end{aligned}$$

The speed of the plane in still air is 150 mph.

46. Let $x =$ the time to complete the job working together.
Make a table.

	Rate	Time Working Together	Fractional Part of the Job Done
Machine A	$\frac{1}{2}$	x	$\frac{x}{2}$
Machine B	$\frac{1}{3}$	x	$\frac{x}{3}$

Part done by A	plus	part done by B	equals	1 whole job.
$\frac{x}{2}$	$+$	$\frac{x}{3}$	$=$	1

Multiply by the LCD, 6.

$$\begin{aligned} 6\left(\frac{x}{2} + \frac{x}{3}\right) &= 6 \cdot 1 \\ 3x + 2x &= 6 \\ 5x &= 6 \\ x &= \frac{6}{5} \text{ or } 1\frac{1}{5} \end{aligned}$$

Working together, the machines can complete the job in $\frac{6}{5}$ or $1\frac{1}{5}$ hours.

CHAPTER 8 ROOTS, RADICALS, AND ROOT FUNCTIONS

Section 8.1

1. $-\sqrt{16} = -(4) = -4$ **(E)**

3. $\sqrt[3]{-27} = -3$ because $(-3)^3 = -27$. **(D)**

5. $\sqrt[4]{81} = 3$ because $3^4 = 81$. **(A)**

7. $\sqrt{123.5} \approx \sqrt{121} = 11$ because $11^2 = 121$. **(C)**

9. The length $\sqrt{98}$ is closer to $\sqrt{100} = 10$ than to $\sqrt{81} = 9$. The width $\sqrt{26}$ is closer to $\sqrt{25} = 5$ than to $\sqrt{36} = 6$. Use the estimates $L = 10$ and $W = 5$ in $A = LW$ to find an estimate of the area.

$$A \approx 10 \cdot 5 = 50$$

Choice **C** is the best estimate.

11. **(a)** If $a > 0$, then $\sqrt{-a}$ is *not a real number.*

(b) If $a < 0$, then $\sqrt{-a}$ is a positive number and then $-\sqrt{-a}$ is a *negative* number.

(c) If $a = 0$, then $\sqrt{-a} = \sqrt{0} = 0$.

13. $\sqrt{81} = 9$ because $9^2 = 81$. Notice that -9 is not an answer because the symbol $\sqrt{81}$ means the *nonnegative* square root of 81.

15. $\sqrt[3]{216} = 6$ because $6^3 = 216$, so $-\sqrt[3]{216} = -6$.

17. $\sqrt[3]{-64} = -4$ because $(-4)^3 = -64$.

19. $\sqrt[3]{512} = 8$ because $8^3 = 512$.

21. $\sqrt[4]{1296} = 6$ because $6^4 = 1296$.

23. $\sqrt[4]{81} = 3$ because $3^4 = 81$, so $-\sqrt[4]{81} = -3$.

25. $\sqrt[4]{-16}$ is not a real number since no real number to the fourth power equals -16. Any real number raised to the fourth power is 0 or positive.

27. $\sqrt[6]{64} = 2$ because $2^6 = 64$.

29. $\sqrt[6]{-32}$ is not a real number because the index, 6, is even and the radicand, -32, is negative.

31. $\sqrt{\dfrac{64}{81}} = \dfrac{8}{9}$ because $\left(\dfrac{8}{9}\right)^2 = \dfrac{64}{81}$.

33. $\sqrt[3]{\dfrac{8}{27}} = \dfrac{2}{3}$ because $\left(\dfrac{2}{3}\right)^3 = \dfrac{8}{27}$.

35. $\sqrt[6]{\dfrac{1}{64}} = \dfrac{1}{2}$ because $\left(\dfrac{1}{2}\right)^6 = \dfrac{1}{64}$.

37. $f(x) = \sqrt{x+3}$
For the radicand to be nonnegative, we must have

$$x + 3 \geq 0 \quad \text{or} \quad x \geq -3.$$

Thus, the domain is $[-3, \infty)$.
The function values are positive or zero (the result of the radical), so the range is $[0, \infty)$.

x	$f(x) = \sqrt{x+3}$
-3	$\sqrt{-3+3} = 0$
-2	$\sqrt{-2+3} = 1$
1	$\sqrt{1+3} = 2$

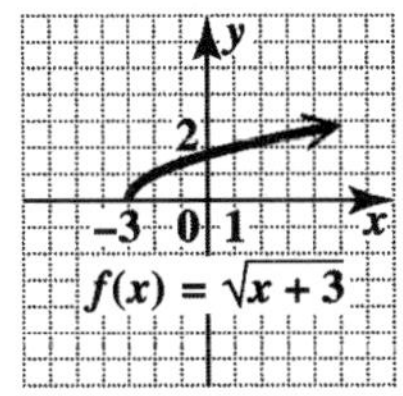

39. $f(x) = \sqrt{x} - 2$
For the radicand to be nonnegative, we must have

$$x \geq 0.$$

Note that the "-2" does not affect the domain, which is $[0, \infty)$.
The result of the radical is positive or zero, but the function values are 2 less than those values, so the range is $[-2, \infty)$.

x	$f(x) = \sqrt{x} \quad 2$
0	$\sqrt{0} - 2 = -2$
1	$\sqrt{1} - 2 = -1$
4	$\sqrt{4} - 2 = 0$

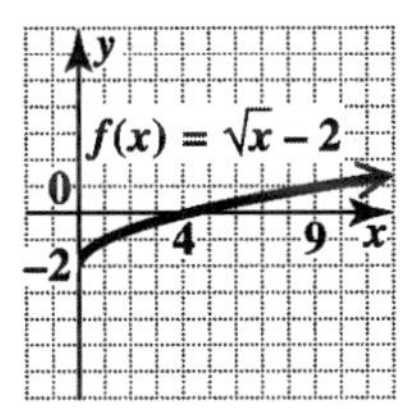

41. $f(x) = \sqrt[3]{x} - 3$
Since we can take the cube root of any real number, the domain is $(-\infty, \infty)$.
The result of a cube root can be any real number, so the range is $(-\infty, \infty)$. (The "-3" does not affect the range.)

x	$f(x) = \sqrt[3]{x} - 3$
-8	$\sqrt[3]{-8} - 3 = -5$
-1	$\sqrt[3]{-1} - 3 = -4$
0	$\sqrt[3]{0} - 3 = -3$
1	$\sqrt[3]{1} - 3 = -2$
8	$\sqrt[3]{8} - 3 = -1$

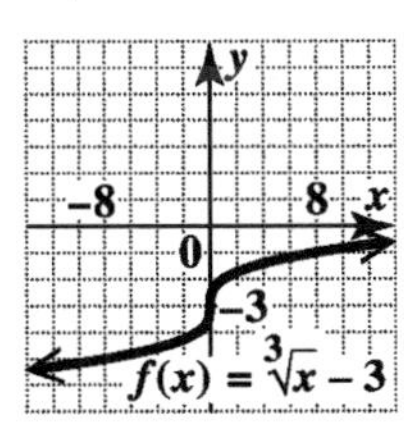

43. $f(x) = \sqrt[3]{x-3}$

Both the domain and range are $(-\infty, \infty)$.

x	$f(x) = \sqrt[3]{x-3}$
-5	$\sqrt[3]{-5-3} = -2$
2	$\sqrt[3]{2-3} = -1$
3	$\sqrt[3]{3-3} = 0$
4	$\sqrt[3]{4-3} = 1$
11	$\sqrt[3]{11-3} = 2$

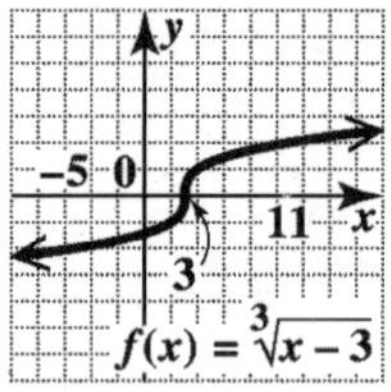

45. $\sqrt{12^2} = |12| = 12$

47. $\sqrt{(-10)^2} = |-10| = -(-10) = 10$

49. Since 6 is an even positive integer, $\sqrt[6]{a^6} = |a|$, so $\sqrt[6]{(-2)^6} = |-2| = 2$.

51. Since 5 is odd, $\sqrt[5]{a^5} = a$, so

$$\sqrt[5]{(-9)^5} = -9.$$

53. Since the index is even, $\sqrt{x^2} = |x|$.

55. Since the index is odd, $\sqrt[3]{x^3} = x$.

57. $\sqrt[3]{x^{15}} = \sqrt[3]{(x^5)^3} = x^5$ (3 is odd)

In Exercises 59–71, use a calculator and round to three decimal places

59. $\sqrt{9483} \approx 97.381$

61. $\sqrt{284.361} \approx 16.863$

63. $\sqrt{7} \approx 2.646$

65. $-\sqrt{82} \approx -9.055$

67. $\sqrt[3]{423} \approx 7.507$

69. $\sqrt[4]{100} \approx 3.162$

71. $\sqrt[5]{23.8} \approx 1.885$

73.
$$\begin{aligned} f &= \frac{1}{2\pi\sqrt{LC}} \\ &= \frac{1}{2\pi\sqrt{(7.237 \times 10^{-5})(2.5 \times 10^{-10})}} \\ &\approx 1,183,235 \end{aligned}$$

or about 1,183,000 cycles per second.

75. Since $H = 44 + 6 = 50$ ft, substitute 50 for H in the formula.

$$D = \sqrt{2H} = \sqrt{2 \cdot 50} = \sqrt{100} = 10$$

She will be able to see about 10 miles.

77. Let $a = 850$, $b = 925$, and $c = 1300$. First find s.

$$\begin{aligned} s &= \frac{1}{2}(a+b+c) \\ &= \frac{1}{2}(850 + 925 + 1300) \\ &= \frac{3075}{2} = 1537.5 \end{aligned}$$

Now find the area using Heron's formula.

$$\begin{aligned} A &= \sqrt{s(s-a)(s-b)(s-c)} \\ &= \sqrt{1537.5(687.5)(612.5)(237.5)} \\ &\approx 392,128.8 \end{aligned}$$

The area of the Bermuda Triangle is about 392,000 square miles.

79.
$$\begin{aligned} I &= \sqrt{\frac{2P}{L}} \\ &= \sqrt{\frac{2(120)}{80}} = \sqrt{3} \approx 1.732 \text{ amps} \end{aligned}$$

Section 8.2

1. $2^{1/2} = \sqrt[2]{2^1} = \sqrt{2}$ **(C)**

3. $-16^{1/2} = -\sqrt{16} = -(4) = -4$ **(A)**

5. $(-32)^{1/5} = \left[(-2)^5\right]^{1/5} = (-2)^1 = -2$ **(H)**

7. $4^{3/2} = \left[(2^2)^{1/2}\right]^3 = 2^3 = 8$ **(B)**

9. $-6^{2/4} = -(6^{1/2}) = -\sqrt{6}$ **(D)**

11. $169^{1/2} = (13^2)^{1/2} = 13^1 = 13$

We could use radical notation as follows:

$$169^{1/2} = \sqrt{169} = 13$$

13. $729^{1/3} = (9^3)^{1/3} = 9$

15. $16^{1/4} = (2^4)^{1/4} = 2$

17. $\left(\frac{64}{81}\right)^{1/2} = \left[\left(\frac{8}{9}\right)^2\right]^{1/2} = \frac{8}{9}$

19. $(-27)^{1/3} = \left[(-3)^3\right]^{1/3} = -3$

21. $100^{3/2} = (100^{1/2})^3 = 10^3 = 1000$

23. $-16^{5/2} = -(16^{1/2})^5 = -(4)^5 = -1024$

25. $(-144)^{1/2}$ is not a real number because no real number squared equals -144.

27. $64^{-3/2} = \frac{1}{64^{3/2}} = \frac{1}{(64^{1/2})^3} = \frac{1}{8^3} = \frac{1}{512}$

29. $\left(-\frac{8}{27}\right)^{-2/3} = \left(-\frac{27}{8}\right)^{2/3}$
$$= \left(\left[\left(-\frac{3}{2}\right)^3\right]^{1/3}\right)^2$$
$$= \left(-\frac{3}{2}\right)^2 = \frac{9}{4}$$

31. $(-64)^{1/2}$ is an even root of a negative number. No real number squared will give -64. On the other hand, $-64^{1/2} = -\sqrt{64} = -8$, which is a real number. ($-64^{1/2}$ is the opposite of $64^{1/2}$.)

33. $12^{1/2} = \sqrt{12}$

35. $8^{3/4} = \left(8^{1/4}\right)^3 = \left(\sqrt[4]{8}\right)^3$

37. $(9q)^{5/8} - (2x)^{2/3}$
$$= \left[(9q)^{1/8}\right]^5 - \left[(2x)^{1/3}\right]^2$$
$$= \left(\sqrt[8]{9q}\right)^5 - \left(\sqrt[3]{2x}\right)^2$$

39. $(2m)^{-3/2} = \left[(2m)^{1/2}\right]^{-3}$
$$= \left(\sqrt{2m}\right)^{-3}$$
$$= \frac{1}{\left(\sqrt{2m}\right)^3}$$

41. $(2y+x)^{2/3} = \left[(2y+x)^{1/3}\right]^2$
$$= \left(\sqrt[3]{2y+x}\right)^2$$

43. $\left(3m^4 + 2k^2\right)^{-2/3}$
$$= \frac{1}{(3m^4+2k^2)^{2/3}}$$
$$= \frac{1}{\left[(3m^4+2k^2)^{1/3}\right]^2}$$
$$= \frac{1}{\left(\sqrt[3]{3m^4+2k^2}\right)^2}$$

45. We are to show that, in general,
$$\sqrt{a^2+b^2} \neq a+b.$$
When $a = 3$ and $b = 4$,
$$\sqrt{a^2+b^2} = \sqrt{3^2+4^2}$$
$$= \sqrt{9+16}$$
$$= \sqrt{25} = 5,$$
but
$$a + b = 3 + 4 = 7.$$
Since $5 \neq 7$, $\sqrt{a^2+b^2} \neq a+b$.

47. $\sqrt{2^{12}} = \left(2^{12}\right)^{1/2} = 2^{12/2} = 2^6 = 64$

49. $\sqrt[3]{4^9} = 4^{9/3} = 4^3 = 64$

51. $\sqrt{x^{20}} = x^{20/2} = x^{10}$

53. $\sqrt[3]{x} \cdot \sqrt{x} = x^{1/3} \cdot x^{1/2} = x^{1/3+1/2} = x^{2/6+3/6}$
$$= x^{5/6} = \sqrt[6]{x^5}$$

55. $\frac{\sqrt[3]{t^4}}{\sqrt[5]{t^4}} = \frac{t^{4/3}}{t^{4/5}} = \frac{t^{20/15}}{t^{12/15}} = t^{20/15-12/15}$
$$= t^{8/15} = \sqrt[15]{t^8}$$

57. $3^{1/2} \cdot 3^{3/2} = 3^{1/2+3/2} = 3^{4/2} = 3^2 = 9$

59. $\frac{64^{5/3}}{64^{4/3}} = 64^{5/3-4/3} = 64^{1/3} = \sqrt[3]{64} = 4$

61. $y^{7/3} \cdot y^{-4/3} = y^{7/3+(-4/3)} = y^{3/3} = y$

63. $\frac{k^{1/3}}{k^{2/3} \cdot k^{-1}} = \frac{k^{1/3}}{k^{-1/3}} = k^{1/3-(-1/3)} = k^{2/3}$

65. $\frac{\left(x^{1/4}y^{2/5}\right)^{20}}{x^2} = \frac{x^{(1/4)\cdot 20}y^{(2/5)\cdot 20}}{x^2}$
$$= \frac{x^5y^8}{x^2}$$
$$= x^3y^8$$

67. $\frac{\left(x^{2/3}\right)^2}{\left(x^2\right)^{7/3}} = \frac{x^{4/3}}{x^{14/3}} = x^{4/3-14/3}$
$$= x^{-10/3} = \frac{1}{x^{10/3}}$$

69. $\frac{m^{3/4}n^{-1/4}}{(m^2n)^{1/2}} = \frac{m^{3/4}n^{-1/4}}{m^1n^{1/2}}$
$$= m^{3/4-1}n^{-1/4-1/2}$$
$$= m^{3/4-4/4}n^{-1/4-2/4}$$
$$= m^{-1/4}n^{-3/4}$$
$$= \frac{1}{m^{1/4}n^{3/4}}$$

71. $\frac{p^{1/5}p^{7/10}p^{1/2}}{\left(p^3\right)^{-1/5}} = \frac{p^{2/10+7/10+5/10}}{p^{-3/5}}$
$$= \frac{p^{14/10}}{p^{-6/10}}$$
$$= p^{14/10-(-6/10)}$$
$$= p^{20/10} = p^2$$

73. $\left(\frac{b^{-3/2}}{c^{-5/3}}\right)^2 \left(b^{-1/4}c^{-1/3}\right)^{-1}$
$$= \left(\frac{c^{5/3}}{b^{3/2}}\right)^2 \left(b^{1/4}c^{1/3}\right)$$
$$= \frac{c^{10/3}}{b^3}\left(b^{1/4}c^{1/3}\right)$$
$$= \frac{c^{10/3}b^{1/4}c^{1/3}}{b^3}$$

continued

$$= c^{10/3+1/3}b^{1/4-3}$$
$$= c^{11/3}b^{-11/4}$$
$$= \frac{c^{11/3}}{b^{11/4}}$$

75. $\left(\frac{p^{-1/4}q^{-3/2}}{3^{-1}p^{-2}q^{-2/3}}\right)^{-2}$
$$= \frac{p^{1/2}q^3}{3^2p^4q^{4/3}}$$
$$= \frac{q^{3-4/3}}{9p^{4-1/2}}$$
$$= \frac{q^{5/3}}{9p^{7/2}}$$

77. $p^{2/3}(p^{1/3} + 2p^{4/3})$
$$= p^{2/3}p^{1/3} + p^{2/3}(2p^{4/3})$$
$$= p^{2/3+1/3} + 2p^{2/3+4/3}$$
$$= p^{3/3} + 2p^{6/3}$$
$$= p^1 + 2p^2$$
$$= p + 2p^2$$

79. $k^{1/4}(k^{3/2} - k^{1/2})$
$$= k^{1/4+3/2} - k^{1/4+1/2}$$
$$= k^{1/4+6/4} - k^{1/4+2/4}$$
$$= k^{7/4} - k^{3/4}$$

81. $6a^{7/4}(a^{-7/4} + 3a^{-3/4})$
$$= 6a^{7/4+(-7/4)} + 18a^{7/4+(-3/4)}$$
$$= 6a^0 + 18a^{4/4}$$
$$= 6(1) + 18a^1 = 6 + 18a$$

83. $\sqrt[5]{x^3} \cdot \sqrt[4]{x} = x^{3/5} \cdot x^{1/4}$
$$= x^{3/5+1/4}$$
$$= x^{12/20+5/20}$$
$$= x^{17/20}$$

85. $\frac{\sqrt{x^5}}{\sqrt{x^8}} = \frac{(x^5)^{1/2}}{(x^8)^{1/2}}$
$$= x^{5/2-(8/2)}$$
$$= x^{-3/2} = \frac{1}{x^{3/2}}$$

87. $\sqrt{y} \cdot \sqrt[3]{yz} = y^{1/2} \cdot (yz)^{1/3}$
$$= y^{1/2}y^{1/3}z^{1/3}$$
$$= y^{1/2+1/3}z^{1/3}$$
$$= y^{3/6+2/6}z^{1/3}$$
$$= y^{5/6}z^{1/3}$$

89. $\sqrt[4]{\sqrt[3]{m}} = \sqrt[4]{m^{1/3}} = (m^{1/3})^{1/4} = m^{1/12}$

91. $\sqrt{\sqrt[3]{\sqrt[4]{x}}} = \sqrt{\sqrt[3]{x^{1/4}}}$
$$= \sqrt{(x^{1/4})^{1/3}}$$
$$= (x^{1/12})^{1/2}$$
$$= x^{1/24}$$

93. Use $T(D) = .07D^{3/2}$ with $D = 16$.

$$T(16) = .07(16)^{3/2} = .07(16^{1/2})^3$$
$$= .07(4)^3 = .07(64) = 4.48$$

To the nearest tenth of an hour, time T is 4.5 hours.

95. $3x^{-1/2} - 4x^{1/2} = x^{-1/2}(3 - 4x^{(1/2)-(-1/2)})$
$$= x^{-1/2}(3 - 4x^1)$$
$$= x^{-1/2}(3 - 4x)$$

Note that we are *subtracting* exponents.

96. $m^3 - 3m^{5/2} = m^{5/2}(m^{3-(5/2)} - 3)$
$$= m^{5/2}(m^{1/2} - 3)$$

97. $4t^{-1/2} + 7t^{3/2} = t^{-1/2}(4 + 7t^{(3/2)-(-1/2)})$
$$= t^{-1/2}(4 + 7t^2)$$

98. $8x^{2/3} + 5x^{-1/3} = x^{-1/3}(8x^{(2/3)-(-1/3)} + 5)$
$$= x^{-1/3}(8x^1 + 5)$$
$$= x^{-1/3}(8x + 5)$$

99. $4p - p^{3/4} = p^{3/4}(4p^{1-(3/4)} - 1)$
$$= p^{3/4}(4p^{1/4} - 1)$$

100. $2m^{1/8} - m^{5/8} = m^{1/8}(2 - m^{(5/8)-(1/8)})$
$$= m^{1/8}(2 - m^{4/8})$$
$$= m^{1/8}(2 - m^{1/2})$$

101. $9k^{-3/4} - 2k^{-1/4} = k^{-3/4}(9 - 2k^{(-1/4)-(-3/4)})$
$$= k^{-3/4}(9 - 2k^{2/4})$$
$$= k^{-3/4}(9 - 2k^{1/2})$$

102. $7z^{-5/8} - z^{-3/4} = z^{-3/4}(7z^{(-5/8)-(-3/4)} - 1)$
$$= z^{-3/4}(7z^{(-5/8)+(6/8)} - 1)$$
$$= z^{-3/4}(7z^{1/8} - 1)$$

Section 8.3

1. Does $2\sqrt{12} = \sqrt{48}$?
$$2\sqrt{12} = 2\sqrt{4 \cdot 3} = 2\sqrt{4} \cdot \sqrt{3} = 2 \cdot 2 \cdot \sqrt{3}$$
$$= 4\sqrt{3}$$
$$\sqrt{48} = \sqrt{16 \cdot 3} = \sqrt{16} \cdot \sqrt{3} = 4\sqrt{3}$$
The calculator approximation for each expression is 6.92820323. The statement is true.

3. Does $3\sqrt{8} = 2\sqrt{18}$?
$3\sqrt{8} = 3\sqrt{4 \cdot 2} = 3\sqrt{4}\sqrt{2} = 3 \cdot 2\sqrt{2} = 6\sqrt{2}$
$2\sqrt{18} = 2\sqrt{9 \cdot 2} = 2\sqrt{9}\sqrt{2} = 2 \cdot 3\sqrt{2} = 6\sqrt{2}$
The calculator approximation for each expression is 8.485281374. The statement is true.

5. **A.** $.5 = \frac{1}{2}$ so $\sqrt{.5} = \sqrt{\frac{1}{2}}$

B. $\frac{2}{4} = \frac{1}{2}$ so $\sqrt{\frac{2}{4}} = \sqrt{\frac{1}{2}}$

C. $\frac{3}{6} = \frac{1}{2}$ so $\sqrt{\frac{3}{6}} = \sqrt{\frac{1}{2}}$

D. $\frac{\sqrt{4}}{\sqrt{16}} = \sqrt{\frac{4}{16}} = \sqrt{\frac{1}{4}} \neq \sqrt{\frac{1}{2}}$

Choice **D** is not equal to $\sqrt{\frac{1}{2}}$.

7. $\sqrt{5} \cdot \sqrt{6} = \sqrt{5 \cdot 6} = \sqrt{30}$

9. $\sqrt{14} \cdot \sqrt{x} = \sqrt{14 \cdot x} = \sqrt{14x}$

11. $\sqrt{14} \cdot \sqrt{3pqr} = \sqrt{14 \cdot 3pqr} = \sqrt{42pqr}$

13. $\sqrt[3]{7x} \cdot \sqrt[3]{2y} = \sqrt[3]{7x \cdot 2y} = \sqrt[3]{14xy}$

15. $\sqrt[4]{11} \cdot \sqrt[4]{3} = \sqrt[4]{11 \cdot 3} = \sqrt[4]{33}$

17. $\sqrt[4]{2x} \cdot \sqrt[4]{3y^2} = \sqrt[4]{2x \cdot 3y^2} = \sqrt[4]{6xy^2}$

19. $\sqrt[3]{7} \cdot \sqrt[4]{3}$ cannot be multiplied using the product rule, because the indexes (3 and 4) are different.

21. To multiply two radical expressions with the same index, multiply the radicands and keep the same index. For example, $\sqrt[3]{3} \cdot \sqrt[3]{5} = \sqrt[3]{15}$.

23. $\sqrt{\frac{64}{121}} = \frac{\sqrt{64}}{\sqrt{121}} = \frac{8}{11}$

25. $\sqrt{\frac{3}{25}} = \frac{\sqrt{3}}{\sqrt{25}} = \frac{\sqrt{3}}{5}$

27. $\sqrt{\frac{x}{25}} = \frac{\sqrt{x}}{\sqrt{25}} = \frac{\sqrt{x}}{5}$

29. $\sqrt{\frac{p^6}{81}} = \frac{\sqrt{p^6}}{\sqrt{81}} = \frac{\sqrt{(p^3)^2}}{9} = \frac{p^3}{9}$

31. $\sqrt[3]{\frac{27}{64}} = \frac{\sqrt[3]{27}}{\sqrt[3]{64}} = \frac{3}{4}$

33. $\sqrt[3]{-\frac{r^2}{8}} = \frac{\sqrt[3]{r^2}}{\sqrt[3]{-8}} = \frac{\sqrt[3]{r^2}}{-2} = -\frac{\sqrt[3]{r^2}}{2}$

35. $-\sqrt[4]{\frac{81}{x^4}} = -\frac{\sqrt[4]{3^4}}{\sqrt[4]{x^4}} = -\frac{3}{x}$

37. $\sqrt[5]{\frac{1}{x^{15}}} = \frac{\sqrt[5]{1}}{\sqrt[5]{(x^3)^5}} = \frac{1}{x^3}$

39. $\sqrt{12} = \sqrt{4 \cdot 3} = \sqrt{4} \cdot \sqrt{3} = 2\sqrt{3}$

41. $\sqrt{288} = \sqrt{144 \cdot 2} = \sqrt{144} \cdot \sqrt{2} = 12\sqrt{2}$

43. $-\sqrt{32} = -\sqrt{16 \cdot 2} = -\sqrt{16} \cdot \sqrt{2} = -4\sqrt{2}$

45. $-\sqrt{28} = -\sqrt{4 \cdot 7} = -\sqrt{4} \cdot \sqrt{7} = -2\sqrt{7}$

47. $\sqrt{-300}$ is not a real number (even index, negative radicand).

49. $\sqrt[3]{128} = \sqrt[3]{64 \cdot 2} = \sqrt[3]{64} \cdot \sqrt[3]{2} = 4\sqrt[3]{2}$

51. $\sqrt[3]{-16} = \sqrt[3]{-8 \cdot 2} = \sqrt[3]{-8} \cdot \sqrt[3]{2} = -2\sqrt[3]{2}$

53. $\sqrt[3]{40} = \sqrt[3]{8 \cdot 5} = \sqrt[3]{8} \cdot \sqrt[3]{5} = 2\sqrt[3]{5}$

55. $-\sqrt[4]{512} = -\sqrt[4]{256 \cdot 2} = -\sqrt[4]{4^4} \cdot \sqrt[4]{2} = -4\sqrt[4]{2}$

57. $\sqrt[5]{64} = \sqrt[5]{32 \cdot 2} = \sqrt[5]{2^5} \cdot \sqrt[5]{2} = 2\sqrt[5]{2}$

59. His reasoning was incorrect. The radicand 14 must be written as a product of two factors (not a sum of two terms) where one of the two factors is a perfect cube.

61. $\sqrt{72k^2} = \sqrt{36k^2 \cdot 2} = \sqrt{36k^2} \cdot \sqrt{2} = 6k\sqrt{2}$

63. $\sqrt{144x^3y^9} = \sqrt{144x^2y^8 \cdot xy}$
$= \sqrt{(12xy^4)^2} \cdot \sqrt{xy}$
$= 12xy^4\sqrt{xy}$

65. $\sqrt{121x^6} = \sqrt{(11x^3)^2} = 11x^3$

67. $-\sqrt[3]{27t^{12}} = -\sqrt[3]{(3t^4)^3} = -3t^4$

69. $-\sqrt{100m^8z^4} = -\sqrt{(10m^4z^2)^2} = -10m^4z^2$

71. $-\sqrt[3]{-125a^6b^9c^{12}} = -\sqrt[3]{(-5a^2b^3c^4)^3}$
$= -(-5a^2b^3c^4)$
$= 5a^2b^3c^4$

73. $\sqrt[4]{\frac{1}{16}r^8t^{20}} = \sqrt[4]{\left(\frac{1}{2}r^2t^5\right)^4} = \frac{1}{2}r^2t^5$

75. $\sqrt{50x^3} = \sqrt{25x^2 \cdot 2x} = \sqrt{(5x)^2} \cdot \sqrt{2x}$
$= 5x\sqrt{2x}$

77. $-\sqrt{500r^{11}} = -\sqrt{100r^{10} \cdot 5r}$
$= -\sqrt{(10r^5)^2} \cdot \sqrt{5r} = -10r^5\sqrt{5r}$

79. $\sqrt{13x^7y^8} = \sqrt{(x^6y^8)(13x)}$
$= \sqrt{x^6y^8} \cdot \sqrt{13x}$
$= x^3y^4\sqrt{13x}$

81. $\sqrt[3]{8z^6w^9} = \sqrt[3]{(2z^2w^3)^3} = 2z^2w^3$

83. $\sqrt[3]{-16z^5t^7} = \sqrt[3]{(-2^3z^3t^6)(2z^2t)}$
$= -2zt^2\sqrt[3]{2z^2t}$

85. $\sqrt[4]{81x^{12}y^{16}} = \sqrt[4]{(3x^3y^4)^4} = 3x^3y^4$

87. $-\sqrt[4]{162r^{15}s^{10}} = -\sqrt[4]{81r^{12}s^8(2r^3s^2)}$
$= -\sqrt[4]{81r^{12}s^8} \cdot \sqrt[4]{2r^3s^2}$
$= -3r^3s^2\sqrt[4]{2r^3s^2}$

89. $\sqrt{\frac{y^{11}}{36}} = \frac{\sqrt{y^{11}}}{\sqrt{36}} = \frac{\sqrt{y^{10} \cdot y}}{6} = \frac{y^5\sqrt{y}}{6}$

91. $\sqrt[3]{\frac{x^{16}}{27}} = \frac{\sqrt[3]{x^{15} \cdot x}}{\sqrt[3]{27}} = \frac{x^5\sqrt[3]{x}}{3}$

93. $\sqrt[4]{48^2} = 48^{2/4} = 48^{1/2} = \sqrt{48} = \sqrt{16 \cdot 3}$
$= \sqrt{16} \cdot \sqrt{3} = 4\sqrt{3}$

95. $\sqrt[4]{25} = 25^{1/4} = (5^2)^{1/4} = 5^{2/4} = 5^{1/2} = \sqrt{5}$

97. $\sqrt[10]{x^{25}} = x^{25/10} = x^{5/2} = \sqrt{x^5}$
$= \sqrt{x^4 \cdot x} = x^2\sqrt{x}$

99. $\sqrt[3]{4} \cdot \sqrt{3}$
The least common index of 3 and 2 is 6. Write each radical as a sixth root.

$$\sqrt[3]{4} = 4^{1/3} = 4^{2/6} = \sqrt[6]{4^2} = \sqrt[6]{16}$$
$$\sqrt{3} = 3^{1/2} = 3^{3/6} = \sqrt[6]{3^3} = \sqrt[6]{27}$$

Therefore,

$$\sqrt[3]{4} \cdot \sqrt{3} = \sqrt[6]{16} \cdot \sqrt[6]{27}$$
$$= \sqrt[6]{16 \cdot 27} = \sqrt[6]{432}.$$

101. $\sqrt[4]{3} \cdot \sqrt[3]{4}$
The least common index of 4 and 3 is 12. Write each radical as a twelfth root.

$$\sqrt[4]{3} = 3^{1/4} = 3^{3/12} = \sqrt[12]{3^3} = \sqrt[12]{27}$$
$$\sqrt[3]{4} = 4^{1/3} = 4^{4/12} = \sqrt[12]{4^4} = \sqrt[12]{256}$$

Therefore,

$$\sqrt[4]{3} \cdot \sqrt[3]{4} = \sqrt[12]{27} \cdot \sqrt[12]{256}$$
$$= \sqrt[12]{27 \cdot 256} = \sqrt[12]{6912}.$$

103. $\sqrt{x} = x^{1/2} = x^{3/6} = \sqrt[6]{x^3}$
$\sqrt[3]{x} = x^{1/3} = x^{2/6} = \sqrt[6]{x^2}$

So $\sqrt{x} \cdot \sqrt[3]{x} = \sqrt[6]{x^3} \cdot \sqrt[6]{x^2}$
$= \sqrt[6]{x^3 \cdot x^2} = \sqrt[6]{x^5}.$

105. $\sqrt{48} = \sqrt{16 \cdot 3} = \sqrt{16}\sqrt{3} = 4\sqrt{3}$
The statement is true so the calculator should return a 1.

107. $\sqrt{5} = 5^{1/2} = 5^{3/6} = \sqrt[6]{5^3} = \sqrt[6]{125}$
$\sqrt[3]{4} = 4^{1/3} = 4^{2/6} = \sqrt[6]{4^2} = \sqrt[6]{16}$
So $\sqrt{5} \cdot \sqrt[3]{4} = \sqrt[6]{125 \cdot 16} = \sqrt[6]{2000}.$
The statement is true so the calculator should return a 1.

109. Substitute 3 for a and 4 for b in the Pythagorean formula to find c.

$$c^2 = a^2 + b^2$$
$$c = \sqrt{a^2 + b^2} = \sqrt{3^2 + 4^2}$$
$$= \sqrt{9 + 16} = \sqrt{25} = 5$$

The length of the hypotenuse is 5.

111. Substitute 12 for c and 4 for a in the Pythagorean formula to find b.

$$a^2 + b^2 = c^2$$
$$b = \sqrt{c^2 - a^2} = \sqrt{12^2 - 4^2}$$
$$= \sqrt{144 - 16} = \sqrt{128}$$
$$= \sqrt{64}\sqrt{2} = 8\sqrt{2}$$

The length of the unknown leg is $8\sqrt{2}$.

113. Use $D = \sqrt{L^2 + W^2 + H^2}$ with $L = 4$, $W = 2$, and $H = 3$ to find D.

$$D = \sqrt{4^2 + 2^2 + 3^2}$$
$$= \sqrt{16 + 4 + 9} = \sqrt{29} \approx 5.4$$

The length of the diagonal is $\sqrt{29}$ feet or about 5.4 feet.

115. Substitute 282 for E, 100 for R, 264 for L, and 120π for ω in the formula to find I.

$$I = \frac{E}{\sqrt{R^2 + \omega^2L^2}}$$
$$= \frac{282}{\sqrt{100^2 + (120\pi)^2(264)^2}}$$
$$\approx \frac{282}{99,526} \approx .003$$

117. Substitute 640 for k and 2 for I in the formula to find d.

$$d = \sqrt{\frac{k}{I}} = \sqrt{\frac{640}{2}} = \sqrt{320} = \sqrt{64 \cdot 5}$$
$$= \sqrt{64} \cdot \sqrt{5} = 8\sqrt{5} \approx 17.9$$

The illumination will be 2 foot-candles at $8\sqrt{5}$ feet or about 17.9 feet from the light source.

In Exercises 119–129, use the distance formula

$$d = \sqrt{(x_2 - x_1)^2 + (y_2 - y_1)^2}.$$

119. $(6, 13)$ and $(1, 1)$

$$\begin{aligned} d &= \sqrt{(6-1)^2 + (13-1)^2} \\ &= \sqrt{5^2 + 12^2} = \sqrt{25 + 144} \\ &= \sqrt{169} = 13 \end{aligned}$$

121. $(-6, 5)$ and $(3, -4)$

$$\begin{aligned} d &= \sqrt{(-6-3)^2 + [5-(-4)]^2} \\ &= \sqrt{(-9)^2 + (9)^2} = \sqrt{81 + 81} \\ &= \sqrt{162} = \sqrt{81} \cdot \sqrt{2} = 9\sqrt{2} \end{aligned}$$

123. $(-8, 2)$ and $(-4, 1)$

$$\begin{aligned} d &= \sqrt{[-8-(-4)]^2 + (2-1)^2} \\ &= \sqrt{(-4)^2 + 1^2} \\ &= \sqrt{16 + 1} = \sqrt{17} \end{aligned}$$

125. $(4.7, 2.3)$ and $(1.7, -1.7)$

$$\begin{aligned} d &= \sqrt{(4.7-1.7)^2 + [2.3-(-1.7)]^2} \\ &= \sqrt{3^2 + 4^2} = \sqrt{9 + 16} \\ &= \sqrt{25} = 5 \end{aligned}$$

127. $\left(\sqrt{2}, \sqrt{6}\right)$ and $\left(-2\sqrt{2}, 4\sqrt{6}\right)$

$$\begin{aligned} d &= \sqrt{\left[\sqrt{2} - \left(-2\sqrt{2}\right)\right]^2 + \left(\sqrt{6} - 4\sqrt{6}\right)^2} \\ &= \sqrt{\left(3\sqrt{2}\right)^2 + \left(-3\sqrt{6}\right)^2} \\ &= \sqrt{9 \cdot 2 + 9 \cdot 6} = \sqrt{18 + 54} \\ &= \sqrt{72} = \sqrt{36} \cdot \sqrt{2} = 6\sqrt{2} \end{aligned}$$

129. $(x + y, y)$ and $(x - y, x)$

$$\begin{aligned} d &= \sqrt{[(x+y)-(x-y)]^2 + (y-x)^2} \\ &= \sqrt{(2y)^2 + (y-x)^2} \\ &= \sqrt{4y^2 + y^2 - 2xy + x^2} \\ &= \sqrt{5y^2 - 2xy + x^2} \end{aligned}$$

131. Since $\sqrt{a} = a^{1/2}$, the distance formula

$$d = \sqrt{(x_2 - x_1)^2 + (y_2 - y_1)^2}$$

may be expressed as

$$d = \left[(x_2 - x_1)^2 + (y_2 - y_1)^2\right]^{1/2}.$$

133. To find the lengths of the three sides of the triangle, use the distance formula to find the distance between each pair of points. Then add the distances to find the perimeter.

$$\begin{aligned} P &= \sqrt{(-3-2)^2 + (-3-6)^2} \\ &\quad + \sqrt{(2-6)^2 + (6-2)^2} \\ &\quad + \sqrt{[6-(-3)]^2 + [2-(-3)]^2} \\ &= \sqrt{(-5)^2 + (-9)^2} + \sqrt{(-4)^2 + 4^2} \\ &\quad + \sqrt{9^2 + 5^2} \\ &= \sqrt{25 + 81} + \sqrt{16 + 16} + \sqrt{81 + 25} \\ &= \sqrt{106} + \sqrt{32} + \sqrt{106} \\ &= 2\sqrt{106} + \sqrt{16} \cdot \sqrt{2} \\ &= 2\sqrt{106} + 4\sqrt{2} \end{aligned}$$

Section 8.4

1. Only choice **B** has like radical factors, so it can be simplified without first simplifying the individual radical expressions.

$$3\sqrt{6} + 9\sqrt{6} = 12\sqrt{6}$$

3. $$\begin{aligned} \sqrt{64} + \sqrt[3]{125} + \sqrt[4]{16} &= \sqrt{8^2} + \sqrt[3]{5^3} + \sqrt[4]{2^4} \\ &= 8 + 5 + 2 = 15 \end{aligned}$$

This sum can be found easily since each radicand has a whole number power corresponding to the index of the radical; that is, each radical expression simplifies to a whole number.

5. Simplify each radical and subtract.

$$\sqrt{36} - \sqrt{100} = 6 - 10 = -4$$

7. $$\begin{aligned} &-2\sqrt{48} + 3\sqrt{75} \\ &= -2\sqrt{16 \cdot 3} + 3\sqrt{25 \cdot 3} \\ &= -2 \cdot 4\sqrt{3} + 3 \cdot 5\sqrt{3} \\ &= -8\sqrt{3} + 15\sqrt{3} = 7\sqrt{3} \end{aligned}$$

9. $$\begin{aligned} &\sqrt[3]{16} + 4\sqrt[3]{54} \\ &= \sqrt[3]{8 \cdot 2} + 4\sqrt[3]{27 \cdot 2} \\ &= \sqrt[3]{8}\sqrt[3]{2} + 4\sqrt[3]{27}\sqrt[3]{2} \\ &= 2\sqrt[3]{2} + 4 \cdot 3\sqrt[3]{2} \\ &= 2\sqrt[3]{2} + 12\sqrt[3]{2} = 14\sqrt[3]{2} \end{aligned}$$

11. $$\begin{aligned} &\sqrt[4]{32} + 3\sqrt[4]{2} \\ &= \sqrt[4]{16 \cdot 2} + 3\sqrt[4]{2} \\ &= \sqrt[4]{16}\sqrt[4]{2} + 3\sqrt[4]{2} \\ &= 2\sqrt[4]{2} + 3\sqrt[4]{2} = 5\sqrt[4]{2} \end{aligned}$$

13. $$\begin{aligned} &6\sqrt{18} - \sqrt{32} + 2\sqrt{50} \\ &= 6\sqrt{9 \cdot 2} - \sqrt{16 \cdot 2} + 2\sqrt{25 \cdot 2} \\ &= 6 \cdot 3\sqrt{2} - 4\sqrt{2} + 2 \cdot 5\sqrt{2} \\ &= 18\sqrt{2} - 4\sqrt{2} + 10\sqrt{2} \\ &= 24\sqrt{2} \end{aligned}$$

15. $5\sqrt{6} + 2\sqrt{10}$

The radicals differ and are already simplified, so the expression cannot be simplified further.

17. $2\sqrt{5}+3\sqrt{20}+4\sqrt{45}$
$= 2\sqrt{5}+3\sqrt{4\cdot 5}+4\sqrt{9\cdot 5}$
$= 2\sqrt{5}+3\cdot 2\sqrt{5}+4\cdot 3\sqrt{5}$
$= 2\sqrt{5}+6\sqrt{5}+12\sqrt{5}$
$= 20\sqrt{5}$

19. $8\sqrt{2x}-\sqrt{8x}+\sqrt{72x}$
$= 8\sqrt{2x}-\sqrt{4\cdot 2x}+\sqrt{36\cdot 2x}$
$= 8\sqrt{2x}-2\sqrt{2x}+6\sqrt{2x}$
$= 12\sqrt{2x}$

21. $3\sqrt{72m^2}-5\sqrt{32m^2}-3\sqrt{18m^2}$
$= 3\sqrt{36m^2\cdot 2}-5\sqrt{16m^2\cdot 2}-3\sqrt{9m^2\cdot 2}$
$= 3\cdot 6m\sqrt{2}-5\cdot 4m\sqrt{2}-3\cdot 3m\sqrt{2}$
$= 18m\sqrt{2}-20m\sqrt{2}-9m\sqrt{2}$
$= (18m-20m-9m)\sqrt{2} = -11m\sqrt{2}$

23. $-\sqrt[3]{54}+2\sqrt[3]{16} = -\sqrt[3]{27\cdot 2}+2\sqrt[3]{8\cdot 2}$
$= -3\sqrt[3]{2}+2\cdot 2\sqrt[3]{2}$
$= -3\sqrt[3]{2}+4\sqrt[3]{2}$
$= \sqrt[3]{2}$

25. $2\sqrt[3]{27x}-2\sqrt[3]{8x}$
$= 2\sqrt[3]{27\cdot x}-2\sqrt[3]{8\cdot x}$
$= 2\cdot 3\sqrt[3]{x}-2\cdot 2\sqrt[3]{x}$
$= 6\sqrt[3]{x}-4\sqrt[3]{x}$
$= 2\sqrt[3]{x}$

27. $\sqrt[3]{x^2y}-\sqrt[3]{8x^2y} = \sqrt[3]{x^2y}-\sqrt[3]{8}\sqrt[3]{x^2y}$
$= 1\sqrt[3]{x^2y}-2\sqrt[3]{x^2y}$
$= (1-2)\sqrt[3]{x^2y}$
$= -\sqrt[3]{x^2y}$

29. $3x\sqrt[3]{xy^2}-2\sqrt[3]{8x^4y^2}$
$= 3x\sqrt[3]{xy^2}-2\sqrt[3]{8x^3}\cdot\sqrt[3]{xy^2}$
$= 3x\sqrt[3]{xy^2}-2\cdot 2x\cdot\sqrt[3]{xy^2}$
$= (3x-4x)\sqrt[3]{xy^2} = -x\sqrt[3]{xy^2}$

31. $5\sqrt[4]{32}+3\sqrt[4]{162} = 5\sqrt[4]{16\cdot 2}+3\sqrt[4]{81\cdot 2}$
$= 5\cdot 2\sqrt[4]{2}+3\cdot 3\sqrt[4]{2}$
$= 10\sqrt[4]{2}+9\sqrt[4]{2}$
$= 19\sqrt[4]{2}$

33. $3\sqrt[4]{x^5y}-2x\sqrt[4]{xy}$
$= 3\sqrt[4]{x^4\cdot xy}-2x\sqrt[4]{xy}$
$= 3x\sqrt[4]{xy}-2x\sqrt[4]{xy}$
$= (3x-2x)\sqrt[4]{xy} = x\sqrt[4]{xy}$

35. $2\sqrt[4]{32a^3}+5\sqrt[4]{2a^3}$
$= 2\sqrt[4]{16}\cdot\sqrt[4]{2a^3}+5\sqrt[4]{2a^3}$
$= 2\cdot 2\cdot\sqrt[4]{2a^3}+5\sqrt[4]{2a^3}$
$= (4+5)\sqrt[4]{2a^3} = 9\sqrt[4]{2a^3}$

37. $\sqrt[3]{64xy^2}+\sqrt[3]{27x^4y^5}$
$= \sqrt[3]{64\cdot xy^2}+\sqrt[3]{27x^3y^3\cdot xy^2}$
$= 4\sqrt[3]{xy^2}+3xy\sqrt[3]{xy^2}$
$= (4+3xy)\sqrt[3]{xy^2}$

39. $\sqrt{8}-\dfrac{\sqrt{64}}{\sqrt{16}} = \sqrt{4\cdot 2}-\dfrac{8}{4}$
$= \sqrt{4}\cdot\sqrt{2}-2$
$= 2\sqrt{2}-2$

41. $\dfrac{2\sqrt{5}}{3}+\dfrac{\sqrt{5}}{6} = \dfrac{4\sqrt{5}}{6}+\dfrac{1\sqrt{5}}{6}$
$= \dfrac{4\sqrt{5}+1\sqrt{5}}{6}$
$= \dfrac{5\sqrt{5}}{6}$

43. $\sqrt{\dfrac{8}{9}}+\sqrt{\dfrac{18}{36}} = \dfrac{\sqrt{8}}{\sqrt{9}}+\dfrac{\sqrt{18}}{\sqrt{36}}$
$= \dfrac{\sqrt{4}\sqrt{2}}{3}+\dfrac{\sqrt{9}\sqrt{2}}{6}$
$= \dfrac{2\sqrt{2}}{3}+\dfrac{3\sqrt{2}}{6}$
$= \dfrac{4\sqrt{2}}{6}+\dfrac{3\sqrt{2}}{6}$
$= \dfrac{4\sqrt{2}+3\sqrt{2}}{6} = \dfrac{7\sqrt{2}}{6}$

45. $\dfrac{\sqrt{32}}{3}+\dfrac{2\sqrt{2}}{3}-\dfrac{\sqrt{2}}{\sqrt{9}}$
$= \dfrac{\sqrt{16}\sqrt{2}}{3}+\dfrac{2\sqrt{2}}{3}-\dfrac{\sqrt{2}}{3}$
$= \dfrac{4\sqrt{2}+2\sqrt{2}-\sqrt{2}}{3} = \dfrac{5\sqrt{2}}{3}$

47. $3\sqrt{\dfrac{50}{9}}+8\dfrac{\sqrt{2}}{\sqrt{8}} = 3\dfrac{\sqrt{50}}{\sqrt{9}}+8\dfrac{\sqrt{2}}{2\sqrt{2}}$
$= 3\cdot\dfrac{5\sqrt{2}}{3}+8\cdot\dfrac{1}{2}$
$= 5\sqrt{2}+4$

49. $\sqrt{\dfrac{25}{x^8}}-\sqrt{\dfrac{9}{x^6}} = \dfrac{\sqrt{25}}{\sqrt{x^8}}-\dfrac{\sqrt{9}}{\sqrt{x^6}}$
$= \dfrac{5}{x^4}-\dfrac{3}{x^3}$
$= \dfrac{5}{x^4}-\dfrac{3\cdot x}{x^3\cdot x}$ $\quad LCD = x^4$
$= \dfrac{5-3x}{x^4}$

51. $3\sqrt[3]{\frac{m^5}{27}} - 2m\sqrt[3]{\frac{m^2}{64}}$

$= \frac{3\sqrt[3]{m^5}}{\sqrt[3]{27}} - \frac{2m\sqrt[3]{m^2}}{\sqrt[3]{64}}$

$= \frac{3\sqrt[3]{m^3}\sqrt[3]{m^2}}{3} - \frac{2m\sqrt[3]{m^2}}{4}$

$= \frac{m\sqrt[3]{m^2}}{1} - \frac{m\sqrt[3]{m^2}}{2}$

$= \frac{2m\sqrt[3]{m^2} - m\sqrt[3]{m^2}}{2} = \frac{m\sqrt[3]{m^2}}{2}$

53. $3\sqrt[3]{\frac{2}{x^6}} - 4\sqrt[3]{\frac{5}{x^9}} = 3\frac{\sqrt[3]{2}}{\sqrt[3]{x^6}} - 4\frac{\sqrt[3]{5}}{\sqrt[3]{x^9}}$

$= 3\frac{\sqrt[3]{2}}{x^2} - 4\frac{\sqrt[3]{5}}{x^3}$

$= \frac{3 \cdot x \cdot \sqrt[3]{2}}{x^2 \cdot x} - \frac{4\sqrt[3]{5}}{x^3}$

$LCD = x^3$

$= \frac{3x\sqrt[3]{2} - 4\sqrt[3]{5}}{x^3}$

55. $3\sqrt{32} - 2\sqrt{8} \approx 11.3137085$

$8\sqrt{2} \approx 11.3137085$

Both calculator approximations are the same, supporting (but not proving) the truth of the statement.

57. Let $L = \sqrt{192} \approx 14$ and $W = \sqrt{48} \approx 7$.
The best estimate is choice **A**.

An estimate of the perimeter is
$2L + 2W = 2(14) + 2(7) = 42$ meters.

59. The perimeter, P, of a triangle is the sum of the measures of the sides.

$P = 3\sqrt{20} + 2\sqrt{45} + \sqrt{75}$

$= 3\sqrt{4 \cdot 5} + 2\sqrt{9 \cdot 5} + \sqrt{25 \cdot 3}$

$= 3 \cdot 2\sqrt{5} + 2 \cdot 3\sqrt{5} + 5\sqrt{3}$

$= 6\sqrt{5} + 6\sqrt{5} + 5\sqrt{3}$

$- 12\sqrt{5} + 5\sqrt{3}$

The perimeter is $\left(12\sqrt{5} + 5\sqrt{3}\right)$ inches.

61. To find the perimeter, add the lengths of the sides.

$4\sqrt{18} + \sqrt{108} + 2\sqrt{72} + 3\sqrt{12}$

$= 4\sqrt{9}\sqrt{2} + \sqrt{36}\sqrt{3} + 2\sqrt{36}\sqrt{2} + 3\sqrt{4}\sqrt{3}$

$= 4 \cdot 3\sqrt{2} + 6\sqrt{3} + 2 \cdot 6\sqrt{2} + 3 \cdot 2\sqrt{3}$

$= 12\sqrt{2} + 6\sqrt{3} + 12\sqrt{2} + 6\sqrt{3}$

$= 24\sqrt{2} + 12\sqrt{3}$

The perimeter is $\left(24\sqrt{2} + 12\sqrt{3}\right)$ inches.

Section 8.5

1. $(x + \sqrt{y})(x - \sqrt{y})$

$= x^2 - x\sqrt{y} + x\sqrt{y} - (\sqrt{y})^2$

$= x^2 - y$ **(E)**

3. $(\sqrt{x} + \sqrt{y})(\sqrt{x} - \sqrt{y})$

$= (\sqrt{x})^2 - (\sqrt{y})^2$

$= x - y$ **(A)**

5. $(\sqrt{x} - \sqrt{y})^2$

$= (\sqrt{x})^2 - 2\sqrt{x}\sqrt{y} + (\sqrt{y})^2$

$= x - 2\sqrt{xy} + y$ **(D)**

7. $\sqrt{6}\left(3 + \sqrt{2}\right) = 3\sqrt{6} + \sqrt{6} \cdot \sqrt{2}$

$= 3\sqrt{6} + \sqrt{12}$

$= 3\sqrt{6} + \sqrt{4} \cdot \sqrt{3}$

$= 3\sqrt{6} + 2\sqrt{3}$

9. $5\left(\sqrt{72} - \sqrt{8}\right) = 5\sqrt{72} - 5\sqrt{8}$

$- 5\sqrt{36} \cdot \sqrt{2} - 5\sqrt{4} \cdot \sqrt{2}$

$= 5 \cdot 6 \cdot \sqrt{2} - 5 \cdot 2 \cdot \sqrt{2}$

$= 30\sqrt{2} - 10\sqrt{2}$

$= 20\sqrt{2}$

11. $\left(\sqrt{7} + 3\right)\left(\sqrt{7} - 3\right) = \left(\sqrt{7}\right)^2 - 3^2$

$= 7 - 9 = -2$

13. $\left(\sqrt{2} - \sqrt{3}\right)\left(\sqrt{2} + \sqrt{3}\right) = \left(\sqrt{2}\right)^2 - \left(\sqrt{3}\right)^2$

$= 2 - 3 = -1$

15. $\left(\sqrt{8} - \sqrt{2}\right)\left(\sqrt{8} + \sqrt{2}\right) = \left(\sqrt{8}\right)^2 - \left(\sqrt{2}\right)^2$

$= 8 - 2 = 6$

17. $\left(\sqrt{2} + 1\right)\left(\sqrt{3} - 1\right)$

F O I L

$= \sqrt{2} \cdot \sqrt{3} - 1\sqrt{2} + 1\sqrt{3} - 1 \cdot 1$

$= \sqrt{6} - \sqrt{2} + \sqrt{3} - 1$

19. $\left(\sqrt{11} - \sqrt{7}\right)\left(\sqrt{2} + \sqrt{5}\right)$

F O I L

$= \sqrt{11} \cdot \sqrt{2} + \sqrt{11} \cdot \sqrt{5} - \sqrt{7} \cdot \sqrt{2} - \sqrt{7} \cdot \sqrt{5}$

$= \sqrt{22} + \sqrt{55} - \sqrt{14} - \sqrt{35}$

21. $\left(2\sqrt{3} + \sqrt{5}\right)\left(3\sqrt{3} - 2\sqrt{5}\right)$

$= \left(2\sqrt{3}\right)\left(3\sqrt{3}\right) + \left(2\sqrt{3}\right)\left(-2\sqrt{5}\right)$

$+ \left(\sqrt{5}\right)\left(3\sqrt{3}\right) + \left(\sqrt{5}\right)\left(-2\sqrt{5}\right)$

$= 2 \cdot 3 \cdot 3 - 2 \cdot 2\sqrt{3 \cdot 5} + 3\sqrt{5 \cdot 3} - 2 \cdot 5$

$= 18 - 4\sqrt{15} + 3\sqrt{15} - 10$

$= 8 - \sqrt{15}$

23. $\left(\sqrt{5}+2\right)^2 = \left(\sqrt{5}\right)^2 + 2\cdot\sqrt{5}\cdot 2 + 2^2$
$= 5 + 4\sqrt{5} + 4$
$= 9 + 4\sqrt{5}$

25. $\left(\sqrt{21}-\sqrt{5}\right)^2$
$= \left(\sqrt{21}\right)^2 - 2\cdot\sqrt{21}\cdot\sqrt{5} + \left(\sqrt{5}\right)^2$
$= 21 - 2\sqrt{105} + 5$
$= 26 - 2\sqrt{105}$

27. $\left(2+\sqrt[3]{6}\right)\left(2-\sqrt[3]{6}\right) = 2^2 - \left(\sqrt[3]{6}\right)^2$
$= 4 - \sqrt[3]{6}\cdot\sqrt[3]{6}$
$= 4 - \sqrt[3]{36}$

29. $\left(2+\sqrt[3]{2}\right)\left(4-2\sqrt[3]{2}+\sqrt[3]{4}\right)$
$= 2\cdot 4 - 2\cdot 2\sqrt[3]{2} + 2\sqrt[3]{4}$
$+ 4\sqrt[3]{2} - 2\sqrt[3]{2}\cdot\sqrt[3]{2} + \sqrt[3]{2}\cdot\sqrt[3]{4}$
$= 8 - 4\sqrt[3]{2} + 2\sqrt[3]{4} + 4\sqrt[3]{2} - 2\sqrt[3]{4} + \sqrt[3]{8}$
$= 8 + 2$
$= 10$

31. $\left(3\sqrt{x}-\sqrt{5}\right)\left(2\sqrt{x}+1\right)$
$= \left(3\sqrt{x}\right)\left(2\sqrt{x}\right) + 3\sqrt{x} - 2\sqrt{5}\sqrt{x} - \sqrt{5}$
$= 6x + 3\sqrt{x} - 2\sqrt{5x} - \sqrt{5}$

33. $\left(3\sqrt{r}-\sqrt{s}\right)\left(3\sqrt{r}+\sqrt{s}\right)$
$= 3\sqrt{r}\cdot 3\sqrt{r} + 3\sqrt{r}\cdot\sqrt{s}$
$- 3\sqrt{r}\cdot\sqrt{s} - \sqrt{s}\cdot\sqrt{s}$
$= 9r - s$

35. $\left(\sqrt[3]{2y}-5\right)\left(4\sqrt[3]{2y}+1\right)$
$= \sqrt[3]{2y}\cdot 4\sqrt[3]{2y} + \sqrt[3]{2y} - 5\cdot 4\sqrt[3]{2y} - 5$
$= 4\sqrt[3]{4y^2} + \sqrt[3]{2y} - 20\sqrt[3]{2y} - 5$
$= 4\sqrt[3]{4y^2} - 19\sqrt[3]{2y} - 5$

37. $\left(\sqrt{3x}+2\right)\left(\sqrt{3x}-2\right) = \left(\sqrt{3x}\right)^2 - 2^2$
$= 3x - 4$

39. $\left(2\sqrt{x}+\sqrt{y}\right)\left(2\sqrt{x}-\sqrt{y}\right)$
$= \left(2\sqrt{x}\right)^2 - \left(\sqrt{y}\right)^2$
$= 2^2\left(\sqrt{x}\right)^2 - y$
$= 4x - y$

41. $\left[\left(\sqrt{2}+\sqrt{3}\right)-\sqrt{6}\right]\left[\left(\sqrt{2}+\sqrt{3}\right)+\sqrt{6}\right]$
$= \left(\sqrt{2}+\sqrt{3}\right)^2 - \left(\sqrt{6}\right)^2$
$= \left[\left(\sqrt{2}\right)^2 + 2\sqrt{2}\sqrt{3} + \left(\sqrt{3}\right)^2\right] - 6$
$= \left(2 + 2\sqrt{6} + 3\right) - 6$
$= 2\sqrt{6} - 1$

43. $\dfrac{7}{\sqrt{7}} = \dfrac{7\cdot\sqrt{7}}{\sqrt{7}\cdot\sqrt{7}} = \dfrac{7\sqrt{7}}{7} = \sqrt{7}$

45. $\dfrac{15}{\sqrt{3}} = \dfrac{15\cdot\sqrt{3}}{\sqrt{3}\cdot\sqrt{3}} = \dfrac{15\sqrt{3}}{3} = 5\sqrt{3}$

47. $\dfrac{\sqrt{3}}{\sqrt{2}} = \dfrac{\sqrt{3}\cdot\sqrt{2}}{\sqrt{2}\cdot\sqrt{2}} = \dfrac{\sqrt{6}}{2}$

49. $\dfrac{9\sqrt{3}}{\sqrt{5}} = \dfrac{9\sqrt{3}\cdot\sqrt{5}}{\sqrt{5}\cdot\sqrt{5}} = \dfrac{9\sqrt{15}}{5}$

51. $\dfrac{-6}{\sqrt{18}} = \dfrac{-6}{\sqrt{9\cdot 2}} = \dfrac{-6}{3\sqrt{2}} = \dfrac{-2}{\sqrt{2}} = \dfrac{-2\cdot\sqrt{2}}{\sqrt{2}\cdot\sqrt{2}}$
$= \dfrac{-2\sqrt{2}}{2} = -\sqrt{2}$

53. $\sqrt{\dfrac{7}{2}} = \dfrac{\sqrt{7}}{\sqrt{2}} = \dfrac{\sqrt{7}\cdot\sqrt{2}}{\sqrt{2}\cdot\sqrt{2}} = \dfrac{\sqrt{14}}{2}$

55. $-\sqrt{\dfrac{7}{50}} = -\dfrac{\sqrt{7}}{\sqrt{25\cdot 2}} = -\dfrac{\sqrt{7}}{5\sqrt{2}}$
$= -\dfrac{\sqrt{7}\cdot\sqrt{2}}{5\sqrt{2}\cdot\sqrt{2}} = -\dfrac{\sqrt{14}}{5\cdot 2} = -\dfrac{\sqrt{14}}{10}$

57. $\sqrt{\dfrac{24}{x}} = \dfrac{\sqrt{24}}{\sqrt{x}} = \dfrac{\sqrt{4\cdot 6}}{\sqrt{x}} = \dfrac{2\sqrt{6}}{\sqrt{x}}$
$= \dfrac{2\sqrt{6}\cdot\sqrt{x}}{\sqrt{x}\cdot\sqrt{x}} = \dfrac{2\sqrt{6x}}{x}$

59. $\dfrac{-8\sqrt{3}}{\sqrt{k}} = \dfrac{-8\sqrt{3}\cdot\sqrt{k}}{\sqrt{k}\cdot\sqrt{k}} = \dfrac{-8\sqrt{3k}}{k}$

61. $-\sqrt{\dfrac{150m^5}{n^3}} = \dfrac{-\sqrt{150m^5}}{\sqrt{n^3}}$
$= \dfrac{-\sqrt{25m^4\cdot 6m}}{\sqrt{n^2\cdot n}} = \dfrac{-5m^2\sqrt{6m}}{n\sqrt{n}}$
$= \dfrac{-5m^2\sqrt{6m}\cdot\sqrt{n}}{n\sqrt{n}\cdot\sqrt{n}}$
$= \dfrac{-5m^2\sqrt{6mn}}{n\cdot n}$
$= \dfrac{-5m^2\sqrt{6mn}}{n^2}$

63. $\sqrt{\dfrac{288x^7}{y^9}} = \dfrac{\sqrt{288x^7}}{\sqrt{y^9}} = \dfrac{\sqrt{144x^6\cdot 2x}}{\sqrt{y^8\cdot y}}$
$= \dfrac{12x^3\sqrt{2x}}{y^4\sqrt{y}} = \dfrac{12x^3\sqrt{2x}\cdot\sqrt{y}}{y^4\sqrt{y}\cdot\sqrt{y}}$
$= \dfrac{12x^3\sqrt{2xy}}{y^4\cdot y} = \dfrac{12x^3\sqrt{2xy}}{y^5}$

65. $\dfrac{5\sqrt{2m}}{\sqrt{y^3}} = \dfrac{5\sqrt{2m}}{\sqrt{y^3}} \cdot \dfrac{\sqrt{y}}{\sqrt{y}}$
$= \dfrac{5\sqrt{2my}}{\sqrt{y^4}}$
$= \dfrac{5\sqrt{2my}}{y^2}$

67. $-\sqrt{\dfrac{48k^2}{z}} = -\dfrac{\sqrt{16k^2} \cdot \sqrt{3}}{\sqrt{z}} \cdot \dfrac{\sqrt{z}}{\sqrt{z}}$
$= -\dfrac{4k\sqrt{3z}}{z}$

69. $\sqrt[3]{\dfrac{2}{3}} = \dfrac{\sqrt[3]{2} \cdot \sqrt[3]{9}}{\sqrt[3]{3} \cdot \sqrt[3]{9}} = \dfrac{\sqrt[3]{18}}{\sqrt[3]{27}} = \dfrac{\sqrt[3]{18}}{3}$

71. $\sqrt[3]{\dfrac{4}{9}} = \dfrac{\sqrt[3]{4}}{\sqrt[3]{9}} = \dfrac{\sqrt[3]{4}}{\sqrt[3]{3^2}} = \dfrac{\sqrt[3]{4} \cdot \sqrt[3]{3}}{\sqrt[3]{3^2} \cdot \sqrt[3]{3}}$
$= \dfrac{\sqrt[3]{12}}{\sqrt[3]{3^3}} = \dfrac{\sqrt[3]{12}}{3}$

73. $\sqrt[3]{\dfrac{9}{32}} = \dfrac{\sqrt[3]{9}}{\sqrt[3]{32}} = \dfrac{\sqrt[3]{9}}{\sqrt[3]{8} \cdot \sqrt[3]{4}} \cdot \dfrac{\sqrt[3]{2}}{\sqrt[3]{2}}$
$= \dfrac{\sqrt[3]{9 \cdot 2}}{2 \cdot \sqrt[3]{8}}$
$= \dfrac{\sqrt[3]{18}}{4}$

75. $-\sqrt[3]{\dfrac{2p}{r^2}} = -\dfrac{\sqrt[3]{2p}}{\sqrt[3]{r^2}} = -\dfrac{\sqrt[3]{2p} \cdot \sqrt[3]{r}}{\sqrt[3]{r^2} \cdot \sqrt[3]{r}}$
$= -\dfrac{\sqrt[3]{2pr}}{\sqrt[3]{r^3}} = -\dfrac{\sqrt[3]{2pr}}{r}$

77. $\sqrt[3]{\dfrac{x^6}{y}} = \dfrac{\sqrt[3]{x^6}}{\sqrt[3]{y}} = \dfrac{x^2}{\sqrt[3]{y}} \cdot \dfrac{\sqrt[3]{y^2}}{\sqrt[3]{y^2}}$
$= \dfrac{x^2\sqrt[3]{y^2}}{\sqrt[3]{y^3}}$
$= \dfrac{x^2\sqrt[3]{y^2}}{y}$

79. $\sqrt[4]{\dfrac{16}{x}} = \dfrac{\sqrt[4]{16}}{\sqrt[4]{x}} = \dfrac{2}{\sqrt[4]{x}} = \dfrac{2 \cdot \sqrt[4]{x^3}}{\sqrt[4]{x} \cdot \sqrt[4]{x^3}}$
$= \dfrac{2\sqrt[4]{x^3}}{\sqrt[4]{x^4}} = \dfrac{2\sqrt[4]{x^3}}{x}$

81. $\sqrt[4]{\dfrac{2y}{z}} = \dfrac{\sqrt[4]{2y}}{\sqrt[4]{z}} \cdot \dfrac{\sqrt[4]{z^3}}{\sqrt[4]{z^3}}$
$= \dfrac{\sqrt[4]{2yz^3}}{z}$

83. To rationalize the denominator in $\dfrac{3}{4+\sqrt{5}}$, multiply both the numerator and the denominator by the conjugate of the denominator, $4-\sqrt{5}$.

Multiply both numerator and denominator by $4+\sqrt{5}$:

$$\frac{3}{4+\sqrt{5}} = \frac{3\left(4+\sqrt{5}\right)}{\left(4+\sqrt{5}\right)\left(4+\sqrt{5}\right)}$$
$$= \frac{3\left(4+\sqrt{5}\right)}{16+8\sqrt{5}+5}$$
$$= \frac{3\left(4+\sqrt{5}\right)}{21+8\sqrt{5}}.$$

The denominator is not rational because we did not multiply it by an expression that would eliminate the radical.

85. $\dfrac{3}{4+\sqrt{5}}$

Multiply both the numerator and denominator by the conjugate of the denominator, $4-\sqrt{5}$.

$$= \frac{3\left(4-\sqrt{5}\right)}{\left(4+\sqrt{5}\right)\left(4-\sqrt{5}\right)}$$
$$= \frac{3\left(4-\sqrt{5}\right)}{16-5} = \frac{3\left(4-\sqrt{5}\right)}{11}$$

87. $\dfrac{\sqrt{8}}{3-\sqrt{2}}$

Multiply both the numerator and denominator by the conjugate of the denominator, $3+\sqrt{2}$.

$$= \frac{\sqrt{4}\sqrt{2}\left(3+\sqrt{2}\right)}{\left(3-\sqrt{2}\right)\left(3+\sqrt{2}\right)}$$
$$= \frac{2\sqrt{2}\left(3+\sqrt{2}\right)}{3^2-2}$$
$$= \frac{2 \cdot 3\sqrt{2}+2 \cdot 2}{7}$$
$$= \frac{6\sqrt{2}+4}{7}$$

89. $\dfrac{2}{3\sqrt{5}+2\sqrt{3}}$

Multiply both the numerator and denominator by the conjugate of the denominator, $3\sqrt{5}-2\sqrt{3}$.

$$= \frac{2\left(3\sqrt{5}-2\sqrt{3}\right)}{\left(3\sqrt{5}+2\sqrt{3}\right)\left(3\sqrt{5}-2\sqrt{3}\right)}$$
$$= \frac{2\left(3\sqrt{5}-2\sqrt{3}\right)}{3^2 \cdot 5 - 2^2 \cdot 3}$$
$$= \frac{2\left(3\sqrt{5}-2\sqrt{3}\right)}{45-12} = \frac{2\left(3\sqrt{5}-2\sqrt{3}\right)}{33}$$

91. $\dfrac{\sqrt{2}-\sqrt{3}}{\sqrt{6}-\sqrt{5}}$

Multiply both the numerator and denominator by the conjugate of the denominator, $\sqrt{6}+\sqrt{5}$.

$$=\frac{\left(\sqrt{2}-\sqrt{3}\right)\left(\sqrt{6}+\sqrt{5}\right)}{\left(\sqrt{6}-\sqrt{5}\right)\left(\sqrt{6}+\sqrt{5}\right)}$$
$$=\frac{\sqrt{12}+\sqrt{10}-\sqrt{18}-\sqrt{15}}{\left(\sqrt{6}\right)^2-\left(\sqrt{5}\right)^2}$$
$$=\frac{\sqrt{4}\cdot\sqrt{3}+\sqrt{10}-\sqrt{9}\cdot\sqrt{2}-\sqrt{15}}{6-5}$$
$$=2\sqrt{3}+\sqrt{10}-3\sqrt{2}-\sqrt{15}$$

93. $\dfrac{m-4}{\sqrt{m}+2}$

Multiply both the numerator and denominator by the conjugate of the denominator, $\sqrt{m}-2$.

$$=\frac{(m-4)(\sqrt{m}-2)}{(\sqrt{m}+2)(\sqrt{m}-2)}$$
$$=\frac{(m-4)(\sqrt{m}-2)}{m-4}$$
$$=\sqrt{m}-2$$

95. $\dfrac{4\sqrt{x}}{\sqrt{x}-2\sqrt{y}}$

Multiply both the numerator and denominator by the conjugate of the denominator, $\sqrt{x}+2\sqrt{y}$.

$$=\frac{4\sqrt{x}(\sqrt{x}+2\sqrt{y})}{(\sqrt{x}-2\sqrt{y})(\sqrt{x}+2\sqrt{y})}$$
$$=\frac{4\sqrt{x}(\sqrt{x}+2\sqrt{y})}{x-4y}$$

97. $\dfrac{\sqrt{x}-\sqrt{y}}{\sqrt{x}+\sqrt{y}}$

Multiply both the numerator and denominator by the conjugate of the denominator, $\sqrt{x}-\sqrt{y}$.

$$=\frac{(\sqrt{x}-\sqrt{y})(\sqrt{x}-\sqrt{y})}{(\sqrt{x}+\sqrt{y})(\sqrt{x}-\sqrt{y})}$$
$$=\frac{(\sqrt{x})^2-2\sqrt{x}\sqrt{y}+(\sqrt{y})^2}{(\sqrt{x})^2-(\sqrt{y})^2}$$
$$=\frac{x-2\sqrt{xy}+y}{x-y}$$

99. $\dfrac{3\sqrt{x}}{\sqrt{x}-2\sqrt{y}}$

Multiply both the numerator and denominator by the conjugate of the denominator, $\sqrt{x}+2\sqrt{y}$.

$$=\frac{3\sqrt{x}(\sqrt{x}+2\sqrt{y})}{(\sqrt{x}-2\sqrt{y})(\sqrt{x}+2\sqrt{y})}$$
$$=\frac{3\sqrt{x}(\sqrt{x}+2\sqrt{y})}{x-4y}$$

101. Let $a=\dfrac{\sqrt{6}-\sqrt{2}}{4}$ and $b=\dfrac{\sqrt{2-\sqrt{3}}}{2}$.

Since the numerators and denominators are all positive, a and b are both positive.

$$a^2=\left(\frac{\sqrt{6}-\sqrt{2}}{4}\right)^2$$
$$=\frac{\left(\sqrt{6}-\sqrt{2}\right)\left(\sqrt{6}-\sqrt{2}\right)}{4^2}$$
$$=\frac{6-2\sqrt{12}+2}{16}=\frac{8-2\sqrt{12}}{16}$$
$$=\frac{8-2\sqrt{4\cdot 3}}{16}=\frac{8-4\sqrt{3}}{16}$$
$$=\frac{4\left(2-\sqrt{3}\right)}{4\cdot 4}=\frac{2-\sqrt{3}}{4}$$
$$b^2=\left(\frac{\sqrt{2-\sqrt{3}}}{2}\right)^2$$
$$=\frac{\sqrt{2-\sqrt{3}}\cdot\sqrt{2-\sqrt{3}}}{2^2}$$
$$=\frac{2-\sqrt{3}}{4}$$

Since $a^2=b^2$ and a and b are both positive, $a=b$.

103. $\dfrac{30-20\sqrt{6}}{10}=\dfrac{10\left(3-2\sqrt{6}\right)}{10}=3-2\sqrt{6}$

105. $\dfrac{3-3\sqrt{5}}{3}=\dfrac{3\left(1-\sqrt{5}\right)}{3}=1-\sqrt{5}$

107. $\dfrac{16-4\sqrt{8}}{12}=\dfrac{16-4\left(2\sqrt{2}\right)}{12}=\dfrac{16-8\sqrt{2}}{12}$

$$=\frac{4\left(4-2\sqrt{2}\right)}{4\cdot 3}=\frac{4-2\sqrt{2}}{3}$$

109. $\dfrac{6p+\sqrt{24p^3}}{3p}$

$$=\frac{6p+\sqrt{4p^2\cdot 6p}}{3p}=\frac{6p+2p\sqrt{6p}}{3p}$$
$$=\frac{p(6+2\sqrt{6p})}{3p}=\frac{6+2\sqrt{6p}}{3}$$

111. $\dfrac{1}{\sqrt{x+y}} = \dfrac{1}{\sqrt{x+y}} \cdot \dfrac{\sqrt{x+y}}{\sqrt{x+y}}$
$= \dfrac{1 \cdot \sqrt{x+y}}{\left(\sqrt{x+y}\right)^2}$
$= \dfrac{\sqrt{x+y}}{x+y}$

113. $\dfrac{p}{\sqrt{p+2}} = \dfrac{p}{\sqrt{p+2}} \cdot \dfrac{\sqrt{p+2}}{\sqrt{p+2}}$
$= \dfrac{p\sqrt{p+2}}{p+2}$

115. $\dfrac{1}{\sqrt{2}} \cdot \dfrac{\sqrt{3}}{2} - \dfrac{1}{\sqrt{2}} \cdot \dfrac{1}{2} = \dfrac{\sqrt{3}}{2\sqrt{2}} - \dfrac{1}{2\sqrt{2}}$
$= \dfrac{\sqrt{3}-1}{2\sqrt{2}} = \dfrac{\left(\sqrt{3}-1\right)\sqrt{2}}{\left(2\sqrt{2}\right)\sqrt{2}}$
$= \dfrac{\sqrt{6}-\sqrt{2}}{2 \cdot 2} = \dfrac{\sqrt{6}-\sqrt{2}}{4}$

Using a calculator,

$\dfrac{1}{\sqrt{2}} \cdot \dfrac{\sqrt{3}}{2} - \dfrac{1}{\sqrt{2}} \cdot \dfrac{1}{2} \approx .2588190451$ and

$\dfrac{\sqrt{6}-\sqrt{2}}{4} \approx .2588190451.$

117. $x - 7 = \left(\sqrt{x}\right)^2 - \left(\sqrt{7}\right)^2$
$= \left(\sqrt{x}+\sqrt{7}\right)\left(\sqrt{x}-\sqrt{7}\right)$

118. $x - 7 = \left(\sqrt[3]{x}\right)^3 - \left(\sqrt[3]{7}\right)^3$
$= \left(\sqrt[3]{x} - \sqrt[3]{7}\right)$
$\cdot \left[\left(\sqrt[3]{x}\right)^2 + \sqrt[3]{x} \cdot \sqrt[3]{7} + \left(\sqrt[3]{7}\right)^2\right]$
$= \left(\sqrt[3]{x} - \sqrt[3]{7}\right)\left(\sqrt[3]{x^2} + \sqrt[3]{7x} + \sqrt[3]{49}\right)$

119. $x + 7 = \left(\sqrt[3]{x}\right)^3 + \left(\sqrt[3]{7}\right)^3$
$= \left(\sqrt[3]{x} + \sqrt[3]{7}\right)$
$\cdot \left[\left(\sqrt[3]{x}\right)^2 - \sqrt[3]{x} \cdot \sqrt[3]{7} + \left(\sqrt[3]{7}\right)^2\right]$
$= \left(\sqrt[3]{x} + \sqrt[3]{7}\right)\left(\sqrt[3]{x^2} - \sqrt[3]{7x} + \sqrt[3]{49}\right)$

120. $\dfrac{x+3}{\sqrt{x}-\sqrt{7}} = \dfrac{(x+3)\left(\sqrt{x}+\sqrt{7}\right)}{\left(\sqrt{x}-\sqrt{7}\right)\left(\sqrt{x}+\sqrt{7}\right)}$
$= \dfrac{(x+3)\left(\sqrt{x}+\sqrt{7}\right)}{x-7}$

121. $\dfrac{x+3}{\sqrt[3]{x}-\sqrt[3]{7}}$
$= \dfrac{(x+3)\left(\sqrt[3]{x^2}+\sqrt[3]{7x}+\sqrt[3]{49}\right)}{\left(\sqrt[3]{x}-\sqrt[3]{7}\right)\left(\sqrt[3]{x^2}+\sqrt[3]{7x}+\sqrt[3]{49}\right)}$
$= \dfrac{(x+3)\left(\sqrt[3]{x^2}+\sqrt[3]{7x}+\sqrt[3]{49}\right)}{x-7}$

122. $\dfrac{x+3}{\sqrt[3]{x^2}-\sqrt[3]{7x}+\sqrt[3]{49}}$
$= \dfrac{(x+3)\left(\sqrt[3]{x}+\sqrt[3]{7}\right)}{\left(\sqrt[3]{x^2}-\sqrt[3]{7x}+\sqrt[3]{49}\right)\left(\sqrt[3]{x}+\sqrt[3]{7}\right)}$
$= \dfrac{(x+3)\left(\sqrt[3]{x}+\sqrt[3]{7}\right)}{x+7}$

123. $2 = 5 - 3 = \left(\sqrt[3]{5}\right)^3 - \left(\sqrt[3]{3}\right)^3$
$= \left(\sqrt[3]{5} - \sqrt[3]{3}\right)$
$\cdot \left[\left(\sqrt[3]{5}\right)^2 + \sqrt[3]{5} \cdot \sqrt[3]{3} + \left(\sqrt[3]{3}\right)^2\right]$
$= \left(\sqrt[3]{5} - \sqrt[3]{3}\right)\left(\sqrt[3]{25} + \sqrt[3]{15} + \sqrt[3]{9}\right)$

124. $\dfrac{2}{\sqrt[3]{5}-\sqrt[3]{3}}$
$= \dfrac{2\left(\sqrt[3]{25}+\sqrt[3]{15}+\sqrt[3]{9}\right)}{\left(\sqrt[3]{5}-\sqrt[3]{3}\right)\left(\sqrt[3]{25}+\sqrt[3]{15}+\sqrt[3]{9}\right)}$
$= \dfrac{2\left(\sqrt[3]{25}+\sqrt[3]{15}+\sqrt[3]{9}\right)}{5-3}$
$= \sqrt[3]{25}+\sqrt[3]{15}+\sqrt[3]{9}$

125. $\dfrac{6-\sqrt{2}}{4} = \dfrac{6-\sqrt{2}}{4} \cdot \dfrac{6+\sqrt{2}}{6+\sqrt{2}}$
$= \dfrac{6^2 - \left(\sqrt{2}\right)^2}{4\left(6+\sqrt{2}\right)}$
$= \dfrac{36-2}{4\left(6+\sqrt{2}\right)}$
$= \dfrac{34}{4\left(6+\sqrt{2}\right)}$
$= \dfrac{17 \cdot 2}{2 \cdot 2\left(6+\sqrt{2}\right)}$
$= \dfrac{17}{2\left(6+\sqrt{2}\right)}$

127. $$\frac{3\sqrt{a}+\sqrt{b}}{b}=\frac{3\sqrt{a}+\sqrt{b}}{b}\cdot\frac{3\sqrt{a}-\sqrt{b}}{3\sqrt{a}-\sqrt{b}}$$
$$=\frac{(3\sqrt{a})^2-(\sqrt{b})^2}{b(3\sqrt{a}-\sqrt{b})}$$
$$=\frac{9a-b}{b(3\sqrt{a}-\sqrt{b})}$$

Summary Exercises on Operations with Radicals

1. $$6\sqrt{10}-12\sqrt{10}=(6-12)\sqrt{10}$$
$$=-6\sqrt{10}$$

3. $$(1-\sqrt{3})(2+\sqrt{6})$$
F O I L
$$=2+\sqrt{6}-2\sqrt{3}-\sqrt{18}$$
$$=2+\sqrt{6}-2\sqrt{3}-\sqrt{9}\cdot\sqrt{2}$$
$$=2+\sqrt{6}-2\sqrt{3}-3\sqrt{2}$$

5. $$(3\sqrt{5}+2\sqrt{7})^2$$
$$=(3\sqrt{5})^2+2(3\sqrt{5})(2\sqrt{7})+(2\sqrt{7})^2$$
$$=9\cdot5+12\sqrt{35}+4\cdot7$$
$$=45+12\sqrt{35}+28$$
$$=73+12\sqrt{35}$$

7. $$\frac{8}{\sqrt{7}+\sqrt{5}}=\frac{8}{\sqrt{7}+\sqrt{5}}\cdot\frac{\sqrt{7}-\sqrt{5}}{\sqrt{7}-\sqrt{5}}$$
$$=\frac{8(\sqrt{7}-\sqrt{5})}{7-5}$$
$$=\frac{8(\sqrt{7}-\sqrt{5})}{2}$$
$$=4(\sqrt{7}-\sqrt{5})$$

9. $$\frac{1-\sqrt{2}}{1+\sqrt{2}}=\frac{1-\sqrt{2}}{1+\sqrt{2}}\cdot\frac{1-\sqrt{2}}{1-\sqrt{2}}$$
$$=\frac{(1-\sqrt{2})^2}{1^2-(\sqrt{2})^2}$$
$$=\frac{1-2\sqrt{2}+2}{1-2}$$
$$=\frac{3-2\sqrt{2}}{-1}$$
$$=-3+2\sqrt{2}$$

11. $$(\sqrt{5}+7)(\sqrt{5}-7)=(\sqrt{5})^2-7^2$$
$$=5-49$$
$$=-44$$

13. $$\sqrt[3]{8a^3b^5c^9}=\sqrt[3]{8a^3b^3c^9}\cdot\sqrt[3]{b^2}$$
$$=2abc^3\sqrt[3]{b^2}$$

15. $$\frac{3}{\sqrt{5}+2}=\frac{3}{\sqrt{5}+2}\cdot\frac{\sqrt{5}-2}{\sqrt{5}-2}$$
$$=\frac{3(\sqrt{5}-2)}{5-4}$$
$$=3(\sqrt{5}-2)$$

17. $$\frac{16\sqrt{3}}{5\sqrt{12}}=\frac{16\sqrt{3}}{5\cdot\sqrt{4}\cdot\sqrt{3}}$$
$$=\frac{16}{5\cdot2}$$
$$=\frac{8}{5}$$

19. $$\frac{-10}{\sqrt[3]{10}}=\frac{-10}{\sqrt[3]{10}}\cdot\frac{\sqrt[3]{100}}{\sqrt[3]{100}}$$
$$=\frac{-10\sqrt[3]{100}}{\sqrt[3]{1000}}$$
$$=\frac{-10\sqrt[3]{100}}{10}=-\sqrt[3]{100}$$

21. $$\sqrt{12x}-\sqrt{75x}=\sqrt{4}\cdot\sqrt{3x}-\sqrt{25}\cdot\sqrt{3x}$$
$$=2\sqrt{3x}-5\sqrt{3x}$$
$$=-3\sqrt{3x}$$

23. $$(\sqrt{74}-\sqrt{73})(\sqrt{74}+\sqrt{73})=74-73=1$$

25. $$-t^2\sqrt[4]{t}+3\sqrt[4]{t^9}-t\sqrt[4]{t^5}$$
$$=-t^2\sqrt[4]{t}+3\sqrt[4]{t^8}\cdot\sqrt[4]{t}-t\cdot\sqrt[4]{t^4}\cdot\sqrt[4]{t}$$
$$=-t^2\sqrt[4]{t}+3t^2\sqrt[4]{t}-t\cdot t\cdot\sqrt[4]{t}$$
$$=-t^2\sqrt[4]{t}+3t^2\sqrt[4]{t}-t^2\sqrt[4]{t}$$
$$=(-t^2+3t^2-t^2)\sqrt[4]{t}=t^2\sqrt[4]{t}$$

Section 8.6

1. $\sqrt{3x+18}=x$

(a) Check $x=6$.

$$\sqrt{3(6)+18}\stackrel{?}{=}6$$
$$\sqrt{18+18}\stackrel{?}{=}6$$
$$\sqrt{36}\stackrel{?}{=}6$$
$$6=6 \quad \textit{True}$$

The number 6 is a solution.

(b) Check $x=-3$.

$\sqrt{3(-3)+18}=-3$ is a false statement since the principal square root of a number is nonnegative. The number -3 is not a solution.

3. $\sqrt{x+2} = \sqrt{9x-2} - 2\sqrt{x-1}$

(a) Check $x = 2$.

$$\sqrt{2+2} \stackrel{?}{=} \sqrt{9(2)-2} - 2\sqrt{2-1}$$
$$\sqrt{4} \stackrel{?}{=} \sqrt{16} - 2\sqrt{1}$$
$$2 \stackrel{?}{=} 4 - 2$$
$$2 = 2 \quad \textit{True}$$

The number 2 is a solution.

(b) Check $x = 7$.

$$\sqrt{7+2} \stackrel{?}{=} \sqrt{9(7)-2} - 2\sqrt{7-1}$$
$$\sqrt{9} \stackrel{?}{=} \sqrt{61} - 2\sqrt{6}$$
$$3 = \sqrt{61} - 2\sqrt{6} \quad \textit{False}$$

The number 7 is not a solution.

5. $\sqrt{9} = 3$, not -3. There is no solution of $\sqrt{x} = -3$ since the value of a principal square root cannot equal a negative number.

In Exercises 7–33, check each solution in the original equation.

7. $\sqrt{r-2} = 3$

Square both sides.

$$\left(\sqrt{r-2}\right)^2 = 3^2$$
$$r - 2 = 9$$
$$r = 11$$

Check the potential solution, 11.

Check $r = 11$: $\sqrt{9} \stackrel{?}{=} 3$ *True*

Solution set: $\{11\}$

9. $\sqrt{6k-1} = 1$

Square both sides.

$$\left(\sqrt{6k-1}\right)^2 = 1^2$$
$$6k - 1 = 1$$
$$6k = 2$$
$$k = \frac{2}{6} = \frac{1}{3}$$

Check the potential solution, $\frac{1}{3}$.

Check $k = \frac{1}{3}$: $\sqrt{1} \stackrel{?}{=} 1$ *True*

Solution set: $\left\{\frac{1}{3}\right\}$

11. $\sqrt{4r+3} + 1 = 0$

Isolate the radical.

$$\sqrt{4r+3} = -1$$

This equation has no solution because $\sqrt{4r+3}$ cannot be negative.

Solution set: $\emptyset$

13. $\sqrt{3k+1} - 4 = 0$

Isolate the radical.

$$\sqrt{3k+1} = 4$$

Square both sides.

$$\left(\sqrt{3k+1}\right)^2 = 4^2$$
$$3k + 1 = 16$$
$$3k = 15$$
$$k = 5$$

Check $k = 5$: $\sqrt{16} - 4 \stackrel{?}{=} 0$ *True*

Solution set: $\{5\}$

15. $4 - \sqrt{x-2} = 0$

$$4 = \sqrt{x-2} \quad \textit{Isolate}$$
$$4^2 = \left(\sqrt{x-2}\right)^2 \quad \textit{Square}$$
$$16 = x - 2$$
$$18 = x$$

Check $x = 18$: $4 - \sqrt{16} \stackrel{?}{=} 0$ *True*

Solution set: $\{18\}$

17. $\sqrt{9a-4} = \sqrt{8a+1}$

$$\left(\sqrt{9a-4}\right)^2 = \left(\sqrt{8a+1}\right)^2 \quad \textit{Square}$$
$$9a - 4 = 8a + 1$$
$$a = 5$$

Check $a = 5$: $\sqrt{41} \stackrel{?}{=} \sqrt{41}$ *True*

Solution set: $\{5\}$

19. $2\sqrt{x} = \sqrt{3x+4}$

$$\left(2\sqrt{x}\right)^2 = \left(\sqrt{3x+4}\right)^2 \quad \textit{Square}$$
$$4x = 3x + 4$$
$$x = 4$$

Check $x = 4$: $4 \stackrel{?}{=} \sqrt{16}$ *True*

Solution set: $\{4\}$

21. $3\sqrt{z-1} = 2\sqrt{2z+2}$

$$\left(3\sqrt{z-1}\right)^2 = \left(2\sqrt{2z+2}\right)^2 \quad \textit{Square}$$
$$9(z-1) = 4(2z+2)$$
$$9z - 9 = 8z + 8$$
$$z = 17$$

Check $z = 17$: $3(4) \stackrel{?}{=} 2(6)$ *True*

Solution set: $\{17\}$

23. $k = \sqrt{k^2 + 4k - 20}$

$k^2 = \left(\sqrt{k^2 + 4k - 20}\right)^2$ *Square*

$k^2 = k^2 + 4k - 20$

$20 = 4k$

$5 = k$

Check $k = 5$: $5 \stackrel{?}{=} \sqrt{25}$ *True*

Solution set: $\{5\}$

25. $a = \sqrt{a^2 + 3a + 9}$

$a^2 = \left(\sqrt{a^2 + 3a + 9}\right)^2$ *Square*

$a^2 = a^2 + 3a + 9$

$-3a = 9$

$a = -3$

Substituting -3 for a makes the left side of the original equation negative, but the right side is nonnegative, so the solution set is $\emptyset$.

27. $\sqrt{9 - x} = x + 3$

$\left(\sqrt{9 - x}\right)^2 = (x + 3)^2$ *Square*

$9 - x = x^2 + 6x + 9$

$0 = x^2 + 7x$

$0 = x(x + 7)$

$x = 0$ or $x + 7 = 0$

$x = -7$

Check $x = -7$: $\sqrt{16} \stackrel{?}{=} -4$ *False*

Check $x = 0$: $\sqrt{9} \stackrel{?}{=} 3$ *True*

Solution set: $\{0\}$

29. $\sqrt{k^2 + 2k + 9} = k + 3$

$\left(\sqrt{k^2 + 2k + 9}\right)^2 = (k + 3)^2$ *Square*

$k^2 + 2k + 9 = k^2 + 6k + 9$

$0 = 4k$

$0 = k$

Check $k = 0$: $\sqrt{9} \stackrel{?}{=} 3$ *True*

Solution set: $\{0\}$

31. $\sqrt{r^2 + 9r + 3} = -r$

$\left(\sqrt{r^2 + 9r + 3}\right)^2 = (-r)^2$ *Square*

$r^2 + 9r + 3 = r^2$

$9r = -3$

$r = -\frac{1}{3}$

Check $r = -\frac{1}{3}$: $\sqrt{\frac{1}{9}} \stackrel{?}{=} \frac{1}{3}$ *True*

Solution set: $\left\{-\frac{1}{3}\right\}$

33. $\sqrt{z^2 + 12z - 4} + 4 - z = 0$

$\sqrt{z^2 + 12z - 4} = z - 4$ *Isolate*

$\left(\sqrt{z^2 + 12z - 4}\right)^2 = (z - 4)^2$ *Square*

$z^2 + 12z - 4 = z^2 - 8z + 16$

$20z = 20$

$z = 1$

Substituting 1 for z makes the left side of the original equation positive, but the right side is zero, so the solution set is $\emptyset$.

35. $\sqrt{3x + 4} = 8 - x$

$(8 - x)^2$ equals $64 - 16x + x^2$, not $64 + x^2$. The first step should be

$3x + 4 = 64 - 16x + x^2$

$0 = x^2 - 19x + 60$

$0 = (x - 4)(x - 15)$

$x - 4 = 0$ or $x - 15 = 0$

$x = 4$ or $x = 15$

Check $x = 4$: $\sqrt{16} \stackrel{?}{=} 8 - 4$ *True*

Check $x = 15$: $\sqrt{49} \stackrel{?}{=} 8 - 15$ *False*

Solution set: $\{4\}$

37. $\sqrt[3]{2x + 5} = \sqrt[3]{6x + 1}$

Cube both sides.

$\left(\sqrt[3]{2x + 5}\right)^3 = \left(\sqrt[3]{6x + 1}\right)^3$

$2x + 5 = 6x + 1$

$4 = 4x$

$1 = x$

Check $x = 1$: $\sqrt[3]{7} \stackrel{?}{=} \sqrt[3]{7}$ *True*

Solution set: $\{1\}$

39. $\sqrt[3]{a^2 + 5a + 1} = \sqrt[3]{a^2 + 4a}$

$\left(\sqrt[3]{a^2 + 5a + 1}\right)^3 = \left(\sqrt[3]{a^2 + 4a}\right)^3$ *Cube*

$a^2 + 5a + 1 = a^2 + 4a$

$a = -1$

Check $a = -1$: $\sqrt[3]{-3} \stackrel{?}{=} \sqrt[3]{-3}$ *True*

Solution set: $\{-1\}$

41. $\sqrt[3]{2m - 1} = \sqrt[3]{m + 13}$

$\left(\sqrt[3]{2m - 1}\right)^3 = \left(\sqrt[3]{m + 13}\right)^3$ *Cube*

$2m - 1 = m + 13$

$m = 14$

Check $m = 14$: $\sqrt[3]{27} \stackrel{?}{=} \sqrt[3]{27}$ *True*

Solution set: $\{14\}$

43. $\sqrt[4]{a+8} = \sqrt[4]{2a}$

Raise each side to the fourth power.

$$\left(\sqrt[4]{a+8}\right)^4 = \left(\sqrt[4]{2a}\right)^4$$
$$a+8 = 2a$$
$$8 = a$$

Check $a = 8$: $\sqrt[4]{16} \stackrel{?}{=} \sqrt[4]{16}$ *True*

Solution set: $\{8\}$

45. $\sqrt[3]{x-8} + 2 = 0$

$$\sqrt[3]{x-8} = -2 \quad \textit{Isolate}$$
$$\left(\sqrt[3]{x-8}\right)^3 = (-2)^3 \quad \textit{Cube}$$
$$x - 8 = -8$$
$$x = 0$$

Check $x = 0$: $\sqrt[3]{-8} + 2 \stackrel{?}{=} 0$ *True*

Solution set: $\{0\}$

47. $\sqrt[4]{2k-5} + 4 = 0$

$$\sqrt[4]{2k-5} = -4 \quad \textit{Isolate}$$

This equation has no solution because $\sqrt[4]{2k-5}$ cannot be negative.

Solution set: $\emptyset$

49. $\sqrt{k+2} - \sqrt{k-3} = 1$

Get one radical on each side of the equals sign.

$$\sqrt{k+2} = 1 + \sqrt{k-3}$$
$$\left(\sqrt{k+2}\right)^2 = \left(1 + \sqrt{k-3}\right)^2 \quad \textit{Square}$$
$$k + 2 = 1 + 2\sqrt{k-3} + k - 3$$
$$4 = 2\sqrt{k-3} \quad \textit{Isolate}$$
$$2 = \sqrt{k-3} \quad \textit{Divide by 2.}$$
$$2^2 = \left(\sqrt{k-3}\right)^2 \quad \textit{Square again}$$
$$4 = k - 3$$
$$7 = k$$

Check $k = 7$: $\sqrt{9} - \sqrt{4} \stackrel{?}{=} 1$ *True*

Solution set: $\{7\}$

51. $\sqrt{2r+11} - \sqrt{5r+1} = -1$

Get one radical on each side of the equals sign.

$$\sqrt{2r+11} = -1 + \sqrt{5r+1}$$

Square both sides.

$$\left(\sqrt{2r+11}\right)^2 = \left(-1 + \sqrt{5r+1}\right)^2$$
$$2r + 11 = 1 - 2\sqrt{5r+1} + 5r + 1$$

Isolate the remaining radical.

$$2\sqrt{5r+1} = 3r - 9$$

Square both sides again.

$$\left(2\sqrt{5r+1}\right)^2 = (3r-9)^2$$
$$4(5r+1) = 9r^2 - 54r + 81$$
$$20r + 4 = 9r^2 - 54r + 81$$
$$0 = 9r^2 - 74r + 77$$
$$0 = (9r - 11)(r - 7)$$

$9r - 11 = 0$ or $r - 7 = 0$

$r = \frac{11}{9}$ $\qquad$ $r = 7$

Check $r = \frac{11}{9}$: $\frac{11}{3} - \frac{8}{3} \stackrel{?}{=} -1$ *False*

Check $r = 7$: $5 - 6 \stackrel{?}{=} -1$ *True*

Solution set: $\{7\}$

53. $\sqrt{3p+4} - \sqrt{2p-4} = 2$

Get one radical on each side of the equals sign.

$$\sqrt{3p+4} = 2 + \sqrt{2p-4}$$

Square both sides.

$$\left(\sqrt{3p+4}\right)^2 = \left(2 + \sqrt{2p-4}\right)^2$$
$$3p + 4 = 4 + 4\sqrt{2p-4} + 2p - 4$$

Isolate the remaining radical.

$$p + 4 = 4\sqrt{2p-4}$$

Square both sides again.

$$(p+4)^2 = \left(4\sqrt{2p-4}\right)^2$$
$$p^2 + 8p + 16 = 16(2p-4)$$
$$p^2 + 8p + 16 = 32p - 64$$
$$p^2 - 24p + 80 = 0$$
$$(p-4)(p-20) = 0$$

$p - 4 = 0$ or $p - 20 = 0$

$p = 4$ $\qquad$ $p = 20$

Check $p = 4$: $\sqrt{16} - \sqrt{4} \stackrel{?}{=} 2$ *True*

Check $p = 20$: $\sqrt{64} - \sqrt{36} \stackrel{?}{=} 2$ *True*

Solution set: $\{4, 20\}$

55. $\sqrt{3-3p} - 3 = \sqrt{3p+2}$

Square both sides.

$$\left(\sqrt{3-3p} - 3\right)^2 = \left(\sqrt{3p+2}\right)^2$$
$$3 - 3p - 6\sqrt{3-3p} + 9 = 3p + 2$$

Isolate the remaining radical.

$$-6\sqrt{3-3p} = 6p - 10$$
$$-3\sqrt{3-3p} = 3p - 5$$

Square both sides again.

$$\left(-3\sqrt{3-3p}\right)^2 = (3p-5)^2$$
$$9(3-3p) = 9p^2 - 30p + 25$$
$$27 - 27p = 9p^2 - 30p + 25$$
$$0 = 9p^2 - 3p - 2$$
$$0 = (3p+1)(3p-2)$$

continued

$$3p+1=0 \quad \text{or} \quad 3p-2=0$$
$$p=-\frac{1}{3} \qquad\qquad p=\frac{2}{3}$$

Check $p=-\frac{1}{3}$: $\quad \sqrt{4}-3 \stackrel{?}{=} \sqrt{1}$ *False*

Check $p=\frac{2}{3}$: $\quad \sqrt{1}-3 \stackrel{?}{=} \sqrt{4}$ *False*

Solution set: $\emptyset$

57.
$$\begin{aligned} \sqrt{2\sqrt{x+11}} &= \sqrt{4x+2} & \\ 2\sqrt{x+11} &= 4x+2 & \textit{Square} \\ \left(2\sqrt{x+11}\right)^2 &= (4x+2)^2 & \textit{Square again} \\ 4(x+11) &= 16x^2+16x+4 & \\ 4x+44 &= 16x^2+16x+4 & \\ 0 &= 16x^2+12x-40 & \\ 0 &= 4x^2+3x-10 & \\ 0 &= (x+2)(4x-5) & \end{aligned}$$

$$x+2=0 \quad \text{or} \quad 4x-5=0$$
$$x=-2 \qquad\qquad x=\frac{5}{4}$$

Check $x=-2$: $\quad \sqrt{6} \stackrel{?}{=} \sqrt{-6}$ *False*

Check $x=\frac{5}{4}$: $\quad \sqrt{7} \stackrel{?}{=} \sqrt{7}$ *True*

Solution set: $\left\{\frac{5}{4}\right\}$

59. Write $\sqrt{x+3}=\sqrt[3]{5+4x}$ as

$$(x+3)^{1/2}=(5+4x)^{1/3}.$$

We wish to raise both sides to the smallest power n that will eliminate the radicals.

$$\begin{aligned} \left[(x+3)^{1/2}\right]^n &= \left[(5+4x)^{1/3}\right]^n \\ (x+3)^{n/2} &= (5+4x)^{n/3} \end{aligned}$$

To eliminate radicals, $n/2$ and $n/3$ must be natural numbers.
So we are looking for the lowest common multiple of 2 and 3, or the least common denominator of $1/2$ and $1/3$. That number is 6.
Test 6 as follows.

$$\begin{aligned} \left[(x+3)^{1/2}\right]^6 &= \left[(5+4x)^{1/3}\right]^6 \\ (x+3)^3 &= (5+4x)^2 \end{aligned}$$

The radicals are eliminated.

61. Graph the functions

$$Y_1=\sqrt{3-3x} \quad \text{and} \quad Y_2=3+\sqrt{3x+2}.$$

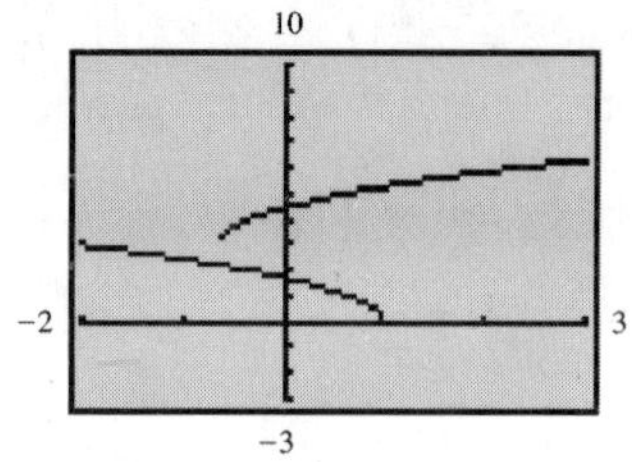

The graphs do not intersect, so the solution set is $\emptyset$.
To find the domain of

$$y=\sqrt{3-3x}-3-\sqrt{3x+2}$$

we must have

$$\begin{aligned} 3-3x &\ge 0 & \text{and} \quad 3x+2 &\ge 0 \\ -3x &\ge -3 & 3x &\ge -2 \\ x &\le 1 & \text{and} \quad x &\ge -\frac{2}{3}. \end{aligned}$$

The domain is $\left[-\frac{2}{3}, 1\right]$. This is the region where the graphs of Y_1 and Y_2 vertically overlap.

63.
$$\begin{aligned} (2x-9)^{1/2} &= 2+(x-8)^{1/2} \\ \sqrt{2x-9} &= 2+\sqrt{x-8} \\ \left(\sqrt{2x-9}\right)^2 &= \left(2+\sqrt{x-8}\right)^2 \\ 2x-9 &= 4+4\sqrt{x-8}+x-8 \\ x-5 &= 4\sqrt{x-8} \\ (x-5)^2 &= \left(4\sqrt{x-8}\right)^2 \\ x^2-10x+25 &= 16(x-8) \\ x^2-10x+25 &= 16x-128 \\ x^2-26x+153 &= 0 \\ (x-9)(x-17) &= 0 \end{aligned}$$

Check $x=9$: $\quad 9^{1/2} \stackrel{?}{=} 2+1^{1/2}$

$$3 \stackrel{?}{=} 2+1 \quad \textit{True}$$

Check $x=17$: $\quad 25^{1/2} \stackrel{?}{=} 2+9^{1/2}$

$$5 \stackrel{?}{=} 2+3 \quad \textit{True}$$

Solution set: $\{9, 17\}$

65.
$$\begin{aligned} (2w-1)^{2/3}-w^{1/3} &= 0 \\ \sqrt[3]{(2w-1)^2} &= \sqrt[3]{w} \\ \left[\sqrt[3]{(2w-1)^2}\right]^3 &= \left(\sqrt[3]{w}\right)^3 \\ (2w-1)^2 &= w \\ 4w^2-4w+1 &= w \\ 4w^2-5w+1 &= 0 \\ (4w-1)(w-1) &= 0 \end{aligned}$$

$$4w - 1 = 0 \quad \text{or} \quad w - 1 = 0$$
$$w = \frac{1}{4} \quad \text{or} \quad w = 1$$

Check $w = \frac{1}{4}$: $\left(-\frac{1}{2}\right)^{2/3} - \left(\frac{1}{4}\right)^{1/3} \stackrel{?}{=} 0$

$\left(\frac{1}{4}\right)^{1/3} - \left(\frac{1}{4}\right)^{1/3} \stackrel{?}{=} 0$ *True*

Check $w = 1$: $1 - 1 \stackrel{?}{=} 0$ *True*

Solution set: $\left\{\frac{1}{4}, 1\right\}$

67. Solve $V = \sqrt{\frac{2K}{m}}$ for K.

$(V)^2 = \left(\sqrt{\frac{2K}{m}}\right)^2$ *Square*

$V^2 = \frac{2K}{m}$ *Multiply by* $\frac{m}{2}$.

$\frac{V^2 m}{2} = K$

69. Solve $f = \frac{1}{2\pi\sqrt{LC}}$ for L.

$$2\pi f\sqrt{LC} = 1$$
$$\left(2\pi f\sqrt{LC}\right)^2 = 1^2$$
$$4\pi^2 f^2 LC = 1$$
$$L = \frac{1}{4\pi^2 f^2 C}$$

71. **(a)** Solve $N = \frac{1}{2\pi}\sqrt{\frac{a}{r}}$ for r.

$$2\pi N = \sqrt{\frac{a}{r}}$$
$$(2\pi N)^2 = \left(\sqrt{\frac{a}{r}}\right)^2$$
$$4\pi^2 N^2 = \frac{a}{r}$$
$$4\pi^2 N^2 r = a$$
$$r = \frac{a}{4\pi^2 N^2}$$

(b) $r = \frac{9.8}{4\pi^2(.063)^2}$

≈ 62.5 meters

(c) $r = \frac{9.8}{4\pi^2(.04)^2}$

≈ 155.1 meters

73. From the figure, the timber grown in each year shown is as follows:

1920: About 7 billion ft^3
1952: About 14 billion ft^3
1976: About 22.5 billion ft^3
1991: About 22.5 billion ft^3.

75. From the figure, the timber harvested in each year shown is as follows:

1920: About 12 billion ft^3
1952: About 12 billion ft^3
1976: About 14 billion ft^3
1991: About 17.5 billion ft^3.

77. The values obtained from the equations are fairly close to the estimates obtained from the figure with the exception of the year 1920. In 1920 the equation yielded 6 billion ft^3 while the estimate was about 12 billion ft^3.

Section 8.7

1. $\sqrt{-1} = i$

3. $i^2 = -1$

5. $\frac{1}{i} = \frac{1}{i} \cdot \frac{-i}{-i}$ *$-i$ is the conjugate of i*

$= \frac{-i}{-i^2} = \frac{i}{i^2} = \frac{i}{-1} = -i$

7. $\sqrt{-169} = i\sqrt{169} = 13i$

9. $-\sqrt{-144} = -i\sqrt{144} = -12i$

11. $\sqrt{-5} = i\sqrt{5}$

13. $\sqrt{-48} = i\sqrt{48} = i\sqrt{16 \cdot 3} = 4i\sqrt{3}$

15. $\sqrt{-7} \cdot \sqrt{-15} = i\sqrt{7} \cdot i\sqrt{15} = i^2\sqrt{7 \cdot 15}$

$= -1\sqrt{105} = -\sqrt{105}$

17. $\sqrt{-4} \cdot \sqrt{-25} = i\sqrt{4} \cdot i\sqrt{25} = 2i \cdot 5i = 10i^2$

$= 10(-1) = -10$

19. $\sqrt{-3} \cdot \sqrt{11} = i\sqrt{3} \cdot \sqrt{11}$

$= i\sqrt{33}$

21. $\frac{\sqrt{-300}}{\sqrt{-100}} = \frac{i\sqrt{300}}{i\sqrt{100}} = \sqrt{\frac{300}{100}} = \sqrt{3}$

23. $\frac{\sqrt{-75}}{\sqrt{3}} = \frac{i\sqrt{75}}{\sqrt{3}} = i\sqrt{\frac{75}{3}} = i\sqrt{25} = 5i$

25. **(a)** Since any real number can be written as $a + bi$, where $b = 0$, every real number is also a complex number.

(b) Not every complex number is a real number. For example, any number $a + bi$, $b \neq 0$, such as $3 + 7i$, is an imaginary number, which is a complex number that is not a real number.

27. $(3 + 2i) + (-4 + 5i)$

$= [(3 + (-4)] + (2 + 5)i$

$= -1 + 7i$

29. $(5 - i) + (-5 + i)$
$= (5 - 5) + (-1 + 1)i$
$= 0$

31. $(4 + i) - (-3 - 2i)$
$= [(4 - (-3)] + [(1 - (-2)]i$
$= 7 + 3i$

33. $(-3 - 4i) - (-1 - 4i)$
$= [-3 - (-1)] + [-4 - (-4)]i$
$= -2$

35. $(-4 + 11i) + (-2 - 4i) + (7 + 6i)$
$= (-4 - 2 + 7) + (11 - 4 + 6)i$
$= 1 + 13i$

37. $[(7 + 3i) - (4 - 2i)] + (3 + i)$
Work inside the brackets first.
$= [(7 - 4) + (3 + 2)i] + (3 + i)$
$= (3 + 5i) + (3 + i)$
$= (3 + 3) + (5 + 1)i$
$= 6 + 6i$

39. If $a - c = b$, then $b + c = a$.
So, $(4 + 2i) - (3 + i) = 1 + i$ implies that

$$(1 + i) + (3 + i) = \underline{4 + 2i}.$$

41. $(3i)(27i) = 81i^2 = 81(-1) = -81$

43. $(-8i)(-2i) = 16i^2 = 16(-1) = -16$

45. $5i(-6 + 2i) = (5i)(-6) + (5i)(2i)$
$= -30i + 10i^2$
$= -30i + 10(-1)$
$= -10 - 30i$

47. $(4 + 3i)(1 - 2i)$

F O I L

$= (4)(1) + 4(-2i) + (3i)(1) + (3i)(-2i)$
$= 4 - 8i + 3i - 6i^2$
$= 4 - 5i - 6(-1)$
$= 4 - 5i + 6 = 10 - 5i$

49. $(4 + 5i)^2 = 4^2 + 2(4)(5i) + (5i)^2$
$= 16 + 40i + 25i^2$
$= 16 + 40i + 25(-1)$
$= 16 + 40i - 25$
$= -9 + 40i$

51. $2i(-4 - i)^2$
$= 2i\left[(-4)^2 - 2(-4)(i) + i^2\right]$
$= 2i(16 + 8i - 1)$
$= 2i(15 + 8i) = 30i + 16i^2$
$= 30i + 16(-1) = -16 + 30i$

53. $(12 + 3i)(12 - 3i)$
$= 12^2 - (3i)^2 = 144 - 9i^2$
$= 144 - 9(-1) = 144 + 9 = 153$

55. $(4 + 9i)(4 - 9i)$
$= 4^2 - (9i)^2 = 16 - 81i^2$
$= 16 - 81(-1) = 16 + 81 = 97$

57. The conjugate of $a + bi$ is $a - bi$.

59. $\dfrac{2}{1 - i}$
Multiply the numerator and the denominator by the conjugate of the denominator, $1 + i$.
$$= \frac{2(1 + i)}{(1 - i)(1 + i)} = \frac{2(1 + i)}{1^2 - i^2}$$
$$= \frac{2(1 + i)}{1 - (-1)} = \frac{2(1 + i)}{2} = 1 + i$$

61. $\dfrac{-7 + 4i}{3 + 2i}$
Multiply the numerator and the denominator by the conjugate of the denominator, $3 - 2i$.
$$= \frac{(-7 + 4i)(3 - 2i)}{(3 + 2i)(3 - 2i)}$$
In the denominator, we make use of the fact that $(a + bi)(a - bi) = a^2 + b^2$.
$$= \frac{-21 + 14i + 12i + 8}{3^2 + 2^2}$$
$$= \frac{-13 + 26i}{13} = \frac{13(-1 + 2i)}{13} = -1 + 2i$$

63. $\dfrac{8i}{2 + 2i}$
Write in lowest terms.
$$= \frac{2 \cdot 4i}{2(1 + i)} = \frac{4i}{1 + i}$$
Multiply the numerator and the denominator by the conjugate of the denominator, $1 - i$.
$$= \frac{4i(1 - i)}{(1 + i)(1 - i)} = \frac{4(i - i^2)}{1^2 + 1^2}$$
$$= \frac{4(i + 1)}{2} = 2(i + 1) = 2 + 2i$$

65. $\dfrac{2 - 3i}{2 + 3i}$
Multiply the numerator and the denominator by the conjugate of the denominator, $2 - 3i$.
$$= \frac{(2 - 3i)(2 - 3i)}{(2 + 3i)(2 - 3i)} = \frac{2^2 - 2(2)(3i) + (3i)^2}{2^2 + 3^2}$$
$$= \frac{4 - 12i + 9i^2}{4 + 9} = \frac{4 - 12i - 9}{13}$$
$$= \frac{-5 - 12i}{13} = -\frac{5}{13} - \frac{12}{13}i$$

67. $\dfrac{3+i}{i}$

Multiply the numerator and the denominator by the conjugate of the denominator, $-i$.

$$= \frac{(3+i)(-i)}{i(-i)} = \frac{-3i-i^2}{-i^2}$$
$$= \frac{-3i-(-1)}{-(-1)} = \frac{-3i+1}{1}$$
$$= 1-3i$$

69. $(8-5i)+2(11+6i)-(1+i)^2$
$$= 8-5i+22+12i-(1+2i-1)$$
$$= 30+7i-(2i)$$
$$= 30+5i$$

71. The reciprocal of $5-4i$ is $\dfrac{1}{5-4i}$.

$$\frac{1}{5-4i} = \frac{1\cdot(5+4i)}{(5-4i)(5+4i)}$$
$$= \frac{5+4i}{25-16i^2} = \frac{5+4i}{25-16(-1)}$$
$$= \frac{5+4i}{25+16} = \frac{5+4i}{41} = \frac{5}{41}+\frac{4}{41}i$$

73. $i^{18} = i^{16}\cdot i^2 = \left(i^4\right)^4\cdot i^2$
$$= 1^4\cdot(-1) = 1\cdot(-1) = -1$$

75. $i^{89} = i^{88}\cdot i = \left(i^4\right)^{22}\cdot i = 1^{22}\cdot i$
$$= 1\cdot i = i$$

77. $i^{38} = i^{36}\cdot i^2$
$$= \left(i^4\right)^9\cdot i^2 = 1^9(-1) = -1$$

79. $i^{43} = i^{40}\cdot i^2\cdot i$
$$= \left(i^4\right)^{10}\cdot(-1)\cdot i$$
$$= 1^{10}\cdot(-i) = -i$$

81. $i^{-5} = \dfrac{1}{i^5} = \dfrac{1}{i^4\cdot i} = \dfrac{1}{1\cdot i} = \dfrac{1}{i}$

From Exercise 5, $\dfrac{1}{i} = -i$.

83. Since $i^{20} = \left(i^4\right)^5 = 1^5 = 1$, the student multiplied by 1, which is justified by the identity property for multiplication.

85. $I = \dfrac{E}{R+(X_L-X_c)i}$

Substitute $2+3i$ for E, 5 for R, 4 for X_L, and 3 for X_c.

$$I = \frac{2+3i}{5+(4-3)i} = \frac{2+3i}{5+i}$$
$$= \frac{(2+3i)(5-i)}{(5+i)(5-i)} = \frac{10-2i+15i-3i^2}{5^2+1^2}$$
$$= \frac{10+3+13i}{25+1} = \frac{13+13i}{26}$$
$$= \frac{13(1+i)}{13\cdot 2} = \frac{1+i}{2} = \frac{1}{2}+\frac{1}{2}i$$

87. To check that $1+5i$ is a solution of the equation, substitute $1+5i$ for x.

$$x^2-2x+26 = 0$$
$$(1+5i)^2-2(1+5i)+26 \stackrel{?}{=} 0$$
$$\left(1+10i+25i^2\right)-2-10i+26 \stackrel{?}{=} 0$$
$$1+10i-25-2-10i+26 \stackrel{?}{=} 0$$
$$(1-25-2+26)+(10-10)i \stackrel{?}{=} 0$$
$$0 \stackrel{?}{=} 0$$

True

Thus, $1+5i$ is a solution of

$$x^2-2x+26 = 0.$$

Now substitute $1-5i$ for x.

$$(1-5i)^2-2(1-5i)+26 \stackrel{?}{=} 0$$
$$\left(1-10i+25i^2\right)-2+10i+26 \stackrel{?}{=} 0$$
$$1-10i-25-2+10i+26 \stackrel{?}{=} 0$$
$$(1-25-2+26)+(-10+10)i \stackrel{?}{=} 0$$
$$0 \stackrel{?}{=} 0$$

True

Thus, $1-5i$ is also a solution of the given equation.

89. **(a)** $(x+2)+(3x-1)$
$$= (1+3)x+(2-1)$$
$$= 4x+1$$

(b) $(1+2i)+(3-i)$
$$= (1+3)+(2-1)i$$
$$= 4+i$$

90. **(a)** $(x+2)-(3x-1)$
$$= (x-3x)+[2-(-1)]$$
$$= (1-3)x+(2+1)$$
$$= -2x+3$$

(b) $(1+2i)-(3-i)$
$$= (1-3)+[2-(-1)]i$$
$$= -2+3i$$

91. **(a)** $(x+2)(3x-1)$
$$= 3x^2-x+6x-2$$
$$= 3x^2+5x-2$$

(b) $(1+2i)(3-i)$
$$= 3-i+6i-2i^2$$
$$= 3+5i-2(-1)$$
$$= 3+5i+2$$
$$= 5+5i$$

92. **(a)** $$\frac{\sqrt{3}-1}{1+\sqrt{2}}=\frac{\left(\sqrt{3}-1\right)\left(1-\sqrt{2}\right)}{\left(1+\sqrt{2}\right)\left(1-\sqrt{2}\right)}$$
$$=\frac{\sqrt{3}-\sqrt{6}-1+\sqrt{2}}{1^2-\left(\sqrt{2}\right)^2}$$
$$=\frac{\sqrt{3}-\sqrt{6}-1+\sqrt{2}}{1-2}$$
$$=\frac{\sqrt{3}-\sqrt{6}-1+\sqrt{2}}{-1}$$
$$=-\left(\sqrt{3}-\sqrt{6}-1+\sqrt{2}\right)$$
$$=-\sqrt{3}+\sqrt{6}+1-\sqrt{2}$$

(b) $$\frac{3-i}{1+2i}=\frac{(3-i)(1-2i)}{(1+2i)(1-2i)}$$
$$=\frac{3-6i-i+2i^2}{1-(2i)^2}$$
$$=\frac{3-7i+2(-1)}{1-4(-1)}$$
$$=\frac{3-7i-2}{1+4}$$
$$=\frac{1-7i}{5}$$
$$=\frac{1}{5}-\frac{7}{5}i$$

93. In parts (a) and (b) of Exercises 89 and 90, real and imaginary parts are added, just like coefficients of similar terms in the binomials, and the answers correspond. In Exercise 91, introducing $i^2=-1$ when a product is found leads to answers that do not correspond.

94. In parts (a) and (b) of Exercises 89 and 90, real and imaginary parts are added, just like coefficients of similar terms in the binomials, and the answers correspond. In Exercise 92, introducing $i^2=-1$ when performing the division leads to answers that do not correspond.

95. $$\frac{3}{2-i}+\frac{5}{1+i}$$
$$=\frac{3(2+i)}{(2-i)(2+i)}+\frac{5(1-i)}{(1+i)(1-i)}$$
$$=\frac{6+3i}{4+1}+\frac{5-5i}{1+1}$$
$$=\frac{2(6+3i)}{2\cdot 5}+\frac{5(5-5i)}{5\cdot 2}$$
$$=\frac{12+6i}{10}+\frac{25-25i}{10}$$
$$=\frac{37-19i}{10}=\frac{37}{10}-\frac{19}{10}i$$

97. $$\frac{2+i}{2-i}+\frac{i}{1+i}$$
$$=\frac{(2+i)(2+i)}{(2-i)(2+i)}+\frac{i(1-i)}{(1+i)(1-i)}$$
$$=\frac{4+4i+i^2}{4+1}+\frac{i-i^2}{1+1}$$
$$=\frac{3+4i}{5}+\frac{1+i}{2}$$
$$=\frac{2(3+4i)}{2\cdot 5}+\frac{5(1+i)}{5\cdot 2}$$
$$=\frac{6+8i}{10}+\frac{5+5i}{10}$$
$$=\frac{11+13i}{10}=\frac{11}{10}+\frac{13}{10}i$$

Thus, $$\left(\frac{2+i}{2-i}+\frac{i}{1+i}\right)i$$
$$=\left(\frac{11}{10}+\frac{13}{10}i\right)i$$
$$=\frac{11}{10}i+\frac{13}{10}i^2$$
$$=-\frac{13}{10}+\frac{11}{10}i.$$

Chapter 8 Review Exercises

1. $\sqrt{1764}=42$ because $42^2=1764$.
2. $-\sqrt{289}=-(17)=-17$, since $17^2=289$.
3. $\sqrt[3]{216}=6$ because $6^3=216$.
4. $\sqrt[3]{-125}=-5$ because $(-5)^3=-125$.
5. $-\sqrt[3]{27}=-(3)=-3$, since $3^3=27$.
6. $\sqrt[5]{-32}=-2$ because $(-2)^5=-32$.
7. $\sqrt[n]{a}$ is not a real number if n is even and a is negative.
8. **(a)** $\sqrt{x^2}=|x|$

   (b) $-\sqrt{x^2}=-|x|$

   (c) $\sqrt[3]{x^3}=x$
9. $-\sqrt{47}\approx -6.856$
10. $\sqrt[3]{-129}\approx -5.053$
11. $\sqrt[4]{605}\approx 4.960$
12. $500^{-3/4}\approx .009$
13. $-500^{4/3}\approx -3968.503$
14. $-28^{-1/2}\approx -.189$
15. $f(x)=\sqrt{x-1}$

    For the radicand to be nonnegative, we must have
    $$x-1\geq 0 \quad\text{or}\quad x\geq 1.$$

Thus, the domain is $[1, \infty)$.
The function values are positive or zero (the result of the radical), so the range is $[0, \infty)$.

x	$f(x) = \sqrt{x-1}$
1	$\sqrt{1-1} = 0$
2	$\sqrt{2-1} = 1$
5	$\sqrt{5-1} = 2$

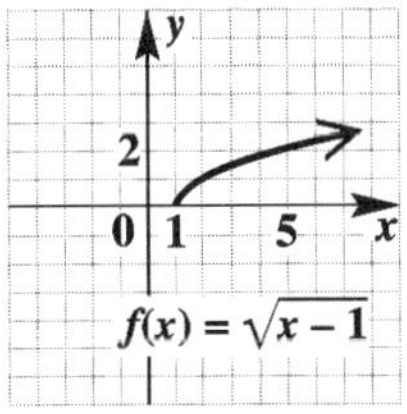

16. $f(x) = \sqrt[3]{x} + 4$
Since we can take the cube root of any real number, the domain is $(-\infty, \infty)$.
The result of a cube root can be any real number, so the range is $(-\infty, \infty)$. (The "+4" does not affect that range.)

x	$f(x) = \sqrt[3]{x} + 4$
-8	$\sqrt[3]{-8} + 4 = 2$
-1	$\sqrt[3]{-1} + 4 = 3$
0	$\sqrt[3]{0} + 4 = 4$
1	$\sqrt[3]{1} + 4 = 5$
8	$\sqrt[3]{8} + 4 = 6$

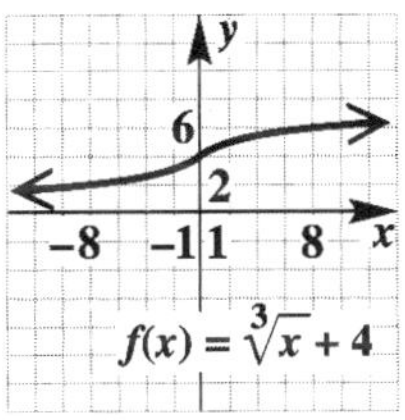

17. The base $\sqrt{38}$ is closest to $\sqrt{36} = 6$. The height $\sqrt{99}$ is closest to $\sqrt{100} = 10$. Use the estimates $b = 6$ and $h = 10$ in $A = \frac{1}{2}bh$ to find an estimate of the area.

$$A \approx \frac{1}{2}(6)(10) = 30$$

Choice **B** is the best estimate.

18. One way to evaluate $8^{2/3}$ is to first find the *cube (or third)* root of *8*, which is *2*. Then raise that result to the *second* power, to get an answer of *4*. Therefore, $8^{2/3} = $ *4*.

19. **A.** $(-27)^{2/3} = \left[(-27)^{1/3}\right]^2$
This number is a square, so it is a positive number.

B. $(-64)^{5/3} = \left[(-64)^{1/3}\right]^5$
This number is the odd power of an odd root of a negative number, so it is a negative number.

C. $(-100)^{1/2}$
This number is the square root of a negative number, so it is not a real number.

D. $(-32)^{1/5}$
This number is an odd root of a negative number, so it is a negative number.
The only positive number is choice **A**.

20. $a^{m/n} = \sqrt[n]{a^m}$
Since n is odd, $\sqrt[n]{a^m}$ is positive if a^m is positive and negative if a^m is negative. Since a is negative, a^m is positive if m is even and negative if m is odd.

(a) If a is negative and n is odd, then $a^{m/n}$ is positive if m is even.

(b) If a is negative and n is odd, then $a^{m/n}$ is negative if m is odd.

21. If a is negative and n is even, then $a^{1/n}$ is not a real number. An example is $(-4)^{1/2}$, not a real number.

22. $49^{1/2} = \left(7^2\right)^{1/2} = 7$

23. $-121^{1/2} = -\left(11^2\right)^{1/2} = -11$

24. $16^{5/4} = \left[\left(2^4\right)^{1/4}\right]^5 = 2^5 = 32$

25. $-8^{2/3} = -\left[\left(2^3\right)^{1/3}\right]^2 = -2^2 = -4$

26. $-\left(\frac{36}{25}\right)^{3/2} = -\left(\left[\left(\frac{6}{5}\right)^2\right]^{1/2}\right)^3$
$= -\left(\frac{6}{5}\right)^3 = -\frac{216}{125}$

27. $\left(-\frac{1}{8}\right)^{-5/3} = (-8)^{5/3} = \left(\left[(-2)^3\right]^{1/3}\right)^5$
$= (-2)^5 = -32$

28. $\left(\frac{81}{10,000}\right)^{-3/4} = \left(\frac{10,000}{81}\right)^{3/4}$
$= \left(\left[\left(\frac{10}{3}\right)^4\right]^{1/4}\right)^3$
$= \left(\frac{10}{3}\right)^3 = \frac{1000}{27}$

29. The base, -16, is negative and the index, 4, is even, so $(-16)^{3/4}$ is not a real number.

30. Solve $a^2 + b^2 = c^2$ for b. $(b > 0)$
$$b^2 = c^2 - a^2$$
$$b = \sqrt{c^2 - a^2}$$

31. The expression with fractional exponents, $a^{m/n}$, is equivalent to the radical expression, $\sqrt[n]{a^m}$. The denominator of the exponent is the index of the radical. For example, $\sqrt[3]{8^2} = \sqrt[3]{64} = 4$, and $8^{2/3} = \left(8^{1/3}\right)^2 = 2^2 = 4$.

32. $(m+3n)^{1/2} = \sqrt{m+3n}$

33. $(3a+b)^{-5/3} = \dfrac{1}{(3a+b)^{5/3}} = \dfrac{1}{\left((3a+b)^{1/3}\right)^5} = \dfrac{1}{\left(\sqrt[3]{3a+b}\right)^5}$ or $\dfrac{1}{\sqrt[3]{(3a+b)^5}}$

34. $\sqrt{7^9} = \left(7^9\right)^{1/2} = 7^{9/2}$

35. $\sqrt[5]{p^4} = \left(p^4\right)^{1/5} = p^{4/5}$

36. $5^{1/4} \cdot 5^{7/4} = 5^{1/4+7/4} = 5^{8/4} = 5^2$ or 25

37. $\dfrac{96^{2/3}}{96^{-1/3}} = 96^{2/3-(-1/3)} = 96^1 = 96$

38. $\dfrac{\left(a^{1/3}\right)^4}{a^{2/3}} = \dfrac{a^{4/3}}{a^{2/3}} = a^{4/3-2/3} = a^{2/3}$

39. $\dfrac{y^{-1/3} \cdot y^{5/6}}{y} = \dfrac{y^{-2/6}y^{5/6}}{y^{6/6}} = y^{-2/6+5/6-6/6} = y^{-3/6} = y^{-1/2} = \dfrac{1}{y^{1/2}}$

40. $\left(\dfrac{z^{-1}x^{-3/5}}{2^{-2}z^{-1/2}x}\right)^{-1} = \dfrac{z^1x^{3/5}}{2^2z^{1/2}x^{-1}} = \dfrac{1}{4}z^{1-1/2}x^{3/5-(-1)} = \dfrac{z^{1/2}x^{8/5}}{4}$

41. $r^{-1/2}\left(r + r^{3/2}\right)$
$= r^{-1/2}(r) + r^{-1/2}\left(r^{3/2}\right)$
$= r^{-1/2+1} + r^{-1/2+3/2}$
$= r^{1/2} + r^{2/2}$
$= r^{1/2} + r$

42. $\sqrt[8]{s^4} = \left(s^4\right)^{1/8} = s^{4/8} = s^{1/2}$

43. $\sqrt[6]{r^9} = \left(r^9\right)^{1/6} = r^{9/6} = r^{3/2}$

44. $\dfrac{\sqrt{p^5}}{p^2} = \dfrac{p^{5/2}}{p^2} = p^{5/2-2} = p^{5/2-4/2} = p^{1/2}$

45. $\sqrt[4]{k^3} \cdot \sqrt{k^3} = \left(k^3\right)^{1/4}\left(k^3\right)^{1/2} = k^{3/4}k^{3/2} = k^{3/4+3/2} = k^{3/4+6/4} = k^{9/4}$

46. $\sqrt[3]{m^5} \cdot \sqrt[3]{m^8} = \left(m^5\right)^{1/3}\left(m^8\right)^{1/3} = m^{5/3}m^{8/3} = m^{5/3+8/3} = m^{13/3}$

47. $\sqrt[4]{\sqrt[3]{z}} = \sqrt[4]{z^{1/3}} = \left(z^{1/3}\right)^{1/4} = z^{1/12}$

48. $\sqrt{\sqrt{\sqrt{x}}} = \sqrt{\sqrt{x^{1/2}}} = \sqrt{\left(x^{1/2}\right)^{1/2}} = \sqrt{x^{1/4}} = \left(x^{1/4}\right)^{1/2} = x^{1/8}$

49. $\sqrt[3]{\sqrt[5]{x}} = \sqrt[3]{x^{1/5}} = \left(x^{1/5}\right)^{1/3} = x^{1/15}$

50. $\sqrt{\sqrt[6]{\sqrt[3]{x}}} = \sqrt{\sqrt[6]{x^{1/3}}} = \sqrt{\left(x^{1/3}\right)^{1/6}} = \sqrt{x^{1/18}} = \left(x^{1/18}\right)^{1/2} = x^{1/36}$

51. The product rule for exponents applies only if the bases are the same.

52. $\sqrt{6} \cdot \sqrt{11} = \sqrt{6 \cdot 11} = \sqrt{66}$

53. $\sqrt{5} \cdot \sqrt{r} = \sqrt{5 \cdot r} = \sqrt{5r}$

54. $\sqrt[3]{6} \cdot \sqrt[3]{5} = \sqrt[3]{6 \cdot 5} = \sqrt[3]{30}$

55. $\sqrt[4]{7} \cdot \sqrt[4]{3} = \sqrt[4]{7 \cdot 3} = \sqrt[4]{21}$

56. $\sqrt{20} = \sqrt{4 \cdot 5} = \sqrt{4}\sqrt{5} = 2\sqrt{5}$

57. $\sqrt{75} = \sqrt{25 \cdot 3} = \sqrt{25}\sqrt{3} = 5\sqrt{3}$

58. $-\sqrt{125} = -\sqrt{25 \cdot 5} = -5\sqrt{5}$

59. $\sqrt[3]{-108} = \sqrt[3]{-27 \cdot 4} = -3\sqrt[3]{4}$

60. $\sqrt{100y^7} = \sqrt{100y^6 \cdot y} = 10y^3\sqrt{y}$

61. $\sqrt[3]{64p^4q^6} = \sqrt[3]{64p^3q^6 \cdot p} = 4pq^2\sqrt[3]{p}$

62. $\sqrt[3]{108a^8b^5} = \sqrt[3]{27a^6b^3 \cdot 4a^2b^2} = 3a^2b\sqrt[3]{4a^2b^2}$

63. $\sqrt[3]{632r^8t^4} = \sqrt[3]{8r^6t^3 \cdot 79r^2t} = 2r^2t\sqrt[3]{79r^2t}$

64. $\sqrt{\dfrac{y^3}{144}} = \dfrac{\sqrt{y^3}}{\sqrt{144}} = \dfrac{\sqrt{y^2 \cdot y}}{12} = \dfrac{y\sqrt{y}}{12}$

65. $\sqrt[3]{\dfrac{m^{15}}{27}} = \dfrac{\sqrt[3]{m^{15}}}{\sqrt[3]{27}} = \dfrac{\sqrt[3]{(m^5)^3}}{\sqrt[3]{3^3}} = \dfrac{m^5}{3}$

66. $\sqrt[3]{\dfrac{r^2}{8}} = \dfrac{\sqrt[3]{r^2}}{\sqrt[3]{8}} = \dfrac{\sqrt[3]{r^2}}{2}$

67. $\sqrt[4]{\dfrac{a^9}{81}} = \dfrac{\sqrt[4]{a^9}}{\sqrt[4]{81}} = \dfrac{\sqrt[4]{a^8 \cdot a}}{3} = \dfrac{a^2\sqrt[4]{a}}{3}$

68. $\sqrt[6]{15^3} = 15^{3/6} = 15^{1/2} = \sqrt{15}$

69. $\sqrt[4]{p^6} = \left(p^6\right)^{1/4} = p^{6/4} = p^{3/2}$
$= p^{2/2}p^{1/2} = p\sqrt{p}$

70. $\sqrt[3]{2} \cdot \sqrt[4]{5} = 2^{1/3} \cdot 5^{1/4}$
$= 2^{4/12} \cdot 5^{3/12}$
$= \left(2^4 \cdot 5^3\right)^{1/12}$
$= \sqrt[12]{16 \cdot 125} = \sqrt[12]{2000}$

71. $\sqrt{x} \cdot \sqrt[5]{x} = x^{1/2} \cdot x^{1/5}$
$= x^{5/10} \cdot x^{2/10}$
$= x^{7/10} = \sqrt[10]{x^7}$

72. Substitute 8 for a and 6 for b in the Pythagorean formula to find the hypotenuse, c.

$c^2 = a^2 + b^2$
$c = \sqrt{a^2 + b^2} = \sqrt{8^2 + 6^2}$
$= \sqrt{64 + 36} = \sqrt{100} = 10$

The length of the hypotenuse is 10.

73. $(-4, 7)$ and $(10, 6)$

$d = \sqrt{(x_2 - x_1)^2 + (y_2 - y_1)^2}$
$= \sqrt{[10 - (-4)]^2 + (6 - 7)^2}$
$= \sqrt{14^2 + (-1)^2}$
$= \sqrt{196 + 1} = \sqrt{197}$

74. $2\sqrt{8} - 3\sqrt{50} = 2\sqrt{4 \cdot 2} - 3\sqrt{25 \cdot 2}$
$= 2 \cdot 2\sqrt{2} - 3 \cdot 5\sqrt{2}$
$= 4\sqrt{2} - 15\sqrt{2} = -11\sqrt{2}$

75. $8\sqrt{80} - 3\sqrt{45} = 8\sqrt{16 \cdot 5} - 3\sqrt{9 \cdot 5}$
$= 8 \cdot 4\sqrt{5} - 3 \cdot 3\sqrt{5}$
$= 32\sqrt{5} - 9\sqrt{5} - 23\sqrt{5}$

76. $-\sqrt{27y} + 2\sqrt{75y} = -\sqrt{9 \cdot 3y} + 2\sqrt{25 \cdot 3y}$
$= -3\sqrt{3y} + 2 \cdot 5\sqrt{3y}$
$= -3\sqrt{3y} + 10\sqrt{3y}$
$= 7\sqrt{3y}$

77. $2\sqrt{54m^3} + 5\sqrt{96m^3}$
$= 2\sqrt{9m^2 \cdot 6m} + 5\sqrt{16m^2 \cdot 6m}$
$= 2 \cdot 3m\sqrt{6m} + 5 \cdot 4m\sqrt{6m}$
$= 6m\sqrt{6m} + 20m\sqrt{6m} = 26m\sqrt{6m}$

78. $3\sqrt[3]{54} + 5\sqrt[3]{16} = 3\sqrt[3]{27 \cdot 2} + 5\sqrt[3]{8 \cdot 2}$
$= 3 \cdot 3\sqrt[3]{2} + 5 \cdot 2\sqrt[3]{2}$
$= 9\sqrt[3]{2} + 10\sqrt[3]{2} = 19\sqrt[3]{2}$

79. $-6\sqrt[4]{32} + \sqrt[4]{512} = -6\sqrt[4]{16 \cdot 2} + \sqrt[4]{256 \cdot 2}$
$= -6 \cdot 2\sqrt[4]{2} + 4\sqrt[4]{2}$
$= -12\sqrt[4]{2} + 4\sqrt[4]{2} = -8\sqrt[4]{2}$

80. $\dfrac{3}{\sqrt{16}} - \dfrac{\sqrt{5}}{2} = \dfrac{3}{4} - \dfrac{2\sqrt{5}}{2 \cdot 2} = \dfrac{3 - 2\sqrt{5}}{4}$

81. $\dfrac{4}{\sqrt{25}} + \dfrac{\sqrt{5}}{4} = \dfrac{4}{5} + \dfrac{\sqrt{5}}{4}$
$= \dfrac{16}{20} + \dfrac{5\sqrt{5}}{20} = \dfrac{16 + 5\sqrt{5}}{20}$

82. Add the measures of the sides.

$P = a + b + c + d$
$P = 4\sqrt{8} + 6\sqrt{12} + 8\sqrt{2} + 3\sqrt{48}$
$= 4\sqrt{4 \cdot 2} + 6\sqrt{4 \cdot 3} + 8\sqrt{2} + 3\sqrt{16 \cdot 3}$
$= 4 \cdot 2\sqrt{2} + 6 \cdot 2\sqrt{3} + 8\sqrt{2} + 3 \cdot 4\sqrt{3}$
$= 8\sqrt{2} + 12\sqrt{3} + 8\sqrt{2} + 12\sqrt{3}$
$= 16\sqrt{2} + 24\sqrt{3}$

The perimeter is $16\sqrt{2} + 24\sqrt{3}$ feet.

83. Add the measures of the sides.

$P = a + b + c$
$P = 2\sqrt{27} + \sqrt{108} + \sqrt{50}$
$= 2\sqrt{9 \cdot 3} + \sqrt{36 \cdot 3} + \sqrt{25 \cdot 2}$
$= 2 \cdot 3\sqrt{3} + 6\sqrt{3} + 5\sqrt{2}$
$= 6\sqrt{3} + 6\sqrt{3} + 5\sqrt{2}$
$- 12\sqrt{3} + 5\sqrt{2}$

The perimeter is $12\sqrt{3} + 5\sqrt{2}$ feet.

84. $\left(\sqrt{3} + 1\right)\left(\sqrt{3} - 2\right) = 3 - 2\sqrt{3} + \sqrt{3} - 2$
$- 1 - \sqrt{3}$

85. $\left(\sqrt{7} + \sqrt{5}\right)\left(\sqrt{7} - \sqrt{5}\right) = \left(\sqrt{7}\right)^2 - \left(\sqrt{5}\right)^2$
$= 7 - 5 = 2$

86. $\left(3\sqrt{2} + 1\right)\left(2\sqrt{2} - 3\right)$
$= 6 \cdot 2 - 9\sqrt{2} + 2\sqrt{2} - 3$
$= 12 - 7\sqrt{2} - 3 = 9 - 7\sqrt{2}$

87. $\left(\sqrt{13}-\sqrt{2}\right)^2$
$= \left(\sqrt{13}\right)^2 - 2\cdot\sqrt{13}\cdot\sqrt{2} + \left(\sqrt{2}\right)^2$
$= 13 - 2\sqrt{26} + 2 = 15 - 2\sqrt{26}$

88. $\left(\sqrt[3]{2}+3\right)\left(\sqrt[3]{4}-3\sqrt[3]{2}+9\right)$
$= \sqrt[3]{2}\cdot\sqrt[3]{4} - \sqrt[3]{2}\cdot 3\sqrt[3]{2} + 9\sqrt[3]{2}$
$+ 3\sqrt[3]{4} - 3\cdot 3\sqrt[3]{2} + 27$
$= \sqrt[3]{8} - 3\sqrt[3]{4} + 9\sqrt[3]{2} + 3\sqrt[3]{4} - 9\sqrt[3]{2} + 27$
$= 2 + 27 = 29$

89. $(\sqrt[3]{4y}-1)(\sqrt[3]{4y}+3)$
$= \sqrt[3]{16y^2} + 3\sqrt[3]{4y} - \sqrt[3]{4y} - 3$
$= \sqrt[3]{8\cdot 2y^2} + 2\sqrt[3]{4y} - 3$
$= 2\sqrt[3]{2y^2} + 2\sqrt[3]{4y} - 3$

90. Show that $15 - 2\sqrt{26} \neq 13\sqrt{26}$.

Find a calculator approximation of each term.

$$4.801960973 \neq 66.28725368$$

Therefore, $15 - 2\sqrt{26} \neq 13\sqrt{26}$.

91. Multiplying by $\sqrt[3]{6}$ still results in an expression with a radical in the denominator.

$$\frac{5\sqrt[3]{6}}{\sqrt[3]{6}\cdot\sqrt[3]{6}} = \frac{5\sqrt[3]{6}}{\sqrt[3]{36}}$$

To rationalize the denominator, multiply by $\sqrt[3]{6^2}$ or $\sqrt[3]{36}$.

$$\frac{5\sqrt[3]{6^2}}{\sqrt[3]{6}\cdot\sqrt[3]{6^2}} = \frac{5\sqrt[3]{36}}{\sqrt[3]{6^3}} = \frac{5\sqrt[3]{36}}{6}$$

92. $\dfrac{\sqrt{6}}{\sqrt{5}} = \dfrac{\sqrt{6}\cdot\sqrt{5}}{\sqrt{5}\cdot\sqrt{5}} = \dfrac{\sqrt{30}}{5}$

93. $\dfrac{-6\sqrt{3}}{\sqrt{2}} = \dfrac{-6\sqrt{3}\cdot\sqrt{2}}{\sqrt{2}\cdot\sqrt{2}} = \dfrac{-6\sqrt{6}}{2} = -3\sqrt{6}$

94. $\dfrac{3\sqrt{7p}}{\sqrt{y}} = \dfrac{3\sqrt{7p}\cdot\sqrt{y}}{\sqrt{y}\cdot\sqrt{y}} = \dfrac{3\sqrt{7py}}{y}$

95. $\sqrt{\dfrac{11}{8}} = \dfrac{\sqrt{11}}{\sqrt{8}} = \dfrac{\sqrt{11}}{\sqrt{4\cdot 2}} = \dfrac{\sqrt{11}}{2\sqrt{2}} = \dfrac{\sqrt{11}\cdot\sqrt{2}}{2\sqrt{2}\cdot\sqrt{2}}$
$= \dfrac{\sqrt{22}}{2\cdot 2} = \dfrac{\sqrt{22}}{4}$

96. $-\sqrt[3]{\dfrac{9}{25}} = -\dfrac{\sqrt[3]{9}}{\sqrt[3]{5^2}} = -\dfrac{\sqrt[3]{9}\cdot\sqrt[3]{5}}{\sqrt[3]{5^2}\cdot\sqrt[3]{5}}$
$= -\dfrac{\sqrt[3]{45}}{\sqrt[3]{5^3}} = -\dfrac{\sqrt[3]{45}}{5}$

97. $\sqrt[3]{\dfrac{108m^3}{n^5}} = \dfrac{\sqrt[3]{108m^3}}{\sqrt[3]{n^5}} = \dfrac{\sqrt[3]{27m^3\cdot 4}}{\sqrt[3]{n^3\cdot n^2}}$
$= \dfrac{3m\sqrt[3]{4}}{n\sqrt[3]{n^2}} = \dfrac{3m\sqrt[3]{4}\cdot\sqrt[3]{n}}{n\sqrt[3]{n^2}\cdot\sqrt[3]{n}}$
$= \dfrac{3m\sqrt[3]{4n}}{n\cdot n} = \dfrac{3m\sqrt[3]{4n}}{n^2}$

98. $\dfrac{1}{\sqrt{2}+\sqrt{7}}$

Multiply the numerator and denominator by the conjugate of the denominator, $\sqrt{2}-\sqrt{7}$.

$= \dfrac{1\left(\sqrt{2}-\sqrt{7}\right)}{\left(\sqrt{2}+\sqrt{7}\right)\left(\sqrt{2}-\sqrt{7}\right)}$
$= \dfrac{\sqrt{2}-\sqrt{7}}{2-7} = \dfrac{\sqrt{2}-\sqrt{7}}{-5}$

99. $\dfrac{-5}{\sqrt{6}-3}$

Multiply the numerator and denominator by the conjugate of the denominator, $\sqrt{6}+3$.

$= \dfrac{-5\left(\sqrt{6}+3\right)}{\left(\sqrt{6}-3\right)\left(\sqrt{6}+3\right)} = \dfrac{-5\left(\sqrt{6}+3\right)}{6-9}$
$= \dfrac{-5\left(\sqrt{6}+3\right)}{-3} = \dfrac{5\left(\sqrt{6}+3\right)}{3}$

100. $\dfrac{2-2\sqrt{5}}{8} = \dfrac{2\left(1-\sqrt{5}\right)}{2\cdot 4} = \dfrac{1-\sqrt{5}}{4}$

101. $\dfrac{4-8\sqrt{8}}{12} = \dfrac{4\left(1-2\sqrt{8}\right)}{3\cdot 4} = \dfrac{1-2\sqrt{8}}{3}$
$= \dfrac{1-2\sqrt{4\cdot 2}}{3} = \dfrac{1-4\sqrt{2}}{3}$

102. $\dfrac{-18+\sqrt{27}}{6} = \dfrac{-18+\sqrt{9}\cdot\sqrt{3}}{6} = \dfrac{-18+3\sqrt{3}}{6}$
$= \dfrac{3\left(-6+\sqrt{3}\right)}{3\cdot 2} = \dfrac{-6+\sqrt{3}}{2}$

103. $\sqrt{8x+9} = 5$
$\left(\sqrt{8x+9}\right)^2 = 5^2$ *Square*
$8x+9 = 25$
$8x = 16$
$x = 2$

Check $x=2$: $\sqrt{25} \stackrel{?}{=} 5$ *True*
Solution set: $\{2\}$

104. $\sqrt{2z-3}-3=0$

$$\sqrt{2z-3}=3 \quad \textit{Isolate}$$
$$\left(\sqrt{2z-3}\right)^2=3^2 \quad \textit{Square}$$
$$2z-3=9$$
$$2z=12$$
$$z=6$$

Check $z=6$: $\sqrt{9}-3\stackrel{?}{=}0$ *True*

Solution set: $\{6\}$

105. $\sqrt{3m+1}-2=-3$

$$\sqrt{3m+1}=-1 \quad \textit{Isolate}$$

This equation has no solution because $\sqrt{3m+1}$ cannot be negative.

Solution set: $\emptyset$

106. $\sqrt{7z+1}=z+1$

$$\left(\sqrt{7z+1}\right)^2=(z+1)^2 \quad \textit{Square}$$
$$7z+1=z^2+2z+1$$
$$0=z^2-5z$$
$$0=z(z-5)$$

$z=0$ or $z=5$

Check $z=0$: $\sqrt{1}\stackrel{?}{=}1$ *True*

Check $z=5$: $\sqrt{36}\stackrel{?}{=}6$ *True*

Solution set: $\{0,5\}$

107. $3\sqrt{m}=\sqrt{10m-9}$

$$\left(3\sqrt{m}\right)^2=\left(\sqrt{10m-9}\right)^2 \quad \textit{Square}$$
$$9m=10m-9$$
$$9=m$$

Check $m=9$: $3\sqrt{9}\stackrel{?}{=}\sqrt{81}$ *True*

Solution set: $\{9\}$

108. $\sqrt{p^2+3p+7}=p+2$

$$\left(\sqrt{p^2+3p+7}\right)^2=(p+2)^2 \quad \textit{Square}$$
$$p^2+3p+7=p^2+4p+4$$
$$3=p$$

Check $p=3$: $\sqrt{25}\stackrel{?}{=}5$ *True*

Solution set: $\{3\}$

109. $\sqrt{a+2}-\sqrt{a-3}=1$

Get one radical on each side of the equals sign.

$$\sqrt{a+2}=1+\sqrt{a-3}$$

Square both sides.

$$\left(\sqrt{a+2}\right)^2=\left(1+\sqrt{a-3}\right)^2$$
$$a+2=1+2\sqrt{a-3}+a-3$$
$$4=2\sqrt{a-3}$$
$$2=\sqrt{a-3}$$

Square both sides again.

$$2^2=\left(\sqrt{a-3}\right)^2$$
$$4=a-3$$
$$7=a$$

Check $a=7$: $\sqrt{9}-\sqrt{4}\stackrel{?}{=}1$ *True*

Solution set: $\{7\}$

110. $\sqrt[3]{5m-1}=\sqrt[3]{3m-2}$

Cube both sides.

$$\left(\sqrt[3]{5m-1}\right)^3=\left(\sqrt[3]{3m-2}\right)^3$$
$$5m-1=3m-2$$
$$2m=-1$$
$$m=-\frac{1}{2}$$

Check $m=-\frac{1}{2}$: $\sqrt[3]{-\frac{7}{2}}\stackrel{?}{=}\sqrt[3]{-\frac{7}{2}}$ *True*

Solution set: $\left\{-\frac{1}{2}\right\}$

111. $\sqrt[3]{2x^2+3x-7}=\sqrt[3]{2x^2+4x+6}$

Cube both sides.

$$\left(\sqrt[3]{2x^2+3x-7}\right)^3=\left(\sqrt[3]{2x^2+4x+6}\right)^3$$
$$2x^2+3x-7=2x^2+4x+6$$
$$-13=x$$

Check $x=-13$: $\sqrt[3]{292}\stackrel{?}{=}\sqrt[3]{292}$ *True*

Solution set: $\{-13\}$

112. $\sqrt[3]{3y^2-4y+6}=\sqrt[3]{3y^2-2y+8}$

$$\left(\sqrt[3]{3y^2-4y+6}\right)^3=\left(\sqrt[3]{3y^2-2y+8}\right)^3$$
$$3y^2-4y+6=3y^2-2y+8$$
$$-2=2y$$
$$-1=y$$

Check $y=-1$: $\sqrt[3]{13}\stackrel{?}{=}\sqrt[3]{13}$ *True*

Solution set: $\{-1\}$

113. $\sqrt[3]{1-2k}-\sqrt[3]{-k-13}=0$

$$\sqrt[3]{1-2k}=\sqrt[3]{-k-13}$$

Cube both sides.

$$\left(\sqrt[3]{1-2k}\right)^3=\left(\sqrt[3]{-k-13}\right)^3$$
$$1-2k=-k-13$$
$$14=k$$

Check $k=14$: $\sqrt[3]{-27}-\sqrt[3]{-27}\stackrel{?}{=}0$ *True*

Solution set: $\{14\}$

114. $\sqrt[3]{11-2t}-\sqrt[3]{-1-5t}=0$
$$\sqrt[3]{11-2t}=\sqrt[3]{-1-5t}$$
$$\left(\sqrt[3]{11-2t}\right)^3=\left(\sqrt[3]{-1-5t}\right)^3$$
$$11-2t=-1-5t$$
$$3t=-12$$
$$t=-4$$
Check $t=-4$: $\sqrt[3]{19}-\sqrt[3]{19}\overset{?}{=}0$ *True*
Solution set: $\{-4\}$

115. $\sqrt[4]{x-1}+2=0$
$$\sqrt[4]{x-1}=-2$$
This equation has no solution because $\sqrt[4]{x-1}$ cannot be negative.
Solution set: $\emptyset$

116. $\sqrt[4]{2k+3}+1=0$
$$\sqrt[4]{2k+3}=-1$$
This equation has no solution because $\sqrt[4]{2k+3}$ cannot be negative.
Solution set: $\emptyset$

117. $\sqrt[4]{x+7}=\sqrt[4]{2x}$
Raise each side to the fourth power.
$$\left(\sqrt[4]{x+7}\right)^4=\left(\sqrt[4]{2x}\right)^4$$
$$x+7=2x$$
$$7=x$$
Check $x=7$: $\sqrt[4]{14}\overset{?}{=}\sqrt[4]{14}$ *True*
Solution set: $\{7\}$

118. $\sqrt[4]{x+8}=\sqrt[4]{3x}$
Raise each side to the fourth power.
$$\left(\sqrt[4]{x+8}\right)^4=\left(\sqrt[4]{3x}\right)^4$$
$$x+8=3x$$
$$8=2x$$
$$4=x$$
Check $x=4$: $\sqrt[4]{12}\overset{?}{=}\sqrt[4]{12}$ *True*
Solution set: $\{4\}$

119. $x=3$
Solution set: $\{3\}$

120. $x=-3$
Solution set: $\{-3\}$

121. $x^2=9$
$3^2=9$ or $(-3)^2=9$
Solution set: $\{-3,3\}$ or $\{\pm 3\}$

122. $x^3=27$
$3^3=27$
Solution set: $\{3\}$

123. $x^4=81$
$3^4=81$ or $(-3)^4=81$
Solution set: $\{-3,3\}$ or $\{\pm 3\}$

124. $x^5=-243$
$(-3)^5=-243$
Solution set: $\{-3\}$

125. Suppose both sides of $x=k$ are raised to the nth power.

(a) If n is even, the number of solutions of the new equation is *more than* the number of solutions of the original equation.

(b) If n is odd, the number of solutions of the new equation, is *the same as* the number of solutions of the original equation.

126. (a) Solve $L=\sqrt{H^2+W^2}$ for H.
$$L^2=H^2+W^2$$
$$L^2-W^2=H^2$$
$$\sqrt{L^2-W^2}=H$$

(b) Substitute 12 for L and 9 for W in $H=\sqrt{L^2-W^2}$.
$$H=\sqrt{12^2-9^2}$$
$$=\sqrt{144-81}=\sqrt{63}$$
To the nearest tenth of a foot, the height is approximately 7.9 feet.

127. $\sqrt{-25}=i\sqrt{25}=5i$

128. $\sqrt{-200}=i\sqrt{100\cdot 2}=10i\sqrt{2}$

129. If a is a positive real number, then $-a$ is negative. So, $\sqrt{-a}$ is not a real number. Therefore, $-\sqrt{-a}$ is not a real number either.

130. $(-2+5i)+(-8-7i)$
$$=[-2+(-8)]+[5+(-7)]i$$
$$=-10-2i$$

131. $(5+4i)-(-9-3i)$
$$=[(5-(-9)]+[(4-(-3)]i$$
$$=14+7i$$

132. $\sqrt{-5}\cdot\sqrt{-7}=i\sqrt{5}\cdot i\sqrt{7}=i^2\sqrt{35}$
$$=-1\left(\sqrt{35}\right)=-\sqrt{35}$$

133. $\sqrt{-25}\cdot\sqrt{-81}=5i\cdot 9i=45i^2$
$$=45(-1)=-45$$

134. $\dfrac{\sqrt{-72}}{\sqrt{-8}}=\dfrac{i\sqrt{72}}{i\sqrt{8}}=\sqrt{\dfrac{72}{8}}=\sqrt{9}=3$

135. $(2+3i)(1-i) = 2-2i+3i-3i^2$
$= 2+i-3(-1)$
$= 2+i+3$
$= 5+i$

136. $(6-2i)^2 = 6^2 - 2\cdot 6\cdot 2i + (2i)^2$
$= 36 - 24i + 4i^2$
$= 36 - 24i + 4(-1)$
$= 36 - 24i - 4$
$= 32 - 24i$

137. $\dfrac{3-i}{2+i}$
Multiply by the conjugate of the denominator, $2-i$.
$= \dfrac{(3-i)(2-i)}{(2+i)(2-i)}$
$= \dfrac{6-3i-2i+i^2}{4-i^2}$
$= \dfrac{6-5i-1}{4-(-1)} = \dfrac{5-5i}{5}$
$= \dfrac{5(1-i)}{5} = 1-i$

138. $\dfrac{5+14i}{2+3i}$
Multiply by the conjugate of the denominator, $2-3i$.
$= \dfrac{(5+14i)(2-3i)}{(2+3i)(2-3i)}$
$= \dfrac{10-15i+28i-42i^2}{4-9i^2}$
$= \dfrac{10+13i-42(-1)}{4-9(-1)} = \dfrac{52+13i}{13}$
$= \dfrac{13(4+i)}{13} = 4+i$

139. $i^{11} = i^8\cdot i^3 = \left(i^4\right)^2\cdot i^2\cdot i$
$= 1^2\cdot(-1)\cdot i = -i$

140. $i^{36} = \left(i^4\right)^9 = 1^9 = 1$

141. $i^{-10} = \dfrac{1}{i^{10}} = \dfrac{1}{i^8\cdot i^2} = \dfrac{1}{(i^4)^2\cdot(-1)}$
$= \dfrac{1}{1^2\cdot(-1)} = \dfrac{1}{-1} = -1$

Another method:
$i^{-10} = i^{-10}\cdot i^{12} = i^2 = -1$

Note that $i^{12} = \left(i^4\right)^3 = 1^3 = 1$.

142. $i^{-8} = \dfrac{1}{i^8} = \dfrac{1}{(i^4)^2} = \dfrac{1}{1^2} = \dfrac{1}{1} = 1$

143. $-\sqrt[4]{256} = -\left(\sqrt[4]{4^4}\right) = -(4) = -4$

144. $1000^{-2/3} = \dfrac{1}{1000^{2/3}} = \dfrac{1}{\left[(10^3)^{1/3}\right]^2}$
$= \dfrac{1}{10^2} = \dfrac{1}{100}$

145. $\dfrac{z^{-1/5}\cdot z^{3/10}}{z^{7/10}} = z^{-2/10+3/10-7/10}$
$= z^{-6/10} = \dfrac{1}{z^{6/10}} = \dfrac{1}{z^{3/5}}$

146. $\sqrt[4]{k^{24}} = k^{24/4} = k^6$

147. $\sqrt[3]{54z^9t^8} = \sqrt[3]{27z^9t^6\cdot 2t^2} = 3z^3t^2\sqrt[3]{2t^2}$

148. $-5\sqrt{18} + 12\sqrt{72} = -5\sqrt{9\cdot 2} + 12\sqrt{36\cdot 2}$
$= -5\cdot 3\sqrt{2} + 12\cdot 6\sqrt{2}$
$= -15\sqrt{2} + 72\sqrt{2} = 57\sqrt{2}$

149. $8\sqrt[3]{x^3y^2} - 2x\sqrt[3]{y^2} = 8x\sqrt[3]{y^2} - 2x\sqrt[3]{y^2}$
$= 6x\sqrt[3]{y^2}$

150. $\left(\sqrt{5}-\sqrt{3}\right)\left(\sqrt{7}+\sqrt{3}\right)$
$= \sqrt{35} + \sqrt{15} - \sqrt{21} - 3$

151. $\dfrac{-1}{\sqrt{12}} = \dfrac{-1}{\sqrt{4\cdot 3}} = \dfrac{-1}{2\sqrt{3}} = \dfrac{-1\cdot\sqrt{3}}{2\sqrt{3}\cdot\sqrt{3}}$
$= \dfrac{-\sqrt{3}}{2\cdot 3} = \dfrac{-\sqrt{3}}{6}$

152. $\sqrt[3]{\dfrac{12}{25}} = \dfrac{\sqrt[3]{12}}{\sqrt[3]{25}} = \dfrac{\sqrt[3]{12}}{\sqrt[3]{5^2}}$
$= \dfrac{\sqrt[3]{12}\cdot\sqrt[3]{5}}{\sqrt[3]{5^2}\cdot\sqrt[3]{5}} = \dfrac{\sqrt[3]{60}}{\sqrt[3]{5^3}} = \dfrac{\sqrt[3]{60}}{5}$

153. $i^{-1000} = \dfrac{1}{i^{1000}} = \dfrac{1}{\left(i^4\right)^{250}}$
$= \dfrac{1}{1^{250}} = \dfrac{1}{1} = 1$

154. $\sqrt{-49} = i\sqrt{49} = 7i$

155. $(4-9i)+(-1+2i)$
$= (4-1)+(-9+2)i$
$= 3-7i$

156. $\dfrac{\sqrt{50}}{\sqrt{-2}} = \dfrac{\sqrt{25\cdot 2}}{i\sqrt{2}} = \dfrac{5\sqrt{2}}{i\sqrt{2}} = \dfrac{5}{i}$
The conjugate of i is $-i$.
$\dfrac{5(-i)}{i(-i)} = \dfrac{-5i}{-i^2} = \dfrac{-5i}{-(-1)} = -5i$

157. $\dfrac{3+\sqrt{54}}{6} = \dfrac{3+\sqrt{9\cdot 6}}{6} = \dfrac{3+3\sqrt{6}}{6}$
$= \dfrac{3\left(1+\sqrt{6}\right)}{2\cdot 3} = \dfrac{1+\sqrt{6}}{2}$

158. $(3+2i)^2 = 3^2 + 2\cdot 3\cdot 2i + (2i)^2$
$= 9 + 12i + 4i^2$
$= 9 + 12i + 4(-1)$
$= 5 + 12i$

159. $\sqrt{x+4} = x-2$
$\left(\sqrt{x+4}\right)^2 = (x-2)^2$ *Square*
$x + 4 = x^2 - 4x + 4$
$0 = x^2 - 5x$
$0 = x(x-5)$

$x = 0$ or $x = 5$

Check $x = 0$: $\sqrt{4} \stackrel{?}{=} -2$ *False*

Check $x = 5$: $\sqrt{9} \stackrel{?}{=} 3$ *True*

Solution set: $\{5\}$

160. $\sqrt[3]{2x-9} = \sqrt[3]{5x+3}$

Cube both sides.

$\left(\sqrt[3]{2x-9}\right)^3 = \left(\sqrt[3]{5x+3}\right)^3$
$2x - 9 = 5x + 3$
$-3x = 12$
$x = -4$

Check $x = -4$: $\sqrt[3]{-17} \stackrel{?}{=} \sqrt[3]{-17}$ *True*

Solution set: $\{-4\}$

161. $\sqrt{6+2x} - 1 = \sqrt{7-2x}$

Square both sides.

$\left(\sqrt{6+2x} - 1\right)^2 = \left(\sqrt{7-2x}\right)^2$
$6 + 2x - 2\sqrt{6+2x} + 1 = 7 - 2x$
$4x = 2\sqrt{6+2x}$
$2x = \sqrt{6+2x}$

Square both sides again.

$(2x)^2 = \left(\sqrt{6+2x}\right)^2$
$4x^2 = 6 + 2x$
$4x^2 - 2x - 6 = 0$
$2x^2 - x - 3 = 0$
$(2x-3)(x+1) = 0$

$2x - 3 = 0$ or $x + 1 = 0$

$x = \frac{3}{2}$ or $x = -1$

Check $x = \frac{3}{2}$: $\sqrt{9} - 1 \stackrel{?}{=} \sqrt{4}$ *True*

Check $x = -1$: $\sqrt{4} - 1 \stackrel{?}{=} \sqrt{9}$ *False*

Solution set: $\left\{\frac{3}{2}\right\}$

162. $\sqrt{7x+11} - 5 = 0$
$\sqrt{7x+11} = 5$
$\left(\sqrt{7x+11}\right)^2 = 5^2$
$7x + 11 = 25$
$7x = 14$
$x = 2$

Check $x = 2$: $\sqrt{25} - 5 \stackrel{?}{=} 0$ *True*

Solution set: $\{2\}$

163. $\sqrt{6x+2} - \sqrt{5x+3} = 0$

Get one radical on each side of the equals sign.

$\sqrt{6x+2} = \sqrt{5x+3}$

Square both sides.

$\left(\sqrt{6x+2}\right)^2 = \left(\sqrt{5x+3}\right)^2$
$6x + 2 = 5x + 3$
$x = 1$

Check $x = 1$: $\sqrt{8} - \sqrt{8} \stackrel{?}{=} 0$ *True*

Solution set: $\{1\}$

164. $\sqrt{3+5x} - \sqrt{x+11} = 0$
$\sqrt{3+5x} = \sqrt{x+11}$
$\left(\sqrt{3+5x}\right)^2 = \left(\sqrt{x+11}\right)^2$
$3 + 5x = x + 11$
$4x = 8$
$x = 2$

Check $x = 2$: $\sqrt{13} - \sqrt{13} \stackrel{?}{=} 0$ *True*

Solution set: $\{2\}$

165. $3\sqrt{x} = \sqrt{8x+9}$
$\left(3\sqrt{x}\right)^2 = \left(\sqrt{8x+9}\right)^2$ *Square*
$9x = 8x + 9$
$x = 9$

Check $x = 9$: $3\sqrt{9} \stackrel{?}{=} \sqrt{81}$ *True*

Solution set: $\{9\}$

166. $6\sqrt{p} = \sqrt{30p+24}$
$\left(6\sqrt{p}\right)^2 = \left(\sqrt{30p+24}\right)^2$
$36p = 30p + 24$
$6p = 24$
$p = 4$

Check $p = 4$: $6\sqrt{4} \stackrel{?}{=} \sqrt{144}$ *True*

Solution set: $\{4\}$

167. $\sqrt{11+2x}+1=\sqrt{5x+1}$

Square both sides.

$$\left(\sqrt{11+2x}+1\right)^2=\left(\sqrt{5x+1}\right)^2$$

$$\left(\sqrt{11+2x}\right)^2+2\cdot\sqrt{11+2x}+1$$
$$=5x+1$$

$$11+2x+2\sqrt{11+2x}+1=5x+1$$

Isolate the remaining radical.

$$2\sqrt{11+2x}=3x-11$$

Square both sides again.

$$\left(2\sqrt{11+2x}\right)^2=(3x-11)^2$$
$$4(11+2x)=9x^2-66x+121$$
$$44+8x=9x^2-66x+121$$
$$0=9x^2-74x+77$$
$$0=(9x-11)(x-7)$$

$9x-11=0$ or $x-7=0$

$x=\frac{11}{9}$ or $x=7$

Check $x=\frac{11}{9}$: $\sqrt{\frac{121}{9}}+1\stackrel{?}{=}\sqrt{\frac{64}{9}}$ *False*

Check $x=7$: $\sqrt{25}+1\stackrel{?}{=}\sqrt{36}$ *True*

Solution set: $\{7\}$

168. $\sqrt{5x+6}-\sqrt{x+3}=3$

Get one radical on each side.

$$\sqrt{5x+6}=3+\sqrt{x+3}$$
$$\left(\sqrt{5x+6}\right)^2=\left(3+\sqrt{x+3}\right)^2$$
$$5x+6=9+6\sqrt{x+3}+x+3$$

Isolate the remaining radical.

$$4x-6=6\sqrt{x+3}$$
$$2x-3=3\sqrt{x+3}$$
$$(2x-3)^2=\left(3\sqrt{x+3}\right)^2$$
$$4x^2-12x+9=9(x+3)$$
$$4x^2-12x+9-9x+27$$
$$4x^2-21x-18=0$$
$$(4x+3)(x-6)=0$$

$4x+3=0$ or $x-6=0$

$x = -\frac{3}{4}$ $\quad x-6$

Check $x=-\frac{3}{4}$: $\sqrt{\frac{9}{4}}-\sqrt{\frac{9}{4}}\stackrel{?}{=}3$ *False*

Check $x=6$: $\sqrt{36}-\sqrt{9}\stackrel{?}{=}3$ *True*

Solution set: $\{6\}$

Chapter 8 Test

1. $\sqrt{841}=29$ because $29^2=841$, so $-\sqrt{841}=-29$.

2. $\sqrt[3]{-512}=-8$ because $(-8)^3=-512$.

3. $125^{1/3}=\sqrt[3]{125}=5$ because $5^3=125$.

4. $\sqrt{146.25}\approx\sqrt{144}=12$ because $12^2=144$, so choice **C** is the best estimate.

5. $\sqrt{478}\approx 21.863$

6. $\sqrt[3]{-832}\approx -9.405$

7. $f(x)=\sqrt{x+6}$

For the radicand to be nonnegative, we must have

$$x+6\ge 0 \quad\text{or}\quad x\ge -6.$$

Thus, the domain is $[-6,\infty)$.

The function values are positive or zero (the result of the radical), so the range is $[0,\infty)$.

x	$f(x)=\sqrt{x+6}$
-6	$\sqrt{-6+6}=0$
-5	$\sqrt{-5+6}=1$
-2	$\sqrt{-2+6}=2$

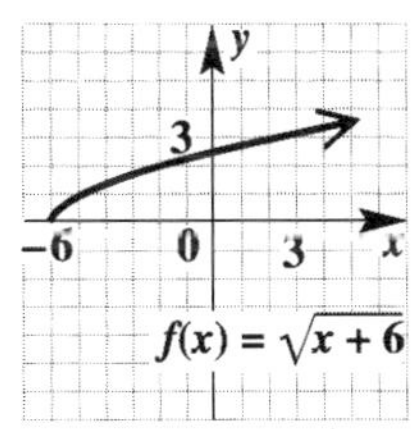

8. $\left(\frac{16}{25}\right)^{-3/2}=\left(\frac{25}{16}\right)^{3/2}=\frac{25^{3/2}}{16^{3/2}}$

$$=\frac{\left[(5^2)^{1/2}\right]^3}{\left[(4^2)^{1/2}\right]^3}=\frac{5^3}{4^3}=\frac{125}{64}$$

9. $(-64)^{-4/3}=\frac{1}{(-64)^{4/3}}=\frac{1}{\left(\left[(-4)^3\right]^{1/3}\right)^4}$

$$=\frac{1}{(-4)^4}=\frac{1}{256}$$

10. $\frac{3^{2/5}x^{-1/4}y^{2/5}}{3^{-8/5}x^{7/4}y^{1/10}}$

$$=3^{2/5-(-8/5)}x^{-1/4-7/4}y^{2/5-1/10}$$
$$=3^{10/5}x^{-8/4}y^{4/10-1/10}$$
$$=3^2x^{-2}y^{3/10}=\frac{9y^{3/10}}{x^2}$$

11. $\left(\frac{x^{-4}y^{-6}}{x^{-2}y^3}\right)^{-2/3}=\left(\frac{x^2}{x^4y^6y^3}\right)^{-2/3}$

$$=\left(\frac{1}{x^2y^9}\right)^{-2/3}=\left(x^2y^9\right)^{2/3}$$
$$=\left(x^2\right)^{2/3}\left(y^9\right)^{2/3}=x^{4/3}y^6$$

12. $7^{3/4}\cdot 7^{-1/4}=7^{3/4+(-1/4)}$

$$=7^{1/2}\text{ or }\sqrt{7}$$

13. $\sqrt[3]{a^4} \cdot \sqrt[3]{a^7} = a^{4/3} \cdot a^{7/3} = a^{4/3+7/3} = a^{11/3}$
$= \sqrt[3]{a^{11}} = \sqrt[3]{(a^3)^3 \cdot a^2} = a^3\sqrt[3]{a^2}$

14. $a^2 + b^2 = c^2$
$12^2 + b^2 = 17^2$ *Let a=12, c=17*
$144 + b^2 = 289$
$b^2 = 145$
$b = \sqrt{145}$

15. $(-4, 2)$ and $(2, 10)$

$d = \sqrt{(x_2 - x_1)^2 + (y_2 - y_1)^2}$
$= \sqrt{[2 - (-4)]^2 + (10 - 2)^2}$
$= \sqrt{6^2 + 8^2} = \sqrt{36 + 64} = \sqrt{100} = 10$

16. $\sqrt{54x^5y^6} = \sqrt{9x^4y^6 \cdot 6x} = 3x^2y^3\sqrt{6x}$

17. $\sqrt[4]{32a^7b^{13}} = \sqrt[4]{16a^4b^{12} \cdot 2a^3b}$
$= 2ab^3\sqrt[4]{2a^3b}$

18. $\sqrt{2} \cdot \sqrt[3]{5} = 2^{1/2} \cdot 5^{1/3} = 2^{3/6} \cdot 5^{2/6}$
$= \left(2^3 \cdot 5^2\right)^{1/6} = \sqrt[6]{2^3 \cdot 5^2}$
$= \sqrt[6]{8 \cdot 25} = \sqrt[6]{200}$

19. $3\sqrt{20} - 5\sqrt{80} + 4\sqrt{500}$
$= 3\sqrt{4 \cdot 5} - 5\sqrt{16 \cdot 5} + 4\sqrt{100 \cdot 5}$
$= 3 \cdot 2\sqrt{5} - 5 \cdot 4\sqrt{5} + 4 \cdot 10\sqrt{5}$
$= 6\sqrt{5} - 20\sqrt{5} + 40\sqrt{5} = 26\sqrt{5}$

20. $\sqrt[3]{16t^3s^5} - \sqrt[3]{54t^6s^2}$
$= \sqrt[3]{8t^3s^3} \cdot \sqrt[3]{2s^2} - \sqrt[3]{27t^6} \cdot \sqrt[3]{2s^2}$
$= 2ts\sqrt[3]{2s^2} - 3t^2\sqrt[3]{2s^2}$
$= \left(2ts - 3t^2\right)\sqrt[3]{2s^2}$

21. $\left(7\sqrt{5} + 4\right)\left(2\sqrt{5} - 1\right)$
$= 14 \cdot 5 - 7\sqrt{5} + 8\sqrt{5} - 4$
$= 70 + \sqrt{5} - 4 = 66 + \sqrt{5}$

22. $\left(\sqrt{3} - 2\sqrt{5}\right)^2$
$= \left(\sqrt{3}\right)^2 - 2 \cdot \sqrt{3} \cdot 2\sqrt{5} + \left(2\sqrt{5}\right)^2$
$= 3 - 4\sqrt{15} + 4 \cdot 5$
$= 3 - 4\sqrt{15} + 20 = 23 - 4\sqrt{15}$

23. $\dfrac{-5}{\sqrt{40}} = \dfrac{-5}{\sqrt{4 \cdot 10}} = \dfrac{-5}{2\sqrt{10}} = \dfrac{-5 \cdot \sqrt{10}}{2\sqrt{10} \cdot \sqrt{10}}$
$= \dfrac{-5\sqrt{10}}{2 \cdot 10} = \dfrac{-5\sqrt{10}}{20} = -\dfrac{\sqrt{10}}{4}$

24. $\dfrac{2}{\sqrt[3]{5}} = \dfrac{2 \cdot \sqrt[3]{5^2}}{\sqrt[3]{5}\sqrt[3]{5^2}} = \dfrac{2\sqrt[3]{25}}{5}$

25. $\dfrac{-4}{\sqrt{7} + \sqrt{5}}$

Multiply the numerator and denominator by the conjugate of the denominator, $\sqrt{7} - \sqrt{5}$.

$= \dfrac{-4\left(\sqrt{7} - \sqrt{5}\right)}{\left(\sqrt{7} + \sqrt{5}\right)\left(\sqrt{7} - \sqrt{5}\right)}$
$= \dfrac{-4\left(\sqrt{7} - \sqrt{5}\right)}{7 - 5}$
$= \dfrac{-4\left(\sqrt{7} - \sqrt{5}\right)}{2} = -2\left(\sqrt{7} - \sqrt{5}\right)$

26. $\dfrac{6 + \sqrt{24}}{2} = \dfrac{6 + \sqrt{4 \cdot 6}}{2} = \dfrac{6 + 2\sqrt{6}}{2}$
$= \dfrac{2\left(3 + \sqrt{6}\right)}{2} = 3 + \sqrt{6}$

27. (a) Substitute 50 for V_0, .01 for k, and 30 for T in the formula.

$V = \dfrac{V_0}{\sqrt{1 - kT}}$
$V = \dfrac{50}{\sqrt{1 - (.01)(30)}}$
$= \dfrac{50}{\sqrt{1 - .3}} = \dfrac{50}{\sqrt{.7}} \approx 59.8$

The velocity is about 59.8.

(b) $V = \dfrac{V_0}{\sqrt{1 - kT}}$
$V^2 = \dfrac{V_0^2}{1 - kT}$ *Square*
$V^2(1 - kT) = V_0^2$
$V^2 - V^2kT = V_0^2$
$V^2 - V_0^2 = V^2kT$
$T = \dfrac{V^2 - V_0^2}{V^2k}$
or $T = \dfrac{V_0^2 - V^2}{-V^2k}$

28. $\sqrt[3]{5x} = \sqrt[3]{2x - 3}$
$\left(\sqrt[3]{5x}\right)^3 = \left(\sqrt[3]{2x - 3}\right)^3$
$5x = 2x - 3$
$3x = -3$
$x = -1$

Check $x = -1$: $\sqrt[3]{-5} \stackrel{?}{=} \sqrt[3]{-5}$ *True*

Solution set: $\{-1\}$

29. $x + \sqrt{x+6} = 9 - x$
Isolate the radical.

$$\sqrt{x+6} = 9 - 2x$$
$$\left(\sqrt{x+6}\right)^2 = (9-2x)^2 \quad \textit{Square}$$
$$x + 6 = 81 - 36x + 4x^2$$
$$0 = 4x^2 - 37x + 75$$
$$0 = (x-3)(4x-25)$$

$x - 3 = 0$ or $4x - 25 = 0$

$x = 3$ $\qquad x = \frac{25}{4}$

Check $x = 3$: $3 + \sqrt{9} \stackrel{?}{=} 6$ *True*

Check $x = \frac{25}{4}$: $\frac{25}{4} + \sqrt{\frac{49}{4}} \stackrel{?}{=} \frac{11}{4}$ *False*

Solution set: $\{3\}$

30. $\sqrt{x+4} - \sqrt{1-x} = -1$

$$\sqrt{x+4} = -1 + \sqrt{1-x}$$
$$\left(\sqrt{x+4}\right)^2 = \left(-1 + \sqrt{1-x}\right)^2$$
$$x + 4 = 1 - 2\sqrt{1-x} + 1 - x$$
$$2x + 2 = -2\sqrt{1-x}$$
$$x + 1 = -\sqrt{1-x}$$
$$(x+1)^2 = \left(-\sqrt{1-x}\right)^2$$
$$x^2 + 2x + 1 = 1 - x$$
$$x^2 + 3x = 0$$
$$x(x+3) = 0$$

$x = 0$ or $x + 3 = 0$

$\qquad x = -3$

Check $x = -3$: $1 - 2 \stackrel{?}{=} -1$ *True*

Check $x = 0$: $2 - 1 \stackrel{?}{=} -1$ *False*

Solution set: $\{-3\}$

31. $(-2+5i) - (3+6i) - 7i$
$= (-2-3) + (5-6-7)i$
$= -5 - 8i$

32. $(1+5i)(3+i)$
$= 3 + i + 15i + 5i^2$
$= 3 + 16i + 5(-1)$
$= -2 + 16i$

33. $\frac{7+i}{1-i}$

Multiply the numerator and denominator by the conjugate of the denominator, $1 + i$.

$$= \frac{(7+i)(1+i)}{(1-i)(1+i)}$$
$$= \frac{7 + 7i + i + i^2}{1 - i^2}$$
$$= \frac{7 + 8i - 1}{1 - (-1)}$$
$$= \frac{6+8i}{2} = \frac{2(3+4i)}{2} = 3 + 4i$$

34. $i^{37} = i^{36} \cdot i = \left(i^4\right)^9 \cdot i = 1^9 \cdot i = 1 \cdot i = i$

35. **(a)** $i^2 = -1$ is *true.*

(b) $i = \sqrt{-1}$ is *true.*

(c) $i = -1$ is *false;* $i = \sqrt{-1}$.

(d) $\sqrt{-3} = i\sqrt{3}$ is *true.*

Cumulative Review Exercises Chapters 1–8

In Exercises 1 and 2, $a = -3$, $b = 5$, and $c = -4$.

1. $\left|2a^2 - 3b + c\right|$
$= \left|2(-3)^2 - 3(5) + (-4)\right|$
$= |2(9) - 15 - 4|$
$= |18 - 19| = |-1| = 1$

2. $\frac{(a+b)(a+c)}{3b-6}$
$= \frac{(-3+5)[(-3+(-4)]}{3(5)-6}$
$= \frac{(2)(-7)}{15-6} = -\frac{14}{9}$

3. $3(x+2) - 4(2x+3) = -3x + 2$

$$3x + 6 - 8x - 12 = -3x + 2$$
$$-5x - 6 = -3x + 2$$
$$-2x = 8$$
$$x = -4$$

Check $x = -4$: $-6 + 20 \stackrel{?}{=} 14$ *True*

Solution set: $\{-4\}$

4. $\frac{1}{3}x + \frac{1}{4}(x+8) = x + 7$

Multiply by the LCD, 12.

$$12\left[\frac{1}{3}x + \frac{1}{4}(x+8)\right] = 12(x+7)$$
$$4x + 3(x+8) = 12x + 84$$
$$4x + 3x + 24 = 12x + 84$$
$$7x + 24 = 12x + 84$$
$$-5x = 60$$
$$x = -12$$

Check $x = -12$: $-4 - 1 \stackrel{?}{=} -5$ *True*

Solution set: $\{-12\}$

5. $.04x + .06(100 - x) = 5.88$

Multiply both sides by 100 to clear decimals.

$$4x + 6(100 - x) = 588$$
$$4x + 600 - 6x = 588$$
$$-2x = -12$$
$$x = 6$$

Check $x = 6$: $.24 + 5.64 \stackrel{?}{=} 5.88$ *True*

Solution set: $\{6\}$

6. $|6x + 7| = 13$

$$6x + 7 = 13 \quad \text{or} \quad 6x + 7 = -13$$
$$6x = 6 \qquad 6x = -20$$
$$x = 1 \quad \text{or} \quad x = -\frac{20}{6} = -\frac{10}{3}$$

Check $x = 1$: $13 \stackrel{?}{=} 13$ *True*

Check $x = -\frac{10}{3}$: $13 \stackrel{?}{=} 13$ *True*

Solution set: $\left\{-\frac{10}{3}, 1\right\}$

7. $|-2x + 4| = |-2x - 3|$

$$-2x + 4 = -2x - 3 \quad \text{or} \quad -2x + 4 = -(-2x - 3)$$
$$4 = -3 \qquad -2x + 4 = 2x + 3$$
$$\textit{False} \qquad -4x = -1$$
$$x = \frac{1}{4}$$

Check $x = \frac{1}{4}$: $\frac{7}{2} \stackrel{?}{=} \frac{7}{2}$ *True*

Solution set: $\left\{\frac{1}{4}\right\}$

8. $-5 - 3(m - 2) < 11 - 2(m + 2)$

$$-5 - 3m + 6 < 11 - 2m - 4$$
$$1 - 3m < 7 - 2m$$
$$-m < 6$$

Multiply by -1; reverse the inequality.

$$m > -6$$

Solution set: $(-6, \infty)$

9. The two angles have the same measure, so

$$10x - 70 = 7x - 25$$
$$3x = 45$$
$$x = 15.$$

Then,

$$10x - 70 = 10(15) - 70$$
$$= 150 - 70 = 80.$$

Each angle measures 80°.

10. Let $x =$ the number of nickels and then $50 - x =$ the number of quarters.

Value of nickels	+	value of quarters	=	total value.
$.05x$	+	$.25(50 - x)$	=	8.90

Multiply by 100 to clear the decimals.

$$5x + 25(50 - x) = 890$$
$$5x + 1250 - 25x = 890$$
$$-20x = -360$$
$$x = 18$$

Since $x = 18$, $50 - x = 50 - 18 = 32$.

There are 18 nickels and 32 quarters.

11. Let $x =$ the amount of pure alcohol.
Make a table.

Number of Liters	Percent (as a decimal)	Liters of Pure Alcohol
x	$100\% = 1$	$1 \cdot x = x$
40	$18\% = .18$	$.18(40) = 7.2$
$x + 40$	$22\% = .22$	$.22(x + 40)$

From the last column:

$$x + 7.2 = .22x + 8.8$$
$$.78x = 1.6$$
$$x = \frac{1.6}{.78} = \frac{160}{78} = \frac{80}{39} \quad \text{or} \quad 2\frac{2}{39}$$

The required amount is $\frac{80}{39}$ or $2\frac{2}{39}$ L of pure alcohol.

12. $4x - 3y = 12$
Let $x = 0$ to find the y-intercept, $(0, -4)$.
Let $y = 0$ to find the x-intercept, $(3, 0)$.
Draw a line through the intercepts.

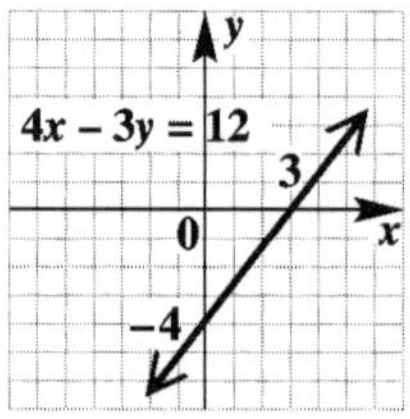

13. $(-4, 6)$ and $(2, -3)$

$$m = \frac{-3 - 6}{2 - (-4)} = \frac{-9}{6} = -\frac{3}{2}$$

Use the point-slope form:

$$y - 6 = -\frac{3}{2}[x - (-4)]$$
$$y - 6 = -\frac{3}{2}(x + 4)$$
$$y - 6 = -\frac{3}{2}x - 6$$
$$y = -\frac{3}{2}x$$

14. $f(x) = 3x - 7$
$f(-10) = 3(-10) - 7 = -30 - 7 = -37$

15. $3x - y = 23$ (1)
$2x + 3y = 8$ (2)
To eliminate y, multiply equation (1) by 3 and add the result to equation (2).

$$\begin{aligned} 9x - 3y &= 69 \quad 3 \times (1) \\ 2x + 3y &= 8 \quad (2) \\ \hline 11x &= 77 \\ x &= 7 \end{aligned}$$

Substitute 7 for x in (2),

$$\begin{aligned} 2(7)+3y&=8\\ 14+3y&=8\\ 3y&=-6\\ y&=-2 \end{aligned}$$

Solution set: $\{(7,-2)\}$

16. $$\begin{aligned} x+y+z&=1\\ x-y-z&=-3\\ x+y-z&=-1 \end{aligned}$$

Write the augmented matrix.

$$\left[\begin{array}{rrr|r} 1 & 1 & 1 & 1\\ 1 & -1 & -1 & -3\\ 1 & 1 & -1 & -1 \end{array}\right]$$

$$\left[\begin{array}{rrr|r} 1 & 1 & 1 & 1\\ 0 & -2 & -2 & -4\\ 0 & 0 & -2 & -2 \end{array}\right] \quad \begin{array}{l} \\ -R_1+R_2\\ -R_1+R_3 \end{array}$$

$$\left[\begin{array}{rrr|r} 1 & 1 & 1 & 1\\ 0 & 1 & 1 & 2\\ 0 & 0 & 1 & 1 \end{array}\right] \quad \begin{array}{l} \\ -\frac{1}{2}R_2\\ -\frac{1}{2}R_3 \end{array}$$

This matrix gives the system

$$\begin{aligned} x+y+z&=1 &\quad (1)\\ y+z&=2 &\quad (2)\\ z&=1. \end{aligned}$$

Substitute 1 for z in (2).

$$\begin{aligned} y+1&=2\\ y&=1 \end{aligned}$$

Substitute 1 for y and 1 for z in (1).

$$\begin{aligned} x+1+1&=1\\ x&=-1 \end{aligned}$$

Solution set: $\{(-1,1,1)\}$

17. Let $x=$ the number of 2-ounce letters and $y=$ the number of 3-ounce letters.

$$\begin{aligned} 5x+3y&=5.09 &\quad (1)\\ 3x+5y&=5.55 &\quad (2) \end{aligned}$$

To eliminate x, multiply (1) by -3 and (2) by 5 and add the results.

$$\begin{array}{rcrl} -15x-9y &=& -15.27 & -3\times(1)\\ 15x+25y &=& 27.75 & 5\times(2)\\ \hline 16y &=& 12.48 & \\ y &=& .78 & \end{array}$$

Substitute $y=.78$ in (1).

$$\begin{aligned} 5x+3y&=5.09 \quad (1)\\ 5x+3(.78)&=5.09\\ 5x+2.34&=5.09\\ 5x&=2.75\\ x&=.55 \end{aligned}$$

The 1997 postage rate for a 2-ounce letter was \$.55 and for a 3-ounce letter, \$.78.

18. $$\begin{aligned} &(3k^3-5k^2+8k-2)-(4k^3+11k+7)\\ &+(2k^2-5k)\\ &=3k^3-4k^3-5k^2+2k^2\\ &\quad+8k-11k-5k-2-7\\ &=-k^3-3k^2-8k-9 \end{aligned}$$

19. $$\begin{aligned} &(8x-7)(x+3)\\ &=8x^2+24x-7x-21\\ &=8x^2+17x-21 \end{aligned}$$

20. $$\begin{aligned} \frac{8z^3-16z^2+24z}{8z^2}&=\frac{8z^3}{8z^2}-\frac{16z^2}{8z^2}+\frac{24z}{8z^2}\\ &=z-2+\frac{3}{z} \end{aligned}$$

21. $$\begin{array}{r} 3y^3-3y^2+4y+1\\ 2y+1\,\overline{)\,6y^4-3y^3+5y^2+6y-9}\\ \underline{6y^4+3y^3\qquad\qquad\qquad}\\ -6y^3+5y^2\qquad\qquad\\ \underline{-6y^3-3y^2\qquad\qquad}\\ 8y^2+6y\qquad\\ \underline{8y^2+4y\qquad}\\ 2y-9\\ \underline{2y+1}\\ -10 \end{array}$$

The answer is

$$3y^3-3y^2+4y+1+\frac{-10}{2y+1}.$$

22. $2p^2-5pq+3q^2=(2p-3q)(p-q)$

23. $$\begin{aligned} 3k^4+k^2-4&=(3k^2+4)(k^2-1)\\ &=(3k^2+4)(k+1)(k-1) \end{aligned}$$

24. $$\begin{aligned} x^3+512&=x^3+8^3\\ &=(x+8)(x^2-8x+64) \end{aligned}$$

25. $$\begin{aligned} 2x^2+11x+15&=0\\ (x+3)(2x+5)&=0 \end{aligned}$$

$x+3=0$ or $2x+5=0$

$x=-3$ or $x=-\frac{5}{2}$

Solution set: $\left\{-3,-\frac{5}{2}\right\}$

26. $$\begin{aligned} 5t(t-1)&=2(1-t)\\ 5t^2-5t&=2-2t\\ 5t^2-3t-2&=0\\ (5t+2)(t-1)&=0 \end{aligned}$$

$5t+2=0$ or $t-1=0$

$t=-\frac{2}{5}$ or $t=1$

Solution set: $\left\{-\frac{2}{5},1\right\}$

27. $f(x)=\frac{2}{x^2-9}=\frac{2}{(x+3)(x-3)}$

The numbers −3 and 3 make the denominator 0 so they must be excluded from the set of all real numbers for the domain of f.

Domain: $\{x|x\neq\pm 3\}$

28. $\frac{y^2+y-12}{y^3+9y^2+20y}\div\frac{y^2-9}{y^3+3y^2}$

$$=\frac{y^2+y-12}{y(y^2+9y+20)}\cdot\frac{y^3+3y^2}{y^2-9}$$
$$=\frac{(y+4)(y-3)}{y(y+4)(y+5)}\cdot\frac{y^2(y+3)}{(y+3)(y-3)}$$
$$=\frac{y}{y+5}$$

29. $\frac{1}{x+y}+\frac{3}{x-y}$ *The LCD is (x+y)(x–y).*

$$=\frac{1(x-y)}{(x+y)(x-y)}+\frac{3(x+y)}{(x-y)(x+y)}$$
$$=\frac{(x-y)+3(x+y)}{(x+y)(x-y)}$$
$$=\frac{x-y+3x+3y}{(x+y)(x-y)}$$
$$=\frac{4x+2y}{(x+y)(x-y)}$$

30. $\frac{\frac{-6}{x-2}}{\frac{8}{3x-6}}=\frac{-6}{x-2}\div\frac{8}{3x-6}$

$$=\frac{-6}{x-2}\cdot\frac{3x-6}{8}$$
$$=\frac{-6}{x-2}\cdot\frac{3(x-2)}{8}$$
$$=\frac{-2\cdot3\cdot3}{2\cdot4}=-\frac{9}{4}$$

31. $\frac{\frac{1}{a}-\frac{1}{b}}{\frac{a}{b}-\frac{b}{a}}$ *The LCD in both numerator and denominator is ab.*

$$=\frac{\frac{b-a}{ab}}{\frac{a^2-b^2}{ab}}$$
$$=\frac{b-a}{ab}\div\frac{a^2-b^2}{ab}$$
$$=\frac{b-a}{ab}\cdot\frac{ab}{a^2-b^2}$$
$$=\frac{b-a}{a^2-b^2}$$
$$=\frac{-(a-b)}{(a-b)(a+b)}=\frac{-1}{a+b}$$

32. $\frac{x^{-1}}{y-x^{-1}}=\frac{\frac{1}{x}}{y-\frac{1}{x}}$

Multiply the numerator and denominator by x.

$$=\frac{x\left(\frac{1}{x}\right)}{x\left(y-\frac{1}{x}\right)}=\frac{1}{xy-1}$$

33. Let $x=$ Chuck's speed.
Then $x+4=$ Natalie's speed.

Use $d=rt$, or $t=\frac{d}{r}$, to make a table.

	Distance	Rate	Time
Chuck	24	x	$\frac{24}{x}$
Natalie	48	$x+4$	$\frac{48}{x+4}$

Since the times are the same,

$$\frac{24}{x}=\frac{48}{x+4}.$$

Multiply by the LCD, $x(x+4)$.

$$x(x+4)\left(\frac{24}{x}\right)=x(x+4)\left(\frac{48}{x+4}\right)$$
$$24(x+4)=48x$$
$$24x+96=48x$$
$$-24x=-96$$
$$x=4$$

Since $x=4$, $x+4=4+4=8$.
Chuck's speed is 4 mph; Natalie's speed is 8 mph.

34. $\frac{p+1}{p-3}=\frac{4}{p-3}+6$

Multiply by the LCD, $p-3$.

$$(p-3)\left(\frac{p+1}{p-3}\right)=(p-3)\left(\frac{4}{p-3}+6\right)$$
$$p+1=4+6(p-3)$$
$$p+1=4+6p-18$$
$$p+1=-14+6p$$
$$-5p=-15$$
$$p=3$$

Substituting 3 in the original equation results in division by zero, so 3 cannot be a solution. The solution set is $\emptyset$.

35. $27^{-2/3}=\frac{1}{27^{2/3}}=\frac{1}{\left[(3^3)^{1/3}\right]^2}=\frac{1}{3^2}=\frac{1}{9}$

36. $\sqrt{200x^4} = \sqrt{100x^4 \cdot 2}$
$= \sqrt{10^2(x^2)^2 \cdot 2} = 10x^2\sqrt{2}$

37. $\sqrt[3]{16x^2y} \cdot \sqrt[3]{3x^3y}$
$= \sqrt[3]{8} \cdot \sqrt[3]{2x^2y} \cdot \sqrt[3]{x^3} \cdot \sqrt[3]{3y}$
$= 2\sqrt[3]{2x^2y} \cdot x\sqrt[3]{3y}$
$= 2x\sqrt[3]{6x^2y^2}$

38. $\sqrt{50} + \sqrt{8} = \sqrt{25 \cdot 2} + \sqrt{4 \cdot 2}$
$= 5\sqrt{2} + 2\sqrt{2} = 7\sqrt{2}$

39. $\dfrac{1}{\sqrt{10} - \sqrt{8}}$
Multiply the numerator and denominator by the conjugate of the denominator, $\sqrt{10} + \sqrt{8}$.
$= \dfrac{1\left(\sqrt{10} + \sqrt{8}\right)}{\left(\sqrt{10} - \sqrt{8}\right)\left(\sqrt{10} + \sqrt{8}\right)}$
$= \dfrac{\sqrt{10} + \sqrt{8}}{10 - 8} = \dfrac{\sqrt{10} + \sqrt{4 \cdot 2}}{2}$
$= \dfrac{\sqrt{10} + 2\sqrt{2}}{2}$

40. $(2\sqrt{x} + \sqrt{y})(-3\sqrt{x} - 4\sqrt{y})$
$= -6x - 8\sqrt{xy} - 3\sqrt{xy} - 4y$
$= -6x - 11\sqrt{xy} - 4y$

41. $(-4, 4)$ and $(-2, 9)$
$d = \sqrt{[(-2) - (-4)]^2 + (9 - 4)^2}$
$= \sqrt{2^2 + 5^2}$
$= \sqrt{4 + 25} = \sqrt{29}$

42. $\sqrt{3r - 8} = r - 2$
Square both sides.
$\left(\sqrt{3r - 8}\right)^2 = (r - 2)^2$
$3r - 8 = r^2 - 4r + 4$
$0 = r^2 - 7r + 12$
$0 = (r - 3)(r - 4)$

$r - 3 = 0$ or $r - 4 = 0$
$r = 3$ or $r = 4$
Check $r = 3$: $\sqrt{1} = 1$ *True*
Check $r = 4$: $\sqrt{4} = 2$ *True*
Solution set: $\{3, 4\}$

43. Substitute 32 for D and 5 for h in the given formula.

$$S = \frac{2.74D}{\sqrt{h}} = \frac{2.74(32)}{\sqrt{5}} = \frac{87.68}{\sqrt{5}} \approx 39.2$$

The fall speed is about 39.2 mph.

44. $(5 + 7i) - (3 - 2i)$
$= (5 - 3) + [7 - (-2)]i$
$= 2 + 9i$

45. $\dfrac{6 - 2i}{1 - i}$
Multiply the numerator and denominator by the conjugate of the denominator, $1 + i$.
$= \dfrac{(6 - 2i)(1 + i)}{(1 - i)(1 + i)}$
$= \dfrac{6 + 6i - 2i - 2i^2}{1^2 + 1^2}$
$= \dfrac{6 + 4i + 2}{1 + 1}$
$= \dfrac{8 + 4i}{2} = \dfrac{2(4 + 2i)}{2} = 4 + 2i$

CHAPTER 9 QUADRATIC EQUATIONS, INEQUALITIES, AND FUNCTIONS

Section 9.1

1. By the square root property, if $x^2 = 16$, then

$$x = +\sqrt{16} \quad \text{or} \quad x = -\sqrt{16}$$

Thus, the equation is also true for $x = -4$.

3. **(a)** A quadratic equation in standard form has a second-degree polynomial in decreasing powers equal to 0.

(b) The zero-factor property states that if a product equals 0, then at least one of the factors equals 0.

(c) The square root property states that if the square of a quantity equals a number, then the quantity equals the positive or negative square root of the number.

5. $x^2 = 81$
$x = 9 \quad \text{or} \quad x = -9$
Solution set: $\{9, -9\}$

7. $t^2 = 17$
$t = \sqrt{17} \quad \text{or} \quad t = -\sqrt{17}$
Solution set: $\left\{\sqrt{17}, -\sqrt{17}\right\}$

9. $m^2 = 32$
$m = \sqrt{32} \quad \text{or} \quad m = -\sqrt{32}$
$m = 4\sqrt{2} \quad \text{or} \quad m = -4\sqrt{2}$
Solution set: $\left\{4\sqrt{2}, -4\sqrt{2}\right\}$

11. $t^2 - 20 = 0$
$t^2 = 20$
$t = \sqrt{20} \quad \text{or} \quad t = -\sqrt{20}$
$t = 2\sqrt{5} \quad \text{or} \quad t = -2\sqrt{5}$
Solution set: $\left\{2\sqrt{5}, -2\sqrt{5}\right\}$

13. $3n^2 - 72 = 0$
$3n^2 = 72$
$n = 24$
$n = \sqrt{24} \quad \text{or} \quad n = -\sqrt{24}$
$n = 2\sqrt{6} \quad \text{or} \quad n = -2\sqrt{6}$
Solution set: $\left\{2\sqrt{6}, -2\sqrt{6}\right\}$

15. $(x+2)^2 = 25$
$x + 2 = \sqrt{25} \quad \text{or} \quad x + 2 = -\sqrt{25}$
$x + 2 = 5 \qquad x + 2 = -5$
$x = 3 \quad \text{or} \quad x = -7$
Solution set: $\{-7, 3\}$

17. $(x-4)^2 = 3$
$x - 4 = \sqrt{3} \quad \text{or} \quad x - 4 = -\sqrt{3}$
$x = 4 + \sqrt{3} \quad \text{or} \quad x = 4 - \sqrt{3}$
Solution set: $\left\{4 + \sqrt{3}, 4 - \sqrt{3}\right\}$

19. $(t+5)^2 = 48$
$t + 5 = \sqrt{48} \quad \text{or} \quad t + 5 = -\sqrt{48}$
$t + 5 = 4\sqrt{3} \qquad t + 5 = -4\sqrt{3}$
$t = -5 + 4\sqrt{3} \quad \text{or} \quad t = -5 - 4\sqrt{3}$
Solution set: $\left\{-5 + 4\sqrt{3}, -5 - 4\sqrt{3}\right\}$

21. $(3x-1)^2 = 7$
$3x - 1 = \sqrt{7} \quad \text{or} \quad 3x - 1 = -\sqrt{7}$
$3x = 1 + \sqrt{7} \qquad 3x = 1 - \sqrt{7}$
$x = \frac{1 + \sqrt{7}}{3} \quad \text{or} \quad x = \frac{1 - \sqrt{7}}{3}$
Solution set: $\left\{\frac{1 + \sqrt{7}}{3}, \frac{1 - \sqrt{7}}{3}\right\}$

23. $(4p+1)^2 = 24$
$4p + 1 = \sqrt{24} \quad \text{or} \quad 4p + 1 = -\sqrt{24}$
$4p + 1 = 2\sqrt{6} \qquad 4p + 1 = -2\sqrt{6}$
$4p = -1 + 2\sqrt{6} \qquad 4p = -1 - 2\sqrt{6}$
$p = \frac{-1 + 2\sqrt{6}}{4} \quad \text{or} \quad p = \frac{-1 - 2\sqrt{6}}{4}$
Solution set: $\left\{\frac{-1 + 2\sqrt{6}}{4}, \frac{-1 - 2\sqrt{6}}{4}\right\}$

25. Use $d = 16t^2$ with $d = 630$.

$630 = 16t^2$
$t^2 = \frac{630}{16} = 39.375$
$t = \sqrt{39.375} \quad \text{or} \quad t = -\sqrt{39.375}$
$t \approx 6.3 \quad \text{or} \quad t \approx -6.3$

Since time cannot be negative, -6.3 is discarded. Therefore, it takes the object about 6.3 seconds to fall 630 feet.

27. To solve $(2x+1)^2 = 5$, we use the square root property and find that $2x + 1 = \sqrt{5}$ or $2x + 1 = -\sqrt{5}$. Then we solve for x in both equations.
To solve $x^2 + 4x = 12$, we find that

$$x^2 + 4x + 4 = 12 + 4$$
$$(x+2)^2 = 16$$

by completing the square.

29. **(a)** We need to add the square of half the coefficient of x to get a perfect square trinomial.

$$\frac{1}{2}(6) = 3 \text{ and } 3^2 = 9$$

Add 9 to $x^2 + 6x$ to get a perfect square trinomial.

(b) $\frac{1}{2}(14) = 7$ and $7^2 = 49$

(c) $\frac{1}{2}(-12) = -6$ and $(-6)^2 = 36$

(d) $\frac{1}{2}(3) = \frac{3}{2}$ and $\left(\frac{3}{2}\right)^2 = \frac{9}{4}$

(e) $\frac{1}{2}(-9) = -\frac{9}{2}$ and $\left(-\frac{9}{2}\right)^2 = \frac{81}{4}$

(f) $\frac{1}{2}\left(-\frac{1}{2}\right) = -\frac{1}{4}$ and $\left(-\frac{1}{4}\right)^2 = \frac{1}{16}$

31. $x^2 + 4x - 2 = 0$

$x^2 + 4x = 2$

$$\left[\frac{1}{2}(4)\right]^2 = 2^2 = 4$$

33. $x^2 + 10x + 18 = 0$

$x^2 + 10x = -18$

$$\left[\frac{1}{2}(10)\right]^2 = 5^2 = 25$$

35. $3w^2 - w - 24 = 0$

$w^2 - \frac{1}{3}w - 8 = 0$ *Divide by 3.*

$w^2 - \frac{1}{3}w = 8$

$$\left[\frac{1}{2}\left(-\frac{1}{3}\right)\right]^2 = \left(-\frac{1}{6}\right)^2 = \frac{1}{36}$$

37. $x^2 - 2x - 24 = 0$

Get the variable terms alone on the left side.

$$x^2 - 2x = 24$$

Complete the square by taking half of -2, the coefficient of x, and squaring the result.

$$\left[\frac{1}{2}(-2)\right]^2 = (-1)^2 = 1$$

Add 1 to each side.

$$x^2 - 2x + 1 = 24 + 1$$

Factor the left side.

$$(x-1)^2 = 25$$

Use the square root property.

$x - 1 = \sqrt{25}$ or $x - 1 = -\sqrt{25}$

$x - 1 = 5$ or $x - 1 = -5$

$x = 6$ or $x = -4$

Solution set: $\{-4, 6\}$

39. $x^2 + 4x - 2 = 0$

$x^2 + 4x = 2$

$x^2 + 4x + 4 = 2 + 4$ $\quad\left[\frac{1}{2}(4)\right]^2 = 4$

$(x+2)^2 = 6$

$x + 2 = \sqrt{6}$ or $x + 2 = -\sqrt{6}$

$x = -2 + \sqrt{6}$ or $x = -2 - \sqrt{6}$

Solution set: $\left\{-2 + \sqrt{6}, -2 - \sqrt{6}\right\}$

41. $x^2 + 10x + 18 = 0$

$x^2 + 10x = -18$

$x^2 + 10x + 25 = -18 + 25$ $\quad\left[\frac{1}{2}(10)\right]^2 = 25$

$(x+5)^2 = 7$

$x + 5 = \sqrt{7}$ or $x + 5 = -\sqrt{7}$

$x = -5 + \sqrt{7}$ or $x = -5 - \sqrt{7}$

Solution set: $\left\{-5 + \sqrt{7}, -5 - \sqrt{7}\right\}$

43. $3w^2 - w = 24$

$w^2 - \frac{1}{3}w = 8$ *Divide by 3.*

Complete the square by taking half of $\frac{1}{3}$, the coefficient of w, and squaring the result.

$$\left[\frac{1}{2}\left(-\frac{1}{3}\right)\right]^2 = \left(-\frac{1}{6}\right)^2 = \frac{1}{36}$$

Add $\frac{1}{36}$ to each side.

$$w^2 - \frac{1}{3}w + \frac{1}{36} = 8 + \frac{1}{36}$$

$$\left(w - \frac{1}{6}\right)^2 = \frac{288}{36} + \frac{1}{36}$$

$$\left(w - \frac{1}{6}\right)^2 = \frac{289}{36}$$

$w - \frac{1}{6} = \sqrt{\frac{289}{36}}$ or $w - \frac{1}{6} = -\sqrt{\frac{289}{36}}$

$w = \frac{1}{6} + \frac{\sqrt{289}}{\sqrt{36}}$ $\quad$ $w = \frac{1}{6} - \frac{\sqrt{289}}{\sqrt{36}}$

$w = \frac{1}{6} + \frac{17}{6}$ $\quad$ $w = \frac{1}{6} - \frac{17}{6}$

$w = \frac{18}{6}$ $\quad$ $w = -\frac{16}{6}$

$w = 3$ or $w = -\frac{8}{3}$

Solution set: $\left\{-\frac{8}{3}, 3\right\}$

45. $2k^2 + 5k - 2 = 0$

$2k^2 + 5k = 2$

$k^2 + \frac{5}{2}k = 1$ *Divide by 2.*

Complete the square.

$$\left(\frac{1}{2} \cdot \frac{5}{2}\right)^2 = \left(\frac{5}{4}\right)^2 = \frac{25}{16}$$

Add $\frac{25}{16}$ to each side.

$$k^2 + \frac{5}{2}k + \frac{25}{16} = 1 + \frac{25}{16}$$

$$\left(k + \frac{5}{4}\right)^2 = \frac{41}{16}$$

$k + \frac{5}{4} = \sqrt{\frac{41}{16}}$ or $k + \frac{5}{4} = -\sqrt{\frac{41}{16}}$

$k = -\frac{5}{4} + \frac{\sqrt{41}}{4}$ $\qquad k = -\frac{5}{4} - \frac{\sqrt{41}}{4}$

$k = \frac{-5 + \sqrt{41}}{4}$ or $k = \frac{-5 - \sqrt{41}}{4}$

Solution set: $\left\{\frac{-5 + \sqrt{41}}{4}, \frac{-5 - \sqrt{41}}{4}\right\}$

47. $5x^2 - 10x + 2 = 0$

$5x^2 - 10x = -2$

$x^2 - 2x = -\frac{2}{5}$ *Divide by 5.*

Complete the square.

$$\left[\frac{1}{2}(-2)\right]^2 = (-1)^2 = 1$$

Add 1 to each side.

$$x^2 - 2x + 1 = -\frac{2}{5} + 1$$

$$(x - 1)^2 = \frac{3}{5}$$

$x - 1 = \sqrt{\frac{3}{5}}$ or $x - 1 = -\sqrt{\frac{3}{5}}$

$x - 1 = \frac{\sqrt{3}}{\sqrt{5}} \cdot \frac{\sqrt{5}}{\sqrt{5}}$ $\qquad x - 1 = -\frac{\sqrt{3}}{\sqrt{5}} \cdot \frac{\sqrt{5}}{\sqrt{5}}$

$x = \frac{5}{5} + \frac{\sqrt{15}}{5}$ $\qquad x = \frac{5}{5} - \frac{\sqrt{15}}{5}$

$x = \frac{5 + \sqrt{15}}{5}$ or $x = \frac{5 - \sqrt{15}}{5}$

Solution set: $\left\{\frac{5 + \sqrt{15}}{5}, \frac{5 - \sqrt{15}}{5}\right\}$

49. $9x^2 - 24x = -13$

$x^2 - \frac{24}{9}x = \frac{-13}{9}$ *Divide by 9.*

$x^2 - \frac{8}{3}x = \frac{-13}{9}$

Complete the square.

$$\left[\frac{1}{2}\left(-\frac{8}{3}\right)\right]^2 = \left(-\frac{4}{3}\right)^2 = \frac{16}{9}$$

Add $\frac{16}{9}$ to each side.

$$x^2 - \frac{8}{3}x + \frac{16}{9} = \frac{-13}{9} + \frac{16}{9}$$

$$\left(x - \frac{4}{3}\right)^2 = \frac{3}{9}$$

$x - \frac{4}{3} = \sqrt{\frac{3}{9}}$ or $x - \frac{4}{3} = -\sqrt{\frac{3}{9}}$

$x = \frac{4}{3} + \frac{\sqrt{3}}{3}$ $\qquad x = \frac{4}{3} - \frac{\sqrt{3}}{3}$

$x = \frac{4 + \sqrt{3}}{3}$ or $x = \frac{4 - \sqrt{3}}{3}$

Solution set: $\left\{\frac{4 + \sqrt{3}}{3}, \frac{4 - \sqrt{3}}{3}\right\}$

51. $z^2 - \frac{4}{3}z = -\frac{1}{9}$

Complete the square.

$$\left[\frac{1}{2}\left(-\frac{4}{3}\right)\right]^2 = \left(-\frac{2}{3}\right)^2 = \frac{4}{9}$$

Add $\frac{4}{9}$ to each side.

$$z^2 - \frac{4}{3}z + \frac{4}{9} = -\frac{1}{9} + \frac{4}{9}$$

$$\left(z - \frac{2}{3}\right)^2 = \frac{3}{9}$$

$z - \frac{2}{3} = \sqrt{\frac{3}{9}}$ or $z - \frac{2}{3} = -\sqrt{\frac{3}{9}}$

$z = \frac{2}{3} + \frac{\sqrt{3}}{3}$ $\qquad z = \frac{2}{3} - \frac{\sqrt{3}}{3}$

$z = \frac{2 + \sqrt{3}}{3}$ or $z = \frac{2 - \sqrt{3}}{3}$

Solution set: $\left\{\frac{2 + \sqrt{3}}{3}, \frac{2 - \sqrt{3}}{3}\right\}$

53. $.1x^2 - .2x - .1 = 0$

Multiply each side by 10 to clear the decimals.

$x^2 - 2x - 1 = 0$

$x^2 - 2x = 1$

Complete the square.

$$\left[\frac{1}{2}(-2)\right]^2 = (-1)^2 = 1$$

Add 1 to each side.

$$x^2 - 2x + 1 = 1 + 1$$

$$(x - 1)^2 = 2$$

$x - 1 = \sqrt{2}$ or $x - 1 = -\sqrt{2}$

$x = 1 + \sqrt{2}$ or $x = 1 - \sqrt{2}$

Solution set: $\left\{1 + \sqrt{2}, 1 - \sqrt{2}\right\}$

55. $x^2 = -12$

$x = \sqrt{-12}$ or $x = -\sqrt{-12}$

$x = i\sqrt{12}$ $\quad x = -i\sqrt{12}$

$x = 2i\sqrt{3}$ or $x = -2i\sqrt{3}$

Solution set: $\left\{2i\sqrt{3}, -2i\sqrt{3}\right\}$

57. $(r-5)^2 = -3$

$r - 5 = \sqrt{-3}$ or $r - 5 = -\sqrt{-3}$

$r = 5 + i\sqrt{3}$ or $r = 5 - i\sqrt{3}$

Solution set: $\left\{5 + i\sqrt{3}, 5 - i\sqrt{3}\right\}$

59. $(6k-1)^2 = -8$

$6k - 1 = \sqrt{-8}$ or $6k - 1 = -\sqrt{-8}$

$6k - 1 = i\sqrt{8}$ $\quad 6k - 1 = -i\sqrt{8}$

$6k - 1 = 2i\sqrt{2}$ $\quad 6k - 1 = -2i\sqrt{2}$

$6k = 1 + 2i\sqrt{2}$ $\quad 6k = 1 - 2i\sqrt{2}$

$k = \dfrac{1 + 2i\sqrt{2}}{6}$ or $k = \dfrac{1 - 2i\sqrt{2}}{6}$

Solution set: $\left\{\dfrac{1 + 2i\sqrt{2}}{6}, \dfrac{1 - 2i\sqrt{2}}{6}\right\}$

61. $m^2 + 4m + 13 = 0$

$m^2 + 4m = -13$

Complete the square.

$$\left(\frac{1}{2}\cdot 4\right)^2 = 2^2 = 4$$

Add 4 to each side.

$m^2 + 4m + 4 = -13 + 4$

$(m+2)^2 = -9$

$m + 2 = \sqrt{-9}$ or $m + 2 = -\sqrt{-9}$

$m = -2 + 3i$ or $m = -2 - 3i$

Solution set: $\{-2 + 3i, -2 - 3i\}$

63. $3r^2 + 4r + 4 = 0$

$3r^2 + 4r = -4$

$r^2 + \dfrac{4}{3}r = \dfrac{-4}{3}$ *Divide by 3.*

Complete the square.

$$\left(\frac{1}{2}\cdot\frac{4}{3}\right)^2 = \left(\frac{2}{3}\right)^2 = \frac{4}{9}$$

Add $\dfrac{4}{9}$ to each side.

$r^2 + \dfrac{4}{3}r + \dfrac{4}{9} = \dfrac{-4}{3} + \dfrac{4}{9}$

$\left(r + \dfrac{2}{3}\right)^2 = \dfrac{-8}{9}$

$r + \dfrac{2}{3} = \dfrac{\sqrt{-8}}{\sqrt{9}}$ or $r + \dfrac{2}{3} = -\dfrac{\sqrt{-8}}{\sqrt{9}}$

$r = -\dfrac{2}{3} + \dfrac{2i\sqrt{2}}{3}$ $\quad r = -\dfrac{2}{3} - \dfrac{2i\sqrt{2}}{3}$

$r = \dfrac{-2 + 2i\sqrt{2}}{3}$ or $r = \dfrac{-2 - 2i\sqrt{2}}{3}$

Solution set: $\left\{\dfrac{-2 + 2i\sqrt{2}}{3}, \dfrac{-2 - 2i\sqrt{2}}{3}\right\}$

65. $-m^2 - 6m - 12 = 0$

Multiply each side by -1.

$m^2 + 6m + 12 = 0$

$m^2 + 6m = -12$

Complete the square.

$$\left(\frac{1}{2}\cdot 6\right)^2 = 3^2 = 9$$

Add 9 to each side.

$m^2 + 6m + 9 = -12 + 9$

$(m+3)^2 = -3$

$m + 3 = \sqrt{-3}$ or $m + 3 = -\sqrt{-3}$

$m = -3 + i\sqrt{3}$ or $m = -3 - i\sqrt{3}$

Solution set: $\left\{-3 + i\sqrt{3}, -3 - i\sqrt{3}\right\}$

67. $x^2 - b = 0$

$x^2 = b$

$x = \sqrt{b}$ or $x = -\sqrt{b}$

Solution set: $\left\{\sqrt{b}, -\sqrt{b}\right\}$

69. $4x^2 = b^2 + 16$

$x^2 = \dfrac{b^2 + 16}{4}$

$x = \sqrt{\dfrac{b^2 + 16}{4}}$ or $x = -\sqrt{\dfrac{b^2 + 16}{4}}$

$x = \dfrac{\sqrt{b^2 + 16}}{2}$ or $x = -\dfrac{\sqrt{b^2 + 16}}{2}$

Solution set: $\left\{\dfrac{\sqrt{b^2 + 16}}{2}, -\dfrac{\sqrt{b^2 + 16}}{2}\right\}$

71. $(5x - 2b)^2 = 3a$

$5x - 2b = \sqrt{3a}$ or $5x - 2b = -\sqrt{3a}$

$5x = 2b + \sqrt{3a}$ $\quad 5x = 2b - \sqrt{3a}$

$x = \dfrac{2b + \sqrt{3a}}{5}$ or $x = \dfrac{2b - \sqrt{3a}}{5}$

Solution set: $\left\{\dfrac{2b + \sqrt{3a}}{5}, \dfrac{2b - \sqrt{3a}}{5}\right\}$

73. The area of the original square is $x \cdot x$, or x^2.

74. Each rectangular strip has length x and width 1, so each strip has an area of $x \cdot 1$, or x.

75. From Exercise 74, the area of a rectangular strip is x. The area of 6 rectangular strips is $6x$.

76. These are 1 by 1 squares, so each has an area of $1 \cdot 1$, or 1.

77. There are 9 small squares, each with area 1 (from Exercise 76), so the total area is $9 \cdot 1$, or 9.

78. The area of the larger square is $(x+3)^2$. Using the results from Exercises 73–77,

$$(x+3)^2 = x^2 + 6x + 9.$$

79. As in the example, we have $x^2 = 17$ or $x = \pm\sqrt{17}$.

Section 9.2

1. If the expression $\frac{-b \pm \sqrt{b^2-4ac}}{2a}$ were written as two terms, we would have $-\frac{b}{2a} \pm \frac{\sqrt{b^2-4ac}}{2a}$. Therefore,

$$\frac{-b \pm \sqrt{b^2-4ac}}{2a} \neq -b \pm \frac{\sqrt{b^2-4ac}}{2a}.$$

3. The last step is wrong. Because 5 is not a common factor in the numerator, the fraction cannot be reduced. The solutions are $\frac{5 \pm \sqrt{5}}{10}$.

5. $m^2 - 8m + 15 = 0$

Here $a = 1$, $b = -8$, and $c = 15$.

$$m = \frac{-b \pm \sqrt{b^2-4ac}}{2a}$$

$$m = \frac{-(-8) \pm \sqrt{(-8)^2 - 4(1)(15)}}{2(1)}$$

$$= \frac{8 \pm \sqrt{64-60}}{2}$$

$$= \frac{8 \pm \sqrt{4}}{2} = \frac{8 \pm 2}{2}$$

$$m = \frac{8+2}{2} = \frac{10}{2} = 5 \text{ or}$$

$$m = \frac{8-2}{2} = \frac{6}{2} = 3$$

Solution set: $\{3, 5\}$

7. $2k^2 + 4k + 1 = 0$

Here $a = 2$, $b = 4$, and $c = 1$.

$$k = \frac{-b \pm \sqrt{b^2-4ac}}{2a}$$

$$k = \frac{-4 \pm \sqrt{4^2 - 4(2)(1)}}{2(2)}$$

$$= \frac{-4 \pm \sqrt{16-8}}{4}$$

$$= \frac{-4 \pm \sqrt{8}}{4} = \frac{-4 \pm 2\sqrt{2}}{4}$$

$$= \frac{2\left(-2 \pm \sqrt{2}\right)}{2 \cdot 2} = \frac{-2 \pm \sqrt{2}}{2}$$

Solution set: $\left\{\frac{-2+\sqrt{2}}{2}, \frac{-2-\sqrt{2}}{2}\right\}$

9. $2x^2 - 2x = 1$

$2x^2 - 2x - 1 = 0$

Here $a = 2$, $b = -2$, and $c = -1$.

$$x = \frac{-b \pm \sqrt{b^2-4ac}}{2a}$$

$$x = \frac{-(-2) \pm \sqrt{(-2)^2 - 4(2)(-1)}}{2(2)}$$

$$= \frac{2 \pm \sqrt{4+8}}{4} = \frac{2 \pm \sqrt{12}}{4} = \frac{2 \pm 2\sqrt{3}}{4}$$

$$= \frac{2\left(1 \pm \sqrt{3}\right)}{2 \cdot 2} = \frac{1 \pm \sqrt{3}}{2}$$

Solution set: $\left\{\frac{1+\sqrt{3}}{2}, \frac{1-\sqrt{3}}{2}\right\}$

11. $x^2 + 18 = 10x$

$x^2 - 10x + 18 = 0$

Here $a = 1$, $b = -10$, and $c = 18$.

$$x = \frac{-b \pm \sqrt{b^2-4ac}}{2a}$$

$$x = \frac{-(-10) \pm \sqrt{(-10)^2 - 4(1)(18)}}{2(1)}$$

$$= \frac{10 \pm \sqrt{100-72}}{2} = \frac{10 \pm \sqrt{28}}{2}$$

$$= \frac{10 \pm 2\sqrt{7}}{2} = \frac{2\left(5 \pm \sqrt{7}\right)}{2} = 5 \pm \sqrt{7}$$

Solution set: $\left\{5+\sqrt{7}, 5-\sqrt{7}\right\}$

13. $4k^2 + 4k - 1 = 0$

Here $a = 4$, $b = 4$, and $c = -1$.

$$k = \frac{-b \pm \sqrt{b^2-4ac}}{2a}$$

$$k = \frac{-4 \pm \sqrt{4^2 - 4(4)(-1)}}{2(4)}$$

$$= \frac{-4 \pm \sqrt{16+16}}{8} = \frac{-4 \pm \sqrt{32}}{8}$$

$$= \frac{-4 \pm 4\sqrt{2}}{8} = \frac{4\left(-1 \pm \sqrt{2}\right)}{2 \cdot 4} = \frac{-1 \pm \sqrt{2}}{2}$$

Solution set: $\left\{\frac{-1+\sqrt{2}}{2}, \frac{-1-\sqrt{2}}{2}\right\}$

15. $2 - 2x = 3x^2$

$0 = 3x^2 + 2x - 2$

Here $a = 3$, $b = 2$, and $c = -2$.

$$x = \frac{-b \pm \sqrt{b^2 - 4ac}}{2a}$$

$$x = \frac{-2 \pm \sqrt{2^2 - 4(3)(-2)}}{2(3)}$$

$$= \frac{-2 \pm \sqrt{4 + 24}}{6} = \frac{-2 \pm \sqrt{28}}{6}$$

$$= \frac{-2 \pm 2\sqrt{7}}{6} = \frac{2\left(-1 \pm \sqrt{7}\right)}{2 \cdot 3} = \frac{-1 \pm \sqrt{7}}{3}$$

Solution set: $\left\{\frac{-1 + \sqrt{7}}{3}, \frac{-1 - \sqrt{7}}{3}\right\}$

17. $\frac{x^2}{4} - \frac{x}{2} = 1$

$\frac{x^2}{4} - \frac{x}{2} - 1 = 0$

$x^2 - 2x - 4 = 0$ *Multiply by 4.*

Here $a = 1$, $b = -2$, and $c = -4$.

$$x = \frac{-b \pm \sqrt{b^2 - 4ac}}{2a}$$

$$x = \frac{-(-2) \pm \sqrt{(-2)^2 - 4(1)(-4)}}{2(1)}$$

$$= \frac{2 \pm \sqrt{4 + 16}}{2} = \frac{2 \pm \sqrt{20}}{2}$$

$$= \frac{2 \pm 2\sqrt{5}}{2} = 1 \pm \sqrt{5}$$

Solution set: $\left\{1 + \sqrt{5}, 1 - \sqrt{5}\right\}$

19. $-2t(t + 2) = -3$

$-2t^2 - 4t = -3$

$-2t^2 - 4t + 3 = 0$

Here $a = -2$, $b = -4$, and $c = 3$.

$$t = \frac{-b \pm \sqrt{b^2 - 4ac}}{2a}$$

$$t = \frac{-(-4) \pm \sqrt{(-4)^2 - 4(-2)(3)}}{2(-2)}$$

$$= \frac{4 \pm \sqrt{16 + 24}}{-4} = \frac{4 \pm \sqrt{40}}{-4}$$

$$= \frac{4 \pm 2\sqrt{10}}{-4} = \frac{2\left(2 \pm \sqrt{10}\right)}{-2 \cdot 2}$$

$$= \frac{2 \pm \sqrt{10}}{-2} \cdot \frac{-1}{-1} = \frac{-2 \mp \sqrt{10}}{2}$$

$$= \frac{-2 \pm \sqrt{10}}{2}$$

Solution set: $\left\{\frac{-2 + \sqrt{10}}{2}, \frac{-2 - \sqrt{10}}{2}\right\}$

21. $(r - 3)(r + 5) = 2$

$r^2 + 2r - 15 = 2$

$r^2 + 2r - 17 = 0$

Here $a = 1$, $b = 2$, and $c = -17$.

$$r = \frac{-b \pm \sqrt{b^2 - 4ac}}{2a}$$

$$r = \frac{-2 \pm \sqrt{2^2 - 4(1)(-17)}}{2(1)}$$

$$= \frac{-2 \pm \sqrt{4 + 68}}{2} = \frac{-2 \pm \sqrt{72}}{2}$$

$$= \frac{-2 \pm 6\sqrt{2}}{2} = \frac{-2\left(1 \pm 3\sqrt{2}\right)}{2}$$

$$= -1\left(1 \pm 3\sqrt{2}\right) = -1 \mp 3\sqrt{2} = -1 \pm 3\sqrt{2}$$

Solution set: $\left\{-1 + 3\sqrt{2}, -1 - 3\sqrt{2}\right\}$

23. $(g + 2)(g - 3) = 1$

$g^2 - g - 6 = 1$

$g^2 - g - 7 = 0$

Here $a = 1$, $b = -1$, and $c = -7$.

$$g = \frac{-b \pm \sqrt{b^2 - 4ac}}{2a}$$

$$g = \frac{-(-1) \pm \sqrt{(-1)^2 - 4(1)(-7)}}{2(1)}$$

$$= \frac{1 \pm \sqrt{1 + 28}}{2} = \frac{1 \pm \sqrt{29}}{2}$$

Solution set: $\left\{\frac{1 + \sqrt{29}}{2}, \frac{1 - \sqrt{29}}{2}\right\}$

25. $p = \frac{5(5 - p)}{3(p + 1)}$

$3p(p + 1) = 5(5 - p)$

$3p^2 + 3p = 25 - 5p$

$3p^2 + 8p - 25 = 0$

Here $a = 3$, $b = 8$, and $c = -25$.

$$p = \frac{-b \pm \sqrt{b^2 - 4ac}}{2a}$$

$$p = \frac{-8 \pm \sqrt{8^2 - 4(3)(-25)}}{2(3)}$$

$$= \frac{-8 \pm \sqrt{64 + 300}}{6} = \frac{-8 \pm \sqrt{364}}{6}$$

$$= \frac{-8 \pm 2\sqrt{91}}{6} = \frac{2\left(-4 \pm \sqrt{91}\right)}{2 \cdot 3} = \frac{-4 \pm \sqrt{91}}{3}$$

Solution set: $\left\{\frac{-4 + \sqrt{91}}{3}, \frac{-4 - \sqrt{91}}{3}\right\}$

27. $x^2 - 3x + 6 = 0$

Here $a = 1$, $b = -3$, and $c = 6$.

$$x = \frac{-b \pm \sqrt{b^2 - 4ac}}{2a}$$

$$x = \frac{-(-3) \pm \sqrt{(-3)^2 - 4(1)(6)}}{2(1)}$$

$$= \frac{3 \pm \sqrt{9 - 24}}{2}$$

$$= \frac{3 \pm \sqrt{-15}}{2}$$

$$= \frac{3 \pm i\sqrt{15}}{2} = \frac{3}{2} \pm \frac{\sqrt{15}}{2}i$$

Solution set: $\left\{\frac{3}{2} + \frac{\sqrt{15}}{2}i, \frac{3}{2} - \frac{\sqrt{15}}{2}i\right\}$

29. $r^2 - 6r + 14 = 0$

Here $a = 1$, $b = -6$, and $c = 14$.

$$r = \frac{-b \pm \sqrt{b^2 - 4ac}}{2a}$$

$$r = \frac{-(-6) \pm \sqrt{(-6)^2 - 4(1)(14)}}{2(1)}$$

$$= \frac{6 \pm \sqrt{36 - 56}}{2}$$

$$= \frac{6 \pm \sqrt{-20}}{2} = \frac{6 \pm 2i\sqrt{5}}{2}$$

$$= \frac{2\left(3 \pm i\sqrt{5}\right)}{2} = 3 \pm i\sqrt{5}$$

Solution set: $\left\{3 + i\sqrt{5}, 3 - i\sqrt{5}\right\}$

31. $4x^2 - 4x = -7$

$4x^2 - 4x + 7 = 0$

Here $a = 4$, $b = -4$, and $c = 7$.

$$x = \frac{-b \pm \sqrt{b^2 - 4ac}}{2a}$$

$$x = \frac{-(-4) \pm \sqrt{(-4)^2 - 4(4)(7)}}{2(4)}$$

$$= \frac{4 \pm \sqrt{16 - 112}}{8} = \frac{4 \pm \sqrt{-96}}{8}$$

$$= \frac{4 \pm 4i\sqrt{6}}{8} = \frac{4\left(1 \pm i\sqrt{6}\right)}{2 \cdot 4}$$

$$= \frac{1 \pm i\sqrt{6}}{2} = \frac{1}{2} \pm \frac{\sqrt{6}}{2}i$$

Solution set: $\left\{\frac{1}{2} + \frac{\sqrt{6}}{2}i, \frac{1}{2} - \frac{\sqrt{6}}{2}i\right\}$

33. $x(3x + 4) = -2$

$3x^2 + 4x = -2$

$3x^2 + 4x + 2 = 0$

Here $a = 3$, $b = 4$, and $c = 2$.

$$x = \frac{-b \pm \sqrt{b^2 - 4ac}}{2a}$$

$$x = \frac{-4 \pm \sqrt{4^2 - 4(3)(2)}}{2(3)}$$

$$= \frac{-4 \pm \sqrt{16 - 24}}{6} = \frac{-4 \pm \sqrt{-8}}{6}$$

$$= \frac{-4 \pm 2i\sqrt{2}}{6} = \frac{2\left(-2 \pm i\sqrt{2}\right)}{2 \cdot 3}$$

$$= \frac{-2 \pm i\sqrt{2}}{3} = -\frac{2}{3} \pm \frac{\sqrt{2}}{3}i$$

Solution set: $\left\{-\frac{2}{3} + \frac{\sqrt{2}}{3}i, -\frac{2}{3} - \frac{\sqrt{2}}{3}i\right\}$

35. $(x + 5)(x - 6) = (2x - 1)(x - 4)$

$x^2 - x - 30 = 2x^2 - 9x + 4$

$0 = x^2 - 8x + 34$

Here $a = 1$, $b = -8$, and $c = 34$.

$$x = \frac{-b \pm \sqrt{b^2 - 4ac}}{2a}$$

$$x = \frac{-(-8) \pm \sqrt{(-8)^2 - 4(1)(34)}}{2(1)}$$

$$= \frac{8 \pm \sqrt{64 - 136}}{2} = \frac{8 \pm \sqrt{-72}}{2}$$

$$= \frac{8 \pm 6i\sqrt{2}}{2} = \frac{2\left(4 \pm 3i\sqrt{2}\right)}{2} = 4 \pm 3i\sqrt{2}$$

Solution set: $\left\{4 + 3i\sqrt{2}, 4 - 3i\sqrt{2}\right\}$

37. $25x^2 + 70x + 49 = 0$

Here $a = 25$, $b = 70$, and $c = 49$, so the discriminant is

$$b^2 - 4ac = 70^2 - 4(25)(49)$$
$$= 4900 - 4900$$
$$= 0.$$

Since the discriminant is 0, the quantity under the radical in the quadratic formula is 0, and there is only one rational solution. The answer is **B**.

39. $x^2 + 4x + 2 = 0$

Here $a = 1$, $b = 4$, and $c = 2$, so the discriminant is

$$b^2 - 4ac = 4^2 - 4(1)(2)$$
$$= 16 - 8$$
$$= 8.$$

Since the discriminant is positive, but not a perfect square, there are two distinct irrational number solutions. The answer is **C**.

41. $$3x^2 = 5x + 2$$
$$3x^2 - 5x - 2 = 0$$

Here $a = 3$, $b = -5$, and $c = -2$, so the discriminant is

$$\begin{aligned} b^2 - 4ac &= (-5)^2 - 4(3)(-2) \\ &= 25 + 24 \\ &= 49. \end{aligned}$$

Since the discriminant is a perfect square, there are two distinct rational solutions. The answer is **A**.

43. $3m^2 - 10m + 15 = 0$

Here $a = 3$, $b = -10$, and $c = 15$, so the discriminant is

$$\begin{aligned} b^2 - 4ac &= (-10)^2 - 4(3)(15) \\ &= 100 - 180 \\ &= -80. \end{aligned}$$

Since the discriminant is negative, there are two distinct imaginary number solutions. The answer is **D**.

45. The equations in Exercises 37, 38, 41, and 42 can be solved by factoring because they have only rational solutions.

47. $$\begin{aligned} 4k^2 - 28k + 49 &= 0 \\ (2k - 7)^2 &= 0 \\ 2k - 7 &= 0 \\ k &= \frac{7}{2} \end{aligned}$$

Solution set: $\left\{\frac{7}{2}\right\}$

49. $$\begin{aligned} 4x^2 &= 4x + 3 \\ 4x^2 - 4x - 3 &= 0 \\ (2x + 1)(2x - 3) &= 0 \end{aligned}$$

$$2x + 1 = 0 \quad \text{or} \quad 2x - 3 = 0$$
$$x = -\frac{1}{2} \quad \text{or} \quad x = \frac{3}{2}$$

Solution set: $\left\{-\frac{1}{2}, \frac{3}{2}\right\}$

51. No, because an irrational solution occurs only if the discriminant is positive, but not the square of an integer. In that case, there will be two irrational solutions.

53. $p^2 + bp + 25 = 0$
For there to be only one rational solution, $b^2 - 4ac$ must equal zero.
Since $a = 1$ and $c = 25$,

$$\begin{aligned} b^2 - 4(1)(25) &= 0 \\ b^2 - 100 &= 0 \\ b^2 &= 100 \\ b &= \pm\sqrt{100} \end{aligned}$$
$$b = 10 \quad \text{or} \quad b = -10.$$

55. $am^2 + 8m + 1 = 0$
For there to be only one rational solution, $b^2 - 4ac$ must equal zero.
Since $b = 8$ and $c = 1$,

$$\begin{aligned} 8^2 - 4(a)(1) &= 0 \\ 64 - 4a &= 0 \\ -4a &= -64 \\ a &= 16. \end{aligned}$$

57. $9x^2 - 30x + c = 0$
Again, $b^2 - 4ac$ must equal zero.
Since $a = 9$ and $b = -30$,

$$\begin{aligned} (-30)^2 - 4(9)(c) &= 0 \\ 900 - 36c &= 0 \\ -36c &= -900 \\ c &= 25. \end{aligned}$$

59. Substitute $-\frac{5}{2}$ for x and solve for b.

$$\begin{aligned} 4x^2 + bx - 3 &= 0 \\ 4\left(-\frac{5}{2}\right)^2 + b\left(-\frac{5}{2}\right) - 3 &= 0 \\ 25 - \frac{5}{2}b - 3 &= 0 \\ 22 &= \frac{5}{2}b \\ \frac{44}{5} &= b \end{aligned}$$

So the equation is

$$4x^2 + \frac{44}{5}x - 3 = 0.$$

Now multiply by 5.

$$20x^2 + 44x - 15 = 0$$

Since $-\frac{5}{2}$ is a solution, $2x + 5$ must be one factor.

$$(2x + 5)(10x - 3) = 0$$

So the other solution is $\frac{3}{10}$.

Section 9.3

1. $(2x + 3)^2 = 4$

Since the equation has the form $(ax + b)^2 = c$, use the *square root property*.

3. $z^2 + 5z - 8 = 0$

The discriminant is

$$\begin{aligned} b^2 - 4ac &= 5^2 - 4(1)(-8) \\ &= 25 + 32 = 57. \end{aligned}$$

Since the discriminant is not a perfect square, use the *quadratic formula*.

5.
$$3m^2 = 2 - 5m$$
$$3m^2 + 5m - 2 = 0$$

The discriminant is

$$b^2 - 4ac = 5^2 - 4(3)(-2) = 25 + 24 = 49.$$

Since the discriminant is a perfect square, use *factoring.*

7. $\dfrac{14}{x} = x - 5$

This is a rational equation, so multiply both sides by the LCD, x.

9. $(r^2 + r)^2 - 8(r^2 + r) + 12 = 0$

This is quadratic in form, so substitute a variable for $r^2 + r$.

11. The potential solution -1 does not check. The solution set is $\{4\}$.

13.
$$\frac{14}{x} = x - 5$$

To clear the fraction, multiply each term by the LCD, x.

$$x\left(\frac{14}{x}\right) = x(x) - 5x$$
$$14 = x^2 - 5x$$
$$-x^2 + 5x + 14 = 0$$
$$x^2 - 5x - 14 = 0$$
$$(x + 2)(x - 7) = 0$$

$x + 2 = 0$ or $x - 7 = 0$

$x = -2$ or $x = 7$

Check $x = -2$: $-7 \stackrel{?}{=} -7$ *True*

Check $x = 7$: $2 \stackrel{?}{=} 2$ *True*

Solution set: $\{-2, 7\}$

15.
$$1 - \frac{3}{x} - \frac{28}{x^2} = 0$$

Multiply by the LCD, x^2.

$$x^2(1) - x^2\left(\frac{3}{x}\right) - x^2\left(\frac{28}{x^2}\right) = x^2 \cdot 0$$
$$x^2 - 3x - 28 = 0$$
$$(x + 4)(x - 7) = 0$$

$x + 4 = 0$ or $x - 7 = 0$

$x = -4$ or $x = 7$

Check $x = -4$: $1 + \frac{3}{4} - \frac{7}{4} \stackrel{?}{=} 0$ *True*

Check $x = 7$: $1 - \frac{3}{7} - \frac{4}{7} \stackrel{?}{=} 0$ *True*

Solution set: $\{-4, 7\}$

17. $3 - \dfrac{1}{t} = \dfrac{2}{t^2}$

Multiply each term by the LCD, t^2.

$$t^2(3) - t^2\left(\frac{1}{t}\right) = t^2\left(\frac{2}{t^2}\right)$$
$$3t^2 - t = 2$$
$$3t^2 - t - 2 = 0$$
$$(3t + 2)(t - 1) = 0$$

$3t + 2 = 0$ or $t - 1 = 0$

$t = -\frac{2}{3}$ or $t = 1$

Check $t = \frac{2}{3}$: $3 + \frac{3}{2} \stackrel{?}{=} \frac{9}{2}$ *True*

Check $t = 1$: $3 - 1 \stackrel{?}{=} 2$ *True*

Solution set: $\left\{-\frac{2}{3}, 1\right\}$

19.
$$\frac{1}{x} + \frac{2}{x + 2} = \frac{17}{35}$$

Multiply by the LCD, $35x(x + 2)$.

$$35x(x + 2)\left(\frac{1}{x}\right) + 35x(x + 2)\left(\frac{2}{x + 2}\right) = 35x(x + 2)\left(\frac{17}{35}\right)$$
$$35(x + 2) + 35x(2) = 17x(x + 2)$$
$$35x + 70 + 70x = 17x^2 + 34x$$
$$70 + 105x = 17x^2 + 34x$$
$$0 = 17x^2 - 71x - 70$$
$$0 = (17x + 14)(x - 5)$$

$17x + 14 = 0$ or $x - 5 = 0$

$x = -\frac{14}{17}$ or $x = 5$

Check $x = -\frac{14}{17}$: $-\frac{17}{14} + \frac{17}{10} \stackrel{?}{=} \frac{17}{35}$ *True*

Check $x = 5$: $\frac{1}{5} + \frac{2}{7} \stackrel{?}{=} \frac{17}{35}$ *True*

Solution set: $\left\{-\frac{14}{17}, 5\right\}$

21. $\dfrac{2}{x + 1} + \dfrac{3}{x + 2} = \dfrac{7}{2}$

Multiply by the LCD, $2(x + 1)(x + 2)$.

$$2(x + 1)(x + 2)\left(\frac{2}{x + 1} + \frac{3}{x + 2}\right) = 2(x + 1)(x + 2)\left(\frac{7}{2}\right)$$
$$2(x + 2)(2) + 2(x + 1)(3) = (x + 1)(x + 2)(7)$$
$$4x + 8 + 6x + 6 = (x^2 + 3x + 2)(7)$$
$$10x + 14 = 7x^2 + 21x + 14$$
$$7x^2 + 11x = 0$$
$$x(7x + 11) = 0$$

continued

$x = 0$ or $7x + 11 = 0$

$x = -\frac{11}{7}$

Check $x = -\frac{11}{7}$: $-\frac{7}{2} + 7 \stackrel{?}{=} \frac{7}{2}$ *True*

Check $x = 0$: $2 + \frac{3}{2} \stackrel{?}{=} \frac{7}{2}$ *True*

Solution set: $\left\{-\frac{11}{7}, 0\right\}$

23. $\frac{3}{2x} - \frac{1}{2(x+2)} = 1$

Multiply by the LCD, $2x(x+2)$.

$$2x(x+2)\left(\frac{3}{2x} - \frac{1}{2(x+2)}\right) = 2x(x+2) \cdot 1$$
$$3(x+2) - x(1) = 2x(x+2)$$
$$3x + 6 - x = 2x^2 + 4x$$
$$0 = 2x^2 + 2x - 6$$
$$0 = x^2 + x - 3$$

Use $a = 1$, $b = 1$, $c = -3$ in the quadratic formula.

$$x = \frac{-b \pm \sqrt{b^2 - 4ac}}{2a}$$
$$x = \frac{-1 \pm \sqrt{1^2 - 4(1)(-3)}}{2(1)}$$
$$= \frac{-1 \pm \sqrt{1+12}}{2} = \frac{-1 \pm \sqrt{13}}{2}$$

Use a calculator to check both potential solutions. Both solutions check.

Solution set: $\left\{\frac{-1+\sqrt{13}}{2}, \frac{-1-\sqrt{13}}{2}\right\}$

25. $3 = \frac{1}{t+2} + \frac{2}{(t+2)^2}$

Multiply by the LCD, $(t+2)^2$.

$$3(t+2)^2 = 1(t+2) + 2$$
$$3(t^2 + 4t + 4) = t + 2 + 2$$
$$3t^2 + 12t + 12 = t + 4$$
$$3t^2 + 11t + 8 = 0$$
$$(3t + 8)(t + 1) = 0$$

$3t + 8 = 0$ or $t + 1 = 0$

$t = -\frac{8}{3}$ $\quad t = -1$

Check $t = -\frac{8}{3}$: $3 \stackrel{?}{=} -\frac{3}{2} + \frac{9}{2}$ *True*

Check $t = -1$: $3 \stackrel{?}{=} 1 + 2$ *True*

Solution set: $\left\{-\frac{8}{3}, -1\right\}$

27. $\frac{6}{p} = 2 + \frac{p}{p+1}$

Multiply by the LCD, $p(p+1)$.

$$6(p+1) = 2p(p+1) + p \cdot p$$
$$6p + 6 = 2p^2 + 2p + p^2$$
$$0 = 3p^2 - 4p - 6$$

Use $a = 3$, $b = -4$, and $c = -6$ in the quadratic formula.

$$p = \frac{-b \pm \sqrt{b^2 - 4ac}}{2a}$$
$$p = \frac{-(-4) \pm \sqrt{(-4)^2 - 4(3)(-6)}}{2(3)}$$
$$= \frac{4 \pm \sqrt{16+72}}{2(3)} = \frac{4 \pm \sqrt{88}}{2(3)}$$
$$= \frac{4 \pm 2\sqrt{22}}{2(3)} = \frac{2 \pm \sqrt{22}}{3}$$

Use a calculator to check both potential solutions. Both solutions check.

Solution set: $\left\{\frac{2+\sqrt{22}}{3}, \frac{2-\sqrt{22}}{3}\right\}$

29. Rate in still water: 20 mph
Rate of current: t mph

(a) When the boat travels upstream, the current works against the rate of the boat in still water, so the rate is $(20 - t)$ mph.

(b) When the boat travels downstream, the current works with the rate of the boat in still water, so the rate is $(20 + t)$ mph.

31. Let x = rate of the boat in still water.
With the speed of the current at 15 mph, then
$x - 15$ = rate going upstream and
$x + 15$ = rate going downstream.
Complete a table using the information in the problem, the rates given above, and the formula $d = rt$ or $t = \frac{d}{r}$.

	d	r	t
Upstream	4	$x - 15$	$\frac{4}{x-15}$
Downstream	16	$x + 15$	$\frac{16}{x+15}$

The time, 48 min, is written as $\frac{48}{60} = \frac{4}{5}$ hr. The time upstream plus the time downstream equals $\frac{4}{5}$. So, from the table, the equation is written as

$$\frac{4}{x-15} + \frac{16}{x+15} = \frac{4}{5}.$$

Multiply by the LCD, $5(x-15)(x+15)$.

$$5(x-15)(x+15)\left(\frac{4}{x-15}+\frac{16}{x+15}\right)$$
$$=5(x-15)(x+15)\bullet\frac{4}{5}$$
$$20(x+15)+80(x-15)$$
$$=4(x-15)(x+15)$$
$$20x+300+80x-1200$$
$$=4\left(x^2-225\right)$$
$$100x-900=4x^2-900$$
$$0=4x^2-100x$$
$$0=4x(x-25)$$

$4x=0$ or $x-25=0$
$x=0$ or $x=25$

Reject $x=0$ mph as a possible boat speed. Yoshiaki's boat had a top speed of 25 mph.

33. Let $x=$ Harry's average speed.
Then $x-20=$ Yoshi's average speed.

	d	***r***	***t***
Harry	300	x	$\frac{300}{x}$
Yoshi	300	$x-20$	$\frac{300}{x-20}$

It takes Harry $1\frac{1}{4}$ or $\frac{5}{4}$ hours less time than Yoshi.

$$\frac{300}{x}=\frac{300}{x-20}-\frac{5}{4}$$

Multiply by the LCD, $4x(x-20)$.

$$4x(x-20)\left(\frac{300}{x}\right)=4x(x-20)\left(\frac{300}{x-20}-\frac{5}{4}\right)$$
$$1200(x-20)=4x(300)-x(x-20)\bullet 5$$
$$1200x-24,000=1200x-5x^2+100x$$
$$5x^2-100x-24,000=0$$
$$x^2-20x-4800=0$$
$$(x-80)(x+60)=0$$

$x-80=0$ or $x+60=0$
$x=80$ or $x=-60$

Reject $x=-60$. Harry's average speed is 80 km/hr.

35. Let x be the time in hours required for the faster person to cut the lawn. Then the slower person requires $x+1$ hours.
Complete the chart.

	Rate	**Time Working Together**	**Fractional Part of the Job Done**
Faster Worker	$\frac{1}{x}$	2	$\frac{2}{x}$
Slower Worker	$\frac{1}{x+1}$	2	$\frac{2}{x+1}$

Part done by faster person + Part done by slower person = one whole job.

$$\frac{2}{x}+\frac{2}{x+1}=1$$

Multiply each side by the LCD, $x(x+1)$.

$$x(x+1)\left(\frac{2}{x}+\frac{2}{x+1}\right)=x(x+1)\bullet 1$$
$$2(x+1)+2x=x^2+x$$
$$2x+2+2x=x^2+x$$
$$0=x^2-3x-2$$

Solve for x using the quadratic formula with $a=1$, $b=-3$, and $c=-2$.

$$x=\frac{-(-3)\pm\sqrt{(-3)^2-4(1)(-2)}}{2(1)}$$
$$=\frac{3\pm\sqrt{9+8}}{2}=\frac{3\pm\sqrt{17}}{2}$$

$x=\frac{3+\sqrt{17}}{2}$ or $x=\frac{3-\sqrt{17}}{2}$
$x\approx 3.6$ or $x\approx -.6$

Discard $-.6$ as a solution since time cannot be negative.
It would take the faster person approximately 3.6 hours.

37. Let x represent the time in hours it takes Nancy to plant the flowers. Then $x+2$ is the time it takes Rusty.
Organize the information in a chart.

Worker	**Rate**	**Time Working Together**	**Fractional Part of the Job Done**
Nancy	$\frac{1}{x}$	12	$\frac{12}{x}$
Rusty	$\frac{1}{x+2}$	12	$\frac{12}{x+2}$

Part done by Nancy + part done by Rusty = one whole job.

$$\frac{12}{x}+\frac{12}{x+2}=1$$

Multiply each side by the LCD, $x(x+2)$.

$$x(x+2)\left(\frac{12}{x}+\frac{12}{x+2}\right)=x(x+2)\bullet 1$$
$$12(x+2)+12x=x^2+2x$$
$$12x+24+12x=x^2+2x$$
$$0=x^2-22x-24$$

Solve for x using the quadratic formula with $a=1$, $b=-22$, and $c=-24$.

continued

$$x = \frac{-(-22) \pm \sqrt{(-22)^2 - 4(1)(-24)}}{2(1)}$$
$$= \frac{22 \pm \sqrt{580}}{2}$$
$$x = \frac{22 + \sqrt{580}}{2} \approx 23.0 \text{ or}$$
$$x = \frac{22 - \sqrt{580}}{2} \approx -1.0$$

Since x represents time, discard the negative solution.
Nancy takes about 23.0 hours planting flowers alone while Rusty takes about 25.0 hours planting alone.

39. Let $x =$ the number of minutes it takes for the cold water tap alone to fill the washer.
$x + 9 =$ the number of minutes it takes for the hot water tap alone to fill the washer.

Working together, both taps can fill the washer in 6 minutes. Complete a chart using the above information.

Tap	Rate	Time	Fractional Part of Washer Filled
Cold	$\frac{1}{x}$	6	$\frac{1}{x}(6)$
Hot	$\frac{1}{x+9}$	6	$\frac{1}{x+9}(6)$

Since together the hot and cold taps fill one washer, the sum of their fractional parts is 1; that is,

$$\frac{6}{x} + \frac{6}{x+9} = 1.$$

Multiply by the LCD, $x(x+9)$.

$$x(x+9)\left(\frac{6}{x} + \frac{6}{x+9}\right) = x(x+9) \cdot 1$$
$$6(x+9) + 6x = x(x+9)$$
$$6x + 54 + 6x = x^2 + 9x$$
$$0 = x^2 - 3x - 54$$
$$0 = (x-9)(x+6)$$

$x - 9 = 0$ or $x + 6 = 0$
$x = 9$ or $x = -6$

Reject -6 as a possible time. The cold water tap can fill the washer in 9 minutes.

41.
$$x = \sqrt{7x - 10}$$
$$(x)^2 = \left(\sqrt{7x-10}\right)^2$$
$$x^2 = 7x - 10$$
$$x^2 - 7x + 10 = 0$$
$$(x-2)(x-5) = 0$$

$x - 2 = 0$ or $x - 5 = 0$
$x = 2$ or $x = 5$

Check $x = 2$: $2 \stackrel{?}{=} \sqrt{4}$ *True*
Check $x = 5$: $5 \stackrel{?}{=} \sqrt{25}$ *True*
Solution set: $\{2, 5\}$

43.
$$2x = \sqrt{11x + 3}$$
$$(2x)^2 = \left(\sqrt{11x+3}\right)^2$$
$$4x^2 = 11x + 3$$
$$4x^2 - 11x - 3 = 0$$
$$(4x+1)(x-3) = 0$$

$4x + 1 = 0$ or $x - 3 = 0$
$x = -\frac{1}{4}$ or $x = 3$

Check $x = -\frac{1}{4}$: $-\frac{1}{2} \stackrel{?}{=} \sqrt{\frac{1}{4}}$ *False*
Check $x = 3$: $6 \stackrel{?}{=} \sqrt{36}$ *True*

Solution set: $\{3\}$

45.
$$3x = \sqrt{16 - 10x}$$
$$(3x)^2 = \left(\sqrt{16-10x}\right)^2$$
$$9x^2 = 16 - 10x$$
$$9x^2 + 10x - 16 = 0$$
$$(9x-8)(x+2) = 0$$

$9x - 8 = 0$ or $x + 2 = 0$
$x = \frac{8}{9}$ or $x = -2$

Check $x = \frac{8}{9}$: $\frac{8}{3} \stackrel{?}{=} \sqrt{\frac{64}{9}}$ *True*
Check $x = -2$: $-6 \stackrel{?}{=} \sqrt{36}$ *False*

Solution set: $\left\{\frac{8}{9}\right\}$

47.
$$p - 2\sqrt{p} = 8$$
$$p - 8 = 2\sqrt{p}$$
$$(p-8)^2 = \left(2\sqrt{p}\right)^2$$
$$p^2 - 16p + 64 = 4p$$
$$p^2 - 20p + 64 = 0$$
$$(p-4)(p-16) = 0$$

$p - 4 = 0$ or $p - 16 = 0$
$p = 4$ or $p = 16$

Check $p = 4$: $4 - 4 \stackrel{?}{=} 8$ *False*
Check $p = 16$: $16 - 8 \stackrel{?}{=} 8$ *True*

Solution set: $\{16\}$

49.
$$m = \sqrt{\frac{6-13m}{5}}$$
$$m^2 = \frac{6-13m}{5}$$
$$5m^2 = 6-13m$$
$$5m^2+13m-6=0$$
$$(5m-2)(m+3)=0$$
$5m-2=0$ or $m+3=0$

$m=\frac{2}{5}$ or $m=-3$

Check $m=\frac{2}{5}$: $\frac{2}{5} \stackrel{?}{=} \sqrt{\frac{4}{25}}$ *True*

Check $m=-3$: $-3 \stackrel{?}{=} \sqrt{9}$ *False*

Solution set: $\left\{\frac{2}{5}\right\}$

51. $t^4-18t^2+81=0$

Let $u=t^2$, so $u^2=t^4$. The equation becomes
$$u^2-18u+81=0.$$
$$(u-9)^2=0$$
$$u-9=0$$
$$u=9$$
To find t, substitute t^2 for u.
$$t^2=9$$
$t=3$ or $t=-3$

Check $t=\pm 3$: $81-162+81 \stackrel{?}{=} 0$ *True*

Solution set: $\{-3,3\}$

53. $4k^4-13k^2+9=0$

Let $u=k^2$ and $u^2=k^4$ to get
$$4u^2-13u+9=0$$
$$(4u-9)(u-1)=0$$
$4u-9=0$ or $u-1=0$

$u=\frac{9}{4}$ or $u=1.$

To find k, substitute k^2 for u.

$k^2=\frac{9}{4}$ or $k^2=1$

$k=\pm\frac{3}{2}$ or $k=\pm 1$

Check $k=\pm\frac{3}{2}$: $\frac{81}{4}-\frac{117}{4}+9 \stackrel{?}{=} 0$ *True*

Check $k=\pm 1$: $4-13+9 \stackrel{?}{=} 0$ *True*

Solution set: $\left\{-\frac{3}{2},-1,1,\frac{3}{2}\right\}$

55.
$$x^4+48=16x^2$$
$$x^4-16x^2+48=0$$
Let $u=x^2$, so $u^2=x^4$. The equation becomes
$$u^2-16u+48=0.$$
$$(u-4)(u-12)=0$$
$u-4=0$ or $u-12=0$

$u=4$ or $u=12$

To find x substitute x^2 for u.

$x^2=4$ $\quad$ $x^2=12$

$x=\pm\sqrt{4}$ $\quad$ $x=\pm\sqrt{12}$

$x=\pm 2$ or $x=\pm 2\sqrt{3}$

Check $x=\pm 2$: $16+48 \stackrel{?}{=} 64$ *True*

Check $x=\pm 2\sqrt{3}$: $144+48 \stackrel{?}{=} 192$ *True*

Solution set: $\left\{-2\sqrt{3},-2,2,2\sqrt{3}\right\}$

57. $(x+3)^2+5(x+3)+6=0$

Let $u=x+3$, so $u^2=(x+3)^2$.
$$u^2+5u+6=0$$
$$(u+3)(u+2)=0$$
$u+3=0$ or $u+2=0$

$u=-3$ or $u=-2$

To find x, substitute $x+3$ for u.

$x+3=-3$ or $x+3=-2$

$x=-6$ or $x=-5$

Check $x=-6$: $9-15+6 \stackrel{?}{=} 0$ *True*

Check $x=-5$: $4-10+6 \stackrel{?}{=} 0$ *True*

Solution set: $\{-6,-5\}$

59. $3(m+4)^2-8=2(m+4)$

Let $x=m+4$, so $x^2=(m+4)^2$.
$$3x^2-8=2x$$
$$3x^2-2x-8=0$$
$$(3x+4)(x-2)=0$$
$3x+4=0$ or $x-2=0$

$x=-\frac{4}{3}$ $\quad$ $x=2$

$m+4=-\frac{4}{3}$ or $m+4=2$

$m=-\frac{16}{3}$ $\quad$ $m=-2$

Check $m=-\frac{16}{3}$: $\frac{16}{3}-8 \stackrel{?}{=} -\frac{8}{3}$ *True*

Check $m=-2$: $12-8 \stackrel{?}{=} 4$ *True*

Solution set: $\left\{-\frac{16}{3},-2\right\}$

61. $$2+\frac{5}{3k-1}=\frac{-2}{(3k-1)^2}$$

Let $u=3k-1$, so $u^2=(3k-1)^2$.

$$2+\frac{5}{u}=-\frac{2}{u^2}$$

Multiply by the LCD, u^2.

$$u^2\left(2+\frac{5}{u}\right)=u^2\left(-\frac{2}{u^2}\right)$$
$$2u^2+5u=-2$$
$$2u^2+5u+2=0$$
$$(2u+1)(u+2)=0$$
$$2u+1=0 \quad \text{or} \quad u+2=0$$
$$u=-\frac{1}{2} \qquad u=-2$$

To find k, substitute $3k-1$ for u.

$$3k-1=-\frac{1}{2} \quad \text{or} \quad 3k-1=-2$$
$$3k=\frac{1}{2} \qquad 3k=-1$$
$$k=\frac{1}{6} \quad \text{or} \quad k=-\frac{1}{3}$$

Check $k=\frac{1}{6}$: $2-10\stackrel{?}{=}-8$ *True*

Check $k=-\frac{1}{3}$: $2-\frac{5}{2}\stackrel{?}{=}-\frac{1}{2}$ *True*

Solution set: $\left\{-\frac{1}{3},\frac{1}{6}\right\}$

63. $2-6(m-1)^{-2}=(m-1)^{-1}$

Let $u=m-1$ to get

$$2-6u^{-2}=u^{-1}$$

or $$2-\frac{6}{u^2}=\frac{1}{u}.$$

Multiply by the LCD, u^2.

$$2u^2-6=u$$
$$2u^2-u-6=0$$
$$(2u+3)(u-2)=0$$
$$2u+3=0 \quad \text{or} \quad u-2=0$$
$$u=-\frac{3}{2} \quad \text{or} \quad u=2$$

To find m, substitute $m-1$ for u.

$$m-1=-\frac{3}{2} \quad \text{or} \quad m-1=2$$
$$m=-\frac{1}{2} \quad \text{or} \quad m=3$$

Check $m=-\frac{1}{2}$: $2-\frac{8}{3}\stackrel{?}{=}-\frac{2}{3}$ *True*

Check $m=3$: $2-\frac{3}{2}\stackrel{?}{=}\frac{1}{2}$ *True*

Solution set: $\left\{-\frac{1}{2},3\right\}$

65. $x^{2/3}+x^{1/3}-2=0$

Let $u=x^{1/3}$, so $u^2=x^{2/3}$.

$$u^2+u-2=0$$
$$(u+2)(u-1)=0$$
$$u+2=0 \quad \text{or} \quad u-1=0$$
$$u=-2 \quad \text{or} \quad u=1$$

To find x, substitute $x^{1/3}$ for u.

$x^{1/3}=-2$ or $x^{1/3}=1$

Cube both sides of each equation.

$$\left(x^{1/3}\right)^3=(-2)^3 \qquad \left(x^{1/3}\right)^3=1^3$$
$$x=-8 \quad \text{or} \quad x=1$$

Check $x=-8$: $4-2-2\stackrel{?}{=}0$ *True*

Check $x=1$: $1+1-2\stackrel{?}{=}0$ *True*

Solution set: $\{-8,1\}$

67. $r^{2/3}+r^{1/3}-12=0$

Let $u=r^{1/3}$, so $u^2=r^{2/3}$. The equation becomes

$$u^2+u-12=0.$$
$$(u+4)(u-3)=0$$
$$u+4=0 \quad \text{or} \quad u-3=0$$
$$u=-4 \quad \text{or} \quad u=3$$

To find r, substitute $r^{1/3}$ for u.

$$r^{1/3}=-4 \quad \text{or} \quad r^{1/3}=3$$
$$\left(r^{1/3}\right)^3=(-4)^3 \qquad \left(r^{1/3}\right)^3=3^3$$
$$r=-64 \quad \text{or} \quad r=27$$

Check $r=-64$: $16-4-12\stackrel{?}{=}0$ *True*

Check $r=27$: $9+3-12\stackrel{?}{=}0$ *True*

Solution set: $\{-64,27\}$

69. $4k^{4/3}-13k^{2/3}+9=0$

Let $x=k^{2/3}$, so $x^2=k^{4/3}$.

$$4x^2-13x+9=0$$
$$(4x-9)(x-1)=0$$
$$4x-9=0 \quad \text{or} \quad x-1=0$$
$$x=\frac{9}{4} \quad \text{or} \quad x=1$$
$$k^{2/3}=\frac{9}{4} \quad \text{or} \quad k^{2/3}=1$$
$$\left(k^{2/3}\right)^3=\left(\frac{9}{4}\right)^3 \quad \text{or} \quad \left(k^{2/3}\right)^3=1^3$$
$$k^2=\frac{729}{64} \quad \text{or} \quad k^2=1$$
$$k=\pm\sqrt{\frac{729}{64}} \quad \text{or} \quad k=\pm\sqrt{1}$$
$$k=\pm\frac{27}{8} \quad \text{or} \quad k=\pm 1$$

Check $k = \pm\frac{27}{8}$: $\frac{81}{4} - \frac{117}{4} + 9 \stackrel{?}{=} 0$ *True*

Check $k = \pm 1$: $4 - 13 + 9 \stackrel{?}{=} 0$ *True*

Solution set: $\left\{-\frac{27}{8}, -1, 1, \frac{27}{8}\right\}$

71. $2(1+\sqrt{r})^2 = 13(1+\sqrt{r}) - 6$

Let $u = 1 + \sqrt{r}$.

$$2u^2 = 13u - 6$$
$$2u^2 - 13u + 6 = 0$$
$$(2u-1)(u-6) = 0$$

$2u - 1 = 0$ or $u - 6 = 0$

$u = \frac{1}{2}$ or $u = 6$

Replace u with $1 + \sqrt{r}$.

$1 + \sqrt{r} = \frac{1}{2}$ or $1 + \sqrt{r} = 6$

$\sqrt{r} = -\frac{1}{2}$ $\quad$ $\sqrt{r} = 5$

Not possible, $\quad$ $r = 25$

since $\sqrt{r} \geq 0$.

Check $r = 25$: $72 \stackrel{?}{=} 78 - 6$ *True*

Solution set: $\{25\}$

73. $2x^4 + x^2 - 3 = 0$

Let $m = x^2$, so $m^2 = x^4$.

$$2m^2 + m - 3 = 0$$
$$(2m+3)(m-1) = 0$$

$2m + 3 = 0$ or $m - 1 = 0$

$m = -\frac{3}{2}$ or $m = 1$

To find x, substitute x^2 for m.

$x^2 = -\frac{3}{2}$ or $x^2 = 1$

$x = \pm\sqrt{-\frac{3}{2}}$ $\quad$ $x = \pm\sqrt{1}$

$x = \pm\frac{\sqrt{3}}{\sqrt{2}} \cdot \frac{\sqrt{2}}{\sqrt{2}} i$ $\quad$ $x = \pm 1$

$x = \pm\frac{\sqrt{6}}{2} i$

Check $x = \pm\frac{\sqrt{6}}{2} i$: $\frac{9}{2} - \frac{3}{2} - 3 \stackrel{?}{=} 0$ *True*

Check $x = \pm 1$: $2 + 1 - 3 \stackrel{?}{=} 0$ *True*

Solution set: $\left\{-1, 1, -\frac{\sqrt{6}}{2} i, \frac{\sqrt{6}}{2} i\right\}$

75. $12x^4 - 11x^2 + 2 = 0$

Let $u = x^2$, so $u^2 = x^4$.

$$12u^2 - 11u + 2 = 0$$
$$(4u-1)(3u-2) = 0$$

$4u - 1 = 0$ or $3u - 2 = 0$

$u = \frac{1}{4}$ or $u = \frac{2}{3}$

$x^2 = \frac{1}{4}$ $\quad$ $x^2 = \frac{2}{3}$

$x = \pm\frac{1}{2}$ $\quad$ $x = \pm\sqrt{\frac{2}{3}}$

Note: $x = \pm\sqrt{\frac{2}{3}} = \pm\frac{\sqrt{2}}{\sqrt{3}} \cdot \frac{\sqrt{3}}{\sqrt{3}} = \pm\frac{\sqrt{6}}{3}$

Check $x = \pm\frac{1}{2}$: $\frac{3}{4} - \frac{11}{4} + 2 \stackrel{?}{=} 0$ *True*

Check $x = \pm\frac{\sqrt{6}}{3}$: $\frac{16}{3} - \frac{22}{3} + 2 \stackrel{?}{=} 0$ *True*

Solution set: $\left\{-\frac{\sqrt{6}}{3}, -\frac{1}{2}, \frac{1}{2}, \frac{\sqrt{6}}{3}\right\}$

77. $\sqrt{2x+3} = 2 + \sqrt{x-2}$

Square both sides.

$$\left(\sqrt{2x+3}\right)^2 = \left(2 + \sqrt{x-2}\right)^2$$
$$2x + 3 = 4 + 4\sqrt{x-2} + (x-2)$$
$$2x + 3 = x + 2 + 4\sqrt{x-2}$$

Isolate the radical term on one side.

$$x + 1 = 4\sqrt{x-2}$$

Square both sides again.

$$(x+1)^2 = \left(4\sqrt{x-2}\right)^2$$
$$x^2 + 2x + 1 = 16(x-2)$$
$$x^2 + 2x + 1 = 16x - 32$$
$$x^2 - 14x + 33 = 0$$
$$(x-11)(x-3) = 0$$

$x - 11 = 0$ or $x - 3 = 0$

$x = 11$ or $x = 3$

Check $x = 11$: $\sqrt{25} \stackrel{?}{=} 2 + \sqrt{9}$ *True*

Check $x = 3$: $\sqrt{9} \stackrel{?}{=} 2 + \sqrt{1}$ *True*

Solution set: $\{3, 11\}$

79. $2m^6 + 11m^3 + 5 = 0$

Let $y = m^3$, so $y^2 = m^6$.

$$2y^2 + 11y + 5 = 0$$
$$(2y+1)(y+5) = 0$$

$2y + 1 = 0$ or $y + 5 = 0$

$y = -\frac{1}{2}$ or $y = -5$

To find m, substitute m^3 for y.

$m^3 = -\frac{1}{2}$ or $m^3 = -5$

continued

Take the cube root of both sides of each equation.

$$m = \sqrt[3]{-\frac{1}{2}} \quad \text{or} \quad m = \sqrt[3]{-5}$$

$$m = -\sqrt[3]{\frac{1}{2}} \qquad m = -\sqrt[3]{5}$$

$$= -\frac{\sqrt[3]{1}}{\sqrt[3]{2}} \cdot \frac{\sqrt[3]{2^2}}{\sqrt[3]{2^2}}$$

$$= -\frac{\sqrt[3]{4}}{2}$$

Check $m = -\frac{\sqrt[3]{4}}{2}$: $\frac{1}{2} - \frac{11}{2} + 5 \stackrel{?}{=} 0$ *True*

Check $m = -\sqrt[3]{5}$: $50 - 55 + 5 \stackrel{?}{=} 0$ *True*

Solution set: $\left\{-\sqrt[3]{5}, -\frac{\sqrt[3]{4}}{2}\right\}$

81. $6 = 7(2w-3)^{-1} + 3(2w-3)^{-2}$

Let $x = (2w-3)^{-1}$, so $x^2 = (2w-3)^{-2}$.

$$6 = 7x + 3x^2$$

$$0 = 3x^2 + 7x - 6$$

$$0 = (3x-2)(x+3)$$

$$3x - 2 = 0 \quad \text{or} \quad x + 3 = 0$$

$$x = \frac{2}{3} \quad \text{or} \quad x = -3$$

$$(2w-3)^{-1} = \frac{2}{3} \qquad (2w-3)^{-1} = -3$$

$$\frac{1}{2w-3} = \frac{2}{3} \qquad \frac{1}{2w-3} = -3$$

$$3 = 2(2w-3) \qquad 1 = -3(2w-3)$$

$$3 = 4w - 6 \qquad 1 = -6w + 9$$

$$9 = 4w \qquad -8 = -6w$$

$$\frac{9}{4} = w \qquad \frac{4}{3} = w$$

Check $w = \frac{4}{3}$: $6 \stackrel{?}{=} -21 + 27$ *True*

Check $w = \frac{9}{4}$: $6 \stackrel{?}{=} \frac{14}{3} + \frac{4}{3}$ *True*

Solution set: $\left\{\frac{4}{3}, \frac{9}{4}\right\}$

83. $2x^4 - 9x^2 = -2$

$$2x^4 - 9x^2 + 2 = 0$$

Let $u = x^2$, so $u^2 = x^4$.

$$2u^2 - 9u + 2 = 0$$

Use $a = 2$, $b = -9$, and $c = 2$ in the quadratic formula.

$$u = \frac{-b \pm \sqrt{b^2 - 4ac}}{2a}$$

$$u = \frac{-(-9) \pm \sqrt{(-9)^2 - 4(2)(2)}}{2(2)}$$

$$= \frac{9 \pm \sqrt{81 - 16}}{4}$$

$$= \frac{9 \pm \sqrt{65}}{4}$$

To find x, substitute x^2 for u.

$$x^2 = \frac{9 \pm \sqrt{65}}{4}$$

$$x = \pm\sqrt{\frac{9 \pm \sqrt{65}}{4}}$$

$$= \pm\frac{\sqrt{9 \pm \sqrt{65}}}{2}$$

Note: the last expression represents four numbers. All four potential solutions check.

Solution set: $\left\{\frac{\sqrt{9+\sqrt{65}}}{2}, -\frac{\sqrt{9+\sqrt{65}}}{2}, \frac{\sqrt{9-\sqrt{65}}}{2}, -\frac{\sqrt{9-\sqrt{65}}}{2}\right\}$

For Exercises 85–90, use the equation

$$\frac{x^2}{(x-3)^2} + \frac{3x}{x-3} - 4 = 0.$$

85. Substituting 3 for x would cause both denominators to be 0, and division by 0 is undefined.

86. $(x-3)^2\left(\frac{x^2}{(x-3)^2}\right) + (x-3)^2\left(\frac{3x}{x-3}\right)$

$$-(x-3)^2 \cdot 4 = (x-3)^2 \cdot 0$$

$$x^2 + 3x(x-3) - 4(x^2 - 6x + 9) = 0$$

$$x^2 + 3x^2 - 9x - 4x^2 + 24x - 36 = 0$$

$$15x - 36 = 0$$

$$15x = 36$$

$$x = \frac{36}{15} = \frac{12}{5}$$

The solution is $\frac{12}{5}$.

87. $\frac{x^2}{(x-3)^2} + \frac{3x}{x-3} - 4 = 0$

$$\left(\frac{x}{x-3}\right)^2 + 3\left(\frac{x}{x-3}\right) - 4 = 0$$

88. If a fraction is equal to 1, then the numerator must be equal to the denominator. But the numerator can never equal the denominator, since the denominator is 3 less than the numerator.

89. From Exercise 87,

$$\left(\frac{x}{x-3}\right)^2 + 3\left(\frac{x}{x-3}\right) - 4 = 0.$$

Let $t = \frac{x}{x-3}$, so $t^2 = \left(\frac{x}{x-3}\right)^2$.

$$t^2 + 3t - 4 = 0$$

$$(t-1)(t+4) = 0$$

$t-1=0$ or $t+4=0$

$t=1$ or $t=-4$

To find x, substitute $\frac{x}{x-3}$ for t.

$\frac{x}{x-3}=1$ or $\frac{x}{x-3}=-4$

The equation $\frac{x}{x-3}=1$ has no solution since there is no value of x for which $x=x-3$. (See Exercise 88.) Therefore, $t=1$ is impossible.

$$\begin{aligned}\frac{x}{x-3}&=-4\\ x&=-4(x-3)\\ x&=-4x+12\\ 5x&=12\\ x&=\frac{12}{5}\end{aligned}$$

Solution set: $\left\{\frac{12}{5}\right\}$

90. $x^2(x-3)^{-2}+3x(x-3)^{-1}-4=0$

Let $s=(x-3)^{-1}$, so $s^2=(x-3)^{-2}$.

$$x^2s^2+3xs-4=0$$
$$(xs+4)(xs-1)=0$$

$xs+4=0$ or $xs-1=0$

$s=-\frac{4}{x}$ or $s=\frac{1}{x}$

To find x, substitute $\frac{1}{x-3}$ for s.

$\frac{1}{x-3}=-\frac{4}{x}$ or $\frac{1}{x-3}=\frac{1}{x}$

$x=-4(x-3)$

$x=-4x+12$ $\qquad$ $x=x-3$

$5x=12$ $\qquad$ $0=-3$ *False*

$x=\frac{12}{5}$ $\qquad$ Thus, $s=\frac{1}{x}$ is impossible.

Solution set: $\left\{\frac{12}{5}\right\}$

Summary Exercises on Solving Quadratic Equations

1. $p^2=7$

$p=\sqrt{7}$ or $p=-\sqrt{7}$

Solution set: $\left\{\sqrt{7},-\sqrt{7}\right\}$

3. $n^2+6n+4=0$

$n^2+6n=-4$

$n^2+6n+9=-4+9$ $\quad$ $\left[\frac{1}{2}(6)\right]^2=9$

$(n+3)^2=5$

$n+3=\sqrt{5}$ or $n+3=-\sqrt{5}$

$n=-3+\sqrt{5}$ or $n=-3-\sqrt{5}$

Solution set: $\left\{-3+\sqrt{5},-3-\sqrt{5}\right\}$

5. $\frac{5}{m}+\frac{12}{m^2}=2$

Multiply by the LCD, m^2.

$$\begin{aligned}5m+12&=2m^2\\ 0&=2m^2-5m-12\\ 0&=(2m+3)(m-4)\end{aligned}$$

$2m+3=0$ or $m-4=0$

$m=-\frac{3}{2}$ or $m=4$

Solution set: $\left\{-\frac{3}{2},4\right\}$

7. $2r^2-4r+1=0$

Use $a=2$, $b=-4$, and $c=1$ in the quadratic formula.

$$\begin{aligned}r&=\frac{-b\pm\sqrt{b^2-4ac}}{2a}\\ r&=\frac{-(-4)\pm\sqrt{(-4)^2-4(2)(1)}}{2(2)}\\ &=\frac{4\pm\sqrt{16-8}}{2(2)}\\ &=\frac{4\pm\sqrt{8}}{2(2)}=\frac{4\pm2\sqrt{2}}{2(2)}\\ &=\frac{2\pm\sqrt{2}}{2}\end{aligned}$$

Solution set: $\left\{\frac{2+\sqrt{2}}{2},\frac{2-\sqrt{2}}{2}\right\}$

9.

$$\begin{aligned}x\sqrt{2}&=\sqrt{5x-2}\\ \left(x\sqrt{2}\right)^2&=\left(\sqrt{5x-2}\right)^2\\ x^2\cdot2&=5x-2\\ 2x^2-5x+2&=0\\ (2x-1)(x-2)&=0\end{aligned}$$

$2x-1=0$ or $x-2=0$

$x=\frac{1}{2}$ $\qquad$ $x=2$

Check $x=\frac{1}{2}$: $\quad\frac{1}{2}\sqrt{2}\stackrel{?}{=}\sqrt{\frac{1}{2}}$ $\quad$ *True*

Check $x=2$: $\quad 2\sqrt{2}\stackrel{?}{=}\sqrt{8}$ $\quad$ *True*

Solution set: $\left\{\frac{1}{2},2\right\}$

11. $(2k+3)^2 = 8$

$$2k+3 = \sqrt{8} \quad \text{or} \quad 2k+3 = -\sqrt{8}$$
$$2k = -3+2\sqrt{2} \qquad 2k = -3-2\sqrt{2}$$
$$k = \frac{-3+2\sqrt{2}}{2} \quad \text{or} \quad k = \frac{-3-2\sqrt{2}}{2}$$

Solution set: $\left\{\frac{-3+2\sqrt{2}}{2}, \frac{-3-2\sqrt{2}}{2}\right\}$

13.
$$t^4 + 14 = 9t^2$$
$$t^4 - 9t^2 + 14 = 0$$
$$\left(t^2-2\right)\left(t^2-7\right) = 0$$
$$t^2 - 2 = 0 \quad \text{or} \quad t^2 - 7 = 0$$
$$t^2 = 2 \qquad t^2 = 7$$
$$t = \pm\sqrt{2} \quad \text{or} \quad t = \pm\sqrt{7}$$

Solution set: $\left\{-\sqrt{7}, -\sqrt{2}, \sqrt{2}, \sqrt{7}\right\}$

15. $z^2 + z + 1 = 0$

Use $a = 1$, $b = 1$, and $c = 1$ in the quadratic formula.

$$z = \frac{-b \pm \sqrt{b^2-4ac}}{2a}$$
$$z = \frac{-1 \pm \sqrt{1^2 - 4(1)(1)}}{2(1)}$$
$$= \frac{-1 \pm \sqrt{1-4}}{2} = \frac{-1 \pm \sqrt{-3}}{2}$$
$$= \frac{-1 \pm i\sqrt{3}}{2} = -\frac{1}{2} \pm \frac{\sqrt{3}}{2}i$$

Solution set: $\left\{-\frac{1}{2} + \frac{\sqrt{3}}{2}i, -\frac{1}{2} - \frac{\sqrt{3}}{2}i\right\}$

17.
$$4t^2 - 12t + 9 = 0$$
$$(2t-3)(2t-3) = 0$$
$$(2t-3)^2 = 0$$
$$2t - 3 = 0$$
$$t = \frac{3}{2}$$

Solution set: $\left\{\frac{3}{2}\right\}$

19. $r^2 - 72 = 0$
$$r^2 = 72$$
$$r = \pm\sqrt{72} = \pm 6\sqrt{2}$$

Solution set: $\left\{6\sqrt{2}, -6\sqrt{2}\right\}$

21. $x^2 - 5x - 36 = 0$
$$(x+4)(x-9) = 0$$
$$x + 4 = 0 \quad \text{or} \quad x - 9 = 0$$
$$x = -4 \quad \text{or} \quad x = 9$$

Solution set: $\{-4, 9\}$

23.
$$3p^2 = 6p - 4$$
$$3p^2 - 6p + 4 = 0$$

Use $a = 3$, $b = -6$, and $c = 4$ in the quadratic formula.

$$p = \frac{-b \pm \sqrt{b^2-4ac}}{2a}$$
$$p = \frac{-(-6) \pm \sqrt{(-6)^2 - 4(3)(4)}}{2(3)}$$
$$= \frac{6 \pm \sqrt{36-48}}{2(3)} = \frac{6 \pm \sqrt{-12}}{2(3)}$$
$$= \frac{6 \pm 2i\sqrt{3}}{2(3)} = \frac{3 \pm i\sqrt{3}}{3}$$
$$= 1 \pm \frac{\sqrt{3}}{3}i$$

Solution set: $\left\{1 + \frac{\sqrt{3}}{3}i, 1 - \frac{\sqrt{3}}{3}i\right\}$

25. $2(3k-1)^2 + 5(3k-1) = -2$

Let $u = 3k - 1$.
$$2u^2 + 5u + 2 = 0$$
$$(2u+1)(u+2) = 0$$
$$2u + 1 = 0 \quad \text{or} \quad u + 2 = 0$$
$$u = -\frac{1}{2} \quad \text{or} \quad u = -2$$
$$3k - 1 = -\frac{1}{2} \qquad 3k - 1 = -2$$
$$3k = \frac{1}{2} \qquad 3k = -1$$
$$k = \frac{1}{6} \qquad k = -\frac{1}{3}$$

Solution set: $\left\{-\frac{1}{3}, \frac{1}{6}\right\}$

Section 9.4

1. The first step in solving a formula that has the specified variable in the denominator is to multiply both sides by the LCD to clear the equation of fractions.

3. We must recognize that a formula like
$$gw^2 = kw + 24$$
is quadratic in w. So the first step is to write the formula in standard form (with 0 on one side, in decreasing powers of w). This allows us to apply the quadratic formula to solve for w.

5. Since the triangle is a right triangle, use the Pythagorean formula with legs m and n and hypotenuse p.

$$m^2 + n^2 = p^2$$
$$m^2 = p^2 - n^2$$
$$m = \sqrt{p^2 - n^2}$$

Only the positive square root is given since m represents the side of a triangle.

7. Solve $d = kt^2$ for t.

$kt^2 = d$

$t^2 = \dfrac{d}{k}$ *Divide by k.*

$t = \pm\sqrt{\dfrac{d}{k}}$ *Use square root property.*

$= \dfrac{\pm\sqrt{d}}{\sqrt{k}} \cdot \dfrac{\sqrt{k}}{\sqrt{k}}$ *Rationalize denominator.*

$t = \dfrac{\pm\sqrt{dk}}{k}$ *Simplify*

9. Solve $I = \dfrac{ks}{d^2}$ for d.

$Id^2 = ks$ *Multiply by d^2.*

$d^2 = \dfrac{ks}{I}$ *Divide by I.*

$d = \pm\sqrt{\dfrac{ks}{I}}$ *Use square root property.*

$= \pm\dfrac{\sqrt{ks}}{\sqrt{I}} \cdot \dfrac{\sqrt{I}}{\sqrt{I}}$ *Rationalize denominator.*

$d = \dfrac{\pm\sqrt{ksI}}{I}$ *Simplify*

11. Solve $F = \dfrac{kA}{v^2}$ for v.

$v^2F = kA$ *Multiply by v^2.*

$v^2 = \dfrac{kA}{F}$ *Divide by F.*

$v = \pm\sqrt{\dfrac{kA}{F}}$ *Use square root property.*

$= \dfrac{\pm\sqrt{kA}}{\sqrt{F}} \cdot \dfrac{\sqrt{F}}{\sqrt{F}}$ *Rationalize denominator.*

$v = \dfrac{\pm\sqrt{kAF}}{F}$ *Simplify*

13. Solve $V = \dfrac{1}{3}\pi r^2 h$ for r.

$3V = \pi r^2 h$ *Multiply by 3.*

$\dfrac{3V}{\pi h} = r^2$ *Divide by πh.*

$r = \pm\sqrt{\dfrac{3V}{\pi h}}$ *Use square root property.*

$= \dfrac{\pm\sqrt{3V} \cdot \sqrt{\pi h}}{\sqrt{\pi h} \cdot \sqrt{\pi h}}$ *Rationalize denominator.*

$r = \dfrac{\pm\sqrt{3\pi V h}}{\pi h}$ *Simplify*

15. Solve $At^2 + Bt = -C$ for t.

$At^2 + Bt + C = 0$

Use the quadratic formula.

$$t = \frac{-B \pm \sqrt{B^2 - 4AC}}{2A}$$

17. Solve $D = \sqrt{kh}$ for h.

$D^2 = kh$ *Square both sides.*

$\dfrac{D^2}{k} = h$ *Divide by k.*

19. Solve $p = \sqrt{\dfrac{k\ell}{g}}$ for ℓ.

$p^2 = \dfrac{k\ell}{g}$ *Square both sides.*

$p^2 g = k\ell$ *Multiply by g.*

$\dfrac{p^2 g}{k} = \ell$ *Divide by k.*

21. If g is positive, the only way to have a real value for p is to have $k\ell$ positive, since the quotient of two positive numbers is positive. If k and ℓ have different signs, their product is negative, leading to a negative radicand.

23. Solve $p = \dfrac{E^2 R}{(r + R)^2}$ $(E > 0)$ for R.

$$p(r + R)^2 = E^2 R$$
$$p(r^2 + 2rR + R^2) = E^2 R$$
$$pr^2 + 2prR + pR^2 = E^2 R$$
$$pR^2 + 2prR - E^2 R + pr^2 = 0$$
$$pR^2 + (2pr - E^2)R + pr^2 = 0$$

Here $a = p$, $b = 2pr - E^2$, and $c = pr^2$.

$$R = \frac{-(2pr - E^2) \pm \sqrt{(2pr - E^2)^2 - 4p \cdot pr^2}}{2p}$$
$$= \frac{E^2 - 2pr \pm \sqrt{4p^2r^2 - 4prE^2 + E^4 - 4p^2r^2}}{2p}$$
$$= \frac{E^2 - 2pr \pm \sqrt{E^4 - 4prE^2}}{2p}$$
$$= \frac{E^2 - 2pr \pm \sqrt{E^2(E^2 - 4pr)}}{2p}$$
$$R = \frac{E^2 - 2pr \pm E\sqrt{E^2 - 4pr}}{2p}$$

25. Solve $10p^2c^2 + 7pcr = 12r^2$ for r.

$0 = 12r^2 - 7pcr - 10p^2c^2$

Here $a = 12$, $b = -7pc$, and $c = -10p^2c^2$.

$$r = \frac{-(-7pc) \pm \sqrt{(-7pc)^2 - 4(12)(-10p^2c^2)}}{2(12)}$$
$$= \frac{7pc \pm \sqrt{49p^2c^2 + 480p^2c^2}}{24}$$
$$= \frac{7pc \pm \sqrt{529p^2c^2}}{24} = \frac{7pc \pm 23pc}{24}$$
$$r = \frac{7pc + 23pc}{24} = \frac{30pc}{24} = \frac{5pc}{4} \text{ or}$$
$$r = \frac{7pc - 23pc}{24} = \frac{-16pc}{24} = -\frac{2pc}{3}$$

27. Solve $LI^2 + RI + \frac{1}{c} = 0$ for I.

$cLI^2 + cRI + 1 = 0$ *Multiply by c.*

Here $a = cL$, $b = cR$, and $c = 1$.

$$I = \frac{-cR \pm \sqrt{(cR)^2 - 4(cL)(1)}}{2(cL)}$$
$$= \frac{-cR \pm \sqrt{c^2R^2 - 4cL}}{2cL}$$

29. Apply the Pythagorean formula.

$$(5m)^2 = (2m)^2 + (2m+3)^2$$
$$25m^2 = 4m^2 + 4m^2 + 12m + 9$$

$17m^2 - 12m - 9 = 0$

Here $a = 17$, $b = -12$, and $c = -9$.

$$m = \frac{-(-12) \pm \sqrt{(-12)^2 - 4(17)(-9)}}{2(17)}$$
$$= \frac{12 \pm \sqrt{144 + 612}}{34} = \frac{12 \pm \sqrt{756}}{34}$$
$$m = \frac{12 + \sqrt{756}}{34} \approx 1.16 \text{ or}$$
$$m = \frac{12 - \sqrt{756}}{34} \approx -.46$$

Reject the negative solution.
If $m = 1.16$, then

$$5m = 5(1.16) = 5.80,$$
$$2m = 2(1.16) = 2.32, \text{ and}$$
$$2m + 3 = 2(1.16) + 3 = 5.32.$$

The lengths of the sides of the triangle are approximately 2.3, 5.3, and 5.8.

31. Let $x =$ the distance traveled by the eastbound ship. Then $x + 70 =$ the distance traveled by the southbound ship.

Since the ships are traveling at right angles to one another, the distance d between them can be found using the Pythagorean formula.

$$c^2 = a^2 + b^2$$
$$d^2 = x^2 + (x+70)^2$$

Let $d = 170$, and solve for x.

$$170^2 = x^2 + (x+70)^2$$
$$28,900 = x^2 + x^2 + 140x + 4900$$
$$0 = 2x^2 + 140x - 24,000$$
$$0 = x^2 + 70x - 12,000$$
$$0 = (x + 150)(x - 80)$$

$x + 150 = 0$ or $x - 80 = 0$

$x = -150$ or $x = 80$

Distance cannot be negative, so reject -150. If $x = 80$, then $x + 70 = 150$. The eastbound ship traveled 80 miles, and the southbound ship traveled 150 miles.

33. Let $x =$ length of the shorter leg;
$2x + 2 =$ length of the longer leg;
$2x + 2 + 1 =$ length of the hypotenuse.

Use the Pythagorean formula.

$$x^2 + (2x+2)^2 = (2x+3)^2$$
$$x^2 + 4x^2 + 8x + 4 = 4x^2 + 12x + 9$$
$$5x^2 + 8x + 4 = 4x^2 + 12x + 9$$
$$x^2 - 4x - 5 = 0$$
$$(x - 5)(x + 1) = 0$$

$x - 5 = 0$ or $x + 1 = 0$

$x = 5$ $\quad$ $x = -1$

Since x represents length, discard -1 as a solution. If $x = 5$, then

$2x + 2 = 2(5) + 2 = 12$ and
$2x + 2 + 1 = 2(5) + 3 = 13.$

The lengths should be 5 cm, 12 cm, and 13 cm.

35. Let $x =$ the width of the rectangle.
Then $2x - 1 =$ the length of the rectangle.

Use the Pythagorean formula.

$$x^2 + (2x-1)^2 = (2.5)^2$$
$$x^2 + 4x^2 - 4x + 1 = 6.25$$
$$5x^2 - 4x - 5.25 = 0$$

Multiply by 4.

$$20x^2 - 16x - 21 = 0$$
$$(2x - 3)(10x + 7) = 0$$

$2x - 3 = 0$ or $10x + 7 = 0$

$x = \frac{3}{2}$ or $x = -\frac{7}{10}$

Discard the negative solution.

If $x = \frac{3}{2}$, then

$$2x - 1 = 2\left(\frac{3}{2}\right) - 1 = 2.$$

The width of the rectangle is 1.5 cm, and the length is 2 cm.

37. Let $x =$ the width of the uncovered strip of flooring.

From the problem,
(length of the rug) • (width of the rug) $= 234$.

The rug is centered in the room a distance x from the walls (width of the strip x), so the length of the rug

$= \text{length of the room} - 2 \bullet (\text{width of the strip})$
$= 20 - 2x,$

and the width of the rug

$= \text{width of the room} - 2 \bullet (\text{width of the strip})$
$= 15 - 2x.$

The equation

(length of rug) • (width of rug) $= 234$

becomes

$$(20 - 2x)(15 - 2x) = 234.$$
$$300 - 70x + 4x^2 = 234$$
$$4x^2 - 70x + 66 = 0$$

Divide by 2.

$$2x^2 - 35x + 33 = 0$$
$$(2x - 33)(x - 1) = 0$$

$2x - 33 = 0 \quad$ or $\quad x - 1 = 0$

$x = \frac{33}{2} \quad$ or $\quad x = 1$

Reject $\frac{33}{2} = 16\frac{1}{2}$ since $16\frac{1}{2}$ is wider than the room itself. The width of the uncovered strip is 1 foot.

39. Let x be the width of grass.

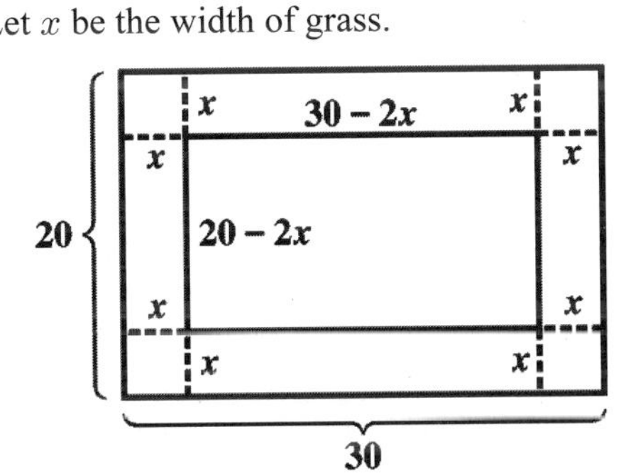

The area of the flower garden is $(20 - 2x)(30 - 2x)$.

Area of back yard	less	area of flower garden	is	area of grass.
$20 \bullet 30$	$-$	$(20 - 2x)(30 - 2x)$	$=$	184

$$600 - (600 - 100x + 4x^2) = 184$$
$$600 - 600 + 100x - 4x^2 = 184$$
$$100x - 4x^2 = 184$$
$$4x^2 - 100x + 184 = 0$$

Divide by 4.

$$x^2 - 25x + 46 = 0$$
$$(x - 23)(x - 2) = 0$$

$x - 23 = 0 \quad$ or $\quad x - 2 = 0$

$x = 23 \quad$ or $\quad x = 2$

Discard $x = 23$ since the width of the garden cannot be

$20 - 2x = 20 - 2(23) = -26.$

Thus, if $x = 2$, then

$30 - 2x = 30 - 2(2) = 26,$

and

$20 - 2x = 20 - 2(2) = 16.$

The flower garden is 26 m by 16 m.

41. Let x be the width of the sheet metal. Then the length is $2x - 4$.

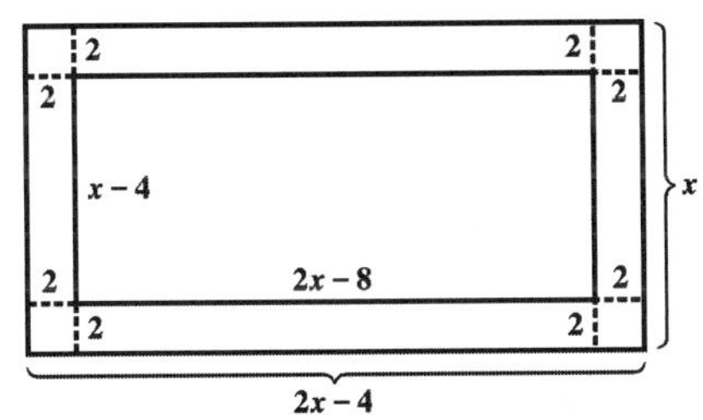

By cutting out 2-inch squares from each corner we get a rectangle with width $x - 4$ and length $(2x - 4) - 4 = 2x - 8$. The uncovered box then has height 2 inches, length $2x - 8$ inches, and width $x - 4$ inches.
Use the formula $V = LWH$ or $V = HLW$.

$256 = 2(2x - 8)(x - 4)$
$256 = 4(x - 4)(x - 4) \qquad$ *Factor out 2.*
$64 = (x - 4)^2 \qquad$ *Divide by 4.*

Use the square root property.
$\pm 8 = x - 4$

$x - 4 = 8 \quad$ or $\quad x - 4 = -8$

$x = 12 \quad$ or $\quad x = -4$

Since x represents width, discard the negative solution.
The width is 12 inches, and the length is $2(12) - 4 = 20$ inches.

43.

$$\begin{aligned} s &= 144t - 16t^2 \\ 128 &= 144t - 16t^2 && \textit{Let } s = 128. \\ 0 &= -16t^2 + 144t - 128 \\ 0 &= t^2 - 9t + 8 && \textit{Divide by } -16. \\ 0 &= (t-8)(t-1) \end{aligned}$$

$$t - 8 = 0 \quad \text{or} \quad t - 1 = 0$$
$$t = 8 \quad \text{or} \quad t = 1$$

The object will be 128 feet above the ground at two times, going up and coming down, or at 1 second and at 8 seconds.

45. Let $s = 213$ in the equation.

$$\begin{aligned} s &= -16t^2 + 128t \\ 213 &= -16t^2 + 128t \\ 0 &= -16t^2 + 128t - 213 \end{aligned}$$

Here $a = -16$, $b = 128$, and $c = -213$.

$$\begin{aligned} t &= \frac{-b \pm \sqrt{b^2 - 4ac}}{2a} \\ t &= \frac{-128 \pm \sqrt{128^2 - 4(-16)(-213)}}{2(-16)} \\ &= \frac{-128 \pm \sqrt{16,384 - 13,632}}{-32} \\ &= \frac{-128 \pm \sqrt{2752}}{-32} \end{aligned}$$

$$t = \frac{-128 + \sqrt{2752}}{-32} \approx 2.4 \text{ or}$$
$$t = \frac{-128 - \sqrt{2752}}{-32} \approx 5.6$$

The ball will be 213 feet from the ground after 2.4 seconds and again after 5.6 seconds.

47.

$$\begin{aligned} D(t) &= 13t^2 - 100t \\ 180 &= 13t^2 - 100t && \textit{Let } D(t) = 180. \\ 0 &= 13t^2 - 100t - 180 \end{aligned}$$

Here $a = 13$, $b = -100$, and $c = -180$.

$$\begin{aligned} t &= \frac{-b \pm \sqrt{b^2 - 4ac}}{2a} \\ t &= \frac{-(-100) \pm \sqrt{(-100)^2 - 4(13)(-180)}}{2(13)} \\ &= \frac{100 \pm \sqrt{10{,}000 + 9360}}{2(13)} \\ &= \frac{100 \pm \sqrt{19{,}360}}{2(13)} \\ &= \frac{100 \pm 44\sqrt{10}}{2(13)} = \frac{50 \pm 22\sqrt{10}}{13} \end{aligned}$$

$$t = \frac{50 + 22\sqrt{10}}{13} \approx 9.2 \text{ or}$$
$$t = \frac{50 - 22\sqrt{10}}{13} \approx -1.5$$

Discard the negative solution. The car will skid 180 feet in approximately 9.2 seconds.

49.

$$\begin{aligned} s(t) &= -16t^2 + 160t \\ 400 &= -16t^2 + 160t && \textit{Let } s(t) = 400. \\ 0 &= -16t^2 + 160t - 400 \\ 0 &= t^2 - 10t + 25 && \textit{Divide by } -16. \\ 0 &= (t-5)(t-5) \\ 0 &= (t-5)^2 \\ 0 &= t - 5 \\ 5 &= t \end{aligned}$$

The rock reaches a height of 400 feet after 5 seconds. This is its maximum height since this is the only time it reaches 400 feet.

51. Supply and demand are equal when

$$3p - 200 = \frac{3200}{p}.$$

Solve for p.

$$\begin{aligned} 3p^2 - 200p &= 3200 \\ 3p^2 - 200p - 3200 &= 0 \end{aligned}$$

Use the quadratic formula with $a = 3$, $b = -200$, and $c = -3200$.

$$\begin{aligned} p &= \frac{-(-200) \pm \sqrt{(-200)^2 - 4(3)(-3200)}}{2(3)} \\ &= \frac{200 \pm \sqrt{40,000 + 38,400}}{6} \\ &= \frac{200 \pm \sqrt{78,400}}{6} = \frac{200 \pm 280}{6} \end{aligned}$$

$$p = \frac{480}{6} = 80 \quad \text{or} \quad p = \frac{-80}{6} = -\frac{40}{3}$$

Discard the negative solution. The supply and demand are equal when the price is 80 cents or $.80.

53. Let $r =$ the interest rate.

$$A = P(1+r)^2$$

Let $A = 2142.25$ and $P = 2000$. Solve for r.

$$\begin{aligned} 2142.25 &= 2000(1+r)^2 \\ 1.071125 &= (1+r)^2 \end{aligned}$$

Use the square root property.

$$1 + r \approx 1.035 \quad \text{or} \quad 1 + r \approx -1.035$$
$$r \approx .035 \quad \text{or} \quad r \approx -2.035$$

Since the interest rate cannot be negative, reject -2.035. The interest rate is about 3.5%.

55. **(a)** 2.4 million

(b) Use $f(x) = .016x^2 + .124x + .787$ with $x = 7$.

$$f(7) = .016(7)^2 + .124(7) + .787$$
$$= 2.439$$

This answer agrees with part (a).

57. Use $f(x) = .016x^2 + .124x + .787$ with $f(x) = 2$.

$$2 = .016x^2 + .124x + .787$$
Multiply each side by 1000.
$$2000 = 16x^2 + 124x + 787$$
$$0 = 16x^2 + 124x - 1213$$

Here $a = 16$, $b = 124$, and $c = -1213$.

$$x = \frac{-b \pm \sqrt{b^2 - 4ac}}{2a}$$
$$x = \frac{-124 \pm \sqrt{124^2 - 4(16)(-1213)}}{2(16)}$$
$$= \frac{-124 \pm \sqrt{15{,}376 + 77{,}632}}{32}$$
$$= \frac{-124 \pm \sqrt{93{,}008}}{32}$$
$$x = \frac{-124 + \sqrt{93{,}008}}{32} \approx 5.7 \quad \text{or}$$
$$x = \frac{-124 - \sqrt{93{,}008}}{32} \approx -13.4$$

Since time cannot be negative, sales reached 2 million in 1995 ($x \approx 5$, rounded down). The graph indicates that sales reached 2 million in 1996.

59. Let F denote the Froude number. Solve

$$F = \frac{v^2}{g\ell}$$

for v.

$$v^2 = Fg\ell$$
$$v = \pm\sqrt{Fg\ell}$$

v is positive, so

$$V = \sqrt{Fg\ell}.$$

For the rhinoceros, $\ell = 1.2$ and $F = 2.57$.

$$V = \sqrt{(2.57)(9.8)(1.2)} \approx 5.5$$

or 5.5 meters per second.

61. Write a proportion.

$$\frac{x-4}{3x-19} = \frac{4}{x-3}$$

Multiply by the LCD, $(3x - 19)(x - 3)$.

$$(3x-19)(x-3)\left(\frac{x-4}{3x-19}\right) = (3x-19)(x-3)\left(\frac{4}{x-3}\right)$$
$$(x-3)(x-4) = (3x-19)4$$
$$x^2 - 7x + 12 = 12x - 76$$
$$x^2 - 19x + 88 = 0$$
$$(x-8)(x-11) = 0$$
$$x - 8 = 0 \quad \text{or} \quad x - 11 = 0$$
$$x = 8 \quad \text{or} \quad x = 11$$

If $x = 8$, then

$$3x - 19 = 3(8) - 19 = 5.$$

If $x = 11$, then

$$3x - 19 = 3(11) - 19 = 14.$$

Thus, $AC = 5$ or $AC = 14$.

Section 9.5

1. A parabola with equation $f(x) = a(x-h)^2 + k$ has vertex $V(h, k)$. We'll identify the vertex for each quadratic function.

(a) $f(x) = (x+2)^2 - 1$

$V(-2, -1)$, choice **B**

(b) $f(x) = (x+2)^2 + 1$

$V(-2, 1)$, choice **C**

(c) $f(x) = (x-2)^2 - 1$

$V(2, -1)$, choice **A**

(d) $f(x) = (x-2)^2 + 1$

$V(2, 1)$, choice **D**

For Exercises 3–9, we write $f(x)$ in the form $f(x) = a(x-h)^2 + k$ and then list the vertex (h, k).

3. $f(x) = -3x^2 = -3(x-0)^2 + 0$
The vertex (h, k) is $(0, 0)$.

5. $f(x) = x^2 + 4 = 1(x-0)^2 + 4$
The vertex (h, k) is $(0, 4)$.

7. $f(x) = (x-1)^2 = 1(x-1)^2 + 0$
The vertex (h, k) is $(1, 0)$.

9. $f(x) = (x+3)^2 - 4 = 1[x-(-3)]^2 - 4$
The vertex (h, k) is $(-3, -4)$.

11. In Exercise 9, the parabola is shifted 3 units to the left and 4 units down. The parabola in Exercise 10 is shifted 5 units to the right and 8 units down.

13. $f(x) = -\frac{2}{5}x^2$

Since $a = -\frac{2}{5} < 0$, the graph opens down. Since $|a| = \left|-\frac{2}{5}\right| = \frac{2}{5} < 1$, the graph is wider than the graph of $f(x) = x^2$.

15. $f(x) = 3x^2 + 1$

Since $a = 3 > 0$, the graph opens up.
Since $|a| = |3| = 3 > 1$, the graph is narrower than the graph of $f(x) = x^2$.

17. Consider $f(x) = a(x-h)^2 + k$.

(a) If $h > 0$ and $k > 0$ in $f(x) = a(x-h)^2 + k$, the shift is to the right and upward, so the vertex is in quadrant I.

(b) If $h > 0$ and $k < 0$, the shift is to the right and downward, so the vertex is in quadrant IV.

(c) If $h < 0$ and $k > 0$, the shift is to the left and upward, so the vertex is in quadrant II.

(d) If $h < 0$ and $k < 0$, the shift is to the left and downward, so the vertex is in quadrant III.

19. **(a)** $f(x) = (x-4)^2 - 2 = 1(x-4)^2 - 2$ has vertex $(4, -2)$. Because $a = 1 > 0$, the graph opens up. The correct answer is **D**.

(b) $f(x) = (x-2)^2 - 4 = 1(x-2)^2 - 4$ has vertex $(2, -4)$. Because $a = 1 > 0$, the graph opens up. The correct answer is **B**.

(c) $f(x) = -(x-4)^2 - 2 = -1(x-4)^2 - 2$ has vertex $(4, -2)$. Because $a = -1 < 0$, the graph opens down. The correct answer is **C**.

(d) $f(x) = -(x-2)^2 - 4 = -1(x-2)^2 - 4$ has vertex $(2, -4)$. Because $a = -1 < 0$, the graph opens down. The correct answer is **A**.

21. $f(x) = -2x^2$ written in the form
$f(x) = a(x-h)^2 + k$ is
$f(x) = -2(x-0)^2 + 0$.

Here, $h = 0$ and $k = 0$, so the vertex (h, k) is $(0, 0)$. Since $a = -2 < 0$, the graph opens down. Since $|a| = |-2| = 2 > 1$, the graph is narrower than the graph of $f(x) = x^2$. By evaluating the function with $x = 2$ and $x = -2$, we see that the points $(2, -8)$ and $(-2, -8)$ are on the graph.

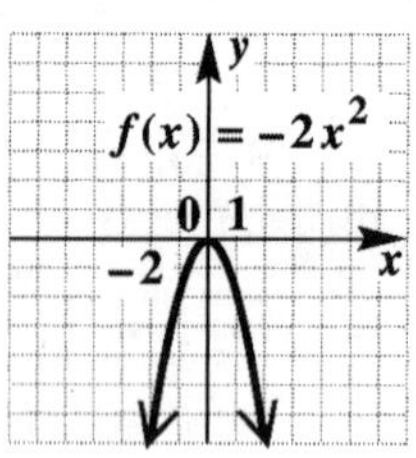

23. $f(x) = x^2 - 1$ written in the form
$f(x) = a(x-h)^2 + k$ is
$f(x) = 1(x-0)^2 + (-1)$.

Here, $h = 0$ and $k = -1$, so the vertex is $(0, -1)$. The graph opens up and has the same shape as $f(x) = x^2$ because $a = 1$. Two other points on the graph are $(-2, 3)$ and $(2, 3)$.

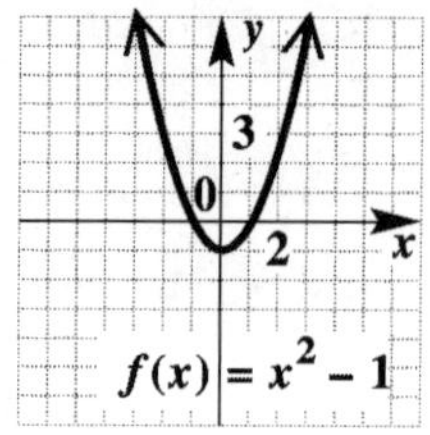

25. $f(x) = -x^2 + 2$ written in the form
$f(x) = a(x-h)^2 + k$ is
$f(x) = -1(x-0)^2 + 2$.

Here, $h = 0$ and $k = 2$, so the vertex (h, k) is $(0, 2)$. Since $a = -1 < 0$, the graph opens down. Since $|a| = |-1| = 1$, the graph has the same shape as $f(x) = x^2$. The points $(2, -2)$ and $(-2, -2)$ are on the graph.

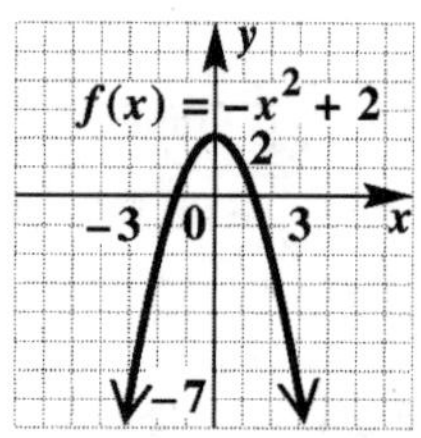

27. $f(x) = (x-4)^2$ written in the form
$f(x) = a(x-h)^2 + k$ is
$f(x) = 1(x-4)^2 + 0$.

Here, $h = 4$ and $k = 0$, so the vertex (h, k) is $(4, 0)$ and the axis is $x = 4$. The graph opens up since a is positive and has the same shape as $f(x) = x^2$ because $|a| = 1$. Two other points on the graph are $(2, 4)$ and $(6, 4)$. We can substitute any value for x, so the domain is $(-\infty, \infty)$. The range is $[0, \infty)$ since the smallest y-value is 0.

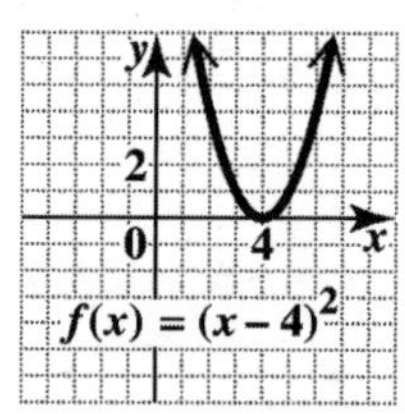

29. $f(x) = (x+2)^2 - 1$ written in the form
$f(x) = a(x-h)^2 + k$ is
$f(x) = 1[x - (-2)]^2 + (-1)$.

Since $h = -2$ and $k = -1$, the vertex (h, k) is $(-2, -1)$ and the axis is $x = -2$. Here, $a = 1$, so the graph opens up and has the same shape as $f(x) = x^2$. The points $(-1, 0)$ and $(-3, 0)$ are on the graph. The domain is $(-\infty, \infty)$. The range is $[-1, \infty)$ since the smallest y-value is -1.

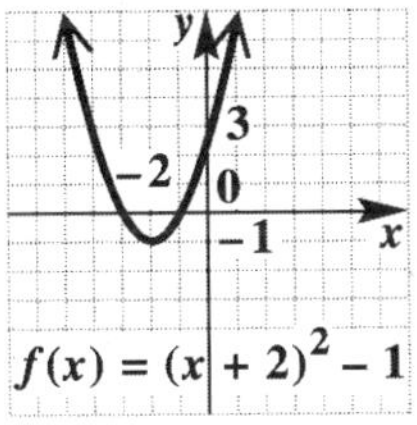

31. $f(x) = 2(x-2)^2 - 4$ written in the form
$f(x) = a(x-h)^2 + k$ is
$f(x) = 2(x-2)^2 + (-4)$.

Here, $h = 2$ and $k = -4$, so the vertex (h, k) is $(2, -4)$ and the axis is $x = 2$. The graph opens up and is narrower than $f(x) = x^2$ because $|a| = |2| > 1$. Two other points on the graph are $(0, 4)$ and $(4, 4)$.
We can substitute any value for x, so the domain is $(-\infty, \infty)$. The value of y is greater than or equal to -4, so the range is $[-4, \infty)$.

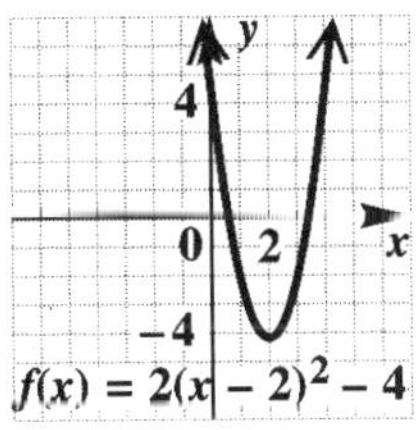

33. $f(x) = -\dfrac{1}{2}(x+1)^2 + 2$ written in the form
$f(x) = a(x-h)^2 + k$ is
$f(x) = -\dfrac{1}{2}[x-(-1)]^2 + 2$.

Since $h = -1$ and $k = 2$, the vertex (h, k) is $(-1, 2)$ and the axis is $x = -1$. Here, $a = -.5 < 0$, so the graph opens down. Also, $|a| = |-.5| = .5 < 1$, so the graph is wider than the graph of $f(x) = x^2$. The points $(1, 0)$ and $(-3, 0)$ are on the graph.
We can substitute any value for x, so the domain is $(-\infty, \infty)$. The value of y is less than or equal to 2, so the range is $(-\infty, 2]$.

$f(x) = -\frac{1}{2}(x+1)^2 + 2$

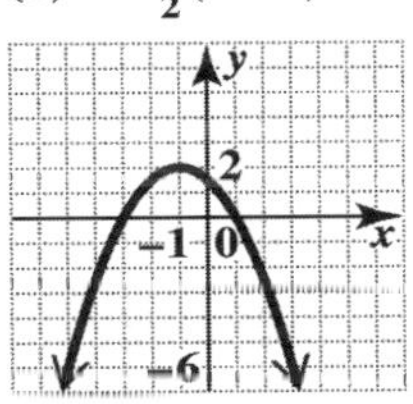

35. $f(x) = 2(x-2)^2 - 3$ written in the form
$f(x) = a(x-h)^2 + k$ is
$f(x) = 2(x-2)^2 + (-3)$.

Here, $h = 2$ and $k = -3$, so the vertex (h, k) is $(2, -3)$ and the axis is $x = 2$. The graph opens up and is narrower than $f(x) = x^2$ because $|a| = |2| > 1$. Two other points on the graph are $(3, -1)$ and $(1, -1)$.

We can substitute any value for x, so the domain is $(-\infty, \infty)$. The value of y is greater than or equal to -3, so the range is $[-3, \infty)$.

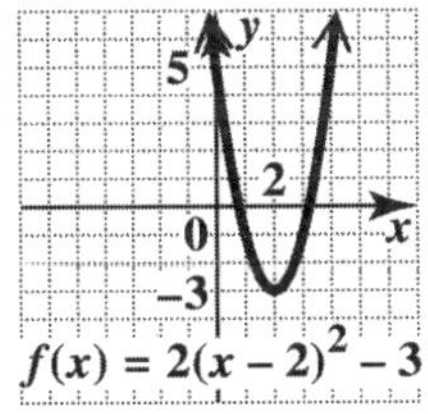

37. The graph of $F(x) = x^2 + 6$ would be shifted 6 units upward from the graph of $f(x) = x^2$.

38. To graph $G(x) = x + 6$, plot the intercepts $(-6, 0)$ and $(0, 6)$, and draw the line through them.

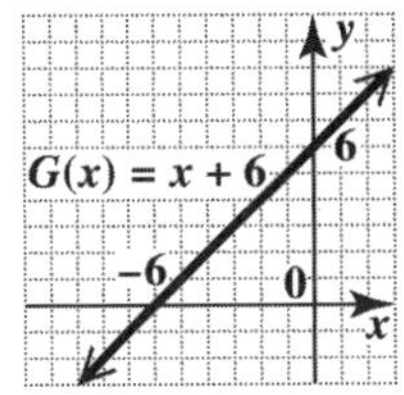

39. When considering the graph of $G(x) = x + 6$, the y-intercept is 6. The graph of $g(x) = x$ has y-intercept 0. Therefore, the graph of $G(x) = x + 6$ is shifted 6 units upward compared to the graph of $g(x) = x$.

40. The graph of $F(x) = (x-6)^2$ is shifted 6 units to the right compared to the graph of $f(x) = x^2$.

41. To graph $G(x) = x - 6$, plot the intercepts $(6, 0)$ and $(0, -6)$, and draw the line through them.

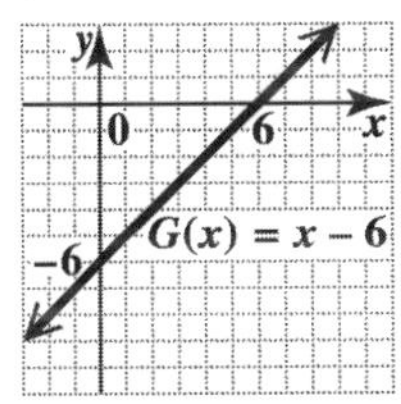

42. When considering the graph of $G(x) = x - 6$, its x-intercept is 6 as compared to the graph of $g(x) = x$ with x-intercept 0. The graph of $G(x) = x - 6$ is shifted 6 units to the right compared to the graph of $g(x) = x$.

43. Since the arrangement of the data points is approximately parabolic, a quadratic function would be the more appropriate model for the data set. The coefficient of x^2 should be positive, since the roughly parabolic shape of the graphed data set opens upward.

45. Since the arrangement of the data points is approximately parabolic, a quadratic function would be the more appropriate model for the data set. The coefficient of x^2 should be negative, since the roughly parabolic shape of the graphed data set opens downward.

47. Since the points lie approximately along a straight line, a linear function would be the more appropriate model for the data set. As the x-values of the points increase, the y-values increase, so the slope should be positive.

49. **(a)**

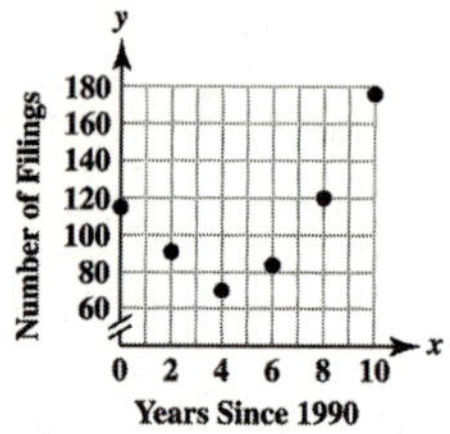

(b) Since the arrangement of the data points is approximately parabolic, a quadratic function would be the more appropriate model for the data set. The coefficient of x^2 should be positive, since the roughly parabolic shape of the graphed data set opens upward.

(c) Use $ax^2 + bx + c = y$ with $(0, 115)$, $(4, 70)$, and $(8, 120)$.

$$\begin{aligned} 0a + 0b + c &= 115 \quad (1) \\ 16a + 4b + c &= 70 \quad (2) \\ 64a + 8b + c &= 120 \quad (3) \end{aligned}$$

From (1), $c = 115$, so the system becomes

$$\begin{aligned} 16a + 4b &= -45 \quad (4) \\ 64a + 8b &= 5 \quad (5) \end{aligned}$$

Now eliminate b.

$$\begin{aligned} -32a - 8b &= 90 \qquad -2 \times (4) \\ 64a + 8b &= 5 \\ \hline 32a &= 95 \\ a &= \frac{95}{32} = 2.96875 \approx 2.969 \end{aligned}$$

From (5) with $a = \frac{95}{32}$,

$$\begin{aligned} 64\left(\frac{95}{32}\right) + 8b &= 5 \\ 190 + 8b &= 5 \\ 8b &= -185 \\ b &= -\frac{185}{8} = -23.125 \end{aligned}$$

The quadratic function is approximately

$$y = f(x) = 2.969x^2 - 23.125x + 115.$$

(d) $x = 2002 - 1990 = 12$ and $f(12) \approx 265$.

(e) No. About 16 companies filed for bankruptcy each month, so at this rate, filings for 2002 would be about 192. The approximation from the model seems high.

51. **(a)** $y = f(x) = .2455x^2 - 1.856x + 30.7105$
$x = 1999 - 1970 = 29$ and
$f(29) = 183.352 \approx 183.4$

(b) The approximation using the model is quite close.

53. $x^2 + 9x + 14 = 0$
From the screens, we see that the x-values of the x-intercepts are -7 and -2, so the solution set is $\{-7, -2\}$.

55. $-8x^2 + 6x + 5 = 0$
From the screens, we see that the x-values of the x-intercepts are $-.5$ and 1.25, so the solution set is $\{-.5, 1.25\}$.

Section 9.6

1. If there is an x^2-term in the equation, the axis is vertical. If there is a y^2-term, the axis is horizontal.

3. Use the discriminant, $b^2 - 4ac$, of the function. If it is positive, there are two x-intercepts. If it is zero, there is one x-intercept (at the vertex), and if it is negative, there is no x-intercept.

5. As in Example 1, we'll complete the square to find the vertex.

$$\begin{aligned} f(x) &= x^2 + 8x + 10 \\ &= x^2 + 8x + \underline{16} + 10 - \underline{16} \qquad \left[\tfrac{1}{2}(8)\right]^2 = 16 \\ &= (x + 4)^2 - 6 \end{aligned}$$

The vertex is $(-4, -6)$.

7. As in Example 2, we'll complete the square to find the vertex.

$$\begin{aligned} f(x) &= -2x^2 + 4x - 5 \\ &= -2(x^2 - 2x) - 5 \\ &= -2(x^2 - 2x + 1 - 1) - 5 \\ &= -2(x^2 - 2x + 1) + (-2)(-1) - 5 \\ &= -2(x - 1)^2 - 3 \end{aligned}$$

The vertex is $(1, -3)$.

9. As in Example 3, we'll use the vertex formula to find the vertex.

$$f(x) = -\tfrac{1}{2}x^2 + 2x - 3$$

The x-coordinate of the vertex is

$$-\frac{b}{2a} = -\frac{2}{2(-\frac{1}{2})} = -\frac{2}{-1} = 2.$$

The y-coordinate of the vertex is

$$f(2) = -2 + 4 - 3 = -1.$$

The vertex is $(2, -1)$.

11. As in Example 2, we'll complete the square to find the vertex.

$$\begin{aligned} f(x) &= 2x^2 + 4x + 5 \\ &= 2(x^2 + 2x) + 5 \\ &= 2(x^2 + 2x + 1 - 1) + 5 \end{aligned}$$

$$= 2(x^2 + 2x + 1) + 2(-1) + 5$$
$$= 2(x + 1)^2 - 2 + 5$$
$$f(x) = 2(x + 1)^2 + 3$$

The vertex is $(-1, 3)$.
Because $a = 2 > 1$, the graph opens up and is narrower than the graph of $y = x^2$.

For $y = f(x) = 2x^2 + 4x + 5$, $a = 2$, $b = 4$, and $c = 5$. The discriminant is

$$b^2 - 4ac = 4^2 - 4(2)(5)$$
$$= 16 - 40 = -24.$$

The discriminant is negative, so the parabola has no x-intercepts.

13. $f(x) = -x^2 + 5x + 3$

Use the vertex formula with $a = -1$ and $b = 5$.

The x-coordinate of the vertex is

$$\frac{-b}{2a} = \frac{-5}{2(-1)} = \frac{5}{2}.$$

The y-coordinate of the vertex is

$$f\left(\frac{-b}{2a}\right) = f\left(\frac{5}{2}\right)$$
$$= -\left(\frac{5}{2}\right)^2 + 5\left(\frac{5}{2}\right) + 3$$
$$= -\frac{25}{4} + \frac{25}{2} + 3$$
$$= \frac{-25 + 50 + 12}{4} = \frac{37}{4}.$$

The vertex is

$$\left(\frac{-b}{2a}, f\left(\frac{-b}{2a}\right)\right) = \left(\frac{5}{2}, \frac{37}{4}\right).$$

Because $a = -1$, the parabola opens down and has the same shape as the graph of $y = x^2$.

$$b^2 - 4ac = 5^2 - 4(-1)(3)$$
$$= 25 + 12 = 37$$

The discriminant is positive, so the parabola has two x-intercepts.

15. Complete the square on the y-terms to find the vertex.

$$x = \frac{1}{3}y^2 + 6y + 24$$
$$= \frac{1}{3}(y^2 + 18y) + 24$$
$$= \frac{1}{3}(y^2 + 18y + 81 - 81) + 24$$
$$= \frac{1}{3}(y^2 + 18y + 81) + \frac{1}{3}(-81) + 24$$
$$= \frac{1}{3}(y + 9)^2 - 27 + 24$$
$$x = \frac{1}{3}(y + 9)^2 - 3$$

The vertex is $(-3, -9)$.
The graph is a horizontal parabola. The graph opens to the right since $a = .\overline{3} > 0$ and is wider than the graph of $y = x^2$ since $|a| = |.\overline{3}| < 1$.

17. The graph of $y = 2x^2 + 4x - 3$ is a vertical parabola opening up, so choice F is correct. **(F)**

19. The graph of $y = -\frac{1}{2}x^2 - x + 1$ is a vertical parabola opening down, so choices A and C are possibilities. The graph in C is wider than the graph in A, so it must correspond to $a = -\frac{1}{2}$ while the graph in A must correspond to $a = -1$. **(C)**

21. The graph of $x = -y^2 - 2y + 4$ is a horizontal parabola opening to the left, so choice D is correct. **(D)**

23. $y = f(x) = x^2 + 8x + 10$

Step 1
Since $a = 1 > 0$, the graph opens up and is the same shape as the graph of $y = x^2$.

Step 2
From Exercise 5, the vertex is $(-4, -6)$. Since the graph opens up, the axis goes through the x-coordinate of the vertex—its equation is $x = -4$.

Step 3
To find the y-intercept, let $x = 0$.
$f(0) = 10$, so the y-intercept is $(0, 10)$.
To find the x-intercepts, let $y = 0$.

$$0 = x^2 + 8x + 10$$
$$x = \frac{-8 \pm \sqrt{64 - 40}}{2} = \frac{-8 \pm \sqrt{24}}{2}$$
$$= \frac{-8 \pm 2\sqrt{6}}{2} = -4 \pm \sqrt{6}$$

The x-intercepts are approximately $(-6.45, 0)$ and $(-1.55, 0)$.

Step 4
For an additional point on the graph, let $x = -2$ (two units to the right of the axis) to get $f(-2) = -2$. So the point $(-2, -2)$ is on the graph. By symmetry, the point $(-6, -2)$ (two units to the left of the axis) is on the graph.

continued

$f(x) = x^2 + 8x + 10$

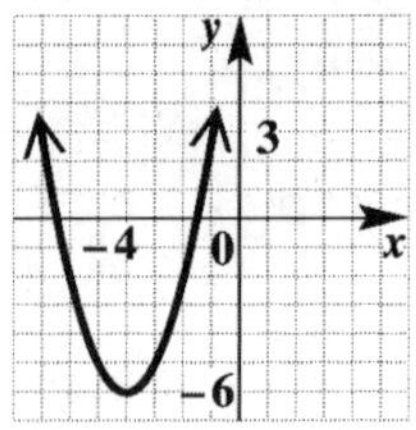

From the graph, we see that the domain is $(-\infty, \infty)$ and the range is $[-6, \infty)$.

25. $y = f(x) = -2x^2 + 4x - 5$

Step 1
Since $a = -2$, the graph opens down and is narrower than the graph of $y = x^2$.

Step 2
From Exercise 7, the vertex is $(1, -3)$. Since the graph opens down, the axis goes through the x-coordinate of the vertex—its equation is $x = 1$.

Step 3
If $x = 0$, $y = -5$, so the y-intercept is $(0, -5)$. To find the x-intercepts, let $y = 0$.

$$0 = -2x^2 + 4x - 5$$
$$x = \frac{-4 \pm \sqrt{16 - 40}}{2(-2)}$$

The discriminant is negative, so there are no x-intercepts.

Step 4
By symmetry, $(2, -5)$ is also on the graph.

$f(x) = -2x^2 + 4x - 5$

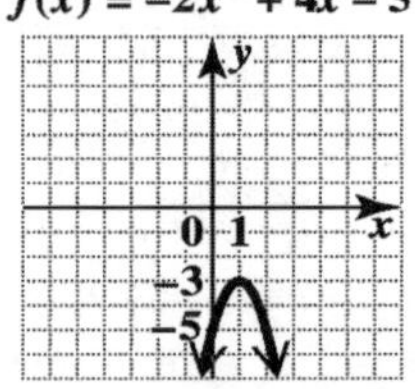

From the graph, we see that the domain is $(-\infty, \infty)$ and the range is $(-\infty, -3]$.

27. $x = (y + 2)^2 + 1 = y^2 + 4y + 5$

The roles of x and y are reversed, so this is a horizontal parabola.

Step 1
The coefficient of y^2 is $1 > 0$, so the graph opens to the right.

Step 2
The vertex can be identified from the given form of the equation. When $y = -2$, $x = 1$, so the vertex is $(1, -2)$. Since the graph opens right, the axis goes through the y-coordinate of the vertex—its equation is $y = -2$.

Step 3
To find the x-intercept, let $y = 0$.
$x = 0^2 + 4(0) + 5 = 5$, so the x-intercept is $(5, 0)$.

To find the y-intercepts, let $x = 0$.

$$0 = (y + 2)^2 + 1$$
$$-1 = (y + 2)^2$$

Since $(y + 2)^2$ cannot be negative, there are no y-intercepts.

Step 4
For an additional point on the graph, let $y = -4$ (two units below the axis) to get
$x = (-4 + 2)^2 + 1 = 5$.

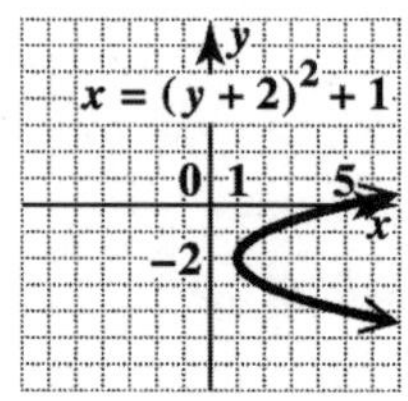

From the graph, we see the domain is $[1, \infty)$ and the range is $(-\infty, \infty)$.

29. $x = -\frac{1}{5}y^2 + 2y - 4$

The roles of x and y are reversed, so this is a horizontal parabola.

Step 1
Since $a = -\frac{1}{5} < 0$, the graph opens to the left and is wider than the graph of $y = x^2$.

Step 2
The y-coordinate of the vertex is

$$\frac{-b}{2a} = \frac{-2}{2(-\frac{1}{5})} = \frac{-2}{-\frac{2}{5}} = 5.$$

The x-coordinate of the vertex is

$$-\frac{1}{5}(5)^2 + 2(5) - 4 = -5 + 10 - 4 = 1.$$

Thus, the vertex is $(1, 5)$. Since the graph opens left, the axis goes through the y-coordinate of the vertex—its equation is $y = 5$.

Step 3
To find the x-intercept, let $y = 0$.
If $y = 0$, $x = -4$, so the x-intercept is $(-4, 0)$.
To find the y-intercepts, let $x = 0$.

$$0 = -\frac{1}{5}y^2 + 2y - 4$$
$$0 = y^2 - 10y + 20 \quad \textit{Multiply by } -5.$$
$$y = \frac{10 \pm \sqrt{100 - 80}}{2} = \frac{10 \pm \sqrt{20}}{2}$$
$$= \frac{10 \pm 2\sqrt{5}}{2} = 5 \pm \sqrt{5}$$

The y-intercepts are approximately $(0, 7.2)$ and $(0, 2.8)$.

Step 4
For an additional point on the graph, let $y = 7$ (two units above the axis) to get $x = \frac{1}{5}$. So the point $\left(\frac{1}{5}, 7\right)$ is on the graph. By symmetry, the point $\left(\frac{1}{5}, 3\right)$ (two units below the axis) is on the graph.

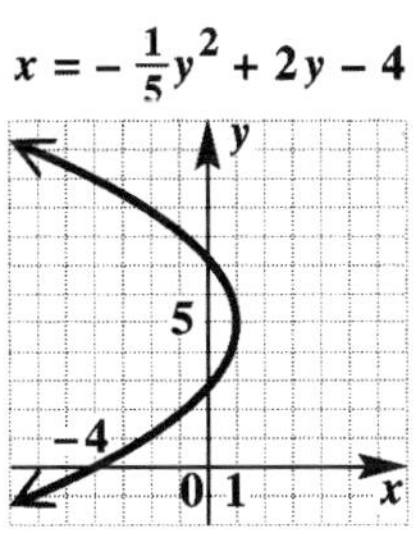

From the graph, we see that the domain is $(-\infty, 1]$ and the range is $(-\infty, \infty)$.

31. $x = 3y^2 + 12y + 5$

The roles of x and y are reversed, so this is a horizontal parabola.

Step 1
Since $a = 3 > 0$, the graph opens to the right and is narrower than the graph of $y = x^2$.

Step 2
Use the formula to find the y-value of the vertex.

$$\frac{-b}{2a} = \frac{-12}{2(3)} = -2$$

If $y = -2$, $x = -7$, so the vertex is $(-7, -2)$. Since the graph opens right, the axis goes through the y-coordinate of the vertex—its equation is $y = -2$.

Step 3
If $y = 0$, $x = 5$, so the x-intercept is $(5, 0)$. To find the y-intercepts, let $x = 0$.

$$0 = 3y^2 + 12y + 5$$
$$y = \frac{-12 \pm \sqrt{144 - 60}}{6} = \frac{-12 \pm \sqrt{84}}{6}$$
$$= \frac{-12 \pm 2\sqrt{21}}{6} = -2 \pm \frac{1}{3}\sqrt{21}$$

The y-intercepts are approximately $(0, -.5)$ and $(0, -3.5)$.

Step 4
By symmetry, $(5, -4)$ is also on the graph.

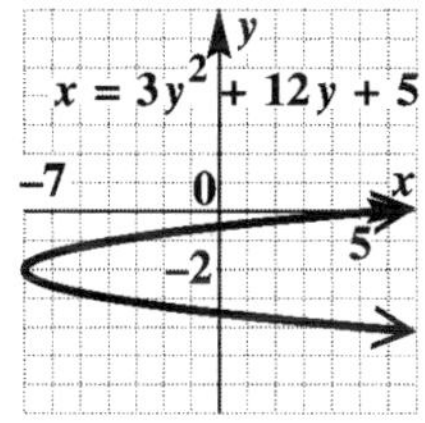

From the graph, we see that the domain is $[-7, \infty)$ and the range is $(-\infty, \infty)$.

33. Let x = one number, $60 - x$ = the other number, and P = the product.

$$P = x(60 - x)$$
$$= 60x - x^2 \quad \text{or} \quad -x^2 + 60x$$

This parabola opens down so the maximum occurs at the vertex.

Here $a = -1$, $b = 60$, and $c = 0$.

$$\frac{-b}{2a} = \frac{-60}{2(-1)} = 30$$

$x = 30$ when the product is a maximum.

Since $x = 30$, $60 - x = 30$, and the two numbers are 30 and 30.

35. Let x represent the length of the two equal sides, and let $280 - 2x$ represent the length of the remaining side (the side parallel to the highway). Then substitute x for L and $280 - 2x$ for W in the formula for the area of a rectangle, $A = LW$.

$$A = x(280 - 2x)$$
$$= 280x - 2x^2 \text{ or } -2x^2 + 280x$$

The maximum area will occur at the vertex.

$$x = \frac{-b}{2a} = \frac{-280}{2(-2)} = 70$$

When $x = 70$, $280 - 2x = 140$, and $A = 9800$. Thus, the dimensions of the lot with maximum area are 140 feet by 70 feet and the maximum area is 9800 square feet.

37. $h(t) = 32t - 16t^2$ or $h(t) = -16t^2 + 32t$

Here, $a = -16 < 0$, so the parabola opens down. The time it takes to reach the maximum height and the maximum height are given by the vertex of the parabola. Use the vertex formula to find that

$$t = \frac{b}{2a} = \frac{32}{2(-16)} = \frac{-32}{-32} = 1,$$

and $$h(t) = -16(1)^2 + 32(1)$$
$$= -16 + 32 = 16.$$

The vertex is $(1, 16)$, so the maximum height is 16 feet which occurs when the time is 1 second. The object hits the ground when $h = 0$.

continued

$$0 = -16t^2 + 32t$$
$$0 = -16t(t-2)$$
$$-16t = 0 \quad \text{or} \quad t - 2 = 0$$
$$t = 0 \quad \text{or} \quad t = 2$$

It takes 2 seconds for the object to hit the ground.

39. The graph of the height of the projectile,

$$s(t) = -16t^2 + 64t + 3,$$

is a parabola that opens down since $a = -16 < 0$. The time at which the cork reaches its maximum height and the maximum height are the t and s coordinates of the vertex.

$$t = \frac{-b}{2a} = \frac{-64}{2(-16)} = 2$$

$$s(2) = -16(2)^2 + 64(2) + 3$$
$$= -64 + 128 + 3 = 67$$

The cork reaches a maximum height of 67 feet after 2 seconds.

41. **(a)** Since the parabola opens up, the y-value of the vertex of this graph is a minimum.

(b) $f(x) = .228x^2 - 2.57x + 8.97$

To find the vertex, use $\left(-\frac{b}{2a}, f\left(-\frac{b}{2a}\right)\right)$,

where $a = .228$, $b = -2.57$, and $c = 8.97$.

$$-\frac{b}{2a} = -\frac{-2.57}{2(.228)} \approx 5.6$$

Rounding down to the nearest year, the minimum percent increase occurred in $1990 + 5 = 1995$.

$$f(5.6) = .228(5.6)^2 - 2.57(5.6) + 8.97 \approx 1.73$$

To the nearest tenth, the percent increase in 1995 was 1.7%.

43. $f(x) = -20.57x^2 + 758.9x - 3140$

(a) The coefficient of x^2 is negative because the parabola opens down.

(b) Use the vertex formula.

$$x = \frac{-b}{2a} = \frac{-758.9}{2(-20.57)} \approx 18.45$$
$$f(18.45) \approx 3860$$

The vertex is approximately $(18.45, 3860)$.

(c) 18 corresponds to 2018, so in 2018 social security assets will reach their maximum value of $3860 billion.

45. The number of people on the plane is $100 - x$ since x is the number of unsold seats. The price per seat is $200 + 4x$.

(a) The total revenue received for the flight is found by multiplying the number of seats by the price per seat. Thus, the revenue is

$$R(x) = (100 - x)(200 + 4x)$$
$$= 20,000 + 200x - 4x^2.$$

(b) Use the formula for the vertex.

$$x = \frac{-b}{2a} = \frac{-200}{2(-4)} = 25$$

$R(25) = 22,500$, so the vertex is $(25,\ 22,500)$. $R(0) = 20,000$, so the R-intercept is $(0,\ 20,000)$. From the factored form for R, we see that the positive x-intercept is $(100, 0)$. (The factor $200 + 4x$ leads to a negative x-intercept, meaningless in this problem.)

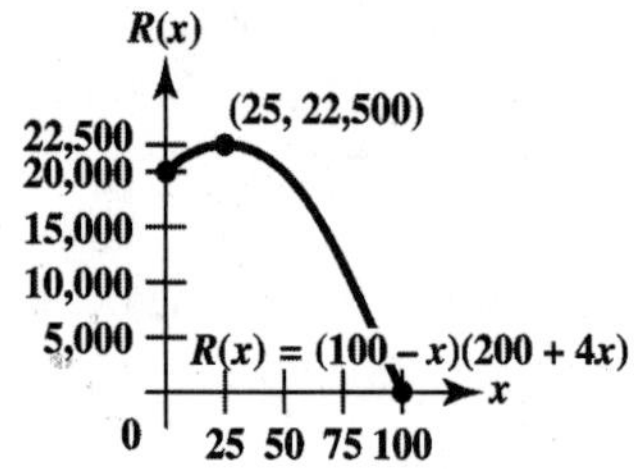

(c) The number of unsold seats x that produce the maximum revenue is 25, the x-value of the vertex.

(d) The maximum revenue is $22,500, the y-value of the vertex.

47. $f(x) = x^2 - 8x + 18$

$$\frac{-b}{2a} = \frac{-(-8)}{2(1)} = 4$$

$f(4) = 2$, so the vertex is $(4, 2)$, which matches choice **B**.

49. $f(x) = x^2 - 8x + 14$

$$\frac{-b}{2a} = \frac{-(-8)}{2(1)} = 4$$

$f(4) = -2$, so the vertex is $(4, -2)$, which matches choice **A**.

51. **(a)** $|x - (-p)| = |x + p|$

(b) The focus should have coordinates $(p, 0)$ because the distance from the focus to the origin should equal the distance from the directrix to the origin.

(c) The distance from (x, y) to $(p, 0)$ is

$$\sqrt{(x-p)^2 + (y-0)^2} = \sqrt{(x-p)^2 + y^2}.$$

(d) Using the results from parts (a) and (c), these distances should be equal.

$$\sqrt{(x-p)^2+y^2} = |x+p|$$

Square both sides.

$$(x-p)^2+y^2 = (x+p)^2$$
$$x^2-2px+p^2+y^2 = x^2+2px+p^2$$
$$y^2 = 4px$$

Section 9.7

1. **(a)** The x-intercepts determine the solutions of the equation $x^2-4x+3=0$. From the graph, the solution set is $\{1,3\}$.

(b) The x-values of the points on the graph that are *above* the x-axis form the solution set of the inequality $x^2-4x+3>0$. From the graph, the solution set is $(-\infty,1)\cup(3,\infty)$.

(c) The x-values of the points on the graph that are *below* the x-axis form the solution set of the inequality $x^2-4x+3<0$. From the graph, the solution set is $(1,3)$.

3. **(a)** The x-intercepts determine the solutions of the equation $-2x^2-x+15=0$. From the graph, the solution set is $\left\{-3,\frac{5}{2}\right\}$.

(b) The x-values of the points on the graph that are *above* the x-axis form the solution set of the inequality $-2x^2-x+15>0$. From the graph, the solution set for $-2x^2-x+15\geq 0$ is $\left[-3,\frac{5}{2}\right]$.

(c) The x-values of the points on the graph that are *below* the x-axis form the solution set of the inequality $-2x^2-x+15<0$. From the graph, the solution set for $-2x^2-x+15\leq 0$ is $(-\infty,-3]\cup\left[\frac{5}{2},\infty\right)$.

5. Include the endpoints if the symbol is $\leq$ or $\geq$. Exclude the endpoints if the symbol is $<$ or $>$.

7. $(x+1)(x-5)>0$

Solve the equation $(x+1)(x-5)=0$.

$x+1=0$ or $x-5=0$
$x=-1$ or $x=5$

The numbers -1 and 5 divide a number line into three intervals: A, B, and C.

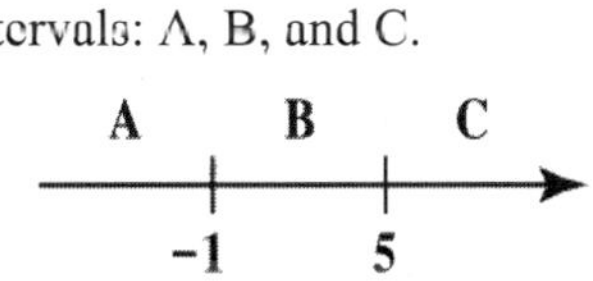

Test a number from each interval in the original inequality.

Interval A: Let $x=-2$.

$$(x+1)(x-5)>0$$
$$(-2+1)(-2-5)>0 \quad ?$$
$$-1(-7)>0 \quad ?$$
$$7>0 \quad \textit{True}$$

Interval B: Let $x=0$.

$$(0+1)(0-5)>0 \quad ?$$
$$-5>0 \quad \textit{False}$$

Interval C: Let $x=6$.

$$(6+1)(6-5)>0 \quad ?$$
$$7>0 \quad \textit{True}$$

The solution set includes the numbers in Intervals A and C, excluding -1 and 5 because of $>$.

Solution set: $(-\infty,-1)\cup(5,\infty)$

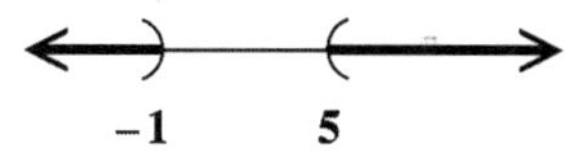

9. $(r+4)(r-6)<0$

Solve the equation $(r+4)(r-6)=0$.

$r+4=0$ or $r-6=0$
$r=-4$ or $r=6$

These numbers divide a number line into three intervals: A, B, and C.

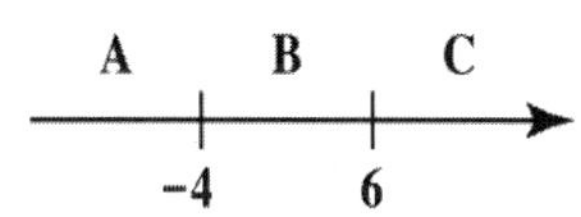

Test a number from each interval in the original inequality.

Interval A: Let $r=-5$.

$$(r+4)(r-6)<0$$
$$(-5+4)(-5-6)<0 \quad ?$$
$$-1(-11)<0 \quad ?$$
$$11<0 \quad \textit{False}$$

Interval B: Let $r=0$.

$$4(-6)<0 \quad ?$$
$$-24<0 \quad \textit{True}$$

Interval C: Let $r=7$.

$$(7+4)(7-6)<0 \quad ?$$
$$11(1)<0 \quad ?$$
$$11<0 \quad \textit{False}$$

The solution set includes Interval B, where the expression is negative.

Solution set: $(-4,6)$

11. $x^2 - 4x + 3 \geq 0$

Solve the equation

$x^2 - 4x + 3 = 0.$

$(x - 1)(x - 3) = 0$

$x - 1 = 0 \quad \text{or} \quad x - 3 = 0$

$x = 1 \quad \text{or} \quad x = 3$

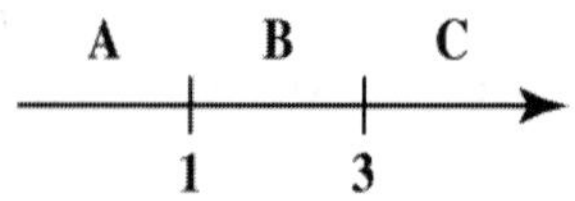

Interval A: Let $x = 0$.

$3 \geq 0$ *True*

Interval B: Let $x = 2$.

$2^2 - 4(2) + 3 \geq 0$?

$-1 \geq 0$ *False*

Interval C: Let $x = 4$.

$4^2 - 4(4) + 3 \geq 0$?

$3 \geq 0$ *True*

The solution set includes the numbers in Intervals A and C, including 1 and 3 because of $\geq$.

Solution set: $(-\infty, 1] \cup [3, \infty)$

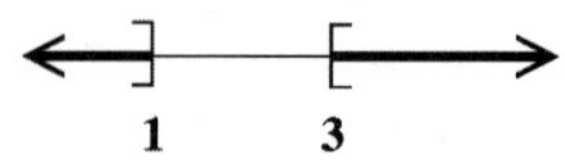

13. $10t^2 + 9t \geq 9$

$10t^2 + 9t - 9 \geq 0$

Solve the equation

$10t^2 + 9t - 9 = 0.$

$(2t + 3)(5t - 3) = 0$

$2t + 3 = 0 \quad \text{or} \quad 5t - 3 = 0$

$t = -\frac{3}{2} \quad \text{or} \quad t = \frac{3}{5}$

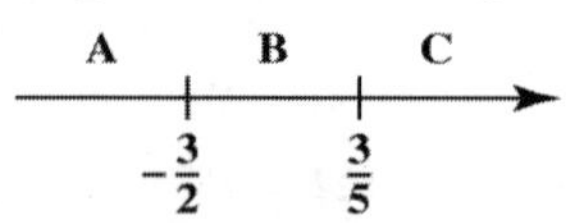

Test a number from each interval in the original inequality.

Interval A: Let $t = -2$.

$10(-2)^2 + 9(-2) \geq 9$?

$40 - 18 \geq 9$?

$22 \geq 9$ *True*

Interval B: Let $t = 0$.

$0 \geq 9$ *False*

Interval C: Let $t = 1$.

$10(1)^2 + 9(1) \geq 9$?

$10 + 9 \geq 9$?

$19 \geq 9$ *True*

The solution set includes the numbers in Intervals A and C, including $-\frac{3}{2}$ and $\frac{3}{5}$ because of $\geq$.

Solution set: $\left(-\infty, -\frac{3}{2}\right] \cup \left[\frac{3}{5}, \infty\right)$

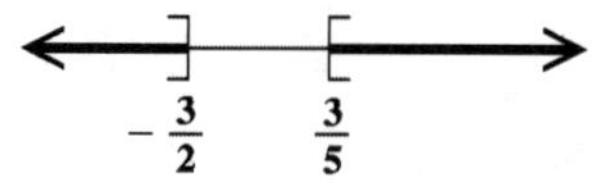

15. $4x^2 - 9 \leq 0$

Solve the equation

$4x^2 - 9 = 0.$

$(2x + 3)(2x - 3) = 0$

$2x + 3 = 0 \quad \text{or} \quad 2x - 3 = 0$

$x = -\frac{3}{2} \quad \text{or} \quad x = \frac{3}{2}$

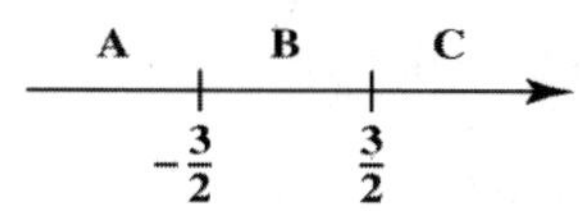

Test a number from each interval in the inequality

$4x^2 - 9 \leq 0.$

Interval A: Let $x = -2$.

$4(-2)^2 - 9 \leq 0$?

$7 \leq 0$ *False*

Interval B: Let $x = 0$.

$-9 \leq 0$ *True*

Interval C: Let $x = 2$.

$4(2)^2 - 9 \leq 0$?

$7 \leq 0$ *False*

The solution set includes Interval B, including the endpoints.

Solution set: $\left[-\frac{3}{2}, \frac{3}{2}\right]$

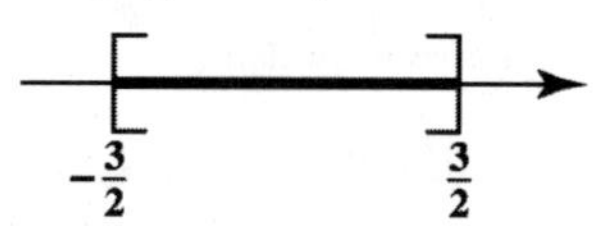

17. $6x^2 + x \geq 1$

$6x^2 + x - 1 \geq 0$

Solve the equation

$6x^2 + x - 1 = 0.$

$(2x + 1)(3x - 1) = 0$

$2x + 1 = 0 \quad \text{or} \quad 3x - 1 = 0$

$x = -\frac{1}{2} \quad \text{or} \quad x = \frac{1}{3}$

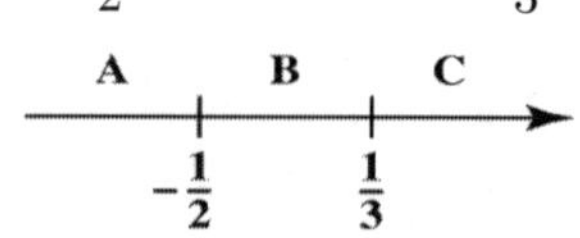

Test a number from each interval in the inequality.

$$6x^2 + x \geq 1.$$

Interval A: Let $x = -1$.

$$6(-1)^2 + (-1) \geq 1 \quad ?$$
$$5 \geq 1 \quad \textit{True}$$

Interval B: Let $x = 0$.

$$0 \geq 1 \quad \textit{False}$$

Interval C: Let $x = 1$.

$$6(1)^2 + 1 \geq 1 \quad ?$$
$$7 \geq 1 \quad \textit{True}$$

The solution set includes the numbers in Intervals A and C, including $-\frac{1}{2}$ and $\frac{1}{3}$ because of $\geq$.

Solution set: $\left(-\infty, -\frac{1}{2}\right] \cup \left[\frac{1}{3}, \infty\right)$

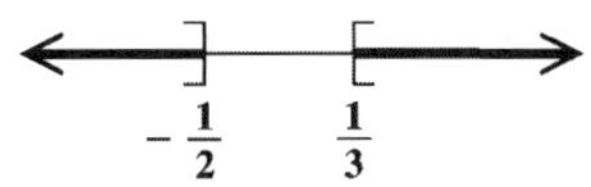

19. $z^2 - 4z \geq 0$

Solve the equation

$z^2 - 4z = 0.$

$z(z-4) = 0$

$z = 0$ or $z - 4 = 0$

$z = 4$

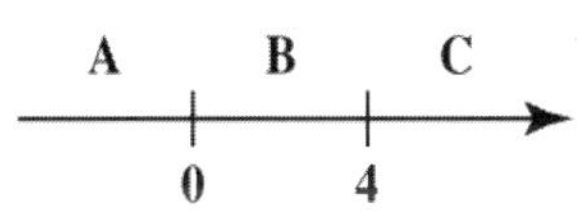

Interval A: Let $z = -1$.

$$(-1)^2 - 4(-1) \geq 0 \quad ?$$
$$5 \geq 0 \quad \textit{True}$$

Interval B: Let $z = 2$.

$$2^2 - 4(2) \geq 0 \quad ?$$
$$-4 \geq 0 \quad \textit{False}$$

Interval C: Let $z = 5$.

$$5^2 - 4(5) \geq 0 \quad ?$$
$$5 \geq 0 \quad \textit{True}$$

The solution set includes the numbers in Intervals A and C, including 0 and 4 because of $\geq$.

Solution set: $(-\infty, 0] \cup [4, \infty)$

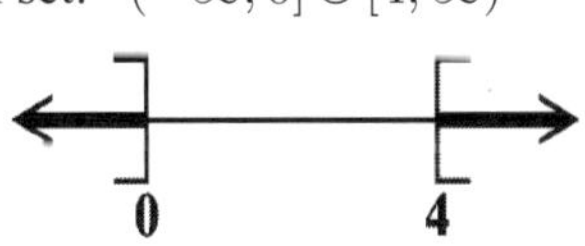

21. $3k^2 - 5k \leq 0$

Solve the equation

$3k^2 - 5k = 0.$

$k(3k - 5) = 0$

$k = 0$ or $3k - 5 = 0$

$k = \frac{5}{3}$

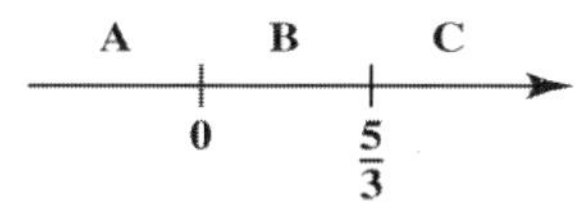

Interval A: Let $k = -1$.

$$3(-1)^2 - 5(-1) \leq 0 \quad ?$$
$$8 \leq 0 \quad \textit{False}$$

Interval B: Let $k = 1$.

$$3(1)^2 - 5(1) \leq 0 \quad ?$$
$$-2 \leq 0 \quad \textit{True}$$

Interval C: Let $k = 2$.

$$3(2)^2 - 5(2) \leq 0 \quad ?$$
$$2 \leq 0 \quad \textit{False}$$

The solution set includes the numbers in Interval B, including the endpoints.

Solution set: $\left[0, \frac{5}{3}\right]$

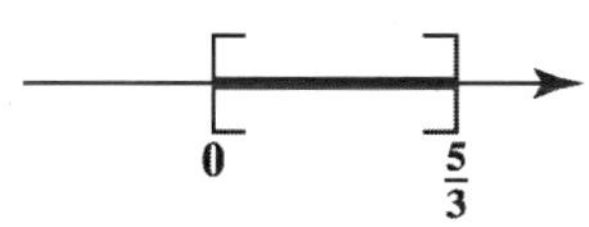

23. $x^2 - 6x + 6 \geq 0$

Solve the equation

$$x^2 - 6x + 6 = 0.$$

Since $x^2 - 6x + 6$ does not factor, let $a = 1$, $b = -6$, and $c = 6$ in the quadratic formula.

$$x = \frac{-(-6) \pm \sqrt{(-6)^2 - 4(1)(6)}}{2(1)}$$
$$= \frac{6 \pm \sqrt{12}}{2} = \frac{6 \pm 2\sqrt{3}}{2}$$
$$= \frac{2\left(3 \pm \sqrt{3}\right)}{2} = 3 \pm \sqrt{3}$$

$x = 3 + \sqrt{3} \approx 4.7$ or

$x = 3 - \sqrt{3} \approx 1.3$

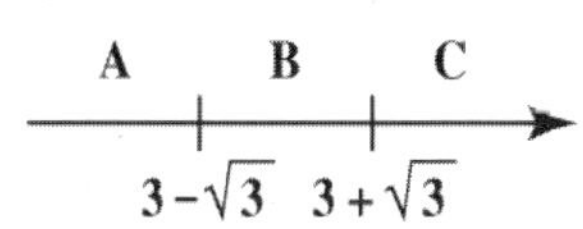

Test a number from each interval in the inequality

$$x^2 - 6x + 6 \geq 0.$$

continued

Interval A: Let $x = 0$.

$6 \geq 0$ *True*

Interval B: Let $x = 3$.

$3^2 - 6(3) + 6 \geq 0$?

$-3 \geq 0$ *False*

Interval C: Let $x = 5$.

$5^2 - 6(5) + 6 \geq 0$?

$1 \geq 0$ *True*

The solution set includes the numbers in Intervals A and C, including $3 - \sqrt{3}$ and $3 + \sqrt{3}$ because of $\geq$.

Solution set: $\left(-\infty, 3 - \sqrt{3}\right] \cup \left[3 + \sqrt{3}, \infty\right)$

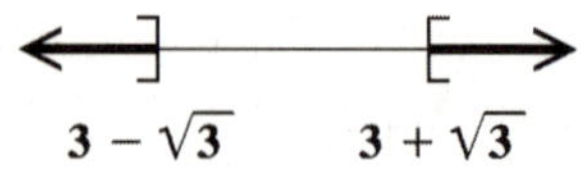

25. $(4 - 3x)^2 \geq -2$

Since $(4 - 3x)^2$ is either 0 or positive, $(4 - 3x)^2$ will always be greater than -2. Therefore, the solution set is $(-\infty, \infty)$.

27. $(3x + 5)^2 \leq -4$

Since $(3x + 5)^2$ is never negative, $(3x + 5)^2$ will never be less than or equal to a negative number. Therefore, the solution set is $\emptyset$.

29. $(p - 1)(p - 2)(p - 4) < 0$

The numbers 1, 2, and 4 are solutions of the cubic equation

$$(p - 1)(p - 2)(p - 4) = 0.$$

These numbers divide a number line into four intervals.

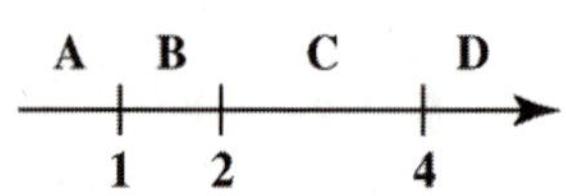

Test a number from each interval in the inequality

$$(p - 1)(p - 2)(p - 4) < 0.$$

Interval A: Let $p = 0$.

$-1(-2)(-4) < 0$?

$-8 < 0$ *True*

Interval B: Let $p = 1.5$.

$(1.5 - 1)(1.5 - 2)(1.5 - 4) < 0$?

$.5(-.5)(-2.5) < 0$?

$.625 < 0$ *False*

Interval C: Let $p = 3$.

$(3 - 1)(3 - 2)(3 - 4) < 0$?

$2(1)(-1) < 0$?

$-2 < 0$ *True*

Interval D: Let $p = 5$.

$(5 - 1)(5 - 2)(5 - 4) < 0$?

$4(3)(1) < 0$?

$12 < 0$ *False*

The numbers in Intervals A and C, not including 1, 2, or 4, are solutions.

Solution set: $(-\infty, 1) \cup (2, 4)$

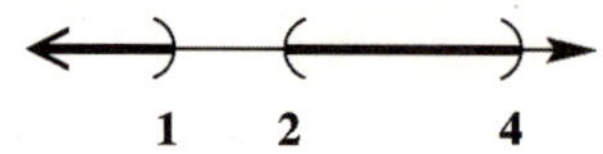

31. $(x - 4)(2x + 3)(3x - 1) \geq 0$

The numbers 4, $-\frac{3}{2}$, and $\frac{1}{3}$ are solutions of the cubic equation

$$(x - 4)(2x + 3)(3x - 1) = 0.$$

These numbers divide a number line into 4 intervals.

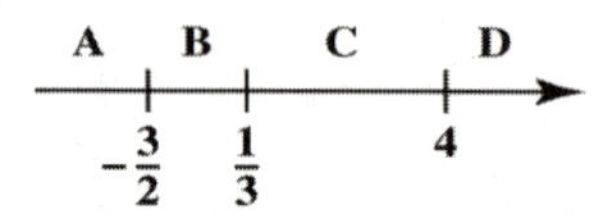

Interval A: Let $x = -2$.

$-6(-1)(-7) \geq 0$?

$-42 \geq 0$ *False*

Interval B: Let $x = 0$.

$-4(3)(-1) \geq 0$?

$12 \geq 0$ *True*

Interval C: Let $x = 1$.

$-3(5)(2) \geq 0$?

$-30 \geq 0$ *False*

Interval D: Let $x = 5$.

$1(13)(14) \geq 0$?

$182 \geq 0$ *True*

The solution set includes numbers in Intervals B and D, including the endpoints.

Solution set: $\left[-\frac{3}{2}, \frac{1}{3}\right] \cup [4, \infty)$

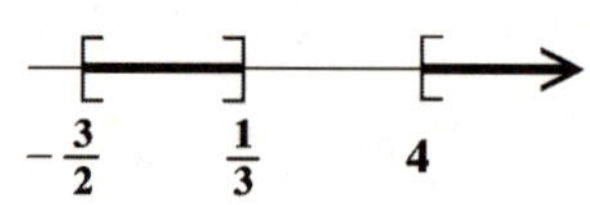

33. $\frac{x - 1}{x - 4} > 0$

The number 1 makes the numerator 0, and 4 makes the denominator 0. These two numbers determine three intervals.

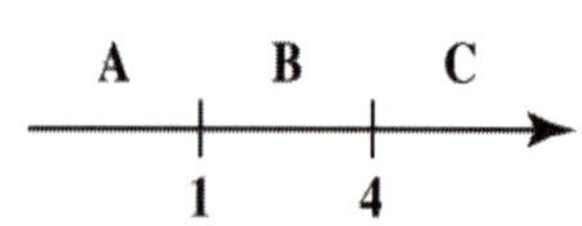

Test a number from each interval in the inequality

$$\frac{x - 1}{x - 4} > 0.$$

Interval A: Let $x = 0$.

$$\frac{0-1}{0-4} > 0 \quad ?$$

$$\frac{1}{4} > 0 \qquad \textit{True}$$

Interval B: Let $x = 2$.

$$\frac{2-1}{2-4} > 0 \quad ?$$

$$\frac{1}{-2} > 0 \qquad \textit{False}$$

Interval C: Let $x = 5$.

$$\frac{5-1}{5-4} > 0 \quad ?$$

$$4 > 0 \qquad \textit{True}$$

The solution set includes numbers in Intervals A and C, excluding endpoints.
Solution set: $(-\infty, 1) \cup (4, \infty)$

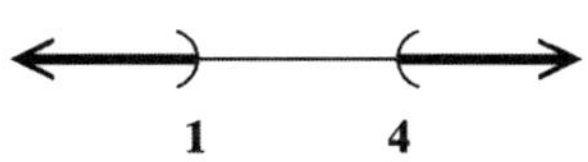

35. $\dfrac{2n+3}{n-5} \le 0$

The number $-\frac{3}{2}$ makes the numerator 0, and 5 makes the denominator 0. These two numbers determine three intervals.

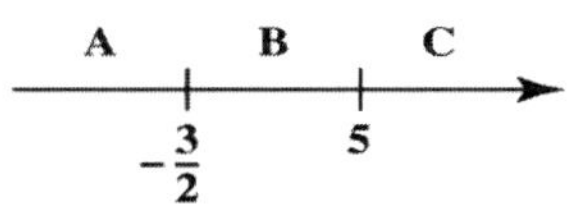

Test a number from each interval in the inequality

$$\frac{2n+3}{n-5} \le 0.$$

Interval A: Let $n = -2$.

$$\frac{2(-2)+3}{(-2)-5} \le 0 \quad ?$$

$$\frac{1}{7} \le 0 \qquad \textit{False}$$

Interval B: Let $n = 0$.

$$\frac{2(0)+3}{0-5} \le 0 \quad ?$$

$$-\frac{3}{5} \le 0 \qquad \textit{True}$$

Interval C: Let $n = 6$.

$$\frac{2(6)+3}{6-5} \le 0 \quad ?$$

$$15 \le 0 \qquad \textit{False}$$

The solution set includes the points in Interval B. The endpoint 5 is not included since it makes the left side undefined. The endpoint $-\dfrac{3}{2}$ is included because it makes the left side equal to 0.

Solution set: $\left[-\dfrac{3}{2}, 5\right)$

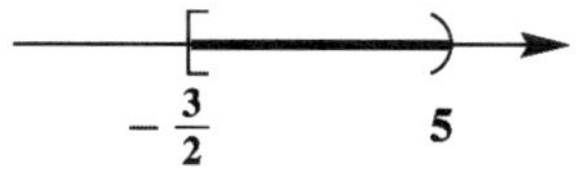

37. $\dfrac{8}{x-2} \ge 2$

Write the inequality so that 0 is on one side.

$$\frac{8}{x-2} - 2 \ge 0$$

$$\frac{8}{x-2} - \frac{2(x-2)}{x-2} \ge 0$$

$$\frac{8-2x+4}{x-2} \ge 0$$

$$\frac{-2x+12}{x-2} \ge 0$$

The number 6 makes the numerator 0, and 2 makes the denominator 0. These two numbers determine three intervals.

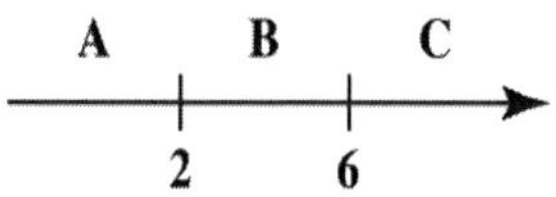

Test a number from each interval in the inequality

$$\frac{8}{x-2} \ge 2.$$

Interval A: Let $x = 0$.

$$\frac{8}{0-2} \ge 2 \quad ?$$

$$-4 \ge 2 \qquad \textit{False}$$

Interval B: Let $x = 3$.

$$\frac{8}{3-2} \ge 2 \quad ?$$

$$8 \ge 2 \qquad \textit{True}$$

Interval C: Let $x = 7$.

$$\frac{8}{7-2} \ge 2 \quad ?$$

$$\frac{8}{5} \ge 2 \qquad \textit{False}$$

The solution set includes numbers in Interval B, including 6 but excluding 2, which makes the fraction undefined.
Solution set: $(2, 6]$

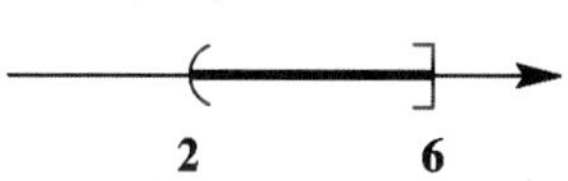

39. $\dfrac{3}{2t-1} < 2$

Write the inequality so that 0 is on one side.

$$\frac{3}{2t-1} - 2 < 0$$

$$\frac{3}{2t-1} - \frac{2(2t-1)}{2t-1} < 0$$

$$\frac{3-4t+2}{2t-1} < 0$$

continued

$$\frac{-4t+5}{2t-1} < 0$$

The number $\frac{5}{4}$ makes the numerator 0, and $\frac{1}{2}$ makes the denominator 0. These two numbers determine three intervals.

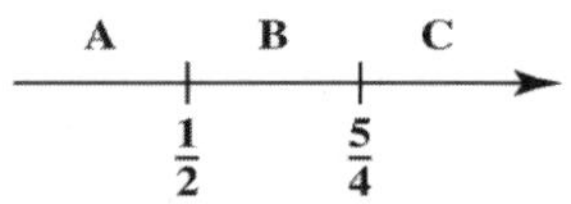

Test a number from each interval in the inequality

$$\frac{3}{2t-1} < 2.$$

Interval A: Let $t = 0$.

$$\frac{3}{2(0)-1} < 2 \quad ?$$
$$-3 < 2 \qquad \textit{True}$$

Interval B: Let $t = 1$.

$$\frac{3}{2(1)-1} < 2 \quad ?$$
$$3 < 2 \qquad \textit{False}$$

Interval C: Let $t = 2$.

$$\frac{3}{2(2)-1} < 2 \quad ?$$
$$1 < 2 \qquad \textit{True}$$

The solution set includes numbers in Intervals A and C, excluding endpoints.

Solution set: $\left(-\infty, \frac{1}{2}\right) \cup \left(\frac{5}{4}, \infty\right)$

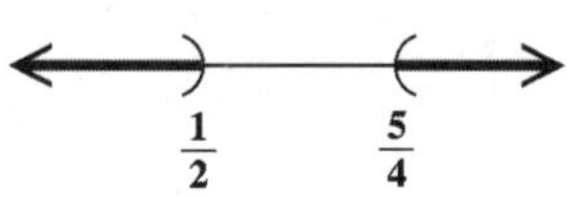

41. $$\frac{g-3}{g+2} \geq 2$$

Write the inequality so that 0 is on one side.

$$\frac{g-3}{g+2} - 2 \geq 0$$
$$\frac{g-3}{g+2} - \frac{2(g+2)}{g+2} \geq 0$$
$$\frac{g-3-2g-4}{g+2} \geq 0$$
$$\frac{-g-7}{g+2} \geq 0$$

The number -7 makes the numerator 0, and -2 makes the denominator 0. These two numbers determine three intervals.

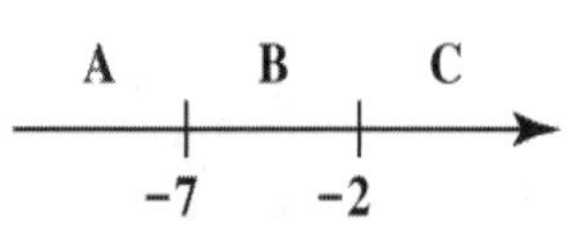

Test a number from each interval in the inequality

$$\frac{g-3}{g+2} \geq 2.$$

Interval A: Let $g = -8$.

$$\frac{-11}{-6} \geq 2 \quad ?$$
$$\frac{11}{6} \geq 2 \qquad \textit{False}$$

Interval B: Let $g = -4$.

$$\frac{-7}{-2} \geq 2 \quad ?$$
$$\frac{7}{2} \geq 2 \qquad \textit{True}$$

Interval C: Let $g = 0$.

$$\frac{-3}{2} \geq 2 \qquad \textit{False}$$

The solution set includes numbers in Interval B, including -7 but excluding -2, which makes the fraction undefined.

Solution set: $[-7, -2)$

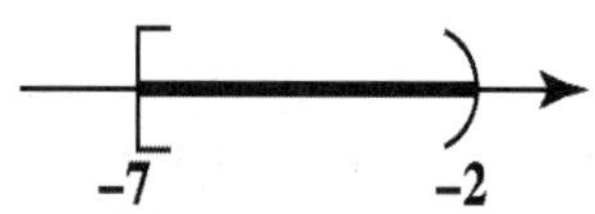

43. $$\frac{x-8}{x-4} < 3$$

Write the inequality so that 0 is on one side.

$$\frac{x-8}{x-4} - 3 < 0$$
$$\frac{x-8}{x-4} - \frac{3(x-4)}{x-4} < 0$$
$$\frac{x-8-3x+12}{x-4} < 0$$
$$\frac{-2x+4}{x-4} < 0$$

The number 2 makes the numerator 0, and 4 makes the denominator 0. These two numbers determine three intervals.

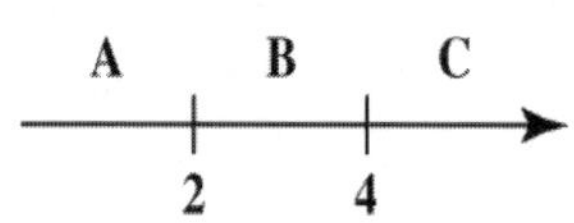

Test a number from each interval in the inequality

$$\frac{x-8}{x-4} < 3.$$

Interval A: Let $x = 0$.

$$\frac{-8}{-4} < 3 \quad ?$$
$$2 < 3 \qquad \textit{True}$$

Interval B: Let $x = 3$.

$$\frac{-5}{-1} < 3 \quad ?$$
$$5 < 3 \qquad \textit{False}$$

Interval C: Let $x = 5$.

$$\frac{-3}{1} < 3 \qquad \textit{True}$$

The solution set includes numbers in Intervals A and C, excluding endpoints.

Solution set: $(-\infty, 2) \cup (4, \infty)$

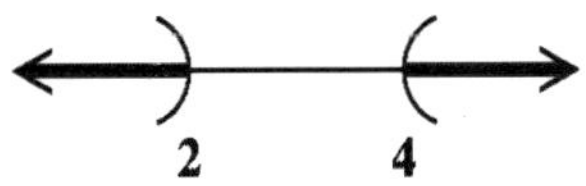

45. $$\frac{4k}{2k-1} < k$$

Write the inequality so that 0 is on one side.

$$\frac{4k}{2k-1} - k < 0$$

$$\frac{4k}{2k-1} - \frac{k(2k-1)}{2k-1} < 0$$

$$\frac{4k - 2k^2 + k}{2k-1} < 0$$

$$\frac{-2k^2 + 5k}{2k-1} < 0$$

$$\frac{k(-2k+5)}{2k-1} < 0$$

The numbers 0 and $\frac{5}{2}$ make the numerator 0, and $\frac{1}{2}$ makes the denominator 0. These three numbers determine four intervals.

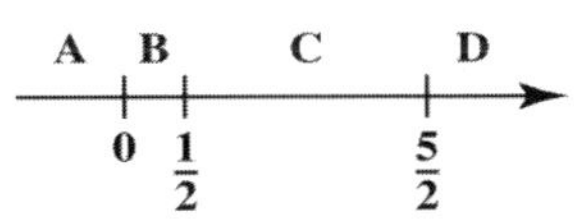

Test a number from each interval in the inequality

$$\frac{4k}{2k-1} < k.$$

Interval A: Let $k = -1$.

$$\frac{4(-1)}{2(-1)-1} < -1 \quad ?$$

$$\frac{4}{3} < -1 \qquad \textit{False}$$

Interval B: Let $k = \frac{1}{4}$.

$$\frac{4\left(\frac{1}{4}\right)}{2\left(\frac{1}{4}\right) - 1} < \frac{1}{4} \quad ?$$

$$-2 < \frac{1}{4} \qquad \textit{True}$$

Interval C: Let $k = 1$.

$$\frac{4(1)}{2(1)-1} < 1 \quad ?$$

$$4 < 1 \qquad \textit{False}$$

Interval D: Let $k = 3$.

$$\frac{4(3)}{2(3)-1} < 3 \quad ?$$

$$\frac{12}{5} < 3 \qquad \textit{True}$$

The solution set includes numbers in Intervals B and D. None of the endpoints are included.

Solution set: $\left(0, \frac{1}{2}\right) \cup \left(\frac{5}{2}, \infty\right)$

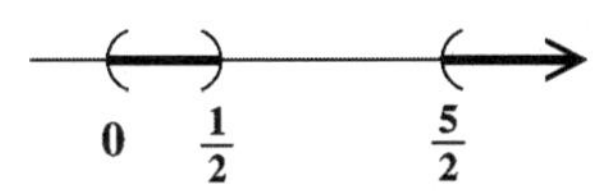

47. $$\frac{2x-3}{x^2+1} \geq 0$$

The denominator is positive for all real numbers x, so it has no effect on the solution set for the inequality.

$$2x - 3 \geq 0$$

$$2x \geq 3$$

$$x \geq \frac{3}{2}$$

Solution set: $\left[\frac{3}{2}, \infty\right)$

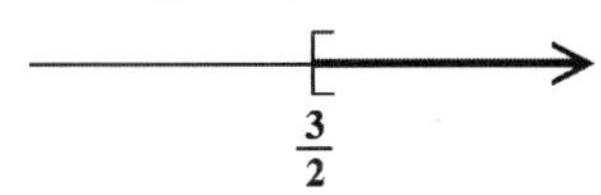

49. $$\frac{(3x-5)^2}{x+2} > 0$$

The numerator is positive for all real numbers x except $x = \frac{5}{3}$, which makes it equal to 0. If we solve the inequality $x + 2 > 0$, then we only have to be sure to exclude $\frac{5}{3}$ from that solution set to determine the solution set of the original inequality.

$$x + 2 > 0$$

$$x > -2$$

Solution set: $\left(-2, \frac{5}{3}\right) \cup \left(\frac{5}{3}, \infty\right)$

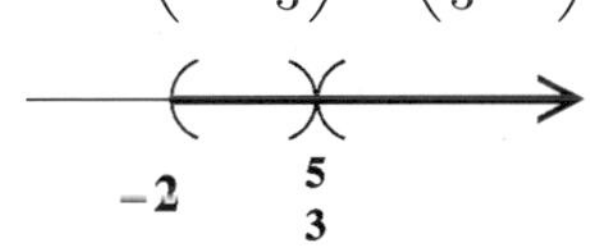

51.

$$s(t) = 624$$

$$-16t^2 + 256t = 624$$

$$t^2 - 16t = -39 \quad \textit{Divide by } -16.$$

$$t^2 - 16t + 39 = 0$$

$$(t-3)(t-13) = 0$$

continued

$t-3=0$ or $t-13=0$
$t=3$ or $t=13$

The rock will be 624 feet above the ground after 3 seconds and after 13 seconds.

52. See Exercise 51 for the solution to the equation $s(t)=624$.

For this problem, our common sense tells us that $s(t)>624$ between 3 seconds and 13 seconds.

53.
$$s(t)=0$$
$$-16t^2+256t=0$$
$$t^2-16t=0 \quad \textit{Divide by } -16.$$
$$t(t-16)=0$$
$$t=0 \quad \text{or} \quad t-16=0$$
$$t=16$$

The rock is at ground level at 0 seconds (the time when it is initially projected) and at 16 seconds (the time when it hits the ground).

54. Based on the answers for Exercises 51 and 53, we deduce that $s(t)<624$ between 0 and 3 seconds and between 13 and 16 seconds.

Chapter 9 Review Exercises

1. $t^2=121$
$t=11$ or $t=-11$
Solution set: $\{-11,11\}$

2. $p^2=3$
$p=\sqrt{3}$ or $p=-\sqrt{3}$
Solution set: $\left\{-\sqrt{3},\sqrt{3}\right\}$

3. $(2x+5)^2=100$
$2x+5=10$ or $2x+5=-10$
$2x=5$ $\qquad 2x=-15$
$x=\frac{5}{2}$ or $x=-\frac{15}{2}$

Solution set: $\left\{-\frac{15}{2},\frac{5}{2}\right\}$

4. $(3k-2)^2=-25$
$3k-2=\sqrt{-25}$ or $3k-2=-\sqrt{-25}$
$3k-2=5i$ $\qquad 3k-2=-5i$
$3k=2+5i$ $\qquad 3k=2-5i$
$k=\frac{2+5i}{3}$ $\qquad k=\frac{2-5i}{3}$
$k=\frac{2}{3}+\frac{5}{3}i$ or $k=\frac{2}{3}-\frac{5}{3}i$

Solution set: $\left\{\frac{2}{3}+\frac{5}{3}i,\frac{2}{3}-\frac{5}{3}i\right\}$

5. $x^2+4x=15$
Complete the square.
$$\left(\frac{1}{2}\cdot 4\right)^2=2^2=4$$
Add 4 to each side.
$$x^2+4x+4=15+4$$
$$(x+2)^2=19$$
$x+2=\sqrt{19}$ or $x+2=-\sqrt{19}$
$x=-2+\sqrt{19}$ or $x=-2-\sqrt{19}$
Solution set: $\left\{-2+\sqrt{19},-2-\sqrt{19}\right\}$

6. $2m^2-3m=-1$
$$m^2-\frac{3}{2}m=-\frac{1}{2} \qquad \textit{Divide by 2.}$$
Complete the square.
$$\left[\frac{1}{2}\left(-\frac{3}{2}\right)\right]^2=\left(-\frac{3}{4}\right)^2=\frac{9}{16}$$
Add $\frac{9}{16}$ to each side.
$$m^2-\frac{3}{2}m+\frac{9}{16}=-\frac{1}{2}+\frac{9}{16}$$
$$\left(m-\frac{3}{4}\right)^2=-\frac{8}{16}+\frac{9}{16}$$
$$\left(m-\frac{3}{4}\right)^2=\frac{1}{16}$$
$m-\frac{3}{4}=\sqrt{\frac{1}{16}}$ or $m-\frac{3}{4}=-\sqrt{\frac{1}{16}}$
$m-\frac{3}{4}=\frac{1}{4}$ $\qquad m-\frac{3}{4}=-\frac{1}{4}$
$m=\frac{3}{4}+\frac{1}{4}$ $\qquad m=\frac{3}{4}-\frac{1}{4}$
$m=1$ or $m=\frac{1}{2}$

Solution set: $\left\{\frac{1}{2},1\right\}$

7. By the square root property, the first step should be
$$x=\sqrt{12} \text{ or } x=-\sqrt{12}.$$

8. $16t^2=d$
$16t^2=150 \qquad$ *Let $d=150$.*
$$t^2=\frac{150}{16}$$
$$t=\sqrt{\frac{150}{16}} \qquad t\geq 0$$
$$t\approx 3.1 \text{ seconds}$$

It would take about 3.1 seconds for the wallet to fall 150 feet.

9. $2x^2 + x - 21 = 0$

Here $a = 2$, $b = 1$, and $c = -21$.

$$x = \frac{-b \pm \sqrt{b^2 - 4ac}}{2a}$$
$$x = \frac{-1 \pm \sqrt{1^2 - 4(2)(-21)}}{2(2)}$$
$$= \frac{-1 \pm \sqrt{1 + 168}}{4}$$
$$= \frac{-1 \pm \sqrt{169}}{4} = \frac{-1 \pm 13}{4}$$
$$x = \frac{-1 + 13}{4} = \frac{12}{4} = 3 \text{ or}$$
$$x = \frac{-1 - 13}{4} = -\frac{14}{4} = -\frac{7}{2}$$

Solution set: $\left\{-\frac{7}{2}, 3\right\}$

10. $k^2 + 5k = 7$

$k^2 + 5k - 7 = 0$

Here $a = 1$, $b = 5$, and $c = -7$.

$$k = \frac{-b \pm \sqrt{b^2 - 4ac}}{2a}$$
$$k = \frac{-5 \pm \sqrt{5^2 - 4(1)(-7)}}{2(1)}$$
$$= \frac{-5 \pm \sqrt{25 + 28}}{2}$$
$$= \frac{-5 \pm \sqrt{53}}{2}$$

Solution set: $\left\{\frac{-5 + \sqrt{53}}{2}, \frac{-5 - \sqrt{53}}{2}\right\}$

11. $(t + 3)(t - 4) = -2$

$t^2 - t - 12 = -2$

$t^2 - t - 10 = 0$

Here $a = 1$, $b = -1$, and $c = -10$.

$$t = \frac{-b \pm \sqrt{b^2 - 4ac}}{2a}$$
$$t = \frac{-(-1) \pm \sqrt{(-1)^2 - 4(1)(-10)}}{2(1)}$$
$$= \frac{1 \pm \sqrt{1 + 40}}{2} = \frac{1 \pm \sqrt{41}}{2}$$

Solution set: $\left\{\frac{1 + \sqrt{41}}{2}, \frac{1 - \sqrt{41}}{2}\right\}$

12. $2x^2 + 3x + 4 = 0$

Here $a = 2$, $b = 3$, and $c = 4$.

$$x = \frac{-b \pm \sqrt{b^2 - 4ac}}{2a}$$
$$x = \frac{-3 \pm \sqrt{3^2 - 4(2)(4)}}{2(2)}$$
$$= \frac{-3 \pm \sqrt{9 - 32}}{4}$$
$$= \frac{-3 \pm \sqrt{-23}}{4}$$
$$= \frac{-3 \pm i\sqrt{23}}{4}$$

Solution set: $\left\{\frac{-3 + i\sqrt{23}}{4}, \frac{-3 - i\sqrt{23}}{4}\right\}$

13. $3p^2 = 2(2p - 1)$

$3p^2 = 4p - 2$

$3p^2 - 4p + 2 = 0$

Here $a = 3$, $b = -4$, and $c = 2$.

$$p = \frac{-b \pm \sqrt{b^2 - 4ac}}{2a}$$
$$p = \frac{-(-4) + \sqrt{(-4)^2 - 4(3)(2)}}{2(3)}$$
$$= \frac{4 \pm \sqrt{16 - 24}}{6} = \frac{4 \pm \sqrt{-8}}{6}$$
$$= \frac{4 \pm 2i\sqrt{2}}{6} = \frac{2\left(2 \pm i\sqrt{2}\right)}{6}$$
$$= \frac{2 \pm i\sqrt{2}}{3} = \frac{2}{3} \pm \frac{\sqrt{2}}{3}i$$

Solution set: $\left\{\frac{2}{3} + \frac{\sqrt{2}}{3}i, \frac{2}{3} - \frac{\sqrt{2}}{3}i\right\}$

14. $m(2m - 7) = 3m^2 + 3$

$2m^2 - 7m = 3m^2 + 3$

$0 = m^2 + 7m + 3$

Here $a = 1$, $b = 7$, and $c = 3$.

$$m = \frac{-b \pm \sqrt{b^2 - 4ac}}{2a}$$
$$m = \frac{-7 \pm \sqrt{7^2 - 4(1)(3)}}{2(1)}$$
$$= \frac{-7 \pm \sqrt{49 - 12}}{2}$$
$$= \frac{-7 \pm \sqrt{37}}{2}$$

Solution set: $\left\{\frac{-7 + \sqrt{37}}{2}, \frac{-7 - \sqrt{37}}{2}\right\}$

15. $x^2 + 5x + 2 = 0$

Here $a = 1$, $b = 5$, and $c = 2$.

$$\begin{aligned} b^2 - 4ac &= 5^2 - 4(1)(2) \\ &= 25 - 8 = 17 \end{aligned}$$

Since the discriminant is positive, but not a perfect square, there are two distinct irrational number solutions. The answer is **C**.

16. $$4t^2 = 3 - 4t$$

$4t^2 + 4t - 3 = 0$

Here $a = 4$, $b = 4$, and $c = -3$.

$$\begin{aligned} b^2 - 4ac &= 4^2 - 4(4)(-3) \\ &= 16 + 48 \\ &= 64 \text{ or } 8^2 \end{aligned}$$

Since the discriminant is positive, and a perfect square, there are two distinct rational number solutions. The answer is **A**.

17. $$4x^2 = 6x - 8$$

$4x^2 - 6x + 8 = 0$

Here $a = 4$, $b = -6$, and $c = 8$.

$$\begin{aligned} b^2 - 4ac &= (-6)^2 - 4(4)(8) \\ &= 36 - 128 = -92 \end{aligned}$$

Since the discriminant is negative, there are two distinct imaginary number solutions. The answer is **D**.

18. $9z^2 + 30z + 25 = 0$

Here $a = 9$, $b = 30$, and $c = 25$.

$$\begin{aligned} b^2 - 4ac &= 30^2 - 4(9)(25) \\ &= 900 - 900 = 0 \end{aligned}$$

Since the discriminant is zero, there is exactly one rational number solution. The answer is **B**.

19. $$\frac{15}{x} = 2x - 1$$

$$x\left(\frac{15}{x}\right) = x(2x - 1) \qquad \textit{Multiply by the LCD, x.}$$

$$\begin{aligned} 15 &= 2x^2 - x \\ 0 &= 2x^2 - x - 15 \\ 0 &= (2x + 5)(x - 3) \end{aligned}$$

$$2x + 5 = 0 \quad \text{or} \quad x - 3 = 0$$

$$x = -\frac{5}{2} \quad \text{or} \quad x = 3$$

Check $x = -\frac{5}{2}$: $-6 \stackrel{?}{=} -5 - 1$ *True*

Check $x = 3$: $5 \stackrel{?}{=} 6 - 1$ *True*

Solution set: $\left\{-\frac{5}{2}, 3\right\}$

20. $$\frac{1}{n} + \frac{2}{n+1} = 2$$

$$n(n+1)\left(\frac{1}{n} + \frac{2}{n+1}\right) = n(n+1) \cdot 2 \qquad \textit{Multiply by the LCD, n(n+1).}$$

$$\begin{aligned} (n+1) + 2n &= 2n^2 + 2n \\ 0 &= 2n^2 - n - 1 \\ 0 &= (2n+1)(n-1) \end{aligned}$$

$$2n + 1 = 0 \quad \text{or} \quad n - 1 = 0$$

$$n = -\frac{1}{2} \quad \text{or} \quad n = 1$$

Check $n = -\frac{1}{2}$: $-2 + 4 \stackrel{?}{=} 2$ *True*

Check $n = 1$: $1 + 1 \stackrel{?}{=} 2$ *True*

Solution set: $\left\{-\frac{1}{2}, 1\right\}$

21. $$-2r = \sqrt{\frac{48 - 20r}{2}}$$

Square both sides.

$$\begin{aligned} (-2r)^2 &= \left(\sqrt{\frac{48 - 20r}{2}}\right)^2 \\ 4r^2 &= \frac{48 - 20r}{2} \\ 4r^2 &= 24 - 10r \end{aligned}$$

$$\begin{aligned} 4r^2 + 10r - 24 &= 0 \\ 2r^2 + 5r - 12 &= 0 \\ (r + 4)(2r - 3) &= 0 \end{aligned}$$

$$r + 4 = 0 \quad \text{or} \quad 2r - 3 = 0$$

$$r = -4 \quad \text{or} \quad r = \frac{3}{2}$$

Check $r = -4$: $8 \stackrel{?}{=} \sqrt{64}$ *True*

Check $r = \frac{3}{2}$: $-3 \stackrel{?}{=} \sqrt{9}$ *False*

Solution set: $\{-4\}$

22. $8(3x + 5)^2 + 2(3x + 5) - 1 = 0$

Let $u = 3x + 5$. The equation becomes

$$\begin{aligned} 8u^2 + 2u - 1 &= 0. \\ (2u + 1)(4u - 1) &= 0 \end{aligned}$$

$$2u + 1 = 0 \quad \text{or} \quad 4u - 1 = 0$$

$$u = -\frac{1}{2} \quad \text{or} \quad u = \frac{1}{4}$$

To find x, substitute $3x + 5$ for u.

$$3x + 5 = -\frac{1}{2} \quad \text{or} \quad 3x + 5 = \frac{1}{4}$$

$$3x = -\frac{11}{2} \qquad 3x = -\frac{19}{4}$$

$$x = -\frac{11}{6} \quad \text{or} \quad x = -\frac{19}{12}$$

Check $x = -\frac{11}{6}$: $2 - 1 - 1 \stackrel{?}{=} 0$ *True*

Check $x = -\frac{19}{12}$: $.5 + .5 - 1 \stackrel{?}{=} 0$ *True*

Solution set: $\left\{-\frac{11}{6}, -\frac{19}{12}\right\}$

23. $2x^{2/3} - x^{1/3} - 28 = 0$

Let $u = x^{1/3}$, so $u^2 = \left(x^{1/3}\right)^2 = x^{2/3}$.

The equation becomes

$$2u^2 - u - 28 = 0.$$
$$(2u+7)(u-4) = 0$$
$$2u + 7 = 0 \quad \text{or} \quad u - 4 = 0$$
$$u = -\frac{7}{2} \quad \text{or} \quad u = 4$$

To find x, substitute $x^{1/3}$ for u.

$$x^{1/3} = -\frac{7}{2} \quad \text{or} \quad x^{1/3} = 4$$
$$\left(x^{1/3}\right)^3 = \left(-\frac{7}{2}\right)^3 \qquad \left(x^{1/3}\right)^3 = 4^3$$
$$x = -\frac{343}{8} \quad \text{or} \quad x = 64$$

Check $x = -\dfrac{343}{8}$: $24.5 + 3.5 - 28 \stackrel{?}{=} 0$ *True*

Check $x = 64$: $32 - 4 - 28 \stackrel{?}{=} 0$ *True*

Solution set: $\left\{-\dfrac{343}{8}, 64\right\}$

24. $p^4 - 10p^2 + 9 = 0$

Let $x = p^2$, so $x^2 = p^4$.

$$x^2 - 10x + 9 = 0$$
$$(x-1)(x-9) = 0$$
$$x - 1 = 0 \quad \text{or} \quad x - 9 = 0$$
$$x = 1 \quad \text{or} \quad x = 9$$

To find p, substitute p^2 for x.

$$p^2 = 1 \quad \text{or} \quad p^2 = 9$$
$$p = \pm\sqrt{1} \qquad p = \pm\sqrt{9}$$
$$p = \pm 1 \quad \text{or} \quad p = \pm 3$$

Check $p = \pm 1$: $1 - 10 + 9 \stackrel{?}{=} 0$ *True*

Check $p = \pm 3$: $81 - 90 + 9 \stackrel{?}{=} 0$ *True*

Solution set: $\{-3, -1, 1, 3\}$

25. Let $x =$ Phong's speed.

Make a table. Use $d = rt$, or $t = \dfrac{d}{r}$.

	Distance	Rate	Time
Upstream	20	$x-3$	$\dfrac{20}{x-3}$
Downstream	20	$x+3$	$\dfrac{20}{x+3}$

Time upstream plus time downstream equals 7 hr.

$$\frac{20}{x-3} + \frac{20}{x+3} = 7$$

Multiply each side by the LCD, $(x+3)(x-3)$.

$$(x+3)(x-3)\left(\frac{20}{x-3} + \frac{20}{x+3}\right) = (x+3)(x-3)(7)$$
$$20(x+3) + 20(x-3) = 7\left(x^2 - 9\right)$$
$$20x + 60 + 20x - 60 = 7x^2 - 63$$
$$0 = 7x^2 - 40x - 63$$
$$0 = (x-7)(7x+9)$$
$$x - 7 = 0 \quad \text{or} \quad 7x + 9 = 0$$
$$x = 7 \quad \text{or} \quad x = -\frac{9}{7}$$

Reject $-\frac{9}{7}$ since speed can't be negative. Phong's speed was 7 mph.

26. Let $x =$ Maureen's speed on the trip to pick up Laurie.

Make a chart. Use $d = rt$, or $t = \dfrac{d}{r}$.

	Distance	Rate	Time
To Laurie	8	x	$\dfrac{8}{x}$
To the mall	11	$x+15$	$\dfrac{11}{x+15}$

Time to pick up Laurie + time to mall = 24 min (or .4 hr).

$$\frac{8}{x} + \frac{11}{x+15} = .4$$

Multiply each side by the LCD, $x(x+15)$.

$$x(x+15)\left(\frac{8}{x} + \frac{11}{x+15}\right) = x(x+15)(.4)$$
$$8(x+15) + 11x = .4x(x+15)$$
$$8x + 120 + 11x = .4x^2 + 6x$$
$$0 = .4x^2 - 13x - 120$$

Multiply by 5 to clear the decimal.

$$0 = 2x^2 - 65x - 600$$
$$0 = (x-40)(2x+15)$$
$$x - 40 = 0 \quad \text{or} \quad 2x + 15 = 0$$
$$x = 40 \quad \text{or} \quad x = -\frac{15}{2}$$

Speed cannot be negative, so $-\dfrac{15}{2}$ is not a solution. Maureen's speed on the trip to pick up Laurie was 40 mph.

27. Let $x =$ the amount of time for the old machine alone and

$x - 1 =$ the amount of time for the new machine alone.

Make a chart.

continued

Machine	Rate	Time Together	Part of Job Done
Old	$\frac{1}{x}$	2	$\frac{2}{x}$
New	$\frac{1}{x-1}$	2	$\frac{2}{x-1}$

Part done by old machine + Part done by new machine = 1 whole job.

$$\frac{2}{x} + \frac{2}{x-1} = 1$$

Multiply by the LCD, $x(x-1)$.

$$x(x-1)\left(\frac{2}{x}+\frac{2}{x-1}\right) = x(x-1)\cdot 1$$
$$2(x-1)+2x = x^2 - x$$
$$2x-2+2x = x^2 - x$$
$$0 = x^2 - 5x + 2$$

Use the quadratic formula.

$$x = \frac{-b \pm \sqrt{b^2-4ac}}{2a}$$
$$x = \frac{-(-5) \pm \sqrt{(-5)^2 - 4(1)(2)}}{2(1)}$$
$$= \frac{5 \pm \sqrt{25-8}}{2} = \frac{5 \pm \sqrt{17}}{2}$$
$$x = \frac{5+\sqrt{17}}{2} \approx 4.6 \text{ or}$$
$$x = \frac{5-\sqrt{17}}{2} \approx .4$$

Reject .4 as the time for the old machine, because that would yield a negative time for the new machine. Thus, the old machine takes about 4.6 hours.

28. Let $x =$ the time for Carter alone and $x - 1 =$ the time for Greg alone.

Worker	Rate	Time Together	Part of Job Done
Carter	$\frac{1}{x}$	1.5	$\frac{1.5}{x}$
Greg	$\frac{1}{x-1}$	1.5	$\frac{1.5}{x-1}$

Part by Carter + part by Greg = 1 whole job.

$$\frac{1.5}{x} + \frac{1.5}{x-1} = 1$$

Multiply by the LCD, $x(x-1)$.

$$x(x-1)\left(\frac{1.5}{x}+\frac{1.5}{x-1}\right) = x(x-1)\cdot 1$$
$$1.5(x-1)+1.5x = x^2 - x$$
$$1.5x - 1.5 + 1.5x = x^2 - x$$
$$0 = x^2 - 4x + 1.5$$

Use the quadratic formula.

$$x = \frac{-b \pm \sqrt{b^2-4ac}}{2a}$$
$$x = \frac{-(-4) \pm \sqrt{(-4)^2 - 4(1)(1.5)}}{2(1)}$$
$$= \frac{4 \pm \sqrt{16-6}}{2} = \frac{4 \pm \sqrt{10}}{2}$$
$$= \frac{4 \pm \sqrt{10}}{2}$$
$$x = \frac{4+\sqrt{10}}{2} \approx 3.6 \text{ or}$$
$$x = \frac{4-\sqrt{10}}{2} \approx .4$$

Reject .4 as Carter's time, because that would make Greg's time negative. Thus, Carter's time alone is about 3.6 hours and Greg's time alone is about $x - 1 = 2.6$ hours.

29. Solve $k = \frac{rF}{wv^2}$ for v.

Multiply both sides by v^2, then divide by k.

$$v^2 = \frac{rF}{kw}$$
$$v = \pm\sqrt{\frac{rF}{kw}} = \frac{\pm\sqrt{rF}}{\sqrt{kw}}$$
$$= \frac{\pm\sqrt{rF}}{\sqrt{kw}} \cdot \frac{\sqrt{kw}}{\sqrt{kw}}$$
$$v = \frac{\pm\sqrt{rFkw}}{kw}$$

30. Solve $p = \sqrt{\frac{yz}{6}}$ for y.

Square both sides.

$$p^2 = \left(\sqrt{\frac{yz}{6}}\right)^2$$
$$p^2 = \frac{yz}{6}$$
$$\frac{6p^2}{z} = y \quad \text{or} \quad y = \frac{6p^2}{z}$$

31. Solve $mt^2 = 3mt + 6$ for t.

$$mt^2 - 3mt - 6 = 0$$

Use the quadratic formula with $a = m$, $b = -3m$, and $c = -6$.

$$t = \frac{-b \pm \sqrt{b^2-4ac}}{2a}$$
$$t = \frac{3m \pm \sqrt{(-3m)^2 - 4(m)(-6)}}{2m}$$
$$= \frac{3m \pm \sqrt{9m^2 + 24m}}{2m}$$

32. Let $x =$ the length of the longer leg;

$\frac{3}{4}x =$ the length of the shorter leg;

$2x - 9 =$ the length of the hypotenuse.

Use the Pythagorean formula.

$$c^2 = a^2 + b^2$$

$$(2x-9)^2 = x^2 + \left(\frac{3}{4}x\right)^2$$

$$4x^2 - 36x + 81 = x^2 + \frac{9}{16}x^2$$

$$16(4x^2 - 36x + 81) = 16\left(x^2 + \frac{9}{16}x^2\right)$$

$$64x^2 - 576x + 1296 = 16x^2 + 9x^2$$

$$39x^2 - 576x + 1296 = 0$$

$$13x^2 - 192x + 432 = 0 \quad \textit{Divide by 3.}$$

$$(13x - 36)(x - 12) = 0$$

$$13x - 36 = 0 \quad \text{or} \quad x - 12 = 0$$

$$x = \frac{36}{13} \quad \text{or} \quad x = 12$$

Reject $x = \frac{36}{13}$ since $2\left(\frac{36}{13}\right) - 9$ is negative.

If $x = 12$, then

$$\frac{3}{4}x = \frac{3}{4}(12) = 9$$

and

$$2x - 9 = 2(12) - 9 = 15.$$

The lengths of the three sides are 9 feet, 12 feet, and 15 feet.

33. Let $x =$ the amount removed from one dimension.

The area of the square is 256 cm^2, so the length of one side is $\sqrt{256}$ or 16 cm. The dimensions of the new rectangle are $16 + x$ and $16 - x$ cm. The area of the new rectangle is 16 cm^2 less than the area of the square.

$$(16 + x)(16 - x) = 256 - 16$$

$$256 - x^2 = 240$$

$$-x^2 = -16$$

$$x^2 = 16$$

$$x = \pm\sqrt{16} = \pm 4$$

Length cannot be negative, so reject -4. If $x = 4$, then $16 + x = 20$, and $16 - x = 12$. The dimensions are 20 cm by 12 cm.

34. Let $x =$ the width of the border.

Area of mat = length • width

$$352 = (2x + 20)(2x + 14)$$

$$352 = 4x^2 + 68x + 280$$

$$0 = 4x^2 + 68x - 72$$

$$0 = x^2 + 17x - 18$$

$$0 = (x + 18)(x - 1)$$

$$x + 18 = 0 \quad \text{or} \quad x - 1 = 0$$

$$x = -18 \quad \text{or} \quad x = 1$$

Reject the negative answer for length. The mat is 1 inch wide.

35. $f(t) = 100t^2 - 300t$ is the distance of the light from the starting point at t minutes. When the light returns to the starting point, the value of $f(t)$ will be 0.

$$0 = 100t^2 - 300t$$

$$0 = t^2 - 3t \quad \textit{Divide by 100.}$$

$$0 = t(t - 3)$$

$$t = 0 \quad \text{or} \quad t - 3 = 0$$

$$t = 3$$

Since $t = 0$ represents the starting time, the light will return to the starting point in 3 minutes.

36. $f(t) = -16t^2 + 45t + 400$

$$200 = -16t^2 + 45t + 400 \quad \textit{Let } f(t) = 200.$$

$$0 = -16t^2 + 45t + 200$$

$$0 = 16t^2 - 45t - 200$$

Here $a = 16$, $b = -45$, and $c = -200$.

$$t = \frac{-b \pm \sqrt{b^2 - 4ac}}{2a}$$

$$t = \frac{-(-45) \pm \sqrt{(-45)^2 - 4(16)(-200)}}{2(16)}$$

$$= \frac{45 \pm \sqrt{2025 + 12{,}800}}{32}$$

$$= \frac{45 \pm \sqrt{14{,}825}}{32}$$

$$t = \frac{45 + \sqrt{14{,}825}}{32} \approx 5.2 \quad \text{or}$$

$$t = \frac{45 - \sqrt{14{,}825}}{32} \approx -2.4$$

Reject the negative solution since time cannot be negative. The ball will reach a height of 200 ft above the ground after about 5.2 seconds.

37. $s(t) = -16t^2 + 75t + 407$

$450 = -16t^2 + 75t + 407$ *Let s(t) = 450.*

$0 = -16t^2 + 75t - 43$

Here $a = -16$, $b = 75$, and $c = -43$.

$$t = \frac{-b \pm \sqrt{b^2 - 4ac}}{2a}$$

$$t = \frac{-75 \pm \sqrt{75^2 - 4(-16)(-43)}}{2(-16)}$$

$$= \frac{-75 \pm \sqrt{5625 - 2752}}{-32}$$

$$= \frac{-75 \pm \sqrt{2873}}{-32}$$

$$t = \frac{-75 + \sqrt{2873}}{-32} \approx .7 \quad \text{or}$$

$$t = \frac{-75 - \sqrt{2873}}{-32} \approx 4.0$$

The ball will be 450 ft above the ground after .7 second (going up) and again after 4.0 seconds (coming down).

38. Let $p =$ the price.

Demand = Supply

$$\frac{25}{p} = 70p + 15$$

$$p\left(\frac{25}{p}\right) = p(70p + 15) \quad \textit{Multiply by p.}$$

$$25 = 70p^2 + 15p$$

$$0 = 70p^2 + 15p - 25$$

$$0 = 14p^2 + 3p - 5 \quad \textit{Divide by 5.}$$

$$0 = (2p - 1)(7p + 5)$$

$2p - 1 = 0$ or $7p + 5 = 0$

$p = \frac{1}{2}$ or $p = -\frac{5}{7}$

Reject the negative answer for price. Supply and demand are equal when p is $\frac{1}{2}$ dollar or \$.50.

39. Let $A = 10{,}920.25$, $P = 10{,}000$, and solve for r.

$$A = P(1 + r)^2$$

$$10{,}920.25 = 10{,}000(1 + r)^2$$

$$1.092025 = (1 + r)^2$$

$1 + r = \sqrt{1.092025}$ or $1 + r = -\sqrt{1.092025}$

$1 + r = 1.045$ or $1 + r = -1.045$

$r = .045$ or $r = -2.045$

Reject the negative rate. The required interest rate is $.045 = 4.5\%$.

40. **(a)** Use $f(x) = 3.29x^2 - 10.4x + 21.6$ with $x = 11$.

$$f(11) = 3.29(11)^2 - 10.4(11) + 21.6$$

$$f(11) \approx 305$$

The value of 305 is close to the number shown on the graph.

(b) $200 = 3.29x^2 - 10.4x + 21.6$ *Let f(x) = 200.*

$0 = 3.29x^2 - 10.4x - 178.4$

Here $a = 3.29$, $b = -10.4$, and $c = -178.4$.

$$x = \frac{-b \pm \sqrt{b^2 - 4ac}}{2a}$$

$$x = \frac{-(-10.4) \pm \sqrt{(-10.4)^2 - 4(3.29)(-178.4)}}{2(3.29)}$$

$$= \frac{10.4 \pm \sqrt{2455.904}}{6.58}$$

$$x = \frac{10.4 + \sqrt{2455.904}}{6.58} \approx 9.1 \quad \text{or}$$

$$x = \frac{10.4 - \sqrt{2455.904}}{6.58} \approx -6.0$$

Since time cannot be negative, the correct answer is $x \approx 9$, which represents 1999. Based on the graph, the number of e-mail boxes reached 200 million in 2000.

41. $y = 6 - 2x^2$
Write in $y = a(x - h)^2 + k$ form as
$y = -2(x - 0)^2 + 6$. The vertex (h, k) is $(0, 6)$.

42. $f(x) = -(x - 1)^2$
Write in $y = a(x - h)^2 + k$ form as
$y = -1(x - 1)^2 + 0$. The vertex (h, k) is $(1, 0)$.

43. $y = (x - 3)^2 + 7$
The equation is in the form $y = a(x - h)^2 + k$, so the vertex (h, k) is $(3, 7)$.

44. $y = f(x) = -3x^2 + 4x - 2$

Use the vertex formula with $a = -3$ and $b = 4$.

The x-coordinate of the vertex is

$$\frac{-b}{2a} = \frac{-4}{2(-3)} = \frac{2}{3}.$$

The y-coordinate of the vertex is

$$f\left(\frac{-b}{2a}\right) = f\left(\frac{2}{3}\right)$$

$$= -3\left(\frac{2}{3}\right)^2 + 4\left(\frac{2}{3}\right) - 2$$

$$= -\frac{4}{3} + \frac{8}{3} - 2$$

$$= -\frac{2}{3}.$$

The vertex is $\left(\frac{2}{3}, -\frac{2}{3}\right)$.

45. $x = (y - 3)^2 - 4$

When $y = 3$, $x = -4$, so the vertex is $(-4, 3)$.

46. If the discriminant is negative, there are no x-intercepts.

47. $y = 2(x - 2)^2 - 3$

The graph opens up since $a = 2 > 0$.

The vertex is $(2, -3)$ and the axis is $x = 2$. The domain is $(-\infty, \infty)$. The smallest y-value is -3, so the range is $[-3, \infty)$.

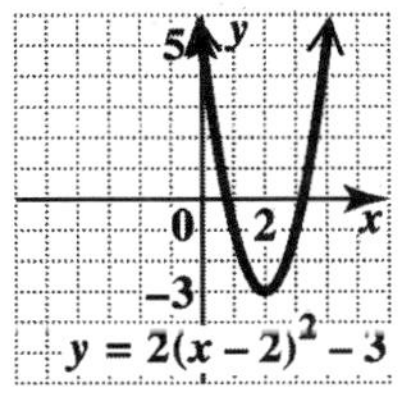

48. $f(x) = -2x^2 + 8x - 5$
Complete the square to find the vertex.

$$\begin{aligned} f(x) &= -2(x^2 - 4x) - 5 \\ &= -2(x^2 - 4x + 4 - 4) - 5 \\ &= -2(x^2 - 4x + 4) - 2(-4) - 5 \\ &= -2(x - 2)^2 + 8 - 5 \\ &= -2(x - 2)^2 + 3 \end{aligned}$$

The equation is in the form $y = a(x - h)^2 + k$, so the vertex (h, k) is $(2, 3)$ and the axis is $x = 2$. Here, $a = -2 < 0$, so the parabola opens down.

Also, $|a| = |-2| = 2 > 1$, so the graph is narrower than the graph of $y = x^2$. The points $(0, -5)$, $(1, 1)$, and $(3, 1)$ are on the graph.

$f(x) = -2x^2 + 8x - 5$

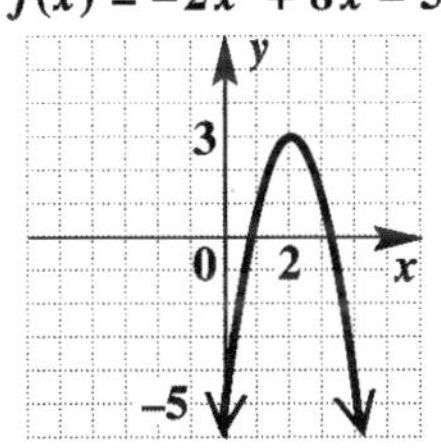

The domain is $(-\infty, \infty)$. The largest y-value is 3, so the range is $(-\infty, 3]$.

49. $x = 2(y + 3)^2 - 4$
Since the roles of x and y are reversed, this is a horizontal parabola.
$x = 2[y - (-3)]^2 + (-4)$
The equation is in the form

$$x = a(y - k)^2 + h,$$

so the vertex (h, k) is $(-4, -3)$ and the axis is $y = -3$. Here, $a = 2 > 0$, so the parabola opens to the right and is narrower than the graph of $y = x^2$.

Two other points on the graph are $(4, -1)$ and $(4, -5)$.

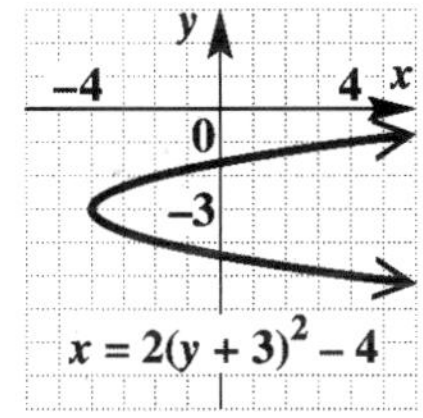

The smallest x-value is -4, so the domain is $[-4, \infty)$. The range is $(-\infty, \infty)$.

50. $x = -\frac{1}{2}y^2 + 6y - 14$

Since the roles of x and y are reversed, this is a horizontal parabola. Complete the square to find the vertex.

$$\begin{aligned} x &= -\frac{1}{2}y^2 + 6y - 14 \\ &= -\frac{1}{2}(y^2 - 12y) - 14 \\ &= -\frac{1}{2}(y^2 - 12y + 36 - 36) - 14 \\ &= -\frac{1}{2}(y^2 - 12y + 36) - \frac{1}{2}(-36) - 14 \\ &= -\frac{1}{2}(y - 6)^2 + 18 - 14 \\ x &= -\frac{1}{2}(y - 6)^2 + 4 \end{aligned}$$

The equation is in the form $x = a(y - k)^2 + h$, so the vertex (h, k) is $(4, 6)$ and the axis is $y = 6$.

Here, $a = -\frac{1}{2} < 0$, so the parabola opens to the left.

Also, $|a| = \left|-\frac{1}{2}\right| = \frac{1}{2} < 1$, so the graph is wider than the graph of $y = x^2$. The points $(-14, 0)$, $(2, 4)$, and $(2, 8)$ are on the graph.

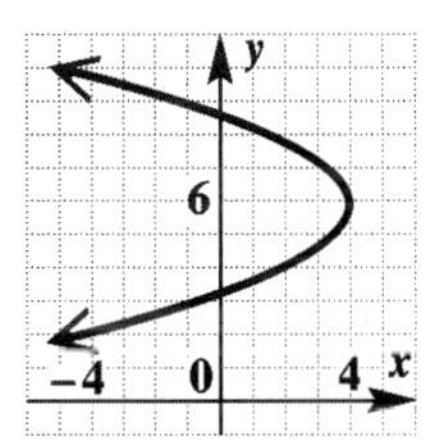

$x = -\frac{1}{2}y^2 + 6y - 14$

The largest x-value is 4, so the domain is $(-\infty, 4]$. The range is $(-\infty, \infty)$.

51. **(a)** Use $ax^2 + bx + c = y$ with $(0, 12.39)$, $(4, 15.78)$, and $(7, 22.71)$.

$$\begin{aligned} c &= 12.39 \quad (1) \\ 16a + 4b + c &= 15.78 \quad (2) \\ 49a + 7b + c &= 22.71 \quad (3) \end{aligned}$$

(b) Rewrite equations (2) and (3) with $c = 12.39$.

$$16a + 4b + 12.39 = 15.78$$
$$49a + 7b + 12.39 = 22.71$$

$$16a + 4b = 3.39 \quad (4)$$
$$49a + 7b = 10.32 \quad (5)$$

Now eliminate b.

$$\begin{array}{rcrl} -112a - 28b & = & -23.73 & -7 \times (4) \\ 196a + 28b & = & 41.28 & 4 \times (5) \\ \hline 84a & = & 17.55 & \\ a & \approx & .2089 & \end{array}$$

Use (4) to approximate b.

$$16\left(\frac{17.55}{84}\right) + 4b = 3.39$$
$$4b \approx .0471$$
$$b \approx .0118$$

Thus, $f(x) = .2089x^2 + .0118x + 12.39$.

(c) Use $f(x) = .2089x^2 + .0118x + 12.39$ with $x = 8$.

$$f(8) = .2089(8)^2 + .0118(8) + 12.39$$
$$\approx \$25.85$$

The result using the model is a little high.

52. $s(t) = -16t^2 + 160t$
The equation represents a parabola. Since $a = -16 < 0$, the parabola opens down. The time and maximum height occur at the vertex (h, k) of the parabola, given by

$$\left(\frac{-b}{2a}, s\left(\frac{-b}{2a}\right)\right).$$

Using the standard form of the equation, $a = -16$ and $b = 160$, so

$$h = \frac{-b}{2a} = \frac{-160}{2(-16)} = 5,$$

and $k = s(h) = -16(5)^2 + 160(5)$
$= -400 + 800 = 400.$

The vertex is $(5, 400)$. The time at which the maximum height is reached is 5 seconds. The maximum height is 400 feet.

53. Let L = the length of the rectangle and
W = the width.

The perimeter of the rectangle is 200 m, so

$$2L + 2W = 200$$
$$2W = 200 - 2L$$
$$W = 100 - L.$$

Since the area is length times width, substitute $100 - L$ for W.

$$A = LW$$
$$= L(100 - L)$$
$$= 100L - L^2 \quad \text{or} \quad -L^2 + 100L$$

Use the vertex formula.

$$L = \frac{-b}{2a} = \frac{-100}{2(-1)} = 50$$

So $L = 50$ meters and
$W = 100 - L = 100 - 50 = 50$ meters.

54. $(x - 4)(2x + 3) > 0$
Solve the equation
$(x - 4)(2x + 3) = 0.$
$x - 4 = 0 \quad \text{or} \quad 2x + 3 = 0$
$x = 4 \quad \text{or} \quad x = -\frac{3}{2}$

The numbers $-\frac{3}{2}$ and 4 divide a number line into three intervals.

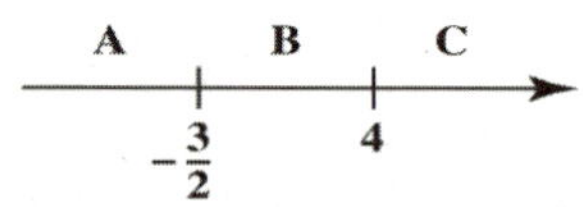

Test a number from each interval in the inequality

$$(x - 4)(2x + 3) > 0.$$

Interval A: Let $x = -2$.
$-6(-1) > 0$?
$6 > 0$ *True*

Interval B: Let $x = 0$.
$-4(3) > 0$?
$-12 > 0$ *False*

Interval C: Let $x = 5$.
$1(13) > 0$?
$13 > 0$ *True*

The solution set includes numbers in Intervals A and C, excluding endpoints.

Solution set: $\left(-\infty, -\frac{3}{2}\right) \cup (4, \infty)$.

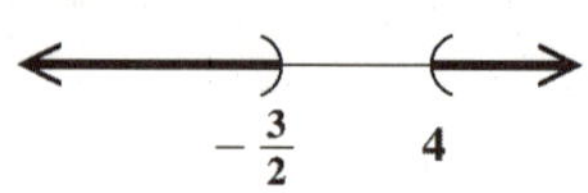

55. $x^2 + x \le 12$
Solve the equation
$x^2 + x = 12.$
$x^2 + x - 12 = 0$
$(x + 4)(x - 3) = 0$
$x + 4 = 0 \quad \text{or} \quad x - 3 = 0$
$x = -4 \quad \text{or} \quad x = 3$
The numbers -4 and 3 divide a number line into three intervals.

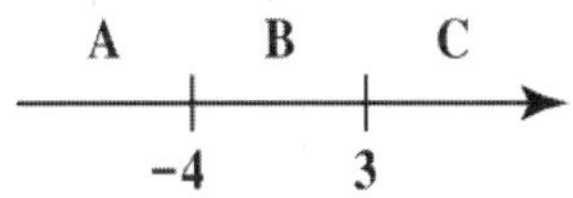

Test a number from each interval in the inequality

$$x^2 + x \leq 12.$$

Interval A: Let $x = -5$.
$25 - 5 \leq 12$?
$20 \leq 12$ *False*

Interval B: Let $x = 0$.
$0 \leq 12$ *True*

Interval C: Let $x = 4$.
$16 + 4 \leq 12$?
$20 \leq 12$ *False*

The numbers in Interval B, including -4 and 3, are solutions.

Solution set: $[-4, 3]$

56. $(x + 2)(x - 3)(x + 5) \leq 0$

The numbers -2, 3, and -5 are solutions of the cubic equation

$$(x + 2)(x - 3)(x + 5) = 0.$$

These numbers divide a number line into four intervals.

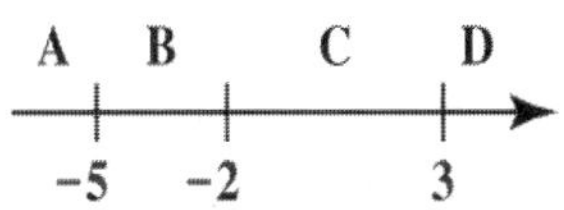

Test a number from each interval in the inequality

$$(x + 2)(x - 3)(x + 5) \leq 0.$$

Interval A: Let $x = -6$.
$(-4)(-9)(-1) \leq 0$ *True*

Interval B: Let $x = -3$.
$(-5)(-6)(2) \leq 0$ *False*

Interval C: Let $x = 0$.
$(2)(-3)(5) \leq 0$ *True*

Interval D: Let $x = 4$.
$(6)(1)(9) \leq 0$ *False*

The numbers in Intervals A and C, including -5, -2, and 3, are solutions.

Solution set: $(-\infty, -5] \cup [-2, 3]$

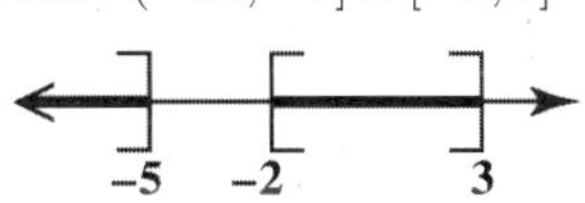

57. $(4m + 3)^2 \leq -4$

The square of a real number is never negative. So, the solution set of this inequality is $\emptyset$.

58. $$\frac{6}{2z - 1} < 2$$

Write the inequality so that 0 is on one side.

$$\frac{6}{2z - 1} - 2 < 0$$
$$\frac{6}{2z - 1} - \frac{2(2z - 1)}{2z - 1} < 0$$
$$\frac{6 - 4z + 2}{2z - 1} < 0$$
$$\frac{-4z + 8}{2z - 1} < 0$$

The number 2 makes the numerator 0, and $\frac{1}{2}$ makes the denominator 0. These two numbers determine three intervals.

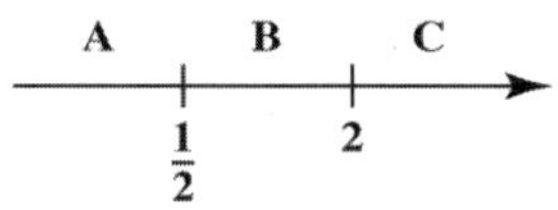

Test a number from each interval in the inequality

$$\frac{6}{2z - 1} < 2.$$

Interval A: Let $z = 0$.
$-6 < 2$ *True*

Interval B: Let $z = 1$.
$6 < 2$ *False*

Interval C: Let $z = 3$.
$\frac{6}{5} < 2$ *True*

The solution set includes numbers in Intervals A and C, excluding endpoints.

Solution set: $\left(-\infty, \frac{1}{2}\right) \cup (2, \infty)$

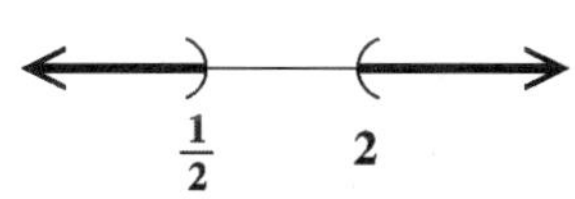

59. $$\frac{3t + 4}{t - 2} \leq 1$$

Write the inequality so that 0 is on one side.

$$\frac{3t + 4}{t - 2} - 1 \leq 0$$
$$\frac{3t + 4}{t - 2} - \frac{1(t - 2)}{t - 2} \leq 0$$
$$\frac{3t + 4 - t + 2}{t - 2} \leq 0$$
$$\frac{2t + 6}{t - 2} \leq 0$$

The number -3 makes the numerator 0, and 2 makes the denominator 0. These two numbers determine three intervals.

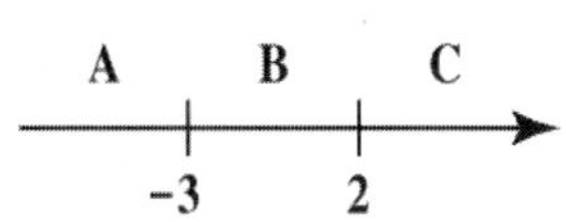

continued

Test a number from each interval in the inequality

$$\frac{3t+4}{t-2} \le 1.$$

Interval A: Let $t = -4$.

$$\frac{-8}{-6} \le 1 \quad ?$$

$$\frac{4}{3} \le 1 \quad \textit{False}$$

Interval B: Let $t = 0$.

$$\frac{4}{-2} \le 1 \quad ?$$

$$-2 \le 1 \quad \textit{True}$$

Interval C: Let $t = 3$.

$$\frac{13}{1} \le 1 \quad ?$$

$$13 \le 1 \quad \textit{False}$$

The numbers in Interval B, including -3 but not 2, are solutions.
Solution set: $[-3, 2)$

60. Solve $V = r^2 + R^2h$ for R.

$$V - r^2 = R^2h$$
$$R^2h = V - r^2$$
$$R^2 = \frac{V - r^2}{h}$$
$$R = \pm\sqrt{\frac{V - r^2}{h}} = \frac{\pm\sqrt{V - r^2}}{\sqrt{h}}$$
$$= \frac{\pm\sqrt{V - r^2}}{\sqrt{h}} \cdot \frac{\sqrt{h}}{\sqrt{h}}$$
$$= \frac{\pm\sqrt{Vh - r^2h}}{h}$$

61. $3t^2 - 6t = -4$

$$3t^2 - 6t + 4 = 0$$

Use the quadratic formula.

$$t = \frac{-b \pm \sqrt{b^2 - 4ac}}{2a}$$
$$t = \frac{-(-6) \pm \sqrt{(-6)^2 - 4(3)(4)}}{2(3)}$$
$$= \frac{6 \pm \sqrt{-12}}{6} = \frac{6 \pm 2i\sqrt{3}}{6}$$
$$= \frac{2\left(3 \pm i\sqrt{3}\right)}{6} = \frac{3 \pm i\sqrt{3}}{3}$$

Solution set: $\left\{\frac{3 + i\sqrt{3}}{3}, \frac{3 - i\sqrt{3}}{3}\right\}$

62.

$$x^4 - 1 = 0$$
$$\left(x^2 + 1\right)\left(x^2 - 1\right) = 0$$

$$x^2 + 1 = 0 \quad \text{or} \quad x^2 - 1 = 0$$
$$x^2 = -1 \qquad x^2 = 1$$
$$x = \pm\sqrt{-1} \qquad x = \pm\sqrt{1}$$
$$x = \pm i \quad \text{or} \quad x = \pm 1$$

Solution set: $\{-i, i, -1, 1\}$

63. $\left(x^2 - 2x\right)^2 = 11\left(x^2 - 2x\right) - 24$

Let $u = x^2 - 2x$. The equation becomes

$$u^2 = 11u - 24.$$
$$u^2 - 11u + 24 = 0$$
$$(u - 8)(u - 3) = 0$$

$$u - 8 = 0 \quad \text{or} \quad u - 3 = 0$$
$$u = 8 \quad \text{or} \quad u = 3$$

To find x, substitute $x^2 - 2x$ for u.

$$x^2 - 2x = 8 \quad \text{or} \quad x^2 - 2x = 3$$
$$x^2 - 2x - 8 = 0 \qquad x^2 - 2x - 3 = 0$$
$$(x - 4)(x + 2) = 0 \qquad (x - 3)(x + 1) = 0$$
$$x - 4 = 0 \text{ or } x + 2 = 0 \qquad x - 3 = 0 \text{ or } x + 1 = 0$$
$$x = 4 \quad \text{or} \quad x = -2 \quad \text{or} \quad x = 3 \quad \text{or} \quad x = -1$$

The potential solutions all check.
Solution set: $\{-2, -1, 3, 4\}$

64. $(r - 1)(2r + 3)(r + 6) < 0$

Solve the equation
$(r - 1)(2r + 3)(r + 6) = 0$.

$$r - 1 = 0 \quad \text{or} \quad 2r + 3 = 0 \quad \text{or} \quad r + 6 = 0$$
$$r = 1 \quad \text{or} \quad r = -\frac{3}{2} \quad \text{or} \quad r = -6$$

The numbers -6, $-\frac{3}{2}$, and 1 divide a number line into four intervals.

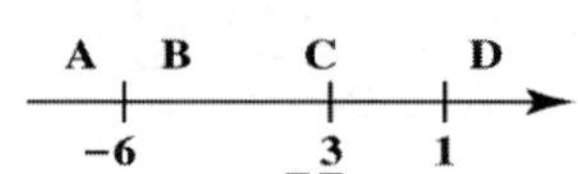

Test a number from each interval in the inequality

$$(r - 1)(2r + 3)(r + 6) < 0.$$

Interval A: Let $r = -7$.

$$-8(-11)(-1) < 0 \quad ?$$
$$-88 < 0 \quad \textit{True}$$

Interval B: Let $r = -2$.

$$-3(-1)(4) < 0 \quad ?$$
$$12 < 0 \quad \textit{False}$$

Interval C: Let $r = 0$.

$$-1(3)(6) < 0 \quad ?$$
$$-18 < 0 \quad \textit{True}$$

Interval D: Let $r = 2$.
$$1(7)(8) < 0 \quad ?$$
$$56 < 0 \qquad \textit{False}$$

The numbers in Intervals A and C, not including $-6, -\frac{3}{2}$, or 1, are solutions.

Solution set: $(-\infty, -6) \cup \left(-\frac{3}{2}, 1\right)$

65. $2x - \sqrt{x} = 6$
$$2x - \sqrt{x} - 6 = 0$$
Let $u = \sqrt{x}$, so $u^2 = x$.
$$2u^2 - u - 6 = 0$$
$$(2u + 3)(u - 2) = 0$$
$$2u + 3 = 0 \quad \text{or} \quad u - 2 = 0$$
$$u = -\frac{3}{2} \quad \text{or} \quad u = 2$$
$$\sqrt{x} = -\frac{3}{2} \quad \text{or} \quad \sqrt{x} = 2$$
Since $\sqrt{x} \geq 0$, we must have $\sqrt{x} = 2$, or $x = 4$.

Check $x = 4$: $8 - 2 \stackrel{?}{=} 6$ *True*

Solution set: $\{4\}$

66. $(3k + 11)^2 = 7$
$$3k + 11 = \sqrt{7} \quad \text{or} \quad 3k + 11 = -\sqrt{7}$$
$$3k = -11 + \sqrt{7} \qquad 3k = -11 - \sqrt{7}$$
$$k = \frac{-11 + \sqrt{7}}{3} \quad \text{or} \quad k = \frac{-11 - \sqrt{7}}{3}$$
Solution set: $\left\{\frac{-11 + \sqrt{7}}{3}, \frac{-11 - \sqrt{7}}{3}\right\}$

67. Solve $S = \frac{Id^2}{k}$ for d.

Multiply both sides by k, then divide by I.
$$\frac{Sk}{I} = d^2$$
$$d = \pm\sqrt{\frac{Sk}{I}} = \frac{\pm\sqrt{Sk}}{\sqrt{I}}$$
$$= \frac{\pm\sqrt{Sk}}{\sqrt{I}} \cdot \frac{\sqrt{I}}{\sqrt{I}}$$
$$d = \frac{\pm\sqrt{SkI}}{I}$$

68. $(8k - 7)^2 \geq -1$

The square of any real number is always greater than or equal to 0, so any real number satisfies this inequality. The solution set is $(-\infty, \infty)$.

69. $6 + \frac{15}{s^2} = -\frac{19}{s}$

Multiply by the LCD, s^2.
$$s^2\left(6 + \frac{15}{s^2}\right) = s^2\left(-\frac{19}{s}\right)$$
$$6s^2 + 15 = -19s$$
$$6s^2 + 19s + 15 = 0$$
$$(3s + 5)(2s + 3) = 0$$
$$3s + 5 = 0 \quad \text{or} \quad 2s + 3 = 0$$
$$s = -\frac{5}{3} \quad \text{or} \quad s = -\frac{3}{2}$$
Check $s = -\frac{5}{3}$: $6 + \frac{27}{5} \stackrel{?}{=} \frac{57}{5}$ *True*

Check $s = -\frac{3}{2}$: $6 + \frac{20}{3} \stackrel{?}{=} \frac{38}{3}$ *True*

Solution set: $\left\{-\frac{5}{3}, -\frac{3}{2}\right\}$

70. $x^4 - 8x^2 = -1$
$$x^4 - 8x^2 + 1 = 0$$
Use the quadratic formula with $a = 1, b = -8$, and $c = 1$ to solve for x^2, not x.
$$x^2 = \frac{-b + \sqrt{b^2 - 4ac}}{2a}$$
$$x^2 = \frac{-(-8) \pm \sqrt{(-8)^2 - 4(1)(1)}}{2(1)}$$
$$= \frac{8 \pm \sqrt{64 - 4}}{2} = \frac{8 \pm \sqrt{60}}{2}$$
$$= \frac{8 \pm 2\sqrt{15}}{2} = 4 \pm \sqrt{15}$$
Both $4 + \sqrt{15}$ and $4 - \sqrt{15}$ are positive and can be set equal to x^2.
$$x^2 = 4 + \sqrt{15} \quad \text{or} \quad x^2 = 4 - \sqrt{15}$$
$$x = \pm\sqrt{4 + \sqrt{15}} \qquad x = \pm\sqrt{4 - \sqrt{15}}$$
The potential solutions all check.

Solution set: $\left\{\sqrt{4 + \sqrt{15}}, -\sqrt{4 + \sqrt{15}}, \sqrt{4 - \sqrt{15}}, -\sqrt{4 - \sqrt{15}}\right\}$

71. $\frac{-2}{x + 5} \leq -5$

Write the inequality so that 0 is on one side.
$$\frac{-2}{x + 5} + 5 \leq 0$$
$$\frac{-2}{x + 5} + \frac{5(x + 5)}{x + 5} \leq 0$$
$$\frac{-2 + 5x + 25}{x + 5} \leq 0$$

continued

$$\frac{5x+23}{x+5} \le 0$$

The number $-\frac{23}{5}$ makes the numerator 0, and -5 makes the denominator 0. These two numbers determine three intervals.

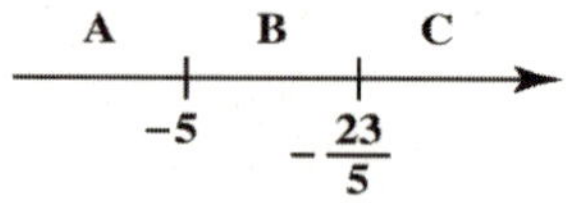

Test a number from each interval in the inequality

$$\frac{-2}{x+5} \le -5.$$

Interval A: Let $x=-6$.

$$\frac{-2}{-1} \le -5 \quad ?$$

$$2 \le -5 \quad \textit{False}$$

Interval B: Let $x=-\frac{24}{5}$.

$$-\frac{2}{\frac{1}{5}} \le -5 \quad ?$$

$$-10 \le -5 \quad \textit{True}$$

Interval C: Let $x=0$.

$$\frac{-2}{5} \le -5 \quad \textit{False}$$

The numbers in Interval B, including $-\frac{23}{5}$ but not -5, are solutions.

Solution set: $\left(-5, -\frac{23}{5}\right]$

72. $g(x)=x^2-5$ written in the form
$g(x)=a(x-h)^2+k$ is
$g(x)=1(x-0)^2+(-5)$.

Here, $h=0$ and $k=-5$, so the vertex (h,k) is $(0,-5)$. The graph is shifted 5 units down from the graph of $f(x)=x^2$. Since $a=1>0$, the graph opens up. The correct figure is **F**.

73. $h(x)=-x^2+4$
Because the coefficient of x^2 is negative, the parabola opens down. Because $k=4$, the shift is upward 4 units. The figure that most closely resembles the graph of $h(x)$ is Choice **B**.

74. $F(x)=(x-1)^2$ written in the form
$F(x)=a(x-h)^2+k$ is
$F(x)=1(x-1)^2+0$.

Here, $h=1$ and $k=0$, so the vertex (h,k) is $(1,0)$. The graph is shifted 1 unit to the right of the graph of $f(x)=x^2$. Since $a=1>0$, the graph opens up. The correct figure is **C**.

75. $G(x)=(x+1)^2$
Because the coefficient of the x^2-term is positive, the graph opens up. Because $h=-1$, the shift is to the left 1 unit. The figure that most closely resembles the graph of $G(x)$ is Choice **A**.

76. $H(x)=(x-1)^2+1$ is written in the form
$H(x)=a(x-h)^2+k$.

Here, $h=1$ and $k=1$, so the vertex (h,k) is $(1,1)$. The graph is shifted 1 unit to the right and 1 unit up from the graph of $f(x)=x^2$. Since $a=1>0$, the graph opens up. The correct figure is **E**.

77. $K(x)=(x+1)^2+1$
Because $h=-1$ and $k=1$, the shift is to the left 1 unit and up 1 unit. The figure that most closely resembles the graph of $K(x)$ is choice **D**.

78. $y=f(x)=4x^2+4x-2$
Complete the square to find the vertex.

$$\begin{aligned}
y &= 4(x^2+x)-2 \\
&= 4\left(x^2+x+\frac{1}{4}-\frac{1}{4}\right)-2 \\
&= 4\left(x^2+x+\frac{1}{4}\right)+4\left(-\frac{1}{4}\right)-2 \\
&= 4\left(x+\frac{1}{2}\right)^2-1-2 \\
&= 4\left(x+\frac{1}{2}\right)^2-3 \\
y &= 4\left(x+\frac{1}{2}\right)^2-3
\end{aligned}$$

The equation is now in the form $y=a(x-h)^2+k$, so the vertex (h,k) is $\left(-\frac{1}{2},-3\right)$ and the axis is $x=-\frac{1}{2}$. Since $a=4>0$, the parabola opens up. Also, $|a|=|4|=4>1$, so the graph is narrower than the graph of $y=x^2$. The points $(-2,6)$, $(0,-2)$, and $(1,6)$ are on the graph.

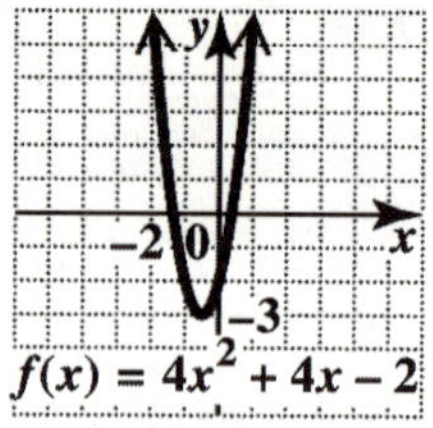

The domain is $(-\infty,\infty)$.
The smallest y-value is -3, so the range is $[-3,\infty)$.

79. **(a)** Use $f(x)=.014x^2-.396x+21.2$ with $x=30$.

$$\begin{aligned}
f(30) &= .014(30)^2-.396(30)+21.2 \\
&= 21.92
\end{aligned}$$

The natural gas use in 2000 was approximately 21.92 trillion ft^3.

(b) $25 = .014x^2 - .396x + 21.2$ *Let f(x) = 25.*

$0 = .014x^2 - .396x - 3.8$

Here $a = .014$, $b = -.396$, and $c = -3.8$.

$$x = \frac{-b \pm \sqrt{b^2 - 4ac}}{2a}$$

$$x = \frac{-(-.396) \pm \sqrt{(-.396)^2 - 4(.014)(-3.8)}}{2(.014)}$$

$$x = \frac{.396 + \sqrt{.369616}}{.028} \approx 35 \text{ (rounding down)} \quad \text{or}$$

$$x = \frac{.396 - \sqrt{.369616}}{.028} \approx -8$$

Since time cannot be negative, the answer is $1970 + 35 = 2005$.

80. Let $x =$ the speed of the boat in still water. Use $t = d/r$ to make a table.

	Distance	Rate	Time
Upstream	15	$x - 5$	$\frac{15}{x-5}$
Downstream	15	$x + 5$	$\frac{15}{x+5}$

The total time is 4 hours.

$$\frac{15}{x-5} + \frac{15}{x+5} = 4$$

Multiply by the LCD, $(x+5)(x-5)$.

$$15(x+5) + 15(x-5) = 4(x+5)(x-5)$$
$$15x + 75 + 15x - 75 = 4x^2 - 100$$
$$0 = 4x^2 - 30x - 100$$
$$0 = 2x^2 - 15x - 50$$
$$0 = (2x+5)(x-10)$$

$2x + 5 = 0$ or $x - 10 = 0$

$x = -\frac{5}{2}$ or $x = 10$

The speed must be positive, so check $x = 10$. The 15-mile trip upstream would take 3 hours at 5 mph. The trip downstream would take 1 hour at 15 mph. Thus, the total time would be 4 hours and we have the speed of the boat in still water equal to 10 mph.

81. Let $x =$ the length of the wire. Use the Pythagorean formula.

$$x^2 = 100^2 + 400^2 \quad \textit{Height=400}$$
$$= 10,000 + 160,000$$
$$= 170,000$$
$$x = \pm\sqrt{170,000} \approx \pm 412.3$$

Reject the negative solution. The length of the wire is about 412.3 feet.

82. **(a)** $3x - (4x + 2) = 0$

$3x - 4x - 2 = 0$

$-x - 2 = 0$

$x = -2$

Solution set: $\{-2\}$

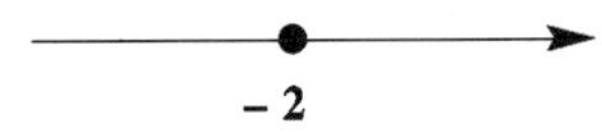

(b) $3x - (4x + 2) > 0$

$3x - 4x - 2 > 0$

$-x - 2 > 0$

$-x > 2$

$x < -2$

Solution set: $(-\infty, -2)$

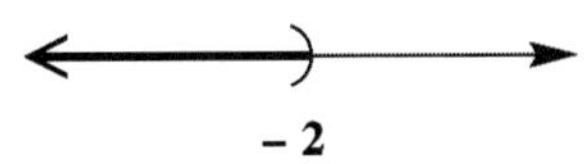

(c) $3x - (4x + 2) < 0$

$3x - 4x - 2 < 0$

$-x - 2 < 0$

$-x < 2$

$x > -2$

Solution set: $(-2, \infty)$

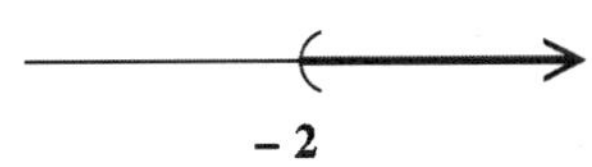

83. **(a)** $x^2 - 6x + 5 = 0$

$(x - 1)(x - 5) = 0$

$x - 1 = 0$ or $x - 5 = 0$

$x = 1$ or $x = 5$

Solution set: $\{1, 5\}$

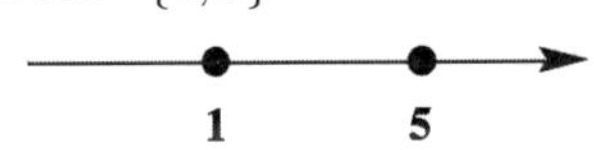

(b) $x^2 - 6x + 5 > 0$

From the equation

$$x^2 - 6x + 5 = 0$$

in part (a), the solutions are $x = 1$ or $x = 5$. These numbers divide the number line into three intervals.

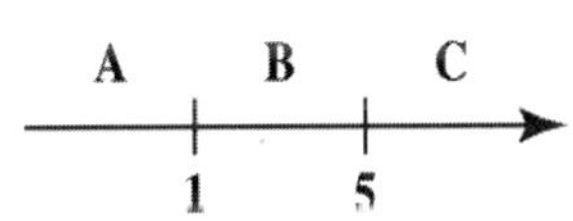

Test a point from each interval in the inequality

$$x^2 - 6x + 5 > 0.$$

Interval A: Let $x = 0$.

$5 > 0$ *True*

Interval B: Let $x = 2$.

$4 - 12 + 5 > 0$?

$-3 > 0$ *False*

continued

Interval C: Let $x = 6$.

$$36 - 36 + 5 > 0 \quad ?$$
$$5 > 0 \qquad \textit{True}$$

The numbers in Intervals A and C, not including the endpoints, are solutions.
Solution set: $(-\infty, 1) \cup (5, \infty)$

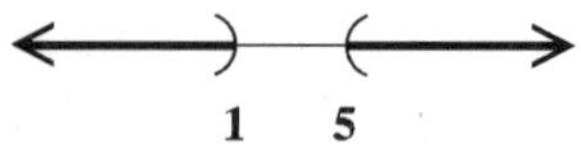

(c) $x^2 - 6x + 5 < 0$
The intervals are the same as in part (b). Solutions to this inequality would be the numbers in Interval B, again not including the endpoints.
Solution set: $(1, 5)$

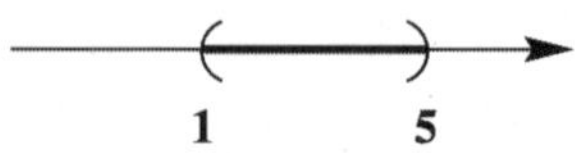

84. (a)

$$\frac{-5x+20}{x-2} = 0$$
$$(x-2)\left(\frac{-5x+20}{x-2}\right) = (x-2) \cdot 0$$
$$-5x + 20 = 0$$
$$-5x = -20$$
$$x = 4$$

Solution set: $\{4\}$

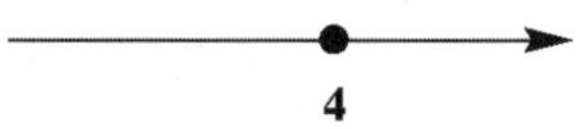

(b) $\dfrac{-5x+20}{x-2} > 0$

The number that makes the numerator 0 is $x = 4$. The number that makes the denominator 0 is $x = 2$.
These numbers divide a number line into three intervals.

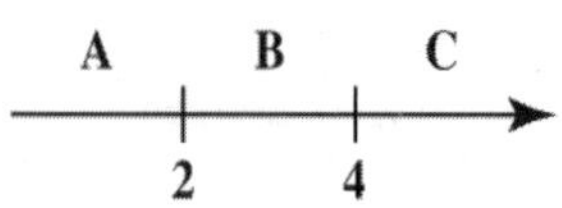

Test a number from each interval in the inequality

$$\frac{-5x+20}{x-2} > 0.$$

Interval A: Let $x = 0$.

$$\frac{20}{-2} > 0 \quad ?$$
$$-10 > 0 \qquad \textit{False}$$

Interval B: Let $x = 3$.

$$\frac{5}{1} > 0 \quad ?$$
$$5 > 0 \qquad \textit{True}$$

Interval C: Let $x = 5$.

$$\frac{-5}{3} > 0 \quad ?$$
$$-\frac{5}{3} > 0 \qquad \textit{False}$$

The numbers in Interval B, not including the endpoints, are solutions.
Solution set: $(2, 4)$

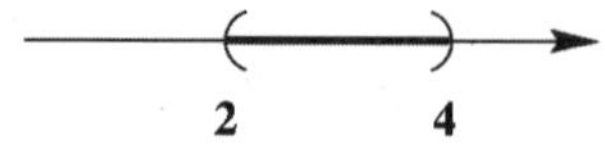

(c) $\dfrac{-5x+20}{x-2} < 0$

The intervals are the same as in part (b). Solutions to this inequality would be the numbers in Intervals A and C, again not including the endpoints.
Solution set: $(-\infty, 2) \cup (4, \infty)$

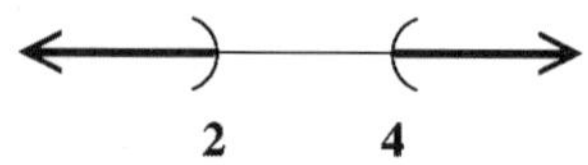

85. If we solve a linear, quadratic, or rational equation and the two inequalities associated with it, the union of the three sets will be $\underline{(-\infty, \infty)}$; the only exception will be in the case of the rational equation and inequalities, where the number or numbers that cause the *denominator* to be 0 will be excluded.

86. If set S is the solution of the other inequality, then

$$S \cup \{-5, 3\} \cup (-\infty, -5) \cup (3, \infty) = (-\infty, \infty).$$

Therefore, $S = (-5, 3)$.

Chapter 9 Test

1. $t^2 = 54$

$$t = \sqrt{54} \quad \text{or} \quad t = -\sqrt{54}$$
$$t = 3\sqrt{6} \quad \text{or} \quad t = -3\sqrt{6}$$

Solution set: $\left\{3\sqrt{6}, -3\sqrt{6}\right\}$

2. $(7x+3)^2 = 25$

$$7x + 3 = 5 \quad \text{or} \quad 7x + 3 = -5$$
$$7x = 2 \qquad\qquad 7x = -8$$
$$x = \frac{2}{7} \quad \text{or} \quad x = -\frac{8}{7}$$

Solution set: $\left\{-\frac{8}{7}, \frac{2}{7}\right\}$

3.

$$2x^2 + 4x = 8$$
$$x^2 + 2x = 4 \qquad \textit{Divide by 2.}$$
$$x^2 + 2x + 1 = 4 + 1$$
$$(x+1)^2 = 5$$

$$x+1=\sqrt{5} \quad \text{or} \quad x+1=-\sqrt{5}$$
$$x=-1+\sqrt{5} \quad \text{or} \quad x=-1-\sqrt{5}$$

Solution set: $\left\{-1+\sqrt{5},-1-\sqrt{5}\right\}$

4. $2x^2-3x-1=0$

Here $a=2$, $b=-3$, and $c=-1$.

$$x=\frac{-b\pm\sqrt{b^2-4ac}}{2a}$$
$$x=\frac{-(-3)\pm\sqrt{(-3)^2-4(2)(-1)}}{2(2)}$$
$$=\frac{3\pm\sqrt{17}}{4}$$

Solution set: $\left\{\frac{3+\sqrt{17}}{4},\frac{3-\sqrt{17}}{4}\right\}$

5. $3t^2-4t=-5$

$3t^2-4t+5=0$

Here $a=3$, $b=-4$, and $c=5$.

$$t=\frac{-b\pm\sqrt{b^2-4ac}}{2a}$$
$$t=\frac{-(-4)\pm\sqrt{(-4)^2-4(3)(5)}}{2(3)}$$
$$=\frac{4\pm\sqrt{-44}}{6}=\frac{4\pm 2i\sqrt{11}}{6}$$
$$=\frac{2\left(2\pm i\sqrt{11}\right)}{6}=\frac{2\pm i\sqrt{11}}{3}$$

Solution set: $\left\{\frac{2+i\sqrt{11}}{3},\frac{2-i\sqrt{11}}{3}\right\}$

6. $3x=\sqrt{\frac{9x+2}{2}}$

Square both sides.

$$9x^2=\frac{9x+2}{2}$$
$$18x^2=9x+2$$
$$18x^2-9x-2=0$$

As directed, use the quadratic formula with $a=18$, $b=-9$, and $c=-2$.

$$x=\frac{-b\pm\sqrt{b^2-4ac}}{2a}$$
$$x=\frac{-(-9)\pm\sqrt{(-9)^2-4(18)(-2)}}{2(18)}$$
$$=\frac{9\pm\sqrt{225}}{36}=\frac{9\pm 15}{36}$$
$$x=\frac{9+15}{36}=\frac{24}{36}=\frac{2}{3} \text{ or}$$
$$x=\frac{9-15}{36}=\frac{-6}{36}=-\frac{1}{6}$$

Check $x=\frac{2}{3}$: $2\stackrel{?}{=}\sqrt{4}$ *True*

Check $x=-\frac{1}{6}$: $-\frac{1}{2}\stackrel{?}{=}\sqrt{\frac{1}{4}}$ *False*

Solution set: $\left\{\frac{2}{3}\right\}$

7. If k is a negative number, then $4k$ is also negative, so the equation $x^2=4k$ will have two imaginary solutions. The answer is **A**.

8. $2x^2-8x-3=0$

$$b^2-4ac=(-8)^2-4(2)(-3)$$
$$=64+24=88$$

The discriminant, 88, is positive but not a perfect square, so there will be two distinct irrational number solutions.

9. $3-\frac{16}{x}-\frac{12}{x^2}=0$

Multiply by the LCD, x^2.

$$x^2\left(3-\frac{16}{x}-\frac{12}{x^2}\right)=x^2\cdot 0$$
$$3x^2-16x-12=0$$
$$(3x+2)(x-6)=0$$
$$3x+2=0 \quad \text{or} \quad x-6=0$$
$$x=-\frac{2}{3} \quad \text{or} \quad x=6$$

Check $x=-\frac{2}{3}$: $3+24-27\stackrel{?}{=}0$ *True*

Check $x=6$: $3-\frac{8}{3}-\frac{1}{3}\stackrel{?}{=}0$ *True*

Solution set: $\left\{-\frac{2}{3},6\right\}$

10. $4x^2+7x-3=0$

Use the quadratic formula with $a=4$, $b=7$, and $c=-3$.

$$x=\frac{-b\pm\sqrt{b^2-4ac}}{2a}$$
$$x=\frac{-7\pm\sqrt{7^2-4(4)(-3)}}{2(4)}$$
$$=\frac{-7\pm\sqrt{97}}{8}$$

Solution set: $\left\{\frac{-7+\sqrt{97}}{8},\frac{-7-\sqrt{97}}{8}\right\}$

11. $9x^4+4=37x^2$

$9x^4-37x^2+4=0$

Let $u=x^2$, so $u^2=\left(x^2\right)^2=x^4$.

The equation becomes

$9u^2-37u+4=0.$

$(9u-1)(u-4)=0$

continued

$9u - 1 = 0$ or $u - 4 = 0$

$u = \frac{1}{9}$ or $u = 4$

To find x, substitute x^2 for u.

$x^2 = \frac{1}{9}$ or $x^2 = 4$

$x = \pm\sqrt{\frac{1}{9}}$ $\quad x = \pm\sqrt{4}$

$x = \pm\frac{1}{3}$ or $x = \pm 2$

Check $x = \pm\frac{1}{3}$: $\frac{1}{9} + 4 \overset{?}{=} \frac{37}{9}$ *True*

Check $x = \pm 2$: $144 + 4 \overset{?}{=} 37(4)$ *True*

Solution set: $\left\{-2, -\frac{1}{3}, \frac{1}{3}, 2\right\}$

12. $12 = (2n+1)^2 + (2n+1)$
Let $u = 2n + 1$. The equation becomes
$12 = u^2 + u$.
$0 = u^2 + u - 12$
$0 = (u+4)(u-3)$

$u + 4 = 0$ or $u - 3 = 0$

$u = -4$ or $u = 3$

To find n, substitute $2n + 1$ for u.

$2n + 1 = -4$ or $2n + 1 = 3$

$2n = -5$ $\quad 2n = 2$

$n = -\frac{5}{2}$ or $n = 1$

Check $n = -\frac{5}{2}$: $12 \overset{?}{=} 16 - 4$ *True*

Check $n = 1$: $12 \overset{?}{=} 9 + 3$ *True*

Solution set: $\left\{-\frac{5}{2}, 1\right\}$

13. Solve $S = 4\pi r^2$ for r.

$$\frac{S}{4\pi} = r^2$$

$$r = \pm\sqrt{\frac{S}{4\pi}} = \frac{\pm\sqrt{S}}{2\sqrt{\pi}}$$

$$= \frac{\pm\sqrt{S}}{2\sqrt{\pi}} \cdot \frac{\sqrt{\pi}}{\sqrt{\pi}}$$

$$r = \frac{\pm\sqrt{\pi S}}{2\pi}$$

14. Let x = Andrew's time alone.
Then $x - 2$ = Kent's time alone.

Make a table.

	Rate	Time Together	Part of Job Done
Andrew	$\frac{1}{x}$	5	$\frac{5}{x}$
Kent	$\frac{1}{x-2}$	5	$\frac{5}{x-2}$

Part done by Andrew plus part done by Kent equals 1 whole job.

$$\frac{5}{x} + \frac{5}{x-2} = 1$$

Multiply both sides by the LCD, $x(x-2)$.

$$x(x-2)\left(\frac{5}{x} + \frac{5}{x-2}\right) = x(x-2) \cdot 1$$

$$5x - 10 + 5x = x^2 - 2x$$

$$0 = x^2 - 12x + 10$$

Use the quadratic formula with $a = 1$, $b = -12$, and $c = 10$.

$$x = \frac{-b \pm \sqrt{b^2 - 4ac}}{2a}$$

$$x = \frac{-(-12) \pm \sqrt{(-12)^2 - 4(1)(10)}}{2(1)}$$

$$= \frac{12 \pm \sqrt{104}}{2} = \frac{12 \pm 2\sqrt{26}}{2}$$

$$= \frac{2\left(6 \pm \sqrt{26}\right)}{2} = 6 \pm \sqrt{26}$$

$x = 6 + \sqrt{26} \approx 11.1$ or

$x = 6 - \sqrt{26} \approx .9$

Reject .9 for Andrew's time, because that would yield a negative time for Kent. Thus, Andrew's time is about 11.1 hours and Kent's time is $x - 2 \approx 9.1$ hours.

15. Let x = Abby's rate.

Make a table. Use $d = rt$, or $t = \frac{d}{r}$.

	Distance	Rate	Time
Upstream	10	$x - 3$	$\frac{10}{x-3}$
Downstream	10	$x + 3$	$\frac{10}{x+3}$

Time upstream plus Time downstream equals 3.5 hr.

$$\frac{10}{x-3} + \frac{10}{x+3} = \frac{7}{2}$$

Multiply both sides by the LCD, $2(x+3)(x-3)$.

$$2(x+3)(x-3)\left(\frac{10}{x-3}+\frac{10}{x+3}\right)=2(x+3)(x-3)\left(\frac{7}{2}\right)$$
$$20(x+3)+20(x-3)=7(x^2-9)$$
$$20x+60+20x-60=7x^2-63$$
$$0=7x^2-40x-63$$
$$0=(x-7)(7x+9)$$
$$x-7=0 \quad \text{or} \quad 7x+9=0$$
$$x=7 \quad \text{or} \quad x=-\frac{9}{7}$$

Reject $-\frac{9}{7}$ since the rate can't be negative. Abby's rate was 7 mph.

16. Let $x =$ the width of the walk.
The area of the walk is equal to the area of the outer figure minus the area of the pool.

$$152=(10+2x)(24+2x)-(24)(10)$$
$$152=240+68x+4x^2-240$$
$$0=4x^2+68x-152$$
$$0=x^2+17x-38$$
$$0=(x+19)(x-2)$$
$$x+19=0 \quad \text{or} \quad x-2=0$$
$$x=-19 \quad \text{or} \quad x=2$$

Reject -19 since width can't be negative. The walk is 2 feet wide.

17. Let $x =$ the height of the tower. Then
$2x+2 =$ the distance from the point to the top.

The distance from the base to the point is 30 m. These three segments form a right triangle, so the Pythagorean formula applies.

$$a^2+b^2=c^2$$
$$x^2+30^2=(2x+2)^2$$
$$x^2+900=4x^2+8x+4$$
$$0=3x^2+8x-896$$
$$0=(x-16)(3x+56)$$
$$x-16=0 \quad \text{or} \quad 3x+56=0$$
$$x=16 \quad \text{or} \quad x=-\frac{56}{3}$$

Reject $-\frac{56}{3}$ since height can't be negative. The tower is 16 m high.

18. $f(x)=a(x-h)^2+k$
Since $a<0$, the parabola opens down. Since $h>0$ and $k<0$, the x-coordinate is positive and the y-coordinate is negative. Therefore, the vertex is in quadrant IV. The correct graph is **A**.

19. $f(x)=\frac{1}{2}x^2-2$

$f(x)=\frac{1}{2}(x-0)^2-2$
The graph is a parabola in $f(x)=a(x-h)^2+k$ form with vertex (h,k) at $(0,-2)$. The axis is $x=0$. Since $a=\frac{1}{2}>0$, the parabola opens up.

Also, $|a|=\left|\frac{1}{2}\right|=\frac{1}{2}<1$, so the graph of the parabola is wider than the graph of $f(x)=x^2$. The points $(2,0)$ and $(-2,0)$ are on the graph.

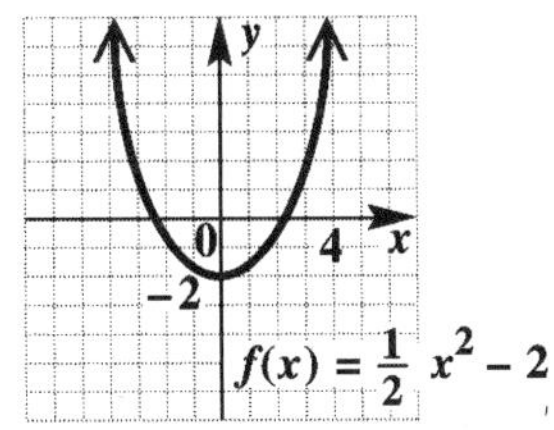

From the graph, we see that the x-values can be any real number, so the domain is $(-\infty,\infty)$. The y-values are greater than or equal to -2, so the range is $[-2,\infty)$.

20. $f(x)=-x^2+4x-1$

The x-coordinate of the vertex is

$$x=\frac{-b}{2a}=\frac{-4}{2(-1)}=2.$$

The y-coordinate of the vertex is

$$f(2)=-4+8-1=3.$$

The graph is a parabola with vertex (h,k) at $(2,3)$ and axis $x=2$. Since $a=-1<0$, the parabola opens down.
Also, $|a|=|-1|=1$, so the graph has the same shape as the graph of $f(x)=x^2$. The points $(0,-1)$ and $(4,-1)$ are on the graph.

$f(x) = -x^2 + 4x - 1$

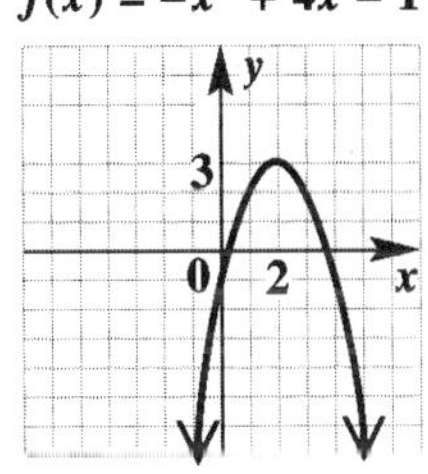

From the graph, we see that the x-values can be any real number, so the domain is $(-\infty,\infty)$.

The y-values are less than or equal to 3, so the range is $(-\infty,3]$.

21. $x = -(y-2)^2 + 2$
The equation is in $x = a(y-k)^2 + h$ form. The graph is a horizontal parabola with vertex (h, k) at $(2, 2)$ and axis $y = 2$. Since $a = -1 < 0$, the graph opens to the left. Also, $|a| = |-1| = 1$, so the graph has the same shape as the graph of $y = x^2$. The points $(-2, 0)$ and $(-2, 4)$ are on the same graph.

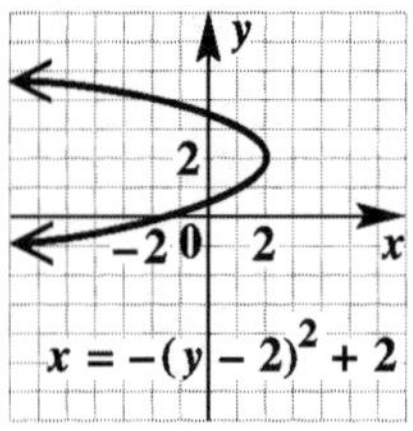

The largest value of x is 2, so the domain is $(-\infty, 2]$.

The y-values can be any real number, so the range is $(-\infty, \infty)$.

22. **(a)** Use $f(x) = .156x^2 - 2.05x + 10.2$ with $x = 11$.

$$\begin{aligned} f(11) &= .156(11)^2 - 2.05(11) + 10.2 \\ &\approx 6.5 \text{ (to the nearest tenth)} \end{aligned}$$

In 2001, there was a 6.5% increase in tuition.

(b) Find the vertex.

$$\frac{-b}{2a} = \frac{-(-2.05)}{2(.156)} \approx 6 \text{ (rounding down)}$$

$$\begin{aligned} f(6) &= .156(6)^2 - 2.05(6) + 10.2 \\ &\approx 3.5 \text{ (to the nearest tenth)} \end{aligned}$$

The minimum tuition increase was 3.5%, which occurred in $1990 + 6 = 1996$.

23. Let $x =$ the width of the lot.
Then $640 - 2x =$ the length of the lot.

Area A is length times width.

$$\begin{aligned} A &= x(640 - 2x) \\ A(x) &= 640x - 2x^2 = -2x^2 + 640x \end{aligned}$$

Use the vertex formula.

$$x = \frac{-b}{2a} = \frac{-640}{2(-2)} = 160$$

$$\begin{aligned} A(160) &= -2(160)^2 + 640(160) \\ &= -51{,}200 + 102{,}400 \\ &= 51{,}200 \end{aligned}$$

The graph is a parabola that opens down, so the maximum occurs at the vertex $(160, 51{,}200)$. The maximum area is 51,200 ft^2 if the width x is 160 ft and the length is $640 - 2x = 640 - 2(160) = 320$ ft.

24.
$$\begin{aligned} 2x^2 + 7x &> 15 \\ 2x^2 + 7x - 15 &> 0 \end{aligned}$$

Solve the equation
$2x^2 + 7x - 15 = 0.$
$(2x-3)(x+5) = 0$

$$2x - 3 = 0 \quad \text{or} \quad x + 5 = 0$$
$$x = \frac{3}{2} \quad \text{or} \quad x = -5$$

The numbers -5 and $\frac{3}{2}$ divide a number line into three intervals.

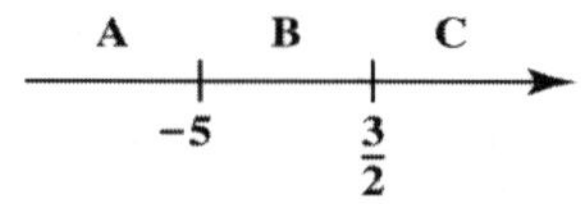

Test a number from each interval in the inequality

$$2x^2 + 7x > 15.$$

Interval A: Let $x = -6$.
$72 - 42 > 15$?
$30 > 15$ *True*

Interval B: Let $x = 0$.
$0 > 15$ *False*

Interval C: Let $x = 2$.
$8 + 14 > 15$?
$22 > 15$ *True*

The numbers in Intervals A and C, not including -5 and $\frac{3}{2}$, are solutions.

Solution set: $(-\infty, -5) \cup \left(\frac{3}{2}, \infty\right)$

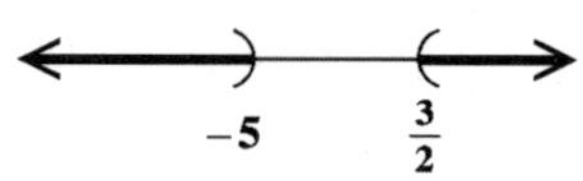

25.
$$\frac{5}{t-4} \le 1$$

Write the inequality so that 0 is on one side.

$$\begin{aligned} \frac{5}{t-4} - 1 &\le 0 \\ \frac{5}{t-4} - \frac{1(t-4)}{t-4} &\le 0 \\ \frac{5 - t + 4}{t-4} &\le 0 \\ \frac{-t+9}{t-4} &\le 0 \end{aligned}$$

The number 9 makes the numerator 0, and 4 makes the denominator 0. These two numbers determine three intervals.

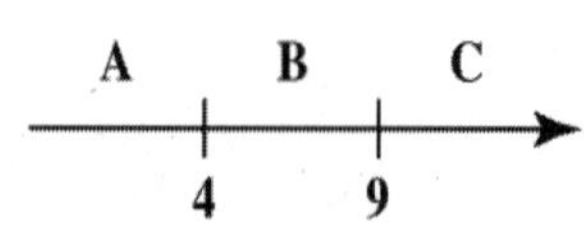

Test a number from each interval in the inequality

$$\frac{5}{t-4} \le 1.$$

Interval A: Let $t = 0$.

$$\frac{5}{-4} \le 1 \quad \textit{True}$$

Interval B: Let $t = 7$.

$$\frac{5}{3} \le 1 \quad \textit{False}$$

Interval C: Let $t = 10$.

$$\frac{5}{6} \le 1 \quad \textit{True}$$

The numbers in Intervals A and C, including 9 but not 4, are solutions.

Solution set: $(-\infty, 4) \cup [9, \infty)$

Cumulative Review Exercises Chapters 1–9

1. $S = \left\{-\frac{7}{3}, -2, -\sqrt{3}, 0, .7, \sqrt{12}, \sqrt{-8}, 7, \frac{32}{3}\right\}$

(a) The elements of S that are integers are -2, 0, and 7.

(b) The elements of S that are rational numbers are $-\frac{7}{3}$, -2, 0, .7, 7, and $\frac{32}{3}$.

(c) All the elements of S except $\sqrt{-8}$ are real numbers.

(d) All the elements of S are complex numbers.

2. $$|-3| + 8 - |-9| - (-7 + 3) = 3 + 8 - 9 - (-4)$$
$$= 3 + 8 - 9 + 4$$
$$= 6$$

3. $$2(-3)^2 + (-8)(-5) + (-17)$$
$$= 2(9) + 40 - 17$$
$$= 18 + 23 = 41$$

4. $$7 - (4 + 3t) + 2t = -6(t - 2) - 5$$
$$7 - 4 - 3t + 2t = -6t + 12 - 5$$
$$3 - t = -6t + 7$$
$$5t = 4$$
$$t = \frac{4}{5}$$

Check $t = \frac{4}{5}$: $\frac{11}{5} = \frac{11}{5}$

Solution set: $\left\{\frac{4}{5}\right\}$

5. $$|6x - 9| = |-4x + 2|$$

$6x - 9 = -4x + 2$ or $6x - 9 = -(-4x + 2)$

$10x = 11$ $\qquad$ $6x - 9 = 4x - 2$

$\qquad\qquad\qquad\quad 2x = 7$

$x = \frac{11}{10}$ or $x = \frac{7}{2}$

Check $x = \frac{11}{10}$: $\left|-\frac{24}{10}\right| \stackrel{?}{=} \left|-\frac{24}{10}\right|$ *True*

Check $x = \frac{7}{2}$: $|12| \stackrel{?}{=} |-12|$ *True*

Solution set: $\left\{\frac{11}{10}, \frac{7}{2}\right\}$

6. $$2x = \sqrt{\frac{5x+2}{3}}$$

Square both sides.

$$(2x)^2 = \left(\sqrt{\frac{5x+2}{3}}\right)^2$$
$$4x^2 = \frac{5x+2}{3}$$
$$12x^2 = 5x + 2$$
$$12x^2 - 5x - 2 = 0$$
$$(3x - 2)(4x + 1) = 0$$

$3x - 2 = 0$ or $4x + 1 = 0$

$x = \frac{2}{3}$ or $x = -\frac{1}{4}$

Check $x = \frac{2}{3}$: $\frac{4}{3} \stackrel{?}{=} \sqrt{\frac{16}{9}}$ *True*

Check $x = -\frac{1}{4}$: $-\frac{1}{2} \stackrel{?}{=} \sqrt{\frac{1}{4}}$ *False*

Solution set: $\left\{\frac{2}{3}\right\}$

7. $$\frac{3}{x-3} - \frac{2}{x-2} = \frac{3}{x^2 - 5x + 6}$$
$$\frac{3}{x-3} - \frac{2}{x-2} = \frac{3}{(x-3)(x-2)}$$

Multiply by the LCD, $(x - 3)(x - 2)$.

$$(x-3)(x-2)\left(\frac{3}{x-3} - \frac{2}{x-2}\right)$$
$$= (x-3)(x-2)\left[\frac{3}{(x-3)(x-2)}\right]$$
$$3(x - 2) - 2(x - 3) = 3$$
$$3x - 6 - 2x + 6 = 3$$
$$x = 3$$

The number 3 is not allowed as a solution since it makes the denominator 0. The solution set is $\emptyset$.

8. $$(r - 5)(2r + 3) = 1$$
$$2r^2 - 7r - 15 = 1$$
$$2r^2 - 7r - 16 = 0$$

Use the quadratic formula.

$$r = \frac{-b \pm \sqrt{b^2 - 4ac}}{2a}$$
$$r = \frac{-(-7) \pm \sqrt{(-7)^2 - 4(2)(-16)}}{2(2)}$$

continued

$$= \frac{7 \pm \sqrt{49+128}}{4} = \frac{7 \pm \sqrt{177}}{4}$$

Solution set: $\left\{\frac{7+\sqrt{177}}{4}, \frac{7-\sqrt{177}}{4}\right\}$

9. $x^4 - 5x^2 + 4 = 0$

Let $u = x^2$, so $u^2 = (x^2)^2 = x^4$.

The equation becomes

$u^2 - 5u + 4 = 0.$

$(u-4)(u-1) = 0$

$u - 4 = 0$ or $u - 1 = 0$

$u = 4$ or $u = 1$

To find x, substitute x^2 for u.

$x^2 = 4$ or $x^2 = 1$

$x = 2$ or $x = -2$ or $x = 1$ or $x = -1$

The potential solutions check.

Solution set: $\{-2, -1, 1, 2\}$

10. $-2x + 4 \leq -x + 3$

$-x \leq -1$

Multiply by -1, and reverse the direction of the inequality.

$x \geq 1$

Solution set: $[1, \infty)$

11. $|3x - 7| \leq 1$

$-1 \leq 3x - 7 \leq 1$

$6 \leq 3x \leq 8$

$2 \leq x \leq \frac{8}{3}$

Solution set: $\left[2, \frac{8}{3}\right]$

12. $x^2 - 4x + 3 < 0$

Solve the equation

$x^2 - 4x + 3 = 0.$

$(x-3)(x-1) = 0$

$x - 3 = 0$ or $x - 1 = 0$

$x = 3$ or $x = 1$

The numbers 1 and 3 divide a number line into three intervals.

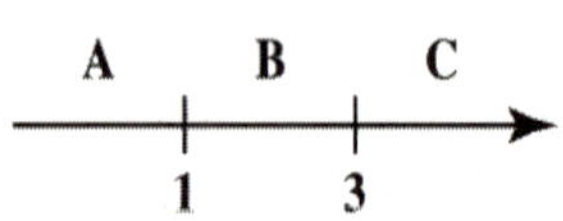

Test a number from each interval in the inequality

$$x^2 - 4x + 3 < 0.$$

Interval A: Let $x = 0$.

$3 < 0$ *False*

Interval B: Let $x = 2$.

$4 - 8 + 3 < 0$?

$-1 < 0$ *True*

Interval C: Let $x = 4$.

$16 - 16 + 3 < 0$?

$3 < 0$ *False*

The numbers from Interval B, not including 1 or 3, are solutions.

Solution set: $(1, 3)$

13. $\frac{3}{p+2} > 1$

Write the inequality so that 0 is on one side.

$$\frac{3}{p+2} - 1 > 0$$

$$\frac{3}{p+2} - \frac{1(p+2)}{p+2} > 0$$

$$\frac{3-p-2}{p+2} > 0$$

$$\frac{-p+1}{p+2} > 0$$

The number 1 makes the numerator 0, and -2 makes the denominator 0. These two numbers determine three intervals.

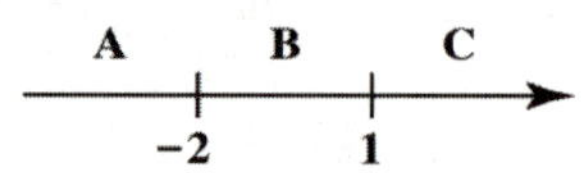

Test a number from each interval in the inequality

$$\frac{3}{p+2} > 1.$$

Interval A: Let $p = -4$.

$\frac{3}{-2} > 1$ *False*

Interval B: Let $p = 0$.

$\frac{3}{2} > 1$ *True*

Interval C: Let $p = 2$.

$\frac{3}{4} > 1$ *False*

The numbers from Interval B, not including -2 or 1, are solutions.

Solution set: $(-2, 1)$

14. $4x - 5y = 15$

Draw the line through its intercepts, $\left(\frac{15}{4}, 0\right)$ and $(0, -3)$. The graph passes the vertical line test, so the relation is a function. As with any line that is not horizontal or vertical, the domain and range are both $(-\infty, \infty)$.

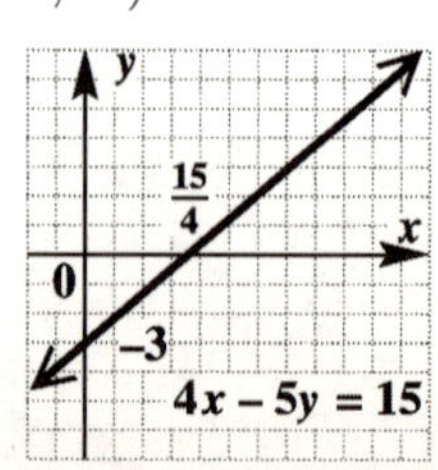

15. $4x - 5y < 15$

Draw a dashed line through the points $\left(\frac{15}{4}, 0\right)$ and $(0, -3)$. Check the origin:

$$4(0) - 5(0) < 15 \quad ?$$
$$0 < 15 \qquad \textit{True}$$

Shade the region that contains the origin.

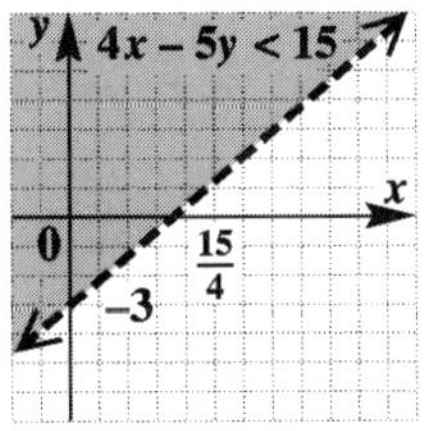

The relation is not a function since for any value of x, there is more than one value of y.

16. $f(x) = -2(x-1)^2 + 3$
The equation is in $f(x) = a(x-h)^2 + k$ form, so the graph is a parabola with vertex (h, k) at $(1, 3)$. Since $a = -2 < 0$, the parabola opens down. Also $|a| = |-2| = 2 > 1$, so the graph is narrower than the graph of $f(x) = x^2$. The points $(0, 1)$ and $(2, 1)$ are on the graph.

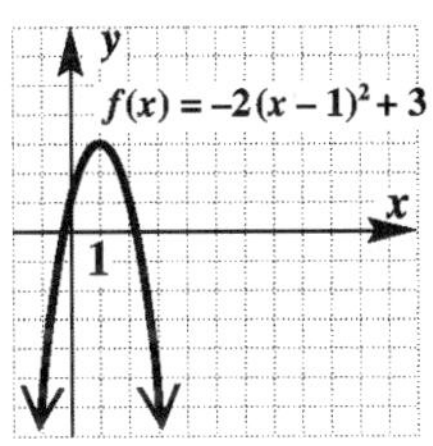

The relation is a function since it passes the vertical line test. The domain is $(-\infty, \infty)$. The largest value of y is 3, so the range is $(-\infty, 3]$.

17. $-2x + 7y = 16$
Solve the equation for y.

$$7y = 2x + 16$$
$$y = \frac{2}{7}x + \frac{16}{7}$$

So the slope is $\frac{2}{7}$ and the y-intercept is $\left(0, \frac{16}{7}\right)$.

Let $y = 0$ in $-2x + 7y = 16$ to find the x-intercept.

$$2x + 7(0) = 16$$
$$-2x = 16$$
$$x = -8$$

The x-intercept is $(-8, 0)$.

18. **(a)** Solve $5x + 2y = 6$ for y.

$$2y = -5x + 6$$
$$y = -\frac{5}{2}x + 3$$

So the slope of the given line and the desired line is $-\frac{5}{2}$. The required form is

$$y = -\frac{5}{2}x + b.$$
$$-3 = -\frac{5}{2}(2) + b \qquad \textit{Let } x=2,\ y=-3.$$
$$-3 = -5 + b$$
$$2 = b$$

The equation is $y = -\frac{5}{2}x + 2$.

(b) The negative reciprocal of the slope in part (a) is

$$-\frac{1}{-\frac{5}{2}} = \frac{2}{5},$$

which is the slope of the line perpendicular to the given line.

$$\text{So } y = \frac{2}{5}x + b.$$
$$1 = \frac{2}{5}(-4) + b \qquad \textit{Let } x=-4,\ y=1.$$
$$1 = -\frac{8}{5} + b$$
$$\frac{13}{5} = b$$

The equation is $y = \frac{2}{5}x + \frac{13}{5}$.

19. **(a)** The points are $(0, 116.26)$, $(10, 132.02)$, $(20, 141.85)$, $(30, 155.38)$, and $(33, 156.36)$.

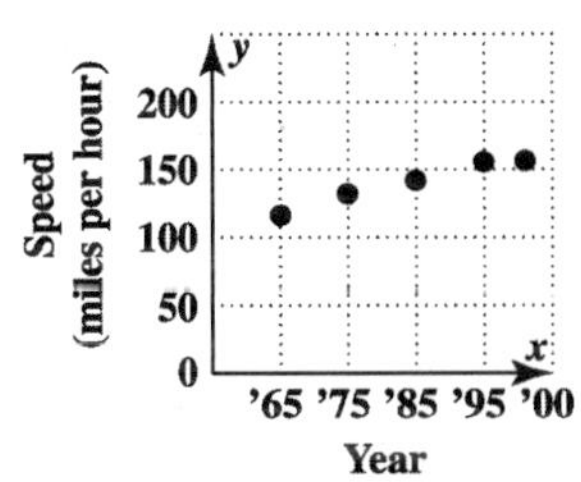

(b) As the years increase, the speeds increase, so the slope of the linear model is positive.

(c) The slope m of the line passing through $(0, 116.26)$ and $(20, 141.85)$ is

$$m = \frac{141.85 - 116.26}{20 - 0} = \frac{25.59}{20} = 1.2795.$$

Note that $x = 0$ represents 1965, so the point $(0, 116.26)$ can be considered to be the y-intercept.

continued

The slope-intercept form of the linear equation that models these data is

$$y = 1.2795x + 116.26.$$

(d) 1998 corresponds to $x = 33$.

$$\begin{aligned} y &= 1.2795(33) + 116.26 \\ &= 158.4835 \approx 158.48, \end{aligned}$$

which is an overestimate of the actual 156.36.

20. The relation $x = 5$ does not define a function because its graph is a vertical line, which is not the graph of a function by the vertical line test.

21. $f(x) = 2(x-1)^2 - 5$

(a) $\begin{aligned} f(-2) &= 2(-2-1)^2 - 5 \\ &= 2(-3)^2 - 5 \\ &= 2(9) - 5 = 13 \end{aligned}$

(b) Any value can be substituted for x, so the domain is $(-\infty, \infty)$. The graph of f is a parabola that opens up with vertex $(1, -5)$. The vertex is a minimum point so the y-values are all greater than or equal to -5. Thus, the range is $[-5, \infty)$.

22. $\begin{aligned} 2x - 4y &= 10 \quad (1) \\ 9x + 3y &= 3 \quad (2) \end{aligned}$

Simplify the equations.

$$\begin{aligned} x - 2y &= 5 \quad (3) \quad \tfrac{1}{2} \times (1) \\ 3x + y &= 1 \quad (4) \quad \tfrac{1}{3} \times (2) \end{aligned}$$

To eliminate y, multiply (4) by 2 and add the result to (3).

$$\begin{array}{rcll} x - 2y &=& 5 & (3) \\ 6x + 2y &=& 2 & 2 \times (4) \\ \hline 7x &=& 7 & \\ x &=& 1 & \end{array}$$

Substitute $x = 1$ into (4).

$$\begin{aligned} 3x + y &= 1 \quad (4) \\ 3(1) + y &= 1 \\ y &= -2 \end{aligned}$$

Solution set: $\{(1, -2)\}$

23. $\begin{aligned} x + y + 2z &= 3 \quad (1) \\ -x + y + z &= -5 \quad (2) \\ 2x + 3y - z &= -8 \quad (3) \end{aligned}$

Eliminate z by adding (2) and (3).

$$\begin{array}{rcll} -x + y + z &=& -5 & (2) \\ 2x + 3y - z &=& -8 & (3) \\ \hline x + 4y &=& -13 & (4) \end{array}$$

To get another equation without z, multiply equation (3) by 2 and add the result to equation (1).

$$\begin{array}{rcll} x + y + 2z &=& 3 & (1) \\ 4x + 6y - 2z &=& -16 & 2 \times (3) \\ \hline 5x + 7y &=& -13 & (5) \end{array}$$

To eliminate x, multiply (4) by -5 and add the result to (5).

$$\begin{array}{rcll} -5x - 20y &=& 65 & -5 \times (4) \\ 5x + 7y &=& -13 & (5) \\ \hline -13y &=& 52 & \\ y &=& -4 & \end{array}$$

Use (4) to find x.

$$\begin{aligned} x + 4y &= -13 \quad (4) \\ x + 4(-4) &= -13 \\ x - 16 &= -13 \\ x &= 3 \end{aligned}$$

Use (2) to find z.

$$\begin{aligned} -x + y + z &= -5 \quad (2) \\ -3 - 4 + z &= -5 \\ -7 + z &= -5 \\ z &= 2 \end{aligned}$$

Solution set: $\{(3, -4, 2)\}$

24. **(a)** Let $x =$ the amount of sales (in billions) for AOL and $y =$ the amount of sales for Time Warner.

The combined sales were \$34.2 billion, so

$$x + y = 34.2. \quad (1)$$

Sales for AOL were \$.3 billion less than 4 times the sales of Time Warner, so

$$x = 4y - .3. \quad (2)$$

(b) Substitute $4y - .3$ for x in (1), then solve for y.

$$\begin{aligned} (4y - .3) + y &= 34.2 \\ 5y &= 34.5 \\ y &= 6.9 \end{aligned}$$

From (2), $x = 4(6.9) - .3 = 27.3$. Thus, AOL had \$27.3 billion in sales and Time Warner had \$6.9 billion in sales.

25. $$\begin{aligned} \left(\frac{x^{-3}y^2}{x^5y^{-2}}\right)^{-1} &= \left(x^{-3-5}y^{2-(-2)}\right)^{-1} \\ &= \left(x^{-8}y^4\right)^{-1} \\ &= x^8y^{-4} \\ &= \frac{x^8}{y^4} \end{aligned}$$

26. $$\begin{aligned} \frac{(4x^{-2})^2(2y^3)}{8x^{-3}y^5} &= \frac{16x^{-4}(2y^3)}{8x^{-3}y^5} \\ &= \frac{4x^{-4}y^3}{x^{-3}y^5} \end{aligned}$$

$$= 4x^{-4-(-3)}y^{3-5}$$
$$= 4x^{-1}y^{-2}$$
$$= \frac{4}{xy^2}$$

27. $(7x + 4)(2x - 3)$

$$\quad \textbf{F} \quad \textbf{O} \quad \textbf{I} \quad \textbf{L}$$
$$= 14x^2 - 21x + 8x - 12$$
$$= 14x^2 - 13x - 12$$

28. $\left(\frac{2}{3}t + 9\right)^2 = \left(\frac{2}{3}t\right)^2 + 2\left(\frac{2}{3}t\right)(9) + 9^2$

$$= \frac{4}{9}t^2 + 12t + 81$$

29. $\left(3t^3 + 5t^2 - 8t + 7\right) - \left(6t^3 + 4t - 8\right)$

$$= 3t^3 + 5t^2 - 8t + 7 - 6t^3 - 4t + 8$$
$$= -3t^3 + 5t^2 - 12t + 15$$

30. Divide $4x^3 + 2x^2 - x + 26$ by $x + 2$.

$$\begin{array}{r|rrrr} & 4x^2 & -\ 6x & +\ 11 & \\ \hline x + 2 & 4x^3 + & 2x^2 & -\ \ x & +\ 26 \\ & 4x^3 + & 8x^2 & & \\ \hline & & -6x^2 & -\ \ x & \\ & & -6x^2 & -\ 12x & \\ \hline & & & 11x & +\ 26 \\ & & & 11x & +\ 22 \\ \hline & & & & 4 \end{array}$$

The answer is

$$4x^2 - 6x + 11 + \frac{4}{x + 2}.$$

31. $16x - x^3 = x\left(16 - x^2\right)$

$$= x(4 + x)(4 - x)$$

32. $24m^2 + 2m - 15$

The two integers whose product is $24(-15) = -360$ and whose sum is 2 are 20 and -18.

$$24m^2 + 2m - 15$$
$$= 24m^2 + 20m - 18m - 15$$
$$= 4m(6m + 5) - 3(6m + 5)$$
$$= (6m + 5)(4m - 3)$$

33. $8x^3 + 27y^3$

Use the sum of cubes formula,

$$a^3 + b^3 = (a + b)\left(a^2 - ab + b^2\right),$$

with $a = 2x$ and $b = 3y$.

$$8x^3 + 27y^3$$
$$= (2x + 3y)\left[(2x)^2 - (2x)(3y) + (3y)^2\right]$$
$$= (2x + 3y)\left(4x^2 - 6xy + 9y^2\right)$$

34. $9x^2 - 30xy + 25y^2$

Use the perfect square formula,

$$a^2 - 2ab + b^2 = (a - b)^2,$$

with $a = 3x$ and $b = 5y$.

$$9x^2 - 30xy + 25y^2$$
$$= \left[(3x)^2 - 2(3x)(5y) + (5y)^2\right]$$
$$= (3x - 5y)^2$$

35. $\frac{x^2 - 3x - 10}{x^2 + 3x + 2} \bullet \frac{x^2 - 2x - 3}{x^2 + 2x - 15}$

$$= \frac{(x - 5)(x + 2)}{(x + 2)(x + 1)} \bullet \frac{(x - 3)(x + 1)}{(x + 5)(x - 3)}$$
$$= \frac{x - 5}{x + 5}$$

36. $\frac{3}{2 - k} - \frac{5}{k} + \frac{6}{k^2 - 2k}$

$$= \frac{3}{2 - k} - \frac{5}{k} + \frac{6}{k(k - 2)}$$
$$= \frac{-3}{k - 2} - \frac{5}{k} + \frac{6}{k(k - 2)}$$

The LCD is $k(k - 2)$.

$$= \frac{-3k}{(k - 2)k} - \frac{5(k - 2)}{k(k - 2)} + \frac{6}{k(k - 2)}$$
$$= \frac{-3k - 5(k - 2) + 6}{k(k - 2)}$$
$$= \frac{-3k - 5k + 10 + 6}{k(k - 2)}$$
$$= \frac{-8k + 16}{k(k - 2)}$$
$$= \frac{-8(k - 2)}{k(k - 2)} = -\frac{8}{k}$$

37. $\dfrac{\frac{r}{s} - \frac{s}{r}}{\frac{r}{s} + 1}$

Multiply the numerator and denominator by the LCD of all the fractions, rs.

$$= \frac{\left(\frac{r}{s} - \frac{s}{r}\right)rs}{\left(\frac{r}{s} + 1\right)rs} = \frac{r^2 - s^2}{r^2 + rs}$$
$$= \frac{(r - s)(r + s)}{r(r + s)} = \frac{r - s}{r}$$

38. $\sqrt[3]{\frac{27}{16}} = \frac{\sqrt[3]{27}}{\sqrt[3]{16}} = \frac{\sqrt[3]{3^3}}{\sqrt[3]{8 \bullet 2}} = \frac{3}{2\sqrt[3]{2}}$

$$= \frac{3 \bullet \sqrt[3]{4}}{2\sqrt[3]{2} \bullet \sqrt[3]{4}} = \frac{3\sqrt[3]{4}}{2\sqrt[3]{8}}$$
$$= \frac{3\sqrt[3]{4}}{2 \bullet 2} = \frac{3\sqrt[3]{4}}{4}$$

39. $$\frac{2}{\sqrt{7}-\sqrt{5}} = \frac{2\left(\sqrt{7}+\sqrt{5}\right)}{\left(\sqrt{7}-\sqrt{5}\right)\left(\sqrt{7}+\sqrt{5}\right)} = \frac{2\left(\sqrt{7}+\sqrt{5}\right)}{7-5} = \frac{2\left(\sqrt{7}+\sqrt{5}\right)}{2} = \sqrt{7}+\sqrt{5}$$

40. Let $x =$ Tri's rate on the bicycle.
Then $x - 10 =$ Tri's rate while walking.

Make a chart. Use $d = rt$, or $t = \frac{d}{r}$.

	Distance	Rate	Time
Bicycle	12	x	$\frac{12}{x}$
Walking	8	$x-10$	$\frac{8}{x-10}$

Tri's time on the bicycle + Tri's time walking = 5 hours.

$$\frac{12}{x} + \frac{8}{x-10} = 5$$

Multiply by the LCD, $x(x-10)$.

$$x(x-10)\left(\frac{12}{x}+\frac{8}{x-10}\right) = x(x-10) \cdot 5$$
$$12(x-10) + 8x = 5x(x-10)$$
$$12x - 120 + 8x = 5x^2 - 50x$$
$$0 = 5x^2 - 70x + 120$$
$$0 = x^2 - 14x + 24$$
$$0 = (x-12)(x-2)$$

$$x - 12 = 0 \quad \text{or} \quad x - 2 = 0$$
$$x = 12 \quad \text{or} \quad x = 2$$

Reject 2 for Tri's bicycle speed, since it would yield a negative walking speed. Thus, his bicycle speed was 12 mph, and his walking speed was $x - 10 = 2$ mph.

41. Let $x =$ the distance traveled by the southbound car and
$2x - 38 =$ the distance traveled by the eastbound car.

Since the cars are traveling at right angles with one another, the Pythagorean formula can be applied.

$$a^2 + b^2 = c^2$$
$$x^2 + (2x-38)^2 = 95^2$$
$$x^2 + 4x^2 - 152x + 1444 = 9025$$
$$5x^2 - 152x - 7581 = 0$$

$$x = \frac{-(-152) \pm \sqrt{(-152)^2 - 4(5)(-7581)}}{2(5)} = \frac{152 \pm \sqrt{174,724}}{10} = \frac{152 \pm 418}{10}$$

Thus, $x = \frac{152+418}{10} = 57$ (the other value is negative). The southbound car traveled 57 miles, and the eastbound car traveled $2x - 38 = 2(57) - 38 = 76$ miles.

42. Since 31% of the people in the United States curse the ATM, the number of people in the United States who curse the ATM in a sample of 3000 is

$$.31(3000) = 930.$$

43. $.24(3000) = 720$

44. $.33(3000) = 990$

45. German cursers: $.53(3000) = 1590$
United States thankers: $.22(3000) = 660$

The difference is

$$1590 - 660 = 930.$$

CHAPTER 10 INVERSE, EXPONENTIAL, AND LOGARITHMIC FUNCTIONS

Section 10.1

1. This function is not one-to-one because both Illinois and Wisconsin are paired with the same range element, 40.

3. The function in the table that pairs a city with a distance is a one-to-one function because for each city there is one distance and each distance has only one city to which it is paired.

If the distance from Indianapolis to Denver had 1 mile added to it, it would be $1058 + 1 = 1059$ mi, the same as the distance from Los Angeles to Denver. In this case, one distance would have two cities to which it is paired, and the function would not be one-to-one.

5. If a function is made up of ordered pairs in such a way that the same y-value appears in a correspondence with two different x-values, then the function is not one-to-one. Choice **B**

7. All of the graphs pass the vertical line test, so they all represent functions. The graph in choice **A** is the only one that passes the horizontal line test, so it is the one-to-one function.

9. $\{(3,6), (2,10), (5,12)\}$ is a one-to-one function, since each x-value corresponds to only one y-value and each y-value corresponds to only one x-value. To find the inverse, interchange x and y in each ordered pair. The inverse is

$$\{(6,3), (10,2), (12,5)\}.$$

11. $\{(-1,3), (2,7), (4,3), (5,8)\}$ is not a one-to-one function. The ordered pairs $(-1,3)$ and $(4,3)$ have the same y-values for two different x-values.

13. The graph of $f(x) = 2x + 4$ is a nonvertical, nonhorizontal line. By the horizontal line test, $f(x)$ is a one-to-one function. To find the inverse, replace $f(x)$ with y.

$$y = 2x + 4$$

Interchange x and y.

$$x = 2y + 4$$

Solve for y.

$$2y = x - 4$$

$$y = \frac{x-4}{2}$$

Replace y with $f^{-1}(x)$.

$$f^{-1}(x) = \frac{x-4}{2}$$

15. Write $g(x) = \sqrt{x-3}$ as $y = \sqrt{x-3}$. Since $x \geq 3$, $y \geq 0$. The graph of g is half of a horizontal parabola that opens to the right. The graph passes the horizontal line test, so g is one-to-one. To find the inverse, interchange x and y to get

$$x = \sqrt{y-3}.$$

Note that now $y \geq 3$, so $x \geq 0$.
Solve for y by squaring both sides.

$$x^2 = y - 3$$

$$x^2 + 3 = y$$

Replace y with $g^{-1}(x)$.

$$g^{-1}(x) = x^2 + 3,\ x \geq 0$$

17. $f(x) = 3x^2 + 2$ is not a one-to-one function because two x-values, such as 1 and -1, both have the same y-value, in this case 5. The graph of this function is a vertical parabola which does not pass the horizontal line test.

19. The graph of $f(x) = x^3 - 4$ is the graph of $g(x) = x^3$ shifted down 4 units. (Recall that $g(x) = x^3$ is the elongated S-shaped curve.) The graph of f passes the horizontal line test, so f is one-to-one.

Replace $f(x)$ with y.

$$y = x^3 - 4$$

Interchange x and y.

$$x = y^3 - 4$$

Solve for y.

$$x + 4 = y^3$$

Take the cube root of each side.

$$\sqrt[3]{x+4} = y$$

Replace y with $f^{-1}(x)$.

$$f^{-1}(x) = \sqrt[3]{x+4}$$

In Exercises 21–23, $f(x) = 2^x$ is a one-to-one function.

21. **(a)** To find $f(3)$, substitute 3 for x.

$f(x) = 2^x$, so $f(3) = 2^3 = 8$.

(b) Since f is one-to-one and $f(3) = 8$, it follows that $f^{-1}(8) = 3$.

23. **(a)** To find $f(0)$, substitute 0 for x.

$f(x) = 2^x$, so $f(0) = 2^0 = 1$.

(b) Since f is one to one and $f(0) = 1$, it follows that $f^{-1}(1) = 0$.

25. **(a)** The function is one-to-one since any horizontal line intersects the graph at most once.

(b) In the graph, the two points marked on the line are $(-1,5)$ and $(2,-1)$. Interchange x and y in each ordered pair to get $(5,-1)$ and $(-1,2)$.

continued

Plot these points, then draw a dashed line through them to obtain the graph of the inverse function.

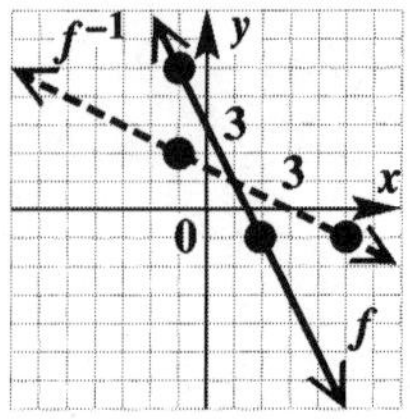

27. **(a)** The function is not one-to-one since there are horizontal lines that intersect the graph more than once. For example, the line $y = 1$ intersects the graph twice.

29. **(a)** The function is one-to-one since any horizontal line intersects the graph at most once.

(b) In the graph, the four points marked on the curve are $(-4, 2)$, $(-1, 1)$, $(1, -1)$, and $(4, -2)$. Interchange x and y in each ordered pair to get $(2, -4)$, $(1, -1)$, $(-1, 1)$, and $(-2, 4)$. Plot these points, then draw a dashed curve (symmetric to the original graph about the line $y = x$) through them to obtain the graph of the inverse.

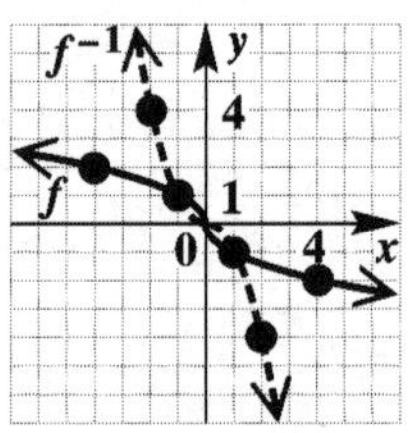

31. $f(x) = 2x - 1$ or $y = 2x - 1$
The graph is a line through $(-2, -5)$, $(0, -1)$, and $(3, 5)$. Plot these points and draw the solid line through them. Then the inverse will be a line through $(-5, -2)$, $(-1, 0)$, and $(5, 3)$. Plot these points and draw the dashed line through them.

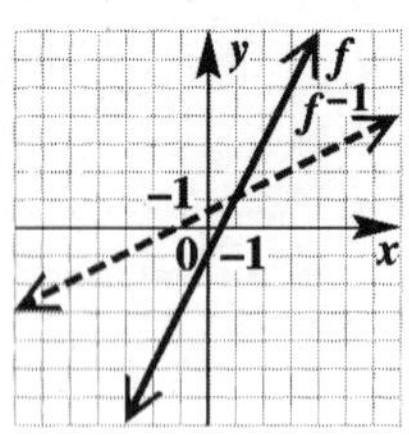

33. $g(x) = -4x$ or $y = -4x$
The graph is a line through $(0, 0)$ and $(1, -4)$. For the inverse, interchange x and y in each ordered pair to get the points $(0, 0)$ and $(-4, 1)$. Draw a dashed line through these points to obtain the graph of the inverse function.

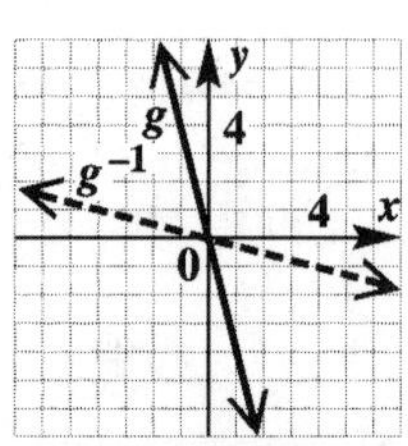

35. $f(x) = \sqrt{x}$, $x \geq 0$
Complete the table of values.

x	$f(x)$
0	0
1	1
4	2

Plot these points and connect them with a solid smooth curve.
Since $f(x)$ is one-to-one, make a table of values for $f^{-1}(x)$ by interchanging x and y.

x	$f^{-1}(x)$
0	0
1	1
2	4

Plot these points and connect them with a dashed smooth curve.

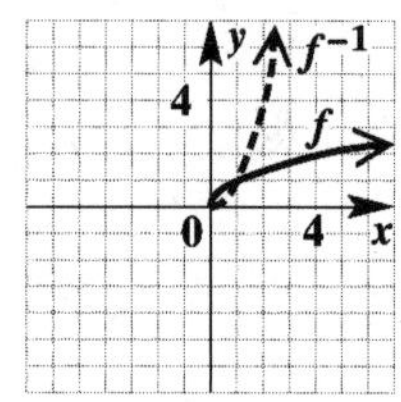

37. $f(x) = x^3 - 2$
Complete the table of values.

x	$f(x)$
-1	-3
0	-2
1	-1
2	6

Plot these points and connect them with a solid smooth curve.
Make a table of values for f^{-1}.

x	$f^{-1}(x)$
-3	-1
-2	0
-1	1
6	2

Plot these points and connect them with a dashed smooth curve.

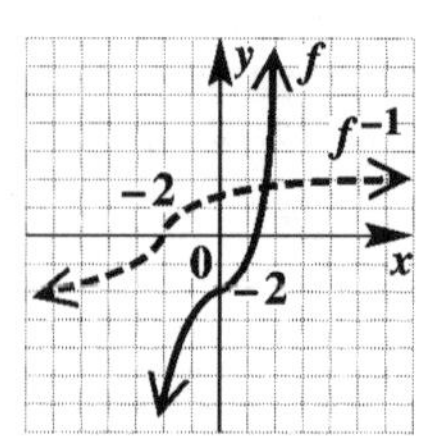

39. $$f(x) = 4x - 5$$
Replace $f(x)$ with y.
$$y = 4x - 5$$
Interchange x and y.
$$x = 4y - 5$$
Solve for y.
$$x + 5 = 4y$$
$$\frac{x+5}{4} = y$$
Replace y with $f^{-1}(x)$.
$$\frac{x+5}{4} = f^{-1}(x)$$

40. Insert each number in the inverse function found in Exercise 39,

$$f^{-1}(x) = \frac{x+5}{4}.$$
$$f^{-1}(47) = \frac{47+5}{4} = \frac{52}{4} = 13 = \text{M},$$
$$f^{-1}(95) = \frac{95+5}{4} = \frac{100}{4} = 25 = \text{Y},$$
and so on.

The decoded message is as follows:
My graphing calculator is the greatest thing since sliced bread.

41. A one-to-one code is essential to this process because if the code is not one-to-one, an encoded number would refer to two different letters.

42. Answers will vary according to the student's name. For example, Jane Doe is encoded as follows:

$$1004 \quad 5 \quad 2748 \quad 129 \quad 68 \quad 3379 \quad 129.$$

43. $\text{Y}_1 = f(x) = 2x - 7$
Replace $f(x)$ with y.
$$y = 2x - 7$$
Interchange x and y.
$$x = 2y - 7$$
Solve for y.
$$x + 7 = 2y$$
$$\frac{x+7}{2} = y$$
Replace y with $f^{-1}(x)$.
$$\frac{x+7}{2} = f^{-1}(x) = \text{Y}_2$$

Now graph Y_1 and Y_2.

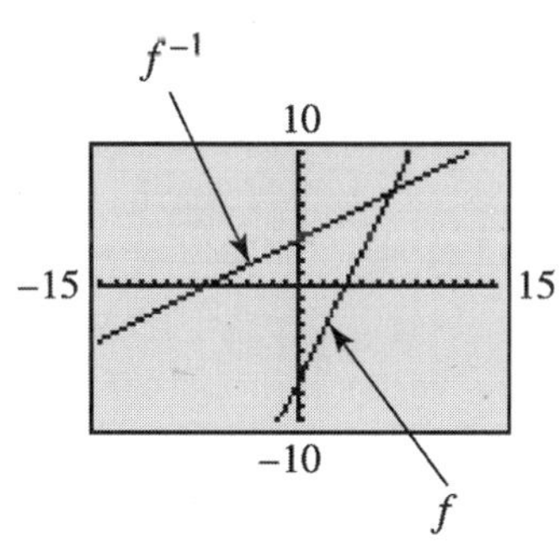

45. $\text{Y}_1 = f(x) = x^3 + 5$
Replace $f(x)$ with y.
$$y = x^3 + 5$$
Interchange x and y.
$$x = y^3 + 5$$
Solve for y.
$$x - 5 = y^3$$
Take the cube root of each side.
$$\sqrt[3]{x-5} = y$$
Replace y with $f^{-1}(x)$.
$$\sqrt[3]{x-5} = f^{-1}(x) = \text{Y}_2$$

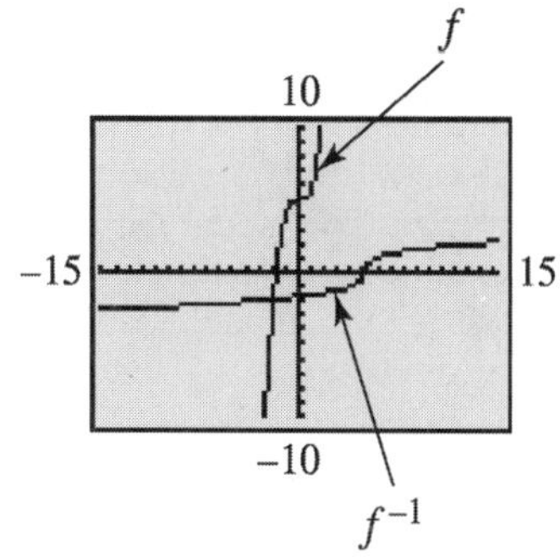

47. $\text{Y}_1 = \text{X}^2 + 3\text{X} + 4$
Graph Y_1 and its inverse in the same square window on a graphing calculator. On a TI-83, graph Y_1 and then enter

DrawInv Y_1

on the home screen. DrawInv is choice 8 under the DRAW menu. Y_1 is choice 1 under VARS, Y-VARS, Function.

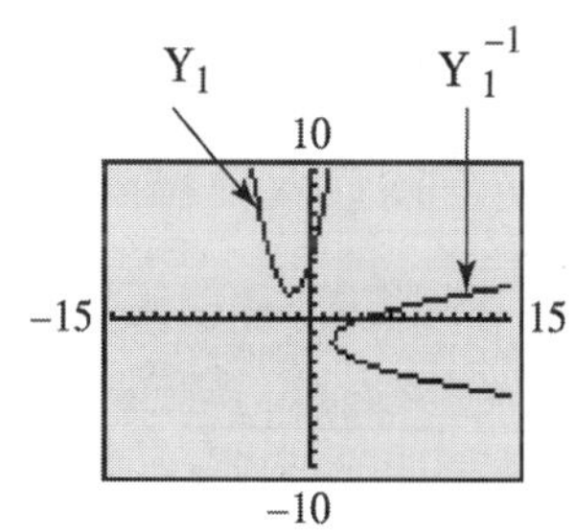

49. The "inverse" of the function in Exercise 47 does not actually satisfy the definition of inverse because the original function is required to be one-to-one. The original function $y = x^2 + 3x + 4$ is not one-to-one as shown by the fact that $(0, 4)$ and $(-3, 4)$ lie on the graph.

Section 10.2

1. Since the graph of $F(x) = a^x$ always contains the point $(0, 1)$, the correct response is **C**.

3. Since the graph of $F(x) = a^x$ always approaches the x-axis, the correct response is **A**.

5. $f(x) = 3^x$
Make a table of values.

$f(-2) = 3^{-2} = \frac{1}{3^2} = \frac{1}{9}$,

$f(-1) = 3^{-1} = \frac{1}{3^1} = \frac{1}{3}$, and so on.

x	-2	-1	0	1	2
$f(x)$	$\frac{1}{9}$	$\frac{1}{3}$	1	3	9

Plot the points from the table and draw a smooth curve through them.

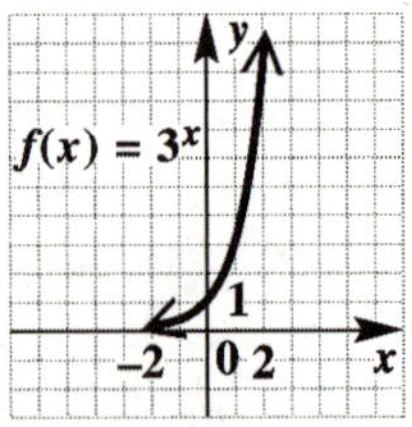

7. $g(x) = \left(\frac{1}{3}\right)^x$
Make a table of values.

$g(-2) = \left(\frac{1}{3}\right)^{-2} = \left(\frac{3}{1}\right)^2 = 9$,

$g(-1) = \left(\frac{1}{3}\right)^{-1} = \left(\frac{3}{1}\right)^1 = 3$, and so on.

x	-2	-1	0	1	2
$g(x)$	9	3	1	$\frac{1}{3}$	$\frac{1}{9}$

Plot the points from the table and draw a smooth curve through them.

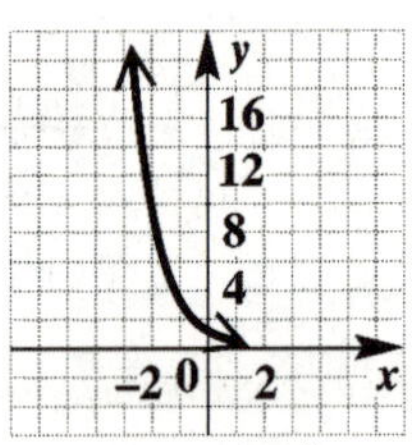

$g(x) = \left(\frac{1}{3}\right)^x$

9. $y = 4^{-x}$
This equation can be rewritten as

$$y = (4^{-1})^x = \left(\frac{1}{4}\right)^x,$$

which shows that it is *falling* from left to right. Make a table of values.

x	-2	-1	0	1	2
y	16	4	1	$\frac{1}{4}$	$\frac{1}{16}$

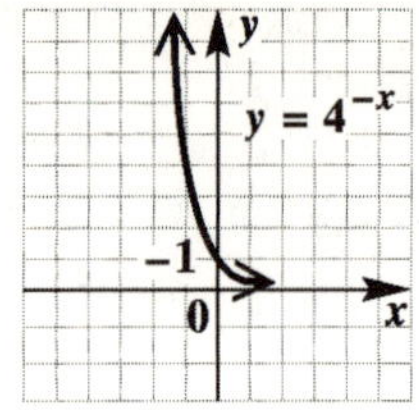

11. $y = 2^{2x-2}$
Make a table of values. It will help to find values for $2x - 2$ before you find y.

x	-2	-1	0	1	2	3
$2x-2$	-6	-4	-2	0	2	4
y	$\frac{1}{64}$	$\frac{1}{16}$	$\frac{1}{4}$	1	4	16

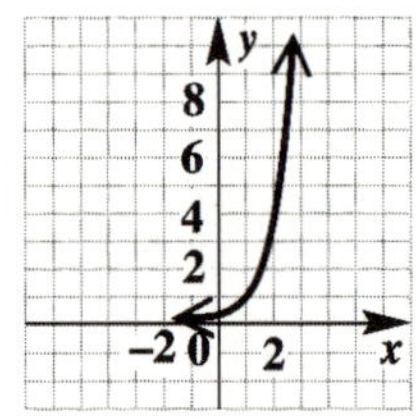

$y = 2^{2x-2}$

13. **(a)** For an exponential function defined by $f(x) = a^x$, if $a > 1$, the graph rises from left to right. (See Example 1, $f(x) = 2^x$, in your text.) If $0 < a < 1$, the graph falls from left to right. (See Example 2, $g(x) = \left(\frac{1}{2}\right)^x = 2^{-x}$, in your text.)

(b) An exponential function defined by $f(x) = a^x$ is one-to-one and has an inverse, since each value of $f(x)$ corresponds to one and only one value of x.

15. $$6^x = 36$$
Write each side as a power of 6.
$$6^x = 6^2$$
For $a > 0$ and $a \neq 1$, if $a^x = a^y$, then $x = y$. Set the exponents equal to each other.
$$x = 2$$

Check $x = 2$: $6^2 \stackrel{?}{=} 36$ *True*

Solution set: $\{2\}$

17. $$100^x = 1000$$
Write each side as a power of 10.
$$(10^2)^x = 10^3$$
$$10^{2x} = 10^3$$
For $a > 0$ and $a \neq 1$, if $a^x = a^y$, then $x = y$. Set the exponents equal to each other.
$$2x = 3$$
$$x = \frac{3}{2}$$

Check $x = \frac{3}{2}$: $100^{3/2} \stackrel{?}{=} 1000$ *True*

Solution set: $\left\{\frac{3}{2}\right\}$

19. $16^{2x+1} = 64^{x+3}$

Write each side as a power of 4.

$$\left(4^2\right)^{2x+1} = \left(4^3\right)^{x+3}$$
$$4^{4x+2} = 4^{3x+9}$$

Set the exponents equal.

$$4x + 2 = 3x + 9$$
$$x = 7$$

Check $x = 7$: $16^{15} \stackrel{?}{=} 64^{10}$ *True*

Solution set: $\{7\}$

21. $5^x = \frac{1}{125}$

$$5^x = \left(\frac{1}{5}\right)^3$$

Write each side as a power of 5.

$$5^x = 5^{-3}$$

Set the exponents equal.

$$x = -3$$

Check $x = -3$: $5^{-3} \stackrel{?}{=} \frac{1}{125}$ *True*

Solution set: $\{-3\}$

23. $5^x = .2$

$$5^x = \frac{2}{10} = \frac{1}{5}$$

Write each side as a power of 5.

$$5^x = 5^{-1}$$

Set the exponents equal.

$$x = -1$$

Check $x = -1$: $5^{-1} \stackrel{?}{=} .2$ *True*

Solution set: $\{-1\}$

25. $\left(\frac{3}{2}\right)^x = \frac{8}{27}$

$$\left(\frac{3}{2}\right)^x = \left(\frac{2}{3}\right)^3$$

Write each side as a power of $\frac{3}{2}$.

$$\left(\frac{3}{2}\right)^x = \left(\frac{3}{2}\right)^{-3}$$

Set the exponents equal.

$$x = -3$$

Check $x = -3$: $\left(\frac{3}{2}\right)^{-3} \stackrel{?}{=} \frac{8}{27}$ *True*

Solution set: $\{-3\}$

27. $12^{2.6} \approx 639.545$

29. $.5^{3.921} \approx .066$

31. $2.718^{2.5} \approx 12.179$

33. The reason many scientific calculators cannot calculate something like $(-2)^4$ is that in an exponential function, the base must be positive. This is necessary so the function can be defined for all real numbers. For example, $f(x) = (-4)^x$ would not be defined for $x = \frac{1}{2}$ since it is impossible to take the square root of a negative number.

35. **(a)** The increase for the exponential-type curve in the year 2000 is about .5°C.

(b) The increase for the linear graph in the year 2000 is about .35°C.

37. **(a)** The increase for the exponential-type curve in the year 2020 is about 1.6°C.

(b) The increase for the linear graph in the year 2020 is about .5°C.

39. $f(x) = 132{,}359(1.0124)^{-x}$

(a) 1970 corresponds to $x = 0$.

$$f(0) = 132{,}359(1.0124)^0$$
$$= 132{,}359(1) = 132{,}359$$

The answer has units in thousands of tons.

(b) 1995 corresponds to $x = 25$.

$$f(25) = 132{,}359(1.0124)^{-25} \approx 97{,}264$$

(c) 1998 corresponds to $x = 28$.

$$f(28) = 132{,}359(1.0124)^{-28} \approx 93{,}733$$

The actual amount, 89,454 thousand tons, is slightly less than the 93,733 thousand tons that the model provides.

41. $V(t) = 5000(2)^{-.15t}$

(a) The original value is found when $t = 0$.

$$V(0) = 5000(2)^{-.15(0)}$$
$$= 5000(2)^0$$
$$= 5000(1) = 5000$$

The original value is \$5000.

(b) The value after 5 years is found when $t = 5$.

$$V(5) = 5000(2)^{-.15(5)}$$
$$= 5000(2)^{-.75} \approx 2973.02$$

The value after 5 years is about \$2973.

(c) The value after 10 years is found when $t = 10$.

$$V(10) = 5000(2)^{-.15(10)}$$
$$= 5000(2)^{-1.5} \approx 1767.77$$

The value after 10 years is about \$1768.

(d) Use the results of parts (a) – (c) to make a table of values.

t	0	5	10
$V(t)$	5000	2973	1768

Plot the points from the table and draw a smooth curve through them.

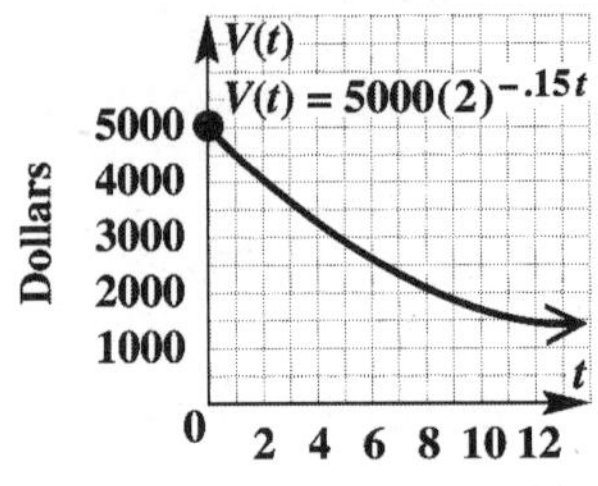

43. $V(t) = 5000(2)^{-.15t}$

$2500 = 5000(2)^{-.15t}$ *Let V(t) = 2500.*

$\frac{1}{2} = (2)^{-.15t}$ *Divide by 5000.*

$2^{-1} = 2^{-.15t}$

$-1 = -.15t$ *Equate exponents*

$t = \frac{-1}{-.15} \approx 6.67$

The value of the machine will be \$2500 in approximately 6.67 years after it was purchased.

45. $16^{3/4} = \left(\sqrt[4]{16}\right)^3 = (2)^3 = 8$

46. $16^{3/4} = \sqrt[4]{16^3} = \sqrt[4]{4096} = 8$

47. $\sqrt{16^3} = 64$ and $\sqrt{64} = 8$.

48. In Exercise 47, we are finding $\sqrt{\sqrt{16^3}}$, that is, taking the square root twice.

$$\sqrt{\sqrt{16^3}} = \sqrt{(16^3)^{1/2}} = \sqrt{16^{3/2}}$$
$$= \left(16^{3/2}\right)^{1/2} = 16^{3/4}$$

49. Since $16^{.5} = \sqrt{16} = 4$ and $16^1 = 16$, a reasonable prediction for $16^{.75}$ must be between 4 and 16. From a calculator,

$$16^{.75} = 8.$$

50. $\sqrt[100]{16^{75}} = 16^{75/100} = 16^{3/4}$

$= \left(\sqrt[4]{16}\right)^3 = (2)^3 = 8$

Section 10.3

1. **(a)** $\log_4 16$ is equal to 2, because 2 is the exponent to which 4 must be raised to obtain 16. **(C)**

(b) $\log_3 81$ is equal to 4, because 4 is the exponent to which 3 must be raised to obtain 81. **(F)**

(c) $\log_3\left(\frac{1}{3}\right)$ is equal to -1, because -1 is the exponent to which 3 must be raised to obtain $\frac{1}{3}$. **(B)**

(d) $\log_{10} .01$ is equal to -2, because -2 is the exponent to which 10 must be raised to obtain .01. **(A)**

(e) $\log_5 \sqrt{5}$ is equal to $\frac{1}{2}$, because $\frac{1}{2}$ is the exponent to which 5 must be raised to obtain $\sqrt{5}$. **(E)**

(f) $\log_{13} 1$ is equal to 0, because 0 is the exponent to which 13 must be raised to obtain 1. **(D)**

3. The base is 4, the exponent (logarithm) is 5, and the number is 1024, so $4^5 = 1024$ becomes $\log_4 1024 = 5$ in logarithmic form.

5. $\frac{1}{2}$ is the base and -3 is the exponent, so $\left(\frac{1}{2}\right)^{-3} = 8$ becomes $\log_{1/2} 8 = -3$ in logarithmic form.

7. The base is 10, the exponent (logarithm) is -3, and the number is .001, so $10^{-3} = .001$ becomes $\log_{10} .001 = -3$ in logarithmic form.

9. $\sqrt[4]{625} = 625^{1/4} = 5$

The base is 625, the exponent (logarithm) is $\frac{1}{4}$, and the number is 5, so $\sqrt[4]{625} = 5$ becomes $\log_{625} 5 = \frac{1}{4}$ in logarithmic form.

11. In $\log_4 64 = 3$, 4 is the base and 3 is the logarithm (exponent), so $\log_4 64 = 3$ becomes $4^3 = 64$ in exponential form.

13. In $\log_{10} \frac{1}{10{,}000} = -4$, the base is 10, the logarithm (exponent) is -4, and the number is $\frac{1}{10{,}000}$, so $\log_{10} \frac{1}{10{,}000} = -4$ becomes $10^{-4} = \frac{1}{10{,}000}$ in exponential form.

15. In $\log_6 1 = 0$, 6 is the base and 0 is the logarithm (exponent), so $\log_6 1 = 0$ becomes $6^0 = 1$ in exponential form.

17. In $\log_9 3 = \frac{1}{2}$, the base is 9, the logarithm (exponent) is $\frac{1}{2}$, and the number is 3, so $\log_9 3 = \frac{1}{2}$ becomes $9^{1/2} = 3$ in exponential form.

19. To evaluate $\log_9 3$, one has to ask "9 raised to what power gives you a result of 3?" We know that the square root (a radical) of 9 is 3. Therefore, the teacher's hint was to see what root of 9 equals 3. The answer is the reciprocal of the root index.

21. $x = \log_{27} 3$
Write in exponential form.
$27^x = 3$
Write each side as a power of 3.
$(3^3)^x = 3$
$3^{3x} = 3^1$
Set the exponents equal.
$3x = 1$
$x = \frac{1}{3}$
Solution set: $\left\{\frac{1}{3}\right\}$

23. $\log_x 9 = \frac{1}{2}$
Change to exponential form.
$x^{1/2} = 9$
$(x^{1/2})^2 = 9^2$ *Square*
$x^1 = 81$
$x = 81$
Solution set: $\{81\}$

25. $\log_x 125 = -3$
Write in exponential form.
$x^{-3} = 125$
$\frac{1}{x^3} = 125$
$1 = 125(x^3)$
$\frac{1}{125} = x^3$
Take the cube root of each side.
$\sqrt[3]{\frac{1}{125}} = \sqrt[3]{x^3}$
$x = \sqrt[3]{\frac{1}{5^3}} = \frac{1}{5}$
Solution set: $\left\{\frac{1}{5}\right\}$

27. $\log_{12} x = 0$
Write in exponential form.
$12^0 = x$
$1 = x$
Solution set: $\{1\}$

29. $\log_x x = 1$
Write in exponential form.
$x^1 = x$
This equation is true for all the numbers x that are allowed as the base of a logarithm; that is, all positive numbers x, $x \neq 1$.
Solution set: $\{x | x > 0,\ x \neq 1\}$

31. $\log_x \frac{1}{25} = -2$
Write in exponential form.
$x^{-2} = \frac{1}{25}$
$\frac{1}{x^2} = \frac{1}{25}$
$x^2 = 25$ *Denominators must be equal.*
$x = \pm 5$
Reject $x = -5$ since the base of a logarithm must be positive and not equal to 1.
Solution set: $\{5\}$

33. $\log_8 32 = x$
$8^x = 32$ *Exponential form*
Write each side as a power of 2.
$(2^3)^x = 2^5$
$2^{3x} = 2^5$
$3x = 5$ *Equate exponents*
$x = \frac{5}{3}$
Solution set: $\left\{\frac{5}{3}\right\}$

35. $\log_\pi \pi^4 = x$
$\pi^x = \pi^4$ *Exponential form*
$x = 4$ *Equate exponents*
Solution set: $\{4\}$

37. $\log_6 \sqrt{216} = x$
$\log_6 216^{1/2} = x$ *Equivalent form*
$6^x = 216^{1/2}$ *Exponential form*
$6^x = (6^3)^{1/2}$ *Same base*
$6^x = 6^{3/2}$
$x = \frac{3}{2}$ *Equate exponents*
Solution set: $\left\{\frac{3}{2}\right\}$

39. $\log_4(2x + 4) = 3$
$2x + 4 = 4^3$ *Exponential form*
$2x = 64 - 4$
$2x = 60$
$x = 30$

Solution set: $\{30\}$

41. $y = \log_3 x$

Change to exponential form.

$3^y = x$

Refer to Section 10.2, Exercise 5, for the graph of $f(x) = 3^x$. Since $y = \log_3 x$ (or $3^y = x$) is the inverse of $f(x) = y = 3^x$, its graph is symmetric about the line $y = x$ to the graph of $f(x) = 3^x$. The graph can be plotted by reversing the ordered pairs in the table of values belonging to $f(x) = 3^x$.

x	$\frac{1}{9}$	$\frac{1}{3}$	1	3	9
y	-2	-1	0	1	2

Plot the points, and draw a smooth curve through them.

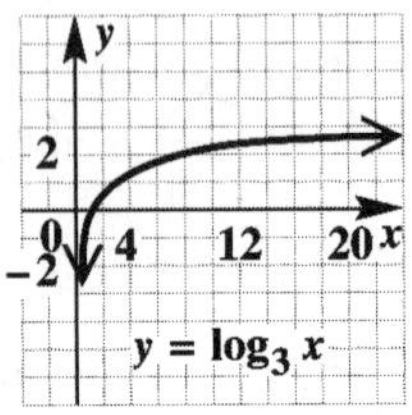

43. $y = \log_{1/3} x$

Change to exponential form.

$$\left(\frac{1}{3}\right)^y = x$$

Refer to Section 10.2, Exercise 7, for the graph of $g(x) = \left(\frac{1}{3}\right)^x$. Since $y = \log_{1/3} x$ $\left(\text{or } \left(\frac{1}{3}\right)^y = x\right)$ is the inverse of $y = \left(\frac{1}{3}\right)^x$, its graph is symmetric about the line $y = x$ to the graph of $y = \left(\frac{1}{3}\right)^x$. The graph can be plotted by reversing the ordered pairs in the table of values belonging to $g(x) = \left(\frac{1}{3}\right)^x$.

x	9	3	1	$\frac{1}{3}$	$\frac{1}{9}$
y	-2	-1	0	1	2

Plot the points, and draw a smooth curve through them.

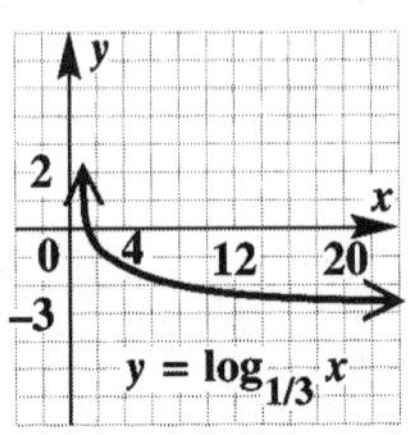

45. The number 1 is not used as a base for a logarithmic function since the function would look like $x = 1^y$ in exponential form. Then, for any real value of y, the statement $1 = 1$ would always be the result since every power of 1 is equal to 1.

47. The range of $F(x) = a^x$ is the domain of $G(x) = \log_a x$, that is, $(0, \infty)$.
The domain of $F(x) = a^x$ is the range of $G(x) = \log_a x$, that is, $(-\infty, \infty)$.

49. The values of t are on the horizontal axis, and the values of $f(t)$ are on the vertical axis. Read the value of $f(t)$ from the graph for the given value of t. At $t = 0$, $f(0) = 8$.

51. To find $f(60)$, find 60 on the t-axis, then go up to the graph and across to the $f(t)$ axis to read the value of $f(60)$. At $t = 60$, $f(60) = 24$.

53. $f(x) = 11.34 + 317.01 \log_2 x$

(a) 1984 corresponds to $x = 4$.

$$\begin{aligned} f(4) &= 11.34 + 317.01 \log_2 4 \\ &= 11.34 + 317.01(2) \\ &= 11.34 + 634.02 = 645.36 \end{aligned}$$

The model gives 645 sites for 1984.

(b) 1988 corresponds to $x = 8$.

$$\begin{aligned} f(8) &= 11.34 + 317.01 \log_2 8 \\ &= 11.34 + 317.01(3) \\ &= 11.34 + 951.03 = 962.37 \end{aligned}$$

The model gives 962 sites for 1988.

(c) 1996 corresponds to $x = 16$.

$$\begin{aligned} f(16) &= 11.34 + 317.01 \log_2 16 \\ &= 11.34 + 317.01(4) \\ &= 11.34 + 1268.04 = 1279.38 \end{aligned}$$

The model gives 1279 sites for 1996.

55. $S(t) = 100 + 30 \log_3(2t + 1)$

(a) $$\begin{aligned} S(1) &= 100 + 30 \log_3(2 \cdot 1 + 1) \\ &= 100 + 30 \log_3(3) \\ &= 100 + 30(1) = 130 \end{aligned}$$

After 1 year, the sales were 130 thousand units.

(b) $$\begin{aligned} S(13) &= 100 + 30 \log_3(2 \cdot 13 + 1) \\ &= 100 + 30 \log_3(27) \\ &= 100 + 30(3) = 190 \end{aligned}$$

After 13 years, the sales were 190 thousand units.

(c) Make a table of values, plot the points they represent, and draw a smooth curve through them.

To make the table, find values of t such that $2t + 1 = 3^k$, where $k = 0, 1, 2, 3, 4$.

k	0	1	2	3	4
3^k	1	3	9	27	81
$2t+1$	1	3	9	27	81
$2t$	0	2	8	26	80
t	0	1	4	13	40
$S(t)$	100	130	160	190	220

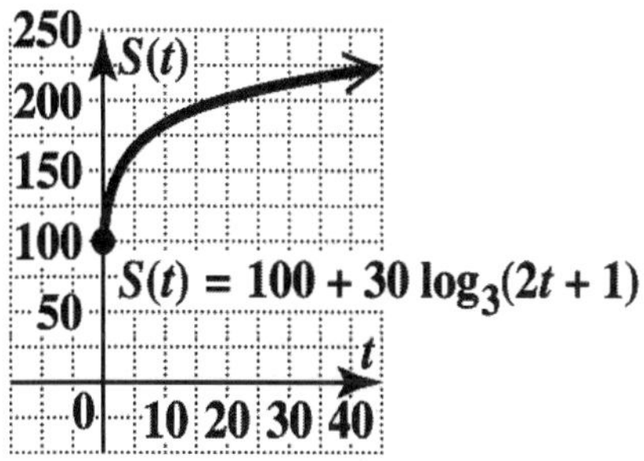

57. $B(t) = 500 \log_3 (2t+3)$

(a) For January 1990, $t = 0$.

$$\begin{aligned} B(0) &= 500 \log_3 (2 \cdot 0 + 3) \\ &= 500 \log_3 3 \\ &= 500(1) = 500 \end{aligned}$$

(b) For January 1993, $t = 3$.

$$\begin{aligned} B(3) &= 500 \log_3 (2 \cdot 3 + 3) \\ &= 500 \log_3 9 \\ &= 500(2) = 1000 \end{aligned}$$

(c) For January 2002, $t = 12$.

$$\begin{aligned} B(12) &= 500 \log_3 (2 \cdot 12 + 3) \\ &= 500 \log_3 27 \\ &= 500(3) = 1500 \end{aligned}$$

(d) Make a table for $B(t)$ when $0 \le t \le 12$ using the results of parts (a) – (c).

t	0	3	12
$B(t)$	500	1000	1500

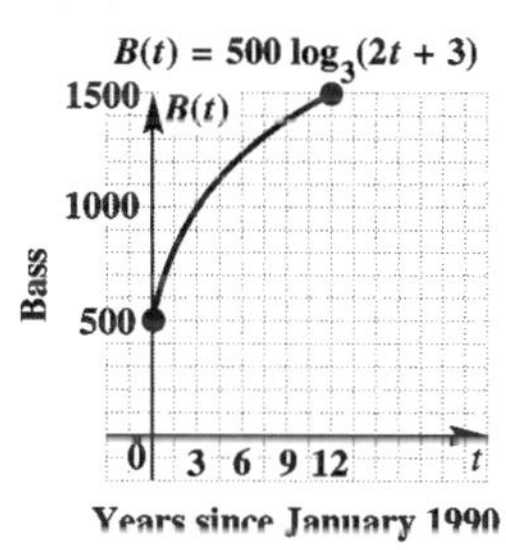

59. $g(x) = \log_3 x$

On a TI-83, assign 3^x to Y_1. Then enter

DrawInv Y_1

on the home screen to obtain the figure that follows. (See Exercise 47 in Section 10.1 for TI-83 specifics.)

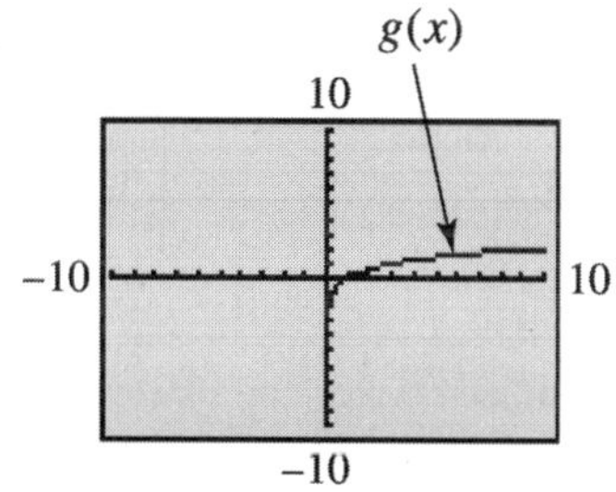

61. $g(x) = \log_{1/3} x$

Assign (1/3)^x to Y_1 and enter DrawInv Y_1.

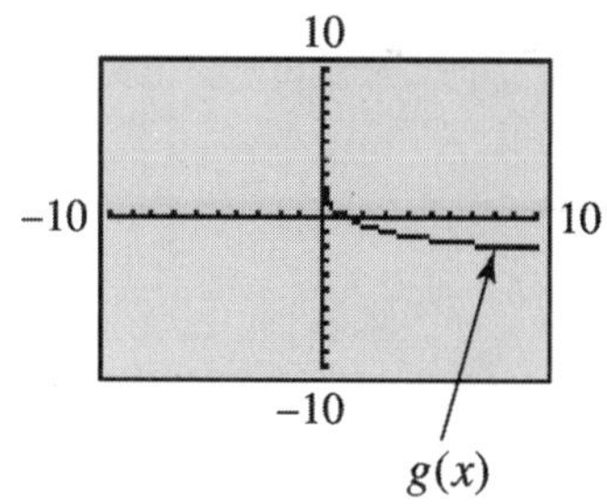

Section 10.4

1. By the product rule,

$$\log_{10} (3 \cdot 4) = \log_{10} 3 + \log_{10} 4.$$

3. By a special property (see page 636 in the text),

$$3^{\log_3 4} = 4.$$

5. By a special property (see page 636 in the text),

$$\log_3 3^4 = 4.$$

7. Use the product rule for logarithms.

$$\log_7 (4 \cdot 5) = \log_7 4 + \log_7 5$$

9. Use the quotient rule for logarithms.

$$\log_5 \frac{8}{3} = \log_5 8 - \log_5 3$$

11. Use the power rule for logarithms.

$$\log_4 6^2 = 2 \log_4 6$$

13. $\log_3 \dfrac{\sqrt[3]{4}}{x^2 y} = \log_3 \dfrac{4^{1/3}}{x^2 y}$

Use the quotient rule for logarithms.

$$= \log_3 4^{1/3} - \log_3 (x^2 y)$$

Use the product rule for logarithms.

$$= \log_3 4^{1/3} - \left(\log_3 x^2 + \log_3 y\right)$$

$$= \log_3 4^{1/3} - \log_3 x^2 - \log_3 y$$

Use the power rule for logarithms.

$$= \frac{1}{3} \log_3 4 - 2 \log_3 x - \log_3 y$$

15. $\log_3 \sqrt{\frac{xy}{5}}$

$= \log_3 \left(\frac{xy}{5}\right)^{1/2}$

$= \frac{1}{2} \log_3 \left(\frac{xy}{5}\right)$ *Power rule*

$= \frac{1}{2}\left[\log_3 (xy) - \log_3 5\right]$ *Quotient rule*

$= \frac{1}{2}\left(\log_3 x + \log_3 y - \log_3 5\right)$ *Product rule*

$= \frac{1}{2} \log_3 x + \frac{1}{2} \log_3 y - \frac{1}{2} \log_3 5$

17. $\log_2 \frac{\sqrt[3]{x} \cdot \sqrt[5]{y}}{r^2}$

$= \log_2 \frac{x^{1/3} y^{1/5}}{r^2}$

$= \log_2 \left(x^{1/3} y^{1/5}\right) - \log_2 r^2$ *Quotient rule*

$= \log_2 x^{1/3} + \log_2 y^{1/5} - \log_2 r^2$ *Product rule*

$= \frac{1}{3} \log_2 x + \frac{1}{5} \log_2 y - 2 \log_2 r$ *Power rule*

19. The distributive property tells us that the *product* $a(x + y)$ equals the sum $ax + ay$. In the notation $\log_a (x + y)$, the parentheses do not indicate multiplication. They indicate that $x + y$ is the result of raising a to some power.

21. By the product rule for logarithms,

$$\log_b x + \log_b y = \log_b xy.$$

23. By the quotient rule for logarithms,

$$\log_a m - \log_a n = \log_a \frac{m}{n}.$$

25. $\left(\log_a r - \log_a s\right) + 3 \log_a t$

Use the quotient and power rules for logarithms.

$= \log_a \frac{r}{s} + \log_a t^3$

$= \log_a \frac{rt^3}{s}$ *Product rule*

27. $3 \log_a 5 - 4 \log_a 3$

$= \log_a 5^3 - \log_a 3^4$ *Power rule*

$= \log_a \frac{5^3}{3^4}$ *Quotient rule*

$= \log_a \frac{125}{81}$

29. $\log_{10} (x + 3) + \log_{10} (x - 3)$

$= \log_{10} (x + 3)(x - 3)$ *Product rule*

$= \log_{10} (x^2 - 9)$

31. By the power rule for logarithms,

$3 \log_p x + \frac{1}{2} \log_p y - \frac{3}{2} \log_p z - 3 \log_p a$

$= \log_p x^3 + \log_p y^{1/2} - \log_p z^{3/2} - \log_p a^3$

Group the terms into sums.

$= \left(\log_p x^3 + \log_p y^{1/2}\right)$

$- \left(\log_p z^{3/2} + \log_p a^3\right)$

$= \log_p x^3 y^{1/2} - \log_p z^{3/2} a^3$ *Product rule*

$= \log_p \frac{x^3 y^{1/2}}{z^{3/2} a^3}$ *Quotient rule*

In Exercises 33–39, $\log_{10} 2 \approx .3010$ and $\log_{10} 9 \approx .9542$.

33. By the product rule for logarithms,

$$\begin{aligned} \log_{10} 18 &= \log_{10} (2 \cdot 9) \\ &= \log_{10} 2 + \log_{10} 9 \\ &\approx .3010 + .9542 \\ &= 1.2552. \end{aligned}$$

35. By the quotient rule for logarithms,

$$\begin{aligned} \log_{10} \frac{2}{9} &= \log_{10} 2 - \log_{10} 9 \\ &\approx .3010 - .9542 \\ &= -.6532. \end{aligned}$$

37. By the product and power rules for logarithms,

$$\begin{aligned} \log_{10} 36 &= \log_{10} 2^2 \cdot 9 \\ &= 2 \log_{10} 2 + \log_{10} 9 \\ &\approx 2(.3010) + .9542 \\ &= 1.5562. \end{aligned}$$

39. $\log_{10} 3 = \log_{10} 9^{1/2}$ *Rename 3*

$= \frac{1}{2} \log_{10} 9$ *Power rule*

$\approx \frac{1}{2}(.9542)$

$= .4771$

41. Note: The first printing of the text had a different logarithm.

$\log_{10} \sqrt[4]{9} = \log_{10} 9^{1/4}$

$= \frac{1}{4} \log_{10} 9$ *Power rule*

$\approx \frac{1}{4}(.9542)$

$= .23855 \approx .2386$

43. $\log_{10} 9^5 = 5 \log_{10} 9$ *Power rule*

$\approx 5(.9542)$

$= 4.7710$

45. $LS = \log_2(8+32) = \log_2 40$
$RS = \log_2 8 + \log_2 32 = \log_2(8 \cdot 32)$
$= \log_2 256$

LS $\neq$ RS, so the statement is *false.*

47. $\log_3 7 + \log_3 7^{-1} = \log_3 7 + (-1)\log_3 7$
$= 0$

The statement is *true.*

49. $\log_6 60 - \log_6 10 = \log_6 \frac{60}{10}$
$= \log_6 6 = 1$

The statement is *true.*

51. $\frac{\log_{10} 7}{\log_{10} 14} \stackrel{?}{=} \frac{1}{2}$

$2\log_{10} 7 \stackrel{?}{=} 1\log_{10}(7 \cdot 2)$ *Cross products are equal*

$2\log_{10} 7 \stackrel{?}{=} \log_{10} 7 + \log_{10} 2$

$\log_{10} 7 \stackrel{?}{=} \log_{10} 2$ *Subtract $\log_{10} 7$*

The statement is *false.*

53. The exponent of a quotient is the difference between the exponent of the numerator and the exponent of the denominator.

55. $\log_2 8 - \log_2 4 = \log_2 \frac{8}{4}$
$= \log_2 2 = 1$

57. $\log_3 81 = \log_3 3^4 = 4$

58. $\log_3 81$ is the exponent to which 3 must be raised to obtain 81.

59. Using the result from Exercise 57,

$$3^{\log_3 81} = 3^4 = 81.$$

60. $\log_2 19$ is the exponent to which 2 must be raised to obtain 19.

61. Keeping in mind the result from Exercise 59,

$$2^{\log_2 19} = 19.$$

62. To find $k^{\log_k m}$, first assume $\log_k m = y$. This means, changing to an exponential equation, $k^y = m$. Therefore,

$$k^{\log_k m} = k^y = m.$$

Section 10.5

1. Since $\log x = \log_{10} x$, the base is 10. The correct response is **C**.

3. $10^0 = 1$ and $10^1 = 10$, so $\log 1 = 0$ and $\log 10 = 1$. Thus, the value of $\log 5.6$ must lie between 0 and 1. The correct response is **C**.

5. $\log 10^{19.2} = \log_{10} 10^{19.2} = 19.2$ by the special property, $\log_b b^x = x$.

7. To four decimal places,

$$\log 43 \approx 1.6335.$$

9. $\log 328.4 \approx 2.5164$

11. $\log .0326 \approx -1.4868$

13. $\log(4.76 \times 10^9) \approx 9.6776$
On a TI-83, enter

[LOG] 4.76 [2nd] [EE] 9).

15. $\ln 7.84 \approx 2.0592$

17. $\ln .0556 \approx -2.8896$

19. $\ln 388.1 \approx 5.9613$

21. $\ln(8.59 \times e^2) \approx 4.1506$
On a TI-83, enter

[LN] 8.59 [X] [2nd] [e^x] 2)).

23. $\ln 10 \approx 2.3026$

25. **(a)** $\log 356.8 \approx 2.552424846$

(b) $\log 35.68 \approx 1.552424846$

(c) $\log 3.568 \approx 0.552424846$

(d) The whole number part of the answers (2, 1, or 0) varies, whereas the decimal part (.552424846) remains the same, indicating that the whole number part corresponds to the placement of the decimal point and the decimal part corresponds to the digits 3, 5, 6, and 8.

27. When you try to find $\log(-1)$ on a calculator, an error message is displayed. This is because the domain of the logarithmic function is $(0, \infty)$; -1 is not in the domain.

29. $pH = -\log[H_3O^+]$
$= -\log(2.5 \times 10^{-2})$
≈ 1.6

Since the pH is less than 3.0, the wetland is classified as a *bog.*

31. Ammonia has a hydronium ion concentration of 2.5×10^{-12}.
$pH = -\log[H_3O^+]$
$pH = -\log(2.5 \times 10^{-12}) \approx 11.6$

33. Grapes have a hydronium ion concentration of 5.0×10^{-5}.
$pH = -\log[H_3O^+]$
$pH = -\log(5.0 \times 10^{-5}) \approx 4.3$

35. Human blood plasma has a pH of 7.4.

$$\begin{aligned} pH &= -\log\,[H_3O^+] \\ 7.4 &= -\log\,[H_3O^+] \\ \log_{10}[H_3O^+] &= -7.4 \\ [H_3O^+] &= 10^{-7.4} \approx 4.0 \times 10^{-8} \end{aligned}$$

37. Spinach has a pH value of 5.4.

$$\begin{aligned} pH &= -\log\,[H_3O^+] \\ 5.4 &= -\log\,[H_3O^+] \\ \log_{10}[H_3O^+] &= -5.4 \\ [H_3O^+] &= 10^{-5.4} \approx 4.0 \times 10^{-6} \end{aligned}$$

39. $N(r) = -5000 \ln r$

(a) 85% (or .85)

$N(.85) = -5000 \ln .85 \approx 813 \approx 800$

The number of years since the split for 85% is about 800 years.

(b) 35% (or .35)

$N(.35) = -5000 \ln .35 \approx 5249 \approx 5200$

The number of years since the split for 35% is about 5200 years.

(c) 10% (or .10)

$N(.10) = -5000 \ln .10 \approx 11{,}513 \approx 11{,}500$

The number of years since the split for 10% is about 11,500 years.

41. $D = 10 \log\left(\dfrac{I}{I_0}\right)$

(a) $$\begin{aligned} D &= 10 \log\left(\frac{5.012 \times 10^{10} I_0}{I_0}\right) \\ &= 10 \log\,(5.012 \times 10^{10}) \approx 107 \end{aligned}$$

The average decibel level for *Armageddon* is about 107 dB.

(b) $$\begin{aligned} D &= 10 \log\left(\frac{10^{10} I_0}{I_0}\right) \\ &= 10 \log 10^{10} = 10 \cdot 10 = 100 \end{aligned}$$

The average decibel level for *Godzilla* is 100 dB.

(c) $$\begin{aligned} D &= 10 \log\left(\frac{6{,}310{,}000{,}000\, I_0}{I_0}\right) \\ &= 10 \log\,(6{,}310{,}000{,}000) \approx 98 \end{aligned}$$

The average decibel level for *Saving Private Ryan* is about 98 dB.

43. $f(x) = -1317 + 304 \ln x$

(a) $x = 1998 - 1900 = 98$

$f(98) = -1317 + 304 \ln (98) \approx 77$

The prediction for 1998 is 77%.

(b) $$\begin{aligned} f(x) &= 50 \\ 50 &= -1317 + 304 \ln x \\ 1367 &= 304 \ln x \\ \ln x &= \frac{1367}{304} \\ e^{\ln x} &= e^{1367/304} \\ x &\approx 89.72 \end{aligned}$$

The outpatient surgeries reached 50% in 1989.

45. $T = -.642 - 189 \ln (1 - p)$

(a) $$\begin{aligned} T &= -.642 - 189 \ln (1 - .25) \\ &= -.642 - 189 \ln (.75) \approx 53.73 \end{aligned}$$

About \$54 per ton will reduce emissions by 25%.

(b) If $p = 0$, then $\ln (1 - p) = \ln 1 = 0$, so T would be negative. If $p = 1$, then $\ln (1 - p) = \ln 0$, but the domain of $\ln x$ is $(0, \infty)$.

47. The change-of-base rule is

$$\log_a x = \frac{\log_b x}{\log_b a}.$$

Use common logarithms ($b = 10$).

$$\log_3 12 = \frac{\log_{10} 12}{\log_{10} 3} = \frac{\log 12}{\log 3} \approx 2.2619$$

49. Use natural logarithms ($b = e$).

$$\log_5 3 = \frac{\log_e 3}{\log_e 5} = \frac{\ln 3}{\ln 5} \approx .6826$$

51. $\log_3 \sqrt{2} = \dfrac{\ln \sqrt{2}}{\ln 3} \approx .3155$

53. $\log_\pi e = \dfrac{\ln e}{\ln \pi} = \dfrac{1}{\ln \pi} \approx .8736$

55. $\log_e 12 = \dfrac{\ln 12}{\ln e} = \dfrac{\ln 12}{1} \approx 2.4849$

57. Let $m =$ the number of letters in your first name and $n =$ the number of letters in your last name.

Answers will vary, but suppose the name is Paul Bunyan, with $m = 4$ and $n = 6$.

(a) $\log_m n = \log_4 6$ is the exponent to which 4 must be raised in order to obtain 6.

(b) Use the change-of-base rule.

$$\begin{aligned} \log_4 6 &= \frac{\log 6}{\log 4} \\ &\approx 1.29248125 \end{aligned}$$

(c) Here, $m = 4$. Use the power key (y^x, x^y, ^) on your calculator.

$$4^{1.29248125} \approx 6$$

The result is 6, the value of n.

59. $f(x) = 11.34 + 317.01 \log_2 x$

1998 corresponds to $x = 18$.

$$\begin{aligned} f(18) &= 11.34 + 317.01 \log_2 18 \\ &= 11.34 + 317.01\left(\frac{\log 18}{\log 2}\right) \\ &\approx 11.34 + 317.01(4.1699) \\ &\approx 1333.2 \end{aligned}$$

The model gives about 1333 sites for 1998.

61. To graph $g(x) = \log_3 x$, assign either

$\frac{\log X}{\log 3}$ or $\frac{\ln X}{\ln 3}$ to Y_1.

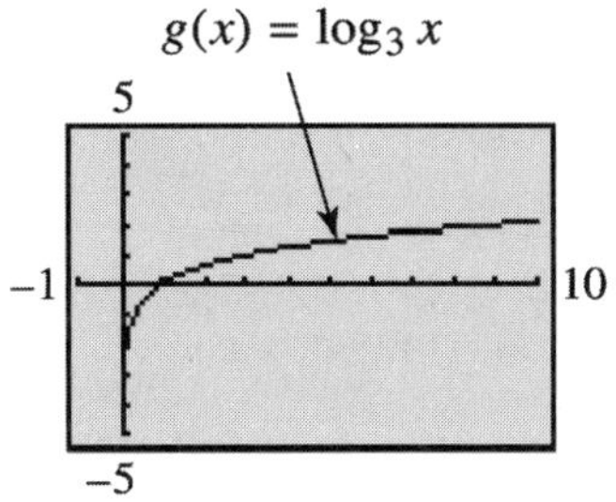

63. To graph $g(x) = \log_{1/3} x$, assign either

$\frac{\log X}{\log \frac{1}{3}}$ or $\frac{\ln X}{\ln \frac{1}{3}}$ to Y_1.

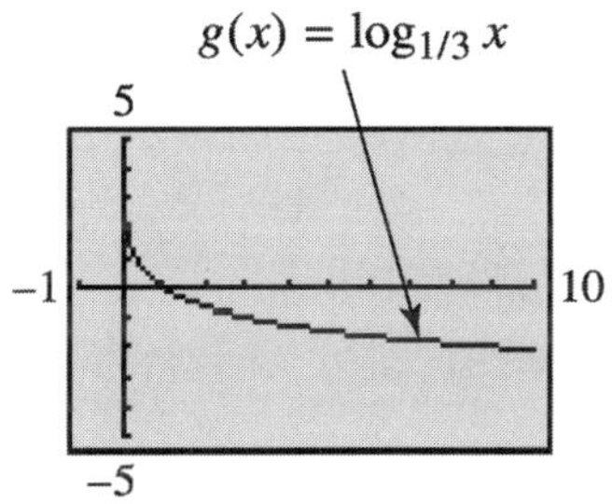

Section 10.6

1. $5^x = 125$

$\log 5^x = \log 125$

2. $x \log 5 = \log 125$

3. $$\frac{x \log 5}{\log 5} = \frac{\log 125}{\log 5}$$

$$x = \frac{\log 125}{\log 5}$$

4. $$\frac{\log 125}{\log 5} = 3 \text{ (from calculator)}$$

Solution set: $\{3\}$

5. $7^x = 5$

Take the logarithm of each side.

$\log 7^x = \log 5$

Use the power rule for logarithms.

$x \log 7 = \log 5$

$$x = \frac{\log 5}{\log 7} \approx .827$$

Solution set: $\{.827\}$

7.

$$\begin{aligned} 9^{-x+2} &= 13 \\ \log 9^{-x+2} &= \log 13 \\ (-x+2)\log 9 &= \log 13 \ (*) \\ -x \log 9 + 2\log 9 &= \log 13 \\ -x \log 9 &= \log 13 - 2\log 9 \\ x \log 9 &= 2\log 9 - \log 13 \\ x &= \frac{2\log 9 - \log 13}{\log 9} \\ &\approx .833 \end{aligned}$$

(*) Alternative solution steps:

$$\begin{aligned} (-x+2)\log 9 &= \log 13 \\ -x + 2 &= \frac{\log 13}{\log 9} \\ 2 - \frac{\log 13}{\log 9} &= x \end{aligned}$$

Solution set: $\{.833\}$

9.

$$\begin{aligned} 3^{2x} &= 14 \\ \log 3^{2x} &= \log 14 \\ 2x \log 3 &= \log 14 \\ x &= \frac{\log 14}{2\log 3} \approx 1.201 \end{aligned}$$

Solution set: $\{1.201\}$

11.

$$\begin{aligned} 2^{x+3} &= 5^x \\ \log 2^{x+3} &= \log 5^x \\ (x+3)\log 2 &= x \log 5 \\ x \log 2 + 3\log 2 &= x \log 5 \end{aligned}$$

Get x-terms on one side.

$$\begin{aligned} x \log 2 - x \log 5 &= -3\log 2 \\ x(\log 2 - \log 5) &= -3\log 2 \quad \textit{Factor out } x. \\ x &= \frac{-3\log 2}{\log 2 - \log 5} \\ &\approx 2.269 \end{aligned}$$

Solution set: $\{2.269\}$

13.
$$\begin{aligned} 2^{x+3} &= 3^{x-4} \\ \log 2^{x+3} &= \log 3^{x-4} \\ (x+3)\log 2 &= (x-4)\log 3 \\ x\log 2 + 3\log 2 &= x\log 3 - 4\log 3 \end{aligned}$$
Distributive property

Get x-terms on one side.

$$x\log 2 - x\log 3 = -3\log 2 - 4\log 3$$

Factor out x.

$$\begin{aligned} x(\log 2 - \log 3) &= -3\log 2 - 4\log 3 \\ x &= \frac{-3\log 2 - 4\log 3}{\log 2 - \log 3} \\ &\approx 15.967 \end{aligned}$$

Solution set: $\{15.967\}$

15.
$$\begin{aligned} e^{.012x} &= 23 \\ \ln e^{.012x} &= \ln 23 \\ .012x(\ln e) &= \ln 23 \\ .012x &= \ln 23 \quad \textit{ln e = 1} \\ x &= \frac{\ln 23}{.012} \approx 261.291 \end{aligned}$$

Solution set: $\{261.291\}$

17.
$$\begin{aligned} e^{-.205x} &= 9 \\ \ln e^{-.205x} &= \ln 9 \\ -.205x(\ln e) &= \ln 9 \\ -.205x &= \ln 9 \quad \textit{ln e = 1} \\ x &= \frac{\ln 9}{-.205} \approx -10.718 \end{aligned}$$

Solution set: $\{-10.718\}$

19.
$$\begin{aligned} \ln e^{3x} &= 9 \\ 3x &= 9 \\ x &= 3 \end{aligned}$$

Solution set: $\{3\}$

21.
$$\begin{aligned} \ln e^{.45x} &= \sqrt{7} \\ .45x &= \sqrt{7} \\ x &= \frac{\sqrt{7}}{.45} \approx 5.879 \end{aligned}$$

Solution set: $\{5.879\}$

23. Let's try Exercise 14.

$$\begin{aligned} e^{.006x} &= 30 \\ \log e^{.006x} &= \log 30 \\ .006x(\log e) &= \log 30 \\ x &= \frac{\log 30}{.006\log e} \approx 566.866 \end{aligned}$$

The natural logarithm is easier because $\ln e = 1$, whereas $\log e$ needs to be calculated.

25.
$$\begin{aligned} \log_3(6x+5) &= 2 \\ 6x+5 &= 3^2 \quad \textit{Exponential form} \\ 6x+5 &= 9 \\ 6x &= 4 \\ x &= \frac{4}{6} = \frac{2}{3} \end{aligned}$$

Solution set: $\left\{\frac{2}{3}\right\}$

27.
$$\begin{aligned} \log_2(2x-1) &= 5 \\ 2x-1 &= 2^5 \quad \textit{Exponential form} \\ 2x-1 &= 32 \\ 2x &= 33 \\ x &= \frac{33}{2} \end{aligned}$$

Solution set: $\left\{\frac{33}{2}\right\}$

29.
$$\begin{aligned} \log_7(x+1)^3 &= 2 \\ (x+1)^3 &= 7^2 \quad \textit{Exponential form} \\ x+1 &= \sqrt[3]{49} \quad \textit{Cube root} \\ x &= -1+\sqrt[3]{49} \end{aligned}$$

Solution set: $\left\{-1+\sqrt[3]{49}\right\}$

31. 2 cannot be a solution because $\log(2-3) = \log(-1)$, and -1 is not in the domain of $\log x$.

33.
$$\begin{aligned} \log(6x+1) &= \log 3 \\ 6x+1 &= 3 \quad \textit{Property 4} \\ 6x &= 2 \\ x &= \frac{2}{6} = \frac{1}{3} \end{aligned}$$

Solution set: $\left\{\frac{1}{3}\right\}$

35.
$$\begin{aligned} \log_5(3t+2) - \log_5 t &= \log_5 4 \\ \log_5 \frac{3t+2}{t} &= \log_5 4 \\ \frac{3t+2}{t} &= 4 \\ 3t+2 &= 4t \\ 2 &= t \end{aligned}$$

Solution set: $\{2\}$

37.
$$\begin{aligned} \log 4x - \log(x-3) &= \log 2 \\ \log\frac{4x}{x-3} &= \log 2 \\ \frac{4x}{x-3} &= 2 \\ 4x &= 2(x-3) \\ 4x &= 2x-6 \\ 2x &= -6 \\ x &= -3 \end{aligned}$$

Reject $x = -3$, because $4x = -12$, which yields an equation in which the logarithm of a negative number must be found.

Solution set: $\emptyset$

39.
$$\begin{aligned}\log_2 x + \log_2 (x-7) &= 3\\ \log_2 [x(x-7)] &= 3\\ x(x-7) &= 2^3 \quad \textit{Exponential form}\\ x^2 - 7x &= 8\\ x^2 - 7x - 8 &= 0\\ (x-8)(x+1) &= 0\end{aligned}$$

$x - 8 = 0$ or $x + 1 = 0$

$x = 8$ or $x = -1$

Reject $x = -1$, because it yields an equation in which the logarithm of a negative number must be found.

Solution set: $\{8\}$

41.
$$\begin{aligned}\log 5x - \log (2x-1) &= \log 4\\ \log \frac{5x}{2x-1} &= \log 4\\ \frac{5x}{2x-1} &= 4\\ 5x &= 8x - 4\\ 4 &= 3x\\ \frac{4}{3} &= x\end{aligned}$$

Solution set: $\left\{\frac{4}{3}\right\}$

43.
$$\begin{aligned}\log_2 x + \log_2 (x-6) &= 4\\ \log_2 [x(x-6)] &= 4\\ x(x-6) &= 2^4 \quad \textit{Exponential form}\\ x^2 - 6x &= 16\\ x^2 - 6x - 16 &= 0\\ (x-8)(x+2) &= 0\end{aligned}$$

$x - 8 = 0$ or $x + 2 = 0$

$x = 8$ or $x = -2$

Reject $x = -2$, because it yields an equation in which the logarithm of a negative number must be found.

Solution set: $\{8\}$

45. **(a)** Use the formula $A = P\left(1 + \frac{r}{n}\right)^{nt}$ with $P = 2000$, $r = .04$, $n = 4$, and $t = 6$.

$$\begin{aligned}A &= 2000\left(1 + \frac{.04}{4}\right)^{4 \cdot 6}\\ &= 2000(1.01)^{24} \approx 2539.47\end{aligned}$$

The account will contain \$2539.47.

(b)
$$\begin{aligned}3000 &= 2000\left(1 + \frac{.04}{4}\right)^{4t}\\ \frac{3000}{2000} &= (1.01)^{4t}\\ \log\left(\frac{3}{2}\right) &= \log (1.01)^{4t}\\ \log\left(\frac{3}{2}\right) &= 4t \log (1.01)\end{aligned}$$

$$t = \frac{\log\left(\frac{3}{2}\right)}{4 \log (1.01)} \approx 10.2$$

It will take about 10.2 years for the account to grow to \$3000.

47. **(a)** Use the formula $A = Pe^{rt}$ with $P = 4000$, $r = .035$, and $t = 6$.

$$\begin{aligned}A &= 4000e^{(.035)(6)}\\ &= 4000e^{.21} \approx 4934.71\end{aligned}$$

There will be \$4934.71 in the account.

(b) If the initial amount doubles, then $A = 2P$, or \$8000.

$$\begin{aligned}8000 &= 4000e^{.035t}\\ 2 &= e^{.035t} \quad \textit{Divide by 4000.}\\ \ln 2 &= \ln e^{.035t}\\ \ln 2 &= .035t\\ t &= \frac{\ln 2}{.035} \approx 19.8\end{aligned}$$

The initial amount will double in about 19.8 years.

49. Use $A = P\left(1 + \frac{r}{n}\right)^{nt}$ with $P = 5000$, $r = .07$, and $t = 12$.

(a) If the interest is compounded annually, $n = 1$.

$$\begin{aligned}A &= 5000\left(1 + \frac{.07}{1}\right)^{1 \cdot 12}\\ &= 5000(1.07)^{12} \approx 11,260.96\end{aligned}$$

There will be \$11,260.96 in the account.

(b) If the interest is compounded semiannually, $n = 2$.

$$\begin{aligned}A &= 5000\left(1 + \frac{.07}{2}\right)^{2 \cdot 12}\\ &= 5000(1.035)^{24} \approx 11,416.64\end{aligned}$$

There will be \$11,416.64 in the account.

(c) If the interest is compounded quarterly, $n = 4$.

$$\begin{aligned}A &= 5000\left(1 + \frac{.07}{4}\right)^{4 \cdot 12}\\ &= 5000(1.0175)^{48} \approx 11,497.99\end{aligned}$$

There will be \$11,497.99 in the account.

(d) If the interest is compounded daily, $n = 365$.

$$\begin{aligned}A &= 5000\left(1 + \frac{.07}{365}\right)^{365 \cdot 12}\\ &\approx 11,580.90\end{aligned}$$

There will be \$11,580.90 in the account.

(e) Use the continuous compound interest formula.

$$A = Pe^{rt}$$
$$A = 5000e^{.07(12)}$$
$$= 5000e^{.84} \approx 11,581.83$$

There will be $11,581.83 in the account.

51. In the continuous compound interest formula, let $A = 1850$, $r = .065$, and $t = 40$.

$$A = Pe^{rt}$$
$$1850 = Pe^{.065(40)}$$
$$1850 = Pe^{2.6}$$
$$P = \frac{1850}{e^{2.6}} \approx 137.41$$

Deposit $137.41 today.

53. $P(x) = 70,967e^{.0526x}$

(a) 1987 corresponds to $x = 2$.

$$P(2) = 70,967e^{.0526(2)}$$
$$\approx 78,839.56$$

The approximate expenditures for 1987 is 78,840 million dollars.

(b) 1990 corresponds to $x = 5$.

$$P(5) \approx 92,316 \text{ million dollars}$$

(c) 1993 corresponds to $x = 8$.

$$P(8) \approx 108,095 \text{ million dollars}$$

(d) 1985 corresponds to $x = 0$.

$$P(0) = 70,967 \text{ million dollars}$$

55. $B(x) = 8768e^{.072x}$
1998 corresponds to $x = 18$.
$B(18) = 8768e^{.072(18)}$
$\approx 32,044$
The approximate consumer expenditures for 1998 is 32,044 million dollars.

57. $A(t) = 2.00e^{-.053t}$

(a) Let $t = 4$.

$$A(4) = 2.00e^{-.053(4)}$$
$$= 2.00e^{-.212}$$
$$\approx 1.62$$

About 1.62 grams would be present.

(b) $A(10) = 2.00e^{-.053(10)}$
≈ 1.18

About 1.18 grams would be present.

(c) $A(20) = 2.00e^{-.053(20)}$
$\approx .69$

About .69 grams would be present.

(d) The initial amount is the amount $A(t)$ present at time $t = 0$.

$$A(0) = 2.00e^{-.053(0)}$$
$$= 2.00e^{0} = 2.00(1) = 2.00$$

Initially, 2.00 grams were present.

59. **(a)** Find $A(t) = 400e^{-.032t}$ when $t = 25$.

$$A(25) = 400e^{-.032(25)}$$
$$= 400e^{-.8} \approx 179.73$$

About 180 grams of lead will be left.

(b) Use $A(t) = 400e^{-.032t}$, with $A(t) = \frac{1}{2}(400) = 200$.

$$200 = 400e^{-.032t}$$
$$\frac{200}{400} = e^{-.032t}$$
$$.5 = e^{-.032t}$$
$$\ln .5 = \ln e^{-.032t}$$
$$\ln .5 = -.032t(\ln e)$$
$$t = \frac{\ln .5}{-.032} \approx 21.66$$

It would take about 21.66 years for the sample to decay to half its original amount.

61.

$$P(x) = 70,967e^{.0526x}$$
$$133,500 = 70,967e^{.0526x}$$
$$e^{.0526x} = \frac{133,500}{70,967}$$
$$\ln e^{.0526x} = \ln \frac{133,500}{70,967}$$
$$.0526x\,(\ln e) = \ln \frac{133,500}{70,967}$$
$$.0526x = \ln \frac{133,500}{70,967}$$
$$x = \frac{\ln \frac{133,500}{70,967}}{.0526} \approx 12.0$$

Since $x = 0$ corresponds to 1985, $x = 12$ corresponds to 1997.

63. $f(t) = 300e^{.4t}$
The original number of ants, at $t = 0$, is 300. Replace $f(t)$ with 600, (double the number of ants), and solve for t.

$$600 = 300e^{.4t}$$
$$2 = e^{.4t}$$
$$\ln 2 = \ln e^{.4t}$$
$$\ln 2 = .4t\,(\ln e)$$
$$\ln 2 = .4t$$
$$t = \frac{\ln 2}{.4} \approx 1.733$$

It will take about 1.733 days for the number of ants to double.

65. The display means that after 250 years, approximately 2.9187487 grams of the original sample of radium-226 remain.

67. **(a)** From the screen:

$$\log e = \log_{10} e$$
$$\ln 10 = \log_e 10$$

So from the given property of logarithms,

$$\log_b x = \frac{1}{\log_x b},$$

we see that

$$\log_e 10 = \frac{1}{\log_{10} e}.$$

Thus, a decimal representation for

$$\frac{1}{.4342944819},$$

which represents $\frac{1}{\log_{10} e}$, must be $\log_e 10$, or

$$2.302585093.$$

(b) As in part (a), $\frac{1}{\log_e 10}$ is represented by

$$\frac{1}{2.302585093},$$

so its value is $\log_{10} e$ or

$$.4342944819.$$

Chapter 10 Review Exercises

1. Since a horizontal line intersects the graph in two points, the function is not one-to-one.

2. Since every horizontal line intersects the graph in no more than one point, the function is one-to-one.

3. This function is not one-to-one because two sodas in the list have 41 mg of caffeine.

4. The function $f(x) = -3x + 7$ is a linear function. By the horizontal line test, it is a one-to-one function. To find the inverse, replace $f(x)$ with y.

$$y = -3x + 7$$

Interchange x and y.

$$x = -3y + 7$$

Solve for y.

$$x - 7 = -3y$$

$$\frac{x-7}{-3} = y \quad \text{or} \quad \frac{7-x}{3} = y$$

Replace y with $f^{-1}(x)$.

$$f^{-1}(x) = \frac{x-7}{-3} \quad \text{or} \quad f^{-1}(x) = \frac{7-x}{3}$$

5. $f(x) = \sqrt[3]{6x-4}$
The cube root causes each value of x to be matched with only one value of $f(x)$. The function is one-to-one.
To find the inverse, replace $f(x)$ with y.

$$y = \sqrt[3]{6x-4}$$

Cube both sides.

$$y^3 = 6x - 4$$

Interchange x and y.

$$x^3 = 6y - 4$$

Solve for y.

$$x^3 + 4 = 6y$$

$$\frac{x^3+4}{6} = y$$

Replace y with $f^{-1}(x)$

$$\frac{x^3+4}{6} = f^{-1}(x)$$

6. $f(x) = -x^2 + 3$
This is an equation of a vertical parabola which opens down.
Since a horizontal line will intersect the graph in two points, the function is not one-to-one.

7. The graph is a linear function through $(0, 1)$ and $(3, 0)$. The graph of $f^{-1}(x)$ will include the points $(1, 0)$ and $(0, 3)$, found by interchanging x and y. Plot these points, and draw a straight line through them.

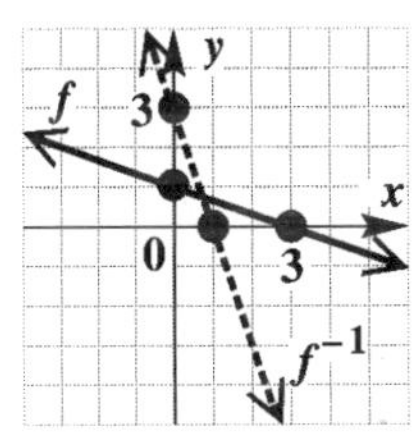

8. The graph is a curve through $(1, 2)$, $(0, 1)$, and $\left(-1, \frac{1}{2}\right)$. Interchange x and y to get $(2, 1)$, $(1, 0)$, and $\left(\frac{1}{2}, -1\right)$, which are on the graph of $f^{-1}(x)$. Plot these points, and draw a smooth curve through them.

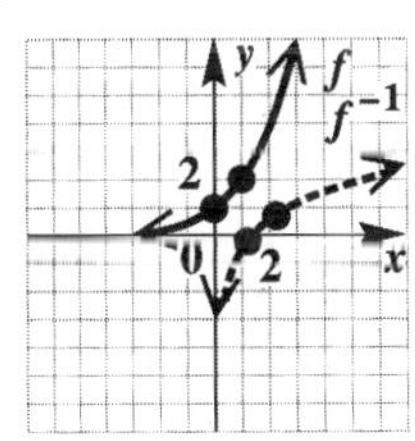

9. $f(x) = 3^x$

Make a table of values.

x	-2	-1	0	1	2
$f(x)$	$\frac{1}{9}$	$\frac{1}{3}$	1	3	9

Plot the points from the table and draw a smooth curve through them.

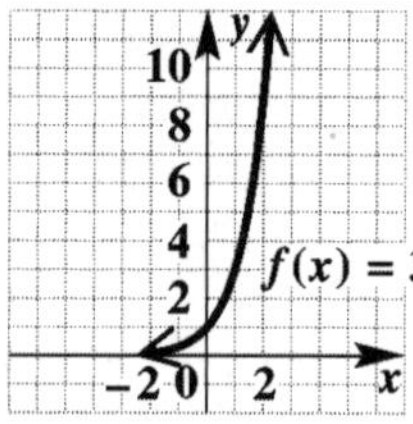

10. $f(x) = \left(\frac{1}{3}\right)^x$

Make a table of values.

x	-2	-1	0	1	2
$f(x)$	9	3	1	$\frac{1}{3}$	$\frac{1}{9}$

Plot the points from the table and draw a smooth curve through them.

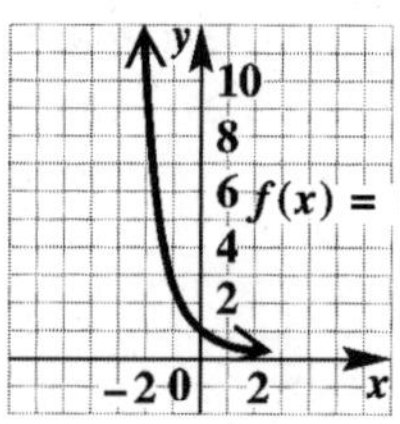

11. $y = 3^{x+1}$

Make a table of values.

x	-3	-2	-1	0	1
y	$\frac{1}{9}$	$\frac{1}{3}$	1	3	9

Plot the points from the table and draw a smooth curve through them.

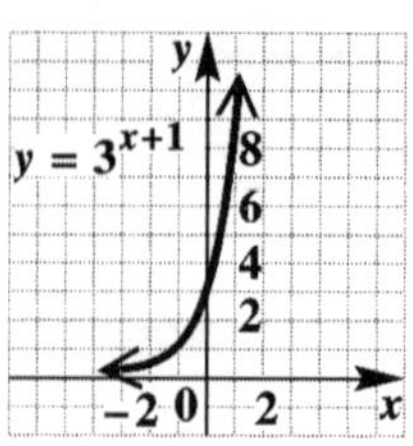

12. $y = 2^{2x+3}$

Make a table of values.

x	-2	$-\frac{3}{2}$	-1	0	$\frac{1}{2}$
y	$\frac{1}{2}$	1	2	8	16

Plot the points from the table and draw a smooth curve through them.

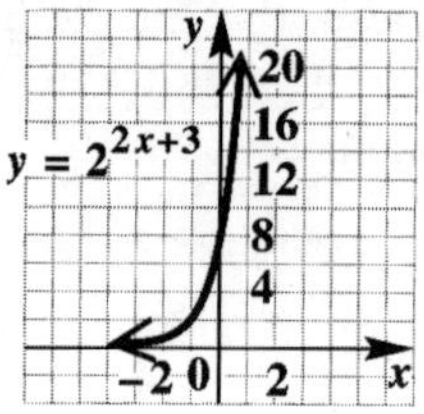

13. $4^{3x} = 8^{x+4}$

Write each side as a power of 2.

$$(2^2)^{3x} = (2^3)^{(x+4)}$$
$$2^{6x} = 2^{(3x+12)}$$
$$6x = 3x + 12 \quad \textit{Equate exponents}$$
$$3x = 12$$
$$x = 4$$

Solution set: $\{4\}$

14. $\left(\frac{1}{27}\right)^{x-1} = 9^{2x}$

$$\left[\left(\frac{1}{3}\right)^3\right]^{x-1} = (3^2)^{2x}$$

Write each side as a power of 3.

$$(3^{-3})^{x-1} = (3^2)^{2x}$$
$$3^{(-3x+3)} = 3^{4x}$$
$$-3x + 3 = 4x \quad \textit{Equate exponents}$$
$$3 = 7x$$
$$\frac{3}{7} = x$$

Solution set: $\left\{\frac{3}{7}\right\}$

15. $W(x) = .67(1.123)^x$

(a) $x = 1965 - 1960 = 5$

$$W(5) = .67(1.123)^5 \approx 1.2$$

In 1965, the approximation is 1.2 million tons.

(b) $x = 1975 - 1960 = 15$

$$W(15) = .67(1.123)^{15} \approx 3.8$$

In 1975, the approximation is 3.8 million tons.

(c) $x = 1990 - 1960 = 30$

$$W(30) = .67(1.123)^{30} \approx 21.8$$

In 1990, the approximation is 21.8 million tons.

16. $g(x) = \log_3 x$

Replace $g(x)$ with y, and write in exponential form.

$$y = \log_3 x$$
$$3^y = x$$

Make a table of values. Since $x = 3^y$ is the inverse of $f(x) = y = 3^x$ in Exercise 9, simply reverse the ordered pairs in the table of values belonging to $f(x) = 3^x$.

x	$\frac{1}{9}$	$\frac{1}{3}$	1	3	9
y	-2	-1	0	1	2

Plot the points from the table and draw a smooth curve through them.

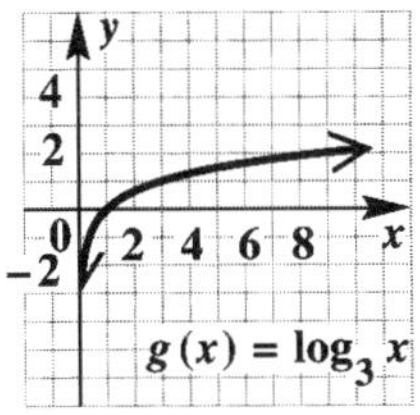

17. $g(x) = \log_{1/3} x$
Replace $g(x)$ with y, and write in exponential form.

$$y = \log_{1/3} x$$
$$\left(\frac{1}{3}\right)^y = x$$

Make a table of values. Since $x = \left(\frac{1}{3}\right)^y$ is the inverse of $f(x) = y = \left(\frac{1}{3}\right)^x$ in Exercise 10, simply reverse the ordered pairs in the table of values belonging to $f(x) = \left(\frac{1}{3}\right)^x$.

x	9	3	1	$\frac{1}{3}$	$\frac{1}{9}$
y	-2	-1	0	1	2

Plot the points from the table and draw a smooth curve through them.

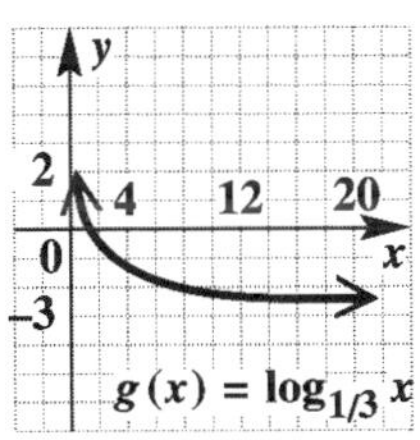

18. $\log_8 64 = x$
$8^x = 64$ *Exponential form*
Write each side as a power of 8.
$8^x = 8^2$
$x = 2$ *Equate exponents*
Solution set: $\{2\}$

19. $\log_2 \sqrt{8} = x$
$2^x = \sqrt{8}$ *Exponential form*
$2^x = 8^{1/2}$
Write each side as a power of 2.
$2^x = (2^3)^{1/2}$
$2^x = 2^{3/2}$
$x = \frac{3}{2}$ *Equate exponents*
Solution set: $\left\{\frac{3}{2}\right\}$

20. $\log_x \left(\frac{1}{49}\right) = -2$
$x^{-2} = \frac{1}{49}$ *Exponential form*
$\frac{1}{x^2} = \frac{1}{49}$
$x^2 = 49$
$x = \pm 7$

Since x is the base, we cannot have a negative number.
Solution set: $\{7\}$

21. $\log_4 x = \frac{3}{2}$
$x = 4^{3/2}$ *Exponential form*
$x = \left(\sqrt{4}\right)^3 = 2^3 = 8$
Solution set: $\{8\}$

22. $\log_k 4 = 1$
$k^1 = 4$ *Exponential form*
$k = 4$
Solution set: $\{4\}$

23. $\log_b b^2 = 2$
$b^2 = b^2$ *Exponential form*
This is an identity. Thus, b can be any real number, $b > 0$ and $b \neq 1$.
Solution set: $\{b | b > 0, b \neq 1\}$

24. $\log_b a$ is the exponent to which b must be raised to obtain a.

25. From Exercise 24,

$$b^{\log_b a} = a.$$

26. $S(x) = 100 \log_2 (x + 2)$
(a) When $x = 6$,

$$S(x) = 100 \log_2 (x + 2)$$
$$= 100(3) = 300.$$

After 6 weeks the sales were 300 thousand dollars or $300, 000.

(b) To graph the function, make a table of values that includes the ordered pair from above.

x	0	2	6
$S(x)$	100	200	300

Plot the ordered pairs and draw the graph through them.

continued

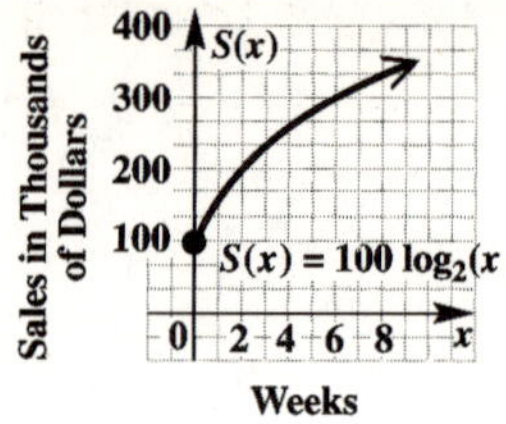

27. $\log_2 3xy^2$

$= \log_2 3 + \log_2 x + \log_2 y^2$ *Product rule*

$= \log_2 3 + \log_2 x + 2\log_2 y$ *Power rule*

28. $\log_4 \dfrac{\sqrt{x} \cdot w^2}{z}$

$= \log_4 (\sqrt{x} \cdot w^2) - \log_4 z$ *Quotient rule*

$= \log_4 x^{1/2} + \log_4 w^2 - \log_4 z$ *Product rule*

$= \dfrac{1}{2}\log_4 x + 2\log_4 w - \log_4 z$ *Power rule*

29. $\log_b 3 + \log_b x - 2\log_b y$

Use the product and power rules for logarithms.

$= \log_b (3 \cdot x) - \log_b y^2$

$= \log_b \dfrac{3x}{y^2}$ *Quotient rule*

30. $\log_3 (x+7) - \log_3 (4x+6)$

$= \log_3 \left(\dfrac{x+7}{4x+6}\right)$ *Quotient rule*

31. $\log 28.9 \approx 1.4609$

32. $\log .257 \approx -.5901$

33. $\ln 28.9 \approx 3.3638$

34. $\ln .257 \approx -1.3587$

35. $\log_{16} 13 = \dfrac{\log 13}{\log 16} \approx .9251$

36. $\log_4 12 = \dfrac{\log 12}{\log 4} \approx 1.7925$

37. Milk has a hydronium ion concentration of 4.0×10^{-7}.

$\text{pH} = -\log [\text{H}_3\text{O}^+]$

$\text{pH} = -\log (4.0 \times 10^{-7}) \approx 6.4$

38. Crackers have a hydronium ion concentration of 3.8×10^{-9}.

$\text{pH} = -\log [\text{H}_3\text{O}^+]$

$\text{pH} = -\log (3.8 \times 10^{-9}) \approx 8.4$

39. Orange juice has a pH of 4.6.

$$\text{pH} = -\log [\text{H}_3\text{O}^+]$$
$$4.6 = -\log [\text{H}_3\text{O}^+]$$
$$\log_{10} [\text{H}_3\text{O}^+] = -4.6$$
$$[\text{H}_3\text{O}^+] = 10^{-4.6} \approx 2.5 \times 10^{-5}$$

40. $Q(t) = 500e^{-.05t}$

(a) Let $t = 0$.

$Q(0) = 500e^{-.05(0)}$

$= 500e^0 = 500(1) = 500$

There are 500 grams.

(b) Let $t = 4$.

$Q(4) = 500e^{-.05(4)}$

$= 500e^{-.2} \approx 409.4$

There will be about 409 grams in 4 days.

41. $t(r) = \dfrac{\ln 2}{\ln (1+r)}$

(a) $4\% = .04$; $t(.04) = \dfrac{\ln 2}{\ln (1+.04)} \approx 18$

At 4%, it would take about 18 years.

(b) $6\% = .06$; $t(.06) = \dfrac{\ln 2}{\ln (1+.06)} \approx 12$

At 6%, it would take about 12 years.

(c) $10\% = .10$; $t(.10) = \dfrac{\ln 2}{\ln (1+.10)} \approx 7$

At 10%, it would take about 7 years.

(d) $12\% = .12$; $t(.12) = \dfrac{\ln 2}{\ln (1+.12)} \approx 6$

At 12%, it would take about 6 years.

(e) Each comparison shows approximately the same number. For example, in part (a) the doubling time is 18 yr (rounded) and $\frac{72}{4} = 18$. Thus, the formula $t = \dfrac{72}{100r}$ (called the *rule of 72*) is an excellent approximation of the doubling time formula. (It is used by bankers for that purpose.)

42. $g(x) = -650 + 143 \ln x$

(a) $g(92) = -650 + 143 \ln 92 \approx -3.4$

$g(99) = -650 + 143 \ln 99 \approx 7.1$

(b) $g(92)$ agrees closely with the y-value from the graph of about -3; $g(99)$ also is reasonably close to the y-value from the graph of about 7.5.

43.
$$\begin{aligned} 3^x &= 9.42 \\ \log 3^x &= \log 9.42 \\ x \log 3 &= \log 9.42 \\ x &= \frac{\log 9.42}{\log 3} \approx 2.042 \end{aligned}$$
Solution set: $\{2.042\}$

44.
$$\begin{aligned} 2^{x-1} &= 15 \\ \log 2^{x-1} &= \log 15 \\ (x-1)\log 2 &= \log 15 \\ x - 1 &= \frac{\log 15}{\log 2} \\ x &= \frac{\log 15}{\log 2} + 1 \approx 4.907 \end{aligned}$$
Solution set: $\{4.907\}$

45.
$$e^{.06x} = 3$$
Take base e logarithms on both sides.
$$\begin{aligned} \ln e^{.06x} &= \ln 3 \\ .06x \ln e &= \ln 3 \\ .06x &= \ln 3 \qquad \textit{ln e = 1} \\ x &= \frac{\ln 3}{.06} \approx 18.310 \end{aligned}$$
Solution set: $\{18.310\}$

46.
$$\begin{aligned} \log_3 (9x + 8) &= 2 \\ 9x + 8 &= 3^2 \qquad \textit{Exponential form} \\ 9x + 8 &= 9 \\ 9x &= 1 \\ x &= \frac{1}{9} \end{aligned}$$
Solution set: $\left\{\frac{1}{9}\right\}$

47. $\log_5 (y + 6)^3 = 2$
Change to exponential form.
$$\begin{aligned} (y+6)^3 &= 5^2 \\ (y+6)^3 &= 25 \end{aligned}$$
Take the cube root of each side.
$$\begin{aligned} y + 6 &= \sqrt[3]{25} \\ y &= \sqrt[3]{25} - 6 \end{aligned}$$
Solution set: $\left\{-6 + \sqrt[3]{25}\right\}$

48.
$$\begin{aligned} \log_3 (p+2) - \log_3 p &= \log_3 2 \\ \log_3 \frac{p+2}{p} &= \log_3 2 \qquad \textit{Quotient rule} \\ \frac{p+2}{p} &= 2 \qquad \textit{Property 4} \\ p + 2 &= 2p \\ 2 &= p \end{aligned}$$
Solution set: $\{2\}$

49.
$$\begin{aligned} \log (2x + 3) &= 1 + \log x \\ \log_{10} (2x+3) - \log_{10} x &= 1 \\ \log_{10} \frac{2x+3}{x} &= 1 \qquad \textit{Quotient rule} \\ 10^1 &= \frac{2x+3}{x} \qquad \textit{Exponential form} \\ 10x &= 2x + 3 \\ 8x &= 3 \\ x &= \frac{3}{8} \end{aligned}$$
Solution set: $\left\{\frac{3}{8}\right\}$

50.
$$\begin{aligned} \log_4 x + \log_4 (8 - x) &= 2 \\ \log_4 [x(8-x)] &= 2 \qquad \textit{Product rule} \\ x(8-x) &= 4^2 \qquad \textit{Exponential form} \\ 8x - x^2 &= 16 \\ x^2 - 8x + 16 &= 0 \\ (x-4)(x-4) &= 0 \\ x - 4 &= 0 \\ x &= 4 \end{aligned}$$
Solution set: $\{4\}$

51.
$$\begin{aligned} \log_2 x + \log_2 (x + 15) &= \log_2 16 \\ \log_2 [x(x+15)] &= \log_2 16 \qquad \textit{Product rule} \\ x^2 + 15x &= 16 \qquad \textit{Property 4} \\ x^2 + 15x - 16 &= 0 \\ (x+16)(x-1) &= 0 \end{aligned}$$
$$\begin{aligned} x + 16 = 0 \quad &\text{or} \quad x - 1 = 0 \\ x = -16 \quad &\text{or} \quad x = 1 \end{aligned}$$
Reject $x = -16$, because it yields an equation in which the logarithm of a negative number must be found.
Solution set: $\{1\}$

52. When the power rule was applied in the second step, the domain was changed from $\{x | x \neq 0\}$ to $\{x | x > 0\}$. Instead of using the power rule for logarithms, we can change the original equation to the exponential form $x^2 = 10^2$ and get $x = \pm 10$. As you can see in the erroneous solution, the valid solution -10 was "lost." The solution set is $\{\pm 10\}$.

53. $A = P\left(1 + \frac{r}{n}\right)^{nt}$
Let $P = 20,000$, $r = .07$, and $t = 5$. For $n = 4$ (quarterly compounding),
$$A = 20,000\left(1 + \frac{.07}{4}\right)^{4 \cdot 5} \approx 28,295.56.$$
There will be $\$28,295.56$ in the account after 5 years.

54. In the continuous compounding formula, let $P = 10,000$, $r = .06$, and $t = 3$.

$$A = Pe^{rt}$$
$$A = 10,000e^{.06(3)} \approx 11,972.17$$

There will be \$11,972.17 in the account after 3 years.

55. Use $A = P\left(1 + \frac{r}{n}\right)^{nt}$.

Plan A:
Let $P = 1000$, $r = .04$, $n = 4$, and $t = 3$.

$$A = 1000\left(1 + \frac{.04}{4}\right)^{4 \cdot 3} \approx 1126.83$$

Plan B:
Let $P = 1000$, $r = .039$, $n = 12$, and $t = 3$.

$$A = 1000\left(1 + \frac{.039}{12}\right)^{12 \cdot 3} \approx 1123.91$$

Plan A is the better plan by \$2.92.

56. Let $Q(t) = \frac{1}{2}(500) = 250$ to find the half-life of the radioactive substance.

$$\begin{aligned} Q(t) &= 500e^{-.05t} \\ 250 &= 500e^{-.05t} \\ .5 &= e^{-.05t} \\ \ln .5 &= \ln e^{-.05t} \\ \ln .5 &= -.05t\,(\ln e) \\ -.05t &= \ln .5 \\ t &= \frac{\ln .5}{-.05} \approx 13.9 \end{aligned}$$

The half-life is about 13.9 days.

57. $S = C(1 - r)^n$

(a) Let $C = 30,000$, $r = .15$, and $n = 12$.

$$\begin{aligned} S &= 30,000(1 - .15)^{12} \\ &= 30,000(.85)^{12} \approx 4267 \end{aligned}$$

The scrap value is about \$4267.

(b) Let $S = \frac{1}{2}C$ and $n = 6$.

$$\begin{aligned} S &= C(1 - r)^n \\ \frac{1}{2}C &= C(1 - r)^6 \\ .5 &= (1 - r)^6 \ (*) \\ \ln .5 &= \ln (1 - r)^6 \\ \ln .5 &= 6 \ln (1 - r) \\ \ln (1 - r) &= \frac{\ln .5}{6} \\ \ln (1 - r) &\approx -.1155 \end{aligned}$$

$$\begin{aligned} 1 - r &= e^{-.1155} \\ 1 - r &\approx .89 \\ r &= .11 \end{aligned}$$

The rate is approximately 11%.

$(*)$ Alternative solution steps without logarithms:

$$\begin{aligned} .5 &= (1 - r)^6 \\ \sqrt[6]{.5} &= 1 - r \\ r &= 1 - \sqrt[6]{.5} \approx .11 \end{aligned}$$

Note that $1 - r$ must be positive, so $\pm \sqrt[6]{.5}$ is not needed.

58. $N(r) = -5000 \ln r$
Replace $N(r)$ by 2000, and solve for r.

$$\begin{aligned} 2000 &= -5000 \ln r \\ \ln r &= -\frac{2000}{5000} \\ \log_e r &= -.4 \end{aligned}$$

Change to exponential form and approximate.

$$r = e^{-.4} \approx .67$$

About 67% of the words are common to both of the evolving languages.

59. **A.** Solve $7^x = 23$ by using the power rule with common logarithms.

$$\begin{aligned} 7^x &= 23 \\ \log 7^x &= \log 23 \\ x \log 7 &= \log 23 \\ x &= \frac{\log 23}{\log 7} \end{aligned}$$

B. Solve $7^x = 23$ by using the power rule with natural logarithms.

$$\begin{aligned} 7^x &= 23 \\ \ln 7^x &= \ln 23 \\ x \ln 7 &= \ln 23 \\ x &= \frac{\ln 23}{\ln 7} \end{aligned}$$

C. Use the change-of-base rule with the solution from **A.**

$$x = \frac{\log 23}{\log 7} = \log_7 23$$

D. $x = \frac{\log 23}{\log 7} \neq \log_{23} 7$

The answer is **D**.

60. **(a)**

$$\begin{aligned} \log (2x + 3) &= \log x + 1 \\ \log (2x + 3) - \log x &= 1 \\ \log_{10} \frac{2x + 3}{x} &= 1 \quad \textit{Quotient rule} \end{aligned}$$

$$\begin{aligned} \frac{2x+3}{x} &= 10^1 \quad \textit{Exponential form} \\ 2x+3 &= 10x \\ 3 &= 8x \\ \frac{3}{8} &= x \end{aligned}$$

Solution set: $\left\{\frac{3}{8}\right\}$

(b) From the graph, the x-value of the x-intercept is .375, the decimal equivalent of $\frac{3}{8}$. Note that the solutions of $Y_1 - Y_2 = 0$ are the same as the solutions of $Y_1 = Y_2$.

61. $f(x) = 2^x$

x	$f(x)$
-2	$2^{-2} = \frac{1}{4}$
-1	$2^{-1} = \frac{1}{2}$
0	$2^0 = 1$
1	$2^1 = 2$
2	$2^2 = 4$
3	$2^3 = 8$

Plot the ordered pairs from the table, and draw a smooth curve through them.

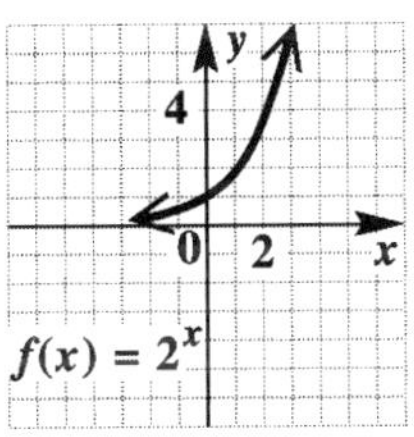

62. $g(x) = \log_2 x$

x	$g(x)$
$\frac{1}{4}$	$\log_2 \frac{1}{4} = \log_2 2^{-2} = -2$
$\frac{1}{2}$	$\log_2 \frac{1}{2} = \log_2 2^{-1} = -1$
1	$\log_2 1 = \log_2 2^0 = 0$
2	$\log_2 2 = \log_2 2^1 = 1$
4	$\log_2 4 = \log_2 2^2 = 2$
8	$\log_2 8 = \log_2 2^3 = 3$

Plot the ordered pairs from the table, and draw a smooth curve through them.

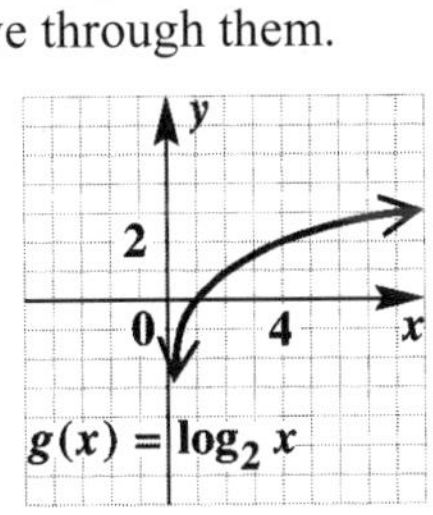

63. The ordered pairs in Exercises 61 and 62 are reverses of each other. In other words, if (x, y) is an ordered pair in Exercise 61, (y, x) is an ordered pair in Exercise 62. Functions f and g are inverses of each other.

64. The graph of f in Exercise 61 has a <u>*horizontal*</u> asymptote, $y = 0$, while the graph of g in Exercise 62 has a <u>*vertical*</u> asymptote, $x = 0$.

65. $2^2 \cdot 2^3 = 2^{\underline{5}}$ because $\underline{2} + \underline{3} = \underline{5}$.

66. It is a fact that $32 = 4 \cdot 8$. Therefore, using properties of logarithms,

$$\log_2 32 = \log_2 \underline{4} + \log_2 \underline{8}.$$

67. $\log_2 13 = \dfrac{\log 13}{\log 2} = 3.700439718$

68. Based on the result of Exercise 67, $\log_2 13$ means that $2^{3.700439718} = 13$. $\log_2 13$ is the exponent to which 2 must be raised to obtain 13.

69. $2^{\log_2 13} = 13$ by the special property of logarithms, $a^{\log_a x} = x$.

70. Using a calculator,

$$2^{3.700439718} = 13.$$

The number in Exercise 67 is the exponent to which 2 must be raised to obtain 13.

71. Based on the result in Exercise 67, the point $(13, \underline{3.700439718})$ lies on the graph of $g(x) = \log_2 x$.

72.
$$\begin{aligned} 2^{x+1} &= 8^{2x+3} \\ 2^{x+1} &= \left(2^3\right)^{2x+3} \quad \textit{Equal bases} \\ 2^{x+1} &= 2^{6x+9} \\ x+1 &= 6x+9 \quad \textit{Equate exponents} \\ -8 &= 5x \\ -\frac{8}{5} &= x \end{aligned}$$

Solution set: $\left\{-\frac{8}{5}\right\}$

73. $\log_2 128 = \log_2 2^7 = 7$, by a special property of logarithms.

74. By a special property of logarithms,

$$5^{\log_5 36} = 36.$$

75. $e^{\ln 4} = 4$ since $e^{\ln x} = x$.

76. $10^{\log e} = e$ since $10^{\log x} = x$.

77. $\log_3 3^{-5} = -5$ since $\log_a a^x = x$.

78. $\ln e^{5.4} = 5.4$ since $\ln e^x = x$.

79. $\log_3(x+9) = 4$

$x + 9 = 3^4$ *Exponential form*

$x + 9 = 81$

$x = 72$

Solution set: $\{72\}$

80. $\ln e^x = 3$

$x = 3$

Solution set: $\{3\}$

81. $\log_x \frac{1}{81} = 2$

$x^2 = \frac{1}{81}$ *Exponential form*

$x^2 = \left(\frac{1}{9}\right)^2$

$x = \pm\frac{1}{9}$ *Square root property*

The base x cannot be negative.

Solution set: $\left\{\frac{1}{9}\right\}$

82. $27^x = 81$

Write each side as a power of 3.

$(3^3)^x = 3^4$

$3^{3x} = 3^4$

$3x = 4$ *Equate exponents*

$x = \frac{4}{3}$

Solution set: $\left\{\frac{4}{3}\right\}$

83. $2^{2x-3} = 8$

Write each side as a power of 2.

$2^{2x-3} = 2^3$

$2x - 3 = 3$ *Equate exponents*

$2x = 6$

$x = 3$

Solution set: $\{3\}$

84. $5^{x+2} = 25^{2x+1}$

$5^{x+2} = (5^2)^{2x+1}$ *Equal bases*

$5^{x+2} = 5^{4x+2}$

$x + 2 = 4x + 2$ *Equate exponents*

$0 = 3x$

$0 = x$

Solution set: $\{0\}$

85. $\log_3(x+1) - \log_3 x = 2$

$\log_3 \frac{x+1}{x} = 2$ *Quotient rule*

$\frac{x+1}{x} = 3^2$ *Exponential form*

$9x = x + 1$

$8x = 1$

$x = \frac{1}{8}$

Solution set: $\left\{\frac{1}{8}\right\}$

86. $\log(3x-1) = \log 10$

$3x - 1 = 10$

$3x = 11$

$x = \frac{11}{3}$

Solution set: $\left\{\frac{11}{3}\right\}$

87. $\ln(x^2+3x+4) = \ln 2$

$x^2 + 3x + 4 = 2$

$x^2 + 3x + 2 = 0$

$(x+2)(x+1) = 0$

$x + 2 = 0$ or $x + 1 = 0$

$x = -2$ or $x = -1$

Both answers check.

Solution set: $\{-2, -1\}$

88. $f(t) = 5000(2)^{-.15t}$

(a) The original value is found when $t = 0$.

$f(0) = 5000(2)^{-.15(0)}$
$= 5000(2)^0 = 5000(1) = 5000$

The original value is \$5000.

(b) The value after 5 years is found when $t = 5$.

$f(5) = 5000(2)^{-.15(5)} \approx 2973.018$

The value after 5 years is about \$2973.

(c) The value after 10 years is found when $t = 10$.

$f(10) = 5000(2)^{-.15(10)} \approx 1767.767$

The value after 10 years is about \$1768.

89. $S = C(1-r)^n$

Let $S = 10{,}000$, $C = 30{,}000$, and $r = 15\% = .15$.

$10{,}000 = 30{,}000(1-.15)^n$

$\frac{1}{3} = (.85)^n$

$\log\left(\frac{1}{3}\right) = \log(.85)^n$

$\log\left(\frac{1}{3}\right) = n\log(.85)$

$n = \frac{\log\left(\frac{1}{3}\right)}{\log(.85)} \approx 6.8$

The useful life is 6.8 years.

90. $R(x) = .0597e^{.0553x}$

1990 corresponds to $x = 20$.

$R(20) = .0597e^{.0553(20)}$

$\approx .1804$

In 1990, about 18.04% of municipal solid waste was recovered.

91. There are 90 of one species and 10 of another, so

$$p_1 = \frac{90}{100} = .9 \quad \text{and} \quad p_2 = \frac{10}{100} = .1.$$

Thus, the index of diversity is

$$-(p_1 \ln p_1 + p_2 \ln p_2)$$
$$= -(.9 \ln .9 + .1 \ln .1)$$
$$\approx .325.$$

92. There are 60 of one species and 40 of another, so

$$p_1 = \frac{60}{100} = .6 \quad \text{and} \quad p_2 = \frac{40}{100} = .4.$$

Thus, the index of diversity is

$$-(p_1 \ln p_1 + p_2 \ln p_2)$$
$$= -(.6 \ln .6 + .4 \ln .4)$$
$$\approx .673.$$

Chapter 10 Test

1. **(a)** $f(x) = x^2 + 9$

This function is not one-to-one. The graph of $f(x)$ is a vertical parabola. A horizontal line will intersect the graph more than once.

(b) This function is one-to-one. A horizontal line will not intersect the graph in more than one point.

2. $f(x) = \sqrt[3]{x+7}$

Replace $f(x)$ with y.

$y = \sqrt[3]{x+7}$

Interchange x and y.

$x = \sqrt[3]{y+7}$

Solve for y.

$x^3 = y + 7$

$x^3 - 7 = y$

Replace y with $f^{-1}(x)$.

$f^{-1}(x) = x^3 - 7$

3. By the horizontal line test, $f(x)$ is a one-to-one function and has an inverse. Choose some points on the graph of $f(x)$, such as $(4, 0)$, $(3, -1)$, and $(0, -2)$. To graph the inverse, interchange the x- and y-values to get $(0, 4)$, $(-1, 3)$, and $(-2, 0)$. Plot these points and draw a smooth curve through them.

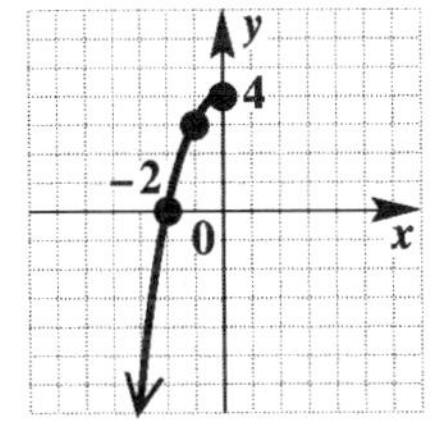

4. $f(x) = 6^x$

Make a table of values.

x	-2	-1	0	1
$f(x)$	$\frac{1}{36}$	$\frac{1}{6}$	1	6

Plot these points and draw a smooth exponential curve through them.

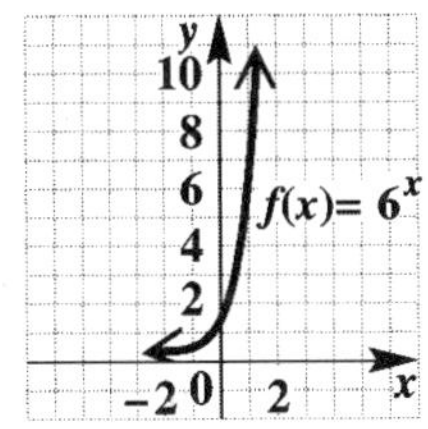

5. $g(x) = \log_6 x$

Make a table of values.

Powers of 6	6^{-2}	6^{-1}	6^0	6^1
x	$\frac{1}{36}$	$\frac{1}{6}$	1	6
$g(x)$	-2	-1	0	1

Plot these points and draw a smooth logarithmic curve through them.

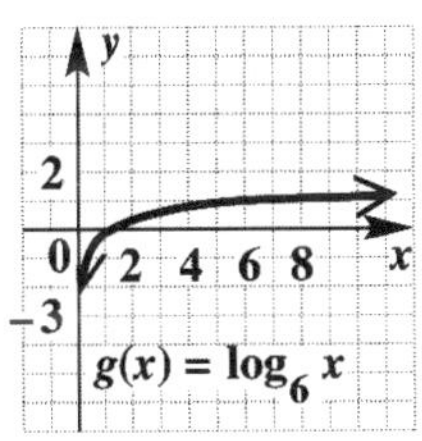

6. $y = 6^x$ and $y = \log_6 x$ are inverse functions. To use the graph from Exercise 4 to obtain the graph of the function in Exercise 5, interchange the x- and y-coordinates of the ordered pairs $\left(-2, \frac{1}{36}\right)$, $\left(-1, \frac{1}{6}\right)$, $(0, 1)$, and $(1, 6)$ to get $\left(\frac{1}{36}, -2\right)$, $\left(\frac{1}{6}, -1\right)$, $(1, 0)$, and $(6, 1)$. Plot these points and draw a smooth logarithmic curve through them.

7. $5^x = \frac{1}{625}$

$5^x = \left(\frac{1}{5}\right)^4$

Write each side as a power of 5.

continued

$5^x = 5^{-4}$

$x = -4$ *Equate exponents*

Solution set: $\{-4\}$

8. $2^{3x-7} = 8^{2x+2}$

Write each side as a power of 2.

$2^{3x-7} = 2^{3(2x+2)}$

$3x - 7 = 3(2x+2)$ *Equate exponents*

$3x - 7 = 6x + 6$

$-13 = 3x$

$-\frac{13}{3} = x$

Solution set: $\left\{-\frac{13}{3}\right\}$

9. $f(x) = 26.7e^{.023x}$

(a) 2000: $x = 2000 - 1995 = 5$

$f(5) = 26.7e^{.023(5)} \approx 30.0$

The U.S. Hispanic population estimate for 2000 is 30.0 million.

(b) 2010: $x = 2010 - 1995 = 15$

$f(15) = 26.7e^{.023(15)} \approx 37.7$

The U.S. Hispanic population estimate for 2010 is 37.7 million.

10. The base is 4, the exponent (logarithm) is -2, and the number is .0625, so $4^{-2} = .0625$ becomes $\log_4 .0625 = -2$ in logarithmic form.

11. The base is 7, the logarithm (exponent) is 2, and the number is 49, so $\log_7 49 = 2$ becomes $7^2 = 49$ in exponential form.

12. $\log_{1/2} x = -5$

$x = \left(\frac{1}{2}\right)^{-5}$ *Exponential form*

$x = \left(\frac{2}{1}\right)^{5} = 32$

Solution set: $\{32\}$

13. $x = \log_9 3$

$9^x = 3$ *Exponential form*

Write each side as a power of 3.

$(3^2)^x = 3$

$3^{2x} = 3^1$

$2x = 1$ *Equate exponents*

$x = \frac{1}{2}$

Solution set: $\left\{\frac{1}{2}\right\}$

14. $\log_x 16 = 4$

$x^4 = 16$ *Exponential form*

$x^2 = \pm 4$ *Square root property*

Reject -4 since $x^2 \geq 0$.

$x^2 = 4$

$x = \pm 2$ *Square root property*

Reject -2 since the base cannot be negative.

Solution set: $\{2\}$

15. The value of $\log_2 32$ is $\underline{5}$. This means that if we raise $\underline{2}$ to the $\underline{\text{5th}}$ power, the result is $\underline{32}$.

16. $\log_3 x^2 y$

$= \log_3 x^2 + \log_3 y$ *Product rule*

$= 2\log_3 x + \log_3 y$ *Power rule*

17. $\log_5 \left(\frac{\sqrt{x}}{yz}\right)$

$= \log_5 \sqrt{x} - \log_5 yz$ *Quotient rule*

$= \log_5 x^{1/2} - (\log_5 y + \log_5 z)$ *Product rule*

$= \frac{1}{2}\log_5 x - \log_5 y - \log_5 z$ *Power rule*

18. $3\log_b s - \log_b t$

$= \log_b s^3 - \log_b t$ *Power rule*

$= \log_b \frac{s^3}{t}$ *Quotient rule*

19. $\frac{1}{4}\log_b r + 2\log_b s - \frac{2}{3}\log_b t$

Use the power rule for logarithms.

$= \log_b r^{1/4} + \log_b s^2 - \log_b t^{2/3}$

Use the product and quotient rules for logarithms.

$= \log_b \frac{r^{1/4}s^2}{t^{2/3}}$

20. **(a)** $\log 23.1 \approx 1.3636$

(b) $\ln .82 \approx -.1985$

21. **(a)** $\log_3 19 = \frac{\log_{10} 19}{\log_{10} 3} = \frac{\log 19}{\log 3}$

(b) $\log_3 19 = \frac{\log_e 19}{\log_e 3} = \frac{\ln 19}{\ln 3}$

(c) The four-decimal-place approximation of either fraction is 2.6801.

22. $3^x = 78$

$\ln 3^x = \ln 78$

$x \ln 3 = \ln 78$ *Power rule*

$x = \frac{\ln 78}{\ln 3} \approx 3.9656$

Solution set: $\{3.9656\}$

23. $\log_8 (x+5) + \log_8 (x-2) = 1$

Use the product rule for logarithms.

$$\log_8 [(x+5)(x-2)] = 1$$
$$(x+5)(x-2) = 8^1 \quad \textit{Exp. form}$$
$$x^2 + 3x - 10 = 8$$
$$x^2 + 3x - 18 = 0$$
$$(x+6)(x-3) = 0$$

$$x + 6 = 0 \quad \text{or} \quad x - 3 = 0$$
$$x = -6 \quad \text{or} \quad x = 3$$

Reject $x = -6$, because $x + 5 = -1$, which yields an equation in which the logarithm of a negative number must be found.

Solution set: $\{3\}$

24. $A = P\left(1 + \frac{r}{n}\right)^{nt}$

$$A = 10,000\left(1 + \frac{.045}{4}\right)^{4 \cdot 5} \approx 12,507.51$$

$\$10,000$ invested at 4.5% annual interest, compounded quarterly, will increase to $\$12,507.51$ in 5 years.

25. $A = Pe^{rt}$

(a) $A = 15,000e^{.05(5)} \approx 19,260.38$

There will be $\$19,260.38$ in the account.

(b) Let $A = 2(15,000) = 30,000$ and solve for t.

$$30,000 = 15,000e^{.05t}$$
$$2 = e^{.05t}$$
$$\ln 2 = \ln e^{.05t}$$
$$\ln 2 = .05t\,(\ln e)$$
$$.05t = \ln 2$$
$$t = \frac{\ln 2}{.05} \approx 13.9$$

The principal will double in about 13.9 years.

Cumulative Review Exercises Chapters 1–10

For Exercises 1–4,

$$S = \left\{-\frac{9}{4}, -2, -\sqrt{2}, 0, .6, \sqrt{11}, \sqrt{-8}, 6, \frac{30}{3}\right\}.$$

1. The integers are $-2, 0, 6$, and $\frac{30}{3}$ (or 10).

2. The rational numbers are $-\frac{9}{4}, -2, 0, .6, 6,$ and $\frac{30}{3}$ (or 10).

Each can be expressed as a quotient of two integers.

3. The irrational numbers are $-\sqrt{2}$ and $\sqrt{11}$.

4. All are real numbers except $\sqrt{-8}$.

5. $|-8| + 6 - |-2| - (-6+2)$
$$= 8 + 6 - 2 - (-4)$$
$$= 14 - 2 + 4 = 16$$

6. $-12 - |-3| - 7 - |-5|$
$$= -12 - 3 - 7 - 5 = -27$$

7. $2(-5) + (-8)(4) - (-3)$
$$= -10 - 32 + 3 = -39$$

8. $7 - (3 + 4a) + 2a = -5(a-1) - 3$
$$7 - 3 - 4a + 2a = -5a + 5 - 3$$
$$4 - 2a = -5a + 2$$
$$3a = -2$$
$$a = -\frac{2}{3}$$

Solution set: $\left\{-\frac{2}{3}\right\}$

9. $2m + 2 \le 5m - 1$
$$-3m \le -3$$

Divide by -3; reverse the inequality.
$$m \ge 1$$

Solution set: $[1, \infty)$

10. $|2x - 5| = 9$

$$2x - 5 = 9 \quad \text{or} \quad 2x - 5 = -9$$
$$2x = 14 \qquad 2x = -4$$
$$x = 7 \quad \text{or} \quad x = -2$$

Solution set: $\{-2, 7\}$

11. $|3p| - 4 = 12$
$$|3p| = 16$$

$$3p = 16 \quad \text{or} \quad 3p = -16$$
$$p = \frac{16}{3} \quad \text{or} \quad p = -\frac{16}{3}$$

Solution set: $\left\{\pm\frac{16}{3}\right\}$

12. $|3k - 8| \le 1$
$$-1 \le 3k - 8 \le 1$$
$$7 \le 3k \le 9$$
$$\frac{7}{3} \le k \le 3$$

Solution set: $\left[\frac{7}{3}, 3\right]$

13. $|4m + 2| > 10$

$$4m + 2 > 10 \quad \text{or} \quad 4m + 2 < -10$$
$$4m > 8 \qquad 4m < -12$$
$$m > 2 \quad \text{or} \quad m < -3$$

Solution set: $(-\infty, -3) \cup (2, \infty)$

14. $5x + 2y = 10$

Find the x- and y-intercepts. To find the x-intercept, let $y = 0$.

$$\begin{aligned} 5x + 2(0) &= 10 \\ 5x &= 10 \\ x &= 2 \end{aligned}$$

The x-intercept is $(2, 0)$.
To find the y-intercept, let $x = 0$.

$$\begin{aligned} 5(0) + 2y &= 10 \\ 2y &= 10 \\ y &= 5 \end{aligned}$$

The y-intercept is $(0, 5)$.
Plot the intercepts and draw the line through them.

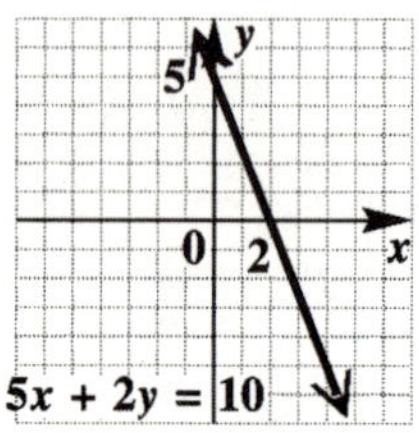

15. $-4x + y \le 5$

Graph the line $-4x + y = 5$, which has intercepts $(0, 5)$ and $\left(-\frac{5}{4}, 0\right)$, as a solid line because the inequality involves $\le$. Test $(0, 0)$, which yields $0 \le 5$, a true statement. Shade the region that includes $(0, 0)$.

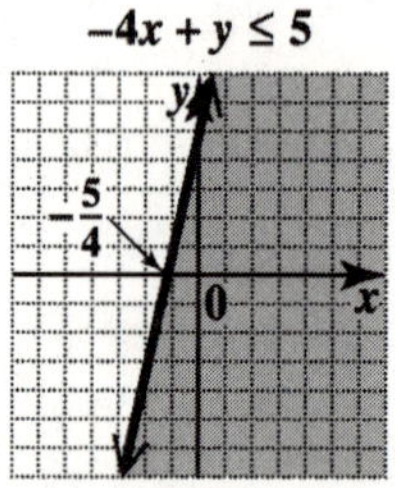

16. **(a)** Yes, this is the graph of a function because it passes the vertical line test.

(b) $(x_1, y_1) = (1997, 17{,}716)$, $(x_2, y_2) = (1999, 9{,}733)$

$$\begin{aligned} m &= \frac{y_2 - y_1}{x_2 - x_1} \\ &= \frac{9{,}733 - 17{,}716}{1999 - 1997} \\ &= \frac{-7983}{2} \approx -4000 \end{aligned}$$

The number of acres harvested decreased by approximately 4000 acres per year during 1997–1999.

17. Through $(5, -1)$; parallel to $3x - 4y = 12$
Find the slope of

$$\begin{aligned} 3x - 4y &= 12 \\ -4y &= -3x + 12 \\ y &= \frac{3}{4}x - 3. \end{aligned}$$

The slope is $\frac{3}{4}$, so a line parallel to it also has slope $\frac{3}{4}$.

Let $m = \frac{3}{4}$ and $(x_1, y_1) = (5, -1)$ in the point-slope form.

$$\begin{aligned} y - y_1 &= m(x - x_1) \\ y - (-1) &= \frac{3}{4}(x - 5) \\ y + 1 &= \frac{3}{4}x - \frac{15}{4} \\ y &= \frac{3}{4}x - \frac{19}{4} \end{aligned}$$

18. $5x - 3y = 14 \quad (1)$
$2x + 5y = 18 \quad (2)$

Multiply equation (1) by 5 and equation (2) by 3. Then add the results.

$$\begin{array}{rcrcrl} 25x & - & 15y & = & 70 & \quad 5 \times (1) \\ 6x & + & 15y & = & 54 & \quad 3 \times (2) \\ \hline 31x & & & = & 124 & \textit{Add} \\ & & x & = & 4 & \end{array}$$

Substitute 4 for x in equation (1) to find y.

$$\begin{aligned} 5x - 3y &= 14 \quad (1) \\ 5(4) - 3y &= 14 \\ 20 - 3y &= 14 \\ -3y &= -6 \\ y &= 2 \end{aligned}$$

Solution set: $\{(4, 2)\}$

19. $2x - 7y = 8 \quad (1)$
$4x - 14y = 3 \quad (2)$

Multiply equation (1) by -2 and add the result to equation (2).

$$\begin{array}{rcrcrl} -4x & + & 14y & = & -16 & \quad -2 \times (1) \\ 4x & - & 14y & = & 3 & \quad (2) \\ \hline & & 0 & = & -13 & \textit{Add} \end{array}$$

The statement $0 = -13$ is false, so the solution set is $\emptyset$.

20.
$$\begin{aligned} x + 2y + 3z &= 11 \quad (1) \\ 3x - y + z &= 8 \quad (2) \\ 2x + 2y - 3z &= -12 \quad (3) \end{aligned}$$

To eliminate z, add equations (1) and (3).

$$\begin{array}{rcrcrcrl} x & + & 2y & + & 3z & = & 11 & (1) \\ 2x & + & 2y & - & 3z & = & -12 & (3) \\ \hline 3x & + & 4y & & & = & -1 & (4) \end{array}$$

To eliminate z again, multiply equation (2) by 3 and add the result to equation (3).

$$\begin{array}{rcrcrcrl} 9x & - & 3y & + & 3z & = & 24 & 3 \times (2) \\ 2x & + & 2y & - & 3z & = & -12 & (3) \\ \hline 11x & - & y & & & = & 12 & (5) \end{array}$$

Multiply equation (5) by 4 and add the result to equation (4).

$$\begin{array}{rcrcrl} 44x & - & 4y & = & 48 & 4 \times (5) \\ 3x & + & 4y & = & -1 & (4) \\ \hline 47x & & & = & 47 & \\ & & x & = & 1 & \end{array}$$

Substitute 1 for x in equation (5) to find y.

$$\begin{aligned} 11x - y &= 12 \quad (5) \\ 11(1) - y &= 12 \\ 11 - y &= 12 \\ -y &= 1 \\ y &= -1 \end{aligned}$$

Substitute 1 for x and -1 for y in equation (2) to find z.

$$\begin{aligned} 3x - y + z &= 8 \quad (2) \\ 3(1) - (-1) + z &= 8 \\ 3 + 1 + z &= 8 \\ 4 + z &= 8 \\ z &= 4 \end{aligned}$$

Solution set: $\{(1, -1, 4)\}$

21. Let $x =$ the amount of candy at \$1.00 per pound.

	Number of Pounds	Price per Pound	Value
First Candy	x	\$1.00	$1.00x$
Second Candy	10	\$1.96	$1.96(10)$
Mixture	$x + 10$	\$1.60	$1.60(x + 10)$

The sum of the values of each candy must equal the value of the mixture.

$$1.00x + 1.96(10) = 1.60(x + 10)$$

Multiply by 10 to clear the decimals.

$$\begin{aligned} 10x + 196 &= 16(x + 10) \\ 10x + 196 &= 16x + 160 \\ 36 &= 6x \\ 6 &= x \end{aligned}$$

Use 6 pounds of the \$1.00 candy.

22. $(2p + 3)(3p - 1) = 6p^2 - 2p + 9p - 3$
$= 6p^2 + 7p - 3$

23. $(4k - 3)^2 = (4k)^2 - 2(4k)(3) + 3^2$
$= 16k^2 - 24k + 9$

24. $(3m^3 + 2m^2 - 5m) - (8m^3 + 2m - 4)$
$= 3m^3 + 2m^2 - 5m - 8m^3 - 2m + 4$
$= 3m^3 - 8m^3 + 2m^2 - 5m - 2m + 4$
$= -5m^3 + 2m^2 - 7m + 4$

25.

$$\begin{array}{r|l} & 2t^3 + 5t^2 - 3t + 4 \\ \hline 3t + 1 & 6t^4 + 17t^3 - 4t^2 + 9t + 4 \\ & 6t^4 + 2t^3 \\ \hline & \quad 15t^3 - 4t^2 \\ & \quad 15t^3 + 5t^2 \\ \hline & \qquad -9t^2 + 9t \\ & \qquad -9t^2 - 3t \\ \hline & \qquad\qquad 12t + 4 \\ & \qquad\qquad 12t + 4 \\ \hline & \qquad\qquad\qquad 0 \end{array}$$

The quotient is

$$2t^3 + 5t^2 - 3t + 4.$$

26. $8x + x^3 = x(8 + x^2)$

27. $24y^2 - 7y - 6 = (8y + 3)(3y - 2)$

28. $5z^3 - 19z^2 - 4z = z(5z^2 - 19z - 4)$
$= z(5z + 1)(z - 4)$

29. $16a^2 - 25b^4$

Use the difference of squares formula,

$$x^2 - y^2 = (x + y)(x - y),$$

where $x = 4a$ and $y = 5b^2$.

$$16a^2 - 25b^4 = (4a + 5b^2)(4a - 5b^2)$$

30. $8c^3 + d^3$

Use the sum of cubes formula,

$$x^3 + y^3 = (x + y)(x^2 - xy + y^2),$$

where $x = 2c$ and $y = d$.

$$8c^3 + d^3 = (2c + d)(4c^2 - 2cd + d^2)$$

31. $16r^2 + 56rq + 49q^2$
$= (4r)^2 + 2(4r)(7q) + (7q)^2$

Use the perfect square formula,

$$x^2 + 2xy + y^2 = (x + y)^2,$$

where $x = 4r$ and $y = 7q$.

$$16r^2 + 56rq + 49q^2 = (4r + 7q)^2$$

32. $\dfrac{(5p^3)^4(-3p^7)}{2p^2(4p^4)} = \dfrac{(5^4p^{12})(-3p^7)}{8p^6}$
$= \dfrac{(625)(-3)p^{19}}{8p^6}$
$= -\dfrac{1875p^{13}}{8}$

33. $\dfrac{x^2-9}{x^2+7x+12} \div \dfrac{x-3}{x+5}$

Multiply by the reciprocal.

$= \dfrac{x^2-9}{x^2+7x+12} \bullet \dfrac{x+5}{x-3}$

$= \dfrac{(x+3)(x-3)}{(x+3)(x+4)} \bullet \dfrac{(x+5)}{(x-3)}$ *Factor*

$= \dfrac{x+5}{x+4}$

34. $\dfrac{2}{k+3} - \dfrac{5}{k-2}$

The LCD is $(k+3)(k-2)$.

$= \dfrac{2(k-2)}{(k+3)(k-2)} - \dfrac{5(k+3)}{(k-2)(k+3)}$

$= \dfrac{2k-4-5k-15}{(k+3)(k-2)}$

$= \dfrac{-3k-19}{(k+3)(k-2)}$

35. $\dfrac{3}{p^2-4p} - \dfrac{4}{p^2+2p}$

$= \dfrac{3}{p(p-4)} - \dfrac{4}{p(p+2)}$

The LCD is $p(p-4)(p+2)$.

$= \dfrac{3(p+2)}{p(p-4)(p+2)} - \dfrac{4(p-4)}{p(p+2)(p-4)}$

$= \dfrac{3p+6-4p+16}{p(p-4)(p+2)}$

$= \dfrac{22-p}{p(p-4)(p+2)}$

36. $\sqrt{288} = \sqrt{144 \bullet 2} = \sqrt{144}\sqrt{2} = 12\sqrt{2}$

37. $2\sqrt{32} - 5\sqrt{98} = 2\sqrt{16 \bullet 2} - 5\sqrt{49 \bullet 2}$
$= 2 \bullet 4\sqrt{2} - 5 \bullet 7\sqrt{2}$
$= 8\sqrt{2} - 35\sqrt{2}$
$= -27\sqrt{2}$

38. $\sqrt{2x+1} - \sqrt{x} = 1$
$\sqrt{2x+1} = 1 + \sqrt{x}$
$\left(\sqrt{2x+1}\right)^2 = (1+\sqrt{x})^2$
$2x+1 = 1 + 2\sqrt{x} + x$
$x = 2\sqrt{x}$
$(x)^2 = (2\sqrt{x})^2$
$x^2 = 4x$
$x^2 - 4x = 0$
$x(x-4) = 0$

$x = 0$ or $x = 4$

Check $x = 0$: $\sqrt{1} - \sqrt{0} \stackrel{?}{=} 1$ *True*

Check $x = 4$: $\sqrt{9} - \sqrt{4} \stackrel{?}{=} 1$ *True*

Solution set: $\{0, 4\}$

39. $(5+4i)(5-4i) = 5^2 - (4i)^2$
$= 25 - 16i^2$
$= 25 - 16(-1)$
$= 25 + 16 = 41$

40. $3x^2 - x - 1 = 0$

Here $a = 3$, $b = -1$, and $c = -1$.

Use the quadratic formula.

$x = \dfrac{-b \pm \sqrt{b^2-4ac}}{2a}$

$x = \dfrac{-(-1) \pm \sqrt{(-1)^2 - 4(3)(-1)}}{2(3)}$

$= \dfrac{1 \pm \sqrt{1+12}}{6} = \dfrac{1 \pm \sqrt{13}}{6}$

Solution set: $\left\{\dfrac{1+\sqrt{13}}{6}, \dfrac{1-\sqrt{13}}{6}\right\}$

41. $k^2 + 2k - 8 > 0$

Solve the equation

$$k^2 + 2k - 8 = 0.$$
$$(k+4)(k-2) = 0$$

$k + 4 = 0$ or $k - 2 = 0$
$k = -4$ or $k = 2$

The numbers -4 and 2 divide a number line into three intervals.

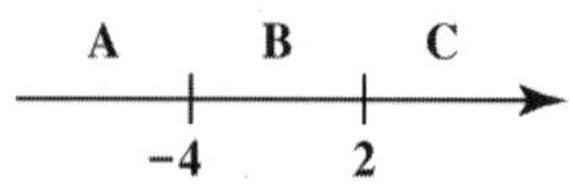

Test a number from each interval in the inequality

$$k^2 + 2k - 8 > 0.$$

Interval A:	Let $k = -5$.		
	$25 - 10 - 8 > 0$	?	
	$7 > 0$		*True*
Interval B:	Let $k = 0$.		
	$-8 > 0$		*False*
Interval C:	Let $k = 3$.		
	$9 + 6 \quad 8 > 0$	?	
	$7 > 0$		*True*

The numbers in Intervals A and C, not including -4 or 2 because of $>$, are solutions.
Solution set: $(-\infty, -4) \cup (2, \infty)$

42. $x^4 - 5x^2 + 4 = 0$

Let $u = x^2$, so $u^2 = (x^2)^2 = x^4$.

$$u^2 - 5u + 4 = 0$$
$$(u-1)(u-4) = 0$$

$u - 1 = 0$ or $u - 4 = 0$
$u = 1$ or $u = 4$

To find x, substitute x^2 for u.

$x^2 = 1$ or $x^2 = 4$
$x = \pm 1$ or $x = \pm 2$
Solution set: $\{\pm 1, \pm 2\}$

43. Let $x =$ one of the numbers.
Then $300 - x =$ the other number.

The product of the two numbers is given by

$$P = x(300 - x).$$

Writing this equation in standard form gives us

$$P = -x^2 + 300x.$$

Finding the maximum of the product is the same as finding the vertex of the graph of P.
The x-value of the vertex is

$$x = -\frac{b}{2a} = -\frac{300}{2(-1)} = 150.$$

If x is 150, then $300 - x$ must also be 150. The two numbers are 150 and 150 and the product is $150 \cdot 150 = 22,500$.

44. $f(x) = \frac{1}{3}(x-1)^2 + 2$ is in $f(x) = a(x-h)^2 + k$ form. The graph is a vertical parabola with vertex (h, k) at $(1, 2)$. Since $a = \frac{1}{3} > 0$, the graph opens up.

Also, $|a| = \left|\frac{1}{3}\right| = \frac{1}{3} < 1$, so the graph is wider than the graph of $f(x) = x^2$. The points $\left(0, 2\frac{1}{3}\right)$, $(-2, 5)$, and $(4, 5)$ are also on the graph.

$f(x) = \frac{1}{3}(x-1)^2 + 2$

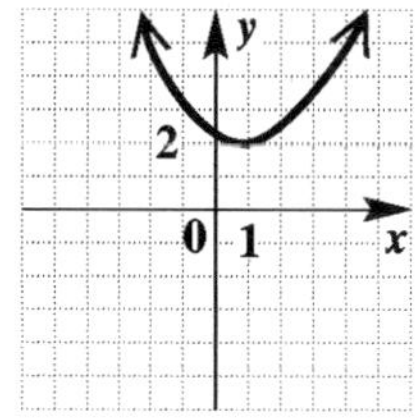

45. $f(x) = 2^x$
Make a table of values.

x	-2	-1	0	1	2
$f(x)$	$\frac{1}{4}$	$\frac{1}{2}$	1	2	4

Plot the ordered pairs from the table, and draw a smooth exponential curve through the points.

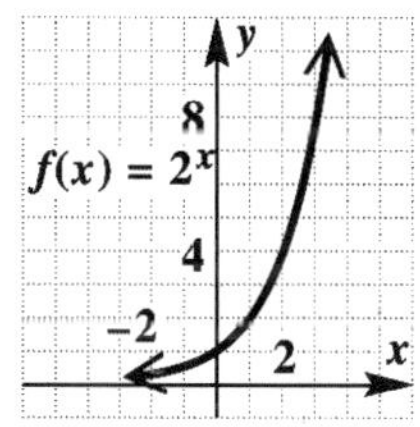

46. $5^{x+3} = \left(\frac{1}{25}\right)^{3x+2}$

$$5^{x+3} = \left[\left(\frac{1}{5}\right)^2\right]^{3x+2}$$

Write each side to the power of 5.

$$5^{x+3} = \left(5^{-2}\right)^{(3x+2)}$$
$$5^{x+3} = 5^{-2(3x+2)}$$
$$x + 3 = -2(3x + 2)$$
$$x + 3 = -6x - 4$$
$$7x = -7$$
$$x = -1$$

Solution set: $\{-1\}$

47. $f(x) = \log_3 x$

Make a table of values.

Powers of 3	3^{-2}	3^{-1}	3^0	3^1	3^2
x	$\frac{1}{9}$	$\frac{1}{3}$	1	3	9
y	-2	-1	0	1	2

Plot the ordered pairs and draw a smooth logarithmic curve through the points.

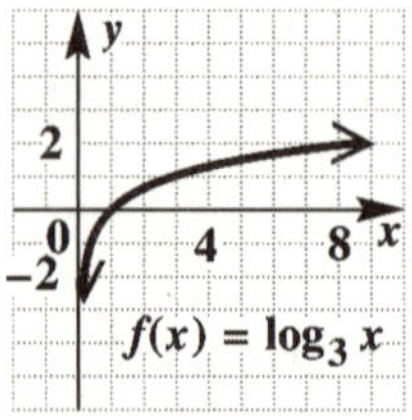

48. $\log_2 81 = \log_2 9^2 = 2\log_2 9$
$\approx 2(3.1699) = 6.3398$

49. $\log \dfrac{x^3\sqrt{y}}{z}$

$= \log \dfrac{x^3 y^{1/2}}{z}$

$= \log\left(x^3 y^{1/2}\right) - \log z$ *Quotient rule*

$= \log x^3 + \log y^{1/2} - \log z$ *Product rule*

$= 3\log x + \dfrac{1}{2}\log y - \log z$ *Power rule*

50. $B(t) = 25{,}000e^{.2t}$

(a) At noon, $t = 0$.

$$B(0) = 25{,}000e^{.2(0)} = 25{,}000e^0 = 25{,}000(1) = 25{,}000$$

25, 000 bacteria are present at noon.

(b) At 1 P.M., $t = 1$.

$$B(1) = 25{,}000e^{.2(1)} \approx 30{,}535$$

About 30, 500 bacteria are present at 1 P.M.

(c) At 2 P.M., $t = 2$.

$$B(2) = 25{,}000e^{.2(2)} \approx 37{,}296$$

About 37, 300 bacteria are present at 2 P.M.

(d) The population doubles when $B(t) = 50{,}000$.

$$50{,}000 = 25{,}000e^{.2t}$$
$$2 = e^{.2t}$$
$$\ln 2 = \ln e^{.2t}$$
$$\ln 2 = .2t \ln e$$
$$t = \frac{\ln 2}{.2} \approx 3.5$$

The population will double in about 3.5 hours, or at about 3:30 P.M.

CHAPTER 11 NONLINEAR FUNCTIONS, CONIC SECTIONS, AND NONLINEAR SYSTEMS

Section 11.1

1. For the reciprocal function defined by $f(x) = 1/x$, $\underline{0}$ is the only real number not in the domain since division by 0 is undefined.

3. The lowest point on the graph of $f(x) = |x|$ has coordinates $(\underline{0}, \underline{0})$.

5. $f(x) = |x - 2| + 2$

The graph of this function has its "vertex" at $(2, 2)$, so the correct graph is **B**.

7. $f(x) = |x - 2| - 2$

The graph of this function has its "vertex" at $(2, -2)$, so the correct graph is **A**.

9. $f(x) = |x + 1|$

Since x can be any real number, the domain is $(-\infty, \infty)$.

The value of y is always greater than or equal to 0, so the range is $[0, \infty)$.

The graph of $y = |x + 1|$ looks like the graph of the absolute value function $y = |x|$, but the graph is translated 1 unit to the left. The x-value of its "vertex" is obtained by setting $x + 1 = 0$ and solving for x:

$$x + 1 = 0$$
$$x = -1.$$

Since the corresponding y-value is 0, the "vertex" is $(-1, 0)$. Some additional points are $(-3, 2)$, $(-2, 1)$, $(0, 1)$, and $(1, 2)$.

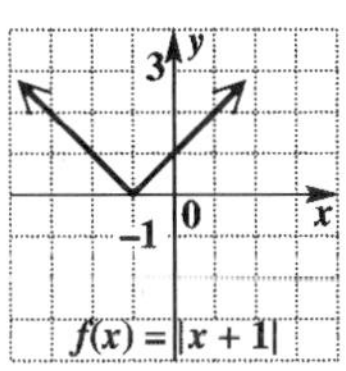

11. $f(x) = \dfrac{1}{x} + 1$

The graph of this function is similar to the graph of $g(x) = \dfrac{1}{x}$, except that each point is translated 1 unit upward. Just as with $g(x) = \dfrac{1}{x}$, $x = 0$ is the vertical asymptote, but this graph has $y = 1$ as its horizontal asymptote.

The domain is all real numbers except 0, that is, $(-\infty, 0) \cup (0, \infty)$.

The range is all real numbers except 1, that is, $(-\infty, 1) \cup (1, \infty)$.

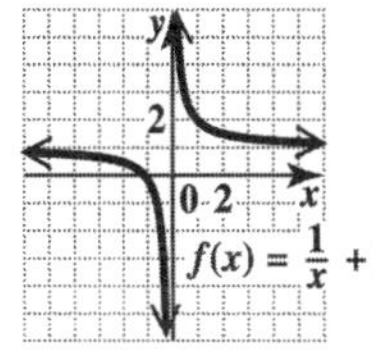

13. $f(x) = \sqrt{x - 2}$

The graph is found by shifting the graph of $y = \sqrt{x}$ two units to the right. The following table of ordered pairs gives some specific points the graph passes through.

x	2	3	6
y	0	1	2

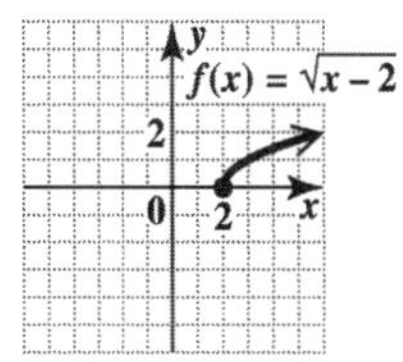

The domain of the function is $[2, \infty)$ and its range is $[0, \infty)$.

15. $f(x) = \dfrac{1}{x - 2}$

This is the graph of the reciprocal function, $g(x) = \dfrac{1}{x}$, shifted 2 units to the right. Since $x \neq 2$ (or a denominator of 0 results), the line $x = 2$ is a vertical asymptote. Since $\dfrac{1}{x-2} \neq 0$, the line $y = 0$ is a horizontal asymptote.

x	$\frac{5}{2}$	3	4	5
y	2	1	$\frac{1}{2}$	$\frac{1}{3}$

x	-1	0	1	$\frac{3}{2}$
y	$-\frac{1}{3}$	$-\frac{1}{2}$	-1	-2

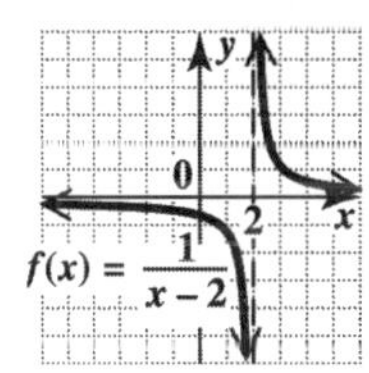

The domain of the function is $(-\infty, 2) \cup (2, \infty)$ and its range is $(-\infty, 0) \cup (0, \infty)$.

17. $f(x) = \sqrt{x + 3} - 3$

The graph is found by shifting the graph of $y = \sqrt{x}$ three units to the left and three units down. The following table of ordered pairs gives some specific points the graph passes through.

continued

x	-3	-2	1
y	-3	-2	-1

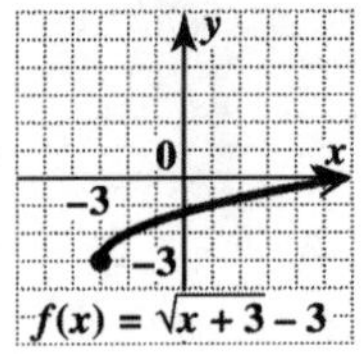

The domain of the function is $[-3, \infty)$ and its range is $[-3, \infty)$.

19. $f(x) = [\![x - 3]\!]$

This is the graph of the greatest integer function, $g(x) = [\![x]\!]$, shifted 3 units to the right.

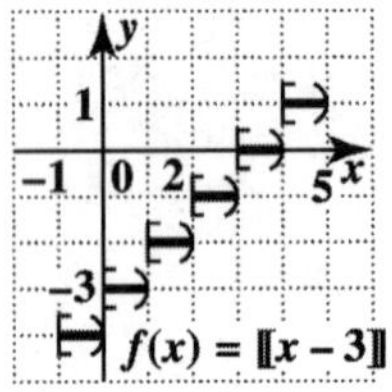

21. For any portion of the first ounce, the cost will be one 37¢ stamp. If the weight exceeds one ounce (up to two ounces), an additional 23¢ stamp is required. The following table summarizes the weight of a letter, x, and the number of stamps required, $p(x)$, on the interval $(0, 5]$.

x	$(0,1]$	$(1,2]$	$(2,3]$	$(3,4]$	$(4,5]$
$p(x)$	1	2	3	4	5

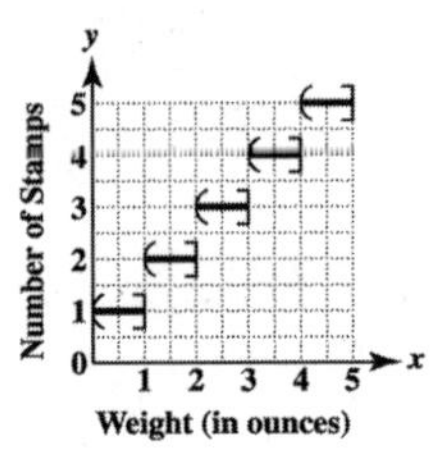

For Exercises 23–33,

$f(x) = x^2 + 4$, $g(x) = 2x + 3$, and $h(x) = x + 5$.

23. $$\begin{aligned}(h \circ g)(4) &= h[g(4)]\\ &= h(2 \cdot 4 + 3)\\ &= h(11)\\ &= 11 + 5\\ &= 16\end{aligned}$$

25. $$\begin{aligned}(g \circ f)(6) &= g[f(6)]\\ &= g(6^2 + 4)\\ &= g(40)\\ &= 2 \cdot 40 + 3\\ &= 83\end{aligned}$$

27. $$\begin{aligned}(f \circ h)(-2) &= f[h(-2)]\\ &= f(-2 + 5)\\ &= f(3)\\ &= 3^2 + 4\\ &= 13\end{aligned}$$

29. $$\begin{aligned}(f \circ g)(x) &= f[g(x)]\\ &= f(2x + 3)\\ &= (2x + 3)^2 + 4\\ &= 4x^2 + 12x + 9 + 4\\ &= 4x^2 + 12x + 13\end{aligned}$$

31. $$\begin{aligned}(f \circ h)(x) &= f[h(x)]\\ &= f(x + 5)\\ &= (x + 5)^2 + 4\\ &= x^2 + 10x + 25 + 4\\ &= x^2 + 10x + 29\end{aligned}$$

33. $$\begin{aligned}(h \circ g)(x) &= h[g(x)]\\ &= h(2x + 3)\\ &= 2x + 3 + 5\\ &= 2x + 8\end{aligned}$$

35. $f(x) = 12x$, $g(x) = 5280x$

$$\begin{aligned}(f \circ g)(x) &= f[g(x)]\\ &= f(5280x)\\ &= 12(5280x)\\ &= 63{,}360x\end{aligned}$$

$(f \circ g)(x)$ computes the number of inches in x mi.

37. $r(t) = 2t$, $A(r) = \pi r^2$

$$\begin{aligned}(A \circ r)(t) &= A[r(t)]\\ &= A(2t)\\ &= \pi(2t)^2\\ &= 4\pi t^2\end{aligned}$$

This is the area of the circular layer as a function of time.

Section 11.2

1. **(a)** $x^2 + y^2 = 25$ can be written in the center-radius form as

$$(x - 0)^2 + (y - 0)^2 = 5^2.$$

The center is the point $(0, 0)$.

(b) The radius is 5.

(c) The x-intercepts are $(5, 0)$ and $(-5, 0)$. The y-intercepts are $(0, 5)$ and $(0, -5)$.

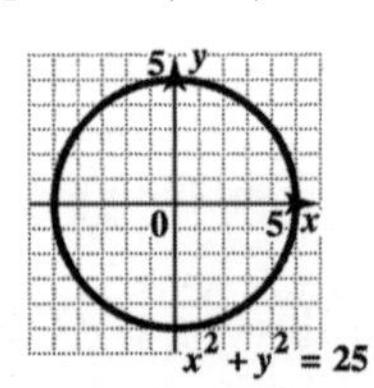

3. $(x-3)^2+(y-2)^2=25$ is an equation of a circle with center $(3,2)$ and radius 5, choice **B**.

5. $(x+3)^2+(y-2)^2=25$ is an equation of a circle with center $(-3,2)$ and radius 5, choice **D**.

7. Center: $(-4,3)$; radius: 2

Substitute $h=-4$, $k=3$, and $r=2$ in the center-radius form of the equation of a circle.

$$(x-h)^2+(y-k)^2=r^2$$
$$[x-(-4)]^2+(y-3)^2=2^2$$
$$(x+4)^2+(y-3)^2=4$$

9. Center: $(-8,-5)$; radius: $\sqrt{5}$

Substitute $h=-8$, $k=-5$, and $r=\sqrt{5}$ in the center-radius form of the equation of a circle.

$$(x-h)^2+(y-k)^2=r^2$$
$$[x-(-8)]^2+[y-(-5)]^2=\left(\sqrt{5}\right)^2$$
$$(x+8)^2+(y+5)^2=5$$

11. $$x^2+y^2+4x+6y+9=0$$

Rewrite the equation keeping only the variable terms on the left and grouping the x-terms and y terms.

$$x^2+4x+y^2+6y=-9$$

Complete both squares on the left, and add the same constants to the right.

$$(x^2+4x+\underline{4})+(y^2+6y+\underline{9})=-9+\underline{4}+\underline{9}$$
$$(x+2)^2+(y+3)^2=4$$

From the form $(x-h)^2+(y-k)^2=r^2$, we have $h=-2$, $k=-3$, and $r=2$. The center is $(-2,-3)$, and the radius r is 2.

13. $x^2+y^2+10x-14y-7=0$

$$(x^2+10x\qquad)+(y^2-14y\qquad)=7$$
$$(x^2+10x+\underline{25})+(y^2-14y+\underline{49})$$
$$=7+\underline{25}+\underline{49}$$
$$(x+5)^2+(y-7)^2=81$$

The center is $(-5,7)$, and the radius is $\sqrt{81}=9$.

15. $$3x^2+3y^2-12x-24y+12=0$$
$$3(x^2-4x\qquad)+3(y^2-8y\qquad)=-12$$

Divide by 3.

$$(x^2-4x\qquad)+(y^2-8y\qquad)=-4$$
$$(x^2-4x+\underline{4})+(y^2-8y+\underline{16})$$
$$=-4+\underline{4}+\underline{16}$$
$$(x-2)^2+(y-4)^2=16$$

The center is $(2,4)$, and the radius is $\sqrt{16}=4$.

17. This method works because the pencil is always the same distance from the fastened end. The fastened end works as the center, and the length of the string from the fastened end to the pencil is the radius.

19. $$x^2+y^2=9$$
$$(x-0)^2+(y-0)^2=3^2$$

Here, $h=0$, $k=0$, and $r=3$, so the graph is a circle with center $(0,0)$ and radius 3.

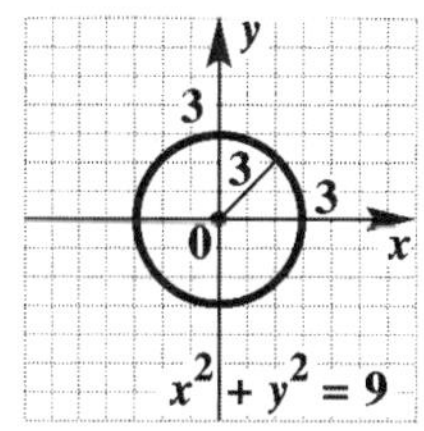

21. $$2y^2=10-2x^2$$
$$2x^2+2y^2=10$$
$$x^2+y^2=5 \qquad \textit{Divide by 2.}$$
$$(x-0)^2+(y-0)^2=\left(\sqrt{5}\right)^2$$

Here, $h=0$, $k=0$, and $r=\sqrt{5}\approx 2.2$, so the graph is a circle with center $(0,0)$ and radius $\sqrt{5}$.

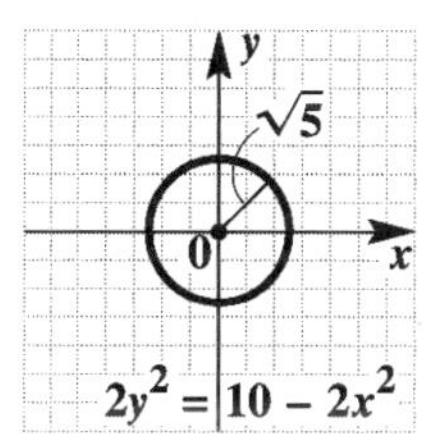

23. $(x+3)^2+(y-2)^2=9$

Here, $h=-3$, $k=2$, and $r=\sqrt{9}=3$. The graph is a circle with center $(-3,2)$ and radius 3.

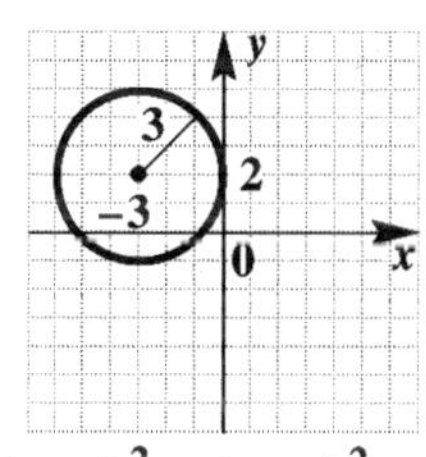

$(x+3)^2+(y-2)^2=9$

25. $$x^2+y^2-4x-6y+9=0$$
$$(x^2-4x\qquad)+(y^2-6y\qquad)=-9$$

Complete the square for each variable.

$$(x^2-4x+\underline{4})+(y^2-6y+\underline{9})$$
$$=-9+\underline{4}+\underline{9}$$
$$(x-2)^2+(y-3)^2=4$$

continued

Here, $h = 2$, $k = 3$, and $r = \sqrt{4} = 2$. The graph is a circle with center $(2, 3)$ and radius 2.

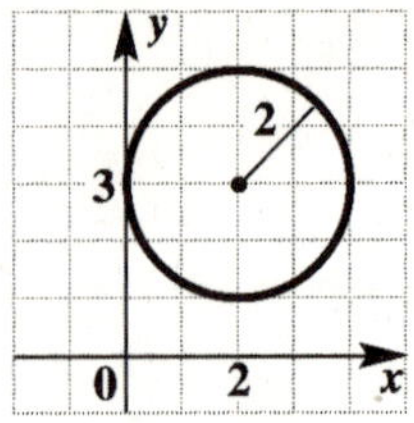

$x^2 + y^2 - 4x - 6y + 9 = 0$

27. The equation $\dfrac{x^2}{9} + \dfrac{y^2}{25} = 1$ is of the form $\dfrac{x^2}{a^2} + \dfrac{y^2}{b^2} = 1$. The graph is an ellipse with $a^2 = 9$ and $b^2 = 25$, so $a = 3$ and $b = 5$. The x-intercepts are $(3, 0)$ and $(-3, 0)$. The y-intercepts are $(0, 5)$ and $(0, -5)$. Plot the intercepts, and draw the ellipse through them.

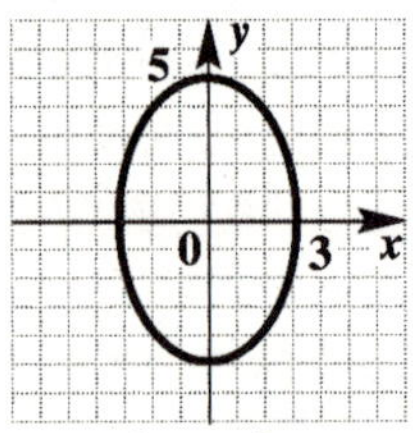

$\dfrac{x^2}{9} + \dfrac{y^2}{25} = 1$

29. $\dfrac{x^2}{36} = 1 - \dfrac{y^2}{16}$

$\dfrac{x^2}{36} + \dfrac{y^2}{16} = 1$ is in the form $\dfrac{x^2}{a^2} + \dfrac{y^2}{b^2} = 1.$

The graph is an ellipse with $a^2 = 36$ and $b^2 = 16$, so $a = 6$ and $b = 4$. The x-intercepts are $(6, 0)$ and $(-6, 0)$. The y-intercepts are $(0, 4)$ and $(0, -4)$. Plot the intercepts, and draw the ellipse through them.

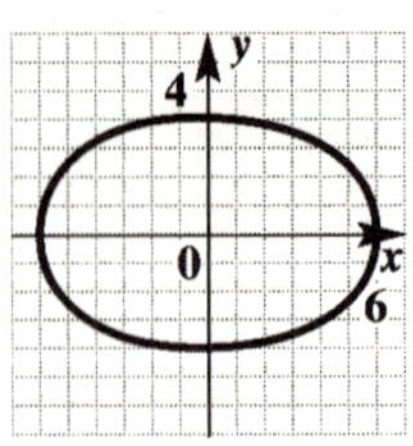

$\dfrac{x^2}{36} = 1 - \dfrac{y^2}{16}$

31. $\dfrac{y^2}{25} = 1 - \dfrac{x^2}{49}$

$\dfrac{x^2}{49} + \dfrac{y^2}{25} = 1$ is in the form $\dfrac{x^2}{a^2} + \dfrac{y^2}{b^2} = 1.$

The graph is an ellipse with $a^2 = 49$ and $b^2 = 25$, so $a = 7$ and $b = 5$. The x-intercepts are $(7, 0)$ and $(-7, 0)$. The y-intercepts are $(0, 5)$ and $(0, -5)$. Plot the intercepts, and draw the ellipse through them.

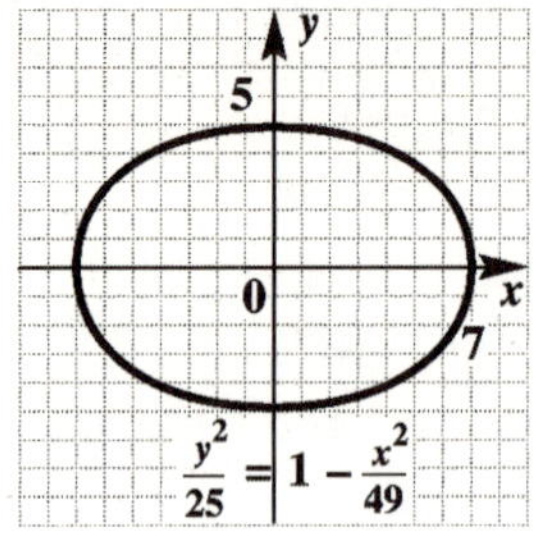

33. $\dfrac{x^2}{16} + \dfrac{y^2}{4} = 1$ is in the form

$$\frac{x^2}{a^2} + \frac{y^2}{b^2} = 1,$$

so $a = 4$ and $b = 2$. Its x-intercepts are $(4, 0)$ and $(-4, 0)$, and its y-intercepts are $(0, 2)$ and $(0, -2)$.

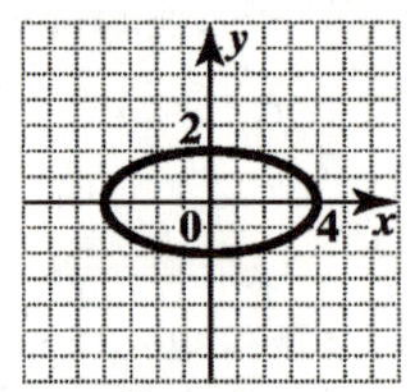

$\dfrac{x^2}{16} + \dfrac{y^2}{4} = 1$

35. $\dfrac{(x+1)^2}{64} + \dfrac{(y-2)^2}{49} = 1$

This equation is of the form

$$\frac{(x-h)^2}{a^2} + \frac{(y-k)^2}{b^2} = 1,$$

so the center of the ellipse is at $(-1, 2)$. Since $a^2 = 64$, $a = 8$. Since $b^2 = 49$, $b = 7$. Add ± 8 to -1, and add ± 7 to 2 to find the points $(7, 2)$, $(-9, 2)$, $(-1, 9)$, and $(-1, -5)$. Plot the points, and draw the ellipse through them.

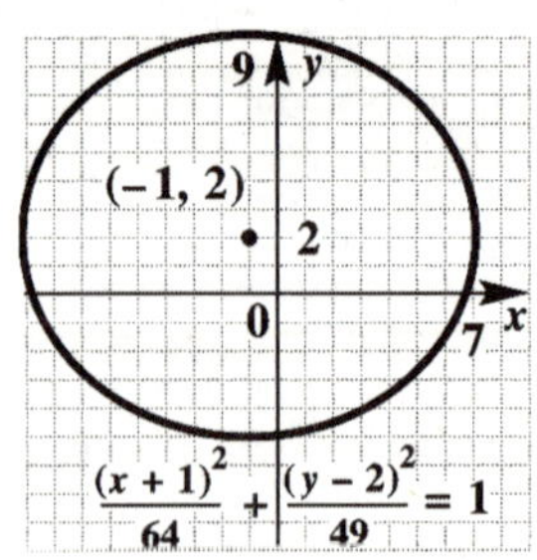

37. $\dfrac{(x-2)^2}{16} + \dfrac{(y-1)^2}{9} = 1$

The center of the ellipse is at $(2, 1)$. Since $a^2 = 16$, $a = 4$. Since $b^2 = 9$, $b = 3$. Add ± 4 to 2, and add ± 3 to 1 to find the points $(6, 1)$, $(-2, 1)$, $(2, 4)$, and $(2, -2)$. Plot the points, and draw the ellipse through them.

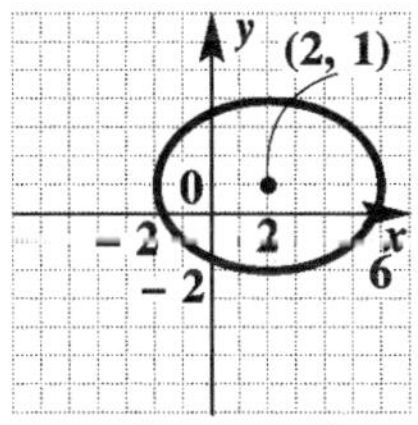

$$\frac{(x-2)^2}{16}+\frac{(y-1)^2}{9}=1$$

39. As the pencil traces the graph along the taut string, notice that the distances from the pencil to the two fastened ends continue to change, but the sum of the two distances is always the same, namely, the length of the string between the two fastened ends. This is exactly the definition of an ellipse—the set of all points in a plane the sum of whose distances from two fixed points (the fastened ends) is constant.

41. By the vertical line test the set is not a function, because a vertical line may intersect the graph of an ellipse in two points.

43.
$$(x+2)^2+(y-4)^2=16$$
$$(y-4)^2=16-(x+2)^2$$

Take the square root of each side.

$$y-4=\pm\sqrt{16-(x+2)^2}$$
$$y=4\pm\sqrt{16-(x+2)^2}$$

Therefore, the two functions used to obtain the graph were

$$y_1=4+\sqrt{16-(x+2)^2} \quad \text{and}$$
$$y_2=4-\sqrt{16-(x+2)^2}.$$

45.
$$x^2+y^2=36$$
$$y^2=36-x^2$$

Take the square root of both sides.

$$y=\pm\sqrt{36-x^2}$$

Therefore,

$$y_1=\sqrt{36-x^2} \text{ and } y_2=-\sqrt{36-x^2}.$$

Use these two functions to obtain the graph.

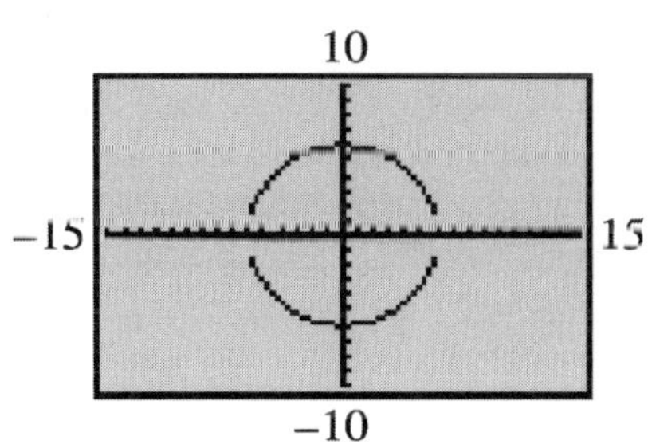

47.
$$\frac{x^2}{16}+\frac{y^2}{4}=1$$
$$\frac{y^2}{4}=1-\frac{x^2}{16}$$
$$y^2=4\left(1-\frac{x^2}{16}\right)$$

Take the square root of both sides.

$$y=\pm 2\sqrt{1-\frac{x^2}{16}}$$

Therefore,

$$y_1=2\sqrt{1-\frac{x^2}{16}} \text{ and } y_2=-2\sqrt{1-\frac{x^2}{16}}.$$

Use these two functions to obtain the graph.

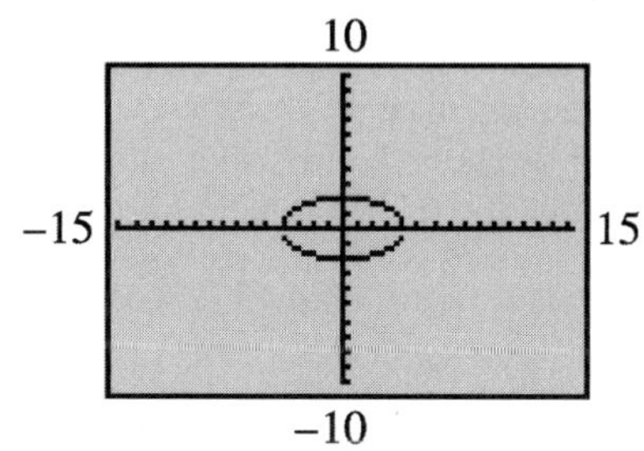

49. **(a)** $100x^2+324y^2=32,400$
Divide by $32,400$.

$$\frac{x^2}{324}+\frac{y^2}{100}=1$$
$$\frac{x^2}{18^2}+\frac{y^2}{10^2}=1$$

The height in the center is the y-coordinate of the positive y-intercept. The height is 10 meters.

(b) The width of the ellipse is the distance between the x-intercepts, $(-18,0)$ and $(18,0)$. The width across the bottom of the arch is $18+18=36$ meters.

51.
$$\frac{x^2}{141.7^2}+\frac{y^2}{141.1^2}=1$$

(a) $c^2=a^2-b^2$, so

$$c=\sqrt{a^2-b^2}=\sqrt{141.7^2-141.1^2}$$
$$=\sqrt{169.68}\approx 13.0$$

From the figure, the *apogee* is $a+c=141.7+13.0=154.7$ million miles.

(b) The *perigee* is $a-c=141.7-13.0=128.7$ million miles.

53. $\dfrac{x^2}{36}+\dfrac{y^2}{9}=1$

$c^2=a^2-b^2=36-9=27,$ so

$c=\sqrt{27}=3\sqrt{3}.$

The kidney stone and the source of the beam must be placed $3\sqrt{3}$ units from the center of the ellipse.

55. **(a)** The foci are $(c,0)$ and $(-c,0)$. The sum of the distances is $2a$. By the definition of an ellipse,

$$\sqrt{(x-c)^2+(y-0)^2} + \sqrt{(x+c)^2+(y-0)^2}=2a$$
$$\sqrt{(x-c)^2+y^2}+\sqrt{(x+c)^2+y^2}=2a$$
$$\sqrt{(x-c)^2+y^2}=2a-\sqrt{(x+c)^2+y^2}.$$

Square both sides.

$$(x-c)^2+y^2=4a^2-4a\sqrt{(x+c)^2+y^2} + (x+c)^2+y^2$$
$$x^2-2xc+c^2+y^2 = 4a^2-4a\sqrt{(x+c)^2+y^2} + x^2+2xc+c^2+y^2$$
$$-4xc=4a^2-4a\sqrt{(x+c)^2+y^2}$$
$$-4xc-4a^2=-4a\sqrt{(x+c)^2+y^2}$$

Divide by -4.

$$xc+a^2=a\sqrt{(x+c)^2+y^2}$$

Square both sides.

$$x^2c^2+2xca^2+a^4=a^2\left[(x+c)^2+y^2\right]$$
$$x^2c^2+2xca^2+a^4=a^2\left(x^2+2xc+c^2+y^2\right)$$
$$x^2c^2+2xca^2+a^4=a^2x^2+2a^2xc+a^2c^2+a^2y^2$$
$$x^2c^2+a^4=a^2x^2+a^2c^2+a^2y^2$$

Isolate the x^2- and y^2-terms on one side of the equals sign.

$$a^4-a^2c^2=a^2x^2-x^2c^2+a^2y^2$$
$$a^2\left(a^2-c^2\right)=x^2\left(a^2-c^2\right)+a^2y^2$$

Divide each term by $a^2(a^2-c^2)$ to force the left side to equal 1.

$$1=\frac{x^2(a^2-c^2)}{a^2(a^2-c^2)}+\frac{a^2y^2}{a^2(a^2-c^2)}$$
$$1=\frac{x^2}{a^2}+\frac{y^2}{a^2-c^2}$$

(b) $\dfrac{x^2}{a^2}+\dfrac{y^2}{a^2-c^2}=1$

To find the x-intercepts, let $y=0$.

$$\frac{x^2}{a^2}+\frac{0^2}{a^2-c^2}=1$$
$$\frac{x^2}{a^2}=1$$
$$x^2=a^2$$
$$x=\pm a$$

The x-intercepts are $(a,0)$ and $(-a,0)$.

(c) Let $b^2=a^2-c^2$.
Therefore, the equation is
$$\frac{x^2}{a^2}+\frac{y^2}{b^2}=1.$$

To find the y-intercepts, let $x=0$.

$$\frac{0^2}{a^2}+\frac{y^2}{b^2}=1$$
$$\frac{y^2}{b^2}=1$$
$$y^2=b^2$$
$$y=\pm b$$

The y-intercepts are $(0,b)$ and $(0,-b)$.

Section 11.3

1. $\dfrac{x^2}{25}+\dfrac{y^2}{9}=1$

This is the standard form for the equation of an ellipse with x-intercepts $(5,0)$ and $(-5,0)$ and y-intercepts $(0,3)$ and $(0,-3)$. This is graph **C**.

3. $\dfrac{x^2}{9}-\dfrac{y^2}{25}=1$

This is the standard form for the equation of a hyperbola that opens left and right. Its x-intercepts are $(3,0)$ and $(-3,0)$. This is graph **D**.

5. If the equation of a hyperbola is in standard form (that is, equal to one), the hyperbola would open to the left and right if the x^2-term was positive. It would open up and down if the y^2-term was positive.

7. The equation $\dfrac{x^2}{16}-\dfrac{y^2}{9}=1$ is in the form $\dfrac{x^2}{a^2}-\dfrac{y^2}{b^2}=1$. The graph is a hyperbola with $a=4$ and $b=3$. The x-intercepts are $(4,0)$ and $(-4,0)$. There are no y-intercepts. The vertices of the fundamental rectangle are $(4,3)$, $(4,-3)$, $(-4,-3)$, and $(-4,3)$. Extend the diagonals of the rectangle through these points to get the asymptotes. Graph a branch of the hyperbola through each intercept and approaching the asymptotes.

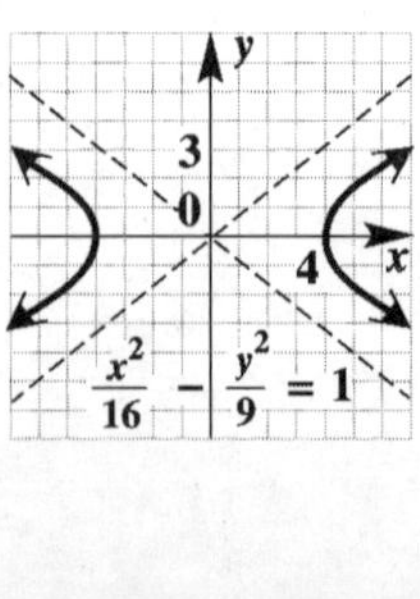

9. $\dfrac{y^2}{9} - \dfrac{x^2}{9} = 1$ is a hyperbola with $a = 3$ and $b = 3$. The y-intercepts are $(0, 3)$ and $(0, -3)$. There are no x-intercepts. One asymptote passes through $(3, 3)$ and $(-3, -3)$. The other asymptote passes through $(-3, 3)$ and $(3, -3)$. Draw the asymptotes and sketch the hyperbola through the intercepts and approaching the asymptotes.

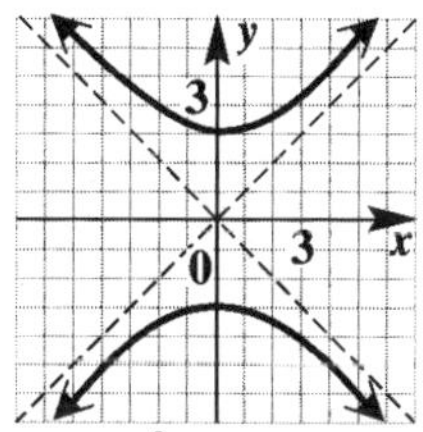

$\dfrac{y^2}{9} - \dfrac{x^2}{9} = 1$

11. $\dfrac{x^2}{25} - \dfrac{y^2}{36} = 1$ is a hyperbola with $a = 5$ and $b = 6$. The x-intercepts are $(5, 0)$ and $(-5, 0)$. There are no y-intercepts. To sketch the graph, draw the diagonals of the fundamental rectangle with vertices $(5, 6)$, $(5, -6)$, $(-5, -6)$ and $(-5, 6)$. Graph a branch of the hyperbola through each intercept and approaching the asymptotes.

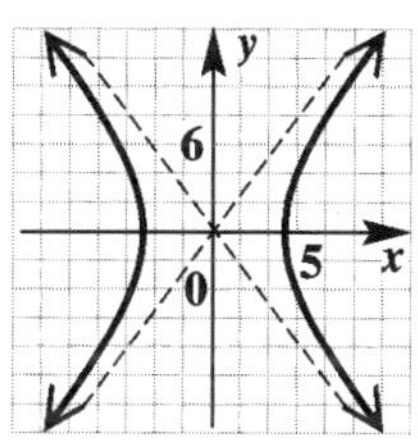

$\dfrac{x^2}{25} - \dfrac{y^2}{36} = 1$

13. $\dfrac{y^2}{16} - \dfrac{x^2}{16} = 1$ is in the form $\dfrac{y^2}{b^2} - \dfrac{x^2}{a^2} = 1$, where $b = 4$ and $a = 4$. The branches open upward and downward, so the hyperbola has y-intercepts $(0, 4)$ and $(0, -4)$. The fundamental rectangle has vertices $(4, 4)$, $(4, -4)$, $(-4, -4)$, and $(-4, 4)$. Sketch the extended diagonals, which are the asymptotes of this hyperbola. Graph a branch of the hyperbola through each intercept and approaching the asymptotes.

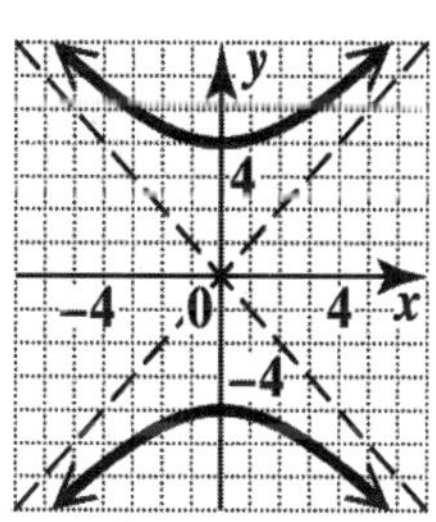

$\dfrac{y^2}{16} - \dfrac{x^2}{16} = 1$

15. $x^2 - y^2 = 16$

$\dfrac{x^2}{16} - \dfrac{y^2}{16} = 1$ *Divide by 16.*

This equation is in the form $\dfrac{x^2}{a^2} - \dfrac{y^2}{b^2} = 1$ with $a = 4$ and $b = 4$. The graph is a hyperbola with x-intercepts $(4, 0)$ and $(-4, 0)$ and no y-intercepts. One asymptote passes through $(4, 4)$ and $(-4, -4)$. The other asymptote passes through $(-4, 4)$ and $(4, -4)$. Sketch the graph through the intercepts and approaching the asymptotes.

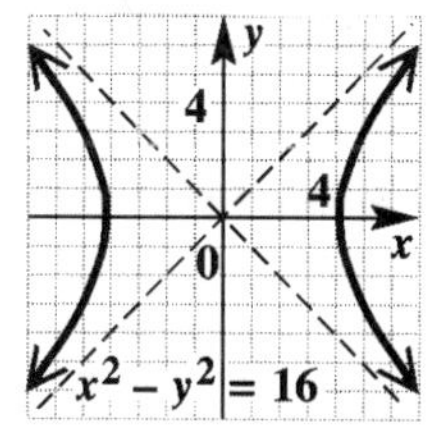

17. $4x^2 + y^2 = 16$

$\dfrac{x^2}{4} + \dfrac{y^2}{16} = 1$ *Divide by 16.*

This equation is in the form $\dfrac{x^2}{a^2} + \dfrac{y^2}{b^2} = 1$ with $a = 2$ and $b = 4$. The graph is an ellipse. The x-intercepts $(2, 0)$ and $(-2, 0)$. The y-intercepts are $(0, 4)$ and $(0, -4)$. Plot the intercepts and draw the ellipse through them.

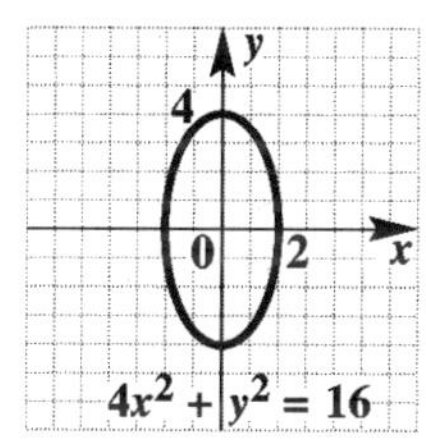

19.
$$y^2 = 36 - x^2$$
$$x^2 + y^2 = 36$$
$$(x - 0)^2 + (y - 0)^2 = 36$$

The graph is a circle with center at $(0, 0)$ and radius $\sqrt{36} = 6$.

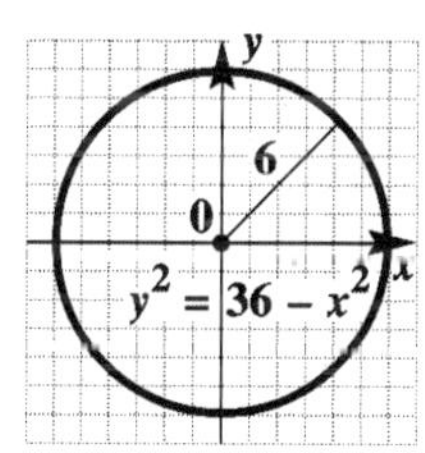

21.
$$9x^2 = 144 + 16y^2$$
$$9x^2 - 16y^2 = 144$$
$$\frac{x^2}{16} - \frac{y^2}{9} = 1 \quad \textit{Divide by 144.}$$

continued

The equation is a hyperbola in the form $\frac{x^2}{a^2} - \frac{y^2}{b^2} = 1$ with $a = 4$ and $b = 3$. The x-intercepts are $(4, 0)$ and $(-4, 0)$. There are no y-intercepts. To sketch the graph, draw the diagonals of the fundamental rectangle with vertices $(4, 3)$, $(4, -3)$, $(-4, -3)$ and $(-4, 3)$. These are the asymptotes. Graph a branch of the hyperbola through each intercept approaching the asymptotes.

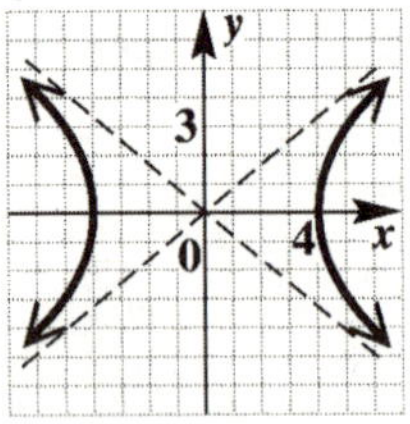

$9x^2 = 144 + 16y^2$

23.

$$\begin{aligned} y^2 &= 4 + x^2 \\ y^2 - x^2 &= 4 \\ \frac{y^2}{4} - \frac{x^2}{4} &= 1 \qquad \textit{Divide by 4.} \end{aligned}$$

The graph is a hyperbola with y-intercepts $(0, 2)$ and $(0, -2)$. One asymptote passes through $(2, 2)$ and $(-2, -2)$. The other asymptote passes through $(-2, 2)$ and $(2, -2)$.

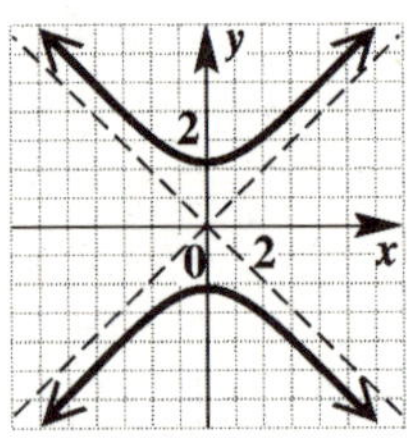

$y^2 = 4 + x^2$

25. $f(x) = \sqrt{16 - x^2}$

Replace $f(x)$ with y and square both sides to get the equation

$$y^2 = 16 - x^2 \quad \text{or} \quad x^2 + y^2 = 16.$$

This is the graph of a circle with center $(0, 0)$ and radius 4. Since $f(x)$, or y, represents a principal square root in the original equation, $f(x)$ must be nonnegative. This restricts the graph to the upper half of the circle.

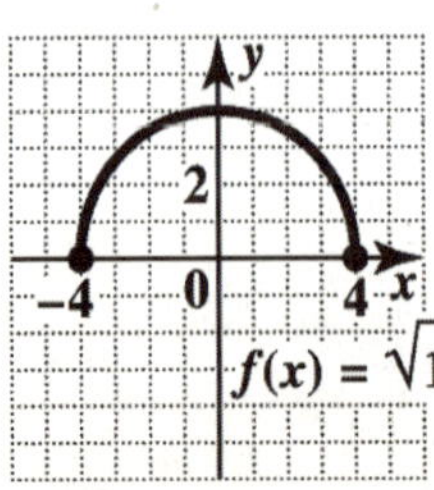

The domain is $[-4, 4]$, and the range is $[0, 4]$.

27.

$$f(x) = -\sqrt{36 - x^2}$$

Replace $f(x)$ with y, and square both sides of the equation.

$$\begin{aligned} y &= -\sqrt{36 - x^2} \\ y^2 &= 36 - x^2 \\ x^2 + y^2 &= 36 \end{aligned}$$

This is a circle centered at the origin with radius $\sqrt{36} = 6$. Since $f(x)$, or y, represents a nonpositive square root in the original equation, $f(x)$ must be nonpositive. This restricts the graph to the bottom half of the circle.

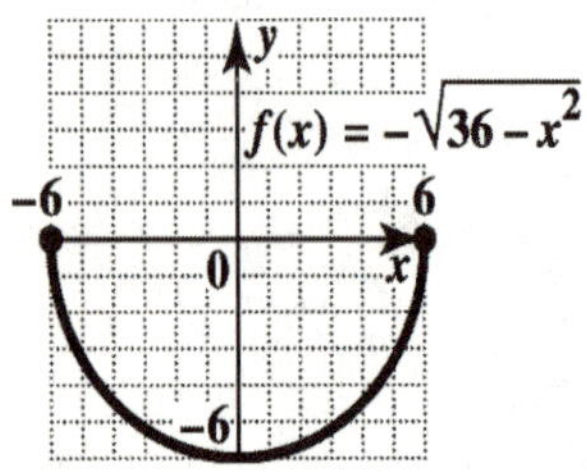

The domain is $[-6, 6]$, and the range is $[-6, 0]$.

29.

$$\frac{y}{3} = \sqrt{1 + \frac{x^2}{9}}$$

Square both sides.

$$\begin{aligned} \frac{y^2}{9} &= 1 + \frac{x^2}{9} \\ \frac{y^2}{9} - \frac{x^2}{9} &= 1 \end{aligned}$$

This is a hyperbola opening up and down with y-intercepts $(0, 3)$ and $(0, -3)$. The four points $(3, 3)$, $(3, -3)$, $(-3, 3)$, and $(-3, -3)$ are the vertices of the rectangle that determine the asymptotes. Since $f(x)$, or y, represents a square root in the original equation, $f(x)$ must be nonnegative. This restricts the graph to the upper half of the hyperbola.

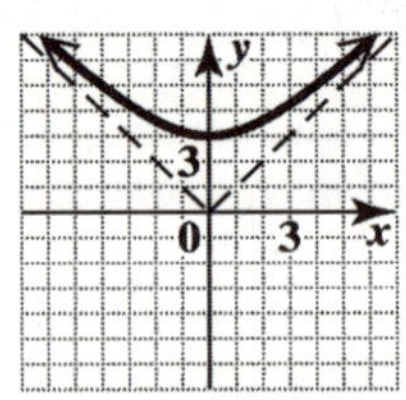

$\frac{y}{3} = \sqrt{1 + \frac{x^2}{9}}$

The domain is $(-\infty, \infty)$, and the range is $[3, \infty)$.

31.

$$\frac{(x-2)^2}{4} - \frac{(y+1)^2}{9} = 1$$

is a hyperbola centered at $(2, -1)$, with $a = 2$ and $b = 3$. The x-intercepts are $(2 \pm 2, -1)$ or $(4, -1)$ and $(0, -1)$. The asymptotes are the extended diagonals of the rectangle with vertices

$(2,3)$, $(2,-3)$, $(-2,-3)$ and $(-2,3)$ shifted 2 units right and 1 unit down, or $(4,2)$, $(4,-4)$, $(0,-4)$ and $(0,2)$. Draw the hyperbola.

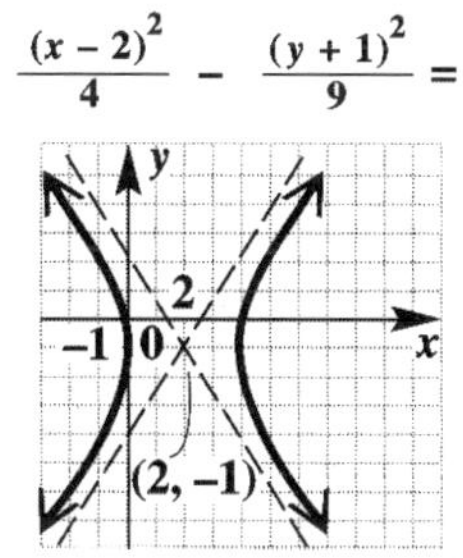

33. $\dfrac{y^2}{36} - \dfrac{(x-2)^2}{49} = 1$

is a hyperbola centered at $(2,0)$ with $a = 7$, and $b = 6$. The asymptotes are the extended diagonals of the rectangle with vertices $(7,6)$, $(7,-6)$, $(-7,-6)$, and $(-7,6)$ shifted right 2 units, or $(9,6)$, $(9,-6)$, $(-5,-6)$, and $(-5,6)$. Draw the hyperbola.

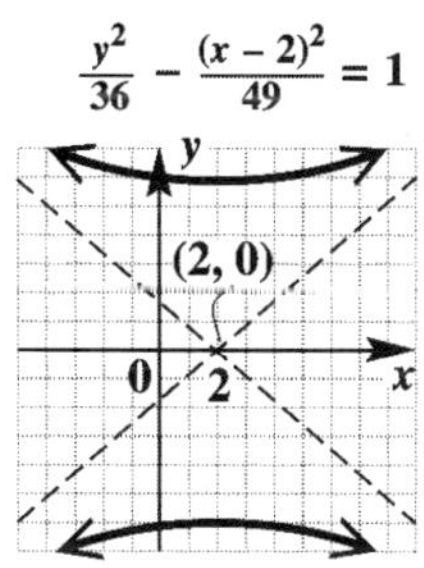

35. **(a)** $400x^2 - 625y^2 = 250,000$

Divide by $250,000$.

$$\frac{x^2}{625} - \frac{y^2}{400} = 1$$

$$\frac{x^2}{25^2} - \frac{y^2}{20^2} = 1$$

The x-intercepts are $(25,0)$ and $(-25,0)$. The distance between the buildings is the distance between the x-intercepts. The buildings are $25 + 25 = 50$ meters apart at their closest point.

(b) At $x = 50$, $y = \dfrac{d}{2}$, so $d = 2y$.

$$400(50)^2 - 625y^2 = 250,000$$
$$1,000,000 - 625y^2 = 250,000$$
$$625y^2 = 750,000$$
$$y^2 = 1200$$
$$y = \sqrt{1200}$$

The distance d is $2\sqrt{1200} \approx 69.3$ meters.

37. $V > \dfrac{2.82 \times 10^7}{\sqrt{D}}$

$$V > \frac{2.82 \times 10^7}{\sqrt{4.25 \times 10^7}} \approx 4325.68$$

The trajectory will be hyperbolic for V greater than 4325.68 meters per second.

39. $\dfrac{x^2}{9} - y^2 = 1$

$$-y^2 = 1 - \frac{x^2}{9}$$

$$y^2 = \frac{x^2}{9} - 1 \qquad \textit{Multiply by } -1.$$

Take the square root of both sides.

$$y = \pm\sqrt{\frac{x^2}{9} - 1}$$

The two functions used to obtain the graph were

$$y_1 = \sqrt{\frac{x^2}{9} - 1} \text{ and } y_2 = -\sqrt{\frac{x^2}{9} - 1}.$$

41. $\dfrac{x^2}{25} - \dfrac{y^2}{49} = 1$

$$-\frac{y^2}{49} = 1 - \frac{x^2}{25}$$

$$\frac{y^2}{49} = \frac{x^2}{25} - 1 \qquad \textit{Multiply by } -1.$$

$$y^2 = 49\left(\frac{x^2}{25} - 1\right)$$

Take the square root of both sides.

$$y = \pm 7\sqrt{\frac{x^2}{25} - 1}$$

To obtain the graph, use the two functions

$$y_1 = 7\sqrt{\frac{x^2}{25} - 1} \text{ and } y_2 = -7\sqrt{\frac{x^2}{25} - 1}.$$

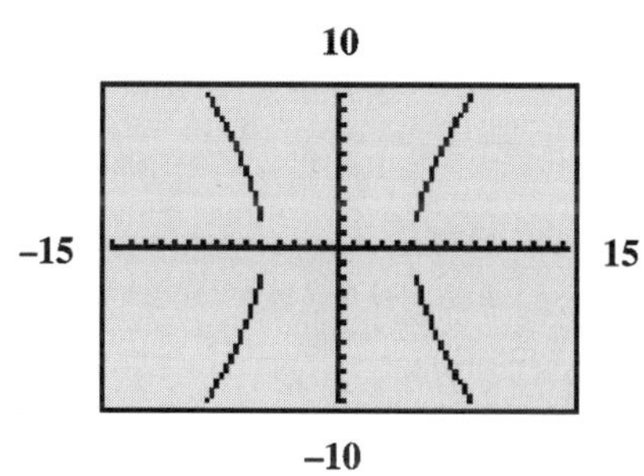

43. $\dfrac{y^2}{9} - x^2 = 1$

$$y^2 - 9x^2 = 9 \qquad \textit{Multiply by } 9.$$

$$y^2 = 9 + 9x^2$$

$$y^2 = 9(1 + x^2)$$

Take the square root of both sides.

$$y = \pm 3\sqrt{1 + x^2}$$

To obtain the graphs, use the two functions

$$y_1 = 3\sqrt{1 + x^2} \text{ and } y_2 = -3\sqrt{1 + x^2}.$$

continued

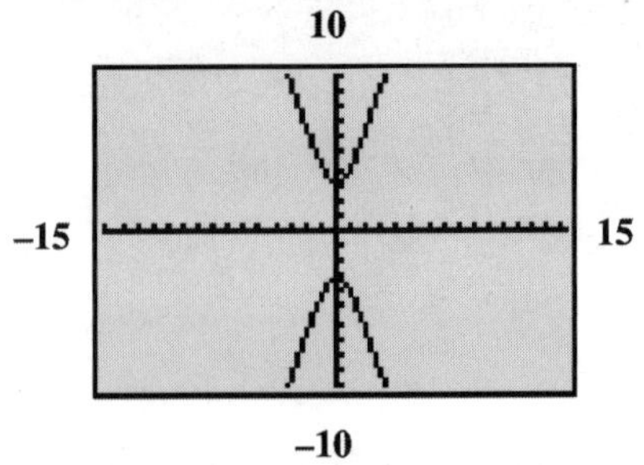

45. $$\sqrt{[x-(-c)]^2+(y-0)^2} - \sqrt{(x-c)^2+(y-0)^2} = 2a$$
$$\sqrt{(x+c)^2+y^2} = 2a + \sqrt{(x-c)^2+y^2}$$

Square both sides.

$$\begin{aligned}(x+c)^2+y^2 &= 4a^2+4a\sqrt{(x-c)^2+y^2} \\ &\quad + (x-c)^2+y^2 \\ x^2+2cx+c^2+y^2 &= 4a^2+4a\sqrt{(x-c)^2+y^2} \\ &\quad + x^2-2cx+c^2+y^2 \\ 4cx-4a^2 &= 4a\sqrt{(x-c)^2+y^2}\end{aligned}$$

Divide by 4.

$$cx-a^2 = a\sqrt{(x-c)^2+y^2}$$

Square both sides again.

$$\begin{aligned}c^2x^2-2a^2cx+a^4 &= a^2\left[(x-c)^2+y^2\right] \\ c^2x^2-2a^2cx+a^4 &= a^2\left(x^2-2cx+c^2+y^2\right) \\ c^2x^2-2a^2cx+a^4 &= a^2x^2-2a^2cx+a^2c^2+a^2y^2 \\ c^2x^2+a^4 &= a^2x^2+a^2c^2+a^2y^2\end{aligned}$$

Isolate the x^2- and y^2-terms on one side of the equals sign.

$$\begin{aligned}c^2x^2-a^2x^2-a^2y^2 &= a^2c^2-a^4 \\ x^2\left(c^2-a^2\right)-a^2y^2 &= a^2\left(c^2-a^2\right)\end{aligned}$$

Since $b^2=c^2-a^2$,

$$\begin{aligned}x^2b^2-a^2y^2 &= a^2b^2 \\ \frac{x^2b^2}{a^2b^2}-\frac{a^2y^2}{a^2b^2} &= \frac{a^2b^2}{a^2b^2} \qquad \textit{Divide by } a^2b^2. \\ \frac{x^2}{a^2}-\frac{y^2}{b^2} &= 1.\end{aligned}$$

Section 11.4

1. Substitute $x-1$ for y in the first equation. Then solve for x. Find the corresponding y-values by substituting back into $y=x-1$. In the first equation, both variables are squared and in the second, both variables are to the first power, so the elimination method is not appropriate.

3. The line intersects the ellipse in exactly one point, so there is *one* point in the solution set of the system.

5. The line does not intersect the hyperbola, so there are no points in the solution set of the system.

7. A line and a circle; no points

Draw any circle, and then draw a line that does not cross the circle.

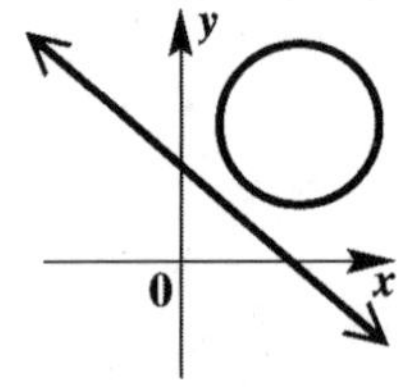

9. A line and a hyperbola; one point

The line is tangent to the hyperbola.

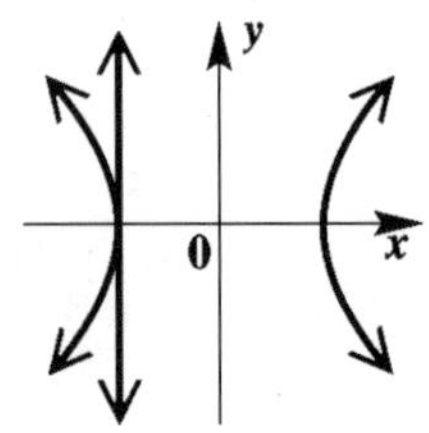

11. A circle and an ellipse; four points

Draw any ellipse, and then draw a circle with the same center whose radius is just large enough so that there are four points of intersection. (If the radius of the circle is too large or too small, there may be fewer points of intersection.)

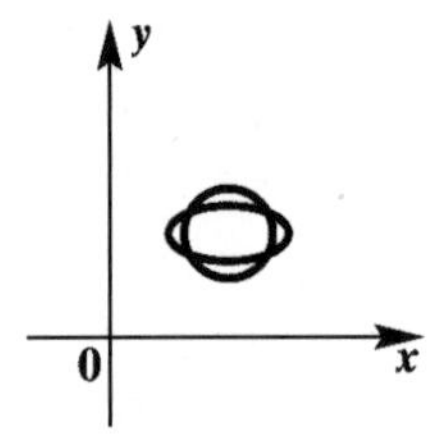

13. A parabola and an ellipse; four points

Draw any parabola, and then draw an ellipse large enough so that there are four points of intersection. (If the ellipse is too large or too small, there may be fewer points of intersection.)

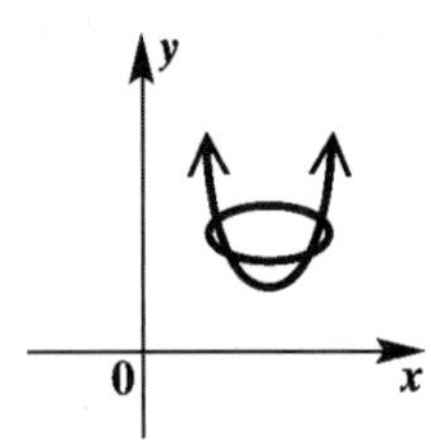

15. $$\begin{aligned}y &= 4x^2-x \qquad (1) \\ y &= x \qquad (2)\end{aligned}$$

Substitute x for y in equation (1).

$$\begin{aligned}y &= 4x^2-x \qquad (1) \\ x &= 4x^2-x \\ 0 &= 4x^2-2x \\ 0 &= 2x(2x-1)\end{aligned}$$

$$2x = 0 \quad \text{or} \quad 2x - 1 = 0$$
$$x = 0 \quad \text{or} \quad x = \frac{1}{2}$$

Use equation (2) to find y for each x-value.

If $x = 0$, then $y = 0$.
If $x = \frac{1}{2}$, then $y = \frac{1}{2}$.

Solution set: $\left\{(0, 0), \left(\frac{1}{2}, \frac{1}{2}\right)\right\}$

17.
$$y = x^2 + 6x + 9 \quad (1)$$
$$x + y = 3 \quad (2)$$

Substitute $x^2 + 6x + 9$ for y in equation (2).

$$x + y = 3 \quad (2)$$
$$x + \left(x^2 + 6x + 9\right) = 3$$
$$x^2 + 7x + 9 = 3$$
$$x^2 + 7x + 6 = 0$$
$$(x + 6)(x + 1) = 0$$

$$x + 6 = 0 \quad \text{or} \quad x + 1 = 0$$
$$x = -6 \quad \text{or} \quad x = -1$$

Substitute these values for x in equation (2) and solve for y.

If $x = -6$, then
$$x + y = 3 \quad (2)$$
$$-6 + y = 3$$
$$y = 9.$$

If $x = -1$, then
$$x + y = 3 \quad (2)$$
$$-1 + y = 3$$
$$y = 4.$$

Solution set: $\{(-6, 9), (-1, 4)\}$

19.
$$x^2 + y^2 = 2 \quad (1)$$
$$2x + y = 1 \quad (2)$$

Solve equation (2) for y.

$$y = 1 - 2x \quad (3)$$

Substitute $1 - 2x$ for y in equation (1).

$$x^2 + y^2 = 2 \quad (1)$$
$$x^2 + (1 - 2x)^2 = 2$$
$$x^2 + 1 - 4x + 4x^2 = 2$$
$$5x^2 - 4x - 1 = 0$$
$$(5x + 1)(x - 1) = 0$$

$$5x + 1 = 0 \quad \text{or} \quad x - 1 = 0$$
$$x = -\frac{1}{5} \quad \text{or} \quad x = 1$$

Use equation (3) to find y for each x-value.

If $x = -\frac{1}{5}$, then

$$y = 1 - 2\left(-\frac{1}{5}\right) = 1 + \frac{2}{5} = \frac{7}{5}.$$

If $x = 1$, then

$$y = 1 - 2(1) = -1.$$

Solution set: $\left\{\left(-\frac{1}{5}, \frac{7}{5}\right), (1, -1)\right\}$

21.
$$xy = 4 \quad (1)$$
$$3x + 2y = -10 \quad (2)$$

Solve equation (1) for y to get $y = \frac{4}{x}$.

Substitute $\frac{4}{x}$ for y in equation (2) to find x.

$$3x + 2y = -10 \quad (2)$$
$$3x + 2\left(\frac{4}{x}\right) = -10$$

Multiply by the LCD, x.

$$3x^2 + 8 = -10x$$
$$3x^2 + 10x + 8 = 0$$
$$(3x + 4)(x + 2) = 0$$

$$3x + 4 = 0 \quad \text{or} \quad x + 2 = 0$$
$$x = -\frac{4}{3} \quad \text{or} \quad x = -2$$

Since $y = \frac{4}{x}$, if $x = -\frac{4}{3}$, then $y = \frac{4}{-\frac{4}{3}} = -3.$

If $x = -2$, then
$$y = \frac{4}{-2} = -2.$$

Solution set: $\left\{(-2, -2), \left(-\frac{4}{3}, -3\right)\right\}$

23.
$$xy = -3 \quad (1)$$
$$x + y = -2 \quad (2)$$

Solve equation (2) for y.

$$y = -x - 2 \quad (3)$$

Substitute $-x - 2$ for y in equation (1).

$$xy = -3 \quad (1)$$
$$x(-x - 2) = -3$$
$$-x^2 - 2x = -3$$
$$-x^2 - 2x + 3 = 0$$
$$x^2 + 2x - 3 = 0 \quad \textit{Multiply by } -1.$$
$$(x + 3)(x - 1) = 0$$

continued

$$x+3=0 \quad \text{or} \quad x-1=0$$
$$x=-3 \quad \text{or} \quad x=1$$

Use equation (3) to find y for each x-value.

If $x=-3$, then
$$y=-(-3)-2=1.$$

If $x=1$, then
$$y=-(1)-2=-3.$$

Solution set: $\{(-3,1),(1,-3)\}$

25. $y=3x^2+6x \quad (1)$
$y=x^2-x-6 \quad (2)$

Substitute x^2-x-6 for y in equation (1) to find x.

$$\begin{aligned} y &= 3x^2+6x \quad (1) \\ x^2-x-6 &= 3x^2+6x \\ 0 &= 2x^2+7x+6 \\ 0 &= (2x+3)(x+2) \end{aligned}$$

$$2x+3=0 \quad \text{or} \quad x+2=0$$
$$x=-\frac{3}{2} \quad \text{or} \quad x=-2$$

Substitute $-\frac{3}{2}$ for x in equation (1) to find y.

$$\begin{aligned} y &= 3x^2+6x \quad (1) \\ y &= 3\left(-\frac{3}{2}\right)^2+6\left(-\frac{3}{2}\right) \\ &= 3\left(\frac{9}{4}\right)+6\left(-\frac{6}{4}\right) \\ &= \frac{27}{4}-\frac{36}{4}=-\frac{9}{4} \end{aligned}$$

Substitute -2 for x in equation (1) to find y.

$$\begin{aligned} y &= 3x^2+6x \quad (1) \\ y &= 3(-2)^2+6(-2) \\ &= 12-12=0 \end{aligned}$$

Solution set: $\left\{\left(-\frac{3}{2},-\frac{9}{4}\right),(-2,0)\right\}$

27. $2x^2-y^2=6 \quad (1)$
$y=x^2-3 \quad (2)$

Substitute x^2-3 for y in equation (1).

$$\begin{aligned} 2x^2-y^2 &= 6 \quad (1) \\ 2x^2-(x^2-3)^2 &= 6 \\ 2x^2-(x^4-6x^2+9) &= 6 \\ -x^4+8x^2-9 &= 6 \\ -x^4+8x^2-15 &= 0 \\ x^4-8x^2+15 &= 0 \quad \textit{Multiply by } -1. \end{aligned}$$

Let $z=x^2$, so $z^2=x^4$.

$$z^2-8z+15=0$$
$$(z-3)(z-5)=0$$

$$z-3=0 \quad \text{or} \quad z-5=0$$
$$z=3 \quad \text{or} \quad z=5$$

Since $z=x^2$,
$$x^2=3 \quad \text{or} \quad x^2=5.$$

Use the square root property.
$$x=\pm\sqrt{3} \quad \text{or} \quad x=\pm\sqrt{5}$$

Use equation (2) to find y for each x-value.

If $x=\sqrt{3}$ or $-\sqrt{3}$, then
$$y=\left(\pm\sqrt{3}\right)^2-3=0.$$

(Note: We could substitute 3 for x^2 in (2) to obtain the same result.)

If $x=\sqrt{5}$ or $-\sqrt{5}$, then
$$y=\left(\pm\sqrt{5}\right)^2-3=2.$$

Solution set:
$$\left\{\left(-\sqrt{3},0\right),\left(\sqrt{3},0\right),\left(-\sqrt{5},2\right),\left(\sqrt{5},2\right)\right\}$$

29. $x^2-xy+y^2=0 \quad (1)$
$x-2y=1 \quad (2)$

Solve equation (2) for x.
$$x=2y+1 \quad (3)$$

Substitute $2y+1$ for x in equation (1).

$$\begin{aligned} (2y+1)^2-(2y+1)y+y^2 &= 0 \\ 4y^2+4y+1-2y^2-y+y^2 &= 0 \\ 3y^2+3y+1 &= 0 \end{aligned}$$

Use the quadratic formula with $a=3$, $b=3$, and $c=1$.

$$\begin{aligned} y &= \frac{-b\pm\sqrt{b^2-4ac}}{2a} \\ y &= \frac{-3\pm\sqrt{9-12}}{6} \\ &= \frac{-3\pm\sqrt{-3}}{6}=\frac{-3\pm i\sqrt{3}}{6} \end{aligned}$$

Use equation (3) to solve for x.

$$\begin{aligned} x &= 2y+1 \quad (3) \\ &= 2\left(\frac{-3\pm i\sqrt{3}}{6}\right)+1 \\ &= \frac{-3\pm i\sqrt{3}}{3}+\frac{3}{3} \end{aligned}$$

$$= \pm\frac{i\sqrt{3}}{3}$$

Solution set:

$$\left\{\left(\frac{i\sqrt{3}}{3}, \frac{-3+i\sqrt{3}}{6}\right), \left(-\frac{i\sqrt{3}}{3}, \frac{-3-i\sqrt{3}}{6}\right)\right\}$$

31. $3x^2 + 2y^2 = 12 \quad (1)$
$x^2 + 2y^2 = 4 \quad (2)$

Multiply equation (2) by -1 and add the result to equation (1).

$$\begin{array}{rcrl} 3x^2 + 2y^2 &=& 12 & (1) \\ -x^2 - 2y^2 &=& -4 & -1\times(2) \\ \hline 2x^2 &=& 8 & \\ x^2 &=& 4 & \\ x &=& \pm 2 & \end{array}$$

Substitute ± 2 for x in equation (2) to find y.

$$\begin{aligned} x^2 + 2y^2 &= 4 \quad (2) \\ (\pm 2)^2 + 2y^2 &= 4 \\ 4 + 2y^2 &= 4 \\ 2y^2 &= 0 \\ y^2 &= 0 \\ y &= 0 \end{aligned}$$

Solution set: $\{(-2, 0), (2, 0)\}$

33. $2x^2 + 3y^2 = 6 \quad (1)$
$x^2 + 3y^2 = 3 \quad (2)$

Multiply equation (2) by -1 and add the result to equation (1).

$$\begin{array}{rcrl} 2x^2 + 3y^2 &=& 6 & (1) \\ -x^2 - 3y^2 &=& -3 & -1\times(2) \\ \hline x^2 &=& 3 & \\ x &=& \pm\sqrt{3} & \end{array}$$

Substitute $\pm\sqrt{3}$ for x in equation (2).

$$\begin{aligned} x^2 + 3y^2 &= 3 \quad (2) \\ \left(\pm\sqrt{3}\right)^2 + 3y^2 &= 3 \\ 3 + 3y^2 &= 3 \\ y^2 &= 0 \\ y &= 0 \end{aligned}$$

Solution set: $\left\{\left(\sqrt{3}, 0\right), \left(-\sqrt{3}, 0\right)\right\}$

35. $5x^2 - 2y^2 = -13 \quad (1)$
$3x^2 + 4y^2 = 39 \quad (2)$

Multiply equation (1) by 2 and add the result to equation (2).

$$\begin{array}{rcrl} 10x^2 - 4y^2 &=& -26 & 2\times(1) \\ 3x^2 + 4y^2 &=& 39 & (2) \\ \hline 13x^2 &=& 13 & \\ x^2 &=& 1 & \\ x &=& \pm 1 & \end{array}$$

Substitute 1 for x^2 in equation (2).

$$\begin{aligned} 3x^2 + 4y^2 &= 39 \quad (2) \\ 3 + 4y^2 &= 39 \\ 4y^2 &= 36 \\ y^2 &= 9 \\ y &= \pm 3 \end{aligned}$$

Since 3 can be paired with either 1 or -1 and -3 can be paired with either 1 or -1, there are four possible solutions.

Solution set: $\{(1, 3), (1, -3), (-1, 3), (-1, -3)\}$

37. $2x^2 = 8 - 2y^2 \quad (1)$
$3x^2 = 24 - 4y^2 \quad (2)$

Multiply equation (1) by -2 and add the result to equation (2).

$$\begin{array}{rcrl} -4x^2 &=& -16 + 4y^2 & -2\times(1) \\ 3x^2 &=& 24 - 4y^2 & (2) \\ \hline -x^2 &=& 8 & \\ x^2 &=& -8 & \\ x = \pm\sqrt{-8} &=& \pm 2i\sqrt{2} & \end{array}$$

Substitute $\pm 2i\sqrt{2}$ for x in equation (1).

$$\begin{aligned} 2x^2 &= 8 - 2y^2 \quad (1) \\ 2\left(\pm 2i\sqrt{2}\right)^2 &= 8 - 2y^2 \\ 2(-8) &= 8 - 2y^2 \\ -16 &= 8 - 2y^2 \\ 2y^2 &= 24 \\ y^2 &= 12 \\ y &= \pm\sqrt{12} = \pm 2\sqrt{3} \end{aligned}$$

Since $2i\sqrt{2}$ can be paired with either $2\sqrt{3}$ or $-2\sqrt{3}$ and $-2i\sqrt{2}$ can be paired with either $2\sqrt{3}$ or $-2\sqrt{3}$, there are four possible solutions.

Solution set:

$$\left\{\left(-2i\sqrt{2}, -2\sqrt{3}\right), \left(-2i\sqrt{2}, 2\sqrt{3}\right), \left(2i\sqrt{2}, -2\sqrt{3}\right), \left(2i\sqrt{2}, 2\sqrt{3}\right)\right\}$$

39. $x^2 + xy + y^2 = 15$ (1)
$x^2 + y^2 = 10$ (2)

Multiply equation (2) by -1 and add the result to equation (1).

$$\begin{array}{rcll} x^2 + xy + y^2 &=& 15 & (1) \\ -x^2 \qquad - y^2 &=& -10 & -1 \times (2) \\ \hline xy &=& 5 & \\ y &=& \frac{5}{x} & \end{array}$$

Substitute $\frac{5}{x}$ for y in equation (2).

$$\begin{aligned} x^2 + y^2 &= 10 \quad (2) \\ x^2 + \left(\frac{5}{x}\right)^2 &= 10 \\ x^2 + \frac{25}{x^2} &= 10 \\ x^4 + 25 &= 10x^2 \quad \textit{Multiply by } x^2. \\ x^4 - 10x^2 + 25 &= 0 \end{aligned}$$

Let $z = x^2$, so $z^2 = x^4$.

$$\begin{aligned} z^2 - 10z + 25 &= 0 \\ (z-5)^2 &= 0 \\ z - 5 &= 0 \\ z &= 5 \end{aligned}$$

Since $z = x^2$,

$$x^2 = 5, \text{ and } x = \pm\sqrt{5}.$$

Using the equation $y = \frac{5}{x}$, we get the following.

If $x = -\sqrt{5}$, then

$$y = \frac{5}{-\sqrt{5}} = \frac{5 \cdot \sqrt{5}}{-\sqrt{5} \cdot \sqrt{5}} = \frac{5\sqrt{5}}{-5} = -\sqrt{5}.$$

Similarly, if $x = \sqrt{5}$, then $y = \sqrt{5}$.

Solution set: $\left\{\left(-\sqrt{5}, -\sqrt{5}\right), \left(\sqrt{5}, \sqrt{5}\right)\right\}$

41. $3x^2 + 2xy - 3y^2 = 5$ (1)
$-x^2 - 3xy + y^2 = 3$ (2)

Multiply equation (2) by 3 and add the result to equation (1).

$$\begin{array}{rcll} 3x^2 + 2xy - 3y^2 &=& 5 & (1) \\ -3x^2 - 9xy + 3y^2 &=& 9 & 3 \times (2) \\ \hline -7xy &=& 14 & \\ x &=& \frac{14}{-7y} = -\frac{2}{y} & \end{array}$$

Substitute $-\frac{2}{y}$ for x in equation (2).

$$\begin{aligned} -x^2 - 3xy + y^2 &= 3 \quad (2) \\ -\left(-\frac{2}{y}\right)^2 - 3\left(-\frac{2}{y}\right)y + y^2 &= 3 \\ -\left(\frac{4}{y^2}\right) + 6 + y^2 &= 3 \\ y^2 + 3 - \frac{4}{y^2} &= 0 \\ y^4 + 3y^2 - 4 &= 0 \quad \textit{Multiply by } y^2. \\ (y^2 + 4)(y^2 - 1) &= 0 \end{aligned}$$

$$\begin{array}{lcl} y^2 + 4 = 0 & \text{or} & y^2 - 1 = 0 \\ y^2 = -4 & & y^2 = 1 \\ y = \pm 2i & \text{or} & y = \pm 1 \end{array}$$

Since $x = -\frac{2}{y}$, substitute these values for y to find the values of x.

If $y = 2i$, then

$$x = -\frac{2}{2i} = -\frac{1}{i} = -\frac{1}{i} \cdot \frac{i}{i} = \frac{-i}{-1} = i.$$

If $y = -2i$, then

$$x = -\frac{2}{-2i} = \frac{1}{i} = \frac{1}{i} \cdot \frac{i}{i} = \frac{i}{-1} = -i.$$

If $y = 1$, then

$$x = -\frac{2}{1} = -2.$$

If $y = -1$, then

$$x = -\frac{2}{-1} = 2.$$

Solution set: $\{(i, 2i), (-i, -2i), (2, -1), (-2, 1)\}$

43. $xy = -6$ (1)
$x + y = -1$ (2)

Solve both equations for y.

$$y = -\frac{6}{x} \quad \text{and} \quad y = -x - 1$$

Graph

$$\mathrm{Y}_1 = -\frac{6}{\mathrm{X}} \quad \text{and} \quad \mathrm{Y}_2 = -\mathrm{X} - 1$$

on a graphing calculator to obtain the solution set $\{(2, -3), (-3, 2)\}$.

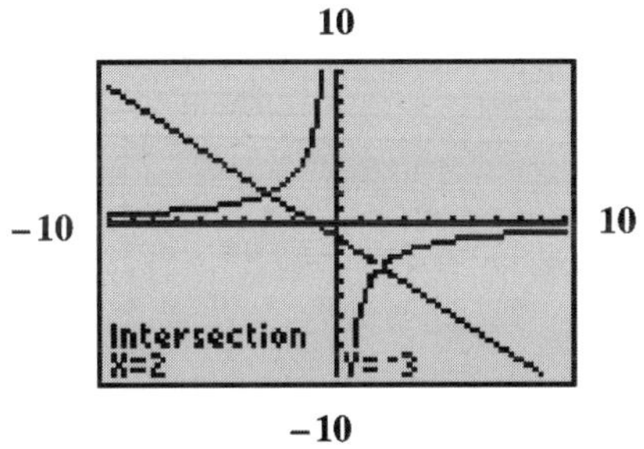

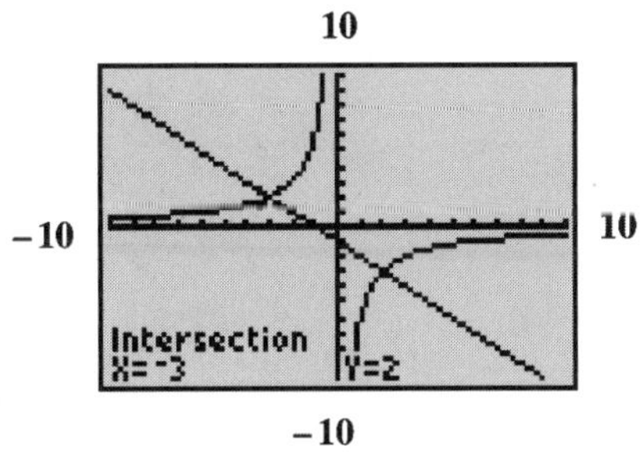

Now solve the system algebraically using substitution.

$$\begin{aligned} xy &= -6 \quad (1) \\ x + y &= -1 \quad (2) \end{aligned}$$

Solve equation (2) for y.

$$y = -1 - x \qquad (3)$$

Substitute $-1 - x$ for y in equation (1).

$$\begin{aligned} x(-1 - x) &= -6 \\ -x - x^2 &= -6 \\ 0 &= x^2 + x - 6 \\ 0 &= (x + 3)(x - 2) \end{aligned}$$

$$\begin{aligned} x + 3 = 0 \quad &\text{or} \quad x - 2 = 0 \\ x = -3 \quad &\text{or} \quad x = 2 \end{aligned}$$

Substitute these values for x in (3) to find y.

If $x = -3$, then $y = -1 - (-3) = 2$.

If $x = 2$, then $y = -1 - 2 = -3$.

The solution set, $\{(2, -3), (-3, 2)\}$, is the same as that obtained using a graphing calculator.

45. Let $W =$ the width, and
$L =$ the length.

The formula for the area of a rectangle is $LW - A$, so

$$LW = 84. \quad (1)$$

The perimeter of a rectangle is given by $2L + 2W = P$, so

$$2L + 2W = 38. \quad (2)$$

Solve equation (2) for L to get

$$L = 19 - W. \quad (3)$$

Substitute $19 - W$ for L in equation (1).

$$\begin{aligned} LW &= 84 \quad (1) \\ (19 - W)W &= 84 \\ 19W - W^2 &= 84 \\ -W^2 + 19W - 84 &= 0 \\ W^2 - 19W + 84 &= 0 \qquad \textit{Multiply by } -1. \\ (W - 7)(W - 12) &= 0 \end{aligned}$$

$$\begin{aligned} W - 7 = 0 \quad &\text{or} \quad W - 12 = 0 \\ W = 7 \quad &\text{or} \quad W = 12 \end{aligned}$$

Using equation (3), with $W = 7$,

$$L = 19 - 7 = 12.$$

If $W = 12$, then $L = 7$, which are the same two numbers. Length must be greater than width, so the length is 12 feet and the width is 7 feet.

47. $px = 16 \qquad (1)$
$p = 10x + 12 \qquad (2)$

Substitute $10x + 12$ for p in equation (1).

$$\begin{aligned} px &= 16 \quad (1) \\ (10x + 12)x &= 16 \\ 10x^2 + 12x &= 16 \\ 10x^2 + 12x \quad 16 &= 0 \\ 5x^2 + 6x - 8 &= 0 \\ (5x - 4)(x + 2) &= 0 \end{aligned}$$

$$\begin{aligned} 5x - 4 = 0 \quad &\text{or} \quad x + 2 = 0 \\ x = \frac{4}{5} \quad &\text{or} \quad x = -2 \end{aligned}$$

Since x cannot be negative, eliminate -2 as a value of x. Substitute $\frac{4}{5}$ for x in equation (2) to find p.

$$\begin{aligned} p &= 10x + 12 \qquad (2) \\ p &= 10\left(\frac{4}{5}\right) + 12 \\ &= 8 + 12 = 20 \end{aligned}$$

Since the demand x is in thousands,

$$\text{demand} = \frac{4}{5}(1000) = 800.$$

The equilibrium price is \$20. The supply/demand at that price is 800 calculators.

49. Men: $y = .138x^2 + .064x + 451 \quad (1)$
Women: $y = 12.1x + 334 \quad (2)$

Substitute $12.1x + 334$ for y in (1).

$$\begin{aligned} 12.1x + 334 &= .138x^2 + .064x + 451 \\ 0 &= .138x^2 - 12.036x + 117 \end{aligned}$$

Use the quadratic formula.

continued

$$x = \frac{-(-12.036) \pm \sqrt{(-12.036)^2 - 4(.138)(117)}}{2(.138)}$$
$$= \frac{12.036 \pm \sqrt{80.28196}}{.276}$$

These values are approximately 76 and 11. $x = 11$ corresponds to 1981 and $x = 76$ is out of the range for these models.

Substituting $x = 11$ into (2) gives us approximately 467 thousand bachelor's degrees (or 470 thousand to the nearest ten thousand) awarded to both men and women in 1981.

Section 11.5

1. $x^2 + y^2 < 25$
$y > -2$

The boundary, $x^2 + y^2 = 25$, is a circle (dashed because of the $<$ sign) with center $(0,0)$ and radius 5. When $(0,0)$ is tested, a true statement, $0 < 25$, results, so the inside of the circle is shaded. The graph of $y = -2$ is a horizontal line through $(0,-2)$ with shading above the dashed line, since $y > -2$. The correct answer is **C**.

3. To graph the solution set of a nonlinear inequality, first graph the corresponding equality. This graph should be a dashed curve for $<$ or $>$ inequalities or a solid curve for $\leq$ or $\geq$ inequalities. Next, decide which region to shade by substituting any point not on the boundary (usually $(0,0)$ is the easiest) into the inequality. If the statement is true, then shade that area. If the statement is false, then shade the other area.

5. $y \geq x^2 + 4$

This is an inequality whose boundary is a solid parabola, opening up, with vertex $(0,4)$. The inside of the parabola is shaded since $(0,0)$ makes the inequality false. This is graph **B**.

7. $y < x^2 + 4$

This is an inequality whose boundary is a dashed parabola, opening up, with vertex $(0,4)$. The outside of the parabola is shaded since $(0,0)$ makes the inequality true. This is graph **A**.

9.
$$y^2 > 4 + x^2$$
$$y^2 - x^2 > 4$$
$$\frac{y^2}{4} - \frac{x^2}{4} > 1$$

The boundary, $\frac{y^2}{4} - \frac{x^2}{4} = 1$, is a hyperbola with y-intercepts $(0,2)$ and $(0,-2)$ and asymptotes formed by the diagonals of the fundamental rectangle with vertices at $(2,2)$, $(2,-2)$, $(-2,-2)$, and $(-2,2)$. The hyperbola has dashed branches because of $>$. Test $(0,0)$.

$$0^2 > 4 + 0^2 \quad ?$$
$$0 > 4 \qquad \textit{False}$$

Shade the sides of the hyperbola that do not contain $(0,0)$. These are the regions inside the branches of the hyperbola.

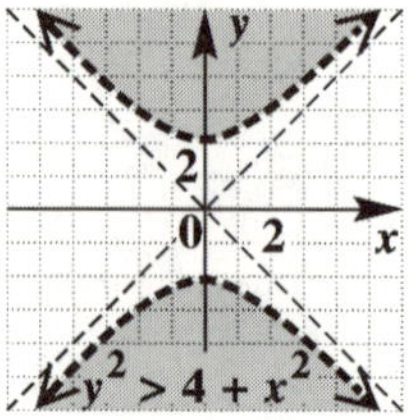

11. $y + 2 \geq x^2$
$y \geq x^2 - 2$
$y \geq (x-0)^2 - 2$

Graph the solid vertical parabola $y = x^2 - 2$ with vertex $(0,-2)$. Two other points on the parabola are $(2,2)$ and $(-2,2)$. Test a point not on the parabola, say $(0,0)$, in $y \geq x^2 - 2$ to get $0 \geq -2$, a true statement. Shade that portion of the graph that contains the point $(0,0)$. This is the region inside the parabola.

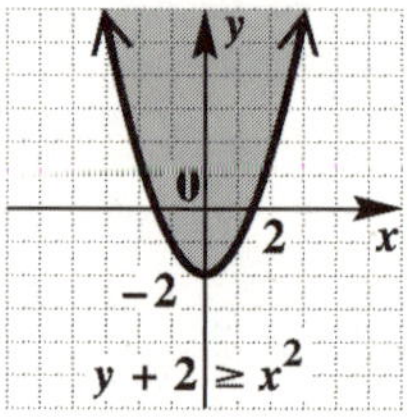

13.
$$2y^2 \geq 8 - x^2$$
$$x^2 + 2y^2 \geq 8$$
$$\frac{x^2}{8} + \frac{y^2}{4} \geq 1$$

The boundary, $\frac{x^2}{8} + \frac{y^2}{4} = 1$, is the ellipse with intercepts $\left(2\sqrt{2},0\right)$, $\left(-2\sqrt{2},0\right)$, $(0,2)$, and $(0,-2)$, drawn as a solid curve because of $\geq$. Test $(0,0)$.

$$2(0)^2 \geq 8 - 0^2 \quad ?$$
$$0 \geq 8 \qquad \textit{False}$$

Shade the region of the ellipse that does not contain $(0,0)$. This is the region outside the ellipse.

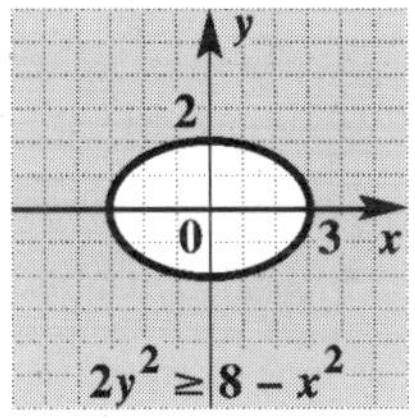

15. $y \le x^2 + 4x + 2$

Graph the solid vertical parabola $y = x^2 + 4x + 2$. Use the vertex formula $x = \frac{-b}{2a}$ to obtain the vertex $(-2, -2)$. Two other points on the parabola are $(0, 2)$ and $(1, 7)$. Test a point not on the parabola, say $(0, 0)$, in $y \le x^2 + 4x + 2$ to get $0 \le 2$, a true statement. Shade outside the parabola, since this region contains $(0, 0)$.

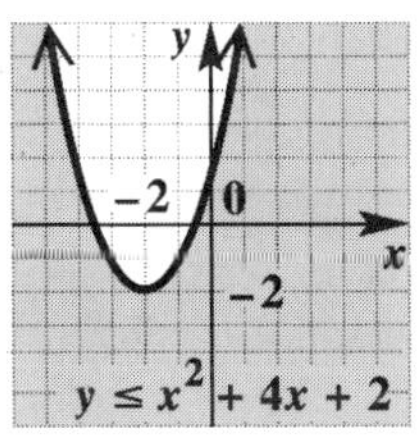

17.
$$9x^2 > 16y^2 + 144$$
$$9x^2 - 16y^2 > 144$$
$$\frac{x^2}{16} - \frac{y^2}{9} > 1$$

The boundary, $\frac{x^2}{16} - \frac{y^2}{9} = 1$, is a hyperbola with x-intercepts $(4, 0)$ and $(-4, 0)$ and asymptotes formed by the diagonals of the fundamental rectangle with vertices at $(4, 3)$, $(4, -3)$, $(-4, -3)$, and $(-4, 3)$. The hyperbola has dashed branches because of $>$. Test $(0, 0)$.

$$9(0)^2 > 16(0)^2 + 144 \quad ?$$
$$0 > 144 \qquad \textit{False}$$

Shade the sides of the hyperbola that do not contain $(0, 0)$. These are the regions inside the branches of the hyperbola.

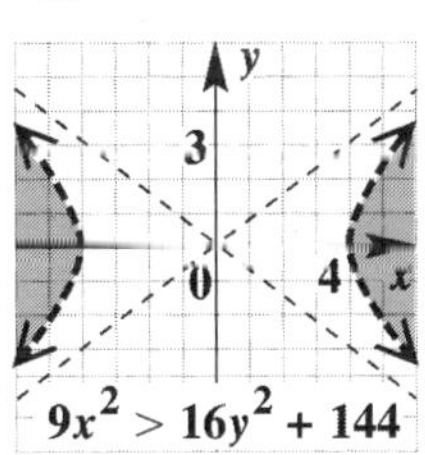

19.
$$x^2 - 4 \ge -4y^2$$
$$x^2 + 4y^2 \ge 4$$
$$\frac{x^2}{4} + \frac{y^2}{1} \ge 1$$

Graph the solid ellipse $\frac{x^2}{4} + \frac{y^2}{1} = 1$ through the x-intercepts $(2, 0)$ and $(-2, 0)$ and y-intercepts $(0, 1)$ and $(0, -1)$. Test a point not on the ellipse, say $(0, 0)$, in $x^2 - 4 \ge -4y^2$ to get $-4 \ge 0$, a false statement. Shade outside the ellipse, since this region does *not* include $(0, 0)$.

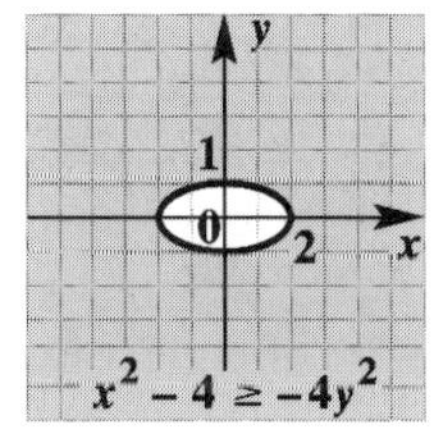

21. $x \le -y^2 + 6y - 7$

Complete the square to find the vertex.

$$\begin{aligned} x &= -y^2 + 6y - 7 \\ &= -(y^2 - 6y) - 7 \\ &= -(y^2 - 6y + 9) + 9 - 7 \\ x &= -(y - 3)^2 + 2 \end{aligned}$$

The boundary, $x = -(y - 3)^2 + 2$, is a solid horizontal parabola with vertex at $(2, 3)$ that opens to the left. Test $(0, 0)$.

$$0 \le -0^2 + 6(0) - 7 \quad ?$$
$$0 \le -7 \qquad \textit{False}$$

Shade the region of the parabola that does not contain $(0, 0)$. This is the region inside the parabola.

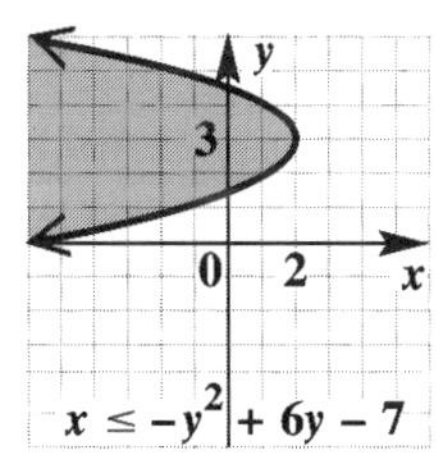

23.
$$\begin{aligned} 2x + 5y &< 10 \\ x - 2y &< 4 \end{aligned}$$

Graph $2x + 5y = 10$ as a dashed line through $(5, 0)$ and $(0, 2)$. Test $(0, 0)$.

$$2x + 5y < 10$$
$$2(0) + 5(0) < 10 \quad ?$$
$$0 < 10 \qquad \textit{True}$$

Shade the region containing $(0, 0)$. Graph $x - 2y = 4$ as a dashed line through $(4, 0)$ and $(0, -2)$. Test $(0, 0)$.

continued

$$\begin{aligned} x - 2y &< 4 \\ 0 - 2(0) &< 4 \quad ? \\ 0 &< 4 \quad \textit{True} \end{aligned}$$

Shade the region containing $(0, 0)$. The graph of the system is the intersection of the two shaded regions.

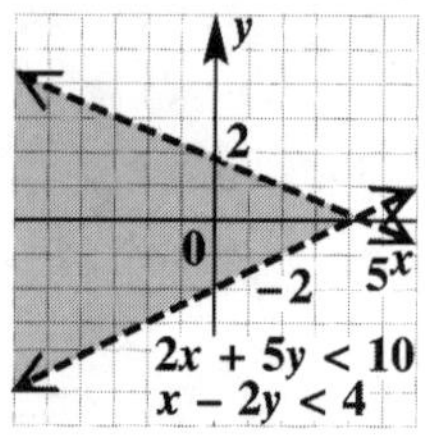

25. $5x - 3y \le 15$
$4x + y \ge 4$

The boundary, $5x - 3y = 15$, is a solid line with intercepts $(3, 0)$ and $(0, -5)$. Test $(0, 0)$.

$$\begin{aligned} 5(0) - 3(0) &\le 15 \quad ? \\ 0 &\le 15 \quad \textit{True} \end{aligned}$$

Shade the side of the line that contains $(0, 0)$.
The boundary, $4x + y = 4$, is a solid line with intercepts $(1, 0)$ and $(0, 4)$. Test $(0, 0)$.

$$\begin{aligned} 4(0) + 0 &\ge 4 \quad ? \\ 0 &\ge 4 \quad \textit{False} \end{aligned}$$

Shade the side of the line that does not contain $(0, 0)$.
The graph of the system is the intersection of the two shaded regions.

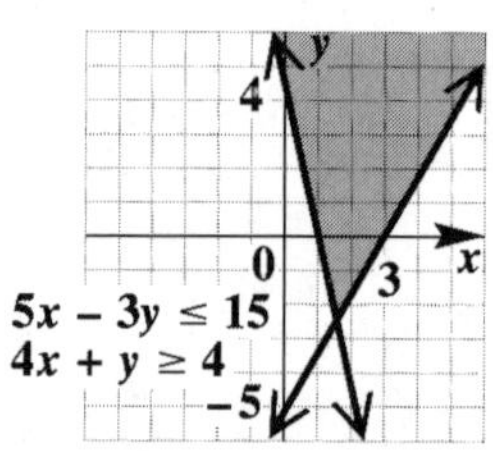

27. $x \le 5$
$y \le 4$

Graph $x = 5$ as a solid vertical line through $(5, 0)$. Since $x \le 5$, shade the left side of the line.
Graph $y = 4$ as a solid horizontal line through $(0, 4)$. Since $y \le 4$, shade below the line.
The graph of the system is the intersection of the two shaded regions.

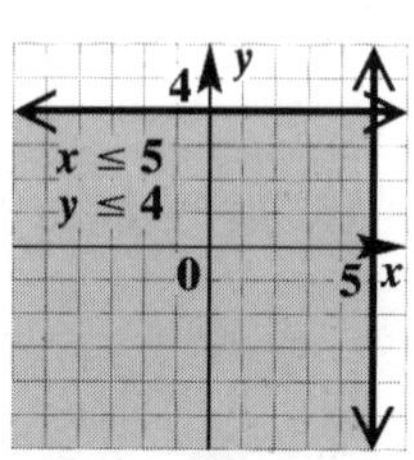

29. $y > x^2 - 4$
$y < -x^2 + 3$

The boundary, $y = x^2 - 4$, is a dashed parabola with vertex $(0, -4)$ that opens up. Test $(0, 0)$.

$$\begin{aligned} 0 &> 0^2 - 4 \quad ? \\ 0 &> -4 \quad \textit{True} \end{aligned}$$

Shade the side of the parabola that contains $(0, 0)$. This is the region inside the parabola.
The boundary, $y = -x^2 + 3$, is a dashed parabola with vertex $(0, 3)$ that opens down. Test $(0, 0)$.

$$\begin{aligned} 0 &< -0^2 + 3 \quad ? \\ 0 &< 3 \quad \textit{True} \end{aligned}$$

Shade the side of the parabola that contains $(0, 0)$. This is the region inside the parabola.
The graph of the system is the intersection of the two shaded regions.

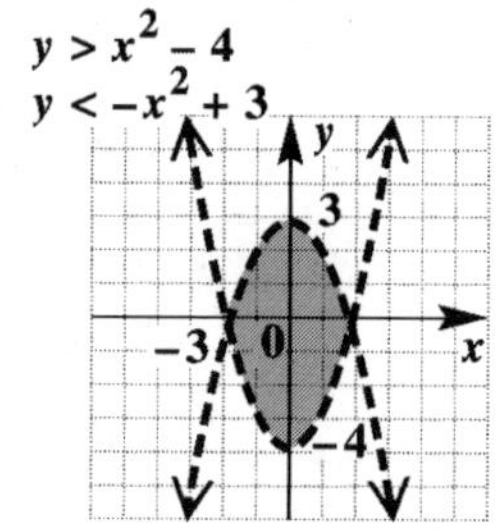

31. $x^2 + y^2 \ge 4$ (1)
$x + y \le 5$ (2)
$x \ge 0$ (3)
$y \ge 0$ (4)

The graph of (1) is the region outside the solid circle boundary $x^2 + y^2 = 4$ with center $(0, 0)$ and radius 2.
The graph of (2) is the region below the solid line $x + y = 5$ with intercepts $(5, 0)$ and $(0, 5)$.
The intersection of the graphs of (3) and (4) is quadrant I where both x and y are positive. The positive x- and y-axes are included.
The graph of the system is the intersection of the graphs of (1) and (2) in quadrant I.

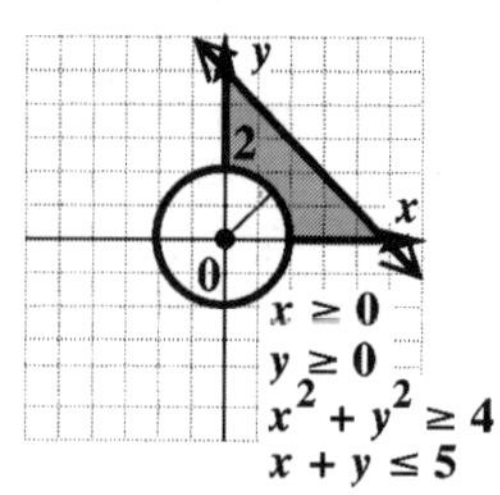

33. $y \le -x^2$
$y \ge x - 3$
$y \le -1$
$x < 1$

The boundary, $y = -x^2$, is a solid parabola with vertex at $(0, 0)$ that opens down. Test $(0, -1)$.

$$-1 \leq -(0)^2 \quad ?$$
$$-1 \leq 0 \qquad \textit{True}$$

Shade the side of the parabola that contains $(0, -1)$. This is the region inside the parabola.
The boundary, $y = x - 3$, is a solid line with intercepts $(3, 0)$ and $(0, -3)$. Test $(0, 0)$.

$$0 \geq 0 - 3 \quad ?$$
$$0 \geq -3 \qquad \textit{True}$$

Shade the side of the line that contains $(0, 0)$.
For $y \leq -1$, shade below the solid horizontal line $y = -1$.
For $x < 1$, shade to the left of the dashed vertical line $x = 1$.
The intersection of the four shaded regions is the graph of the system.

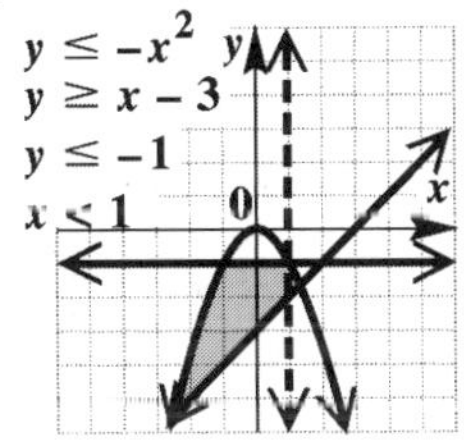

35. $x^2 + y^2 > 36, x \geq 0$

This is a circle of radius 6 centered at the origin. The graph is a dashed curve. Since $(0, 0)$ does not satisfy the inequality, the shading will be outside the circle. By including the restriction $x \geq 0$, we consider only the shading to the right of, and including, the y-axis.

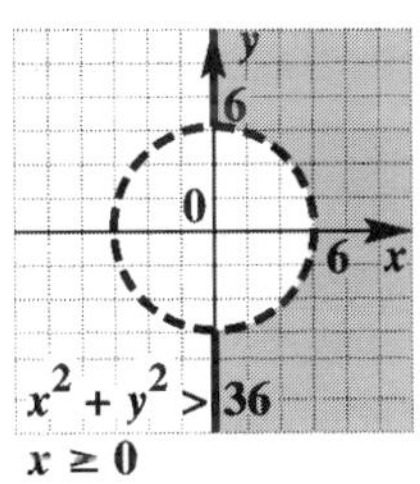

37. $x < y^2 - 3, x < 0$

Consider the equation.

$$x - y^2 - 3, \text{ or}$$
$$x = (y - 0)^2 - 3.$$

This is a parabola with vertex $(-3, 0)$, opening to the right having the same shape as $x = y^2$. The graph is a dashed curve. The shading will be outside the parabola, since $(0, 0)$ does not satisfy the inequality. By including the restriction $x < 0$, we consider only the shading to the left of, but not including, the y-axis. The y-axis is a dashed line.

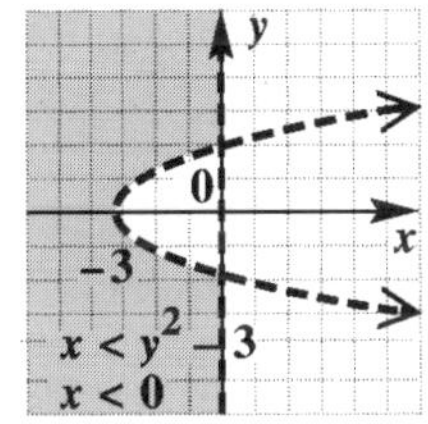

39. $4x^2 - y^2 > 16, x < 0$

Consider the equation

$$4x^2 - y^2 = 16, \text{ or}$$
$$\frac{x^2}{4} - \frac{y^2}{16} = 1.$$

This is a hyperbola with x-intercepts $(2, 0)$ and $(-2, 0)$. The graph is a dashed curve. The shading will be to the left and to the right of the two branches of the hyperbola, since $(0, 0)$ does not satisfy the inequality. By including the restriction $x < 0$, we consider only the shading to the left of the y-axis.

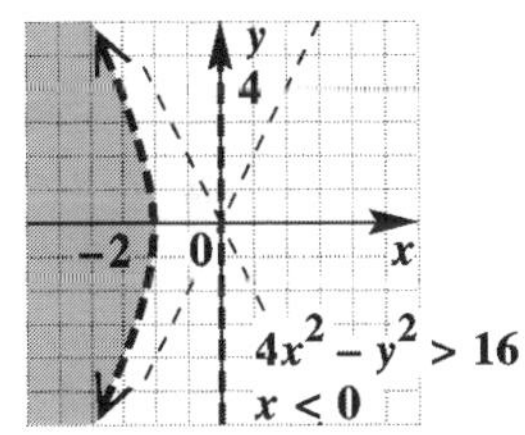

41. $x^2 + 4y^2 \geq 1, x \geq 0, y \geq 0$

Consider the equation

$$x^2 + 4y^2 = 1, \text{ or}$$
$$\frac{x^2}{1} + \frac{y^2}{\frac{1}{4}} = 1.$$

The graph is a solid ellipse with x-intercepts $(1, 0)$ and $(-1, 0)$ and y-intercepts $\left(0, \frac{1}{2}\right)$ and $\left(0, -\frac{1}{2}\right)$. The shading will be outside the ellipse, since $(0, 0)$ does not satisfy the inequality. By including the restrictions $x \geq 0$ and $y \geq 0$, we consider only the shading in quadrant I and portions of the x- and y-axis.

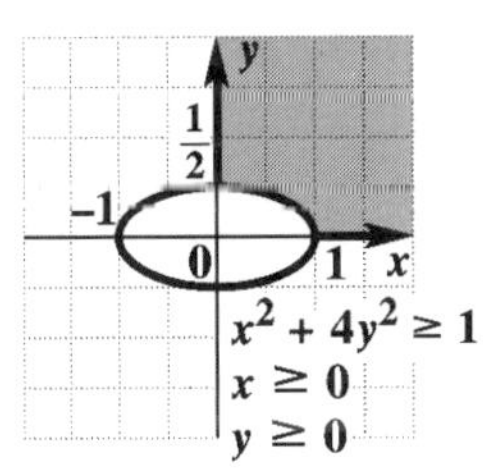

43. $y \geq x - 3$
$y \leq -x + 4$

The graphs of both inequalities include the points on the lines as part of the solution because of $\geq$ and $\leq$.
To produce the graph of the system on the TI-83, make the following Y-assignments:

To get the upper and lower darkened triangles to the left of Y_1 and Y_2, simply place the cursor in that spot and press ENTER to cycle through the graphing choices.

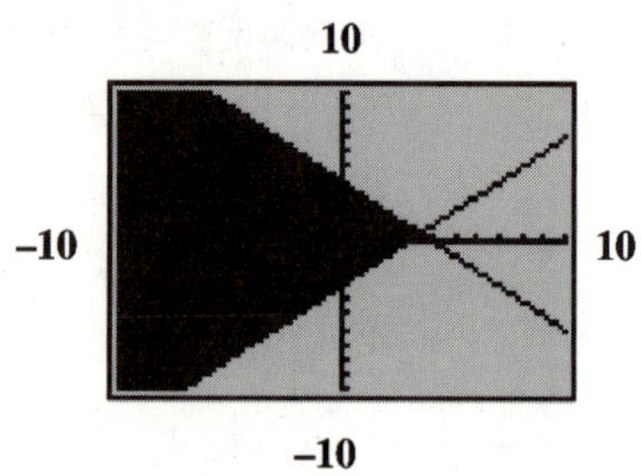

45. $y < x^2 + 4x + 4$
$y > -3$

The graphs do *not* include the points on the parabola or the line as part of the solution because of $<$ and $>$.

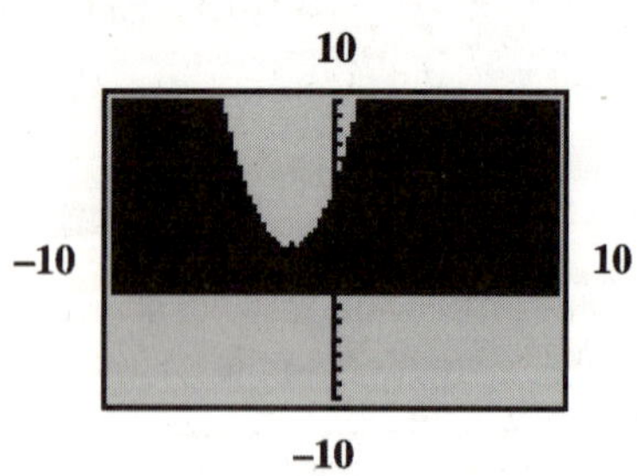

Chapter 11 Review Exercises

1. $f(x) = |x + 4|$

This is the graph of the absolute value function $g(x) = |x|$ shifted 4 units to the left (since $x + 4 = 0$ if $x = -4$).

x	-6	-5	-4	-3	-2
y	2	1	0	1	2

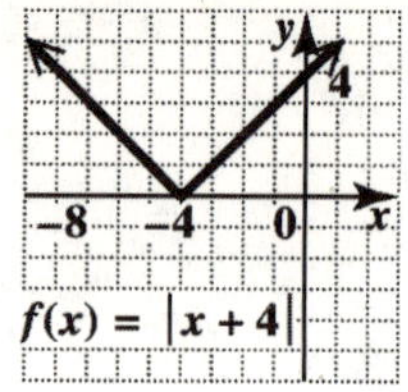

2. $f(x) = \dfrac{1}{x-4}$

This is the graph of the reciprocal function, $g(x) = \dfrac{1}{x}$, shifted 4 units to the right. Since $x \neq 4$ (or a denominator of 0 results), the line $x = 4$ is a vertical asymptote. Since $\dfrac{1}{x-4} \neq 0$, the line $y = 0$ is a horizontal asymptote.

x	$4\frac{1}{2}$	5	6
y	2	1	$\frac{1}{2}$

x	2	3	$3\frac{1}{2}$
y	$-\frac{1}{2}$	-1	-2

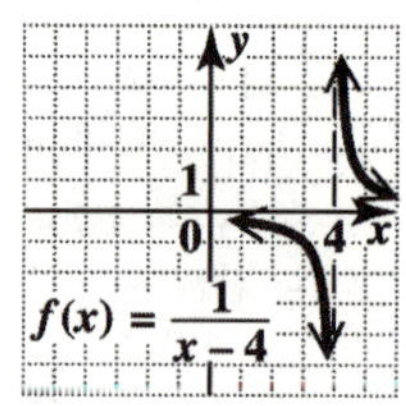

3. $f(x) = \sqrt{x} + 3$

This is the graph of the square root function, $g(x) = \sqrt{x}$, shifted 3 units upward.

x	0	1	4
y	3	4	5

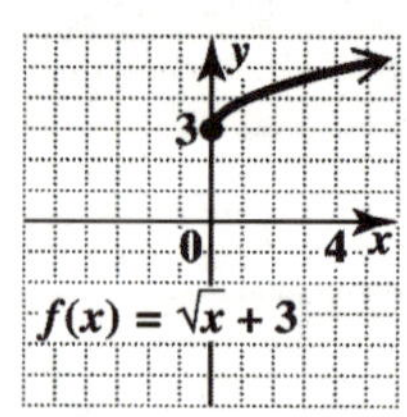

4. $f(x) = [\![x]\!] - 2$

This is the graph of the greatest integer function, $g(x) = [\![x]\!]$, shifted down 2 units.

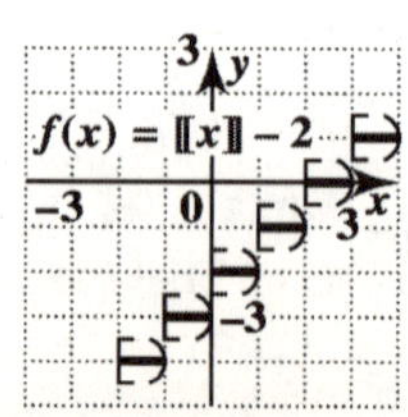

In Exercises 5–10, $f(x)=3x^2+2x-1$ and $g(x)=5x+7$.

5. $$\begin{aligned}(g\circ f)(3)&=g[f(3)]\\&=g(3\cdot 3^2+2\cdot 3-1)\\&=g(32)\\&=5\cdot 32+7\\&=167\end{aligned}$$

6. $$\begin{aligned}(f\circ g)(3)&=f[g(3)]\\&=f(5\cdot 3+7)\\&=f(22)\\&=3\cdot 22^2+2\cdot 22-1\\&=1495\end{aligned}$$

7. $$\begin{aligned}(f\circ g)(-2)&=f[g(-2)]\\&=f[5(-2)+7]\\&=f(-3)\\&=3(-3)^2+2(-3)-1\\&=20\end{aligned}$$

8. $$\begin{aligned}(g\circ f)(-2)&=g[f(-2)]\\&=g[3(-2)^2+2(-2)-1]\\&=g(7)\\&=5\cdot 7+7\\&=42\end{aligned}$$

9. $$\begin{aligned}(f\circ g)(x)&=f[g(x)]\\&=f(5x+7)\\&=3(5x+7)^2+2(5x+7)-1\\&=3(25x^2+70x+49)+10x+14-1\\&=75x^2+210x+147+10x+13\\&=75x^2+220x+160\end{aligned}$$

10. $$\begin{aligned}(g\circ f)(x)&=g[f(x)]\\&=g(3x^2+2x-1)\\&=5(3x^2+2x-1)+7\\&=15x^2+10x-5+7\\&=15x^2+10x+2\end{aligned}$$

11. Center $(-2,4)$, $r=3$
Here $h=-2$, $k=4$, and $r=3$, so an equation of the circle is

$$\begin{aligned}(x-h)^2+(y-k)^2&=r^2\\ [x-(-2)]^2+(y-4)^2&=3^2\\ (x+2)^2+(y-4)^2&=9.\end{aligned}$$

12. Center $(-1,-3)$, $r=5$
Here $h=-1$, $k=-3$, and $r=5$, so an equation of the circle is

$$\begin{aligned}(x-h)^2+(y-k)^2&=r^2\\ [x-(-1)]^2+[y-(-3)]^2&=5^2\\ (x+1)^2+(y+3)^2&=25.\end{aligned}$$

13. Center $(4,2)$, $r=6$
Here $h=4$, $k=2$, and $r=6$, so an equation of the circle is

$$\begin{aligned}(x-h)^2+(y-k)^2&=r^2\\ (x-4)^2+(y-2)^2&=6^2\\ (x-4)^2+(y-2)^2&=36.\end{aligned}$$

14. $x^2+y^2+6x-4y-3=0$
Write the equation in center-radius form,
$$(x-h)^2+(y-k)^2=r^2,$$
by completing the squares on x and y.

$$\begin{aligned}(x^2+6x\quad)+(y^2-4y\quad)&=3\\ (x^2+6x+\underline{9})+(y^2-4y+\underline{4})&\\ &=3+\underline{9}+\underline{4}\\ (x+3)^2+(y-2)^2&=16\\ [x-(-3)]^2+(y-2)^2&=16\end{aligned}$$

The circle has center (h,k) at $(-3,2)$ and radius $\sqrt{16}=4$.

15. $x^2+y^2-8x-2y+13=0$
Write the equation in center-radius form,
$$(x-h)^2+(y-k)^2=r^2,$$
by completing the squares on x and y.

$$\begin{aligned}(x^2-8x\quad)+(y^2-2y\quad)&=-13\\ (x^2-8x+\underline{16})+(y^2-2y+\underline{1})&\\ &=-13+\underline{16}+\underline{1}\\ (x-4)^2+(y-1)^2&=4\end{aligned}$$

The circle has center (h,k) at $(4,1)$ and radius $\sqrt{4}=2$.

16. $$\begin{aligned}2x^2+2y^2+4x+20y&=-34\\ x^2+y^2+2x+10y&=-17\end{aligned}$$
Write the equation in center-radius form,
$$(x-h)^2+(y-k)^2=r^2,$$
by completing the squares on x and y.

$$\begin{aligned}(x^2+2x\quad)+(y^2+10y\quad)&=-17\\ (x^2+2x+\underline{1})+(y^2+10y+\underline{25})&\\ &=-17+\underline{1}+\underline{25}\\ (x+1)^2+(y+5)^2&=9\\ [x-(-1)]^2+[y-(-5)]^2&=9\end{aligned}$$

The circle has center (h,k) at $(-1,-5)$ and radius $\sqrt{9}=3$.

17. $$\begin{aligned}4x^2+4y^2-24x+16y&=48\\ x^2+y^2-6x+4y&=12\end{aligned}$$
Write the equation in center-radius form,
$$(x-h)^2+(y-k)^2=r^2,$$
by completing the squares on x and y.

continued

$$(x^2 - 6x \qquad) + (y^2 + 4y \qquad) = 12$$
$$(x^2 - 6x + \underline{9}) + (y^2 + 4y + \underline{4}) = 12 + \underline{9} + \underline{4}$$
$$(x-3)^2 + (y+2)^2 = 25$$
$$(x-3)^2 + [y - (-2)]^2 = 25$$

The circle has center (h, k) at $(3, -2)$ and radius $\sqrt{25} = 5$.

18.
$$x^2 + y^2 = 16$$
$$(x-0)^2 + (y-0)^2 = 4^2$$

This is a circle with center $(0, 0)$ and radius 4.

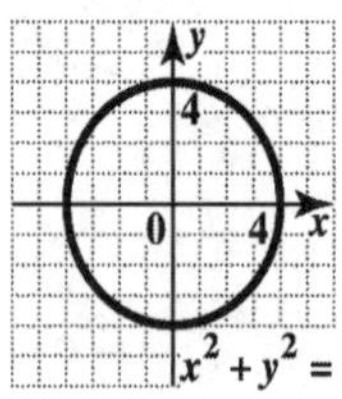

19. $\dfrac{x^2}{16} + \dfrac{y^2}{9} = 1$ is in $\dfrac{x^2}{a^2} + \dfrac{y^2}{b^2} = 1$ form with $a = 4$ and $b = 3$. The graph is an ellipse with x-intercepts $(4, 0)$ and $(-4, 0)$ and y-intercepts $(0, 3)$ and $(0, -3)$. Plot the intercepts, and draw the ellipse through them.

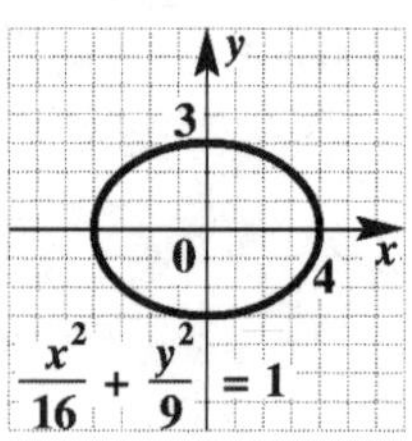

20. $\dfrac{x^2}{49} + \dfrac{y^2}{25} = 1$ is in $\dfrac{x^2}{a^2} + \dfrac{y^2}{b^2} = 1$ form with $a = 7$ and $b = 5$. The graph is an ellipse with x-intercepts $(7, 0)$ and $(-7, 0)$ and y-intercepts $(0, 5)$ and $(0, -5)$. Plot the intercepts, and draw the ellipse through them.

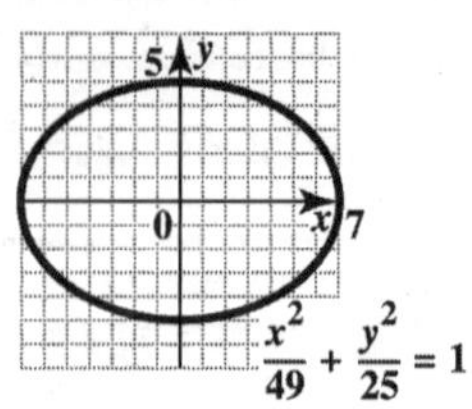

21. The total distance on the horizontal axis is $160 + 16{,}000 = 16{,}160$ km. This represents $2a$, so $a = \frac{1}{2}(16{,}160) = 8080$. The distance from Earth to the center of the ellipse is

$$8080 - 160 = 7920,$$

which is the value of c. From Exercise 51 in Section 11.2, we know that $c^2 = a^2 - b^2$, so

$$b^2 = a^2 - c^2.$$
$$b^2 = 8080^2 - 7920^2 = 2{,}560{,}000$$

Thus, $b = \sqrt{2{,}560{,}000} = 1600$ and the equation is

$$\frac{x^2}{8080^2} + \frac{y^2}{1600^2} = 1$$

or $$\frac{x^2}{65{,}286{,}400} + \frac{y^2}{2{,}560{,}000} = 1.$$

22. **(a)** The distance between the foci is $2c$, where c can be found using the relationship

$$c = \sqrt{a^2 - b^2}.$$
$$= \sqrt{310^2 - (513/2)^2}$$
$$\approx 174.1 \text{ feet}$$

So the distance is about 348.2 feet.

(b) The approximate circumference of the Roman Colosseum is

$$C \approx 2\pi\sqrt{\frac{a^2 + b^2}{2}}.$$
$$= 2\pi\sqrt{\frac{310^2 + (513/2)^2}{2}}$$
$$\approx 1787.6 \text{ feet}$$

23. $\dfrac{x^2}{16} - \dfrac{y^2}{25} = 1$ is in $\dfrac{x^2}{a^2} - \dfrac{y^2}{b^2} = 1$ form with $a = 4$ and $b = 5$. The graph is a hyperbola with x-intercepts $(4, 0)$ and $(-4, 0)$ and asymptotes that are the extended diagonals of the rectangle with vertices $(4, 5)$, $(4, -5)$, $(-4, -5)$, and $(-4, 5)$. Graph a branch of the hyperbola through each intercept approaching the asymptotes.

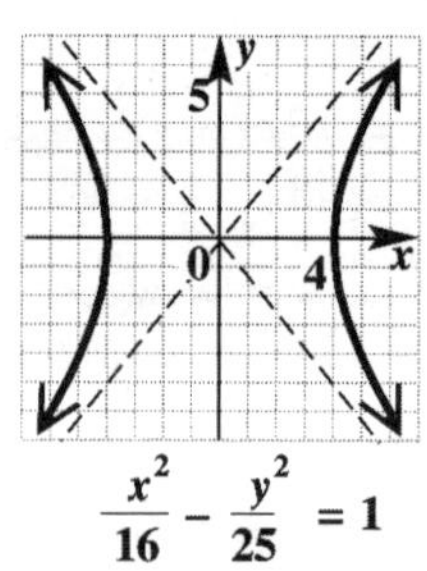

24. $\dfrac{y^2}{25} - \dfrac{x^2}{4} = 1$ is in $\dfrac{y^2}{b^2} - \dfrac{x^2}{a^2} = 1$ form with $a = 2$ and $b = 5$. The graph is a hyperbola with y-intercepts $(0, 5)$ and $(0, -5)$ and asymptotes that are the extended diagonals of the rectangle with vertices $(2, 5)$, $(2, -5)$, $(-2, -5)$, and $(-2, 5)$. Graph a branch of the hyperbola through each intercept approaching the asymptotes.

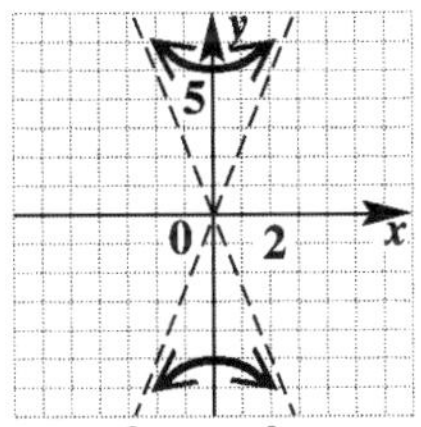

$$\frac{y^2}{25} - \frac{x^2}{4} = 1$$

25. $$f(x) = -\sqrt{16 - x^2}$$

Replace $f(x)$ with y.

$$y = -\sqrt{16 - x^2}$$

Square both sides.

$$y^2 = 16 - x^2$$
$$x^2 + y^2 = 16$$

This equation is the graph of a circle with center $(0, 0)$ and radius 4. Since $f(x)$ represents a nonpositive square root, $f(x)$ is nonpositive and its graph is the lower half of the circle.

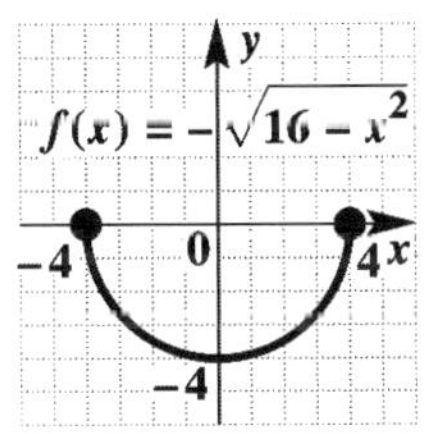

26. $$x^2 + y^2 = 64$$
$$(x - 0)^2 + (y - 0)^2 = 8^2$$

The last equation is in $(x - h)^2 + (y - k)^2 = r^2$ form. The graph is a *circle*.

27. $$y = 2x^2 - 3$$
$$y = 2(x - 0)^2 - 3$$

The last equation is in $y = a(x - h)^2 + k$ form. The graph is a *parabola*.

28. $$y^2 = 2x^2 - 8$$
$$2x^2 - y^2 = 8$$
$$\frac{x^2}{4} - \frac{y^2}{8} = 1$$

The last equation is in $\frac{x^2}{a^2} - \frac{y^2}{b^2} = 1$ form, so the graph is a *hyperbola*.

29. $$y^2 = 8 - 2x^2$$
$$2x^2 + y^2 = 8$$
$$\frac{x^2}{4} + \frac{y^2}{8} = 1$$

The last equation is in $\frac{x^2}{a^2} + \frac{y^2}{b^2} = 1$ form, so the graph is an *ellipse*.

30. $$x = y^2 + 4$$
$$x = (y - 0)^2 + 4$$

The last equation is in $x = a(y - k)^2 + h$ form, so the graph is a *parabola*.

31. $$x^2 - y^2 = 64$$
$$\frac{x^2}{64} - \frac{y^2}{64} = 1$$

The last equation is in $\frac{x^2}{a^2} - \frac{y^2}{b^2} = 1$ form, so the graph is a *hyperbola*.

32. This hyperbola has center at the origin with foci $(-50, 0)$ and $(50, 0)$. The difference between the distances from any point on the hyperbola to the two foci is equal to $80 - 30 = 50$. From Exercise 45 in Section 11.3, the foci are $(-c, 0)$ and $(c, 0)$. Hence, $c = 50$. Since the difference between the distances from any points to the two foci is $2a$, $2a = 50$, or $a = 25$. Since

$$b^2 = c^2 - a^2,$$
$$b^2 = 50^2 - 25^2$$
$$= 2500 - 625 = 1875.$$

Since the equation of a hyperbola is

$$\frac{x^2}{a^2} - \frac{y^2}{b^2} = 1,$$

the equation for this hyperbola is

$$\frac{x^2}{25^2} - \frac{y^2}{1875} = 1 \quad \text{or} \quad \frac{x^2}{625} - \frac{y^2}{1875} = 1.$$

33. $$2y = 3x - x^2 \quad (1)$$
$$x + 2y = -12 \quad (2)$$

Substitute $3x - x^2$ for $2y$ in equation (2).

$$x + 2y = -12 \quad (2)$$
$$x + (3x - x^2) = -12$$
$$-x^2 + 4x + 12 = 0$$
$$x^2 - 4x - 12 = 0$$
$$(x - 6)(x + 2) = 0$$

$$x - 6 = 0 \quad \text{or} \quad x + 2 = 0$$
$$x = 6 \quad \text{or} \quad x = -2$$

Substitute these values for x in equation (2) to find y.

If $x = 6$, then

$$x + 2y = -12 \quad (2)$$
$$6 + 2y = -12$$
$$2y = -18$$
$$y = -9.$$

continued

If $x = -2$ then

$$\begin{aligned} x + 2y &= -12 \quad (2) \\ -2 + 2y &= -12 \\ 2y &= -10 \\ y &= -5 \end{aligned}$$

Solution set: $\{(6, -9), (-2, -5)\}$

34. $y + 1 = x^2 + 2x \quad (1)$
$y + 2x = 4 \quad (2)$

Solve equation (2) for y.

$$y = 4 - 2x \quad (3)$$

Substitute $4 - 2x$ for y in (1).

$$\begin{aligned} (4 - 2x) + 1 &= x^2 + 2x \\ 0 &= x^2 + 4x - 5 \\ 0 &= (x + 5)(x - 1) \end{aligned}$$

$$x + 5 = 0 \quad \text{or} \quad x - 1 = 0$$
$$x = -5 \quad \text{or} \quad x = 1$$

Substitute these values for x in equation (3) to find y.

If $x = -5$, then $y = 4 - 2(-5) = 14$.
If $x = 1$, then $y = 4 - 2(1) = 2$.

Solution set: $\{(1, 2), (-5, 14)\}$

35. $x^2 + 3y^2 = 28 \quad (1)$
$y - x = -2 \quad (2)$

Solve equation (2) for y.

$$y = x - 2$$

Substitute $x - 2$ for y in equation (1).

$$\begin{aligned} x^2 + 3y^2 &= 28 \quad (1) \\ x^2 + 3(x - 2)^2 &= 28 \\ x^2 + 3(x^2 - 4x + 4) - 28 &= 0 \\ x^2 + 3x^2 - 12x + 12 - 28 &= 0 \\ 4x^2 - 12x - 16 &= 0 \\ 4(x^2 - 3x - 4) &= 0 \\ 4(x - 4)(x + 1) &= 0 \end{aligned}$$

$$x - 4 = 0 \quad \text{or} \quad x + 1 = 0$$
$$x = 4 \quad \text{or} \quad x = -1$$

Since $y = x - 2$, if $x = 4$, then $y = 4 - 2 = 2$.

If $x = -1$, then $y = -1 - 2 = -3$.

Solution set: $\{(4, 2), (-1, -3)\}$

36. $xy = 8 \quad (1)$
$x - 2y = 6 \quad (2)$

Solve equation (2) for x.

$$x = 2y + 6 \quad (3)$$

Substitute $2y + 6$ for x in equation (1) to find y.

$$\begin{aligned} xy &= 8 \quad (1) \\ (2y + 6)y &= 8 \\ 2y^2 + 6y - 8 &= 0 \\ 2(y^2 + 3y - 4) &= 0 \\ 2(y + 4)(y - 1) &= 0 \end{aligned}$$

$$y + 4 = 0 \quad \text{or} \quad y - 1 = 0$$
$$y = -4 \quad \text{or} \quad y = 1$$

Substitute these values for y in equation (3) to find x.
If $y = -4$, then $x = 2(-4) + 6 = -2$.
If $y = 1$, then $x = 2(1) + 6 = 8$.

Solution set: $\{(-2, -4), (8, 1)\}$

37. $x^2 + y^2 = 6 \quad (1)$
$x^2 - 2y^2 = -6 \quad (2)$

Multiply equation (2) by -1 and add the result to equation (1).

$$\begin{array}{rcll} x^2 + y^2 &=& 6 & (1) \\ -x^2 + 2y^2 &=& 6 & -1 \times (2) \\ \hline 3y^2 &=& 12 & \\ y^2 &=& 4 & \end{array}$$

$$y = 2 \quad \text{or} \quad y = -2$$

Substitute these values for y in equation (1) to find x.
If $y = \pm 2$, then

$$\begin{aligned} x^2 + y^2 &= 6 \quad (1) \\ x^2 + (\pm 2)^2 &= 6 \\ x^2 + 4 &= 6 \\ x^2 &= 2. \end{aligned}$$

$$x = \sqrt{2} \quad \text{or} \quad x = -\sqrt{2}$$

Since each value of x can be paired with each value of y, there are four points and the solution set is

$$\left\{\left(\sqrt{2}, 2\right), \left(-\sqrt{2}, 2\right), \left(\sqrt{2}, -2\right), \left(-\sqrt{2}, -2\right)\right\}.$$

38. $3x^2 - 2y^2 = 12 \quad (1)$
$x^2 + 4y^2 = 18 \quad (2)$

Multiply equation (1) by 2 and add the result to equation (2).

$$\begin{array}{rcll} 6x^2 - 4y^2 &=& 24 & 2 \times (1) \\ x^2 + 4y^2 &=& 18 & (2) \\ \hline 7x^2 &=& 42 & \\ x^2 &=& 6 & \end{array}$$

$$x = \sqrt{6} \quad \text{or} \quad x = -\sqrt{6}$$

Substitute these values for x in equation (2) to find y.

If $x = \pm\sqrt{6}$, then

$$\begin{aligned} x^2 + 4y^2 &= 18. \quad (2) \\ \left(\pm\sqrt{6}\right)^2 + 4y^2 &= 18 \\ 6 + 4y^2 &= 18 \\ 4y^2 &= 12 \\ y^2 &= 3 \end{aligned}$$

$$y = \sqrt{3} \quad \text{or} \quad y = -\sqrt{3}$$

The solution set is

$$\left\{\left(\sqrt{6}, \sqrt{3}\right), \left(\sqrt{6}, -\sqrt{3}\right), \left(-\sqrt{6}, \sqrt{3}\right), \left(-\sqrt{6}, -\sqrt{3}\right)\right\}.$$

39. A circle and a line can intersect in zero, one, or two points, so zero, one, or two solutions are possible.

40. A parabola and a hyperbola can intersect in zero, one, two, three, or four points, so zero, one, two, three, or four solutions are possible.

41.
$$\begin{aligned} 9x^2 &\geq 16y^2 + 144 \\ 9x^2 - 16y^2 &\geq 144 \\ \frac{x^2}{16} - \frac{y^2}{9} &\geq 1 \end{aligned}$$

The boundary, $\frac{x^2}{16} - \frac{y^2}{9} = 1$, is a solid hyperbola with x-intercepts $(4, 0)$ and $(-4, 0)$. The asymptotes are the extended diagonals of the rectangle with vertices $(4, 3)$, $(4, -3)$, $(-4, -3)$, and $(-4, 3)$. Test $(0, 0)$.

$$\begin{aligned} 9(0)^2 &\geq 16(0)^2 + 144 \quad ? \\ 0 &\geq 144 \quad \textit{False} \end{aligned}$$

Shade the sides of the hyperbola that do not contain $(0, 0)$. These are the regions inside the branches of the hyperbola.

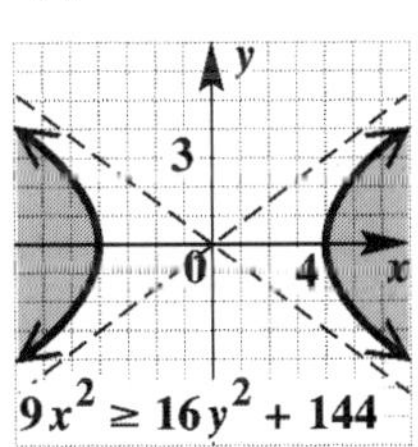

42. $4x^2 + y^2 \geq 16$

$$\frac{x^2}{4} + \frac{y^2}{16} \geq 1$$

The boundary, $\frac{x^2}{4} + \frac{y^2}{16} = 1$, is a solid ellipse with intercepts $(2, 0)$, $(-2, 0)$, $(0, 4)$, and $(0, -4)$. Test $(0, 0)$.

$$\begin{aligned} 4(0)^2 + 0^2 &\geq 16 \quad ? \\ 0 &\geq 16 \quad \textit{False} \end{aligned}$$

Shade the side of ellipse that does not contain $(0, 0)$. This is the region outside the ellipse.

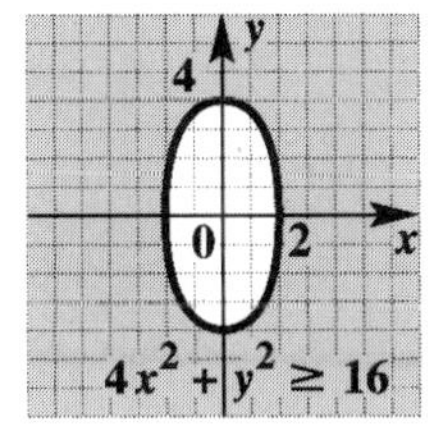

43. $y < -(x + 2)^2 + 1$

The boundary, $y = -(x + 2)^2 + 1$, is a dashed vertical parabola with vertex $(-2, 1)$. Since $a = -1 < 0$, the parabola opens down. Also, $|a| = |-1| = 1$, so the graph has the same shape as the graph of $y = x^2$. Test $(0, 0)$.

$$\begin{aligned} 0 &< -(0 + 2)^2 + 1 \quad ? \\ 0 &< -(4) + 1 \quad ? \\ 0 &< -3 \quad \textit{False} \end{aligned}$$

Shade the side of the parabola that does not contain $(0, 0)$. This is the region inside the parabola.

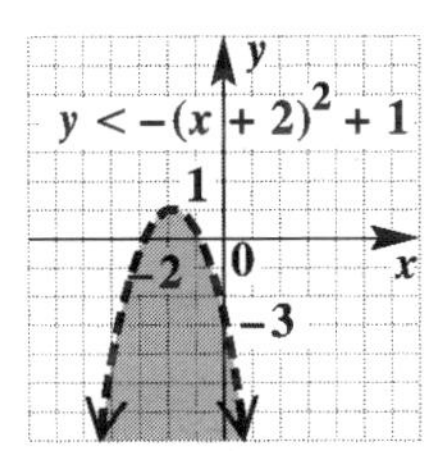

44.
$$\begin{aligned} 2x + 5y &\leq 10 \\ 3x - y &\leq 6 \end{aligned}$$

The boundary, $2x + 5y = 10$ is a solid line with intercepts $(5, 0)$ and $(0, 2)$. Test $(0, 0)$.

$$\begin{aligned} 2(0) + 5(0) &\leq 10 \quad ? \\ 0 &\leq 10 \quad \textit{True} \end{aligned}$$

Shade the side of the line that contains $(0, 0)$. The boundary, $3x - y = 6$, is a solid line with intercepts $(2, 0)$ and $(0, -6)$. Test $(0, 0)$.

$$\begin{aligned} 3(0) - 0 &\leq 6 \quad ? \\ 0 &\leq 6 \quad \textit{True} \end{aligned}$$

Shade the side of the line that contains $(0, 0)$. The graph of the system is the intersection of the two shaded regions.

continued

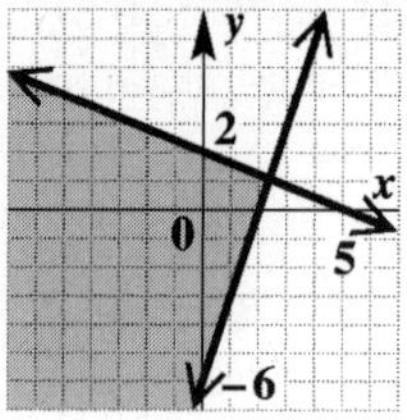

2x + 5y ≤ 10
3x − y ≤ 6

45.
$$|x| \le 2$$
$$|y| > 1$$
$$4x^2 + 9y^2 \le 36$$

The equation of the boundary, $|x| = 2$, can be written as

$$x = -2 \text{ or } x = 2.$$

The graph is these two solid vertical lines. Since $0 \le 2$ is true, the region between the lines, containing $(0, 0)$, is shaded.
The boundary, $|y| = 1$, consists of the two dashed horizontal lines $y = 1$ and $y = -1$. Since $0 > 1$ is false, the regions above and below the lines, not containing $(0, 0)$, are shaded.
The boundary given by

$$4x^2 + 9y^2 = 36$$
$$\text{or } \frac{x^2}{9} + \frac{y^2}{4} = 1$$

is graphed as a solid ellipse with intercepts $(3, 0)$, $(-3, 0)$, $(0, 2)$, and $(0, -2)$. Test $(0, 0)$.

$$4(0)^2 + 9(0)^2 \le 36 \quad ?$$
$$0 \le 36 \quad \textit{True}$$

The region inside the ellipse, containing $(0, 0)$, is shaded.
The graph of the system consists of the regions that include the common points of the three shaded regions.

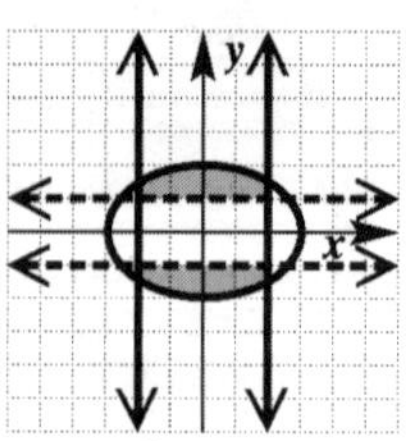

|x| ≤ 2
|y| > 1
4x² + 9y² ≤ 36

46.
$$9x^2 \le 4y^2 + 36$$
$$x^2 + y^2 \le 16$$

The equation of the first boundary is

$$9x^2 = 4y^2 + 36$$
$$9x^2 - 4y^2 = 36$$
$$\frac{x^2}{4} - \frac{y^2}{9} = 1.$$

The graph is a solid hyperbola with x-intercepts $(2, 0)$ and $(-2, 0)$. The asymptotes are the extended diagonals of the rectangle with vertices $(2, 3)$, $(2, -3)$, $(-2, -3)$, and $(-2, 3)$. Test $(0, 0)$.

$$9(0)^2 \le 4(0)^2 + 36 \quad ?$$
$$0 \le 36 \quad \textit{True}$$

Shade the region between the branches of the hyperbola that contains $(0, 0)$.
The equation of the second boundary is $x^2 + y^2 = 16$. This is a solid circle with center $(0, 0)$ and radius 4. Test $(0, 0)$.

$$0^2 + 0^2 \le 16 \quad ?$$
$$0 \le 16 \quad \textit{True}$$

Shade the region inside the circle.
The graph of the system is the intersection of the shaded regions which is between the two branches of the hyperbola and inside the circle.

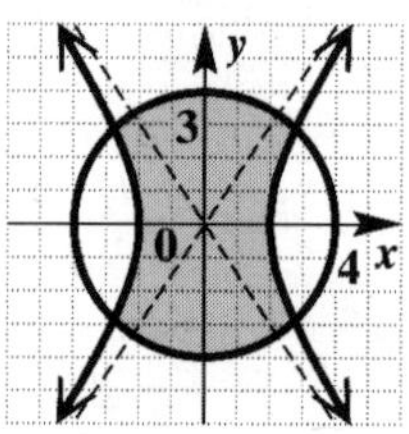

9x² ≤ 4y² + 36
x² + y² ≤ 16

47. $x^2 + y^2 + ax + by + c = 0$

Let $x = 2$ and $y = 4$.

$$2^2 + 4^2 + a(2) + b(4) + c = 0$$
$$4 + 16 + 2a + 4b + c = 0$$
$$2a + 4b + c = -20$$

48. $x^2 + y^2 + ax + by + c = 0$

Let $x = 5$ and $y = 1$.

$$5^2 + 1^2 + a(5) + b(1) + c = 0$$
$$25 + 1 + 5a + b + c = 0$$
$$5a + b + c = -26$$

49. $x^2 + y^2 + ax + by + c = 0$

Let $x = -1$ and $y = 1$.

$$(-1)^2 + 1^2 + a(-1) + b(1) + c = 0$$
$$1 + 1 - a + b + c = 0$$
$$-a + b + c = -2$$

50.
$$\begin{aligned} 2a + 4b + c &= -20 \quad (1)\\ 5a + b + c &= -26 \quad (2)\\ -a + b + c &= -2 \quad (3)\end{aligned}$$

Multiply equation (3) by 2 and add the result to equation (1).

$$\begin{array}{rl} 2a + 4b + c = -20 & (1)\\ -2a + 2b + 2c = -4 & 2 \times (3)\\ \hline 6b + 3c = -24 & (4)\end{array}$$

Multiply equation (3) by 5 and add the result to equation (2).

$$\begin{array}{rl} 5a + b + c = -26 & (2)\\ -5a + 5b + 5c = -10 & 5 \times (3)\\ \hline 6b + 6c = -36 & (5)\end{array}$$

Multiply equation (4) by -1 and add the result to equation (5).

$$\begin{array}{rl} -6b - 3c = 24 & -1 \times (4)\\ 6b + 6c = -36 & (5)\\ \hline 3c = -12 & \\ c = -4 & \end{array}$$

Substitute $c = -4$ into equation (4), and solve for b.

$$\begin{aligned} 6b + 3c &= -24 \quad (4)\\ 6b + 3(-4) &= -24\\ 6b - 12 &= -24\\ 6b &= -12\\ b &= -2\end{aligned}$$

Substitute $c = -4$ and $b = -2$ into equation (3), and solve for a.

$$\begin{aligned} -a + b + c &= -2 \quad (3)\\ -a + (-2) + (-4) &= -2\\ -a - 6 &= -2\\ -a &= 4\\ a &= -4\end{aligned}$$

The solution set is $\{(-4, -2, -4)\}$.

Therefore, the equation of the circle is

$$x^2 + y^2 - 4x - 2y - 4 = 0.$$

51.
$$\begin{aligned} x^2 + y^2 - 4x - 2y - 4 &= 0\\ (x^2 - 4x \quad\quad) + (y^2 - 2y \quad\quad) &= 4\end{aligned}$$

Complete the square for x and y.

$$\begin{aligned} (x^2 - 4x + \underline{4}) + (y^2 - 2y + \underline{1}) &= 4 + \underline{4} + \underline{1}\\ (x-2)^2 + (y-1)^2 &= 9\end{aligned}$$

The center is $(2, 1)$ and the radius is $\sqrt{9} = 3$.

52. $\dfrac{x^2}{64} + \dfrac{y^2}{25} = 1$ is in $\dfrac{x^2}{a^2} + \dfrac{y^2}{b^2} = 1$ form with $a = 8$ and $b = 5$. The graph is an ellipse with intercepts $(8, 0)$, $(-8, 0)$, $(0, 5)$, and $(0, -5)$. Plot the intercepts, and draw the ellipse through them.

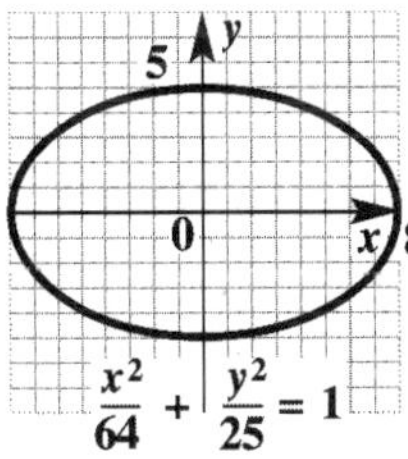

53.
$$\begin{aligned} \frac{y^2}{4} - 1 &= \frac{x^2}{9}\\ \frac{y^2}{4} - \frac{x^2}{9} &= 1\end{aligned}$$

The equation is in $\dfrac{y^2}{b^2} - \dfrac{x^2}{a^2} = 1$ form with $a = 3$ and $b = 2$. The graph is a hyperbola with y-intercepts $(0, 2)$ and $(0, -2)$ and asymptotes that are the extended diagonals of the rectangle with vertices $(3, 2)$, $(3, -2)$, $(-3, -2)$, and $(-3, 2)$. Draw a branch of the hyperbola through each intercept and approaching the asymptotes.

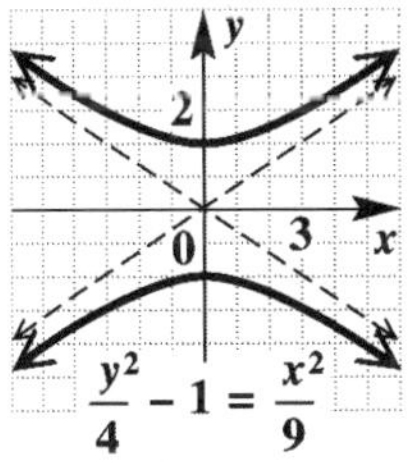

54. $x^2 + y^2 = 25$ is in $(x-h)^2 + (y-k)^2 = r^2$ form. The graph is a circle with center at $(0, 0)$ and radius 5.

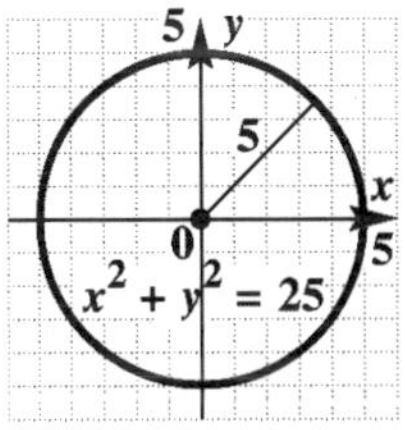

55. $x^2 + 9y^2 = 9$ in $\dfrac{x^2}{a^2} + \dfrac{y^2}{b^2} = 1$ form is $\dfrac{x^2}{9} + \dfrac{y^2}{1} = 1$ with $a = 3$ and $b = 1$. The graph is an ellipse with x-intercepts $(3, 0)$ and $(-3, 0)$ and y-intercepts $(0, 1)$ and $(0, -1)$. Plot the intercepts, and draw the ellipse through them.

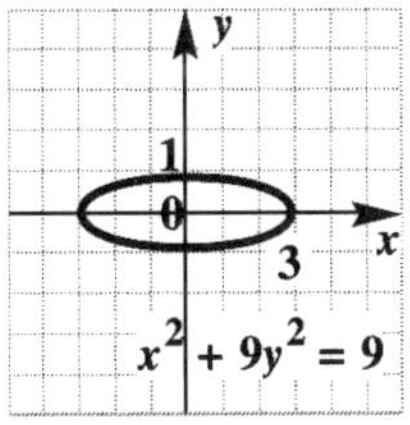

56. $x^2 - 9y^2 = 9$
$$\frac{x^2}{9} - \frac{y^2}{1} = 1$$

The equation is in the $\frac{x^2}{a^2} - \frac{y^2}{b^2} = 1$ form with $a = 3$ and $b = 1$. The graph is a hyperbola with x-intercepts $(3, 0)$ and $(-3, 0)$ and asymptotes that are the extended diagonals of the rectangle with vertices $(3, 1)$, $(3, -1)$, and $(-3, -1)$, and $(-3, 1)$. Graph a branch of the hyperbola through each intercept and approaching the asymptotes.

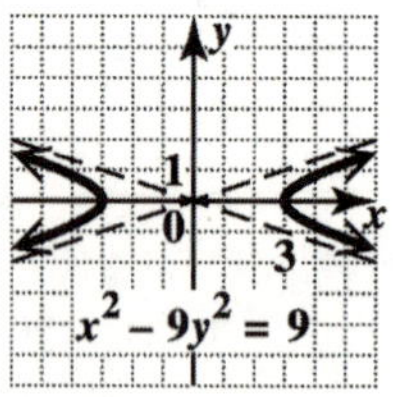

57. $f(x) = \sqrt{4 - x}$

Replace $f(x)$ with y.
$$y = \sqrt{4 - x}$$
Square both sides.
$$y^2 = 4 - x$$
$$x = -y^2 + 4$$
$$x = -1(y - 0)^2 + 4$$

This equation is the graph of a horizontal parabola with vertex $(4, 0)$. Since $a = -1 < 0$, the graph opens to the left. Also, $|a| = |-1| = 1$, so the graph has the same shape as the graph of $y = x^2$. The points $(0, 2)$ and $(3, 1)$ are on the graph. Since $f(x)$ represents a square root, $f(x)$ is nonnegative and its graph is the upper half of the parabola.

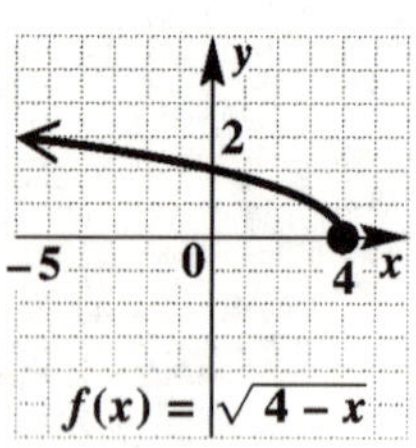

58. $3x + 2y \geq 0$
$$y \leq 4$$
$$x \leq 4$$

The boundary $3x + 2y = 0$ is a solid line through $(0, 0)$ and $(2, -3)$. Test $(0, 1)$.
$$3(0) + 2(1) \geq 0 \quad ?$$
$$2 \geq 0 \quad \textit{True}$$

Shade the side of the line that contains $(0, 1)$. The boundary $y = 4$ is a solid horizontal line through $(0, 4)$. Since $y \leq 4$, shade below the line. The boundary $x = 4$ is a solid vertical line through $(4, 0)$. Since $x \leq 4$, shade the region to the left of the line.

The graph of the system is the intersection of the three shaded regions.

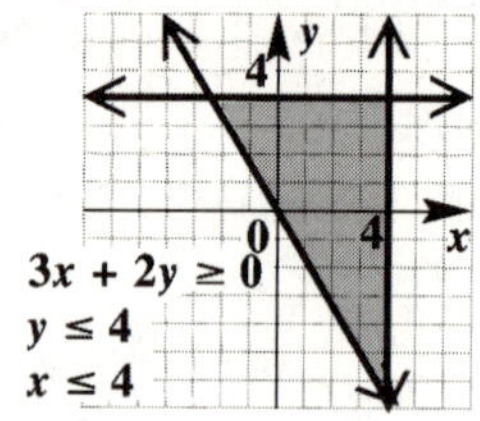

59. $4y > 3x - 12$
$$x^2 < 16 - y^2$$

The boundary $4y = 3x - 12$ is a dashed line with intercepts $(0, -3)$ and $(4, 0)$. Test $(0, 0)$.
$$4(0) > 3(0) - 12 \quad ?$$
$$0 > -12 \quad \textit{True}$$

Shade the side of the line that contains $(0, 0)$. The boundary $x^2 = 16 - y^2$, or $x^2 + y^2 = 16$, is a dashed circle with center at $(0, 0)$ and radius 4. Test $(0, 0)$.
$$0^2 < 16 - 0^2 \quad ?$$
$$0 < 16 \quad \textit{True}$$

Shade the region inside the circle. The graph of the system is the intersection of the two shaded regions.

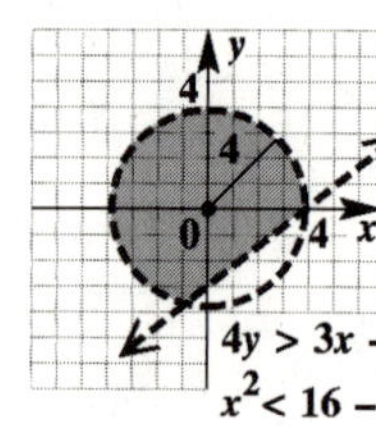

60. There are cases where one x-value will yield two y-values. In a function, every x yields one and only one y.

61. $\frac{x^2}{3352} + \frac{y^2}{3211} = 1$

$c^2 = a^2 - b^2 = 3352 - 3211$,
so $c^2 = 141$ and $c = \sqrt{141}$.

From Exercise 51 in Section 11.2, the apogee is $a + c = \sqrt{3352} + \sqrt{141} \approx 69.8$ million kilometers.

62. The perigee is $a - c = \sqrt{3352} - \sqrt{141} \approx 46.0$ million kilometers.

Chapter 11 Test

1. $f(x) = \sqrt{x - 2}$ is the graph of $g(x) = \sqrt{x}$ shifted 2 units right. The graph is **C**.

2. $f(x)=\sqrt{x+2}$ is the graph of $g(x)=\sqrt{x}$ shifted 2 units left. The graph is **A**.

3. $f(x)=\sqrt{x}+2$ is the graph of $g(x)=\sqrt{x}$ shifted 2 units up. The graph is **D**.

4. $f(x)=\sqrt{x}-2$ is the graph of $g(x)=\sqrt{x}$ shifted 2 units down. The graph is **B**.

5. $f(x)=|x-3|+4$

This is the graph of the absolute value function, $g(x)=|x|$, shifted 3 units to the right (since $x-3=0$ if $x=3$) and 4 units upward (because of the $+4$). Its vertex is at $(3,4)$.

x	0	1	2	3	4	5	6
y	7	6	5	4	5	6	7

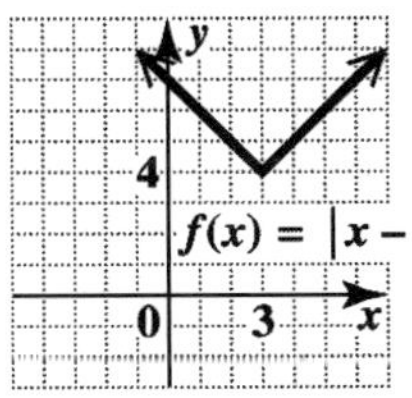

6. $f(x)=3x+5,\ g(x)=x^2+2$

(a) $$\begin{aligned}(f\circ g)(-2)&=f[g(-2)]\\&=f[(-2)^2+2]\\&=f(6)\\&=3\cdot 6+5\\&=23\end{aligned}$$

(b) $$\begin{aligned}(f\circ g)(x)&=f[g(x)]\\&=f(x^2+2)\\&=3(x^2+2)+5\\&=3x^2+6+5\\&=3x^2+11\end{aligned}$$

(c) $$\begin{aligned}(g\circ f)(x)&=g[f(x)]\\&=g(3x+5)\\&=(3x+5)^2+2\\&=9x^2+30x+25+2\\&=9x^2+30x+27\end{aligned}$$

7. $$(x-2)^2+(y+3)^2=16$$
$$(x-2)^2+[y-(-3)]^2=4^2$$

The graph is a circle with center $(2,-3)$ and radius 4.

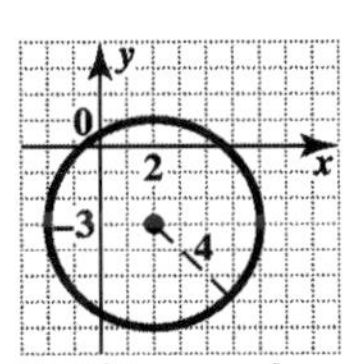

$(x-2)^2+(y+3)^2=16$

8. $x^2+y^2+8x-2y=8$

To find the center and radius, complete the squares on x and y.

$$(x^2+8x\quad\ )+(y^2-2y\quad\ )=8$$
$$(x^2+8x+\underline{16})+(y^2-2y+\underline{1})=8+\underline{16}+\underline{1}$$
$$(x+4)^2+(y-1)^2=25$$

The graph is a circle with center $(-4,1)$ and radius $\sqrt{25}=5$.

9. $$f(x)=\sqrt{9-x^2}$$
Replace $f(x)$ with y.
$$y=\sqrt{9-x^2}$$
Square both sides.
$$y^2=9-x^2$$
$$x^2+y^2=9$$

The graph of $x^2+y^2=9$ is a circle of radius $\sqrt{9}=3$ centered at the origin. Since $f(x)$ is nonnegative, only the top half of the circle is graphed.

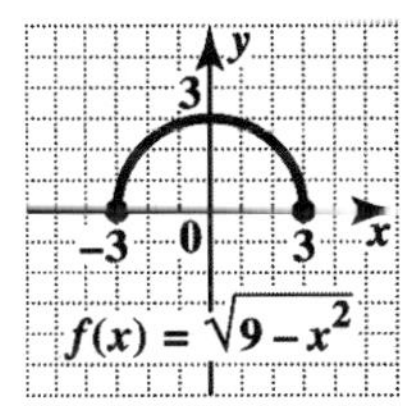

10. $4x^2+9y^2=36$
$$\frac{x^2}{9}+\frac{y^2}{4}=1$$

The equation is in $\frac{x^2}{a^2}+\frac{y^2}{b^2}=1$ form with $a=3$ and $b=2$. The graph is an ellipse with intercepts $(3,0)$, $(-3,0)$, $(0,2)$, and $(0,-2)$. Plot these intercepts, and draw the ellipse through them.

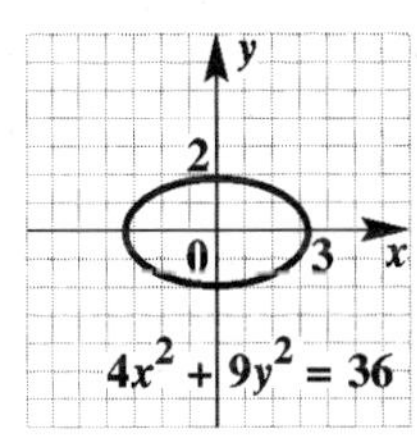

11. $16y^2-4x^2=64$
$$\frac{y^2}{4}-\frac{x^2}{16}=1$$

The equation is in $\frac{y^2}{b^2}-\frac{x^2}{a^2}=1$ form with $a=4$ and $b=2$. The graph is a hyperbola with y-intercepts $(0,2)$ and $(0,-2)$ and asymptotes that are the extended diagonals of the rectangle with vertices $(4,2)$, $(4,-2)$, $(-4,-2)$, and $(-4,2)$.

continued

Draw a branch of the hyperbola through each intercept and approaching the asymptotes.

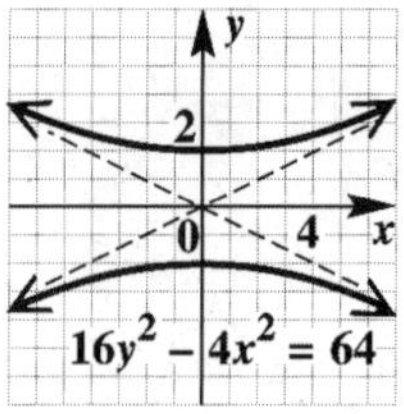

12. $$\frac{y}{2} = -\sqrt{1 - \frac{x^2}{9}}$$

Square both sides.

$$\frac{y^2}{4} = 1 - \frac{x^2}{9}$$
$$\frac{x^2}{9} + \frac{y^2}{4} = 1$$

This is an ellipse with x-intercepts $(3, 0)$ and $(-3, 0)$ and y-intercepts $(0, 2)$ and $(0, -2)$. Since y represents a negative square root in the original equation, y must be nonpositive. This restricts the graph to the lower half of the ellipse.

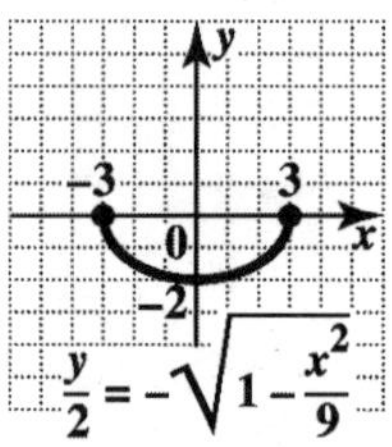

13. $6x^2 + 4y^2 = 12$

We have the *sum* of squares with different coefficients equal to a positive number, so this is an equation of an *ellipse*.

14. $$16x^2 = 144 + 9y^2$$
$$16x^2 - 9y^2 = 144$$

We have the *difference* of squares equal to a positive number, so this is an equation of a *hyperbola.*

15. $$4y^2 + 4x = 9$$
$$4x = -4y^2 + 9$$

We have an x-term and a y^2-term, so this is an equation of a horizontal *parabola*.

16. $2x - y = 9 \quad (1)$
$xy = 5 \quad (2)$

Solve equation (1) for y.

$$y = 2x - 9 \quad (3)$$

Substitute $2x - 9$ for y in equation (2).

$$xy = 5 \quad (2)$$
$$x(2x - 9) = 5$$
$$2x^2 - 9x = 5$$
$$2x^2 - 9x - 5 = 0$$
$$(2x + 1)(x - 5) = 0$$

$$2x + 1 = 0 \quad \text{or} \quad x - 5 = 0$$
$$x = -\frac{1}{2} \quad \text{or} \quad x = 5$$

Substitute these values for x in equation (3) to find y.

If $x = -\frac{1}{2}$, then $y = 2\left(-\frac{1}{2}\right) - 9 = -10$.

If $x = 5$, then $y = 2(5) - 9 = 1$.

Solution set: $\left\{\left(-\frac{1}{2}, -10\right), (5, 1)\right\}$

17. $x - 4 = 3y \quad (1)$
$x^2 + y^2 = 8 \quad (2)$

Solve equation (1) for x.

$$x = 3y + 4$$

Substitute $3y + 4$ for x in equation (2).

$$x^2 + y^2 = 8 \quad (2)$$
$$(3y + 4)^2 + y^2 = 8$$
$$9y^2 + 24y + 16 + y^2 = 8$$
$$10y^2 + 24y + 8 = 0$$
$$2(5y^2 + 12y + 4) = 0$$
$$2(5y + 2)(y + 2) = 0$$

$$5y + 2 = 0 \quad \text{or} \quad y + 2 = 0$$
$$y = -\frac{2}{5} \quad \text{or} \quad y = -2$$

Since $x = 3y + 4$, substitute these values for y to find x.

If $y = -\frac{2}{5}$, then

$$x = 3\left(-\frac{2}{5}\right) + 4 = -\frac{6}{5} + 4 = \frac{14}{5}.$$

If $y = -2$, then $x = 3(-2) + 4 = -2$.

Solution set: $\left\{(-2, -2), \left(\frac{14}{5}, -\frac{2}{5}\right)\right\}$

18. $x^2 + y^2 = 25 \quad (1)$
$x^2 - 2y^2 = 16 \quad (2)$

Multiply equation (1) by 2 and add the result to equation (2).

$$\begin{aligned} 2x^2 + 2y^2 &= 50 \quad 2 \times (1) \\ x^2 - 2y^2 &= 16 \quad (2) \\ \hline 3x^2 &= 66 \\ x^2 &= 22 \end{aligned}$$

$$x = \sqrt{22} \quad \text{or} \quad x = -\sqrt{22}$$

Substitute 22 for x^2 in equation (1).

$$\begin{aligned} x^2 + y^2 &= 25 \quad (1) \\ 22 + y^2 &= 25 \\ y^2 &= 3 \end{aligned}$$

$$y = \sqrt{3} \quad \text{or} \quad y = -\sqrt{3}$$

Solution set: $\left\{\left(\sqrt{22}, \sqrt{3}\right), \left(\sqrt{22}, -\sqrt{3}\right), \left(-\sqrt{22}, \sqrt{3}\right), \left(-\sqrt{22}, -\sqrt{3}\right)\right\}$

19. $y < x^2 - 2$

The boundary, $y = x^2 - 2$, is a dashed parabola in $y = a(x-h)^2 + k$ form with vertex (h, k) at $(0, -2)$. Since $a = 1 > 0$, the parabola opens up. It also has the same shape as $y = x^2$. The points $(2, 2)$ and $(-2, 2)$ are on the graph. Test $(0, 0)$.

$$\begin{aligned} 0 &\le (0)^2 - 2 \quad ? \\ 0 &\le -2 \quad \textit{False} \end{aligned}$$

Shade the side of the parabola that does not contain $(0, 0)$. This is the region outside the parabola.

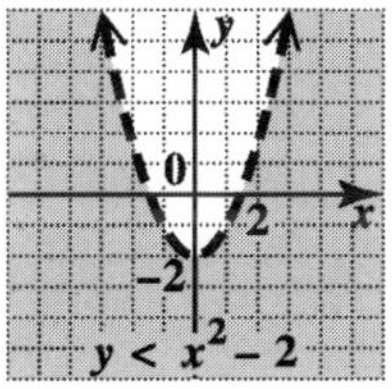

20. $x^2 + 25y^2 \le 25$
$x^2 + y^2 \le 9$

The first boundary, $\frac{x^2}{25} + \frac{y^2}{1} = 1$, is a solid ellipse with intercepts $(5, 0)$, $(-5, 0)$, $(0, 1)$, and $(0, -1)$. Test $(0, 0)$.

$$\begin{aligned} 0^2 + 25 \cdot 0^2 &\le 25 \quad ? \\ 0 &\le 25 \quad \textit{True} \end{aligned}$$

Shade the region inside the ellipse.
The second boundary, $x^2 + y^2 = 9$, is a solid circle with center $(0, 0)$ and radius 3. Test $(0, 0)$.

$$\begin{aligned} 0^2 + 0^2 &\le 9 \quad ? \\ 0 &\le 9 \quad \textit{True} \end{aligned}$$

Shade the region inside the circle. The solution of the system is the intersection of the two shaded regions.

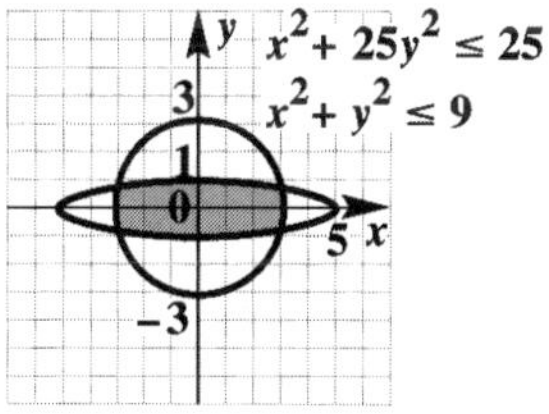

Cumulative Review Exercises Chapters 1–11

1. $$\begin{aligned} &-10 + |-5| - |3| + 4 \\ &= -10 + 5 - 3 + 4 \\ &= -5 - 3 + 4 \\ &= -8 + 4 = -4 \end{aligned}$$

2. $$\begin{aligned} 4 - (2x + 3) + x &= 5x - 3 \\ 4 - 2x - 3 + x &= 5x - 3 \\ -x + 1 &= 5x - 3 \\ -6x &= -4 \\ x &= \frac{2}{3} \end{aligned}$$

Solution set: $\left\{\frac{2}{3}\right\}$

3. $$\begin{aligned} -4k + 7 &\ge 6k + 1 \\ -10k &\ge -6 \end{aligned}$$

Divide by -10; reverse the direction of the inequality.

$$k \le \frac{-6}{-10}$$
$$k \le \frac{3}{5}$$

Solution set: $\left(-\infty, \frac{3}{5}\right]$

4. $$\begin{aligned} |5m| - 6 &= 14 \\ |5m| &= 20 \end{aligned}$$

$$\begin{aligned} 5m &= 20 \quad \text{or} \quad 5m = -20 \\ m &= 4 \quad \text{or} \quad m = -4 \end{aligned}$$

Solution set: $\{-4, 4\}$

5. $|2p - 5| > 15$

$$\begin{aligned} 2p - 5 > 15 &\quad \text{or} \quad 2p - 5 < -15 \\ 2p > 20 &\quad \quad \quad \; 2p < -10 \\ p > 10 &\quad \text{or} \quad p < -5 \end{aligned}$$

Solution set: $(-\infty, -5) \cup (10, \infty)$

6. Let $(x_1, y_1) = (2, 5)$ and $(x_2, y_2) = (-4, 1)$.

$$m = \frac{y_2 - y_1}{x_2 - x_1} = \frac{1-5}{-4-2} = \frac{-4}{-6} = \frac{2}{3}$$

7. Through $(-3, -2)$; perpendicular to $2x - 3y = 7$

Write $2x - 3y = 7$ in slope-intercept form.

$$\begin{aligned} -3y &= -2x + 7 \\ y &= \frac{2}{3}x - \frac{7}{3} \end{aligned}$$

The slope is $\frac{2}{3}$. Perpendicular lines have slopes that are negative reciprocals of each other, so a line perpendicular to the given line will have slope $-\frac{3}{2}$. Let $m = -\frac{3}{2}$ and $(x_1, y_1) = (-3, -2)$ in the point-slope form.

$$\begin{aligned} y - y_1 &= m(x - x_1) \\ y - (-2) &= -\frac{3}{2}[x - (-3)] \\ y + 2 &= -\frac{3}{2}(x + 3) \end{aligned}$$

Multiply by 2 to clear the fraction.

$$\begin{aligned} 2y + 4 &= -3(x + 3) \\ 2y + 4 &= -3x - 9 \\ 3x + 2y &= -13 \end{aligned}$$

8. $$\begin{aligned} 3x - y &= 12 \quad (1) \\ 2x + 3y &= -3 \quad (2) \end{aligned}$$

Multiply equation (1) by 3 and add the result to equation (2).

$$\begin{array}{rcll} 9x - 3y & = & 36 & 3 \times (1) \\ 2x + 3y & = & -3 & (2) \\ \hline 11x & = & 33 & \\ x & = & 3 & \end{array}$$

Substitute 3 for x in equation (1) to find y.

$$\begin{aligned} 3x - y &= 12 \quad (1) \\ 3(3) - y &= 12 \\ 9 - y &= 12 \\ -y &= 3 \\ y &= -3 \end{aligned}$$

Solution set: $\{(3, -3)\}$

9. $$\begin{aligned} x + y - 2z &= 9 \quad (1) \\ 2x + y + z &= 7 \quad (2) \\ 3x - y - z &= 13 \quad (3) \end{aligned}$$

Add equation (2) and equation (3).

$$\begin{array}{rcll} 2x + y + z & = & 7 & (2) \\ 3x - y - z & = & 13 & (3) \\ \hline 5x & = & 20 & \\ x & = & 4 & \end{array}$$

Multiply equation (1) by -1 and add the result to equation (2).

$$\begin{array}{rcll} -x - y + 2z & = & -9 & \\ 2x + y + z & = & 7 & (2) \\ \hline x + 3z & = & -2 & (4) \end{array}$$

Substitute 4 for x in equation (4) to find z.

$$\begin{aligned} x + 3z &= -2 \quad (4) \\ 4 + 3z &= -2 \\ 3z &= -6 \\ z &= -2 \end{aligned}$$

Substitute 4 for x and -2 for z in equation (2) to find y.

$$\begin{aligned} 2x + y + z &= 7 \quad (2) \\ 2(4) + y - 2 &= 7 \\ y + 6 &= 7 \\ y &= 1 \end{aligned}$$

Solution set: $\{(4, 1, -2)\}$

10. $$\begin{aligned} xy &= -5 \quad (1) \\ 2x + y &= 3 \quad (2) \end{aligned}$$

Solve equation (2) for y.

$$y = -2x + 3 \quad (3)$$

Substitute $-2x + 3$ for y in equation (1).

$$\begin{aligned} xy &= -5 \quad (1) \\ x(-2x + 3) &= -5 \\ -2x^2 + 3x &= -5 \\ -2x^2 + 3x + 5 &= 0 \\ 2x^2 - 3x - 5 &= 0 \\ (2x - 5)(x + 1) &= 0 \end{aligned}$$

$$2x - 5 = 0 \quad \text{or} \quad x + 1 = 0$$
$$x = \frac{5}{2} \quad \text{or} \quad x = -1$$

Substitute these values for x in equation (3) to find y.

If $x = \frac{5}{2}$, then $y = -2\left(\frac{5}{2}\right) + 3 = -2$.

If $x = -1$, then $y = -2(-1) + 3 = 5$.

Solution set: $\left\{(-1, 5), \left(\frac{5}{2}, -2\right)\right\}$

11. Let s = Al's speed, $2s$ = Bev's speed, t = Bev's time, and $t + \frac{1}{2}$ = Al's time.

	Distance	Rate	Time
Al	20	s	$t + \frac{1}{2}$
Bev	20	$2s$	t

Since $d = rt$, the system of equations is

$$s\left(t+\frac{1}{2}\right)=20 \qquad (1)$$
$$2st=20. \qquad (2)$$

Solve equation (2) for s.

$$2st=20 \qquad (2)$$
$$s=\frac{20}{2t}=\frac{10}{t}$$

Substitute $\frac{10}{t}$ for s in equation (1) to find t.

$$s\left(t+\frac{1}{2}\right)=20 \qquad (1)$$
$$\frac{10}{t}\left(t+\frac{1}{2}\right)=20$$
$$10+\frac{5}{t}=20$$

Multiply each term by the LCD, t.

$$10t+5=20t$$
$$5=10t$$
$$t=\frac{5}{10}=\frac{1}{2}$$

Since $s=\frac{10}{t}$ and $t=\frac{1}{2}$,

$$s=\frac{10}{\frac{1}{2}}=20.$$

So Al's speed was 20 mph and Bev's speed was $2 \cdot 20=40$ mph.

Another solution:

$$st+\tfrac{1}{2}s=20 \qquad (1)$$
$$2st=20 \qquad (2)$$

Multiply equation (1) by -2 and add the result to equation (2).

$$\begin{array}{rcrl} -2st - s & = & -40 & \quad -2\times(1) \\ 2st & = & 20 & \quad (2) \\ \hline -s & = & -20 & \\ s & = & 20 & \end{array}$$

Substitute 20 for s in equation (2) to find $t=\frac{1}{2}$.

12. $f(x)=.07x+135$
$f(2000)=.07(2000)+135=275$

The weekly fee is \$275 if \$2000 is taken in for the week.

13. $(5y-3)^2=(5y)^2-2(5y)3+3^2$
$=25y^2-30y+9$

14. $(2r+7)(6r-1)$
$=12r^2-2r+42r-7$
$=12r^2+40r-7$

15. $\dfrac{8x^4-4x^3+2x^2+13x+8}{2x+1}$

$$\begin{array}{r|l} & 4x^3 - 4x^2 + 3x + 5 \\ \hline 2x+1 & 8x^4 - 4x^3 + 2x^2 + 13x + 8 \\ & 8x^4 + 4x^3 \\ \hline & -8x^3 + 2x^2 \\ & -8x^3 - 4x^2 \\ \hline & 6x^2 + 13x \\ & 6x^2 + 3x \\ \hline & 10x + 8 \\ & 10x + 5 \\ \hline & 3 \end{array}$$

The answer is

$$4x^3-4x^2+3x+5+\frac{3}{2x+1}.$$

16. $12x^2-7x-10=(4x-5)(3x+2)$

17. $2y^4+5y^2-3$

Let $p=y^2$, so $p^2=y^4$.

$$2y^4+5y^2-3=2p^2+5p-3$$
$$=(2p-1)(p+3)$$

Now substitute y^2 for p.

$$=\left(2y^2-1\right)\left(y^2+3\right)$$

18. $z^4-1=\left(z^2+1\right)\left(z^2-1\right)$
$=\left(z^2+1\right)(z+1)(z-1)$

19. $a^3-27b^3=a^3-(3b)^3$
$=(a-3b)\left(a^2+3ab+9b^2\right)$

20. $\dfrac{5x-15}{24}\cdot\dfrac{64}{3x-9}=\dfrac{5(x-3)}{3\cdot 8}\cdot\dfrac{8\cdot 8}{3(x-3)}$
$=\dfrac{5\cdot 8}{3\cdot 3}=\dfrac{40}{9}$

21. $\dfrac{y^2-4}{y^2-y-6}\div\dfrac{y^2-2y}{y-1}$

Multiply by the reciprocal.

$$=\frac{y^2-4}{y^2-y-6}\cdot\frac{y-1}{y^2-2y}$$

Factor and simplify.

$$=\frac{(y+2)(y-2)}{(y-3)(y+2)}\cdot\frac{(y-1)}{y(y-2)}$$
$$=\frac{y-1}{y(y-3)}$$

22. $\dfrac{5}{c+5}-\dfrac{2}{c+3}$

The LCD is $(c+5)(c+3)$.

$$=\frac{5(c+3)}{(c+5)(c+3)}-\frac{2(c+5)}{(c+3)(c+5)}$$
$$=\frac{5c+15-2c-10}{(c+5)(c+3)}$$
$$=\frac{3c+5}{(c+5)(c+3)}$$

23. $$\frac{p}{p^2+p}+\frac{1}{p^2+p}=\frac{p+1}{p^2+p}$$
$$=\frac{p+1}{p(p+1)}=\frac{1}{p}$$

24. Let $x=$ the time to do the job working together.

Make a chart.

Worker	Rate	Time Together	Part of Job Done
Kareem	$\frac{1}{3}$	x	$\frac{x}{3}$
Jamal	$\frac{1}{2}$	x	$\frac{x}{2}$

Part done by Kareem plus part done by Jamal equals 1 whole job.

$$\frac{x}{3} + \frac{x}{2} = 1$$

Multiply by the LCD, 6.

$$6\left(\frac{x}{3}+\frac{x}{2}\right)=6\cdot 1$$
$$2x+3x=6$$
$$5x=6$$
$$x=\frac{6}{5} \text{ or } 1\frac{1}{5}$$

It takes $\frac{6}{5}$ or $1\frac{1}{5}$ hours to do the job together.

25. $$\left(\frac{4}{3}\right)^{-1}=\left(\frac{3}{4}\right)^{1}=\frac{3}{4}$$

26. $$\frac{(2a)^{-2}a^4}{a^{-3}}=\frac{2^{-2}a^{-2}a^4}{a^{-3}}=\frac{2^{-2}a^2}{a^{-3}}$$
$$=\frac{a^2a^3}{2^2}=\frac{a^5}{4}$$

27. $$4\sqrt[3]{16}-2\sqrt[3]{54}=4\sqrt[3]{8\cdot 2}-2\sqrt[3]{27\cdot 2}$$
$$=4\cdot 2\sqrt[3]{2}-2\cdot 3\sqrt[3]{2}$$
$$=8\sqrt[3]{2}-6\sqrt[3]{2}=2\sqrt[3]{2}$$

28. $$\frac{3\sqrt{5x}}{\sqrt{2x}}=\frac{3\sqrt{5x}\cdot\sqrt{2x}}{\sqrt{2x}\cdot\sqrt{2x}}=\frac{3\sqrt{10x^2}}{2x}$$
$$=\frac{3x\sqrt{10}}{2x}=\frac{3\sqrt{10}}{2}$$

29. $$\frac{5+3i}{2-i}$$

Multiply the numerator and denominator by the conjugate of the denominator.

$$=\frac{(5+3i)(2+i)}{(2-i)(2+i)}$$
$$=\frac{10+5i+6i+3i^2}{4-i^2}$$
$$=\frac{10+11i+3(-1)}{4-(-1)}$$
$$=\frac{7+11i}{5}=\frac{7}{5}+\frac{11}{5}i$$

30. $2\sqrt{k}=\sqrt{5k+3}$

Square both sides.

$$4k=5k+3$$
$$-k=3$$
$$k=-3$$

Since k must be nonnegative so that $\sqrt{k}$ is a real number, -3 cannot be a solution. The solution set is $\emptyset$.

31. $$10q^2+13q=3$$
$$10q^2+13q-3=0$$
$$(5q-1)(2q+3)=0$$
$$5q-1=0 \quad\text{or}\quad 2q+3=0$$
$$q=\frac{1}{5} \quad\text{or}\quad q=-\frac{3}{2}$$

Solution set: $\left\{\frac{1}{5},-\frac{3}{2}\right\}$

32. $(4x-1)^2=8$
$$4x-1=\pm\sqrt{8}$$
$$4x=1\pm 2\sqrt{2}$$
$$x=\frac{1\pm 2\sqrt{2}}{4}$$

Solution set: $\left\{\frac{1+2\sqrt{2}}{4},\frac{1-2\sqrt{2}}{4}\right\}$

33. $3k^2-3k-2=0$

Use the quadratic formula with $a=3$, $b=-3$, and $c=-2$.

$$k=\frac{-b\pm\sqrt{b^2-4ac}}{2a}$$
$$k=\frac{-(-3)\pm\sqrt{(-3)^2-4(3)(-2)}}{2(3)}$$
$$=\frac{3\pm\sqrt{9+24}}{6}=\frac{3\pm\sqrt{33}}{6}$$

Solution set: $\left\{\frac{3+\sqrt{33}}{6},\frac{3-\sqrt{33}}{6}\right\}$

34. $2(x^2-3)^2-5(x^2-3)=12$

Let $u=(x^2-3)$.

$$2u^2-5u=12$$
$$2u^2-5u-12=0$$
$$(2u+3)(u-4)=0$$
$$2u+3=0 \quad\text{or}\quad u-4=0$$
$$u=-\frac{3}{2} \quad\text{or}\quad u=4$$

Substitute x^2-3 for u to find x.

If $u = -\frac{3}{2}$, then

$$x^2 - 3 = -\frac{3}{2}$$
$$x^2 = \frac{3}{2}$$
$$x = \pm\sqrt{\frac{3}{2}} = \pm\frac{\sqrt{3}}{\sqrt{2}} \cdot \frac{\sqrt{2}}{\sqrt{2}} = \pm\frac{\sqrt{6}}{2}.$$

If $u = 4$, then

$$x^2 - 3 = 4$$
$$x^2 = 7$$
$$x = \pm\sqrt{7}.$$

Solution set: $\left\{-\frac{\sqrt{6}}{2}, \frac{\sqrt{6}}{2}, -\sqrt{7}, \sqrt{7}\right\}$

35. Solve $F = \frac{kwv^2}{r}$ for v.

$$Fr = kwv^2$$
$$v^2 = \frac{Fr}{kw}$$

Take the square root of each side.

$$v = \pm\sqrt{\frac{Fr}{kw}} = \frac{\pm\sqrt{Fr}}{\sqrt{kw}} \cdot \frac{\sqrt{kw}}{\sqrt{kw}} = \frac{\pm\sqrt{Frkw}}{kw}$$

36. $f(x) = y = x^3 + 4$

Interchange x and y and solve for y.

$$x = y^3 + 4$$
$$y^3 = x - 4$$
$$y = \sqrt[3]{x - 4}$$
$$f^{-1}(x) = \sqrt[3]{x - 4}$$

37. $a^{\log_a x} = x$, so $3^{\log_3 4} = 4$.

38. $e^{\ln x} = x$, so $e^{\ln 7} = 7$.

39.
$$2\log(3x + 7) - \log 4 = \log(3x + 7)^2 - \log 4 = \log\frac{(3x + 7)^2}{4}$$

40.
$$\log(x + 2) + \log(x - 1) = 1$$
$$\log_{10}[(x + 2)(x - 1)] = 1$$
$$(x + 2)(x - 1) = 10^1$$
$$x^2 + x - 2 = 10$$
$$x^2 + x - 12 = 0$$
$$(x + 4)(x - 3) = 0$$

$x + 4 = 0$ or $x - 3 = 0$
$x = -4$ or $x = 3$

The original equation is undefined when $x = -4$.

Solution set: $\{3\}$

41. $P = 10{,}000$, $r = .05$, $t = 4$

(a)
$$A = P\left(1 + \frac{r}{n}\right)^{nt}$$
$$= 10{,}000\left(1 + \frac{.05}{4}\right)^{4 \cdot 4} \quad \textit{Let } n = 4.$$
$$= 10{,}000(1.0125)^{16}$$
$$\approx 12{,}198.90$$

There will be \$12,198.90 in the account.

(b)
$$A = Pe^{rt}$$
$$= 10{,}000e^{(.05)(4)}$$
$$\approx 12{,}214.03$$

There will be \$12,214.03 in the account.

42.
$$y = 1.38(1.65)^x$$
$$y = 1.38(1.65)^5 \quad \textit{Let } x = 5.$$
$$\approx 16.9 \text{ billion dollars}$$

The sales are estimated to be \$16.9 billion in 2000.

43. $2003 - 1995 = 8$
$$y = 1.38(1.65)^8 \quad \textit{Let } x = 8.$$
$$\approx 75.8 \text{ billion dollars}$$

The sales are estimated to be \$75.8 billion in 2003.

44. $f(x) = x^2 + 2x - 4$, $g(x) = 3x + 2$

(a)
$$(g \circ f)(1) = g[f(1)]$$
$$= g(1^2 + 2 \cdot 1 - 4)$$
$$= g(-1)$$
$$= 3(-1) + 2$$
$$= -1$$

(b)
$$(f \circ g)(x) = f[g(x)]$$
$$= f(3x + 2)$$
$$= (3x + 2)^2 + 2(3x + 2) - 4$$
$$= 9x^2 + 12x + 4 + 6x + 4 - 4$$
$$= 9x^2 + 18x + 4$$

45. $f(x) = -3x + 5$

The equation is in slope-intercept form, so the y-intercept is $(0, 5)$ and $m = -3$ or $\frac{-3}{1}$.

Plot $(0, 5)$. From $(0, 5)$, move down 3 units and right 1 unit. Draw the line through these two points.

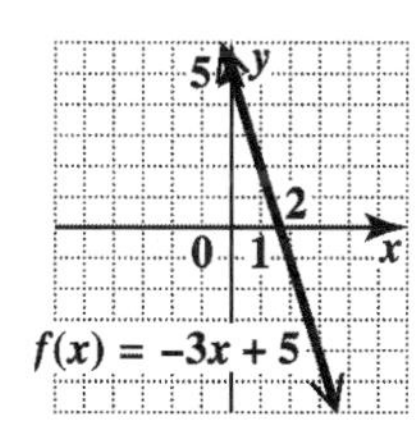

46. $f(x) = -2(x-1)^2 + 3$

The graph is a parabola that has been shifted 1 unit to the right and 3 units upward from $(0, 0)$, so its vertex is at $(1, 3)$. Since $a = -2 < 0$, the parabola opens down. Also $|a| = |-2| = 2 > 1$, so the graph is narrower than the graph of $f(x) = x^2$. The points $(0, 1)$ and $(2, 1)$ are on the graph.

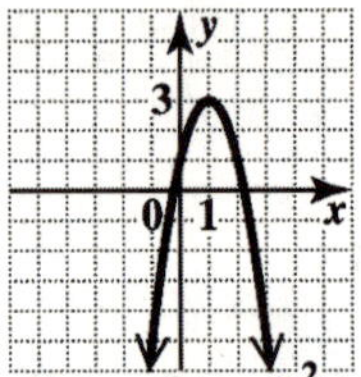

47. $\frac{x^2}{25} + \frac{y^2}{16} \leq 1$

The boundary, $\frac{x^2}{25} + \frac{y^2}{16} = 1$, is a solid ellipse in $\frac{x^2}{a^2} + \frac{y^2}{b^2} = 1$ form with intercepts $(5, 0)$, $(-5, 0)$, $(0, 4)$, and $(0, -4)$. Test $(0, 0)$.

$$\frac{0^2}{25} + \frac{0^2}{16} \leq 1 \quad ?$$

$$0 \leq 1 \quad \textit{True}$$

Shade the region inside the ellipse.

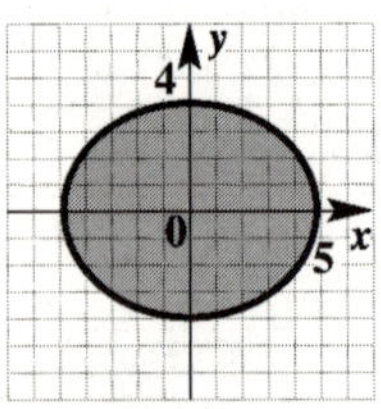

48. $f(x) = \sqrt{x-2}$

This is the graph of $g(x) = \sqrt{x}$ shifted 2 units right.

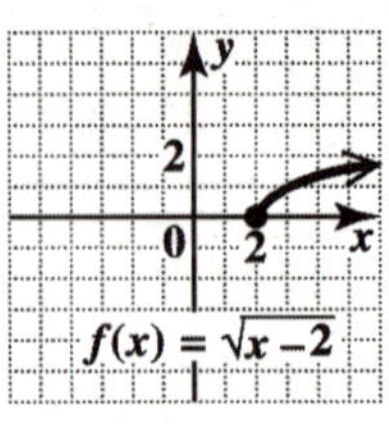

49. $\frac{x^2}{4} - \frac{y^2}{16} = 1$ is in $\frac{x^2}{a^2} - \frac{y^2}{b^2} = 1$ form.

The graph is a hyperbola with x-intercepts $(2, 0)$ and $(-2, 0)$ and asymptotes that are the extended diagonals of the rectangle with vertices $(2, 4)$, $(2, -4)$, $(-2, -4)$, and $(-2, 4)$. Draw a branch of the hyperbola through each intercept approaching the asymptotes.

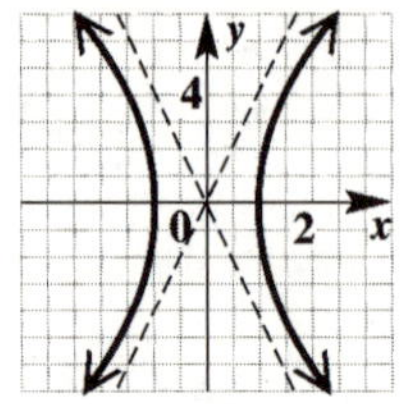

50. $f(x) = 3^x$

The graph of f is an increasing exponential.

x	-1	0	1	2
$f(x)$	$\frac{1}{3}$	1	3	9

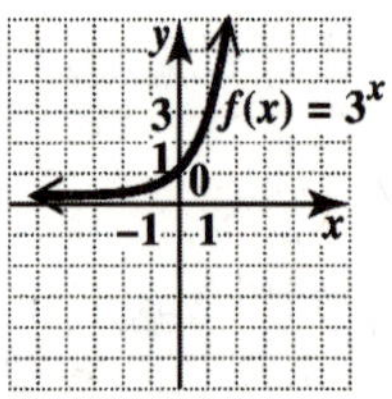

CHAPTER 12 SEQUENCES AND SERIES

Section 12.1

1. $a_n = n + 1$
To get a_1, the first term, replace n with 1.
$a_1 = 1 + 1 = 2$
To get a_2, the second term, replace n with 2.
$a_2 = 2 + 1 = 3$
To get a_3, the third term, replace n with 3.
$a_3 = 3 + 1 = 4$
To get a_4, the fourth term, replace n with 4.
$a_4 = 4 + 1 = 5$
To get a_5, the fifth term, replace n with 5.
$a_5 = 5 + 1 = 6$

Answer: $2, 3, 4, 5, 6$

3. $a_n = \frac{n+3}{n}$
To get a_1, the first term, replace n with 1.
$a_1 = \frac{1+3}{1} = \frac{4}{1} = 4$
To get a_2, the second term, replace n with 2.
$a_2 = \frac{2+3}{2} = \frac{5}{2}$
To get a_3, the third term, replace n with 3.
$a_3 = \frac{3+3}{3} = \frac{6}{3} = 2$
To get a_4, the fourth term, replace n with 4.
$a_4 = \frac{4+3}{4} = \frac{7}{4}$
To get a_5, the fifth term, replace n with 5.
$a_5 = \frac{5+3}{5} = \frac{8}{5}$

Answer: $4, \frac{5}{2}, 2, \frac{7}{4}, \frac{8}{5}$

5. $a_n = 3^n$
$a_1 = 3^1 = 3$
$a_2 = 3^2 = 9$
$a_3 = 3^3 = 27$
$a_4 = 3^4 = 81$
$a_5 = 3^5 = 243$

Answer: $3, 9, 27, 81, 243$

7. $a_n = \frac{1}{n^2}$
$a_1 = \frac{1}{1^2} = 1$
$a_2 = \frac{1}{2^2} = \frac{1}{4}$
$a_3 = \frac{1}{3^2} = \frac{1}{9}$
$a_4 = \frac{1}{4^2} = \frac{1}{16}$
$a_5 = \frac{1}{5^2} = \frac{1}{25}$

Answer: $1, \frac{1}{4}, \frac{1}{9}, \frac{1}{16}, \frac{1}{25}$

9. $a_n = (-1)^n$
$a_1 = (-1)^1 = -1$
$a_2 = (-1)^2 = 1$
$a_3 = (-1)^3 = -1$
$a_4 = (-1)^4 = 1$
$a_5 = (-1)^5 = -1$

Answer: $-1, 1, -1, 1, -1$

11. $a_n = -9n + 2$
To find a_8, replace n with 8.
$a_8 = -9(8) + 2 = -72 + 2 = -70$

13. $a_n = \frac{3n+7}{2n-5}$
$a_{14} = \frac{3(14)+7}{2(14)-5} = \frac{42+7}{28-5} = \frac{49}{23}$

15. $a_n = (n+1)(2n+3)$
$a_8 = (8+1)[2(8)+3]$
$= 9(19) = 171$

17. $4, 8, 12, 16, \ldots$ can be written as
$4 \cdot 1, 4 \cdot 2, 4 \cdot 3, 4 \cdot 4, \ldots$, so $a_n = 4n$.

19. $\frac{1}{3}, \frac{1}{9}, \frac{1}{27}, \frac{1}{81}, \ldots$ can be written as
$\frac{1}{3^1}, \frac{1}{3^2}, \frac{1}{3^3}, \frac{1}{3^4}, \ldots$, so $a_n = \frac{1}{3^n}$.

21. Make a table as follows:

Month	Interest	Payment	Unpaid balance
0			1000
1	$1000(.01) = 10$	$100 + 10 = 110$	$1000 - 100 = 900$
2	$900(.01) = 9$	$100 + 9 = 109$	$900 - 100 = 800$
3	$800(.01) = 8$	$100 + 8 = 108$	$800 - 100 = 700$
4	$700(.01) = 7$	$100 + 7 = 107$	$700 - 100 = 600$
5	$600(.01) = 6$	$100 + 6 = 106$	$600 - 100 = 500$
6	$500(.01) = 5$	$100 + 5 = 105$	$500 - 100 = 400$

The payments are \$110, \$109, \$108, \$107, \$106, and \$105; the unpaid balance is \$400.

23. When new, the car is worth \$20,000.
Let a_n = the value of the car after the nth year.
The car retains $\frac{4}{5}$ of its value each year.

$a_1 = \frac{4}{5}(20{,}000) = \$16{,}000$ (value after the first year)
$a_2 = \frac{4}{5}(16{,}000) = \$12{,}800$
$a_3 = \frac{4}{5}(12{,}800) = \$10{,}240$
$a_4 = \frac{4}{5}(10{,}240) = \$8192$
$a_5 = \frac{4}{5}(8192) = \$6553.60$

The value of the car after 5 years is about \$6554.

25. $\sum_{i=1}^{5}(i+3)$

$= (1+3)+(2+3)+(3+3)+(4+3)+(5+3)$

Let i = 1 to 5 and add terms.

$= 4+5+6+7+8$

$= 30$

27. $\sum_{i=1}^{3}(i^2+2)$

$= (1^2+2)+(2^2+2)+(3^2+2)$

Let i = 1, 2, 3 and add terms.

$= 3+6+11 = 20$

29. $\sum_{i=1}^{6}(-1)^i$

$= (-1)^1+(-1)^2+(-1)^3+(-1)^4$
$+(-1)^5+(-1)^6$

$= -1+1-1+1-1+1 = 0$

31. $\sum_{i=3}^{7}(i-3)(i+2)$

$= (3-3)(3+2)+(4-3)(4+2)$
$+(5-3)(5+2)+(6-3)(6+2)$
$+(7-3)(7+2)$

$= 0(5)+1(6)+2(7)+3(8)+4(9)$

$= 0+6+14+24+36 = 80$

33. $\sum_{i=1}^{5}2x \cdot i$

$= 2x\cdot 1+2x\cdot 2+2x\cdot 3+2x\cdot 4$
$+2x\cdot 5$

$= 2x+4x+6x+8x+10x$

35. $\sum_{i=1}^{5}i \cdot x^i$

$= 1\cdot x^1+2\cdot x^2+3\cdot x^3+4\cdot x^4+5\cdot x^5$

$= x+2x^2+3x^3+4x^4+5x^5$

37. $3+4+5+6+7$

$= (1+2)+(2+2)+(3+2)$
$+(4+2)+(5+2)$

$= \sum_{i=1}^{5}(i+2)$

39. $\frac{1}{2}+\frac{1}{3}+\frac{1}{4}+\frac{1}{5}+\frac{1}{6}$

$= \frac{1}{1+1}+\frac{1}{2+1}+\frac{1}{3+1}+\frac{1}{4+1}$
$+\frac{1}{5+1}$

$= \sum_{i=1}^{5}\frac{1}{i+1}$

41. $1+4+9+16+25$

$= 1^2+2^2+3^2+4^2+5^2$

$= \sum_{i=1}^{5}i^2$

43. The similarities are that both are defined by the same linear expression, and that points satisfying both lie in a straight line. The difference is that the domain of f consists of all real numbers, but the domain of the sequence is $\{1, 2, 3, \ldots\}$. An example of a similarity is $f(1) = 6$ and $a_1 = 6$. An example of a difference is $f\left(\frac{3}{2}\right) = 7$, but $a_{3/2}$ is not allowed.

45. A sequence is a list of terms in a specific order, while a series is the indicated sum of the terms of a sequence.

47. $\overline{x} = \frac{\sum_{i=1}^{n}x_i}{n} = \frac{\sum_{i=1}^{7}x_i}{7}$

$= \frac{8+11+14+9+3+6+8}{7} = \frac{59}{7}$

49. $\overline{x} = \frac{5+9+8+2+4+7+3+2}{8} = \frac{40}{8} = 5$

51. $\overline{x} = \frac{6254+6684+7314+7791+8171}{5}$

$= \frac{36{,}214}{5} = 7242.8$

The average number of funds available for this five-year period was about 7243.

53. $\sum_{i=1}^{6}(i^2+3i+5)$

$= \sum_{i=1}^{6}i^2+\sum_{i=1}^{6}3i+\sum_{i=1}^{6}5$

54. From Exercise 53, the second summation is

$$\sum_{i=1}^{6}3i = 3\sum_{i=1}^{6}i.$$

55. From Exercise 53, the third summation is

$$\sum_{i=1}^{6}5 = 6\cdot 5 = 30.$$

56. $1+2+3+4+\cdots+n = \frac{n(n+1)}{2}$

written in summation notation is

$$\sum_{i=1}^{n}i = \frac{n(n+1)}{2}.$$

57. $1^2+2^2+3^2+4^2+\cdots+n^2$
$= \dfrac{n(n+1)(2n+1)}{6}$
written in summation notation is

$$\sum_{i=1}^{n} i^2 = \frac{n(n+1)(2n+1)}{6}.$$

58. From Exercises 53–55,

$$\sum_{i=1}^{6}(i^2+3i+5)$$
$$= \sum_{i=1}^{6} i^2 + 3\sum_{i=1}^{6} i + 30.$$

Now apply the results of Exercises 56 & 57.

$$= \frac{6(6+1)(2\cdot 6+1)}{6} + 3\cdot\frac{6(6+1)}{2} + 30$$
$$= \frac{6(7)(13)}{6} + 3\cdot\frac{6(7)}{2} + 30$$
$$= 91 + 63 + 30 = 184$$

59. $\displaystyle\sum_{i=1}^{12}(i^2-i)$

$$= \sum_{i=1}^{12} i^2 - \sum_{i=1}^{12} i$$
$$= \frac{12(12+1)(2\cdot 12+1)}{6} - \frac{12(12+1)}{2}$$
$$= \frac{12(13)(25)}{6} - \frac{12(13)}{2}$$
$$= 650 - 78 = 572$$

60. $\displaystyle\sum_{i=1}^{20}(2+i-i^2)$

$$= \sum_{i=1}^{20} 2 + \sum_{i=1}^{20} i - \sum_{i=1}^{20} i^2$$
$$= 20\cdot 2 + \frac{20(20+1)}{2} - \frac{20(20+1)(2\cdot 20+1)}{6}$$
$$= 40 + 210 - 2870 = -2620$$

Section 12.2

1. An arithmetic sequence is a sequence (list) of numbers in a specific order such that there is a common difference between any two successive terms. For example, the sequence $1, 5, 9, 13, \ldots$ is arithmetic with difference $d = 5-1 = 9-5 = 13-9 = 4$. As another example, $2, -1, -4, -7, \ldots$ is an arithmetic sequence with $d = -3$.

3. $1, 2, 3, 4, 5, \ldots$
d is the difference between any two adjacent terms. Choose the terms 3 and 2.

$$d = 3 - 2 = 1$$

The terms 2 and 1 would give

$$d = 2 - 1 = 1,$$

the same result. Therefore, the common difference is $d = 1$.

Note: You should find the difference for all pairs of adjacent terms to determine if the sequence is arithmetic.

5. $2, -4, 6, -8, 10, -12, \ldots$
The difference between the first two terms is $-4-2=-6$, but the difference between the second and third terms is $6-(-4)=10$. The differences are not the same so the sequence is *not arithmetic*.

7. $-10, -5, 0, 5, 10, \ldots$
Choose the terms 10 and 5, and find the difference.

$$d = 10 - 5 = 5$$

The terms -5 and -10 would give

$$d = -5 - (-10) = -5 + 10 = 5,$$

the same result. Therefore, the common difference is $d = 5$.

9. $a_1 = 5$
$a_2 = a_1 + d = 5 + 4 = 9$
$a_3 = a_2 + d = 9 + 4 = 13$
$a_4 = a_3 + d = 13 + 4 = 17$
$a_5 = a_4 + d = 17 + 4 = 21$

11. $a_1 = -2$
$a_2 = a_1 + d = -2 + (-4) = -6$
$a_3 = a_2 + d = -6 + (-4) = -10$
$a_4 = a_3 + d = -10 + (-4) = -14$
$a_5 = a_4 + d = -14 + (-4) = -18$

13. $a_1 = 2,\ d = 5$
$a_n = a_1 + (n-1)d$
$= 2 + (n-1)5$
$= 2 + 5n - 5$
$= 5n - 3$

15. $3, \frac{15}{4}, \frac{9}{2}, \frac{21}{4}, \ldots$
To find d, subtract any two adjacent terms.

$$d = \frac{15}{4} - 3 = \frac{15}{4} - \frac{12}{4} = \frac{3}{4}$$

The first term is $a_1 = 3$. Now find a_n.
$a_n = a_1 + (n-1)d$
$= 3 + (n-1)\left(\dfrac{3}{4}\right)$
$= 3 + \dfrac{3}{4}n - \dfrac{3}{4}$
$= \dfrac{3}{4}n + \dfrac{9}{4}$

17. $-3, 0, 3, \ldots$
To find d, subtract any two adjacent terms.

$$d = 0 - (-3) = 3$$

The first term is $a_1 = -3$. Now find a_n.
$$\begin{aligned} a_n &= a_1 + (n-1)d \\ &= -3 + (n-1)3 \\ &= -3 + 3n - 3 \\ &= 3n - 6 \end{aligned}$$

19. Given $a_1 = 4$ and $d = 3$; find a_{25}.
$$\begin{aligned} a_n &= a_1 + (n-1)d \\ a_{25} &= 4 + (25-1)3 \\ &= 4 + 72 = 76 \end{aligned}$$

21. Given $2, 4, 6, \ldots$; find a_{24}.
Here, $a_1 = 2$ and $d = 4 - 2 = 2$.
Now find a_{24}.
$$\begin{aligned} a_n &= a_1 + (n-1)d \\ a_{24} &= 2 + (24-1)2 \\ &= 2 + 46 = 48 \end{aligned}$$

23. Given $a_{12} = -45$ and $a_{10} = -37$; find a_1.
Use $a_n = a_1 + (n-1)d$ to write a system of equations.
$$\begin{aligned} a_{12} &= a_1 + (12-1)d \\ -45 &= a_1 + 11d \qquad (1) \\ a_{10} &= a_1 + (10-1)d \\ -37 &= a_1 + 9d \qquad (2) \end{aligned}$$
To eliminate d, multiply equation (1) by -9 and equation (2) by 11. Then add the results.
$$\begin{array}{rcrcrl} 405 &=& -9a_1 &-& 99d & \quad -9 \times (1) \\ -407 &=& 11a_1 &+& 99d & \quad 11 \times (2) \\ \hline -2 &=& 2a_1 & & & \\ -1 &=& a_1 & & & \end{array}$$

25. $3, 5, 7, \ldots, 33$
Let n represent the number of terms in the sequence. So, $a_n = 33$, $a_1 = 3$, and $d = 5 - 3 = 2$.
$$\begin{aligned} a_n &= a_1 + (n-1)d \\ 33 &= 3 + (n-1)2 \\ 33 &= 3 + 2n - 2 \\ 33 &= 2n + 1 \\ 32 &= 2n \\ n &= 16 \end{aligned}$$
The sequence has 16 terms.

27. $\frac{3}{4}, 3, \frac{21}{4}, \ldots, 12$
Let n represent the number of terms in the sequence. So, $a_n = 12$,
$a_1 = \frac{3}{4}$, and $d = 3 - \frac{3}{4} = \frac{9}{4}$.
$$\begin{aligned} a_n &= a_1 + (n-1)d \\ 12 &= \frac{3}{4} + (n-1)\left(\frac{9}{4}\right) \\ \frac{45}{4} &= (n-1)\left(\frac{9}{4}\right) \\ 5 &= n - 1 \qquad \textit{Multiply by } \frac{4}{9}. \\ 6 &= n \end{aligned}$$
The sequence has 6 terms.

29. n represents the number of terms.

31. The sum of the left sides is $S + S = \underline{2S}$. For the right sides, $1 + 100 = 101$, $2 + 99 = 101$, $3 + 98 = 101$, and so on. The sum $\underline{101}$ appears $\underline{100}$ times. By multiplying, the sum of the right sides is $101 \cdot 100 = \underline{10,100}$.

32. $2S = 10{,}100$

33. Divide by 2 to obtain $S = \underline{5050}$.

34. The right sides now have a sum of 201 appearing 200 times. Thus, the sum is $201 \cdot 200 = 40,200$.
$$\begin{aligned} 2S &= 40,200 \\ S &= 20,100 \end{aligned}$$
Note that the pattern is $S = \frac{(n+1)(n)}{2}$.

35. Find S_6 given $a_1 = 6$, $d = 3$, and $n = 6$.
$$\begin{aligned} S_n &= \frac{n}{2}[2a_1 + (n-1)d] \\ S_6 &= \frac{6}{2}[2 \cdot 6 + (6-1)3] \\ &= 3(12 + 5 \cdot 3) \\ &= 3(27) = 81 \end{aligned}$$

37. Find S_6 given $a_1 = 7$, $d = -3$, and $n = 6$.
$$\begin{aligned} S_n &= \frac{n}{2}[2a_1 + (n-1)d] \\ S_6 &= \frac{6}{2}[2 \cdot 7 + (6-1)(-3)] \\ &= 3[14 + 5(-3)] \\ &= 3(14 - 15) = 3(-1) = -3 \end{aligned}$$

39. Find S_6 given $a_n = 4 + 3n$.
$$\begin{aligned} a_1 &= 4 + 3(1) = 7 \\ a_6 &= 4 + 3(6) = 22 \\ S_n &= \frac{n}{2}(a_1 + a_n) \\ S_6 &= \frac{6}{2}(7 + 22) \\ &= 3(29) = 87 \end{aligned}$$

41. $\sum_{i=1}^{10} (8i - 5)$
$$\begin{aligned} a_n &= 8n - 5 \\ a_1 &= 8(1) - 5 = 3 \\ a_{10} &= 8(10) - 5 = 75 \end{aligned}$$
Use $S_n = \frac{n}{2}(a_1 + a_n)$ with $n = 10$,

$a_1 = 3$, and $a_{10} = 75$.

$$S_{10} = \frac{10}{2}(3 + 75)$$
$$= 5(78) = 390$$

43. $\sum_{i=1}^{20} (2i - 5)$

$a_n = 2n - 5$
$a_1 = 2(1) - 5 = -3$
$a_{20} = 2(20) - 5 = 35$

Use $S_n = \frac{n}{2}(a_1 + a_n)$ with $n = 20$, $a_1 = -3$, and $a_{20} = 35$.

$$S_{20} = \frac{20}{2}(-3 + 35)$$
$$= 10(32) = 320$$

45. $\sum_{i=1}^{250} i$

$a_n = n$, $a_1 = 1$, $a_{250} = 250$

Use $S_n = \frac{n}{2}(a_1 + a_n)$ with $n = 250$, $a_1 = 1$, and $a_{250} = 250$.

$$S_{250} = \frac{250}{2}(1 + 250)$$
$$= 125(251) = 31,375$$

47. The sequence is $1, 2, 3, \ldots, 30$.

$$S_n = \frac{n}{2}(a_1 + a_n)$$
$$S_{30} = \frac{30}{2}(1 + 30)$$
$$= 15(31) = 465$$

The account will have \$465 deposited in it over the entire month.

49. Your salaries at six-month intervals form an arithmetic sequence with $a_1 = 1600$ and $d = 50$. Since your salary is increased every 6 months, or $2(5) = 10$ times, after 5 years your salary will equal the term a_{11}.

$$a_n = a_1 + (n - 1)d$$
$$a_{11} = 1600 + (11 - 1)50$$
$$= 1600 + 500 = 2100$$

Your salary will be \$2100/month.

51. Given the sequence $20, 22, 24, \ldots$, for 25 terms, find a_{25}. Here, $a_1 = 20$, $d = 22 - 20 = 2$, and $n = 25$.

$$a_n = a_1 + (n - 1)d$$
$$a_{25} = 20 + (25 - 1)2$$
$$= 20 + 48 = 68$$

There are 68 seats in the last row.
Now find S_{25}.

$$S_n = \frac{n}{2}(a_1 + a_n)$$
$$S_{25} = \frac{25}{2}(20 + 68)$$
$$= \frac{25}{2}(88) = 25(44) = 1100$$

There are 1100 seats in the section.

53. Given the sequence $35, 31, 27, \ldots$, can the sequence end in 1? If not, find the last positive value. If the sequence ends in 1, we can find n, a whole number.

$$d = 31 - 35 = -4$$
$$a_n = a_1 + (n - 1)d$$
$$1 = 35 + (n - 1)(-4)$$
$$1 = 35 - 4n + 4$$
$$-38 = -4n$$
$$9.5 = n$$

Since n is not a whole number, the sequence cannot end in 1. The largest n possible is $n = 9$.

$$a_n = a_1 + (n - 1)d$$
$$a_9 = 35 + (9 - 1)(-4)$$
$$= 35 - 32 = 3$$

She can build 9 rows. There are 3 blocks in the last row.

Section 12.3

1. A geometric sequence is an ordered list of numbers such that each term after the first is obtained by multiplying the previous term by a constant, r, called the common ratio. For example, if the first term is 3 and $r = 4$, the sequence is $3, 12, 48, 192, \ldots$. If the first term is 2 and $r = -1$, the sequence is $2, -2, 2, -2, \ldots$.

3. $4, 8, 16, 32, \ldots$

To find r, choose any two adjacent terms and divide the second one by the first one.

$$r = \frac{8}{4} = 2$$

Notice that any two other adjacent terms could have been used with the same result. The common ratio is $r = 2$.

Note: You should find the ratio for all pairs of adjacent terms to determine if the sequence is geometric.

5. $\frac{1}{3}, \frac{2}{3}, \frac{3}{3}, \frac{4}{3}, \frac{5}{3}, \ldots$

Choose any two adjacent terms and divide the second by the first.

$$r = \frac{\frac{2}{3}}{\frac{1}{3}} = \frac{2}{3} \cdot 3 = 2$$

Confirm this result with any two other adjacent terms.

$$r = \frac{\frac{3}{3}}{\frac{2}{3}} = 1 \cdot \frac{3}{2} = \frac{3}{2}$$

continued

Since $2 \neq \frac{3}{2}$, the ratios are not the same. The sequence is *not geometric*.

7. $1, -3, 9, -27, 81, \ldots$

$$r = \frac{-3}{1} = -3$$
$$r = \frac{9}{-3} = -3$$

The common ratio is $r = -3$.

9. $1, -\frac{1}{2}, \frac{1}{4}, -\frac{1}{8}, \frac{1}{16}, \ldots$

$$r = \frac{-\frac{1}{2}}{1} = -\frac{1}{2}$$
$$r = \frac{\frac{1}{4}}{-\frac{1}{2}} = \frac{1}{4} \cdot (-2) = -\frac{1}{2}$$

The common ratio is $r = -\frac{1}{2}$.

11. Find a general term for $5, 10, \ldots$.
First, find r.

$$r = \frac{10}{5} = 2$$

Use $a_1 = 5$ and $r = 2$ to find a_n.

$$a_n = a_1 r^{n-1}$$
$$a_n = 5(2)^{n-1}$$

13. Find a general term for $\frac{1}{9}, \frac{1}{3}, \ldots$.

Here, $a_1 = \frac{1}{9}$. Find r.

$$r = \frac{\frac{1}{3}}{\frac{1}{9}} = \frac{1}{3} \cdot \frac{9}{1} = 3$$

Now find a_n.

$$a_n = a_1 r^{n-1}$$
$$a_n = \frac{1}{9}(3)^{n-1}$$
$$\text{or } a_n = \frac{3^{n-1}}{9}$$

15. Find a general term for $10, -2, \ldots$.
Here, $a_1 = 10$. Find r.

$$r = \frac{-2}{10} = -\frac{1}{5}$$

Now find a_n.

$$a_n = a_1 r^{n-1}$$
$$a_n = 10\left(-\frac{1}{5}\right)^{n-1}$$

17. Substitute $a_1 = 2$, $r = 5$, and $n = 10$ in the nth-term formula to find a_{10}.

$$a_n = a_1 r^{n-1}$$
$$a_{10} = a_1 (r)^{10-1}$$
$$= 2(5)^9 = 3{,}906{,}250$$

19. Given $\frac{1}{2}, \frac{1}{6}, \frac{1}{18}, \ldots$; find a_{12}.
First find the common ratio.

$$r = \frac{\frac{1}{6}}{\frac{1}{2}} = \frac{1}{6} \cdot 2 = \frac{1}{3}$$

Substitute $a_1 = \frac{1}{2}$, $r = \frac{1}{3}$, and $n = 12$ in the nth-term formula.

$$a_n = a_1 r^{n-1}$$
$$a_{12} = a_1 (r)^{12-1}$$
$$= \left(\frac{1}{2}\right)\left(\frac{1}{3}\right)^{11}$$

21. Given $a_3 = \frac{1}{2}$, $a_7 = \frac{1}{32}$; find a_{25}.
Find a_1 and r using the general term $a_n = a_1 r^{n-1}$.

$$a_3 = a_1 r^{3-1}$$
$$\frac{1}{2} = a_1 r^2 \quad (1)$$
$$a_7 = a_1 r^{7-1}$$
$$\frac{1}{32} = a_1 r^6 \quad (2)$$

Solve (1) for a_1.

$$a_1 = \frac{1}{2r^2}$$

Substitute for a_1 in (2).

$$\frac{1}{32} = \frac{1}{2r^2} r^6$$
$$\frac{1}{16} = r^4$$
$$r^2 = \pm\frac{1}{4}$$

Since r^2 is positive,

$$r^2 = \frac{1}{4} \quad \text{and} \quad r = \pm\frac{1}{2}.$$

Substitute $\frac{1}{4}$ for r^2 in (1).

$$\frac{1}{2} = a_1\left(\frac{1}{4}\right)$$
$$2 = a_1$$

Use $a_1 = 2$ and $r = \frac{1}{2}$ $\left(\text{or } -\frac{1}{2}\right)$ to find a_{25}.

$$a_{25} = a_1(r)^{25-1} = 2\left(\frac{1}{2}\right)^{24} = \frac{1}{2^{23}}$$

23. $a_1 = 2, r = 3$; use $a_n = a_1 r^{n-1}$.

$$a_2 = a_1 r^1 = 2(3) = 6$$
$$a_3 = a_1 r^2 = 2(3)^2 = 18$$
$$a_4 = a_1 r^3 = 2(3)^3 = 54$$
$$a_5 = a_1 r^4 = 2(3)^4 = 162$$

Answer: $2, 6, 18, 54, 162$

25. $a_1 = 5, r = -\frac{1}{5}$; use $a_n = a_1 r^{n-1}$.

$$a_2 = a_1 r^1 = 5\left(-\frac{1}{5}\right) = -1$$
$$a_3 = a_1 r^2 = 5\left(-\frac{1}{5}\right)^2 = \frac{1}{5}$$
$$a_4 = a_1 r^3 = 5\left(-\frac{1}{5}\right)^3 = -\frac{1}{25}$$
$$a_5 = a_1 r^4 = 5\left(-\frac{1}{5}\right)^4 = \frac{1}{125}$$

Answer: $5, -1, \frac{1}{5}, -\frac{1}{25}, \frac{1}{125}$

27. $\frac{1}{3}, \frac{1}{9}, \frac{1}{27}, \frac{1}{81}, \frac{1}{243}$

Here, $a_1 = \frac{1}{3}$, $n = 5$, and

$$r = \frac{\frac{1}{9}}{\frac{1}{3}} = \frac{1}{9} \cdot 3 = \frac{1}{3}.$$

$$S_n = \frac{a_1(r^n - 1)}{r - 1}$$

$$S_5 = \frac{\frac{1}{3}\left[\left(\frac{1}{3}\right)^5 - 1\right]}{\frac{1}{3} - 1} = \frac{\frac{1}{3}\left(\frac{1}{243} - 1\right)}{-\frac{2}{3}} = \frac{\frac{1}{3}\left(-\frac{242}{243}\right)}{-\frac{2}{3}} = \frac{121}{243}$$

29. $-\frac{4}{3}, -\frac{4}{9}, -\frac{4}{27}, -\frac{4}{81}, -\frac{4}{243}, -\frac{4}{729}$

Here, $a_1 = -\frac{4}{3}$, $n = 6$, and

$$r = \frac{-\frac{4}{9}}{-\frac{4}{3}} = -\frac{4}{9} \cdot \left(-\frac{3}{4}\right) = \frac{1}{3}.$$

$$S_n = \frac{a_1(r^n - 1)}{r - 1}$$

$$S_6 = \frac{-\frac{4}{3}\left[\left(\frac{1}{3}\right)^6 - 1\right]}{\frac{1}{3} - 1} = \frac{-\frac{4}{3}\left(\frac{1}{729} - 1\right)}{-\frac{2}{3}} = \frac{-\frac{4}{3}\left(-\frac{728}{729}\right)}{-\frac{2}{3}} = -\frac{1456}{729} \approx -1.997$$

31. $\sum_{i=1}^{7} 4\left(\frac{2}{5}\right)^i$

Use $a_1 = 4\left(\frac{2}{5}\right) = \frac{8}{5}$, $n = 7$, and $r = \frac{2}{5}$.

$$S_n = \frac{a_1(r^n - 1)}{r - 1}$$

$$S_7 = \frac{\frac{8}{5}\left[\left(\frac{2}{5}\right)^7 - 1\right]}{\frac{2}{5} - 1} = \frac{\frac{8}{5}\left[\left(\frac{2}{5}\right)^7 - 1\right]}{-\frac{3}{5}} = -\frac{8}{3}\left[\left(\frac{2}{5}\right)^7 - 1\right] \approx 2.662$$

33. $\sum_{i=1}^{10} (-2)\left(\frac{3}{5}\right)^i$

Use $a_1 = (-2)\left(\frac{3}{5}\right) = -\frac{6}{5}$, $n = 10$, and $r = \frac{3}{5}$.

continued

$$S_n = \frac{a_1(r^n - 1)}{r-1}$$

$$S_{10} = \frac{-\frac{6}{5}\left[\left(\frac{3}{5}\right)^{10} - 1\right]}{\frac{3}{5} - 1}$$

$$= \frac{-\frac{6}{5}\left[\left(\frac{3}{5}\right)^{10} - 1\right]}{-\frac{2}{5}}$$

$$= 3\left[\left(\frac{3}{5}\right)^{10} - 1\right] \approx -2.982$$

35. There are 22 deposits, so $n = 22$.

$$S = R\left[\frac{(1+i)^n - 1}{i}\right]$$

$$= 1000\left[\frac{(1+.095)^{22} - 1}{.095}\right]$$

$$= 66,988.91$$

There will be $66,988.91 in the account.

37. Quarterly deposits for 10 years give us $n = 4 \cdot 10 = 40$. The interest rate per period is $i = \frac{.07}{4} = .0175$.

$$S = R\left[\frac{(1+i)^n - 1}{i}\right]$$

$$= 1200\left[\frac{(1+.0175)^{40} - 1}{.0175}\right]$$

$$= 68,680.96$$

We now use the compound interest formula to determine the value of this money after 5 more years.

$$A = P\left(1 + \frac{r}{n}\right)^{nt}$$

$$= 68,680.96\left(1 + \frac{.09}{12}\right)^{12(5)}$$

$$= 107,532.48$$

The woman is also saving $300 per month, so we use the annuity formula to determine that value.

$$S = R\left[\frac{(1+i)^n - 1}{i}\right]$$

$$= 300\left[\frac{\left(1 + \frac{.09}{12}\right)^{12(5)} - 1}{\frac{.09}{12}}\right]$$

$$= 22,627.24$$

Adding $22,627.24 to $107,532.48 gives a total of $130,159.72 in the account.

39. Find the sum if $a_1 = 6$ and $r = \frac{1}{3}$. Since $|r| < 1$, the sum exists.

$$S = \frac{a_1}{1-r} = \frac{6}{1 - \frac{1}{3}} = \frac{6}{\frac{2}{3}} = 6 \cdot \frac{3}{2} = 9$$

41. Find the sum if $a_1 = 1000$ and $r = -\frac{1}{10}$. Since $|r| < 1$, the sum exists.

$$S = \frac{a_1}{1-r} = \frac{1000}{1 - \left(-\frac{1}{10}\right)} = \frac{1000}{\frac{11}{10}}$$

$$= 1000 \cdot \frac{10}{11} = \frac{10,000}{11}$$

43. $\sum_{i=1}^{\infty} \frac{9}{8}\left(-\frac{2}{3}\right)^i$

$a_1 = \frac{9}{8}\left(-\frac{2}{3}\right)^1 = -\frac{3}{4}$ and $r = -\frac{2}{3}$.

Since $|r| < 1$, the sum exists.

$$S = \frac{a_1}{1-r} = \frac{-\frac{3}{4}}{1 - \left(-\frac{2}{3}\right)} = \frac{-\frac{3}{4}}{\frac{5}{3}}$$

$$= -\frac{3}{4} \cdot \frac{3}{5} = -\frac{9}{20}$$

45. $\sum_{i=1}^{\infty} \frac{12}{5}\left(\frac{5}{4}\right)^i$

Since $|r| = \frac{5}{4} > 1$, the sum *does not exist*.

47. The ball is dropped from a height of 10 feet and will rebound $\frac{3}{5}$ of its original height.

Let $a_n =$ the ball's height on the nth rebound.

$a_1 = 10$ and $r = \frac{3}{5}$.

Since we must find the height after the fourth bounce, $n = 5$ (since a_1 is the starting point).

Use $a_n = a_1 r^{n-1}$.

$$a_5 = 10\left(\frac{3}{5}\right)^{5-1} = 10\left(\frac{3}{5}\right)^4 \approx 1.3$$

The ball will rebound 1.3 feet after the fourth bounce.

49. This exercise can be modeled by a geometric sequence with $a_1 = 256$ and $r = \frac{1}{2}$. First we need to find n so that $a_n = 32$.

$$32 = a_1 r^{n-1}$$

$$32 = 256\left(\frac{1}{2}\right)^{n-1}$$

$$\frac{1}{8} = \left(\frac{1}{2}\right)^{n-1}$$
$$\left(\frac{1}{2}\right)^3 = \left(\frac{1}{2}\right)^{n-1}$$
$$3 = n - 1$$
$$4 = n$$

Since n is 4, this means that 32 grams will be present on the day which corresponds to the 4th term of the sequence. That would be on day 3. To find what is left after the tenth day, we need to find a_{11} since we started with a_1.

$$a_{11} = a_1r^{11-1} = 256\left(\frac{1}{2}\right)^{10} = \frac{256}{1024} = \frac{1}{4}$$

There will be $\frac{1}{4}$ gram of the substance after 10 days.

51. **(a)** Here, $a_1 = 1.1$ billion and $r = 106\% = 1.06$. Since we must find the consumption after 5 years, $n = 6$ (since a_1 is the starting point).

$$a_6 = a_1r^{n-1}$$
$$a_6 = 1.1(1.06)^{6-1}$$
$$= 1.1(1.06)^5 \approx 1.5$$

The community will use about 1.5 billion units 5 years from now.

(b) If consumption doubles, then the consumption would be $2a_1$.

$$2a_1 = a_1(1.06)^{n-1}$$
$$2 = (1.06)^{n-1}$$
$$\ln 2 = \ln(1.06)^{n-1}$$
$$\ln 2 = (n-1)\ln(1.06)$$
$$n - 1 = \frac{\ln 2}{\ln 1.06}$$
$$n - 1 \approx 12$$
$$n \approx 13$$

Since n is about 13, that would represent the 13th term of the sequence, which represents about 12 years after the start.

53. Since the machine depreciates by $\frac{1}{4}$ of its value, it retains $1 - \frac{1}{4} = \frac{3}{4}$ of its value. Since the cost of the machine new is $\$50,000$, $a_1 = 50,000$. We want the value after 8 years so since the original cost is a_1, we need to find a_9.

$$a_9 = a_1r^{9-1}$$
$$= 50,000\left(\frac{3}{4}\right)^8 \approx 5006$$

The machine's value after 8 years is about $5000.

55. $\frac{1}{3} = .33333\ldots$

56. $\frac{2}{3} = .66666\ldots$

57.
$$\begin{array}{r} .33333\ldots \\ +\,.66666\ldots \\ \hline .99999\ldots \end{array}$$

58.
$$S = \frac{a_1}{1-r} = \frac{.9}{1-.1} = \frac{.9}{.9} = 1$$

Therefore, $.99999\ldots = 1$.

59. From Exercise 58, $.99999\ldots = 1$, so choice **B** is correct.

60.
$$.49999\ldots = .4 + .09999\ldots = \frac{4}{10} + \frac{1}{10}(.9999\ldots) = \frac{4}{10} + \frac{1}{10}(1) = \frac{5}{10} = \frac{1}{2}$$

Section 12.4

1. $6! = 6 \cdot 5 \cdot 4 \cdot 3 \cdot 2 \cdot 1 = 720$

3. $\dfrac{6!}{4!\,2!} = \dfrac{6 \cdot 5 \cdot 4 \cdot 3 \cdot 2 \cdot 1}{(4 \cdot 3 \cdot 2 \cdot 1)(2 \cdot 1)} = \dfrac{6 \cdot 5}{2 \cdot 1} = 15$

5. ${}_6C_2 = \dfrac{6!}{2!\,(6-2)!} = \dfrac{6!}{2!\,4!} = 15,$ by Exercise 3.

7. $\dfrac{4!}{0!\,4!} = \dfrac{4!}{(1)(4!)} = \dfrac{1}{1} = 1$

9.
$$4! \cdot 5 = (4 \cdot 3 \cdot 2 \cdot 1) \cdot 5 = 24 \cdot 5 = 120$$

11.
$${}_{13}C_{11} = \frac{13!}{11!\,(13-11)!} = \frac{13!}{11!\,2!} = \frac{13 \cdot 12}{2 \cdot 1} \quad \textit{11! cancels}$$
$$= 13 \cdot 6 = 78$$

13. $(m+n)^4$
$$= m^4 + \frac{4!}{3!\,1!}m^3n^1 + \frac{4!}{2!\,2!}m^2n^2 + \frac{4!}{1!\,3!}m^1n^3 + n^4$$
$$= m^4 + 4m^3n + 6m^2n^2 + 4mn^3 + n^4$$

15. $(a-b)^5$
$$= \left[a + (-b)\right]^5$$
$$= a^5 + \frac{5!}{4!\,1!}a^4(-b)^1 + \frac{5!}{3!\,2!}a^3(-b)^2 + \frac{5!}{2!\,3!}a^2(-b)^3 + \frac{5!}{1!\,4!}a^1(-b)^4 + (-b)^5$$
$$= a^5 - 5a^4b + 10a^3b^2 - 10a^2b^3 + 5ab^4 - b^5$$

17. $(2x+3)^3$

$= (2x)^3 + \frac{3!}{2!\,1!}(2x)^2(3)^1 + \frac{3!}{1!\,2!}(2x)(3)^2$
$+ (3)^3$
$= 8x^3 + 36x^2 + 54x + 27$

19. $\left(\frac{x}{2} - y\right)^4$

$= \left[\left(\frac{x}{2}\right) + (-y)\right]^4$
$= \left(\frac{x}{2}\right)^4 + \frac{4!}{3!\,1!}\left(\frac{x}{2}\right)^3(-y)^1 + \frac{4!}{2!\,2!}\left(\frac{x}{2}\right)^2(-y)^2$
$+ \frac{4!}{1!\,3!}\left(\frac{x}{2}\right)^1(-y)^3 + (-y)^4$
$= \frac{x^4}{16} - \frac{x^3y}{2} + \frac{3x^2y^2}{2} - 2xy^3 + y^4$

21. $(mx - n^2)^3$

$= \left[mx + (-n^2)\right]^3$
$= (mx)^3 + \frac{3!}{2!\,1!}(mx)^2(-n^2)^1$
$+ \frac{3!}{1!\,2!}(mx)^1(-n^2)^2 + (-n^2)^3$
$= m^3x^3 - 3m^2n^2x^2 + 3mn^4x - n^6$

23. $(r+2s)^{12}$

$= r^{12} + \frac{12!}{11!\,1!}r^{11}(2s)^1 + \frac{12!}{10!\,2!}r^{10}(2s)^2$
$+ \frac{12!}{9!\,3!}r^9(2s)^3 + \cdots$

The first four terms are
$r^{12} + 24r^{11}s + 264r^{10}s^2 + 1760r^9s^3$.

25. $(3x - y)^{14}$

$= \left[3x + (-y)\right]^{14}$
$= (3x)^{14} + \frac{14!}{13!\,1!}(3x)^{13}(-y)^1$
$+ \frac{14!}{12!\,2!}(3x)^{12}(-y)^2$
$+ \frac{14!}{11!\,3!}(3x)^{11}(-y)^3 + \cdots$

The first four terms are

$3^{14}x^{14} - 14(3^{13})x^{13}y + 91(3^{12})x^{12}y^2$
$- 364(3^{11})x^{11}y^3$.

27. $(t^2 + u^2)^{10}$

$= (t^2)^{10} + \frac{10!}{9!\,1!}(t^2)^9(u^2)^1$
$+ \frac{10!}{8!\,2!}(t^2)^8(u^2)^2$
$+ \frac{10!}{7!\,3!}(t^2)^7(u^2)^3 + \cdots$

The first four terms are

$t^{20} + 10t^{18}u^2 + 45t^{16}u^4 + 120t^{14}u^6$.

29. The rth term of the expansion of $(x+y)^n$ is

$$\frac{n!}{[n-(r-1)]!\,(r-1)!}(x)^{n-(r-1)}(y)^{r-1}.$$

Start with the exponent on y, which is 1 less than the term number r. In this case, we are looking for the fourth term, so $r = 4$ and $r - 1 = 3$. Thus, the fourth term of $(2m+n)^{10}$ is

$\frac{10!}{(10-3)!\,3!}(2m)^{10-3}(n)^3$
$= \frac{10!}{7!\,3!}2^7m^7n^3$
$= 120(2^7)m^7n^3$.

31. The seventh term of $\left(x + \frac{y}{2}\right)^8$ is

$\frac{8!}{(8-6)!\,6!}(x)^{8-6}\left(\frac{y}{2}\right)^6$
$= \frac{8!}{6!\,2!}x^2\frac{y^6}{2^6}$
$= \frac{7x^2y^6}{16}$.

33. The third term of $(k-1)^9$ is

$$\frac{9!}{(9-2)!\,2!}k^{9-2}(-1)^2 = \frac{9!}{7!\,2!}k^7 = 36k^7.$$

35. The expansion of $(x^2 - 2y)^6$ has seven terms, so the middle term is the fourth. The fourth term of $(x^2 - 2y)^6$ is

$\frac{6!}{(6-3)!\,3!}(x^2)^{6-3}(-2y)^3$
$= \frac{6!}{3!\,3!}(x^2)^3(-8y^3)$
$= 20x^6(-8y^3) = -160x^6y^3$.

37. The term of the expansion of $(3x^3 - 4y^2)^5$ with x^9y^4 in it is the term with $(3x^3)^3(-4y^2)^2$, since $(x^3)^3(y^2)^2 = x^9y^4$. The term is

$\frac{5!}{3!\,2!}(3x^3)^3(-4y^2)^2$
$= 10(27x^9)(16y^4) = 4320x^9y^4$.

Chapter 12 Review Exercises

1. $a_n = 2n - 3$

$a_1 = 2(1) - 3 = -1$
$a_2 = 2(2) - 3 = 1$
$a_3 = 2(3) - 3 = 3$
$a_4 = 2(4) - 3 = 5$

Answer: $-1, 1, 3, 5$

2. $a_n = \dfrac{n-1}{n}$

$a_1 = \dfrac{1-1}{1} = 0$

$a_2 = \dfrac{2-1}{2} = \dfrac{1}{2}$

$a_3 = \dfrac{3-1}{3} = \dfrac{2}{3}$

$a_4 = \dfrac{4-1}{4} = \dfrac{3}{4}$

Answer: $0, \dfrac{1}{2}, \dfrac{2}{3}, \dfrac{3}{4}$

3. $a_n = n^2$

$a_1 = (1)^2 = 1$

$a_2 = (2)^2 = 4$

$a_3 = (3)^2 = 9$

$a_4 = (4)^2 = 16$

Answer: $1, 4, 9, 16$

4. $a_n = \left(\dfrac{1}{2}\right)^n$

$a_1 = \left(\dfrac{1}{2}\right)^1 = \dfrac{1}{2}$

$a_2 = \left(\dfrac{1}{2}\right)^2 = \dfrac{1}{4}$

$a_3 = \left(\dfrac{1}{2}\right)^3 = \dfrac{1}{8}$

$a_4 = \left(\dfrac{1}{2}\right)^4 = \dfrac{1}{16}$

Answer: $\dfrac{1}{2}, \dfrac{1}{4}, \dfrac{1}{8}, \dfrac{1}{16}$

5. $a_n = (n+1)(n-1)$

$a_1 = (1+1)(1-1) = 2(0) = 0$

$a_2 = (2+1)(2-1) = 3(1) = 3$

$a_3 = (3+1)(3-1) = 4(2) = 8$

$a_4 = (4+1)(4-1) = 5(3) = 15$

Answer: $0, 3, 8, 15$

6. $\sum_{i=1}^{5} i^2 x$

$= 1^2x + 2^2x + 3^2x + 4^2x + 5^2x$

$= x + 4x + 9x + 16x + 25x$

7. $\sum_{i=1}^{6} (i+1)x^i$

$= (1+1)x^1 + (2+1)x^2 + (3+1)x^3$
$+ (4+1)x^4 + (5+1)x^5 + (6+1)x^6$

$= 2x + 3x^2 + 4x^3 + 5x^4 + 6x^5 + 7x^6$

8. $\sum_{i=1}^{4} (i+2)$

$= (1+2) + (2+2) + (3+2) + (4+2)$

$= 3 + 4 + 5 + 6 = 18$

9. $\sum_{i=1}^{6} 2^i$

$= 2^1 + 2^2 + 2^3 + 2^4 + 2^5 + 2^6$

$= 2 + 4 + 8 + 16 + 32 + 64 = 126$

10. $\sum_{i=4}^{7} \dfrac{i}{i+1}$

$= \dfrac{4}{4+1} + \dfrac{5}{5+1} + \dfrac{6}{6+1} + \dfrac{7}{7+1}$

$= \dfrac{4}{5} + \dfrac{5}{6} + \dfrac{6}{7} + \dfrac{7}{8}$ $\quad$ *LCD*$=2^3 \cdot 3 \cdot 5 \cdot 7 = 840$

$= \dfrac{672}{840} + \dfrac{700}{840} + \dfrac{720}{840} + \dfrac{735}{840} = \dfrac{2827}{840}$

11. $\overline{x} = \dfrac{\text{total}}{5}$, where total $= 1166 + 1509 + 1899$
$+ 2462 + 2408$.

$\overline{x} = \dfrac{9444}{5} = 1888.8$

The average mutual fund retirement assets were 1888.8 billion dollars for the five given years.

12. 2, 5, 8, 11, ... is an *arithmetic* sequence with $d = 5 - 2 = 3$.

13. $-6, -2, 2, 6, 10, \ldots$ is an *arithmetic* sequence with $d = -2 - (-6) = 4$.

14. $\dfrac{2}{3}, -\dfrac{1}{3}, \dfrac{1}{6}, -\dfrac{1}{12}, \ldots$ is a *geometric* sequence with

$r = \dfrac{-\frac{1}{3}}{\frac{2}{3}} = -\dfrac{1}{3} \cdot \dfrac{3}{2} = -\dfrac{1}{2}$.

15. $-1, 1, -1, 1, -1, \ldots$ is a *geometric* sequence with $r = \dfrac{1}{-1} = -1$.

16. $64, 32, 8, \dfrac{1}{2}, \ldots$

$32 - 64 = -32$ and $8 - 32 = -24$, so the sequence is not arithmetic.

$\dfrac{32}{64} \neq \dfrac{8}{32}$, so the sequence is not geometric.

Therefore, the sequence is *neither*.

17. $64, 32, 16, 8, \ldots$ is a *geometric* sequence with

$r = \dfrac{32}{64} = \dfrac{1}{2}$.

18. 10, 8, 6, 4, ... is an *arithmetic* sequence with $d = 8 - 10 = -2$.

19. Given $a_1 = -2$, $d = 5$; find a_{16}.

$a_n = a_1 + (n-1)d$

$a_{16} = -2 + (16-1)5$

$= -2 + 15(5)$

$= -2 + 75 = 73$

20. Given $a_6 = 12$, $a_8 = 18$; find a_{25}.

$a_n = a_1 + (n-1)d$

$a_6 = a_1 + 5d$, so $12 = a_1 + 5d$ (1)

$a_8 = a_1 + 7d$, so $18 = a_1 + 7d$ (2)

Multiply equation (1) by -1 and add the result to equation (2).

$$\begin{aligned} -12 &= -a_1 - 5d \quad && -1 \times (1) \\ 18 &= a_1 + 7d && (2) \\ \hline 6 &= \ 2d \\ 3 &= d \end{aligned}$$

To find a_1, substitute $d = 3$ in equation (1).

$$\begin{aligned} 12 &= a_1 + 5d \quad (1) \\ 12 &= a_1 + 5(3) \\ 12 &= a_1 + 15 \\ -3 &= a_1 \end{aligned}$$

Use $a_1 = -3$ and $d = 3$ to find a_{25}.

$$\begin{aligned} a_n &= a_1 + (n-1)d \\ a_{25} &= -3 + (25-1)3 \\ &= -3 + 24(3) \\ &= -3 + 72 = 69 \end{aligned}$$

21. $a_1 = -4$, $d = -5$

$$\begin{aligned} a_n &= a_1 + (n-1)d \\ a_n &= -4 + (n-1)(-5) \\ &= -4 - 5n + 5 \\ &= -5n + 1 \end{aligned}$$

22. $6, 3, 0, -3, \ldots$

To get the general term, a_n, find d.

$$\begin{aligned} d &= 3 - 6 = -3 \\ a_n &= a_1 + (n-1)d \\ a_n &= 6 + (n-1)(-3) \\ a_n &= 6 - 3n + 3 \\ a_n &= -3n + 9 \end{aligned}$$

23. $7, 10, 13, \ldots, 49$

$a_1 = 7$, $d = 10 - 7 = 3$; find n, the number of terms.

$$\begin{aligned} a_n &= a_1 + (n-1)d \\ 49 &= 7 + (n-1)(3) \\ 42 &= 3(n-1) \\ 14 &= n - 1 \quad \textit{Divide by 3.} \\ 15 &= n \end{aligned}$$

There are 15 terms in this sequence.

24. $5, 1, -3, \ldots, -79$

$a_1 = 5$, $d = 1 - 5 = -4$; find n, the number of terms.

$$\begin{aligned} a_n &= a_1 + (n-1)d \\ -79 &= 5 + (n-1)(-4) \\ -79 &= 9 - 4n \\ -88 &= -4n \\ n &= 22 \end{aligned}$$

There are 22 terms in this sequence.

25. Find S_8 if $a_1 = -2$ and $d = 6$.

$S_n = \frac{n}{2}(a_1 + a_n)$; find a_8 first.

$$\begin{aligned} a_8 &= a_1 + (8-1)d \\ &= -2 + 7(6) \\ &= -2 + 42 = 40 \\ S_8 &= \frac{8}{2}(-2 + 40) \\ &= 4(38) = 152 \end{aligned}$$

26. Find S_8 if $a_n = -2 + 5n$.

$$\begin{aligned} a_1 &= -2 + 5(1) = 3 \\ a_8 &= -2 + 5(8) = 38 \\ S_n &= \frac{n}{2}(a_1 + a_n) \\ S_8 &= \frac{8}{2}(3 + 38) = 4(41) = 164 \end{aligned}$$

27. Find the general term for the geometric sequence $-1, -4, \ldots$.

$a_1 = -1$ and $r = \frac{-4}{-1} = 4.$

$$\begin{aligned} a_n &= a_1 r^{n-1} \\ a_n &= -1(4)^{n-1} \end{aligned}$$

28. $\frac{2}{3}, \frac{2}{15}, \ldots$

$a_1 = \frac{2}{3}$ and $r = \dfrac{\frac{2}{15}}{\frac{2}{3}} = \frac{2}{15} \cdot \frac{3}{2} = \frac{1}{5}.$

$$\begin{aligned} a_n &= a_1 r^{n-1} \\ a_n &= \frac{2}{3}\left(\frac{1}{5}\right)^{n-1} \end{aligned}$$

29. Find a_{11} for $2, -6, 18, \ldots$.

$a_1 = 2$ and $r = \frac{-6}{2} = -3$

$$\begin{aligned} a_n &= a_1 r^{n-1} \\ a_{11} &= 2(-3)^{11-1} \\ &= 2(-3)^{10} = 118{,}098 \end{aligned}$$

30. Given $a_3 = 20$, $a_5 = 80$; find a_{10}.

$a_n = a_1 r^{n-1}$

For a_3, $a_3 = a_1 r^{3-1}$

$20 = a_1 r^2$.

For a_5, $a_5 = a_1 r^{5-1}$

$80 = a_1 r^4$.

The ratio of a_5 to a_3 is

$$\begin{aligned} \frac{80}{20} &= \frac{a_1 r^4}{a_1 r^2} \\ 4 &= r^2 \\ r &= \pm 2. \end{aligned}$$

Since $20 = a_1 r^2$, and $r^2 = 4$,

$$\begin{aligned} 20 &= a_1(4) \\ 5 &= a_1. \end{aligned}$$

Now find a_{10}.

$$a_n = a_1 r^{n-1}$$
$$a_{10} = 5(\pm 2)^{10-1}$$
$$= 5(\pm 2)^9$$

Two answers are possible for a_{10}:

$$5(2)^9 = 2560 \quad \text{or} \quad 5(-2)^9 = -2560.$$

31. $\sum_{i=1}^{5} \left(\frac{1}{4}\right)^i$

$a_1 = \frac{1}{4},\ r = \frac{1}{4},\ n = 5.$

$$S_n = \frac{a_1(1 - r^n)}{1 - r}$$
$$S_5 = \frac{\frac{1}{4}\left[1 - \left(\frac{1}{4}\right)^5\right]}{1 - \frac{1}{4}}$$
$$= \frac{\frac{1}{4}\left(1 - \frac{1}{1024}\right)}{\frac{3}{4}}$$
$$= \frac{1}{3}\left(\frac{1023}{1024}\right) = \frac{341}{1024}$$

32. $\sum_{i=1}^{8} \frac{3}{4}(-1)^i$

$a_1 = -\frac{3}{4}$ and $r = -1$, and $n = 8$.

$$S_n = \frac{a_1(1 - r^n)}{1 - r}$$
$$S_8 = \frac{-\frac{3}{4}\left[1 - (-1)^8\right]}{1 - (-1)}$$
$$= \frac{-\frac{3}{4}(1 - 1)}{2}$$
$$= \frac{-\frac{3}{4}(0)}{2} = 0$$

33. $\sum_{i=1}^{\infty} 4\left(\frac{1}{5}\right)^i$

The terms are the infinite geometric sequence with $a_1 = \frac{4}{5}$ and $r = \frac{1}{5}$.

$$S = \frac{a_1}{1 - r} = \frac{\frac{4}{5}}{1 - \frac{1}{5}} = \frac{\frac{4}{5}}{\frac{4}{5}} = 1$$

34. $\sum_{i=1}^{\infty} 2(3)^i$

The terms are the infinite geometric sequence with $a_1 = 6$ and $r = 3$.

$$S = \frac{a_1}{1 - r} \quad \text{if} \quad |r| < 1,$$

but $r = 3$ so S does not exist, and, thus, the sum does not exist.

35. $(2p - q)^5$

$$= \left[2p + (-q)\right]^5$$
$$= (2p)^5 + \frac{5!}{4!\,1!}(2p)^4(-q)^1$$
$$+ \frac{5!}{3!\,2!}(2p)^3(-q)^2 + \frac{5!}{2!\,3!}(2p)^2(-q)^3$$
$$+ \frac{5!}{1!\,4!}(2p)^1(-q)^4 + (-q)^5$$
$$= 32p^5 + 5(16p^4)(-q) + 10(8p^3)q^2$$
$$+ 10(4p^2)(-q^3) + 5(2p)q^4 - q^5$$
$$= 32p^5 - 80p^4q + 80p^3q^2 - 40p^2q^3$$
$$+ 10pq^4 - q^5$$

36. $(x^2 + 3y)^4$

$$= (x^2)^4 + \frac{4!}{3!\,1!}(x^2)^3(3y)^1$$
$$+ \frac{4!}{2!\,2!}(x^2)^2(3y)^2$$
$$+ \frac{4!}{1!\,3!}(x^2)^1(3y)^3 + (3y)^4$$
$$= x^8 + 4(x^6)(3y) + 6(x^4)(9y^2)$$
$$+ 4(x^2)(27y^3) + 81y^4$$
$$= x^8 + 12x^6y + 54x^4y^2 + 108x^2y^3$$
$$+ 81y^4$$

37. $(\sqrt{m} + \sqrt{n})^4$

$$= (\sqrt{m})^4 + \frac{4!}{3!\,1!}(\sqrt{m})^3(\sqrt{n})^1$$
$$+ \frac{4!}{2!\,2!}(\sqrt{m})^2(\sqrt{n})^2$$
$$+ \frac{4!}{1!\,3!}(\sqrt{m})^1(\sqrt{n})^3 + (\sqrt{n})^4$$
$$= m^2 + 4(m\sqrt{m})(\sqrt{n}) + 6(m)(n)$$
$$+ 4(\sqrt{m})(n\sqrt{n}) + n^2$$
$$= m^2 + 4m\sqrt{mn} + 6mn + 4n\sqrt{mn} + n^2$$

38. The fourth term ($r = 4$, so $r - 1 = 3$) of $(3a + 2b)^{19}$ is

$$\frac{19!}{16!\,3!}(3a)^{16}(2b)^3$$
$$= 969(3)^{16}(a^{16})(8)b^3$$
$$= 7752(3)^{16}a^{16}b^3.$$

39. The twenty-third term ($r = 23$, so $r - 1 = 22$) of $(-2k+3)^{25}$ is

$$= \frac{25!}{3!\,22!}(-2k)^3(3)^{22}$$
$$= -18,400(3)^{22}k^3.$$

40. The arithmetic sequence $1, 7, 13, \ldots$ has $a_1 = 1$ and $d = 7 - 1 = 6$.

$$a_n = a_1 + (n-1)d$$
$$a_{40} = 1 + (40-1)6$$
$$= 1 + (39)6$$
$$= 1 + 234 = 235$$
$$a_{10} = 1 + (10-1)6$$
$$= 1 + (9)6 = 55$$
$$S_n = \frac{n}{2}(a_1 + a_n)$$
$$S_{10} = \frac{10}{2}(1 + 55)$$
$$= 5(56) = 280$$

41. The geometric sequence $-3, 6, -12, \ldots$ has $a_1 = -3$ and $r = \frac{6}{-3} = -2$.

$$a_n = a_1 r^{n-1}$$
$$a_{10} = -3(-2)^9$$
$$= -3(-512) = 1536$$
$$S_n = \frac{a_1(1-r^n)}{1-r}$$
$$S_{10} = \frac{-3\left[1-(-2)^{10}\right]}{1-(-2)}$$
$$= \frac{-3(1-1024)}{3}$$
$$= -(-1023) = 1023$$

42. $a_1 = 1, r = -3$

$$a_n = a_1 r^{n-1}$$
$$a_9 = 1(-3)^{9-1}$$
$$= (-3)^8 = 6561$$
$$S_n = \frac{a_1(r^n - 1)}{r-1}$$
$$S_{10} = \frac{1\left[(-3)^{10} - 1\right]}{-3-1}$$
$$= -\frac{1}{4}(3^{10} - 1)$$
$$= -\frac{1}{4}(59,049 - 1)$$
$$= -14,762$$

43. The arithmetic sequence with $a_1 = -4$ and $d = 3$ is $-4, -1, 2, 5, \ldots$.

$$a_n = a_1 + (n-1)d$$
$$a_{15} = -4 + (15-1)3$$
$$= -4 + 42 = 38$$
$$S_n = \frac{n}{2}\left[2a_1 + (n-1)d\right]$$
$$S_{10} = \frac{10}{2}\left[2(-4) + (10-1)3\right]$$
$$= 5(-8+27)$$
$$= 5(19) = 95$$

44. $2, 7, 12, \ldots$
This is an arithmetic sequence with $a_1 = 2$ and $d = 7 - 2 = 5$.

$$a_n = a_1 + (n-1)d$$
$$a_n = 2 + (n-1)5$$
$$= 2 + 5n - 5$$
$$= 5n - 3$$

45. $2, 8, 32, \ldots$
This is a geometric sequence with $a_1 = 2$ and $r = \frac{8}{2} = 4$.

$$a_n = a_1 r^{n-1}$$
$$a_n = 2(4)^{n-1}$$

46. $27, 9, 3, \ldots$
This is a geometric sequence with $a_1 = 27$ and $r = \frac{9}{27} = \frac{1}{3}$.

$$a_n = a_1 r^{n-1}$$
$$a_n = 27\left(\frac{1}{3}\right)^{n-1}$$

47. $12, 9, 6, \ldots$
This is an arithmetic sequence with $a_1 = 12$ and $d = -3$.

$$a_n = a_1 + (n-1)d$$
$$a_n = 12 + (n-1)(-3)$$
$$= 12 - 3n + 3$$
$$= -3n + 15$$

48. The distances traveled in successive seconds are $3, 7, 11, 15, 19, \ldots$.
This is an arithmetic sequence.

$$S_n = \frac{n}{2}\left[2a_1 + (n-1)d\right]$$
$$S_n = 210, \quad a_1 = 3, \quad d = 4$$
$$210 = \frac{n}{2}\left[2(3) + (n-1)4\right]$$
$$420 = n(6 + 4n - 4)$$
$$420 = 6n + 4n^2 - 4n$$
$$0 = 4n^2 + 2n - 420$$
$$0 = 2n^2 + n - 210$$
$$0 = (2n+21)(n-10)$$
$$2n + 21 = 0 \quad \text{or} \quad n - 10 = 0$$
$$n = -\frac{21}{2} \quad \text{or} \quad n = 10$$

Discard $-\frac{21}{2}$ since time cannot be negative. It takes her 10 seconds.

49. $S = R\left[\frac{(1+i)^n - 1}{i}\right]$

$$S = 672\left[\frac{\left(1+\frac{.08}{4}\right)^{4(7)} - 1}{\frac{.08}{4}}\right]$$

$$= 672\left[\frac{(1.02)^{28} - 1}{.02}\right]$$

$$= 24,898.41$$

The future value of the annuity is \$24,898.41.

50. Since $100\% - 3\% = 97\% = .97$, the population after 1 year is $.97(50,000)$, after 2 years is $.97[.97(50,000)]$ or $(.97)^2(50,000)$, and after n years is $(.97)^n(50,000)$. After 6 years, the population is

$$(.97)^6(50,000) \approx 41,649 \approx 42,000.$$

51. $\left(\frac{1}{2}\right)^n$ is left after n strokes. So

$\left(\frac{1}{2}\right)^7 = \frac{1}{128} = .0078125$ is left after 7 strokes.

52. **(a)** We can write the repeating decimal number .55555... as an infinite geometric sequence as follows:

$$\frac{5}{10} + \frac{5}{10}\left(\frac{1}{10}\right) + \frac{5}{10}\left(\frac{1}{10}\right)^2 + \frac{5}{10}\left(\frac{1}{10}\right)^3 + \cdots$$

(b) The common ratio r is

$$\frac{\frac{5}{10}\left(\frac{1}{10}\right)}{\frac{5}{10}} = \frac{1}{10}.$$

(c) Since $|r| < 1$, the sum exists.

$$S = \frac{a_1}{1-r} = \frac{\frac{5}{10}}{1 - \frac{1}{10}} = \frac{\frac{5}{10}}{\frac{9}{10}} = \frac{5}{9}$$

53. No, the sum cannot be found because $r = 2$, and this value of r does not satisfy $|r| < 1$.

54. No, the terms must be successive, such as the first and second, or the second and third.

Chapter 12 Test

1.
$$a_n = (-1)^n + 1$$
$$a_1 = (-1)^1 + 1 = 0$$
$$a_2 = (-1)^2 + 1 = 1 + 1 = 2$$
$$a_3 = (-1)^3 + 1 = -1 + 1 = 0$$
$$a_4 = (-1)^4 + 1 = 1 + 1 = 2$$
$$a_5 = (-1)^5 + 1 = -1 + 1 = 0$$

Answer: $0, 2, 0, 2, 0$

2.
$$a_1 = 4,\ d = 2$$
$$a_2 = a_1 + d = 4 + 2 = 6$$
$$a_3 = a_2 + d = 6 + 2 = 8$$
$$a_4 = a_3 + d = 8 + 2 = 10$$
$$a_5 = a_4 + d = 10 + 2 = 12$$

Answer: $4, 6, 8, 10, 12$

3.
$$a_4 = 6,\ r = \frac{1}{2}$$
$$a_n = a_1 r^{n-1}$$
$$a_4 = a_1\left(\frac{1}{2}\right)^{4-1}$$
$$6 = a_1\left(\frac{1}{8}\right)$$
$$a_1 = 48$$
$$a_2 = \frac{1}{2}a_1 = \frac{1}{2}(48) = 24$$
$$a_3 = \frac{1}{2}a_2 = \frac{1}{2}(24) = 12$$
$$a_4 = \frac{1}{2}a_3 = \frac{1}{2}(12) = 6$$
$$a_5 = \frac{1}{2}a_4 = \frac{1}{2}(6) = 3$$

Answer: 48, 24, 12, 6, 3

4. Given $a_1 = 6$ and $d = -2$; find a_4.

$$a_n = a_1 + (n-1)d$$
$$a_4 = a_1 + (4-1)d$$
$$= 6 + (3)(-2)$$
$$= 6 - 6 = 0$$

5. Given $a_5 = 16$ and $a_7 = 9$; find a_4.
This is a geometric sequence, so $a_6 = a_5 r$ and $a_7 = a_6 r$. Thus, $a_6 = \frac{a_7}{r}$ and

$$a_5 r = \frac{a_7}{r}$$
$$r^2 = \frac{a_7}{a_5}$$
$$r^2 = \frac{9}{16}$$
$$r = \frac{3}{4} \quad \text{or} \quad r = -\frac{3}{4}.$$
$$a_5 = a_4 r \quad \text{so} \quad a_4 = \frac{a_5}{r}.$$
$$a_4 = \frac{16}{\frac{3}{4}} \quad \text{or} \quad \frac{16}{-\frac{3}{4}}$$
$$a_4 = \frac{64}{3} \quad \text{or} \quad a_4 = -\frac{64}{3}$$

6. Given the arithmetic sequence with $a_2 = 12$ and $a_3 = 15$; find S_5.

$$d = a_3 - a_2 = 15 - 12 = 3$$
$$a_1 = a_2 - 3 = 12 - 3 = 9$$
$$a_5 = a_4 + 3 = a_3 + 6 = 15 + 6 = 21$$

continued

$$S_n = \frac{n}{2}(a_1 + a_n)$$
$$S_5 = \frac{5}{2}(9 + 21)$$
$$= \frac{5}{2}(30) = 75$$

7. Given the geometric sequence with $a_5 = 4$ and $a_7 = 1$; find S_5.

$r^2 = \frac{a_7}{a_5} = \frac{1}{4}$, so $r = \frac{1}{2}$ or $r = -\frac{1}{2}$.

Use $a_7 = a_1 r^6$ to get $1 = a_1\left(\frac{1}{2}\right)^6$, and so $a_1 = 64$.

Use $S_n = \frac{a_1(r^n - 1)}{r - 1}$ or $S_n = \frac{a_1(1 - r^n)}{1 - r}$.

$$S_5 = \frac{64\left[1 - \left(\frac{1}{2}\right)^5\right]}{1 - \frac{1}{2}} \quad \text{or} \quad S_5 = \frac{64\left[1 - \left(-\frac{1}{2}\right)^5\right]}{1 - \left(-\frac{1}{2}\right)}$$
$$= 128\left(1 - \frac{1}{32}\right) \qquad = \frac{128}{3}\left(1 + \frac{1}{32}\right)$$
$$= 128\left(\frac{31}{32}\right) \qquad = \frac{128}{3}\left(\frac{33}{32}\right)$$
$$= 124 \qquad = 44$$

8. $\overline{x} = \frac{\text{total}}{5}$, where total $=$ 66,733 + 69,468 + 70,731 + 72,265 + 72,998.

Thus, $\overline{x} = \frac{352{,}195}{5} = 70{,}439$.

The average number of bands per year for this five-year-period was 70,439.

9. $$S = R\left[\frac{(1+i)^n - 1}{i}\right]$$
$$= 4000\left[\frac{\left(1 + \frac{.06}{4}\right)^{4(7)} - 1}{\frac{.06}{4}}\right]$$
$$= 4000\left[\frac{(1.015)^{28} - 1}{.015}\right]$$
$$\approx 137{,}925.91$$
The account will have \$137,925.91 at the end of this term.

10. An infinite geometric series has a sum if $|r| < 1$, where r is the common ratio.

11. $\sum_{i=1}^{5}(2i + 8)$
$$= [2(1) + 8] + [2(2) + 8] + [2(3) + 8] + [2(4) + 8] + [2(5) + 8]$$
$$= 10 + 12 + 14 + 16 + 18$$
$$= 70$$

12. $\sum_{i=1}^{6}(3i - 5)$

Use the formula $S_6 = \frac{n}{2}(a_1 + a_6)$.
$$a_1 = 3(1) - 5 = -2$$
$$a_6 = 3(6) - 5 = 13$$
$$S_6 = \frac{6}{2}(-2 + 13) = 3(11) = 33$$

13. $\sum_{i=1}^{500} i$

Use the formula $S_{500} = \frac{n}{2}(a_1 + a_{500})$ with $a_1 = 1$ and $a_{500} = 500$.
$$S_{500} = \frac{500}{2}(1 + 500)$$
$$= 250(501) = 125{,}250$$

14. $$\sum_{i=1}^{3}\frac{1}{2}(4^i) = \frac{1}{2}\sum_{i=1}^{3}(4^i)$$
$$= \frac{1}{2}\left[4 + 16 + 64\right] = 42$$

15. $\sum_{i=1}^{\infty}\left(\frac{1}{4}\right)^i$

This is an infinite geometric series with $a_1 = \frac{1}{4}$ and $r = \frac{1}{4}$.

Since $|r| = \frac{1}{4} < 1$, the sum exists.

$$S = \frac{a_1}{1 - r} = \frac{\frac{1}{4}}{1 - \frac{1}{4}} = \frac{\frac{1}{4}}{\frac{3}{4}} = \frac{1}{4} \cdot \frac{4}{3} = \frac{1}{3}$$

16. $\sum_{i=1}^{\infty} 6\left(\frac{3}{2}\right)^i$

This is an infinite geometric series with $a_1 = 6\left(\frac{3}{2}\right)^1 = 9$ and $r = \frac{3}{2}$.

Since $|r| = \frac{3}{2} > 1$, the sum does not exist.

17. $$8! = 8 \cdot 7 \cdot 6 \cdot 5 \cdot 4 \cdot 3 \cdot 2 \cdot 1$$
$$= 40{,}320$$

18. By definition, $0! = 1$.

19. $$\frac{6!}{4!\,2!} = \frac{6 \cdot 5 \cdot 4 \cdot 3 \cdot 2 \cdot 1}{(4 \cdot 3 \cdot 2 \cdot 1)(2 \cdot 1)} = \frac{6 \cdot 5}{2 \cdot 1} = 15$$

20. $${}_{12}C_{10} = \frac{12!}{10!\,(12 - 10)!} = \frac{12 \cdot 11 \cdot 10!}{10!\,(2!)}$$
$$= \frac{12 \cdot 11}{2} = 66$$

21. $(3k-5)^4$

$$= (3k)^4 + \frac{4!}{3!\,1!}(3k)^3(-5)^1 + \frac{4!}{2!\,2!}(3k)^2(-5)^2 + \frac{4!}{1!\,3!}(3k)^1(-5)^3 + (-5)^4$$
$$= 81k^4 - 4(27k^3)(5) + 6(9k^2)(25) - 4(3k)(125) + 625$$
$$= 81k^4 - 540k^3 + 1350k^2 - 1500k + 625$$

22. The fifth term $(r = 5$, so $r - 1 = 4)$ of $\left(2x - \frac{y}{3}\right)^{12}$ is

$$\frac{12!}{(12-4)!\,4!}(2x)^{12-4}\left(-\frac{y}{3}\right)^4$$
$$= \frac{12!}{8!\,4!}(2x)^8\left(-\frac{y}{3}\right)^4$$
$$= \frac{14,080x^8y^4}{9}.$$

23. The amounts of unpaid balance during 15 months form an arithmetic sequence

$$300, 280, 260, \ldots, 40, 20,$$

which is the sequence with $n = 15$, $a_1 = 300$, and $a_{15} = 20$. Find the sum of these balances.

$$S_n = \frac{n}{2}(a_1 + a_n)$$
$$S_{15} = \frac{15}{2}(300 + 20)$$
$$= \frac{15}{2}(320) = 2400$$

Since 1% interest is paid on this total, the interest paid is 1% of \$2400 or \$24. The sewing machine cost \$300 (paid monthly at \$20), so the total cost is \$300 + \$24 = \$324.

24. The weekly populations form a geometric sequence with $a_1 = 20$ and $r = 3$ since the colony begins with 20 insects and triples each week. Find the general term of this geometric sequence.

$$a_n = a_1r^{n-1}$$
$$a_n = 20(3)^{n-1}$$

We're assuming that from the beginning of July to the end of September is 12 weeks, so find a_{12}.

$$a_n = 20(3)^{n-1}$$
$$a_{12} = 20(3)^{11}$$

At the end of September, $20(3)^{11} = 3,542,940$ insects will be present in the colony.

Cumulative Review Exercises Chapters 1–12

In Exercises 1–4, let

$$P = \left\{-\frac{8}{3}, 10, 0, \sqrt{13}, -\sqrt{3}, \frac{45}{15}, \sqrt{-7}, .82, -3\right\}$$

1. The integers are $10, 0, \frac{45}{15}$ (or 3), and -3.

2. The rational numbers are

$$-\frac{8}{3}, 10, 0, \frac{45}{15} \text{ (or 3)}, .82, \text{and } -3.$$

3. The irrational numbers are $\sqrt{13}$ and $-\sqrt{3}$.

4. All are real numbers except $\sqrt{-7}$.

5. $|-7| + 6 - |-10| - (-8+3)$
$= 7 + 6 - 10 - (-5)$
$= 13 - 10 + 5 = 8$

6. $-15 - |-4| - 10 - |-6|$
$= -15 - 4 - 10 - 6 = -35$

7. $4(-6) + (-8)(5) - (-9)$
$= -24 - 40 + 9 = -55$

8.
$$9 - (5 + 3a) + 5a = -4(a-3) - 7$$
$$9 - 5 - 3a + 5a = -4a + 12 - 7$$
$$4 + 2a = -4a + 5$$
$$6a = 1$$
$$a = \frac{1}{6}$$

Solution set: $\left\{\frac{1}{6}\right\}$

9. $7m + 18 \le 9m - 2$
$-2m \le -20$
Divide by -2; reverse the direction of the inequality symbol.
$m \ge 10$
Solution set: $[10, \infty)$

10. $|4x - 3| = 21$
$4x - 3 = 21$ or $4x - 3 = -21$
$4x = 24$ $\qquad$ $4x = -18$
$x = 6$ or $x = -\frac{18}{4} = -\frac{9}{2}$

Solution set: $\left\{-\frac{9}{2}, 6\right\}$

11. $\frac{x+3}{12} - \frac{x-3}{6} = 0$
Multiply by the LCD, 12.
$$12\left(\frac{x+3}{12} - \frac{x-3}{6}\right) = 12(0)$$
$$x + 3 - 2(x-3) = 0$$
$$x + 3 - 2x + 6 = 0$$
$$9 - x = 0$$
$$9 = x$$

Check $x = 9$: $1 - 1 \stackrel{?}{=} 0$ *True*
Solution set: $\{9\}$

12. $2x > 8$ or $-3x > 9$
$x > 4$ or $x < -3$

Solution set: $(-\infty, -3) \cup (4, \infty)$

13. $|2m - 5| \geq 11$

$$2m - 5 \geq 11 \quad \text{or} \quad 2m - 5 \leq -11$$
$$2m \geq 16 \qquad\qquad 2m \leq -6$$
$$m \geq 8 \quad \text{or} \quad m \leq -3$$

Solution set: $(-\infty, -3] \cup [8, \infty)$

14. Let $(x_1, y_1) = (4, -5)$ and $(x_2, y_2) = (-12, -17)$. Then

$$m = \frac{y_2 - y_1}{x_2 - x_1} = \frac{-17 - (-5)}{-12 - 4} = \frac{-12}{-16} = \frac{3}{4}.$$

The slope is $\frac{3}{4}$.

15. To find the equation of the line through $(-2, 10)$ and parallel to $3x + y = 7$, find the slope of

$$3x + y = 7$$
$$y = -3x + 7.$$

The slope is -3, so a line parallel to it also has slope -3. Use $m = -3$ and $(x_1, y_1) = (-2, 10)$ in the point-slope form.

$$y - y_1 = m(x - x_1)$$
$$y - 10 = -3[x - (-2)]$$
$$y - 10 = -3(x + 2)$$

Write in standard form.

$$y - 10 = -3x - 6$$
$$3x + y = 4$$

Alternative solution: The line must be of the form $3x + y = k$ since it is parallel to $3x + y = 7$. Substitute -2 for x and 10 for y to find k.

$$3(-2) + 10 = k$$
$$4 = k$$

The equation is $3x + y = 4$.

16. $x - 3y = 6$

Find the x- and y-intercepts. To find the x-intercept, let $y = 0$.

$$x - 3(0) = 6$$
$$x = 6$$

The x-intercept is $(6, 0)$.
To find the y-intercept, let $x = 0$.

$$0 - 3y = 6$$
$$y = -2$$

The y-intercept is $(0, -2)$.
Plot the intercepts and draw the line through them.

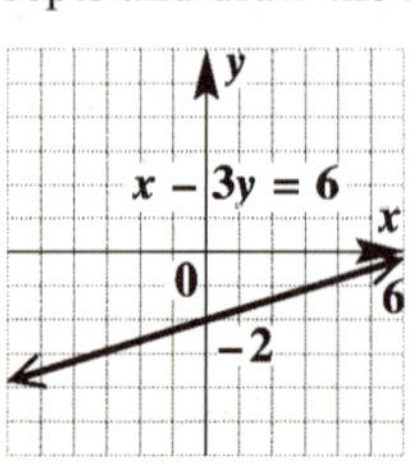

17. $4x - y < 4$
Graph the line $4x - y = 4$, which has intercepts $(0, -4)$ and $(1, 0)$, as a dashed line because the inequality involves $<$. Test $(0, 0)$, which yields $0 < 4$, a true statement. Shade the region on the side of the line that includes $(0, 0)$.

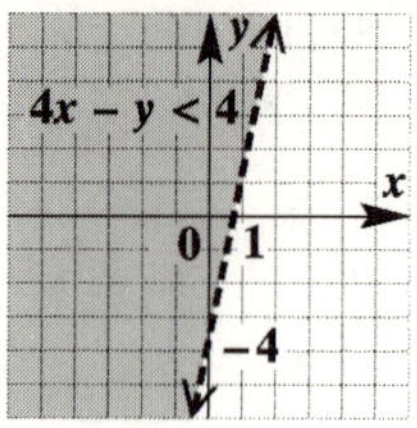

18. $\{(-3, 2), (-2, 6), (0, 4), (1, 2), (2, 6)\}$

(a) The set of ordered pairs is a function since every first coordinate is paired with a unique second coordinate.

(b) The domain is the set of first coordinates, that is,

$$\{-3, -2, 0, 1, 2\}.$$

(c) The range is the set of second coordinates, that is,

$$\{2, 6, 4\}.$$

19.

$$\begin{aligned} 2x + 5y &= -19 \quad (1) \\ -3x + 2y &= -19 \quad (2) \end{aligned}$$

To eliminate x, multiply equation (1) by 3 and equation (2) by 2. Then add the results.

$$\begin{array}{rcrl} 6x + 15y &=& -57 & \quad 3 \times (1) \\ -6x + 4y &=& -38 & \quad 2 \times (2) \\ \hline 19y &=& -95 & \\ y &=& -5 & \end{array}$$

Substitute -5 for y in equation (1) to find x.

$$2x + 5y = -19 \qquad (1)$$
$$2x + 5(-5) = -19$$
$$2x - 25 = -19$$
$$2x = 6$$
$$x = 3$$

Solution set : $\{(3, -5)\}$

20.

$$\begin{aligned} y &= 5x + 3 \quad (1) \\ 2x + 3y &= -8 \quad (2) \end{aligned}$$

From equation (1), substitute $5x + 3$ for y in equation (2). Then solve for x.

$$2x + 3(5x + 3) = -8$$
$$2x + 15x + 9 = -8$$
$$17x = -17$$
$$x = -1$$

From (1), $y = 5(-1) + 3 = -2$.

The solution $(-1, -2)$ checks.

Solution set: $\{(-1, -2)\}$

21.

$$\begin{array}{rcrcrcr} x &+& 2y &+& z &=& 8 \\ 2x &-& y &+& 3z &=& 15 \\ -x &+& 3y &-& 3z &=& -11 \end{array}$$

Write the augmented matrix.

$$\left[\begin{array}{rrr|r} 1 & 2 & 1 & 8 \\ 2 & -1 & 3 & 15 \\ -1 & 3 & -3 & -11 \end{array}\right]$$

$$\left[\begin{array}{rrr|r} 1 & 2 & 1 & 8 \\ 0 & -5 & 1 & -1 \\ 0 & 5 & -2 & -3 \end{array}\right] \quad \begin{array}{r} -2R_1 + R_2 \\ R_1 + R_3 \end{array}$$

$$\left[\begin{array}{rrr|r} 1 & 2 & 1 & 8 \\ 0 & 1 & -\frac{1}{5} & \frac{1}{5} \\ 0 & 5 & -2 & -3 \end{array}\right] \quad -\tfrac{1}{5}R_2$$

$$\left[\begin{array}{rrr|r} 1 & 2 & 1 & 8 \\ 0 & 1 & -\frac{1}{5} & \frac{1}{5} \\ 0 & 0 & -1 & -4 \end{array}\right] \quad -5R_2 + R_3$$

$$\left[\begin{array}{rrr|r} 1 & 2 & 1 & 8 \\ 0 & 1 & -\frac{1}{5} & \frac{1}{5} \\ 0 & 0 & 1 & 4 \end{array}\right] \quad -R_3$$

This matrix gives the system

$$\begin{aligned} x + 2y + z &= 8 \\ y - \frac{1}{5}z &= \frac{1}{5} \\ z &= 4. \end{aligned}$$

Substitute $z = 4$ in the second equation.

$$\begin{aligned} y - \frac{1}{5}(4) &= \frac{1}{5} \\ y &= \frac{1}{5} + \frac{4}{5} = 1 \end{aligned}$$

Substitute $y = 1$ and $z = 4$ in the first equation.

$$\begin{aligned} x + 2(1) + 4 &= 8 \\ x + 6 &= 8 \\ x &= 2 \end{aligned}$$

Solution set: $\{(2, 1, 4)\}$

22. Let $x =$ the number of pounds of \$3 per pound nuts.

	Number of Pounds	Price per Pound	Value
\$3 nuts	x	3	$3x$
\$4.25 nuts	8	4.25	$4.25(8)$
Mixture	$x + 8$	4	$4(x + 8)$

The last column gives the equation.

$$\begin{aligned} 3x + 4.25(8) &= 4(x + 8) \\ 3x + 34 &- 4x + 32 \\ 2 &= x \end{aligned}$$

Use 2 pounds of the \$3 nuts.

23. $(4p + 2)(5p - 3)$

F O I L

$$\begin{aligned} &= 20p^2 - 12p + 10p - 6 \\ &= 20p^2 - 2p - 6 \end{aligned}$$

24. $(3k - 7)^2 = (3k)^2 - 2(3k)(7) + 7^2$
$= 9k^2 - 42k + 49$

25. $(2m^3 - 3m^2 + 8m) - (7m^3 + 5m - 8)$

$$\begin{aligned} &= 2m^3 - 7m^3 - 3m^2 + 8m - 5m + 8 \\ &= -5m^3 - 3m^2 + 3m + 8 \end{aligned}$$

26.

$$\begin{array}{r|l} & 2t^3 + 3t^2 - 4t + 2 \\ \hline 3t - 2 & 6t^4 + 5t^3 - 18t^2 + 14t - 1 \\ & 6t^4 - 4t^3 \\ \hline & \quad 9t^3 - 18t^2 \\ & \quad 9t^3 - 6t^2 \\ \hline & \qquad -12t^2 + 14t \\ & \qquad -12t^2 + 8t \\ \hline & \qquad\qquad 6t - 1 \\ & \qquad\qquad 6t - 4 \\ \hline & \qquad\qquad\quad 3 \end{array}$$

Remainder

Answer: $2t^3 + 3t^2 - 4t + 2 + \dfrac{3}{3t - 2}$

27. $7x + x^3 = x(7 + x^2)$

28. $14y^2 + 13y - 12$

Look for two integers whose product is $(14)(-12) = -168$ and whose sum is 13. The required numbers are 21 and -8.

$$\begin{aligned} &14y^2 + 13y - 12 \\ &= 14y^2 + 21y - 8y - 12 \\ &= 7y(2y + 3) - 4(2y + 3) \\ &= (2y + 3)(7y - 4) \end{aligned}$$

29. $6z^3 + 5z^2 - 4z = z(6z^2 + 5z - 4)$
$= z(3z + 4)(2z - 1)$

30. $49a^4 - 9b^2 = (7a^2)^2 - (3b)^2$
$= (7a^2 + 3b)(7a^2 - 3b)$

31. $c^3 + 27d^3 = c^3 + (3d)^3$
$= (c + 3d)(c^2 - 3cd + 9d^2)$

32. $64r^2 + 48rq + 9q^2$

$$\begin{aligned} &= (8r)^2 + 2(8r)(3q) + (3q)^2 \\ &= (8r + 3q)^2 \end{aligned}$$

33.

$$\begin{aligned} 2x^2 + x &= 10 \\ 2x^2 + x - 10 &= 0 \\ (2x + 5)(x - 2) &= 0 \end{aligned}$$

$2x + 5 = 0$ or $x - 2 = 0$

$x = -\dfrac{5}{2}$ or $x = 2$

Solution set: $\left\{-\frac{5}{2}, 2\right\}$

34. $k^2 - k - 6 \le 0$

Solve the equation

$$\begin{aligned} k^2 - k - 6 &= 0. \\ (k - 3)(k + 2) &= 0 \end{aligned}$$

$k - 3 = 0$ or $k + 2 = 0$

$k = 3$ or $k = -2$

The numbers -2 and 3 divide a number line into three intervals.

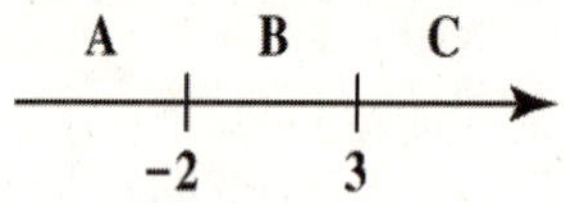

Test a number from each interval in the original inequality.

$$k^2 - k - 6 \le 0$$

Interval A: Let $k = -3$.

$(-3)^2 - (-3) - 6 \le 0$?

$6 \le 0$ *False*

Interval B: Let $k = 0$.

$0^2 - 0 - 6 \le 0$?

$-6 \le 0$ *True*

Interval C: Let $k = 4$.

$4^2 - 4 - 6 \le 0$?

$6 \le 0$ *False*

The numbers in Interval B, including the endpoints -2 and 3, are solutions.

Solution set: $[-2, 3]$

35. $\left(\frac{2}{3}\right)^{-2} = \left(\frac{3}{2}\right)^{2} = \frac{3}{2} \cdot \frac{3}{2} = \frac{9}{4}$

36.
$$\frac{(3p^2)^3(-2p^6)}{4p^3(5p^7)} = \frac{3^3p^6(-2)p^6}{20p^{10}} = \frac{-54p^{12}}{20p^{10}} = -\frac{27}{10}p^{12-10} = -\frac{27p^2}{10}$$

37. $f(x) = \frac{2}{x^2 - 81} = \frac{2}{(x+9)(x-9)}$

The domain of f is the set of all real numbers excluding ± 9 since division by 0 is not defined. In interval notation, the domain is

$$(-\infty, -9) \cup (-9, 9) \cup (9, \infty).$$

38.
$$\frac{x^2 - 16}{x^2 + 2x - 8} \div \frac{x-4}{x+7} = \frac{x^2 - 16}{x^2 + 2x - 8} \cdot \frac{x+7}{x-4} = \frac{(x+4)(x-4)(x+7)}{(x+4)(x-2)(x-4)} = \frac{x+7}{x-2}$$

39.
$$\frac{5}{p^2 + 3p} - \frac{2}{p^2 - 4p} = \frac{5}{p(p+3)} - \frac{2}{p(p-4)}$$

The LCD is $p(p+3)(p-4)$.

$$= \frac{5(p-4)}{p(p+3)(p-4)} - \frac{2(p+3)}{p(p-4)(p+3)} = \frac{5p - 20 - 2p - 6}{p(p+3)(p-4)} = \frac{3p - 26}{p(p+3)(p-4)}$$

40.
$$\frac{4}{x-3} - \frac{6}{x+3} = \frac{24}{x^2 - 9}$$
$$\frac{4}{x-3} - \frac{6}{x+3} = \frac{24}{(x+3)(x-3)}$$

Multiply by the LCD, $(x+3)(x-3)$. $(x \ne \pm 3)$

$$4(x+3) - 6(x-3) = 24$$
$$4x + 12 - 6x + 18 = 24$$
$$-2x + 30 = 24$$
$$-2x = -6$$
$$x = 3$$

But $x \ne 3$.

Solution set: $\emptyset$

41.
$$6x^2 + 5x = 8$$
$$6x^2 + 5x - 8 = 0$$

Use the quadratic formula with $a = 6$, $b = 5$, and $c = -8$.

$$x = \frac{-b \pm \sqrt{b^2 - 4ac}}{2a}$$
$$x = \frac{-5 \pm \sqrt{5^2 - 4(6)(-8)}}{2(6)} = \frac{-5 \pm \sqrt{25 + 192}}{12} = \frac{-5 \pm \sqrt{217}}{12}$$

Solution set: $\left\{\frac{-5 + \sqrt{217}}{12}, \frac{-5 - \sqrt{217}}{12}\right\}$

42.
$$\sqrt{3x - 2} = x$$
$$3x - 2 = x^2 \quad \textit{Square}$$
$$0 = x^2 - 3x + 2$$
$$0 = (x-1)(x-2)$$
$$x - 1 = 0 \quad \text{or} \quad x - 2 = 0$$
$$x = 1 \quad \text{or} \quad x = 2$$

Check $x = 1$: $\sqrt{1} \stackrel{?}{=} 1$ *True*

Check $x = 2$: $\sqrt{4} \stackrel{?}{=} 2$ *True*

Solution set: $\{1, 2\}$

43.
$$5\sqrt{72} - 4\sqrt{50} = 5\sqrt{36 \cdot 2} - 4\sqrt{25 \cdot 2} = 5 \cdot 6\sqrt{2} - 4 \cdot 5\sqrt{2} = 30\sqrt{2} - 20\sqrt{2} = 10\sqrt{2}$$

44. $(8+3i)(8-3i)$
$= 8^2 - (3i)^2 = 64 - 9i^2$
$= 64 - 9(-1) = 64 + 9 = 73$

45. The graph of $f(x) = 9x + 5$ is a line. To find the inverse, replace $f(x)$ with y.

$$y = 9x + 5$$

Interchange x and y.

$$x = 9y + 5$$

Solve for y.

$$x - 5 = 9y$$
$$\frac{x-5}{9} = y$$

Replace y with $f^{-1}(x)$.

$$f^{-1}(x) = \frac{x-5}{9}$$

46. Graph $g(x) = \left(\frac{1}{3}\right)^x$.

Make a table of values.

x	-2	-1	0	1	2
$g(x)$	9	3	1	$\frac{1}{3}$	$\frac{1}{9}$

Plot these points, and draw a smooth decreasing exponential curve through them.

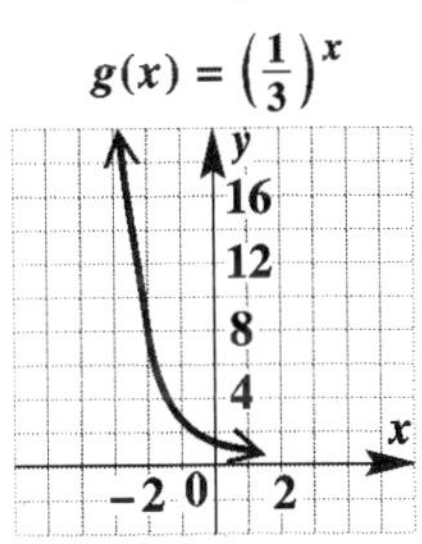

47.
$$3^{2x-1} = 81$$
$$3^{2x-1} = 3^4$$
$$2x - 1 = 4 \qquad \textit{Equate exponents}$$
$$2x = 5$$
$$x = \frac{5}{2}$$

Solution set: $\left\{\frac{5}{2}\right\}$

48. Graph $y = \log_{1/3} x$.
Change to exponential form.

$$\left(\frac{1}{3}\right)^y = x$$

This is the inverse of the graph of

$g(x) = y = \left(\frac{1}{3}\right)^x$ in Exercise 46. To find points on the graph, interchange the x- and y-values in the table.

x	9	3	1	$\frac{1}{3}$	$\frac{1}{9}$
y	-2	-1	0	1	2

Plot these points, and draw a smooth decreasing logarithmic curve through them.

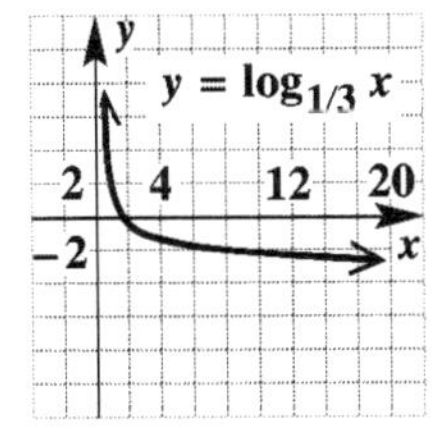

49. $\log_8 x + \log_8 (x+2) = 1$
Use the product rule for logarithms.
$\log_8 x(x+2) = 1$
Change to exponential form.
$x(x+2) = 8^1$
$x^2 + 2x - 8 = 0$
$(x+4)(x-2) = 0$
$x + 4 = 0 \quad$ or $\quad x - 2 = 0$
$x = -4 \quad$ or $\quad x = 2$
$x \neq -4$ because $\log_8(-4)$ does not exist, so the only answer is $x = 2$.

Solution set: $\{2\}$

50. $f(x) = 2(x-2)^2 - 3$ is in $f(x) = a(x-h)^2 + k$ form.

The graph is a vertical parabola with vertex (h, k) at $(2, -3)$. Since $a = 2 > 0$, the graph opens up. Also, $|a| = |2| = 2 > 1$, so the graph is narrower than the graph of $f(x) = x^2$. The points $(0, 5)$ and $(4, 5)$ are on the graph.

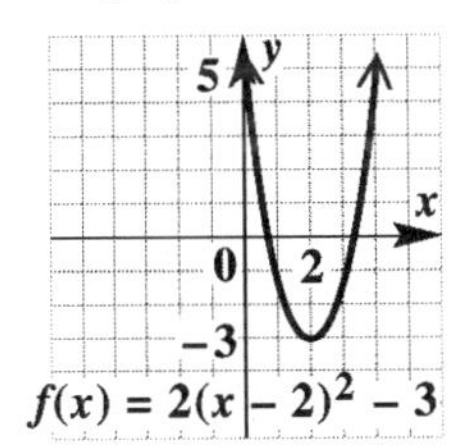

51. $\frac{x^2}{9} + \frac{y^2}{25} = 1$ is in $\frac{x^2}{a^2} + \frac{y^2}{b^2} = 1$ form with $a = 3$ and $b = 5$. The graph is an ellipse centered at $(0, 0)$ with x-intercepts $(3, 0)$ and $(-3, 0)$ and y-intercepts $(0, 5)$ and $(0, -5)$. Plot the intercepts and draw the ellipse through them.

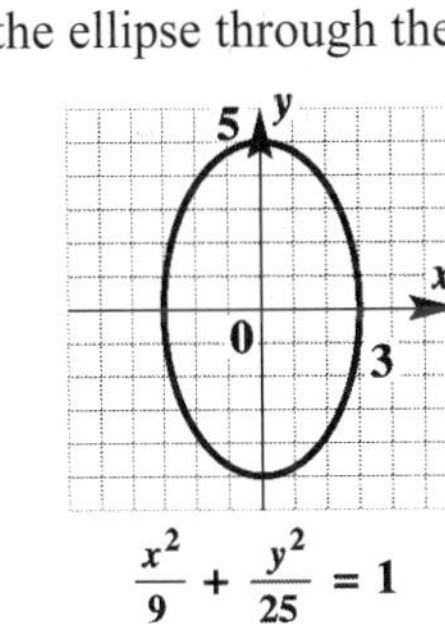

52. $x^2 - y^2 = 9$

$$\frac{x^2}{9} - \frac{y^2}{9} = 1$$

The graph is a hyperbola centered at $(0, 0)$ with x-intercepts $(3, 0)$ and $(-3, 0)$. The asymptotes are $y = \pm x$. Draw the right and left branches through the intercepts and approaching the asymptotes.

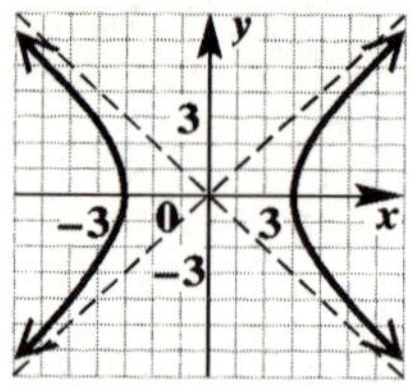

$x^2 - y^2 = 9$

53.
$$xy = -5 \quad (1)$$
$$2x + y = 3 \quad (2)$$

Solve equation (2) for y.

$$y = -2x + 3 \quad (3)$$

Substitute $-2x + 3$ for y in equation (1).

$$\begin{aligned} xy &= -5 \quad (1) \\ x(-2x+3) &= -5 \\ -2x^2 + 3x &= -5 \\ -2x^2 + 3x + 5 &= 0 \\ 2x^2 - 3x - 5 &= 0 \\ (2x-5)(x+1) &= 0 \end{aligned}$$

$$2x - 5 = 0 \quad \text{or} \quad x + 1 = 0$$
$$x = \frac{5}{2} \quad \text{or} \quad x = -1$$

Substitute these values for x in equation (3) to find y.

If $x = \frac{5}{2}$, then $y = -2\left(\frac{5}{2}\right) + 3 = -2$.

If $x = -1$, then $y = -2(-1) + 3 = 5$.

Solution set: $\left\{(-1, 5), \left(\frac{5}{2}, -2\right)\right\}$

54. Center at $(-5, 12)$; radius 9
Use the equation of a circle with $h = -5$, $k = 12$, and $r = 9$.

$$(x-h)^2 + (y-k)^2 = r^2$$
$$\left[x - (-5)\right]^2 + (y-12)^2 = 9^2$$
$$(x+5)^2 + (y-12)^2 = 81$$

55. $a_n = 5n - 12$

$a_1 = 5(1) - 12 = 5 - 12 = -7$
$a_2 = 5(2) - 12 = 10 - 12 = -2$
$a_3 = 5(3) - 12 = 15 - 12 = 3$
$a_4 = 5(4) - 12 = 20 - 12 = 8$
$a_5 = 5(5) - 12 = 25 - 12 = 13$
Answer: $-7, -2, 3, 8, 13$

56. **(a)** $a_1 = 8$, $d = 2$

$$S_n = \frac{n}{2}\left[2a_1 + (n-1)d\right]$$
$$S_6 = \frac{6}{2}\left[2(8) + (6-1)2\right] = 3(16+10) = 3(26) = 78$$

(b) $15 - 6 + \frac{12}{5} - \frac{24}{25} + \ldots$

This is an infinite geometric series with $a_1 = 15$ and $r = \frac{-6}{15} = -\frac{2}{5}$. The sum is

$$S = \frac{a_1}{1-r} = \frac{15}{1 - \left(-\frac{2}{5}\right)} = \frac{15}{\frac{7}{5}} = 15 \cdot \frac{5}{7} = \frac{75}{7}.$$

57. $$\sum_{i=1}^{4} 3i = 3\sum_{i=1}^{4} i = 3\left[\frac{4(4+1)}{2}\right] = 3(10) = 30$$

We used the fact that $\sum_{i=1}^{n} i = \frac{n(n+1)}{2}$.

58. $9! = 9 \cdot 8 \cdot 7 \cdot 6 \cdot 5 \cdot 4 \cdot 3 \cdot 2 \cdot 1 = 362{,}880$

59. $(2a - 1)^5$

$$\begin{aligned} &= (2a)^5 + \frac{5!}{4!\,1!}(2a)^4(-1)^1 \\ &\quad + \frac{5!}{3!\,2!}(2a)^3(-1)^2 + \frac{5!}{2!\,3!}(2a)^2(-1)^3 \\ &\quad + \frac{5!}{1!\,4!}(2a)^1(-1)^4 + (-1)^5 \\ &= 32a^5 + 5(16a^4)(-1) + 10(8a^3)(1) \\ &\quad + 10(4a^2)(-1) + 5(2a)(1) + (-1) \\ &= 32a^5 - 80a^4 + 80a^3 - 40a^2 + 10a - 1 \end{aligned}$$

60. The fourth term ($r = 4$, so $r - 1 = 3$) of $\left(3x^4 - \frac{1}{2}y^2\right)^5$ is

$$\begin{aligned} &\frac{5!}{(5-3)!\,3!}(3x^4)^{5-3}\left(-\frac{1}{2}y^2\right)^3 \\ &= \frac{5!}{2!\,3!}(3x^4)^2\left(-\frac{1}{2}\right)^3(y^2)^3 \\ &= 10(9x^8)\left(-\frac{1}{8}\right)y^6 \\ &= -\frac{45x^8y^6}{4} \end{aligned}$$

APPENDIX B DETERMINANTS AND CRAMER'S RULE

1. $\begin{vmatrix} -2 & -3 \\ 4 & -6 \end{vmatrix} = -2(-6) - (-3)4$

Choice **D** is correct.

3. $\begin{vmatrix} -2 & 5 \\ -1 & 4 \end{vmatrix} = -2(4) - 5(-1)$

$= -8 + 5 = -3$

5. $\begin{vmatrix} 1 & -2 \\ 7 & 0 \end{vmatrix} = 1(0) - (-2)7$

$= 0 + 14 = 14$

7. $\begin{vmatrix} 0 & 4 \\ 0 & 4 \end{vmatrix} = 0(4) - 4(0)$

$= 0 - 0 = 0$

9. $\begin{vmatrix} -1 & 2 & 4 \\ -3 & -2 & -3 \\ 2 & -1 & 5 \end{vmatrix}$ Expand by minors about the first column.

$$= -1\begin{vmatrix} -2 & -3 \\ -1 & 5 \end{vmatrix} - (-3)\begin{vmatrix} 2 & 4 \\ -1 & 5 \end{vmatrix} + 2\begin{vmatrix} 2 & 4 \\ -2 & -3 \end{vmatrix}$$
$$= -1[-2(5) - (-3)(-1)] + 3[2(5) - 4(-1)] + 2[2(-3) - 4(-2)]$$
$$= -1(-13) + 3(14) + 2(2)$$
$$= 13 + 42 + 4 = 59$$

11. $\begin{vmatrix} 1 & 0 & -2 \\ 0 & 2 & 3 \\ 1 & 0 & 5 \end{vmatrix}$ There are two 0s in column 2. We'll expand about that column since there is only 1 minor to evaluate.

$$= -0\begin{vmatrix} 0 & 3 \\ 1 & 5 \end{vmatrix} + 2\begin{vmatrix} 1 & -2 \\ 1 & 5 \end{vmatrix} - 0\begin{vmatrix} 1 & -2 \\ 0 & 3 \end{vmatrix}$$
$$= 0 + 2[1(5) - (-2)(1)] - 0$$
$$= 2[5 + 2] = 2(7) = 14$$

13. Multiply the upper left and lower right entries. Then multiply the upper right and lower left entries. Subtract the second product from the first to obtain the determinant. For example,

$$\begin{vmatrix} 4 & 2 \\ 7 & 1 \end{vmatrix} = 4 \cdot 1 - 2 \cdot 7$$
$$= 4 - 14 = -10.$$

15. $\begin{vmatrix} 4 & 4 & 2 \\ 1 & 1 & 2 \\ 1 & 0 & 2 \end{vmatrix}$ Expand about column 2.

$$= -4\begin{vmatrix} 1 & -2 \\ 1 & 2 \end{vmatrix} + (-1)\begin{vmatrix} 4 & 2 \\ 1 & 2 \end{vmatrix} - 0\begin{vmatrix} 4 & 2 \\ 1 & -2 \end{vmatrix}$$
$$= -4[1(2) - (-2)1] - 1[4(2) - 2(1)] - 0$$
$$= -4(4) - 1(6)$$
$$= -16 - 6 = -22$$

17. $\begin{vmatrix} 3 & 5 & -2 \\ 1 & -4 & 1 \\ 3 & 1 & -2 \end{vmatrix}$ Expand about column 1.

$$= 3\begin{vmatrix} -4 & 1 \\ 1 & -2 \end{vmatrix} - 1\begin{vmatrix} 5 & -2 \\ 1 & -2 \end{vmatrix} + 3\begin{vmatrix} 5 & -2 \\ -4 & 1 \end{vmatrix}$$
$$= 3[-4(-2) - 1(1)] - 1[5(-2) - (-2)1] + 3[5(1) - (-2)(-4)]$$
$$= 3(7) - 1(-8) + 3(-3)$$
$$= 21 + 8 - 9 = 20$$

19. $\begin{vmatrix} 3 & 0 & -2 \\ 1 & -4 & 1 \\ 3 & 1 & -2 \end{vmatrix}$ Expand about row 1.

$$= 3\begin{vmatrix} -4 & 1 \\ 1 & -2 \end{vmatrix} - 0\begin{vmatrix} 1 & 1 \\ 3 & -2 \end{vmatrix} + (-2)\begin{vmatrix} 1 & -4 \\ 3 & 1 \end{vmatrix}$$
$$= 3[-4(-2) - 1(1)] - 0 + (-2)[1(1) - (-4)3]$$
$$= 3(7) - 0 - 2(13)$$
$$= 21 - 26 = -5$$

21. By choosing that row or column to expand about, all terms will have a factor of 0, and so the sum of all these terms will be 0.

23. Enter $\begin{vmatrix} 1.5 & 2.6 & 9.3 \\ 5.2 & -1.4 & 8.6 \\ 0 & .7 & 1.2 \end{vmatrix}$ as [A].

Type the keys for det [A].

The result should be 6.078.

25. **(a)** D is the determinant of coefficients—choice **D**.

(b) D_x is the determinant with the constants replacing the x-coefficients—choice **A**.

(c) D_y is the determinant with the constants replacing the y-coefficients—choice **C**.

(d) D_z is the determinant with the constants replacing the z-coefficients—choice **B**.

27.
$$3x + 5y = -5$$
$$-2x + 3y = 16$$

$$D = \begin{vmatrix} 3 & 5 \\ -2 & 3 \end{vmatrix} = 3(3) - 5(-2) = 19$$
$$D_x = \begin{vmatrix} -5 & 5 \\ 16 & 3 \end{vmatrix} = -5(3) - 5(16) = -95$$
$$D_y = \begin{vmatrix} 3 & -5 \\ -2 & 16 \end{vmatrix} = 3(16) - (-5)(-2) = 38$$

$$x = \frac{D_x}{D} = \frac{-95}{19} = -5;\ y = \frac{D_y}{D} = \frac{38}{19} = 2$$

Solution set: $\{(-5, 2)\}$

29. $8x + 3y = 1$
$6x - 5y = 2$

$D = \begin{vmatrix} 8 & 3 \\ 6 & -5 \end{vmatrix} = 8(-5) - 3(6) = -58$

$D_x = \begin{vmatrix} 1 & 3 \\ 2 & -5 \end{vmatrix} = 1(-5) - 3(2) = -11$

$D_y = \begin{vmatrix} 8 & 1 \\ 6 & 2 \end{vmatrix} = 8(2) - 1(6) = 10$

$x = \frac{D_x}{D} = \frac{-11}{-58} = \frac{11}{58}; y = \frac{D_y}{D} = \frac{10}{-58} = -\frac{5}{29}$

Solution set: $\left\{\left(\frac{11}{58}, -\frac{5}{29}\right)\right\}$

31. $2x + 3y = 4$
$5x + 6y = 7$

$D = \begin{vmatrix} 2 & 3 \\ 5 & 6 \end{vmatrix} = 2(6) - 3(5) = -3$

$D_x = \begin{vmatrix} 4 & 3 \\ 7 & 6 \end{vmatrix} = 4(6) - 3(7) = 3$

$D_y = \begin{vmatrix} 2 & 4 \\ 5 & 7 \end{vmatrix} = 2(7) - 4(5) = -6$

$x = \frac{D_x}{D} = \frac{3}{-3} = -1; y = \frac{D_y}{D} = \frac{-6}{-3} = 2$

Solution set: $\{(-1, 2)\}$

33. $2x + 3y + 2z = 15$
$x - y + 2z = 5$
$x + 2y - 6z = -26$

$D = \begin{vmatrix} 2 & 3 & 2 \\ 1 & -1 & 2 \\ 1 & 2 & -6 \end{vmatrix}$ Expand about row 1.

$= 2\begin{vmatrix} -1 & 2 \\ 2 & -6 \end{vmatrix} - 3\begin{vmatrix} 1 & 2 \\ 1 & -6 \end{vmatrix} + 2\begin{vmatrix} 1 & -1 \\ 1 & 2 \end{vmatrix}$
$= 2(2) - 3(-8) + 2(3)$
$= 4 + 24 + 6 = 34$

$D_x = \begin{vmatrix} 15 & 3 & 2 \\ 5 & -1 & 2 \\ -26 & 2 & -6 \end{vmatrix}$ Expand about row 2.

$= -5\begin{vmatrix} 3 & 2 \\ 2 & -6 \end{vmatrix} + (-1)\begin{vmatrix} 15 & 2 \\ -26 & -6 \end{vmatrix}$
$- 2\begin{vmatrix} 15 & 3 \\ -26 & 2 \end{vmatrix}$
$= -5(-22) - 1(-38) - 2(108)$
$= 110 + 38 - 216 = -68$

$D_y = \begin{vmatrix} 2 & 15 & 2 \\ 1 & 5 & 2 \\ 1 & -26 & -6 \end{vmatrix}$ Expand about column 1.

$= 2\begin{vmatrix} 5 & 2 \\ -26 & -6 \end{vmatrix} - 1\begin{vmatrix} 15 & 2 \\ -26 & -6 \end{vmatrix}$
$+ 1\begin{vmatrix} 15 & 2 \\ 5 & 2 \end{vmatrix}$
$= 2(22) - 1(-38) + 1(20)$
$= 44 + 38 + 20 = 102$

$D_z = \begin{vmatrix} 2 & 3 & 15 \\ 1 & -1 & 5 \\ 1 & 2 & -26 \end{vmatrix}$ Expand about column 1.

$= 2\begin{vmatrix} -1 & 5 \\ 2 & -26 \end{vmatrix} - 1\begin{vmatrix} 3 & 15 \\ 2 & -26 \end{vmatrix} + 1\begin{vmatrix} 3 & 15 \\ -1 & 5 \end{vmatrix}$
$= 2(16) - 1(-108) + 1(30)$
$= 32 + 108 + 30 = 170$

$x = \frac{D_x}{D} = \frac{-68}{34} = -2; y = \frac{D_y}{D} = \frac{102}{34} = 3$

$z = \frac{D_z}{D} = \frac{170}{34} = 5$

Solution set: $\{(-2, 3, 5)\}$

35. $2x - 3y + 4z = 8$
$6x - 9y + 12z = 24$
$-4x + 6y - 8z = -16$

$D = \begin{vmatrix} 2 & -3 & 4 \\ 6 & -9 & 12 \\ -4 & 6 & -8 \end{vmatrix}$ Expand about row 1.

$= 2\begin{vmatrix} -9 & 12 \\ 6 & -8 \end{vmatrix} + 3\begin{vmatrix} 6 & 12 \\ -4 & -8 \end{vmatrix} + 4\begin{vmatrix} 6 & -9 \\ -4 & 6 \end{vmatrix}$
$= 2(0) + 3(0) + 4(0) = 0$

Because $D = 0$, Cramer's rule does not apply.

37. $3x + 5z = 0$
$2x + 3y = 1$
$-y + 2z = -11$

$D = \begin{vmatrix} 3 & 0 & 5 \\ 2 & 3 & 0 \\ 0 & -1 & 2 \end{vmatrix}$ Expand about row 1.

$= 3\begin{vmatrix} 3 & 0 \\ -1 & 2 \end{vmatrix} - 0 + 5\begin{vmatrix} 2 & 3 \\ 0 & -1 \end{vmatrix}$
$= 3(6) + 5(-2)$
$= 18 - 10 = 8$

$D_x = \begin{vmatrix} 0 & 0 & 5 \\ 1 & 3 & 0 \\ -11 & -1 & 2 \end{vmatrix}$ Expand about row 1.

$= 0 - 0 + 5\begin{vmatrix} 1 & 3 \\ -11 & -1 \end{vmatrix}$
$= 5(32) = 160$

$$D_y = \begin{vmatrix} 3 & 0 & 5 \\ 2 & 1 & 0 \\ 0 & -11 & 2 \end{vmatrix} \text{ Expand about row 1.}$$
$$= 3\begin{vmatrix} 1 & 0 \\ -11 & 2 \end{vmatrix} - 0 + 5\begin{vmatrix} 2 & 1 \\ 0 & -11 \end{vmatrix}$$
$$= 3(2) + 5(-22)$$
$$= 6 - 110 = -104$$

$$D_z = \begin{vmatrix} 3 & 0 & 0 \\ 2 & 3 & 1 \\ 0 & -1 & -11 \end{vmatrix} \text{ Expand about row 1.}$$
$$= 3\begin{vmatrix} 3 & 1 \\ -1 & -11 \end{vmatrix} - 0 + 0$$
$$= 3(-32) = -96$$

$$x = \frac{D_x}{D} = \frac{160}{8} = 20;\ y = \frac{D_y}{D} = \frac{-104}{8} = -13$$
$$z = \frac{D_z}{D} = \frac{-96}{8} = -12$$

Solution set: $\{(20, -13, -12)\}$

39.
$$\begin{aligned} x - 3y \quad &= 13 \\ 2y + z &= 5 \\ -x \quad + z &= -7 \end{aligned}$$

$$D - \begin{vmatrix} 1 & -3 & 0 \\ 0 & 2 & 1 \\ -1 & 0 & 1 \end{vmatrix} \text{ Expand about column 1.}$$
$$= 1\begin{vmatrix} 2 & 1 \\ 0 & 1 \end{vmatrix} - 0 + (-1)\begin{vmatrix} -3 & 0 \\ 2 & 1 \end{vmatrix}$$
$$= 1(2) - 1(-3) = 2 + 3 = 5$$

$$D_x = \begin{vmatrix} 13 & -3 & 0 \\ 5 & 2 & 1 \\ -7 & 0 & 1 \end{vmatrix} \text{ Expand about column 3.}$$
$$= 0 - 1\begin{vmatrix} 13 & -3 \\ -7 & 0 \end{vmatrix} + 1\begin{vmatrix} 13 & -3 \\ 5 & 2 \end{vmatrix}$$
$$= -1(-21) + 1(41)$$
$$= 21 + 41 = 62$$

$$D_y = \begin{vmatrix} 1 & 13 & 0 \\ 0 & 5 & 1 \\ -1 & -7 & 1 \end{vmatrix} \text{ Expand about column 1.}$$
$$= 1\begin{vmatrix} 5 & 1 \\ -7 & 1 \end{vmatrix} - 0 + (-1)\begin{vmatrix} 13 & 0 \\ 5 & 1 \end{vmatrix}$$
$$= 1(12) - 1(13) = 12 - 13 = -1$$

$$D_z - \begin{vmatrix} 1 & -3 & 13 \\ 0 & 2 & 5 \\ -1 & 0 & -7 \end{vmatrix} \text{ Expand about column 1.}$$
$$= 1\begin{vmatrix} 2 & 5 \\ 0 & -7 \end{vmatrix} - 0 - 1\begin{vmatrix} -3 & 13 \\ 2 & 5 \end{vmatrix}$$
$$= 1(-14) - 1(-41) = -14 + 41 = 27$$

$$x = \frac{D_x}{D} = \frac{62}{5};\ y = \frac{D_y}{D} = -\frac{1}{5}$$
$$z = \frac{D_z}{D} = \frac{27}{5}$$

Solution set: $\left\{\left(\frac{62}{5}, -\frac{1}{5}, \frac{27}{5}\right)\right\}$

41.
$$\begin{aligned} x + 2y + z &= 10 \\ 2x - y - 3z &= -20 \\ -x + 4y + z &= 18 \end{aligned}$$

Use a calculator to evaluate each determinant.

$$D = \begin{vmatrix} 1 & 2 & 1 \\ 2 & -1 & -3 \\ -1 & 4 & 1 \end{vmatrix} = 20$$
$$D_x = \begin{vmatrix} 10 & 2 & 1 \\ -20 & -1 & -3 \\ 18 & 4 & 1 \end{vmatrix} = -20$$
$$D_y = \begin{vmatrix} 1 & 10 & 1 \\ 2 & -20 & -3 \\ -1 & 18 & 1 \end{vmatrix} = 60$$
$$D_z = \begin{vmatrix} 1 & 2 & 10 \\ 2 & -1 & -20 \\ -1 & 4 & 18 \end{vmatrix} = 100$$

$$x = \frac{D_x}{D} = \frac{-20}{20} = -1;\ y = \frac{D_y}{D} = \frac{60}{20} = 3$$
$$z = \frac{D_z}{D} = \frac{100}{20} = 5$$

Solution set: $\{(-1, 3, 5)\}$

43.
$$\begin{aligned} -8w + 4x - 2y + z &= -28 \\ -w + x - y + z &= -10 \\ w + x + y + z &= -4 \\ 27w + 9x + 3y + z &= 2 \end{aligned}$$

Use a calculator to evaluate each determinant.

$$D = \begin{vmatrix} -8 & 4 & -2 & 1 \\ -1 & 1 & -1 & 1 \\ 1 & 1 & 1 & 1 \\ 27 & 9 & 3 & 1 \end{vmatrix} = 240$$
$$D_w = \begin{vmatrix} 28 & 4 & 2 & 1 \\ -10 & 1 & -1 & 1 \\ -4 & 1 & 1 & 1 \\ 2 & 9 & 3 & 1 \end{vmatrix} = 240$$
$$D_x = \begin{vmatrix} -8 & -28 & -2 & 1 \\ -1 & -10 & -1 & 1 \\ 1 & -4 & 1 & 1 \\ 27 & 2 & 3 & 1 \end{vmatrix} = -720$$
$$D_y = \begin{vmatrix} -8 & 4 & -28 & 1 \\ -1 & 1 & -10 & 1 \\ 1 & 1 & -4 & 1 \\ 27 & 9 & 2 & 1 \end{vmatrix} = 480$$

continued

$$D_z = \begin{vmatrix} -8 & 4 & -2 & -28 \\ -1 & 1 & -1 & -10 \\ 1 & 1 & 1 & -4 \\ 27 & 9 & 3 & 2 \end{vmatrix} = -960$$

$$w = \frac{D_w}{D} = \frac{240}{240} = 1;\ x = \frac{D_x}{D} = \frac{-720}{240} = -3;$$

$$y = \frac{D_y}{D} = \frac{480}{240} = 2;\ z = \frac{D_z}{D} = \frac{-960}{240} = -4$$

Solution set: $\{(1, -3, 2, -4)\}$

45.

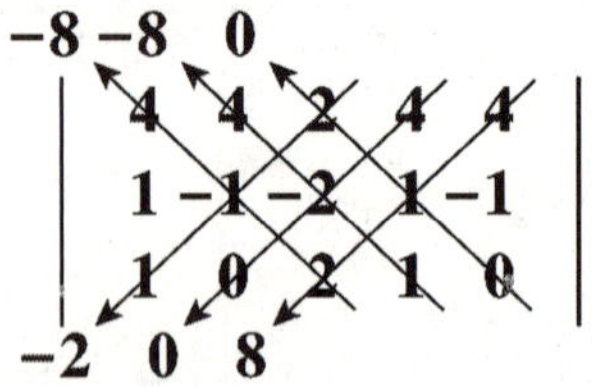

Add the top numbers: $-8 - 8 + 0 = -16$
Add the bottom numbers: $-2 + 0 + 8 = 6$
Find the difference: $(-16) - (6) = -22$

47.

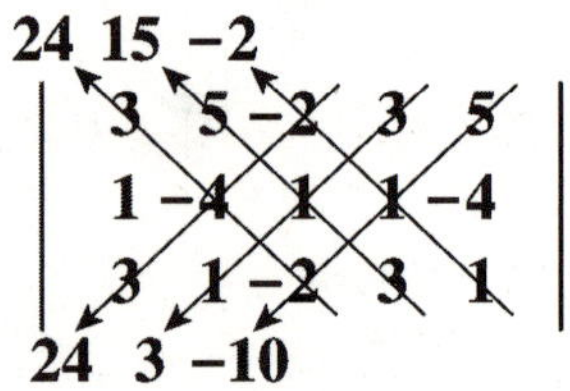

Add the top numbers: $24 + 15 - 2 = 37$
Add the bottom numbers: $24 + 3 - 10 = 17$
Find the difference: $(37) - (17) = 20$

49.

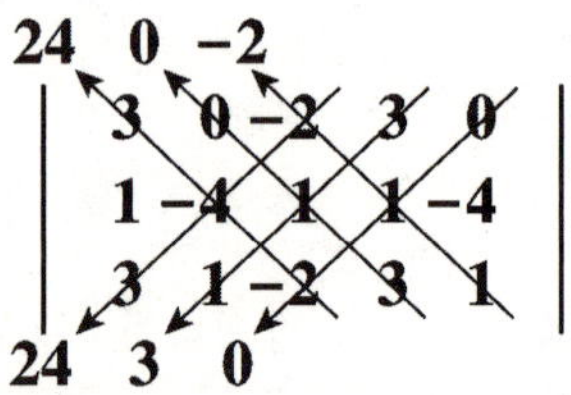

Add the top numbers: $24 + 0 - 2 = 22$
Add the bottom numbers: $24 + 3 + 0 = 27$
Find the difference: $(22) - (27) = -5$

51. $\begin{vmatrix} 4 & x \\ 2 & 3 \end{vmatrix} = 8$

Evaluate the determinant.

$$\begin{vmatrix} 4 & x \\ 2 & 3 \end{vmatrix} = 4(3) - x(2) = 12 - 2x$$

Solve the equation.

$$12 - 2x = 8$$
$$-2x = -4$$
$$x = 2$$

Solution set: $\{2\}$

53. $\begin{vmatrix} x & 4 \\ x & -3 \end{vmatrix} = 0$

Evaluate the determinant.

$$\begin{vmatrix} x & 4 \\ x & -3 \end{vmatrix} = x(-3) - 4x = -3x - 4x = -7x$$

Solve the equation.

$$-7x = 0$$
$$x = 0$$

Solution set: $\{0\}$

55.

$$\begin{aligned} ax + (a+1)y &= a+2 \\ (a+3)x + (a+4)y &= a+5 \end{aligned}$$

$$D = \begin{vmatrix} a & a+1 \\ a+3 & a+4 \end{vmatrix} = a^2 + 4a - (a^2 + 4a + 3) = -3$$

$$D_x = \begin{vmatrix} a+2 & a+1 \\ a+5 & a+4 \end{vmatrix} = a^2 + 6a + 8 - (a^2 + 6a + 5) = 3$$

$$D_y = \begin{vmatrix} a & a+2 \\ a+3 & a+5 \end{vmatrix} = a^2 + 5a - (a^2 + 5a + 6) = -6$$

$$x = \frac{D_x}{D} = \frac{3}{-3} = -1;\ y = \frac{D_y}{D} = \frac{-6}{-3} = 2$$

Solution set: $\{(-1, 2)\}$

57.

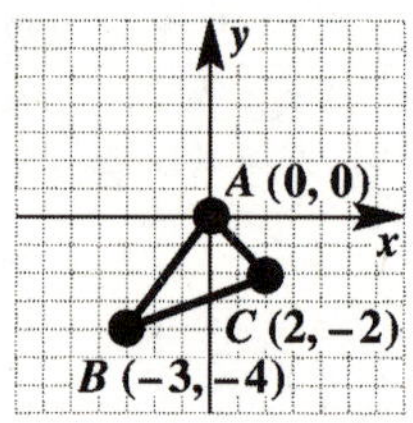

58. Use $(0, 0)$ as (x_1, y_1), $(-3, -4)$ as (x_2, y_2), and $(2, -2)$ as (x_3, y_3). Then

$$\frac{1}{2}\begin{vmatrix} x_1 & y_1 & 1 \\ x_2 & y_2 & 1 \\ x_3 & y_3 & 1 \end{vmatrix} = \frac{1}{2}\begin{vmatrix} 0 & 0 & 1 \\ -3 & -4 & 1 \\ 2 & -2 & 1 \end{vmatrix}.$$

59. $\frac{1}{2}\begin{vmatrix} 0 & 0 & 1 \\ -3 & -4 & 1 \\ 2 & -2 & 1 \end{vmatrix}$ Expand about row 1.

$$= \frac{1}{2}\left(0 - 0 + 1\begin{vmatrix} -3 & -4 \\ 2 & -2 \end{vmatrix}\right)$$
$$= \frac{1}{2}[1(6 + 8)]$$
$$= \frac{1}{2}(14) = 7$$

The area is 7 square units.

60. $\dfrac{1}{2}\begin{vmatrix} 3 & 8 & 1 \\ -1 & 4 & 1 \\ 0 & 1 & 1 \end{vmatrix}$ Expand about row 3.

$= \dfrac{1}{2}\left(0 - 1\begin{vmatrix} 3 & 1 \\ -1 & 1 \end{vmatrix} + 1\begin{vmatrix} 3 & 8 \\ -1 & 4 \end{vmatrix}\right)$

$= \dfrac{1}{2}(-1[3+1] + 1[12+8])$

$= \dfrac{1}{2}(-4+20) = \dfrac{1}{2}(16) = 8$

The area is 8 square units.

61. The slope m through the points (x_1, y_1) and (x_2, y_2) is

$$m = \frac{y_2 - y_1}{x_2 - x_1}.$$

62. $y - y_1 = m(x - x_1)$

Let $m = \dfrac{y_2 - y_1}{x_2 - x_1}$ from Exercise 61.

$y - y_1 = \dfrac{y_2 - y_1}{x_2 - x_1}(x - x_1)$

63. $y - y_1 = \dfrac{y_2 - y_1}{x_2 - x_1}(x - x_1)$

Multiply both sides by $x_2 - x_1$.

$(x_2 - x_1)(y - y_1) = (y_2 - y_1)(x - x_1)$

Subtract $(y_2 - y_1)(x - x_1)$ from both sides.

$(x_2 - x_1)(y - y_1) - (y_2 - y_1)(x - x_1) = 0$

Multiply and collect like terms.

$x_2y - x_2y_1 - x_1y + x_1y_1$
$\quad - (xy_2 - x_1y_2 - xy_1 + x_1y_1) = 0$

$x_2y - x_2y_1 - x_1y + x_1y_1$
$\quad - xy_2 + x_1y_2 + xy_1 - x_1y_1 = 0$

$x_2y - x_2y_1 - x_1y - xy_2 + x_1y_2 + xy_1 = 0$

64. $\begin{vmatrix} x & y & 1 \\ x_1 & y_1 & 1 \\ x_2 & y_2 & 1 \end{vmatrix} = 0$ Expand about row 1.

$x\begin{vmatrix} y_1 & 1 \\ y_2 & 1 \end{vmatrix} - y\begin{vmatrix} x_1 & 1 \\ x_2 & 1 \end{vmatrix} + 1\begin{vmatrix} x_1 & y_1 \\ x_2 & y_2 \end{vmatrix}$

$= x(y_1 - y_2) - y(x_1 - x_2) + 1(x_1y_2 - x_2y_1)$

$= xy_1 - xy_2 - x_1y + x_2y + x_1y_2 - x_2y_1$

Thus,

$xy_1 - xy_2 - x_1y + x_2y + x_1y_2 - x_2y_1 = 0$

or

$x_2y - x_2y_1 - x_1y - xy_2 + x_1y_2 + xy_1 = 0.$

This is the same equation as that found in Exercise 63.

APPENDIX C SYNTHETIC DIVISION

1. Synthetic division provides a quick, easy way to divide a polynomial by a binomial of the form $x-k$.

3. $\dfrac{x^2-6x+5}{x-1}$

$$\begin{array}{r|rrr}1 & 1 & -6 & 5\\ & & 1 & -5\\ \hline & 1 & -5 & 0\end{array}$$

← *Coefficients of numerator*

Write the answer from the bottom row.

$1 \quad -5 \;\to\; x-5$

Answer: $x-5$

5. $\dfrac{4m^2+19m-5}{m+5}$

$m+5=m-(-5)$, so use -5.

$$\begin{array}{r|rrr}-5 & 4 & 19 & -5\\ & & -20 & 5\\ \hline & 4 & -1 & 0\end{array}$$

Answer: $4m-1$

7. $\dfrac{2a^2+8a+13}{a+2}$

$a+2=a-(-2)$, so use -2.

$$\begin{array}{r|rrr}-2 & 2 & 8 & 13\\ & & -4 & -8\\ \hline & 2 & 4 & 5\end{array} \leftarrow \textit{Remainder}$$

Answer: $2a+4+\dfrac{5}{a+2}$

9. $(p^2-3p+5)\div(p+1)$

$$\begin{array}{r|rrr}-1 & 1 & -3 & 5\\ & & -1 & 4\\ \hline & 1 & -4 & 9\end{array}$$

Answer: $p-4+\dfrac{9}{p+1}$

11. $\dfrac{4a^3-3a^2+2a-3}{a-1}$

$$\begin{array}{r|rrrr}1 & 4 & -3 & 2 & -3\\ & & 4 & 1 & 3\\ \hline & 4 & 1 & 3 & 0\end{array}$$

Answer: $4a^2+a+3$

13. $(x^5-2x^3+3x^2-4x-2)\div(x-2)$

Insert 0 for the missing x^4-term.

$$\begin{array}{r|rrrrrr}2 & 1 & 0 & -2 & 3 & -4 & -2\\ & & 2 & 4 & 4 & 14 & 20\\ \hline & 1 & 2 & 2 & 7 & 10 & 18\end{array} \leftarrow \textit{Remainder}$$

Answer: $x^4+2x^3+2x^2+7x+10+\dfrac{18}{x-2}$

15. $\left(-4r^6-3r^5-3r^4+5r^3-6r^2+3r+3\right)\div(r-1)$

$$\begin{array}{r|rrrrrrr}1 & -4 & -3 & -3 & 5 & -6 & 3 & 3\\ & & -4 & -7 & -10 & -5 & -11 & -8\\ \hline & -4 & -7 & -10 & -5 & -11 & -8 & -5\end{array} \leftarrow \textit{Remainder}$$

Answer:

$-4r^5-7r^4-10r^3-5r^2-11r-8+\dfrac{-5}{r-1}$

17. $(-3y^5+2y^4-5y^3-6y^2-1)\div(y+2)$

Insert 0 for the missing y-term.

$$\begin{array}{r|rrrrrr}-2 & -3 & 2 & -5 & -6 & 0 & -1\\ & & 6 & -16 & 42 & -72 & 144\\ \hline & -3 & 8 & -21 & 36 & -72 & 143\end{array} \leftarrow \textit{Remainder}$$

Answer:

$-3y^4+8y^3-21y^2+36y-72+\dfrac{143}{y+2}$

19. $\dfrac{y^3+1}{y-1}=\dfrac{y^3+0y^2+0y+1}{y-1}$

$$\begin{array}{r|rrrr}1 & 1 & 0 & 0 & 1\\ & & 1 & 1 & 1\\ \hline & 1 & 1 & 1 & 2\end{array} \leftarrow \textit{Remainder}$$

Answer: $y^2+y+1+\dfrac{2}{y-1}$

21. $P(x)=2x^3-4x^2+5x-3;\ k=2$

To find $P(2)$, divide the polynomial by $x-2$. $P(2)$ will be the remainder.

$$\begin{array}{r|rrrr}2 & 2 & -4 & 5 & -3\\ & & 4 & 0 & 10\\ \hline & 2 & 0 & 5 & 7\end{array} \leftarrow \textit{Remainder}$$

By the remainder theorem, $P(2)=7$.

23. $P(r)=-r^3-5r^2-4r-2;\ k=-4$

Divide by $r+4$. The remainder is equal to $P(-4)$.

$$\begin{array}{r|rrrr}-4 & -1 & -5 & -4 & -2\\ & & 4 & 4 & 0\\ \hline & -1 & -1 & 0 & -2\end{array} \leftarrow \textit{Remainder}$$

By the remainder theorem, $P(-4)=-2$.

25. $P(y) = 2y^3 - 4y^2 + 5y - 33;\ k = 3$

Divide by $y - 3$. The remainder is equal to $P(3)$.

$$\begin{array}{r|rrrr} 3 & 2 & -4 & 5 & -33 \\ & & 6 & 6 & 33 \\ \hline & 2 & 2 & 11 & 0 \end{array} \leftarrow \textit{Remainder}$$

By the remainder theorem, $P(3) = 0$.

27. By the remainder theorem, a zero remainder means that $P(k) = 0$; that is, k is a number that makes $P(x) = 0$.

29. Is $x = -2$ a solution of

$$x^3 - 2x^2 - 3x + 10 = 0?$$

To decide whether -2 is a solution to the given equation, divide the polynomial by $x + 2$.

$$\begin{array}{r|rrrr} -2 & 1 & -2 & -3 & 10 \\ & & -2 & 8 & -10 \\ \hline & 1 & -4 & 5 & 0 \end{array} \leftarrow \textit{Remainder}$$

Since the remainder is 0, -2 is a solution of the equation.

31. Is $m = -2$ a solution of

$$m^4 + 2m^3 - 3m^2 + 8m - 8 = 0?$$

To decide whether -2 is a solution to the given equation, divide the polynomial by $m + 2$.

$$\begin{array}{r|rrrrr} -2 & 1 & 2 & -3 & 8 & -8 \\ & & -2 & 0 & 6 & -28 \\ \hline & 1 & 0 & -3 & 14 & -36 \end{array} \leftarrow \textit{Remainder}$$

Since the remainder is not 0, -2 is not a solution of the equation.

33. Is $a = -2$ a solution of

$$3a^3 + 2a^2 - 2a + 11 = 0?$$

$$\begin{array}{r|rrrr} -2 & 3 & 2 & -2 & 11 \\ & & -6 & 8 & -12 \\ \hline & 3 & -4 & 6 & -1 \end{array} \leftarrow \textit{Remainder}$$

Since the remainder is not 0, -2 is not a solution of the equation.

35. Is $x = -3$ a solution of

$$2x^3 - x^2 - 13x + 24 = 0?$$

$$\begin{array}{r|rrrr} -3 & 2 & -1 & -13 & 24 \\ & & -6 & 21 & -24 \\ \hline & 2 & -7 & 8 & 0 \end{array} \leftarrow \textit{Remainder}$$

Since the remainder is 0, -3 is a solution of the equation.

In Exercises 37–41,

$$P(x) = 2x^2 + 5x - 12.$$

37. Factor $P(x)$.

$$2x^2 + 5x - 12 = (2x - 3)(x + 4)$$

38. Solve $P(x) = 0$.

$$2x^2 + 5x - 12 = 0$$
$$(2x - 3)(x + 4) = 0$$
$$2x - 3 = 0 \quad \text{or} \quad x + 4 = 0$$
$$2x = 3 \qquad\qquad x = -4$$
$$x = \frac{3}{2}$$

Solution set: $\left\{\frac{3}{2}, -4\right\}$

39.
$$\begin{aligned} P(-4) &= 2(-4)^2 + 5(-4) - 12 \\ &= 2(16) - 20 - 12 \\ &= 32 - 20 - 12 = 0 \end{aligned}$$

$$\begin{aligned} P\left(\frac{3}{2}\right) &= 2\left(\frac{3}{2}\right)^2 + 5\left(\frac{3}{2}\right) - 12 \\ &= 2\left(\frac{9}{4}\right) + \frac{15}{2} - 12 \\ &= \frac{9}{2} + \frac{15}{2} - \frac{24}{2} = 0 \end{aligned}$$

40. If $P(a) = 0$, then $x - \underline{a}$ is a factor of $P(x)$.

41.
$$\begin{aligned} Q(x) &= 3x^3 - 4x^2 - 17x + 6 \\ Q(3) &= 3(3)^3 - 4(3)^2 - 17(3) + 6 \\ &= 81 - 36 - 51 + 6 = 0 \end{aligned}$$

Since $Q(3) = 0$, $x - 3$ is a factor of $Q(x)$. To check, use synthetic division to see if 3 is a solution of the equation.

$$\begin{array}{r|rrrr} 3 & 3 & -4 & -17 & 6 \\ & & 9 & 15 & -6 \\ \hline & 3 & 5 & -2 & 0 \end{array}$$

Therefore, $x - 3$ is a factor of the polynomial and

$$\begin{aligned} 3x^3 - 4x^2 - 17x + 6 &= (x - 3)(3x^2 + 5x - 2) \\ Q(x) &= (x - 3)(3x - 1)(x + 2). \end{aligned}$$

43. From the graph, it appears that $x = 3$ is a solution of the equation

$$x^3 - x^2 - 21x + 45 = 0.$$

Check this with synthetic division.

$$\begin{array}{r|rrrr} 3 & 1 & 1 & -21 & 45 \\ & & 3 & 6 & -45 \\ \hline & 1 & 2 & -15 & 0 \end{array} \leftarrow \textit{Remainder}$$

Since the remainder is 0, 3 is a solution of the equation.

45. From the graph, it appears that $x = -1$ is a solution of the equation

$$x^3 + 3x^2 - 13x - 15 = 0.$$

Check this with synthetic division.

$$\begin{array}{r|rrrr} -1 & 1 & 3 & -13 & -15 \\ & & -1 & -2 & 15 \\ \hline & 1 & 2 & -15 & 0 \end{array} \leftarrow \quad \textit{Remainder}$$

Since the remainder is 0, -1 is a solution of the equation.